An Introduction to Mixed-Signal
IC Test and Measurement

THE OXFORD SERIES IN ELECTRICAL AND COMPUTER ENGINEERING

Adel S. Sedra, Series Editor

Allen and Holberg, *CMOS Analog Circuit Design*, 2nd edition

Bobrow, *Elementary Linear Circuit Analysis*, 2nd edition

Bobrow, *Fundamentals of Electrical Engineering*, 2nd edition

Burns and Roberts, *An Introduction to Mixed-Signal IC Test and Measurement*

Campbell, *Fabrication Engineering at the Micro- and Nanoscale*, 3rd edition

Chen, *Digital Signal Processing*

Chen, *Linear System Theory and Design*, 3rd edition

Chen, *Signals and Systems*, 3rd edition

Comer, *Digital Logic and State Machine Design*, 3rd edition

Comer, *Microprocessor-Based System Design*

Cooper and McGillem, *Probabilistic Methods of Signal and System Analysis*, 3rd edition

Dimitrijev, *Principles of Semiconductor Devices*, 2nd edition

Dimitrijev, *Understanding Semiconductor Devices*

Fortney, *Principles of Electronics: Analog & Digital*

Franco, *Electric Circuits Fundamentals*

Ghausi, *Electronic Devices and Circuits: Discrete and Integrated*

Guru and Hiziroğlu, *Electric Machinery and Transformers*, 3rd edition

Houts, *Signal Analysis in Linear Systems*

Jones, *Introduction to Optical Fiber Communication Systems*

Krein, *Elements of Power Electronics*

Kuo, *Digital Control Systems*, 3rd edition

Lathi, *Linear Systems and Signals*, 2nd edition

Lathi and Ding, *Modern Digital and Analog Communication Systems*, 4th edition

Lathi, *Signal Processing and Linear Systems*

Martin, *Digital Integrated Circuit Design*

Miner, *Lines and Electromagnetic Fields for Engineers*

Parhami, *Computer Architecture*

Parhami, *Computer Arithmetic*

Roberts and Sedra, *SPICE*, 2nd edition

Roulston, *An Introduction to the Physics of Semiconductor Devices*

Sadiku, *Elements of Electromagnetics*, 5th edition

Santina, Stubberud, and Hostetter, *Digital Control System Design*, 2nd edition

Sarma, *Introduction to Electrical Engineering*

Schaumann, Xiao, and Van Valkenburg, *Design of Analog Filters*, 3rd edition

Schwarz and Oldham, *Electrical Engineering: An Introduction*, 2nd edition

Sedra and Smith, *Microelectronic Circuits*, 6th edition

Stefani, Shahian, Savant, and Hostetter, *Design of Feedback Control Systems*, 4th edition

Tsividis, *Operation and Modeling of the MOS Transistor*, 3rd edition

Van Valkenburg, *Analog Filter Design*

Warner and Grung, *Semiconductor Device Electronics*

Wolovich, *Automatic Control Systems*

Yariv and Yeh, *Photonics: Optical Electronics in Modern Communications*, 6th edition

Żak, *Systems and Control*

SECOND EDITION

THE OXFORD SERIES IN ELECTRICAL AND COMPUTER ENGINEERING

An Introduction to Mixed-Signal IC Test and Measurement

GORDON ROBERTS

FRIEDRICH TAENZLER

MARK BURNS

New York   Oxford

OXFORD UNIVERSITY PRESS

Oxford University Press, Inc., publishes works that further Oxford University's
objective of excellence in research, scholarship, and education.

Oxford   New York
Auckland   Cape Town   Dar es Salaam   Hong Kong   Karachi
Kuala Lumpur   Madrid   Melbourne   Mexico City   Nairobi
New Delhi   Shanghai   Taipei   Toronto

With offices in
Argentina   Austria   Brazil   Chile   Czech Republic   France   Greece
Guatemala   Hungary   Italy   Japan   Poland   Portugal   Singapore
South Korea   Switzerland   Thailand   Turkey   Ukraine   Vietnam

For titles covered by Section 112 of the US Higher Education Opportunity Act,
please visit www.oup.com/us/he for the latest information about pricing and
alternate formats.

Published by Oxford University Press, Inc.
198 Madison Avenue, New York, New York 10016
http://www.oup.com

Library of Congress Cataloging-in-Publication Data

Roberts, Gordon W., 1959-
An introduction to mixed-signal IC test and measurement. — 2nd ed. / Gordon Roberts,
Friedrich Taenzler, Mark Burns. — (The Oxford series in electrical and computer engineering)
     p. cm.
   Burns' name appears first on the previous edition.
   Summary: "Textbook for upper-level undergraduate and graduate students studying the testing of digital,
   analog, mixed-signal and radio-frequency circuits and their related devices."— Provided by publisher.
   ISBN 978-0-19-979621-2 (hardback)
   1. Integrated circuits--Testing. 2. Mixed signal circuits—Testing. I. Taenzler, Friedrich. II. Burns, Mark, 1962- III. Title.
TK7874.B825 2011
621.3815--dc23   2011031312

Printing number: 9 8 7 6 5 4 3 2 1

Printed in the United States of America
on acid-free paper

CONTENTS

PREFACE

Since the introduction of the first edition of this textbook in 2001, much change has occurred in the semiconductor industry, especially with the proliferation of complex semiconductor devices containing digital, analog, mixed-signal, and radio-frequency (RF) circuits. The integration of these four circuit types has created many new business opportunities, but at the same time made the economics of test much more significant. Today, product costs are divided among silicon, test, and packaging in various proportions, depending on the maturity of the product as well as the technical skill of the engineering teams. In some market segments, we are seeing packaging costs dominate product costs; in other market segments, device packaging is being done away with entirely, because bare die is mounted directly on the carrier substrate. While this helps to address the packaging costs, it puts greater pressure on test costs, because it now becomes a larger contributor to the overall product cost.

Analog, mixed-signal, and RF IC test and measurement have grown into a highly specialized field of electrical engineering. For the most part, analog and mixed-signal test engineering is handled by one team of specialists, while RF test is handled by another. The skill set required to master both technical areas are quite different, because one involves the particle (electron) perspective of physics whereas the other handles the wave perspective. Nonetheless, these two technical areas have much in common, such as the need to accurately measure a signal in the presence of noise and distortion in a time-sensitive manner, albeit over vastly different dynamic range and power levels.

The goal of the first edition of this textbook was to create a source of information about analog and mixed-signal automated test and measurement as it applies to ICs. At the time, little information was available for the test engineer. One important source was the textbook by M. Mahoney, the pioneer of DSP-based test techniques, however, this was largely limited to the coherency principles of DSP-based testing, and it did not discuss system level tradeoffs of test engineering, nor did it discuss the practical issue related to test interfacing.

Based on the feedback that we received, the first edition of this textbook has been a source of inspiration to many, especially those new to the test field. The industry has seen a great deal of change since the release of the first edition almost a decade ago. RF circuits now play a larger part in many of the devices and systems created today. It is clear to us that engineers need to be fluent in all four-circuit types, digital, analog, mixed-signal and RF, although we limit our discussion in this textbook to the latter three, as digital test is a subject all in itself. We do not believe we could do this topic justice in the amount of space remaining after discussing analog, mixed-signal, and RF test. We encourage our readers to learn as much as possible about digital test as they will most certainly encounter such techniques during their career in test (or design for that matter).

The prerequisite for this book remains at a junior or senior university level; it is assumed that students reading this textbook will have taken courses in linear continuous-time and discrete-time systems, fields, and waves, as well as having had exposure to probability and statistical concepts.

In fact, the three greatest changes made to the second edition of this textbook is a lengthy discussion on RF circuits, high-speed I/Os, and probabilistic reasoning. Over the years, it has become quite clear that test, application, and product engineers make extensive use of probability theory in their day-to-day function, leading us to believe that it is necessary to increase the amount of probability coverage contained in this textbook. By doing so, we could define the concept of noise more rigorously and study its affects on a measurement more concisely. These ideas will be used throughout the textbook to help convey the limitations of a measurement.

OUTLINE TO SECOND EDITION

The book is divided into 16 chapters. The order in which we present these chapters is slightly different from that of the first edition. This reordering is largely an attempt at presenting the material in a more pedagogical order. For instance, DAC and ADC testing are placed before DSP-based testing and after the chapters on DC measurements, because the tests described in these two chapters are largely DC in nature. Data analysis and probability theory follow the chapter on DC measurements as a means to motivate the limitations of performing a simple DC measurement.

A brief breakdown of each chapter goes as follows: Chapter 1 presents an introduction to the context in which mixed-signal testing is performed and why it is necessary. Chapter 2 takes a look at the architecture of a generic mixed-signal automated test equipment (ATE) tester. The generic tester includes instruments such as DC sources, meters, waveform digitizers, arbitrary waveform generators, and digital pattern generators with source and capture functionality. This discussion will also include the source and measurement basics of an RF subsystem.

Chapter 3 introduces basic DC measurement definitions, including continuity, leakage, offset, gain, DC power supply rejection ratio, and many other types of fundamental DC measurements.

Chapter 4 provides the mathematical foundation of data analysis and modeling using probabilistic methods. Probabilities and the related concept of a density function are introduced for various random situations (e.g., Gaussian, sinusoidal, uniform, etc.). One section of this chapter deals with modeling the structure of randomness using Gaussian mixture models. This model is useful for modeling complex random behavior that does not fit into a single Gaussian model. This chapter will also address the expected behavior of sums and differences of different sets of random variables and will look at the central limit theorem of large numbers.

Chapter 5 covers the basics of absolute accuracy, resolution, software calibration, standards traceability, and measurement repeatability. This chapter will also look at the impact of measurement accuracy on manufacturing yield and test time. The topic of guardbanding is also described. Finally, statistical process control (SPC) is explained, including a discussion of process control metrics such as C_p and C_{pk}.

Testing of DACs is covered in Chapter 6. Traditional transfer characteristic based measurements like INL, DNL, and absolute error are discussed. A brief discussion of several dynamic DAC tests such as conversion time, overshoot and undershoot, rise and fall times and skew is also provided. The chapter concludes with a discussion on the types of tests performed on a DAC intended for DC, audio, video or data transmission applications.

Chapter 7 builds upon the concepts in Chapter 6 to show how ADCs are commonly tested. The principles of linear and sinusoidal histogram testing is described and demonstrated with several ADC examples. A brief discussion of several dynamic ADC tests such as conversion and recovery time, sampling frequency, aperture jitter and sparkling effects. The chapter concludes with a discussion on the types of tests performed on a ADC intended for DC, audio, video or data transmission applications.

Chapter 8 presents an introduction to both ADC and DAC sampling theory. DAC sampling theory is applicable to both (a) DAC circuits in the device under test and (b) the arbitrary waveform generators in a mixed-signal tester. ADC sampling theory is applicable to both (a) ADC circuits in the device under test and (b) waveform digitizers in a mixed-signal tester. Coherent multitone sample sets are also introduced as an introduction to DSP-based testing.

Chapter 9 further develops sampling theory concepts and DSP-based testing methodologies, which are at the core of many mixed-signal test and measurement techniques. FFT fundamentals, windowing, frequency domain filtering, and other DSP-based testing fundamentals are covered in Chapters 8 and 9.

Chapter 10 shows how basic AC channel tests can be performed economically using DSP-based testing. This chapter covers only nonsampled channels, consisting of combinations of operational amplifiers, analog filters, PGAs, and other continuous-time circuits.

Chapter 11 explores many of these same tests as described in Chapter 10 as they are applied to sampled channels, which include DACs, ADCs, sample and hold (S/H) amplifiers, and so on. The principle of undersampling under coherent conditions is also discussed in this chapter.

Chapters 12 and 13 are two chapters related to RF testing. They are both new to the second edition of this textbook.

Chapter 12 begins by introducing the reader to the concept of a propagating wave and the various means by which a wave is quantified (e.g., power, wavelength, velocity, etc.). Included in this discussion are the amplitude and phase noise impairments that an RF wave experiences in transmission. A portion of this chapter is devoted to the concept of S-parameters as it applies to an n-port network, such as a two-port network. S-parameters are used to describe the small-signal performance of the network as seen from one port to another. Measures like reflection coefficient, mismatch loss, insertion and transducer loss, and various power gains can easily be defined in terms of these S-parameters. Moreover, the idea of a mismatch uncertainty and its impact on a power measurement is introduced. Mismatch uncertainty is often the most significant contributor to measurement error in an RF test. Several forms of modulation, including analog modulation schemes like AM and PM, followed by several digital modulation schemes, such as ASK, PSK, and QAM used primarily for digital communication systems, are described.

Chapter 13 describes the principles of RF testing of electronic circuits using commercial ATE. These principles are based on the physical concepts related to wave propagation outlined in Chapter 12 combined with the DSP-based sampling principles of Chapters 8–11. This chapter describes the most common types of RF tests using ATE for standard devices, like mixers, VCOs, and power amplifiers. Issues relating to dynamic range, maximum power, noise floor, and phase noise are introduced. Measurement errors introduced by the device interface board due to impedance mismatches, transmission losses, etc., are described, together with de-embedding techniques to compensate for these errors. Also included is a discussion of ATE measurements using directional couplers. These directional couplers enable the measurement of the S-parameters of a device through the direct measurement of the incident and reflected waves at the input and output port of a DUT. Various noise figures measurements (Y-factor and cold noise methods) using an ATE are outlined. Finally, the chapter concludes with a discussion on using an ATE to measure more complex RF system parameters like EVM, ACPR, and BER.

Chapter 14 is also new to the second edition. Chapter 14 outlines the test techniques and metrics used to quantify the behavior of clocks and serial data communication channels. The chapter begins by describing several measures of clock behavior in both the time and frequency domains. In the time domain, measures like instantaneous jitter, period jitter, and cycle-to-cycle jitter are described. In the frequency domain, the concept of phase noise will be derived from a power spectral density description of the clock signal. Chapter 14 will then look at the test attributes associated with communicating serial over a channel. For the most part, this becomes a measure of the bit error rate (BER). However, because BER is a very time-consuming measurement to make, methods to extract estimates of the BER in short period of time are described. These involve a model of channel behavior that is based on parameters known as dependent jitter (DJ) and random jitter (RJ). Finally, the chapter will discuss jitter transmission test such as jitter transfer and jitter tolerance.

Chapter 15 explores the gray art of mixed-signal DIB design. Topics of interest include component selection, power and ground layout, crosstalk, shielding, transmission lines, power matching networks, and tester loading. Chapter 15 also illustrates several common DIB circuits and their use in mixed-signal testing.

Chapter 16 gives a brief introduction to some of the techniques for analog and mixed-signal design for test. There are fewer structured approaches for mixed-signal DfT than for purely digital DfT. The more common ad hoc methods are explained, as well as some of the industry standards such as IEEE Std. 1149.1, 1149.4, and 1500.

ADDITIONS/IMPROVEMENTS TO THE SECOND EDITION

In many ways, the second edition of this textbook has been a large rewrite of the material found in the first edition, together with the addition of three new chapters: two chapters on RF test and one chapter on clocks and serial-data communication channel testing. These three new chapters comprise about one-third of the overall content of this new edition. Hundreds of examples, exercises, and end-of-chapter problems have also been added to the new edition. Both the exercises and end-of-chapter problems are provided with answers. The motivation for the rewrite was largely driven by the need for more probabilistic reasoning throughout the discussions in the textbook, together with the need for new material related to the testing of RF circuits and high-speed I/O devices. Many of today's advancements in test methodologies and techniques are being driven from a probabilistic perspective. It is the authors' belief that tomorrow's test engineer will be better equipped to handle new developments in test if they have a strong understanding of test from a probabilistic perspective. Moreover, a probabilistic perspective goes a long way in describing the reason behind many of the traditional mixed-signal test approaches (e.g., histogram testing of ADCs) used by mixed-signal test engineers.

The arrangement of the chapters in the second edition is quite different from that of the first edition. In this edition, the textbook begins with an introduction into the field of mixed-signal and RF test, as well as a description of some typical ATE hardware. These chapter are attempting to establish the business need behind a set of tests. The next five chapters describe DC measurements and their limitations as it applies to linear and data converter circuits. Process variability and issues of repeatability on a measurement are developed largely from the DC perspective. The next four chapters introduce the reader to DSP-based testing principles. For the most part, this sequence of chapters is identical to that presented in the first edition. However, new material has been added, such as a larger discussion on windowing and its application to noncoherent testing, as well as a detailed look at the principles of undersampling and its application to reconstructing high-frequency signals using the modulo-time shuffling algorithm. The next three chapters are new to this edition. Two chapters are devoted to RF testing, and one describes clock and data communication channel testing. As mentioned above, these three chapters contribute to about a 30% increase in the content of this second edition. While the increase in textbook volume may be considered excessive to some, the evolution of the test engineer is such that their responsibilities now include devices that contain digital, analog, mixed-signal, and RF circuits. Having access to a single volume of test knowledge that ties these circuit types together is believed by the authors to be invaluable. Of course, only time can tell us if our assertion is correct.

The remaining portion of the textbook covers test interfacing issues and design for test (DfT). The test interfacing chapter and DfT chapter was expanded to include RF circuits. Specifically, the test-interfacing chapter expanded on the discussion of transmission lines and added new material related to impedance matching circuits and Smith Charts™. The chapter on DfT was expanded to include a discussion on DFT for RF circuits.

A summary of the changes made are:

- Expanded use of probabilistic approach to problem description.
- Windowing for noncoherent sampling.
- Undersampling and reconstruction using modulo-time shuffling algorithm.
- New chapter on fundamentals of RF testing.
- New chapter on RF test methods.
- New chapter on clock and serial data communications channel measurements.
- Added several sections on RF load board design.
- Hundreds of new examples, exercises, and problems have been added.

A NOTE OF THANKS

First and foremost, a special note of thanks to Mark Burns for starting this project while at Texas Instruments and overseeing the first edition of this book get published. Mark has since retired from the industry and is enjoying life far away from test. It was a great pleasure working with Mark.

We would like to extend our sincere appreciation to the many people who assisted us with this development; beginning with the first edition and subsequently followed by those who contributed to the second edition.

First Edition

The preliminary versions of the first edition were reviewed by a number of practicing test engineers. We would like to thank those who gave us extensive corrections and feedback to improve the textbook:

Steve Lyons, Lucent Technologies
Jim Larson, Teradyne, Inc.
Gary Moraes, Teradyne, Inc.
Justin Ewing, Texas A&M University/Texas Instruments, Inc.
Pramodchandran Variyam, Georgia Tech/Texas Instruments, Inc./Anora LLC.
Geoffrey Zhang, formerly of Texas Instruments, Inc.

We would also like to extend our sincere appreciation to the following for their help in developing this textbook:

Dr. Rainer Fink, Texas A&M University
Dr. Jay Porter, Texas A&M University
Dr. Cajetan Akujuobi, Prairieview A&M University
Dr. Simon Ang, University of Arkansas

Their early adoption of this work at their respective universities has helped to shape the book's content and expose its many weaknesses.

We also thank Juli Boman (Teradyne, Inc.) and Ted Lundquist (Schlumberger Test Equipment) for providing photographs for Chapter 1.

We are extremely grateful to the staff at Oxford University Press, who have helped guide us through the process of writing an enjoyable book. First, we would like to acknowledge the help and constructive feedback of the publishing editor, Peter Gordon. The editorial development help of Karen Shapiro was greatly appreciated.

Finally, on behalf of the test engineering profession, Mark Burns would like to extend his gratitude to

Del Whittaker, formerly of Texas Instruments, Inc.
David VanWinkle, formerly of Texas Instruments, Inc.
Bob Schwartz, formerly of Texas Instruments, Inc.
Ming Chiang, formerly of Texas Instruments, Inc.
Brian Evans, Texas Instruments, Inc.

for allowing him to develop this book as part of his engineering duties for the past three years. It takes great courage and vision for corporate management to expend resources on the production of a work that may ultimately help the competition.

Second Edition

The preliminary versions of the second edition were reviewed by a number of practicing test engineers and professors. We would like to thank those who gave us extensive corrections and feedback to improve the textbook, specifically the following individuals:

Brice Achkir, Cisco Systems Inc.
Rob Aitken, ARM
Benjamin Brown, LTX-Credence Corporation
Cary Champlin, Blue Origin LLC
Ray Clancy, Broadcom Corporation
William DeWilkins, Freescale Semiconductor
Rainer Fink, Texas A&M University
Richard Gale, Texas Tech University
Michael Purtell, Intersil Corporation
Jeff Rearick, Advanced Micro Devices
Tamara Schmitz, Intersil Corporation
Robert J. Weber, Iowa State University

We are extremely grateful to the staff members at Oxford University Press, who have helped guide us through the process of writing the second edition of this book. First, we would like to acknowledge the help and constructive feedback of our editors: Caroline DiTullio, Claire Sullivan, and Rachael Zimmerman.

Gordon Roberts would like to extend his sincere appreciation to all the dedicated staff members and graduate students associated with the Integrated Microsystem Laboratory (formerly the Microelectronics and Computer System Laboratory) at McGill University. Without their desire to learn and ask thought-provoking questions, this textbook would have been less valuable to the students that follow them. For this, I am thankful. More importantly, though, is how the excitement of learning new things has simply made life more enjoyable. For this, I am eternally grateful. A listing of the students that had made some contribution to these works are:

Sadok Aouini
Christopher Taillefer
Mohammad Ali-Bakhshian
Tsung-Yen Tsai
Azhar Chowdhury
Marco Macedo
Kun Chuai
Euisoo. Yoo
Tarek. Alhajj

Dong (Hudson) An
Mouna Safi-Harb
Mourad Oulmane
Ali Ameri
Shudong Lin
George Gal
Michael Guttman
Simon Hong

Gordon Roberts would like to express his love and thanks to his family (Brigid, Sean, and Eileen) for their unequivocal support over the course of this project. Their understanding of the level of commitment that is required to undertake such a large project is essential for the success of this work. Without their support, this project would not have come to completion. For that, all those that learn something from this book owe each one of you some level of gratitude.

In a similar fashion, Friedrich Taenzler would like to express his sincere gratitude to his colleagues at Texas Instruments he had the chance to work with, thereby discussing and learning multiple aspects of test engineering. A special thanks to those colleagues who gave extensive corrections and feedback to improve this textbook: Ganesh Srinivasan, Kausalya Palavesam, and Elida de-Obaldia. Friedrich Taenzler also extends his appreciation to the ATE and test equipment vendors he had the chance to work with while learning the broader view on production tester implementation.

Finally, and most important, Friedrich would like to thank his family, his wife Claudia and his sons Phillip and Ferdinand, for their tremendous support and understanding for the time given to contribute to this book. Needless to say, without their help and tolerance, this book would have never been completed.

Gordon W. Roberts Friedrich Taenzler Mark Burns
McGill University Texas Instruments, Inc. formerly of Texas Instruments, Inc.
Montreal, Quebec, Canada Dallas, Texas, USA Dallas, Texas, USA

CHAPTER 1

Overview of Mixed-Signal Testing

1.1 MIXED-SIGNAL CIRCUITS

1.1.1 Analog, Digital, or Mixed-Signal?

Before delving into the details of mixed-signal IC test and measurement, one might first ask a few pertinent questions. Exactly what are mixed-signal circuits? How are they used in typical applications? Why do we have to test mixed-signal circuits in the first place? What is the role of a test engineer, and how does it differ from that of a design engineer or product engineer? Most training classes offered by mixed-signal tester companies assume that the students already know the answers to these questions. For instance, a typical automated test equipment (ATE) training class shows the students how to program the per-pin current leakage measurement instruments in the tester before the students even know why leakage current is an important parameter to measure. This book will answer many of the what's, when's, and why's of mixed-signal testing, as well as the usual how's. Let's start with a very basic question: What is a mixed-signal circuit?

A mixed-signal circuit can be defined as a circuit consisting of both digital and analog elements. By this definition, a comparator is one of the simplest mixed-signal circuits. It compares two analog voltages and determines if the first voltage is greater than or less than the second voltage. Its digital output changes to one of two states, depending on the outcome of the comparison. In effect, a comparator is a one-bit analog-to-digital converter (ADC). It might also be argued that a simple digital inverter is a mixed-signal circuit, since its digital input controls an "analog" output that swings between two fixed voltages, rising, falling, overshooting, and undershooting according to the laws of analog circuits. In fact, in certain extremely high-frequency applications the outputs of digital circuits have been tested using mixed-signal testing methodologies.[1]

Some mixed-signal experts might argue that a comparator and an inverter are not mixed-signal devices at all. The comparator is typically considered an analog circuit, while an inverter is considered a digital circuit (Figure 1.1). Other examples of borderline mixed-signal devices are analog switches and programmable gain amplifiers. The purist might argue that mixed-signal circuits are those that involve some sort of nontrivial interaction between digital signals and analog signals. Otherwise, the device is simply a combination of digital logic and separate analog

Figure 1.1. Comparator and inverter—analog, digital, or mixed-signal?

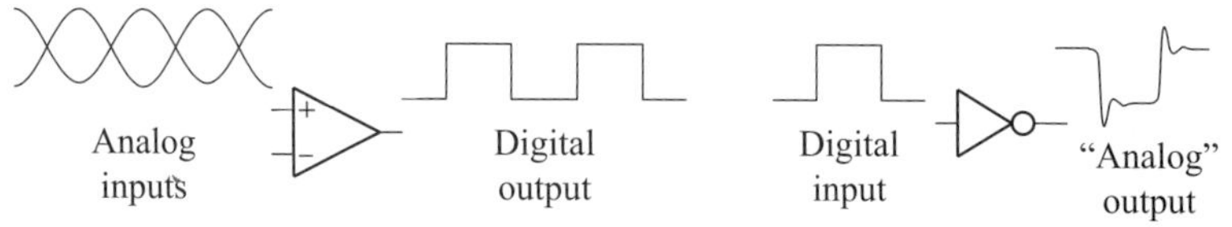

circuitry coexisting on the same die or circuit board. The line between mixed-signal circuits and analog or digital circuits is blurry if one wants to be pedantic.

Fortunately, the blurry lines between digital, analog, and mixed-signal are completely irrelevant in the context of mixed-signal test and measurement. Most complex mixed-signal devices include at least some stand-alone analog circuits that do not interact with digital logic at all. Thus, the testing of op amps, comparators, voltage references, and other purely analog circuits must be included in a comprehensive study of mixed-signal testing. This book encompasses the testing of both analog and mixed-signal circuits, including many of the borderline examples. Digital testing will only be covered superficially, since testing of purely digital circuits has been extensively documented elsewhere.[2–4]

1.1.2 Common Types of Analog and Mixed-Signal Circuits

Analog circuits (also known as *linear circuits*) include operational amplifiers, active or passive filters, comparators, voltage regulators, analog mixers, analog switches, and other specialized functions such as Hall effect transistors. One of the very simplest circuits that can be considered to fall into the mixed-signal realm is the CMOS analog switch. In this circuit, the resistance of a CMOS transistor is varied between high impedance and low impedance under control of a digital signal. The off-resistance may be as high as one megohm (MΩ) or more, while the on-resistance may be 100 Ω or less. Banks of analog switches can be interconnected in a variety of configurations, forming more complex circuits such as analog multiplexers and demultiplexers and analog switch matrices.

Another simple type of mixed-signal circuit is the programmable gain amplifier (PGA). The PGA is often used in the front end of a mixed-signal circuit to allow a wider range of input signal amplitudes. Operating as a digitally adjusted volume control, the PGA is set to high gains for low-amplitude input signals and low gains for high-amplitude input signals. The next circuit following a PGA is thus provided with a consistent signal level. Many circuits require a consistent signal level to achieve optimum performance. These circuits therefore benefit from the use of PGAs.

PGAs and analog switches involve a trivial interaction between the analog and digital circuits. This is why they are not always considered to be mixed-signal circuits at all. The most common circuits that can truly be considered mixed-signal devices are analog-to-digital converters (A/Ds or ADCs) and digital-to-analog converters (D/As or DACs). While the abbreviations A/D and ADC are used interchangeably in the electronics industry, this book will always use the term ADC for consistency. Similarly, the term DAC will be used throughout the book rather than D/A. An ADC is a circuit that samples a continuous analog signal at specific points in time and converts the sampled voltages (or currents) into a digital representation. Each digital representation is called a *sample*. Conversely, a DAC is a circuit that converts digital samples into analog voltages (or currents). ADCs and DACs are the most common mixed-signal components in complex mixed-signal designs, since they form the interface between the physical world and the world of digital logic.

Comprehensive testing of DACs and ADCs is an expansive topic, since there are a wide variety of ADC and DAC designs and a wide variety of techniques to test them. For example, an ADC that is only required to sample once per second may employ a dual slope conversion architecture, whereas a 100-MHz video ADC may have to employ a much faster flash conversion architecture. The weaknesses of these two architectures are totally different. Consequently, the testing of these two converter types is totally different. Similar differences exist between the various types of DACs.

Another common mixed-signal circuit is the phase locked loop, or PLL. PLLs are typically used to generate high-frequency reference clocks or to recover a synchronous clock from an asynchronous data stream. In the former case, the PLL is combined with a digital divider to construct a frequency multiplier. A relatively low-frequency clock, say, 50 MHz, is then multiplied by an integer value to produce a higher-frequency master clock, such as 1 GHz. In the latter case, the recovered clock from the PLL is used to latch the individual bits or bytes of the incoming data stream. Again, depending on the nature of the PLL design and its intended use, the design weaknesses and testing requirements can be very different from one PLL to the next.

1.1.3 Applications of Mixed-Signal Circuits

Many mixed-signal circuits consist of combinations of amplifiers, filters, switches, ADCs, DACs, and other types of specialized analog and digital functions. End-equipment applications such as cellular telephones, hard disk drives, modems, motor controllers, and multimedia audio and video products all employ complex mixed-signal circuits. While it is important to test the individual circuits making up a complex mixed-signal device, it is also important to perform system-level tests. System-level tests guarantee that the circuit as a whole will perform as required in the end-equipment application. Thorough testing of large-scale mixed-signal circuits therefore requires at least a basic understanding of the end-equipment application in which the circuits will be used.

As an example of a mixed-signal application, let us consider a common consumer product using many mixed-signal subcircuits. Figure 1.2 shows a simplified block diagram of a complex mixed-signal application, the digital cellular telephone. It represents an excellent example of a complex mixed-signal system because it employs a variety of mixed-signal components. Since the digital cellular telephone will be used as an example throughout this book, we shall examine its operation in some detail.

A cellular telephone consists of many analog, digital, and mixed-signal circuits working together in a complex fashion. The cellular telephone user interfaces with the keyboard and display to answer incoming calls and to initiate outgoing calls. The control microprocessor handles the interface with the user. It also performs many of the supervisory functions of the telephone, such as helping coordinate the handoff from one base station to the next as the user travels through

Figure 1.2. Digital cellular telephone.

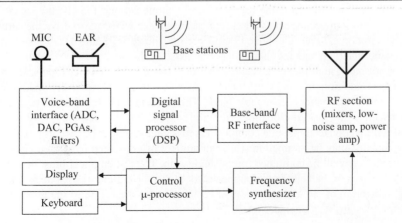

each cellular area. The control microprocessor selects the incoming and outgoing transmission frequencies by sending control signals to the frequency synthesizer. The synthesizer often consists of several PLLs, which control the mixers in the radio-frequency (RF) section of the cellular telephone. The mixers convert the relatively low-frequency signals of the base-band interface to extremely high frequencies that can be transmitted from the cellular telephone's radio antenna. They also convert the very high-frequency incoming signals from the base station into lower-frequency signals that can be processed by the base-band interface.

The voice-band interface, digital signal processor (DSP), and base-band interface perform most of the complex operations. The voice-band interface converts the user's voice into digital samples using an ADC. The volume of the voice signal from the microphone can be adjusted automatically using a programmable gain amplifier (PGA) controlled by either the DSP or the control microprocessor. Alternatively, the PGA may be controlled with a specialized digital circuit built into the voice-band interface itself. Either way, the PGA and automatic adjustment mechanism form an automatic gain control (AGC) circuit. Before the voice signal can be digitized by the voice-band interface ADC, it must first be low-pass filtered to avoid unwanted high-frequency components that might cause aliasing in the transmitted signal. (Aliasing is a type of distortion that can occur in sampled systems, making the speaker's voice difficult to understand.) The digitized samples are sent to the DSP, where they are compressed using a mathematical process called *vocoding*. The vocoding process converts the individual samples of the sound pressure waves into samples that represent the *essence* of the user's speech. The vocoding algorithm calculates a time-varying model of the speaker's vocal tract as each word is spoken. The characteristics of the vocal tract change very slowly compared to the sound pressure waves of the speaker's voice. Therefore, the vocoding algorithm can compress the important characteristics of speech into a much smaller set of data bits than the digitized sound pressure samples. The vocoding process is therefore a type of data compression algorithm that is specifically tailored for speech. The smaller number of transmitted bits frees up airspace for more cellular telephone users. The vocoder's output bits are sent to the base-band interface and RF circuits for modulation and transmission. The base-band interface acts like a modem, converting the digital bits of the vocoder output into modulated analog signals. The RF circuits then transmit the modulated analog waveforms to the base station.

In the receiving direction, the process is reversed. The incoming voice data are received by the RF section and demodulated by the base-band interface to recover the incoming vocoder bit stream. The DSP converts the incoming bit stream back into digitized samples of the incoming speaker's voice. These samples are then passed to the DAC and low pass reconstruction filter of the voice-band interface to reconstruct the voltage samples of the incoming voice. Before the received voice signal is passed to the earpiece, its volume is adjusted using a second PGA. This earpiece PGA is adjusted by signals from the control microprocessor, which monitors the telephone's volume control buttons to determine the user's desired volume setting. Finally, the signal must be passed through a low-impedance buffer to provide the current necessary to drive the earpiece.

Several common cellular telephone circuits are not shown in Figure 1.2. These include DC voltage references and voltage regulators that may exist on the voice-band interface or the base-band processor, analog multiplexers to control the selection of multiple voice inputs, and power-on reset circuits. In addition, a watchdog timer is often included to periodically wake the control microprocessor from its battery-saving idle mode. This allows the microprocessor to receive information such as incoming call notifications from the base station. Clearly, the digital cellular telephone represents a good example of a complex mixed-signal system. The various circuit blocks of a cellular telephone may be grouped into a small number of individual integrated circuits, called a *chipset*, or they may all be combined into a single chip. The test engineer must be ready to test the individual pieces of the cellular telephone and/or to test the cellular telephone as a whole. The increasing integration of circuits into a single semiconductor die is one of the most challenging aspects of mixed-signal test engineering.

1.2 WHY TEST MIXED-SIGNAL DEVICES?

1.2.1 The CMOS Fabrication Process

Integrated circuits (ICs) are fabricated using a series of photolithographic printing, etching, and doping steps. Using a digital CMOS fabrication process as an example, let us look at the idealized IC fabrication process. Some of the steps involved in printing a CMOS transistor pair are illustrated in Figure 1.3a-f. Starting with a lightly doped P⁻ wafer, a layer of silicon dioxide (SiO_2) is deposited on the surface (Figure 1.3a). Next, a negative photoresist is laid down on top of the silicon dioxide. A pattern of ultraviolet light is then projected onto the photoresist using a photographic mask. The photoresist becomes insoluble in the areas where the mask allows the ultraviolet light to pass (Figure 1.3b). An organic solvent is used to dissolve the nonexposed areas of the photoresist (Figure 1.3c). After baking the remaining photoresist, the exposed areas of oxide are removed using an etching process (Figure 1.3d). Next, the exposed areas of silicon are doped to form an N-well using either diffusion or ion implantation (Figure 1.3e).

After many additional steps of printing, masking, etching, implanting, and chemical vapor deposition,[5] a complete integrated circuit can be fabricated as illustrated in Figure 1.3f. The uneven surfaces are exaggerated in the diagram to show that the various layers of oxide, polysilicon, and metal are not at all flat. Even with these exaggerations, this diagram only represents an idealized approximation of actual fabricated circuit structures. The actual circuit structures are not nearly as well-defined as

Figure 1.3. CMOS fabrication steps.

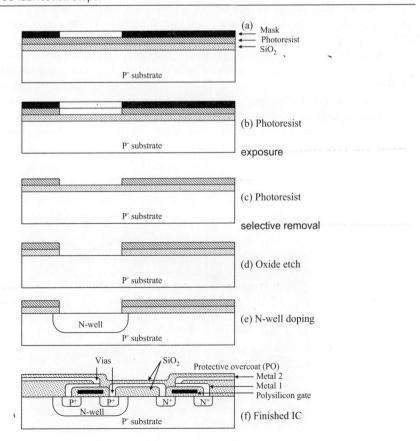

(a) Mask
Photoresist
SiO_2
P⁻ substrate

(b) Photoresist exposure
P⁻ substrate

(c) Photoresist selective removal
P⁻ substrate

(d) Oxide etch
P⁻ substrate

(e) N-well doping
N-well
P⁻ substrate

(f) Finished IC
Vias SiO_2
Protective overcoat (PO)
Metal 2
Metal 1
Polysilicon gate
P⁺ P⁺   N⁺ N⁺
N-well
P⁻ substrate

textbook diagrams would lead us to believe. Cross sections of actual integrated circuits reveal a variety of nonideal physical characteristics that are not entirely under the semiconductor manufacturer's control. Certain characteristics, such as doping profiles that define the boundaries between P and N regions, are not even visible in a cross-section view. Nevertheless, they can have a profound effect on many important analog and mixed-signal circuit characteristics.

1.2.2 Real-World Circuits

Like any photographic printing process, the IC printing process is subject to blemishes and imperfections. These imperfections may cause catastrophic failures in the operation of any individual IC, or they may cause minor variations in performance from one IC to the next. Mixed-signal ICs are often extremely sensitive to tiny imperfections or variations in the printing and doping processes. Many of the fabrication defects that cause problems in mixed-signal devices are difficult to photograph, even with a powerful scanning electron microscope (SEM). For example, a doping error may or may not cause an observable physical defect. However, doping errors can introduce large DC offsets, distortions, and other problems that result in IC performance failures.

In digital circuits, such imperfections in shape may be largely unimportant. However, in mixed-signal circuits, the parasitic capacitance between these traces and surrounding structures may represent significant circuit elements. The exact three-dimensional shape of a metal line and its spacing to adjacent layers may therefore affect the performance of the circuit under test. As circuit geometries continue to shrink, these performance sensitivities will only become more exaggerated. Although a mixed-signal circuit may be essentially functional in the presence of these minor imperfections, it may not meet all its required specifications. For this reason, mixed-signal devices are often tested exhaustively to guard against defects that are not necessarily catastrophic.

Catastrophic defects such as short circuits and open circuits are often easier to detect with test equipment than the subtler ones common in mixed-signal devices. Not surprisingly, the catastrophic defects are often much easier to photograph as well. Several typical defect types are shown in Figures 1.4–1.7. Figure 1.4 shows a defective metal contact, or via, caused by underetching. Figure 1.5 shows a defective via caused by photomask misalignment. A completely defective via usually results in a totally defective circuit, since it represents a complete open circuit. A more

Figure 1.4. Underetched via.

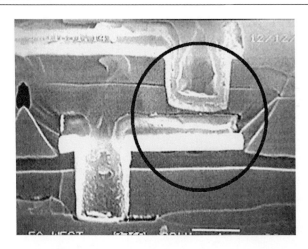

Figure 1.5. Misaligned via.

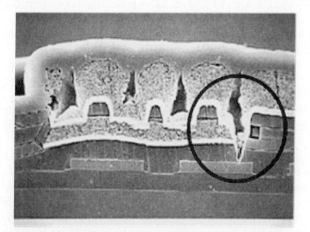

Figure 1.6. Incomplete metal etch.

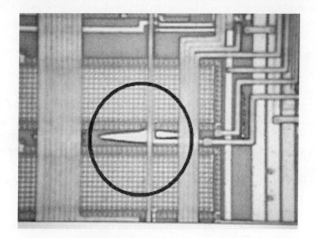

Figure 1.7. Blocked etch (particulate defect).

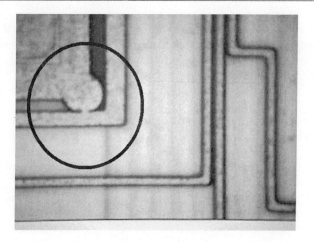

subtle problem is a partially connected via, which may exhibit an abnormally high contact resistance. Depending on the amount of excess resistance, the results of a partially connected via can range from minor DC offset problems to catastrophic distortion problems.

Figure 1.6 shows incomplete etching of the metal surrounding a circuit trace. Incomplete etching can result in catastrophic shorts between circuit nodes. Finally, Figure 1.7 shows a surface defect caused by particulate matter landing on the surface of the wafer or on a photographic mask during one of the processing steps. Again, this type of defect results in a short between circuit nodes. Other catastrophic defects include surface scratches, broken bond wires, and surface explosions caused by electrostatic discharge in a mishandled device. Defects such as these are the reason each semiconductor device must be tested before it can be shipped to the customer.

It has been said that production testing adds no value to the final product. Testing is an expensive process that drives up the cost of integrated circuits without adding any new functionality. Testing cannot change the quality of the individual ICs; it can only *measure* quality if it already exists. However, semiconductor companies would not spend money to test products if the testing process did not add value. This apparent discrepancy is easily explained if we recognize that the product is actually the entire shipment of devices, not just the individual ICs. The quality of the product is certainly improved by testing, since defective devices are not shipped. Therefore, testing does add value to the product, as long as we define the product correctly.

1.2.3 What Is a Test Engineer?

We have mentioned the term *test engineer* several times without actually defining what test engineering is. Perhaps this would be a good time to discuss the traditional roles of test engineers, design engineers, product engineers, and systems engineers. Although each of these engineering professions is involved in the development and production of semiconductor devices, each profession entails its own set of tasks and responsibilities. The various engineering professions are easiest to define if we examine the process by which a new semiconductor product is developed and manufactured.

A new semiconductor product typically begins in one of two ways. Either a customer requests a particular type of product to fill a specific requirement, or a marketing organization realizes an opportunity to produce a product that the market needs. In either case, systems engineers help define the technical requirements of the new product so that it will operate correctly in the end-equipment application. The systems engineers are responsible for defining and documenting the customer's requirements so that the rest of the engineering team can design the product and successfully release it to production.

After the systems engineers have defined the product's technical requirements, design engineers develp the corresponding integrated circuit. Hopefully, the new design meets the technical requirements of the customer's application. Unfortunately, integrated circuits sometimes fail to meet the customer's needs. The failure may be due to a fabrication defect or it may be due to a flaw or weakness in the circuit's design. These failures must be detected before the product is shipped to the customer.

The test engineer's role is to generate hardware and software that will be used by automated test equipment (ATE) to guarantee the performance of each device after it is fabricated. The test software directs the ATE tester to apply a variety of electrical stimuli (such as digital signals and sine waves) to the device under test (DUT). The ATE tester then observes the DUT's response to the various test stimuli to determine whether the device is good or bad (Figure 1.8). A typical mixed-signal DUT must pass hundreds or even thousands of stimulus/response tests before it can be shipped to the customer.

Figure 1.8. Test stimulus and DUT response verification.

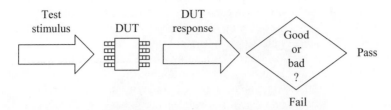

Sometimes the test engineer is also responsible for developing hardware and software that modifies the structure of the semiconductor die to adjust parameters like DC offset and AC gain, or to compensate for grotesque manufacturing defects. Despite claims that production testing adds no value, this is one way in which the testing process can actually enhance the quality of the individual ICs. Circuit modifications can be made in a number of ways, including laser trimming, fuse blowing, and writing to nonvolatile memory cells.

The test engineer is also responsible for reducing the cost of testing through test time reductions and other cost-saving measures. The test cost reduction responsibility is shared with the product engineer. The product engineer's primary role is to support the production of the new device as it matures and proceeds to profitable volume production. The product engineer helps identify and correct process defects, design defects, and tester hardware and software defects.

Sometimes the product engineering function is combined with the test engineering function, forming a single test/product engineering position. The advantage of the combined job function is that the product engineering portion of the job can be performed with a much more thorough understanding of the device and test program details. The disadvantage is that the product engineering responsibilities may interfere with the ability of the engineer to become an expert on the use of the complex test equipment. The choice of combined versus divided job functions is highly dependent on the needs of each organization.

1.3 POST-SILICON PRODUCTION FLOW

1.3.1 Test and Packaging

After silicon wafers have been fabricated, many additional production steps remain before a final packaged device is ready for shipment to the customer. The untested wafers (Figure 1.9) must first be probed using automated test equipment to prevent bad dies from passing on to further production steps. The bad dies can be identified using ink dots, which are applied either after each die is tested or after the whole wafer has been tested. Offline inking is a method used to electronically track bad dies using a computer database. Using pass/fail information from the database, bad dies are inked after the wafer has been removed from the test equipment.

The wafers are then sawed into individual dies and the good ones are attached to lead frames. Lead frames are punched metal holders that eventually become the individual leads of the packaged device. Bond wires are attached from each die's bond pads to the appropriate lead of the lead frame. Then plastic is injection-molded around the dies and lead frame to form packaged devices. Finally, the individual packaged devices are separated from one another by trimming them from the lead frame.

Figure 1.9. Untested wafer.

After the leads have been trimmed and formed, the devices are ready for final testing on a second ATE tester. Final testing guarantees that the performance of the device did not shift during the packaging process. For example, the insertion of plastic over the surface of the die changes the electrical permittivity near the surface of the die. Consequently, trace-to-trace capacitances are increased, which may affect sensitive nodes in the circuit. In addition, the injection-molded plastic introduces mechanical stresses in the silicon, which may consequently introduce DC voltage shifts. Final testing also guarantees that the bond pads are all connected and that the die was not cracked, scratched, or otherwise damaged in the packaging process. After final testing, the devices are ready for shipment to the end-equipment manufacturer.

1.3.2 Characterization Versus Production Testing

When prototype devices are first characterized, the ATE test program is usually very extensive. Tests are performed under many different conditions to evaluate worst-case conditions. For instance, the distortion of an amplifier output may be worse under one loading condition than another. All loading conditions must be tested to identify which one represents the worst-case test. Other examples of exhaustive characterization testing would be DC offset testing using multiple power supply voltages and harmonic distortion testing at multiple signal levels. Characterization testing must be performed over a large number of devices and over several production lots of material before the results can be considered statistically valid and trustworthy.

Characterization testing can be quite time consuming due to the large number of tests involved. Extensive characterization is therefore economically unacceptable in high-volume production testing of mixed-signal devices. Once worst-case test conditions have been established and the design engineers are confident that their circuits meet the required specifications, a more streamlined production test program is needed. The production test program is created from a subset of the characterization tests. The subset must be carefully chosen to guarantee that no bad devices are shipped. Product and test engineers must work very closely to make sure that the reduced test list still catches all manufacturing defects.

1.4 TEST AND DIAGNOSTIC EQUIPMENT

1.4.1 Automated Test Equipment

Automated test equipment is available from a number of commercial vendors, such as Teradyne, LTX-Credence, and Advantest, to name a few. The Teradyne, Inc. Flex mixed-signal tester is shown in Figure 1.10. High-end ATE testers often consist of three major components: a test head, a workstation (not shown), and the mainframe.

The computer workstation serves as the user interface to the tester. The test engineer can debug test programs from the workstation using a variety of software tools from the ATE vendor. Manufacturing personnel can also use the workstation to control the day-to-day operation of the tester as it tests devices in production.

The mainframe contains power supplies, measurement instruments, and one or more computers that control the instruments as the test program is executed. The mainframe may also contain a manipulator to position the test head precisely. It may also contain a refrigeration unit to provide cooled liquid to regulate the temperature of the test head electronics.

Although much of the tester's electronics are contained in the mainframe section, the test head contains the most sensitive measurement electronics. These circuits are the ones that require close proximity to the device under test. For example, high-speed digital signals benefit from short electrical paths between the tester's digital drivers and the pins of the DUT. Therefore, the ATE tester's digital drivers and receivers are located in the test head close to the DUT.

A device interface board (DIB) forms the electrical interface between the ATE tester and the DUT. The DIB is also known as a *performance board*, *swap block*, or *family board*, depending on the ATE vendor's terminology. DIBs come in many shapes and sizes, but their main function is to provide a temporary (socketed) electrical connection between the DUT and the electrical instruments in the tester. The DIB also provides space for DUT-specific

Figure 1.10. Teradyne Flex mixed-signal tester.

Figure 1.11. Octal site device interface board (DIB) showing DUT sockets (left) and local circuits with RF interface (right).

local circuits such as load circuits and buffer amplifiers that are often required for mixed-signal device testing. Figure 1.11 illustrates the top and bottom sides of an octal site DIB. The topside shown on the left displays eight DUT sockets, and the picture on the right shows the local circuits and RF interface.

1.4.2 Wafer Probers

Wafer probers are robotic machines that manipulate wafers as the individual dies are tested by the ATE equipment. The prober moves the wafer underneath a set of tiny electrical probes attached to a probe card. The probes are connected to the electrical resources of the ATE tester through a probe interface board (PIB). The PIB is a specialized type of DIB board that may be connected to the probe card through coaxial cables and/or spring-loaded contacts called *pogo pins*. The PIB and probe card serve the same purpose for the wafer that the DIB board serves for the packaged device. They provide a means of temporarily connecting the DUT to the ATE tester's electrical instrumentation while testing is performed.

The prober informs the tester when it has placed each new die against the probes of the probe card. The ATE tester then executes a series of electrical tests on the die before instructing the prober to move to the next die. The handshaking between tester and prober insures that the tester only begins testing when a die is in position and that the prober does not move the wafer in midtest. Figure 1.12 shows a wafer prober and closeup views of a probe card and its probe tips.

1.4.3 Handlers

Handlers are used to manipulate packaged devices in much the same way that probers are used to manipulate wafers. Most handlers fall into two categories: gravity-fed and robotic. Robotic handlers are also known as *pick-and-place* handlers. Gravity-fed handlers are normally used with dual inline packages, while robotic handlers are used with devices having pins on all four sides or pins on the underside (ball grid array packages, for example).

Either type of handler has one main purpose: to make a temporary electrical connection between the DUT pins and the DIB board. Gravity-fed handlers often perform this task using a contactor assembly that grabs the device pins from either side with metallic contacts that are in turn connected to the DIB board. Robotic handlers usually pick up each device with a suction arm and then plunge the device into a socket on the DIB board.

Figure 1.12. Wafer prober and probe card.

In addition to providing a temporary connection to the DUT, handlers are also responsible for sorting the good DUTs from the bad ones based on test results from the ATE tester. Some handlers also provide a controlled thermal chamber where devices are allowed to "soak" for a few minutes so they can either be cooled or heated before testing. Since many electrical parameters shift with temperature, this is an important handler feature.

1.4.4 E-Beam Probers

Electron beam probers, or e-beam probers as they are often called, are used to probe internal device signals while the device is being stimulated by the tester. These machines are very similar to scanning electron microscopes (SEMs). Unlike an SEM, an e-beam prober is designed to display variations in circuit voltage as the electron beam is swept across the surface of an operating DUT. Variations in the voltage levels on the metal traces in the IC appear as different shades of gray in the e-beam display. E-beam probers are extremely powerful diagnostic tools, since they provide measurement access to internal circuit nodes.

1.4.5 Focused Ion Beam Equipment

Focused ion beam (FIB) equipment is used in conjunction with e-beam probers to modify the device's metal traces and other physical structures. A FIB machine can cut holes in oxide and metal traces and can also lay down new metallic traces on the surface of the device.

Experimental design changes can be implemented without waiting for a complete semiconductor fabrication cycle. The results of the experimental changes can then be observed on the ATE tester to determine the success or failure of the experimental circuit modifications.

1.4.6 Forced-Temperature Systems

As previously mentioned, a handler's thermal chamber allows characterization and testing of large numbers of DUTs at a controlled temperature. When characterizing a small number of DUTs at a variety of temperatures, a less expensive and cumbersome method of temperature control is needed. Portable forced-temperature systems allow DUT performance characterization under a variety of controlled thermal conditions. The nozzle of a forced-temperature system can be seated against the DIB board or bench characterization board, forming a small thermal chamber for the DUT. Many forced-temperature systems are able to raise or lower the DUT's ambient temperature across the full military range (-55 to $+125°C$).

1.5 NEW PRODUCT DEVELOPMENT

1.5.1 Concurrent Engineering

On a poorly managed project, the test engineer might not see the specifications for a device to be tested until after the first prototype devices arrive. The devices must be screened as soon as possible to ship good prototypes to the customer even if they were never designed with testability in mind. In this case, the test engineer's role is completely reactive.

By contrast, the test engineer's role on a well-managed project is proactive. The design engineers and test engineers work together to add testability functions to the design that make the device easier and less expensive to test. The test engineer presents a test plan to the design engineers, explaining all the tests that are to be performed once the device is in production. The design engineers can catch mistakes in the test engineer's understanding of the device operation. They can help eliminate unnecessary tests or point out shortfalls in the proposed test list. This proactive approach is commonly called *concurrent engineering*. True concurrent engineering involves not only design and test engineering personnel, but also systems engineering, product engineering, and manufacturing personnel.

The flow begins with a definition of the device requirements. These include product features, electrical specifications, power consumption requirements, die area estimates, and so on. Once the device requirements are understood, the design team begins to design the individual circuits. In the initial design meetings, test and product engineers work with the design engineers to define the testability features that will make the device less expensive to test and manufacture. Test modes are added to the design to allow access to internal circuit nodes that otherwise would be unobservable in production testing. These observability test modes can be very useful in diagnosing device design flaws.

After the test modes are defined, the test engineer begins working on a test plan while the design process continues. Initially, the main purpose of a test plan is to allow design engineers and test engineers to agree upon a set of tests that will guarantee the quality of a product once it is in production. Eventually, the test plan will serve as documentation for future test and product engineers that may inherit the test program once it is complete. A well-written test plan contains brief background information about the product to be tested, the purpose of each test as it relates to the device specification, setup conditions for each test, and a hardware setup diagram for each test. Once the test plan is complete, all engineers working on the project meet to review the proposed test plan. Last-minute corrections and additions are added at this time. Design engineers point out deficiencies in the proposed test coverage while product engineers point out any problems that may arise on the production floor.

Once the test plan has been approved, the test engineer begins to design the necessary test interface hardware that will connect the automated test equipment to the device under test. Once the initial test hardware has been designed, the test engineer begins writing a test program that will run on the ATE tester. In modern ATE equipment, the test engineer can also debug many of the software routines in the test program before silicon arrives, using an offline simulation environment running on a stand-alone computer workstation.

After the design and layout of the device is complete, the fabrication masks are created from the design database. The database release process is known by various names, such as tape-out or pattern generation. Until pattern generation is complete, the test engineer cannot be certain that the pinout or functionality of the design will not undergo last-minute modifications. The test interface hardware is often fabricated only after the pattern generation step has been completed.

While the silicon wafers and the DIB board are fabricated, the test engineer continues developing the test program. Once the first silicon wafers arrive, the test engineer begins debugging the device, DIB hardware, and software on the ATE tester. Any design problems are reported to the design engineers, who then begin evaluating possible design errors. A second design pass is often required to correct errors and to align the actual circuit performance with specification requirements. Finally, the corrected design is released to production by the product engineer, who then supports the day-to-day manufacturing of the new product.

Of course, the idealized concurrent engineering flow is a simplification of what happens in a typical company doing business in the real world. Concurrent engineering is based on the assumption that adequate personnel and other resources are available to write test plans and generate test hardware and software before the first silicon wafers arrive. It also assumes that only one additional design pass is required to release a device to production. In reality, a high-performance device may require several design passes before it can be successfully manufactured at a profit. This flow also assumes that the market does not demand a change in the device specifications in midstream - a poor assumption in a dynamic world. Nevertheless, concurrent engineering is consistently much more effective than a disjointed development process with poor communication between the various engineering groups.

1.6 MIXED-SIGNAL TESTING CHALLENGES

1.6.1 Time to Market

Time to market is a pressing issue for semiconductor manufacturers. Profit margins for a new product are highest shortly after the product has been released to the market. Margins begin to shrink as competitors introduce similar products at lower prices. The lack of a complete, cost-effective test program can be one of the bottlenecks preventing the release of a new product to profitable volume production.

Mixed-signal test programs are particularly difficult to produce in a short period of time. Surprisingly, the time spent writing test code is often significantly less than the time spent learning about the device under test, defining the test plan, designing the test hardware, and debugging the ATE test solution once silicon is available. Much of the time spent in the debugging phase of test development is actually spent debugging device problems. Mixed-signal test engineers often spend as much time running experiments for design engineers to isolate design errors as they spend debugging their own test code. Perhaps the most aggravating debug time of all is the time spent tracking down problems with the tester itself or the tester's software.

1.6.2 Accuracy, Repeatability, and Correlation

Accuracy is a major concern for mixed-signal test engineers. It is very easy to get an answer from a mixed-signal ATE tester that is simply incorrect. Inaccurate answers are caused by a bewildering number of problems. Electromagnetic interference, improperly calibrated instruments, improperly ranged instruments, and measurements made under incorrect test conditions can all lead to inaccurate test results.

Repeatability is the ability of the test equipment and test program to give the same answer multiple times. Actually, a measurement that never changes at all is suspicious. It sometimes indicates that the tester is improperly configured, giving the same incorrect answer repeatedly. A good measurement typically shows some variability from one test program execution to the next, since electrical noise is present in all electronic circuits. Electrical noise is the source of many repeatability problems.

Another problem facing mixed-signal test engineers is correlation between the answers given by different pieces of measurement hardware. The customer or design engineer often finds that the test program results do not agree with measurements taken using bench equipment in their lab. The test engineer must determine which answer is correct and why there is a discrepancy. It is also common to find that two supposedly identical testers or DIB boards give different answers or that the same tester gives different answers from day to day. These problems frequently result from obscure hardware or software errors that may take days to isolate. Correlation efforts can represent a major portion of the time spent debugging a test program.

1.6.3 Electromechanical Fixturing Challenges

The test head and DIB board must ultimately make contact to the DUT through the handler or prober. There are few mechanical standards in the ATE industry to specify how a tester should be docked to a handler or prober. The test engineer has to design a DIB board that not only meets electrical requirements, but also meets the mechanical docking requirements. These requirements include board thickness, connector locations, DUT socket mechanical holes, and various alignment pins and holes.

Handlers and probers must make a reliable electrical connection between the DUT and the tester. Unfortunately, the metallic contacts between DUT and DIB board are often very inductive and/or capacitive. Stray inductance and capacitance of the contacts can represent a major problem, especially when testing high-impedance or high-frequency circuits. Although several companies have marketed test sockets that reduce these problems, a socketed device will often not perform quite as well as a device soldered directly to a printed circuit board. Performance differences due to sockets are yet another potential source of correlation error and extended time to market.

1.6.4 Economics of Production Testing

Time is money, especially when it comes to production test programs. A high-performance tester may cost two million dollars or more, depending on its configuration. For a specific class of devices developed with design-for-test in mind, the test system can be reduced to as low as two hundred thousand dollars. Probers and handlers may cost five hundred thousand dollars or more. If we also include the cost of providing floor space, electricity, and production personnel, it is easy to understand why testing is an expensive business.

One second of test time can cost a semiconductor manufacturer one to six cents. This may not seem expensive at first glance, but when test costs are multiplied by millions of devices a year the numbers add up quickly. For example, a five-second test program costing four cents per second times one million devices per quarter costs a company eight hundred thousand dollars per year in bottom-line profit. Testing can become the fastest-growing portion of the cost of manufacturing a mixed-signal device if the test engineer does not work on a cost optimized test solution. When testing instead of one device at a time the test solution allows testing eight devices or more with one test insertion, the test cost can be kept under control to be still in the single-digit percentage of the build cost of the device. Continuous process improvements and better photolithography allow the design engineers to add more functions on a single semiconductor chip at little or no additional cost. Unfortunately, test time (especially data collection time) cannot be similarly reduced by simple photolithography. A 100-Hz sine wave takes 10 ms per cycle no matter how small we shrink a transistor.

One feature common in ATE is multisite capability. Multisite testing is a process in which multiple devices are tested on the same test head simultaneously with obvious savings in test cost. The word "site" refers to each socketed DUT. For example, site 0 corresponds to the first DUT, site 1 corresponds to the second DUT, and so on. Multisite testing is primarily a tester operating system feature, although duplicate tester instruments must be added to the tester to allow simultaneous testing on multiple DUT sites.

Clearly, production test economics is an extremely important issue in the field of mixed-signal test engineering. Not only must the test engineer perform accurate measurements of mixed-signal parameters, but the measurements must be performed as quickly as possible to reduce production costs. Since a mixed-signal test program may perform hundreds or even thousands of measurements on each DUT, each measurement must be performed in a small fraction of a second. The conflicting requirements of low test time and high accuracy will be a recurring theme throughout this book.

PROBLEMS

1.1. List four examples of analog circuits.

1.2. List four examples of mixed-signal circuits.

1.3. Problems 1.3–1.6 relate to the cellular telephone in Figure 1.2. Which type of mixed-signal circuit acts as a volume control for the cellular telephone earpiece?

1.4. Which type of mixed-signal circuit converts the digital samples into speaker's voice?

1.5. Which type of mixed-signal circuit converts incoming modulated voice data into digital samples?

1.6. Which type of digital circuit vocodes the speaker's voice samples before they are passed to the base-band interface?

1.7. When a PGA is combined with a digital logic block to keep a signal at a constant level, what is the combined circuit called?

1.8. Assume a particle of dust lands on a photomask during the photolithographic printing process of a metal layer. List at least one possible defect that might occur in the printed IC.

1.9. Why does the cleanliness of the air in a semiconductor fabrication area affect the number of defects in IC manufacturing?

1.10. List at least four production steps after wafers have been fabricated.

1.11. Why would it be improper to draw conclusions about a design based on characterization data from one or two devices?

1.12. List three main components of an ATE tester.

1.13. What is the purpose of a DIB board?

1.14. What type of equipment is used to handle wafers as they are tested by an ATE tester?

1.15. List three advantages of concurrent engineering.

1.16. What is the purpose of a test plan?

1.17. List at least four challenges faced by the mixed-signal test engineer.

1.18. Assume that a test program runs on a tester that costs the company 3 cents per second to operate. This test cost includes tester depreciation, handler depreciation, electricity, floor space, personnel, and so on. How much money can be saved per year by reducing a 5-s test program to 3.5 s, assuming that 5 million devices per year are to be shipped. Assume that only 90% of devices tested are good and that the average time to find a bad device drops to 0.5 s.

1.19. Assume that the profit margin on the device in Problem 1.18 is 20% (i.e., for each $1 worth of devices shipped to the customer, the company makes a profit of 20 cents). How many dollars worth of product would have to be shipped to make a profit equal to the savings offered by the streamlined test program in Problem 1.18? If each device sells for $1.80, how many devices does this represent? What obvious conclusion can we draw about the importance of test time reduction versus the importance of selling and shipping additional devices?

REFERENCES

1. M. Burns, *High Speed Measurements Using Undersampled Delta Modulation*, 1997 Teradyne User's Group proceedings, Teradyne, Inc., Boston.
2. M. Abramovici, M. A. Breuer, and A. D. Friedman, *Digital Systems Testing and Testable Design*, revised printing, IEEE Press, New York, January 1998, ISBN 0780310624.
3. P. K. Lala, *Practical Digital Logic Design and Testing*, Prentice Hall, Upper Saddle River, NJ, 1996, ISBN 0023671718.
4. J. Max Cortner, *Digital Test Engineering*, John Wiley & Sons, New York, 1987, ISBN 0471851353.
5. D. A. Johns and K. Martin, *Analog Integrated Circuit Design*, John Wiley & Sons, New York, 1996, ISBN 0471144487.

CHAPTER 2

Tester Hardware

This chapter explores the architecture of a mixed-signal ATE tester. While we do not focus on any particular ATE from any specific vendor, our intent here is to give a general overview of the common instruments found in such testers. This will include a discussion on DC sources and meters, waveform digitizers, arbitrary waveform generators, an RF measurement subsystem, and digital pattern generators with sources and capture functionality. We shall encounter these instruments again from time to time throughout the remainder of this textbook.

2.1 MIXED-SIGNAL TESTER OVERVIEW

2.1.1 General-Purpose Testers Versus Focused Bench Equipment

General-purpose mixed-signal testers must be capable of testing a variety of dissimilar devices. On any given day, the same mixed-signal tester may be expected to test video palettes, cellular telephone devices, data transceivers, and general-purpose ADCs and DACs. The test requirements for these various devices are very different from one another. For example, the cellular telephone base-band interface shown in Figure 1.2 may require a phase trajectory error test or an error vector magnitude test. Dedicated bench instruments can be purchased that are specifically designed to measure these application-specific parameters. It would be possible to install one of these stand-alone boxes into the tester and communicate with it through an IEEE-488 GPIB bus. However, if every type of DUT required two or three specialized pieces of bolt-on hardware, the tester would soon resemble Frankenstein's monster and would be prohibitively expensive.

The mixed-signal production tester cannot be focused toward a specific type of device if it is to handle a variety of DUTs. Instead of implementing tests like phase trajectory error and error vector magnitude using a dedicated bench instrument, the tester must emulate this type of equipment using a few general-purpose instruments. The instruments are combined with software routines to simulate the operation of the dedicated bench instruments.

2.1.2 Generic Tester Architecture

Mixed-signal testers come in a variety of "flavors" from a variety of vendors. Unlike the ubiquitous PC, testers are not at all standardized in architecture. Each ATE vendor adds special features that they feel will give them a competitive advantage in the marketplace. Consequently, mixed-signal testers from different vendors do not use a common software platform. Furthermore, a test routine implemented on one type of tester is often difficult to translate to another tester type because of subtle differences in the hardware tradeoffs designed into each tester. Nevertheless, most mixed-signal testers share many common features. In this chapter, we will examine these common features without delving too deeply into specific details for any particular brand of tester.

2.2 DC RESOURCES

2.2.1 General-Purpose Multimeters

Most testers incorporate a high-accuracy multimeter that is capable of making fast DC measurements. A tester may also provide a slower, very high-accuracy voltmeter for more demanding measurements such as those needed in focused calibrations. However, this slower instrument may not be usable for production tests because of the longer measurement time. The fast, general-purpose multimeter is used for most of the production tests requiring a nominal level of accuracy.

Figure 2.1. Generic mixed-signal tester architecture.

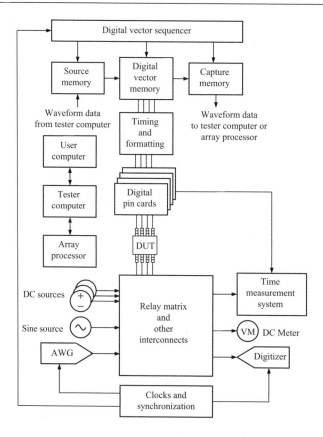

Figure 2.1 shows a generic mixed-signal tester architecture. It includes system computers, DC sources, DC meters, relay control lines, relay matrix lines, time measurement hardware, arbitrary waveform generators, waveform digitizers, clocking and synchronization sources, and a digital subsystem for generating and evaluating digital patterns and signals. This chapter will briefly examine the operation of each of these tester subsystems.

A detailed DC multimeter structure is shown in Figure 2.2. This meter can handle either single-ended or differential inputs. Its architecture includes a high-impedance differential to single-ended converter (instrumentation amplifier), a low-pass filter, a programmable gain amplifier (PGA) for input ranging, a high-linearity ADC, integration hardware, and a sample-and-difference stage. It also includes an input multiplexer stage to select one of several input signals for measurement.

The instrumentation amplifier provides a high-impedance differential input. The high impedance avoids potential DC offset errors caused by bias current leaking into the meter. For single-ended measurements, the low end of the meter may be connected to ground through relays in the input selection multiplexer. The multimeter can also be connected to any of the tester's general-purpose DC voltage sources to measure their output voltage. The meter can also measure current flowing from any of the DC sources. This capability is very useful for measuring power supply currents, impedances, leakage currents, and other common DC parametric values. A PGA placed before the meter's ADC allows proper ranging of the instrument to minimize the effects of the ADC's quantization error (see Section 5.2.4 and Section 5.5).

The meter may also include a low-pass filter in its input path. The low-pass filter removes high-frequency noise from the signal under test, improving the repeatability of DC measurements. This filter can be enabled or bypassed using software commands. It may also have a programmable cutoff frequency so that the test engineer can make tradeoffs between measurement repeatability and test time (see Section 5.6). In addition, some meters may include an integration stage, which acts as a form of hardware averaging circuit to improve measurement repeatability.

Finally, a sample-and-difference stage is included in the front end of many ATE multimeters. The sample-and-difference stage allows highly accurate measurements of small differences between two large DC voltages. During the first phase of the measurement, a hardware sample-and-hold circuit samples a voltage. This first reference voltage is then subtracted from a second voltage (near the first voltage) using an amplifier-based subtractor. The difference between the

Figure 2.2. General-purpose DC multimeter.

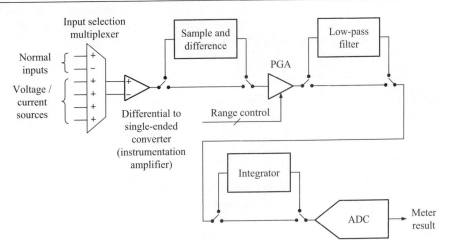

EXAMPLE 2.1

A single-ended DC voltmeter has a resolution of 12 bits. It also features a sample-and-difference front-end circuit. We wish to use this meter to measure the differential offset voltage of a DUT's output buffer. Each of the two outputs is specified to be within a range of 1.35 V ± 10 mV, and the differential offset is specified to be ±5 mV. The meter input can be set to any of the following ranges: ±10 V, ±1 V, ±100 mV, ±10 mV, and ±1 mV. Assuming that all components in the meter are perfectly linear (with the exception of the meter's quantization error), compare the accuracy achieved using two simple DC measurements with the accuracy achieved using the sample-and-difference circuit.

Solution:

The simplest way to measure offset using a single-ended DC voltmeter is to connect the meter to the OUTP output, measure its voltage, connect the meter to the OUTN output, measure its voltage, and subtract the second voltage from the first. Using this approach, we have to set the meter's input range to ±10 V to accommodate the 1.35 V DUT output signals. Thus each measurement may have a quantization error of as much as $\pm\frac{1}{2}$ (20 V/2^{12}–1) = ± 2.44 mV. Therefore the total error might be as high as ±4.88 mV, assuming that the quantization error from the first measurement is positive, while the quantization error from the second measurement is negative. Since the specification limit is ±5 mV, this will be an unacceptable test method.

Using the sample-and-difference circuitry, we could range the meter input to the worst-case difference between the two outputs, which is 5 mV, assuming a good device. The lowest meter range that will accommodate a 5-mV signal is ±10 mV. However, we also need to be able to collect readings from bad devices for purposes of characterization. Therefore, we will choose a range of ±100 mV, giving us a compromise between accuracy and characterization flexibility.

During the first phase of the sample-and-difference measurement, the voltage at the OUTN pin is sampled onto a holding capacitor internal to the meter. Then the meter is connected to the OUTP pin and the second phase of the measurement amplifies the difference between the OUTP voltage and the sampled OUTN voltage. Since the meter is set to a range of ±100 mV, a 100-mV difference between OUTP and OUTN will produce a full-scale 10-V input to the meter's ADC. This serves to reduce the effects of the meter's quantization error. The maximum error is given by $\pm\frac{1}{2}$ (100 mV/2^{12}–1) = ± 12.2 μV. Again, the worst-case error is twice this amount, or ±24.4 μV, which is well within the requirements of our measurement.

two voltages is then amplified and measured by the meter's ADC, resulting in a high-resolution measurement of the difference voltage. This process reduces the quantization error that would otherwise result from a direct measurement of the large voltages using the meter's higher voltage ranges.

2.2.2 General-Purpose Voltage/Current Sources

Most testers include general-purpose DC voltage/current sources, commonly referred to as *V/I sources* or *DC sources*. These programmable power supplies are used to provide the DC voltages and currents necessary to power up the DUT and stimulate its DC inputs. Many general-purpose

Figure 2.3. General-purpose DC source with Kelvin connections (conceptual diagram).

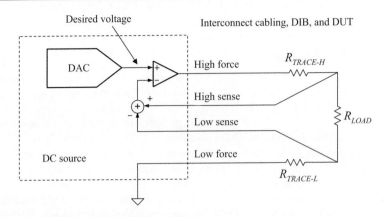

supplies can force either voltage or current, depending on the testing requirements. On most testers, these supplies can be switched to multiple points on the DIB board using the tester's DC matrix (see Subsection 2.2.5). As mentioned in the previous section, the system's general-purpose meter can be connected to any DC source to measure its output voltage or its output current.

Figure 2.3 shows a conceptual block diagram of a DC source having a differential Kelvin connection. A differential Kelvin connection consists of four lines (high force, low force, high sense, and low sense) for forcing highly accurate DC voltages. The Kelvin connection forms a feedback loop that allows the DC source to force an accurate differential voltage through the resistive wires between the source and DUT. Without the Kelvin connection, the small resistance in the force line interconnections ($R_{TRACE-H}$ and $R_{TRACE-L}$) would cause a small *IR* voltage drop. The voltage drop would be proportional to the current through the DUT load (R_{LOAD}). The small *IR* voltage drop would result in errors in the voltage across the DUT load. The sense lines of a Kelvin connection carry no current. Therefore, they are immune to errors caused by *IR* voltage drops.

A sense line is provided on the high side of the DC source and also on the low side of the source. The low-side sense line counteracts the parasitic resistance in the current return path. Since most instruments are referenced to ground, the low sense lines for all the DC instruments in a tester are often lumped into a single ground sense signal called DZ (device zero), DGS (device ground sense), or some other vendor-specific nomenclature. This is one of the most important signals in a mixed-signal tester, since it connects the DUT's ground voltage back to the tester's instruments for use as the entire test system's 0-V "golden zero reference." If any voltage errors are introduced into this ground reference signal relative to the DUT's ground, all the instruments will produce DC voltage offsets.

2.2.3 Precision Voltage References and User Supplies

Mixed-signal testers sometimes include high-accuracy, low-noise voltage references. These voltage sources can be used in place of the general-purpose DC sources when the noise and accuracy characteristics of the standard DC source are inadequate. One common example of a precision voltage reference application is the voltage reference for a high-resolution ADC or DAC. Any noise and DC error on the DC reference of an ADC or DAC translates directly into gain error and increased noise, respectively, in the output of the converter. A precision voltage reference is sometimes used to solve this problem.

Testers may also include nonprogrammable user power supplies with high output current capability. These fixed supplies provide common power supply voltages (±5 V, ±15 V, etc.) for

DIB circuits such as op amps and relay coils. This allows DIB circuits to operate from inexpensive fixed power supplies having high current capability instead of tying up the tester's more expensive programmable DC sources.

2.2.4 Calibration Source

The mixed-signal tester's calibration source is discussed in detail in Section 5.4. The purpose of a calibration source is to provide traceability of standards back to a central agency such as the National Institute of Standards and Technology (NIST). The calibration source must be recalibrated on a periodic basis (six months is a common period). Often, the source is removed from the tester and sent to a certified standards lab for recalibration. The old calibration source is replaced by a freshly calibrated one so that the tester can continue to be used in production. On some testers, the high-accuracy multimeter serves as the calibration source. Also, some testers may have multiple instruments that serve as the calibration sources for various parameters such as voltage or frequency. Clearly, this is a highly tester-specific topic. Calibration and standards traceability is discussed in more detail in Chapter 5, "Yield, Measurement Accuracy, and Test Time."

2.2.5 Relay Matrices

A relay matrix is a bank of electromechanical relays that provides flexible interconnections between many different tester instruments and the DUT. There may be several types of relay matrix in a tester, but they all perform a similar task. At different points in a test program, a particular DUT input may require a DC voltage, an AC waveform, or a connection to a voltmeter. A relay matrix allows each instrument to be connected to a DUT pin at the appropriate time as illustrated in Figure 2.4. General-purpose relay matrices, on the other hand, are used to connect and disconnect various circuit nodes on the DIB board. They have no hardwired connections to tester instruments. Therefore, the purpose and functionality of a general-purpose relay matrix depends on the test engineer's DIB design. It allows flexible interconnections between specific tester instruments and pins of the DUT through connections on the DIB board.

In addition to relay matrices, many other relays and signal paths are distributed throughout a mixed-signal tester to allow flexibility in interconnections without adding unnecessary relays to

Figure 2.4. Instrument relay matrix.

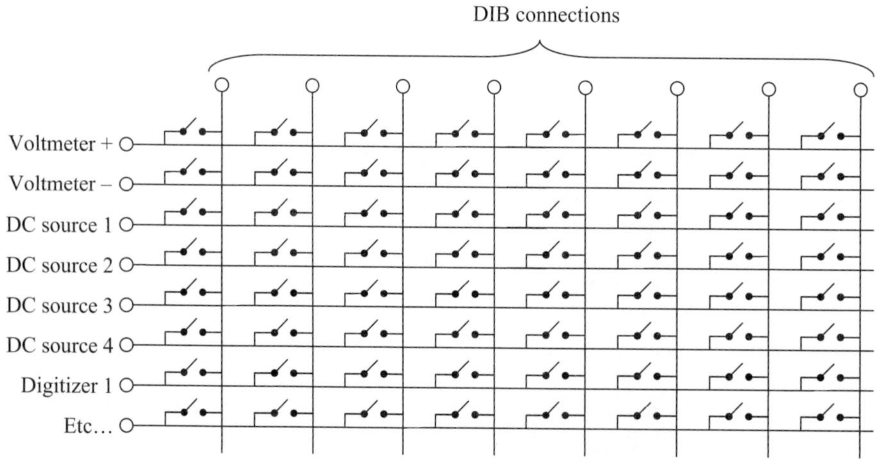

the DIB board. The exact architecture of relays, matrices, and signal paths varies widely from one ATE vendor's tester to the next.

2.2.6 Relay Control Lines

Despite the high degree of interconnection flexibility provided by the general-purpose relay matrix and other instrument interconnect hardware, there are always cases where a local DIB relay (placed near the DUT) is imperative. Usually the need for a local DIB relay is driven by performance of the DUT. For example, there is no better way to get a low-noise ground signal to the input of a DUT than to provide a local relay placed on the DIB directly between the DUT input and the DUT's local ground plane.

Certainly it is possible to feed the local ground through a DIB trace, through a remote relay matrix, and back through another DIB trace, but this connection scheme invariably leads to poor analog performance. The DIB traces are, after all, radio antennae. Many noise problems can be traced to poor layout of ground connections between the DUT and its ground plane. Local DIB relays minimize the radio antenna effect. Local DIB relays are also used to connect device outputs to various passive loads and other DIB circuits.

The test program controls the local DIB relays, opening and closing them at the appropriate time during each test. The relay coils are driven by the tester's relay control lines. A relay control line driver is shown in Figure 2.5. On some testers, the control line is capable of reading back the state of the voltage on the control line through a readback comparator. The readback comparator allows a low-cost method for determining the state of a digital signal.

Relay coils produce an inductive kickback when the current is suddenly changed between the on and off states. The inductive kickback, or flyback as it is known, is induced according to the inductance formula $v(t) = L\, di/dt$. Since high kickback voltages could potentially damage the output circuits of the relay driver, its output circuits contain flyback protection diodes to shunt the excess voltage to a DC source or to ground. Many test engineers also add flyback diodes across the coils of the relay, as shown in Figure 2.5. The extra diode is probably redundant. However, many engineers consider it good practice to add extra flyback diodes even though they ate up quite a bit of DIB board space. To eliminate the board space issue, the test engineer can choose slightly more expensive relays with built-in flyback diodes.

Figure 2.5. Relay coil driver with flyback protection diodes.

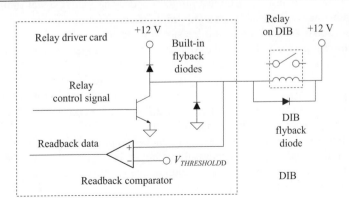

2.3 DIGITAL SUBSYSTEM

2.3.1 Digital Vectors

A mixed-signal tester must test digital circuits as well as mixed-signal and analog circuits. The mixed-signal and digital-only sections of the DUT are exercised using the tester's digital subsystem. The digital subsystem can present high, low, and high-impedance (HIZ) logic levels to the DUT. It can also compare the outputs from the DUT against expected responses to determine whether the digital logic of the DUT has been manufactured without defects. The tester applies a sequence of drive data to the device and simultaneously compares outputs against expected results. Each drive/compare cycle is called a *digital vector*. A series of digital vectors is called a *digital pattern*. Vectors of a digital pattern are usually sourced at a constant frequency, although some testers allow the period of each vector to be set independently. The ability to change digital timing on a vector-by-vector basis is commonly called *timing on the fly*.

2.3.2 Digital Signals

In addition to the simple pass/fail digital pattern tests, the tester must also be capable of sourcing and capturing digital signals. Digital signals are digitized representations of continuous waveforms such as sine waves and multitones. Digital signals are distinct from digital vectors in that they typically carry analog signal information rather than purely digital information. Usually, the samples of a digital signal must be applied to a DUT along with a repetitive digital pattern that keeps the device active and initiates DAC and/or ADC conversions. Each cycle of the repeating digital pattern is called a *frame*.

During a mixed-signal test, the repeating frame vectors must be combined with the non-repeating digital signal sample information to form a repetitive sampling loop. Combining the digital frame vectors with digital signal data, a long sequence of waveform samples can be sent to or captured from the DUT with a very short digital frame pattern. In effect, the sampling frame results in a type of data compression that minimizes the amount of vector memory needed for the tester's digital subsystem.

Looping frames are commonly used when testing DACs and ADCs. A sequence of samples must be loaded into a DAC to produce a continuous sequence of voltages at the DAC's output. In the case of ADC testing, digital signals must be captured and stored into a bank of memory as the looping frame initiates each ADC conversion.

2.3.3 Source Memory

When testing DACs, the digital signal samples representing the desired DAC analog waveform are typically computed in the tester's main test program code. The digital signal samples are stored into a digital subsystem memory block called *source memory* (or *send memory* in some testers). The digital frame data, on the other hand, are stored in vector memory. To generate a repeating frame with a new sample for each loop, the contents of the vector memory and source memory are spliced together in real time as the digital pattern is executed.

A digital signal can be modified quickly without changing the frame loop pattern because its data are generated algorithmically by the main test program. The ability to quickly modify the digital signal data is especially useful during the DUT debug and characterization phase. For example, a DAC may normally be tested using a 1-kHz sine wave digital signal. During the DAC characterization phase, however, the frequency might be swept from 100 Hz to 10 kHz to look for problem areas in the DAC's design. This would be impossibly cumbersome if the digital pattern had to be generated using an expanded, nonlooping sequence of ones and zeros. In fact, some tester architectures attempt to substitute deep, nonlooping vector memory in place of source memory.

This may reduce the cost of tester hardware, but it invariably results in frustrated users. One of the main differences between a mixed-signal tester and a digital tester with bolt-on analog instruments is the presence of source and capture memories in the digital subsystem. Other differences will be pointed out throughout this chapter.

2.3.4 Capture Memory

Devices such as ADCs produce a series of digitized waveform samples that must be captured and stored into a bank of memory called *capture memory* (or *receive memory*). Capture memory serves the opposite function of source memory. Each time the sampling frame is repeated, the digital output from the device is stored into the capture memory. The capture memory address pointer is incremented each time a digital sample is captured. Once a complete set of samples have been collected, they are transferred to an array processor or to the tester computer for analysis.

2.3.5 Pin Card Electronics

The pin card electronics for each digital channel are located inside the test head on most mixed-signal testers. A pin card electronics board may actually contain multiple channels of identical circuitry. Each channel's circuits consist of a programmable driver, a programmable comparator, various relays, dynamic current load circuits, and other circuits necessary to drive and receive signals to and from the DUT. A generic digital pin card is shown in Figure 2.6.

The driver circuitry consists of a fixed impedance driver (typically 50 Ω) with two programmable logic levels, V_{IH} and V_{IL}. These levels are controlled by a pair of driver-level DACs whose voltages are controlled by the test program. The driver can also switch into a high-impedance state (HIZ) at any point in the digital pattern to allow data to come from the DUT into the pin card's comparator. The driver circuits may also include programmable rise and fall times, though fixed rise and fall times are more common. Normally the fixed rise and fall times are designed to be as fast as the ATE vendor can make them. Rise and fall times between 1 ns and 3 ns are typical in today's testers.

The comparator also has two programmable logic levels, V_{OH} and V_{OL}. These are also controlled by another pair of DACs whose voltages are controlled by the test program. The pin card

Figure 2.6. Digital pin card circuits.

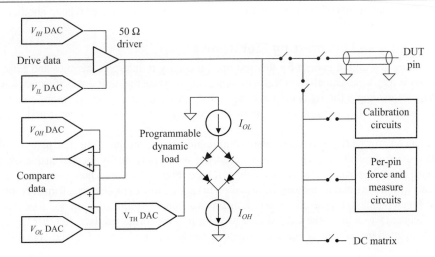

comparator is actually a pair of comparators, one for the V_{OH} level and one for V_{OL}. If the DUT signal is below V_{OL}, then the signal is considered a logic low. If the DUT is above V_{OH}, then it is considered a logic high. If the DUT output is between these thresholds, then the output state is considered a midpoint voltage. If it is outside these thresholds, then it is considered a valid logic level. Comparator results can also be ignored using a mask. Thus there are typically three drive states (HI, LO, and HIZ) and five compare states (HI, LO, and MID, VALID, and MASK).

The usefulness of the valid comparison is not immediately obvious. If we want to test for valid V_{OH} and V_{OL} voltages from the output of a nondeterministic circuit such as an ADC, we cannot set the tester to expect HI or LO. This is because electrical noise in the ADC and tester will produce somewhat unpredictable results at the ADC output. However, we can set the tester to expect valid logic levels during the appropriate digital vectors without specifying whether the ADC should produce a HI or a LO. While the pin card tests for valid logic levels, the samples from the ADC are collected into the digital capture memory for later analysis.

In addition to the drive and compare circuits, digital pin cards may also include dynamic load circuits. A dynamic load is a pair of current sources connected to the DUT output with a diode bridge circuit as shown in Figure 2.6. The diode bridge forces a programmable current into the DUT output whenever its voltage is below a programmable threshold voltage, V_{TH}. It forces current out of the DUT output whenever its voltage is above V_{TH}. The sink and source current settings correspond to the DUT's I_{OH} and I_{OL} specifications (see Section 3.12.4).

Another extremely important function that a digital pin card provides is its per-pin measurement capability. The per-pin measurement circuits of a pin card form a low-resolution, low-current DC voltage/current source for each digital pin. The per-pin circuits also include a relatively low-resolution voltage/current meter. The low-resolution and low-current capabilities are usually adequate for performing certain DC tests like continuity and leakage testing. These DC source and measure circuits can also be used for other types of simple DC tasks like input or output impedance testing.

Some testers may also include overshoot suppression circuits that serve to dampen the overshoot and undershoot characteristics in rapidly rising or falling digital signals. The overshoot and undershoot characteristics are the result of a low-impedance DUT output driving into the DIB traces and coaxial cables leading to the digital pin card electronics. The ringing is minimized as the signal overshoot is shunted to a DC level through a diode.

Digital pin cards also include relays connected to other tester resources such as calibration standards and system DC meters and sources. These connections can be used for a variety of purposes, including calibration of the pin card electronics during the tester's system calibration process. The exact details of these connections vary widely from one tester type to another.

2.3.6 Timing and Formatting Electronics

When looking at a digital pattern for the first time, it is easy to interpret the ones and zeros very literally, as if they represent all the information needed to create the digital waveforms. However, most ATE testers apply timing and formatting to the ones and zeros to create more complicated digital waveforms while minimizing the number of ones and zeros that must be stored in pattern memory.

Timing and formatting is a type of data compression and decompression. The pattern data are formatted using the ATE tester's formatter hardware, which is typically located inside the tester mainframe or on the pin card electronics in the test head. Figure 2.7 shows how the pattern data are combined with timing and formatting information to create more complex waveforms. Notice that the unformatted data in Figure 2.7 require four times as much 1/0 information and four times the bit cell frequency to achieve the same digital waveform as the formatted data. Another key advantage to formatted waveforms is that the formatting hardware in a high-end mixed-signal tester is

Figure 2.7. Drive data compression using formats and timing.

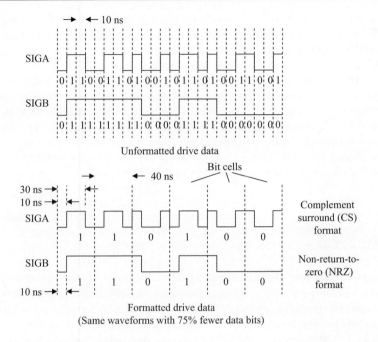

Figure 2.8. Some common digital formats.

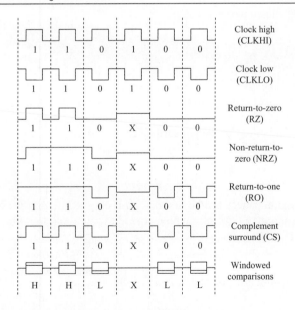

capable of placing the rising and falling edges with an accuracy of a few tens of picoseconds. This gives us better control of edge timing than we could expect to achieve using subgigahertz clocked digital logic.

The programmable drive start and stop times illustrated in Figure 2.7 are generated using digital delay circuitry inside the formatter circuits of the tester. Drive and compare timing is refined during a calibration process called *deskewing*. This allows subnanosecond accuracy in the placement of driven edges and in the placement of compare times (called *strobes* and *windows*). Strobe comparisons are performed at a particular point in time, while window comparisons are performed throughout a period of time. Window timing is typically used when comparing DUT outputs against expected patterns, while strobe timing is typically used when collecting data into capture memory. Again, this depends on the specific tester.

Figure 2.8 shows examples of several different formatting and timing combinations that create many different waveforms from the same digital data stream. In each case, the drive data sequence is 110X00. The compare data sequence is HHLXLL. Notice that certain formats such as Clock High and Clock Low ignore the pattern data altogether. Since digital pin cards can both drive and expect data, a distinction is made between a driven signal (1 or 0) and an expected signal (H or L). This notation is used for clarity in this book, though it is not universally used in the test industry. In fact, some digital pattern standards define H/L as driven data and 1/0 as expected data.

EXAMPLE 2.2

Two digital signals, SIGA and SIGB, are generated by an ATE tester's pattern generator. The pattern generator's vector rate (i.e., its bit cell rate) is set to 4 MHz. SIGA is programmed to RO format, while SIGB is programmed to NRZ format. The start time for SIGA is programmed to 50 ns and the stop time is programmed to 125 ns. Its initial state is programmed to logic high. The start time for SIGB is programmed to 25 ns and the stop time is programmed to 175 ns. Its initial state is programmed to logic low.

The following digital pattern is executed:

SIGA	SIGB
0	1
0	0
1	1
0	1
1	0
1	1

Draw a timing diagram for the two signals SIGA and SIGB produced by this pattern. Show the bit cells in the timing diagram and calculate their period. Assume that we want to produce this same pair of signals using a bank of static random access memory (SRAM) whose address is incremented at a fixed rate (i.e., nonformatted ones and zeros). What SRAM depth would be required to produce this same pair of signals?

Solution:
Figure 2.9 shows the digital waveforms resulting from the specified pattern and timing set. The vector rate is specified to be 4 MHz, so the bit cell period is 250 ns. Also notice that the NRZ format does not have a stop time, so the 175-ns stop time setting is irrelevant. In this example, all

timing edges fall on 25-ns boundaries. If we wanted to generate this same pattern using nonformatted data from a bank of SRAM clocked at a fixed frequency, we would have to source a sequence of 6 × (250 ns / 25 ns) = 60 bits from SRAM memory at a digital vector rate of 1/(25 ns) = 40 MHz.

Figure 2.9. Formatted data using return-to-one and non-return-to-zero formats.

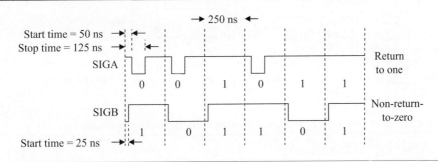

2.4 AC SOURCE AND MEASUREMENT

2.4.1 AC Continuous-Wave Source and AC Meter

The simplest way to apply and measure single-tone AC waveforms is to use a continuous-wave source (CWS) and an RMS voltmeter. The CWS is simply set to the desired frequency and voltage amplitude to stimulate the DUT. The RMS voltmeter is equally simple to use. It is connected to the DUT output, and the RMS output is measured with a single test program command.

But the CWS and RMS voltmeter suffer from a few problems. First, they are only able to measure a single frequency during each measurement. This would be acceptable for bench characterization, but in production testing it would lead to unacceptably long test times. As we will see in Chapters 8 through 11, DSP-based multitone testing is a far more efficient way to test AC performance because multiple frequencies can be tested simultaneously.

Another problem that the RMS voltmeter introduces is that it cannot distinguish the DUT's signal from distortion and noise. Using DSP-based testing, these various signal components can easily be separated from one another. This ability makes DSP-based testing more accurate and reliable than simple RMS-based testing. DSP-based testing is made possible with a more advanced stimulus/measurement pair, the arbitrary waveform generator and the waveform digitizer.

2.4.2 Arbitrary Waveform Generators

An arbitrary waveform generator (AWG) consists of a bank of waveform memory, a DAC that converts the waveform data into stepped analog voltages, and a programmable low-pass filter section, which smoothes the stepped signal into a continuous waveform. An AWG usually includes an output scaling circuit (PGA) to adjust the signal level. It may also include differential outputs and DC offset circuits. Figure 2.10 shows a typical AWG and waveforms that might be seen at each stage in its signal path. (Mathematical signal samples are represented as dots to distinguish them from reconstructed voltages.)

An AWG is capable of creating signals with frequency components below the low-pass filter's cutoff frequency. The frequency components must also be less than one-half the AWG's sampling rate. This so-called Nyquist criterion will be explained in Chapter 8, "Sampling Theory."

Figure 2.10. Arbitrary waveform generator.

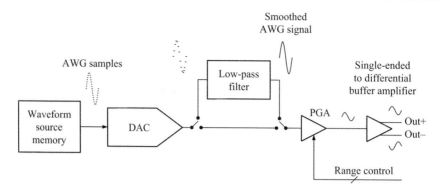

Figure 2.11. Time-domain and frequency-domain views of a three-tone multitone.

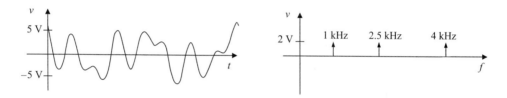

An AWG might create the three-tone multitone illustrated in Figure 2.11. It might also be used to source a sine wave for distortion testing or a triangle wave (up ramp/down ramp) for ADC linearity testing (see Chapter 7, "ADC Testing"). Flexibility in signal creation is the main advantage of AWGs compared to simple sine wave or function generators.

2.4.3 Waveform Digitizers

An AWG converts digital samples from a waveform memory into continuous-time waveforms. A digitizer performs the opposite operation, converting continuous-time analog waveforms into digitized representations. The digitized samples of the continuous waveform are collected into a waveform capture memory. The structure of a typical digitizer is shown in Figure 2.12. A digitizer usually includes a programmable low-pass filter to limit the bandwidth of the incoming signal. The purpose of the bandwidth limitation is to reduce noise and prevent signal aliasing, which we will discuss in Chapter 8, "Sampling Theory."

Like the DC meter, the digitizer has a programmable gain stage at its input to adjust the signal level entering the digitizer's ADC stage. This minimizes the noise effects of quantization error from the digitizer's ADC. Waveform digitizers may also include a differential to single-ended conversion stage for measuring differential outputs from the DUT. Digitizers may also include a sample-and-hold circuit at the front end of the ADC to allow undersampled measurements of very high-frequency signals. Undersampling is explained in more detail in Chapter 11, "Sampled Channel Testing."

2.4.4 Clocking and Synchronization

Many of the subsections and instruments in a mixed-signal tester derive their timing from a central frequency reference. This frequency determines the repetition rate of the sample loop and

Figure 2.12. Waveform digitizer.

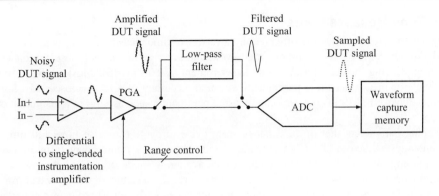

Figure 2.13. Synchronization in a mixed-signal tester.

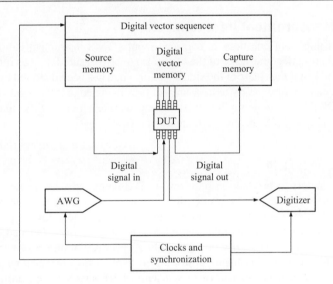

therefore sets the frequency of the DAC or ADC sampling rates. The AWG and digitizer also operate from clock sources that must be synchronized to each other and to the digital pattern's frame loop repetition rate.

Figure 2.13 shows a clock distribution scheme that allows synchronized sampling rates between all the DSP-based measurement instruments. Since the clocking frequency for each instrument is derived from a common source, frequency synchronization is possible. Without precise sampling rate synchronization, the accuracy and repeatability of all the DSP-based measurements in a mixed-signal test program would be degraded.

The reason these clocks must all be synchronized will become more apparent in Chapter 8, "Sampling Theory," and Chapter 9, "DSP-Based Testing." Proper synchronization of sample rates between the various AWGs, digitizers, and digital pattern generators is another of the key distinguishing features of a mixed-signal tester. A digital tester with bolt-on analog instruments often lacks a good clocking and synchronization architecture.

2.5 TIME MEASUREMENT SYSTEM

2.5.1 Time Measurements

Digital and mixed-signal devices often require a variety of time measurements, such as frequency, period, duty cycle, rise and fall times, jitter, skew, and propagation delay. These parameters can be measured using the ATE tester's time measurement system (TMS). Most TMS instruments are capable of measuring these parameters within an accuracy of a few nanoseconds. Some of the more advanced TMS instruments can measure parameters such as jitter to a resolution of less than 1 ps.

Timing parameters that do not change from cycle to cycle (i.e., rise time, fall time, etc.) can sometimes be measured using a very high-bandwidth undersampling waveform digitizer. An undersampling digitizer is similar in nature to the averaging mode of a digitizing oscilloscope. Like digitizing oscilloscopes, undersampling digitizers require a stable, repeating waveform. Thus nonperiodic features such as jitter and random glitches cannot be measured using an undersampling approach. Unfortunately, undersampling digitizers are often considerably slower than dedicated time measurement instruments.

2.5.2 Time Measurement Interconnects

One of the most important questions to consider about a TMS instrument is how its input and interconnection paths affect the shape of the waveform to be measured. It does little good to measure a rise time of 1 ns if the shape of the signal's rising edge has been distorted by a 50-Ω coaxial connection. It is equally futile to try to measure a 100-ps rising edge if the bandwidth of the TMS input is only 300 MHz. Accurate timing measurements require a high-quality signal path between the DUT output and the TMS time measurement circuits.

2.6 RF SUBSYSTEM

Testing RF parameters in production requires an ATE equipped with an RF subsystem. The architecture of the RF subsystem will not only have an impact on the measurements of performance and capabilities, but also has a significant impact on the test cost. A poor implementation might limit the test execution to a single site, whereas a more advanced one will enable octal site testing without necessarily increasing the ATE cost. For an easier understanding, we describe the source and measurement path of a typical RF subsystem below. The interface to the source and measurement path is connected with an RF switch to special RF ports on the test head.

2.6.1 Source Path

Figure 2.14 shows an implementation of the source path of a typical RF subsystem. In this implementation, four RF sources can be switched by a source switch matrix to power a combiner. This is followed by a source multiplexer connecting the RF ports on the test head. This system will be able to combine signals from four separate sources and present it to a single RF port to supply a device with a four-tone signal or supply a single-tone RF signal to four RF ports. The disadvantage of this architecture is that a number of expensive RF sources are required; in addition, the given architecture is limited to testing four devices at a time.

It is essential for all tests that the RF sources are synchronized in frequency. This is typically achieved by using a single source in the ATE as a reference as shown in Figure 2.14. This reference is used for all sources rather than operate from separate independent build-in reference oscillators.

Figure 2.14. Source path of an RF subsystem.

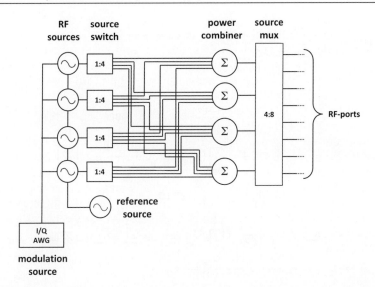

An important option is to have a modulation capability associated with the RF sources. This can be achieved as shown in Figure 2.14 with an AWG, or by a build-in AWG associated with the RF sources. Often the AWG of the mixed signal instrumentation can be routed to the modulation inputs of the RF sources.

RF ATE vendors reduce the cost of their systems by replacing multiple RF sources as shown in Figure 2.14 by a single source, or two sources when a two-tone capability is required. The power level of each RF port can be controlled individually through a gain stage in series with each RF port. For this architecture, the frequency on all RF ports will be the same, which in most cases is not a significant disadvantage. The main advantage of this architecture is that additional RF ports can be realized with only minor additional cost associated with each gain stage. Octal site RF ATE systems can be easily built without adding the high cost of additional RF sources.

2.6.2 Measurement Path

The measurement path of an RF subsystem can be built as simple as a zero-IF system (ZIF) with a mixer mixing the RF DUT signal to a base-band frequency. The mixed signal digitizer of the ATE can then capture this base-band frequency signal. A more advanced architecture is shown in Figure 2.15. The RF signal of the RF port is directed to a directional coupler. This directional coupler enables the measurement of the S-parameter of the DUT. In the architecture shown in Figure 2.15, two ADCs are connected to each port of the directional coupler so that all the S-parameters of a DUT can be obtained during a single measurement. The two IF stages, including the low-pass/bandpass filter cascades, maintains the signal integrity of the signal captured by the ADC by limiting the measuring bandwidth. As well, this architecture enables the signal level to be adjusted through the enabling of either a gain or attenuator stage so that the measured signal exercises the full dynamic range of the ADC.

The measuring bandwidth of the fixed-frequency second IF section can be further reduced. This significantly improves the sensitivity of the system and is especially important when making noise figure measurements. When wide-band signals are measured, the bandpass filter can be bypassed. The pre-amp in the through-pass of the directional coupler is one of the most critical components of the measurement path. To enable the measurement of test signals with large

Figure 2.15. Measurement path with S-parameter measurement capability of an RF subsystem.

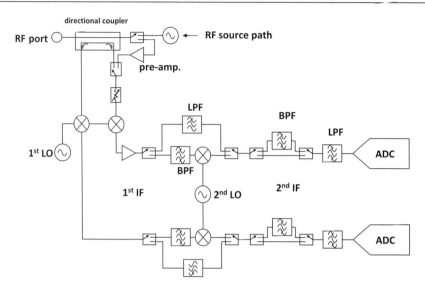

dynamic range, the pre-amp needs to have a low noise figure but at the same time must have a high compression point. This is critical when testing phase noise (see Chapter 13) with a strong carrier and a low phase noise.

2.7 COMPUTING HARDWARE

2.7.1 User Computer

Mixed-signal testers typically contain several computers and signal processors. The test engineer is most familiar with the user computer, since this is the one that is attached to the keyboard. The user computer is responsible for all the editing and compiling processes necessary to debug a test program. It is also responsible for keeping track of the datalogs and other data collection information. On low-cost testers, the user computer may also drive the measurement electronics as well. On more advanced mainframe testers, the execution of the test program, including I/O functions to the tester's measurement electronics, may be delegated to one or more tester computers located inside the tester's mainframe.

2.7.2 Tester Computer

The tester computer executes the compiled test program and interfaces to all the tester's instruments through a high-speed data backplane. By concentrating most of its processing power on the test program itself, the tester computer can execute a test program more efficiently than the user computer. The tester computer also performs all the mathematical operations on the data collected during each test. In some cases, the more advanced digital signal processing (DSP) operations may be handled by a dedicated array processor to further reduce test time. However, computer workstations have become fast enough in recent years that the DSP operations are often handled by the tester computer itself rather than by a dedicated array processor.

2.7.3 Array Processors and Distributed Digital Signal Processors

Many mixed-signal testers include one or more dedicated array processors for performing DSP operations quickly. This is another difference between a mixed-signal tester and a bolted-together digital/analog tester. Some mixed-signal instruments may even include local DSP processors for computing test results before they are transferred to the tester computer. This type of tester architecture and test methodology is called *distributed processing*. Distributed processing can reduce test time by splitting the DSP computation task among several processors throughout the tester. Test time is further reduced by eliminated much of the raw data transfer that would otherwise occur between digitizer instruments and a centralized tester computer or array processor. Unfortunately, distributed processing may have the disadvantage that the resulting test code may be harder to understand and debug.

2.7.4 Network Connectivity

The user computer and/or tester computer are typically connected into a network using Ethernet or similar networking hardware. This allows data and programs to be quickly transferred to the test engineer's desk for offline debugging and data analysis. It also allows for large amounts of production data to be stored and analyzed for characterization purposes.

2.8 SUMMARY

In this chapter we have examined many of the common building blocks of a generic mixed-signal tester. Of course, there are many differences between any two ATE vendors' preferred tester architectures. For example, ATE Vendor A may use a sigma-delta-based digitizer and AWG, while ATE Vendor B may choose to use a more conventional successive approximation architecture for its AWG and digitizer. Each architecture has advantages and disadvantages, which the test engineer must deal with. The test engineer's approach to measuring a given parameter will often be driven by the vendor's architectural choices. In the end, though, each tester has to test the same variety of mixed-signal parameters regardless of its architectural peculiarities. A test engineer's job often involves testing parameters the tester was simply not designed to measure. This can be one of the more challenging and interesting parts of a test engineer's task.

In later chapters we will see how digitizers, AWGs, and digital pattern generators, combined with digital signal processing, can provide greater speed and accuracy than conventional measurement techniques. We will also explain why it is so critical to mixed-signal testing that we achieve precise synchronization of sampling frequencies between all the tester's instruments. Most mixed-signal testing involves DSP-based measurements of one type or another; thus the student will need to devote special attention to these chapters.

PROBLEMS

2.1. Name at least six types of subsystems found in a typical mixed-signal tester.

2.2. What is the purpose of the low-pass filter in a DC multimeter's front end?

2.3. What is the purpose of the PGA in a DC multimeter's front end?

2.4. A single-ended DC voltmeter features a sample-and-difference front-end circuit. We wish to use this meter to measure the differential offset voltage of a DUT's output buffer. Each of the two outputs is specified to be within a range of 3.5 V ± 25 mV, and the differential offset is specified in the device data sheet to be ±15 mV. The meter input can be set to any of the following ranges: ±10 V, ±5 V, ±2 V, and ±1 V. The meter has a maximum error of 0.1% of its programmed range. The error includes all sources of inaccuracy (quantization error, linearity error, gain error, etc.). Compare the accuracy achieved using two simple DC measurements with the accuracy achieved using the sample-and-difference circuit. Assume no errors due to nonrepeatability.

2.5. Why are Kelvin connections used to connect high-current DC power supplies to the DUT?

2.6. Name an instance where a local DIB relay might prove to be a better choice for interconnecting signals than a general-purpose relay matrix.

2.7. What is the purpose of the diodes in the output stage of the relay driver in Figure 2.5?

2.8. What is the difference between a digital pattern and a digital signal?

2.9. What is the purpose of source memory?

2.10. What is the purpose of capture memory?

2.11. Why is formatting and timing information combined with one/zero information to produce digital waveforms?

2.12. A series of digital bits are driven from a digital pin card at a rate of 1 MHz (1-μs period). The series of bits are 10110X1. The format for this pin is set to return-to-zero (RZ) format. Its initial state is set to logic low. The start time for the drive data is set to 500 ns, and the stop time is set to 900 ns. Draw this waveform using the notation in Figure 2.8. Draw the waveform timing approximately to scale. Next, draw the waveform that would result if we set the format to non-return-to-zero (NRZ). To produce these waveforms using clocked digital logic without timing and formatting circuits, what clock rate would be required? If we wanted to be able to set the start and stop times to 500 ns and 901 ns, respectively, at what rate would we have to operate the clocked digital logic?

2.13. Name two reasons that AWGs and digitizers are used in mixed signal testing rather than CW sources and RMS voltmeters.

2.14. What is the purpose of the low-pass filter in the AWG illustrated in Figure 2.10?

2.15. Why is a programmable gain amplifier needed in the front end of the waveform digitizer illustrated in Figure 2.12?

2.16. What is the purpose of distributed digital signal processing hardware?

CHAPTER 3

DC and Parametric Measurements

This chapter introduces the reader to basic DC measurement definitions, including continuity, leakage, impedance, offset, gain, and DC power supply rejection ratio tests. In addition, search techniques used to establish specific DC test conditions are described. The chapter concludes with a brief discussion about the DC tests performed on digital circuits.

3.1 CONTINUITY

3.1.1 Purpose of Continuity Testing

Before a test program can evaluate the quality of a device under test (DUT), the DUT must be connected to the ATE tester using a test fixture such as a device interface board (DIB). A typical interconnection scheme is shown in Figure 3.1. When packaged devices are tested, a socket or handler contactor assembly provides the contact between the DUT and the DIB. When testing a bare die on a wafer, the contact is made through the probe needles of a probe card. The tester's instruments are connected to the DIB through one or more layers of connectors such as spring-loaded pogo pins or edge connectors. The exact connection scheme varies from tester to tester, depending on the mechanical/electrical performance tradeoffs made by the ATE vendor.

In addition to pogo pins and other connectors, electromechanical relays are often used to route signals from the tester electronics to the DUT. A relay is an electrical switch whose position is controlled by an electromagnetic field. The field is created by a current forced through a coil of wire inside the relay. Relays are used extensively in mixed-signal testing to modify the electrical connections to and from the DUT as the test program progresses from test to test.

Any of the electrical connections between a DUT and the tester can be defective, resulting in open circuits or shorts between electrical signals. For example, the wiper of a relay can become stuck in either the open or closed position after millions of open/close cycles. While interconnect problems may not pose a serious problem in a lab environment, defective connections can be a major source of tester down time on the production floor. Continuity tests (also known as *contact tests*) are performed on a device to verify that all the electrical connections are sound. If continuity

Figure 3.1. ATE test head to DUT interconnections.

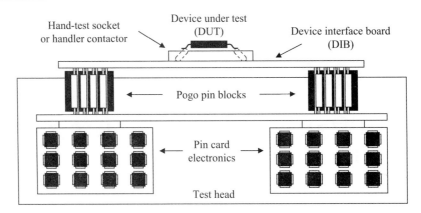

testing is not performed, then the production floor personnel cannot distinguish between bad lots of silicon and defective test hardware connections. Without continuity testing, thousands of good devices could be rejected simply because a pogo pin was bent or because a relay was defective.

3.1.2 Continuity Test Technique

Continuity testing is usually performed by detecting the presence of on-chip protection circuits. These circuits protect each input and output of the device from electrostatic discharge (ESD) and other excessive voltage conditions. The ESD protection circuits prevent the input and output pins from exceeding a small voltage above or below the power supply voltage or ground. Diodes and silicon-controlled rectifiers (SCRs) can be used to short the excess currents from the protected pin to ground or to a power terminal.

An ESD protection diode conducts the excess ESD current to ground or power any time the pin's voltage exceeds one diode drop above (or below) the power or ground voltage. SCRs are similar to ESD protection diodes, but they are triggered by a separate detection circuit. Any of a variety of detection circuits can be used to trigger the SCR when the protected pin's voltage exceeds a safe voltage range. Once triggered, an SCR behaves like a forward-biased diode from the protected pin to power or ground (Figure 3.2). The SCR remains in its triggered state until the excessive voltage is removed. Since an SCR behaves much like a diode when triggered, the term "protection diode" is used to describe ESD protection circuits whether they employ a simple diode or a more elaborate SCR structure. We will use the term "protection diode" throughout the remainder of this book with the understanding that a more complex circuit may actually be employed.

DUT pins may be configured with either one or two protection diodes, connected as shown in Figure 3.3. Notice that the diodes are reverse-biased when the device is powered up, assuming normal input and output voltage levels. This effectively makes them "invisible" to the DUT circuits during normal operation.

To verify that each pin can be connected to the tester without electrical shorts or open circuits, the ATE tester forces a small current across each protection diode in the forward-biased direction. The DUT's power supply pins are set to zero volts to disable all on-chip circuits and to connect the far end of each diode to ground. ESD protection diodes connected to the positive supply are tested by forcing a current I_{CONT} into the pin as shown in Figure 3.4 and measuring the voltage, V_{CONT}, that appears at the pin with respect to ground. If the tester does not see the expected diode drops on each pin, then the continuity test fails and the device is not tested further. Protection

Figure 3.2. SCR-based ESD protection circuit.

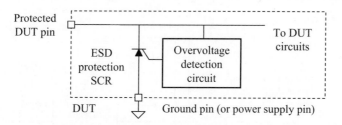

Figure 3.3. Dual and single protection diodes.

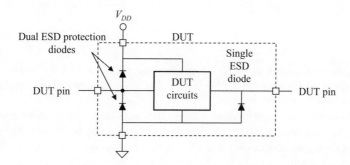

diodes connected to the negative supply or ground are tested by reversing the direction of the forced current.

In the case of an SCR-based protection circuit, the current source initially sees an open circuit. Because the current source output tries to force current into an open circuit, its output voltage rises rapidly. The rising voltage soon triggers the SCR's detection circuit. Once triggered, the SCR accepts current from the current source and the voltage returns to one diode drop above ground. Thus the difference between a diode-based ESD protection circuit and an SCR-based circuit is hardly noticeable during a continuity test.

The amount of current chosen is typically between 100 µA and 1 mA, but the ideal value depends on the characteristics of the protection diodes. Too much current may damage the diodes, while too little current may not fully bias them. The voltage drop across a good protection diode usually measures between 550 and 750 mV. For the purpose of illustration, we shall assume that a conducting diode has voltage drop of 0.7 V. A dead short to ground will result in a reading of 0 V, while an open circuit will cause the tester's current source to reach a programmed clamp voltage.

Many mixed-signal devices have multiple power supply and ground pins. Continuity to these power and ground connections may or may not be testable. If all supply pins or all ground pins are not properly connected to ground, then continuity to some or all of the nonsupply pins will fail. However, if only some of the supply or ground pins are not grounded, the others will provide a continuity path to zero volts. Therefore, the unconnected power supply or ground pins may not be detected. One way to test the power and ground pins individually is to connect them to ground one at a time, using relays to break the connections to the other power and ground pins. Continuity to the power or ground pin can then be verified by looking for the protection diode between it and another DUT pin.

Occasionally, a device pin may not include any protection diodes at all. Continuity to these unprotected pins must be verified by an alternative method, perhaps by detecting a small amount

Figure 3.4. Checking the continuity of the diode connected to the positive supply. The other diode is tested by reversing the direction of the forced current.

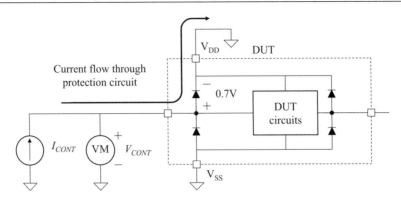

of current leaking into the pin or by detecting the presence of an on-chip component such as a capacitor or resistor. Since unprotected pins are highly vulnerable to ESD damage, they are used only in special cases.

One such example is a high-frequency input requiring very low parasitic capacitance. The space-charge layer in a reverse-biased protection diode might add several picofarads of parasitic capacitance to a device pin. Since even a small amount of stray capacitance presents a low impedance to very high-frequency signals, the protection diode must sometimes be omitted to enhance electrical performance of the DUT.

3.1.3 Serial Versus Parallel Continuity Testing

Continuity can be tested one pin at a time, an approach known as *serial continuity testing*. Unfortunately, serial testing is a time-consuming and costly approach. Modern ATE testers are capable of measuring continuity on all or most pins in parallel rather than measuring the protection diode drops one at a time. These testers accomplish parallel testing using so-called *per-pin measurement instruments* as shown in Figure 3.5a.

Clearly it is more economical to test all pins at once using many current sources and voltage meters. Unfortunately, there are a few potential problems to consider. First, a fully parallel test of pins may not detect pin-to-pin shorts. If two device pins are shorted together for some reason, the net current through each diode does not change. Twice as much current is forced through the parallel combination of two diodes. The shorted circuit configuration will therefore result in the expected voltage drop across each diode, resulting in both pins passing the continuity test. Obviously, the problem can be solved by performing a continuity test on each pin in a serial manner at the cost of extra test time. However, a more economical approach is to test every other pin for continuity on one test pass while grounding the remaining pins. Then the remaining pins can be tested during a second pass while the previously tested pins are grounded. Shorts between adjacent pins would be detected using this dual-pass approach, as illustrated in Figure 3.5b.

A second, subtler problem with parallel continuity testing is related to analog measurement performance. Both analog pins and digital pins must be tested for continuity. On some testers the per-pin continuity test circuitry is limited to digital pins only. The analog pins of the tester may not include per-pin continuity measurement capability. On these testers, continuity testing on analog pins can be performed one pin at a time using a single current source and voltmeter. These two instruments can be connected to each device pin one at a time to measure protection diode drops. Of course, this is a very time-consuming serial test method, which should be avoided if possible.

Figure 3.5. Parallel continuity testing: (a) Full parallel testing with possible adjacent fault masking; (b) Minimizing potential adjacent fault masking by exciting every second pin.

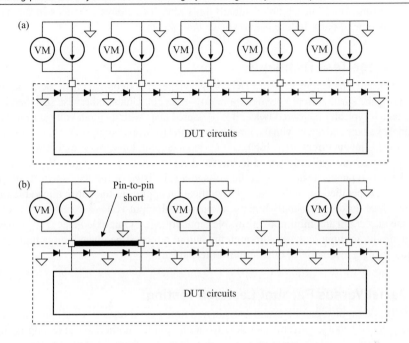

Alternatively, the analog pins can be connected to the per-pin measurement electronics of digital pins. This allows completely parallel testing of continuity. Unfortunately, the digital per-pin electronics may inject noise into sensitive analog signals. Also, the signal trace connecting the DUT to the per-pin continuity electronics adds a complex capacitive and inductive load to the analog pin, which may be unacceptable. The signal trace can also behave as a parasitic radio antenna into which unwanted signals can couple into analog inputs. Clearly, full parallel testing of analog pins should be treated with care. One solution to the noise and parasitic loading problems is to isolate each analog pin from its per-pin continuity circuit using a relay. This complicates the DIB design but gives high performance with minimal test time. Of course, a tester having per-pin continuity measurement circuits on both analog and digital pins represents a superior solution.

3.2 LEAKAGE CURRENTS

3.2.1 Purpose of Leakage Testing

Each input pin and output pin of a DUT exhibits a phenomenon called *leakage*. When a voltage is applied to a high-impedance analog or digital input pin, a small amount of current will typically leak into or out of the pin. This current is called *leakage current*, or simply *leakage*. Leakage can also be measured on output pins that are placed into a nondriving high-impedance mode. A good design and manufacturing process should result in very low leakage currents. Typically the leakage is less than 1 μA, although this can vary from one device design to the next.

One of the main reasons to measure leakage is to detect improperly processed integrated circuits. Leakage can be caused by many physical defects such as metal filaments and particulate matter that forms shorts and leakage paths between layers in the IC. Another reason to measure leakage is that excessive leakage currents can cause improper operation of the customer's end application. Leakage currents can cause DC offsets and other parametric shifts. A third reason to test leakage is that excessive leakage currents can indicate a poorly processed device that initially appears to be functional but which eventually fails after a few days or weeks in the customer's product.[1] This type of early failure is known as *infant mortality*.

3.2.2 Leakage Test Technique

Leakage is measured by simply forcing a DC voltage on the input or output pin of the device under test and measuring the small current flowing into or out of the pin. Unless otherwise specified in the data sheet, leakage is typically measured twice. It is measured once with an input voltage near the positive power supply voltage and again with the input near ground (or negative supply). These two currents are referred to as I_{IH} (input current, logic high) and I_{IL} (input current, logic low), respectively.

Digital inputs are typically tested at the valid input threshold voltages, V_{IH} and V_{IL}. Analog input leakage is typically tested at specific voltage levels listed in the data sheet. If no particular input voltage is specified, then the leakage specification applies to the entire allowable input voltage range. Since leakage is usually highest at one or both input voltage extremes, it is often measured at the maximum and minimum allowable input voltages. Output leakage (I_{OZ}) is measured in a manner similar to input leakage, although the output pin must be placed into a high-impedance (HIZ) state using a test mode or other control mechanism.

3.2.3 Serial Versus Parallel Leakage Testing

Leakage, like continuity, can be tested one pin at a time (serial testing) or all pins at once (parallel testing). Since leakage currents can flow from one pin to another, serial testing is superior to parallel testing from a defect detection perspective. However, from a test time perspective, parallel testing is desired. As in continuity testing, a compromise can be achieved by testing every other pin in a dual-pass approach.

Continuity tests are usually implemented by forcing DC current and measuring voltage. By contrast, leakage tests are implemented by forcing DC voltage and measuring current. Since the tests are similar in nature, tester vendors generally design both capabilities into the per-pin measurement circuits of the ATE tester's pin cards. Analog leakage, like analog continuity, is often measured using the per-pin resources of digital pin cards. Again, a tester with per-pin continuity measurement circuits on both analog and digital pins represents a superior solution, assuming that the extra per-pin circuits are not prohibitively expensive.

3.3 POWER SUPPLY CURRENTS

3.3.1 Importance of Supply Current Tests

One of the fastest ways to detect a device with catastrophic defects is to measure the amount of current it draws from each of its power supplies. Many gross defects such as those illustrated in Figures 1.4–1.7 result in a low-impedance path from one of the power supplies to ground. Supply currents are often tested near the beginning of a test program to screen out completely defective devices quickly and cost effectively.

Of course, the main reason to measure power supply current is to guarantee limited power consumption in the customer's end application. Supply current is an important electrical parameter for the customer who needs to design a system that consumes as little power as possible. Low power consumption is especially important to manufacturers of battery operated equipment like cellular telephones. Even devices that draw large amounts of current by design should draw only as much power as necessary. Therefore, power supply current tests are performed on most if not all devices.

3.3.2 Test Techniques

Most ATE testers are able to measure the current flowing from each voltage source connected to the DUT. Supply currents are therefore very easy to measure in most cases. The power supply is simply set to the desired voltage and the current from its output is measured using one of the tester's ammeters.

When measuring supply currents, the only difficulties arise out of ambiguities in the data sheet. For example, are the analog outputs loaded or unloaded during the supply current test? Is digital block XYZ operating in mode A, mode B, or idle mode? In general, it is safe to assume that the supply currents are to be tested under worst-case conditions.

The test engineer should work with the design engineers to attempt to specify the test conditions that are likely to result in worst-case test conditions. These test conditions should be spelled out clearly in the test plan so that everyone understands the exact conditions used during production testing. Often the actual worst-case conditions are not known until the device has been thoroughly characterized. In these cases, the test program and test plan have to be updated to reflect the characterized worst-case conditions.

Supply currents are often specified under several test conditions, such as power-down mode, standby mode, and normal operational mode. In addition, the digital supply currents are specified separately from the analog supply currents. I_{DD} (CMOS) and I_{CC} (bipolar) are commonly used designations for supply current. I_{DDA}, I_{DDD}, I_{CCA}, and I_{CCD} are the terms used when analog and digital supplies are measured separately.

Many devices have multiple power supply pins that are connected to a common power supply in normal operation. Design engineers often need to know how much current is flowing into each individual power supply pin. Sometimes the test engineer can accommodate this requirement by connecting each power supply pin to its own supply. Other times there are too many DUT supply pins to provide each with its own separate power supply. In these cases, relays can be used to temporarily connect a dedicated power supply to the pin under test.

Another problem that can plague power supply current tests is settling time. The supply current flowing into a DUT must settle to a stable value before it can be measured. The tester and DIB circuits must also settle to a stable value. This normally takes 5–10 ms in normal modes of DUT operation. However, in power-down modes the specified supply current is often less than 100 μA.

Figure 3.6. Arranging different-sized bypass capacitors to minimize power supply current settling behavior.

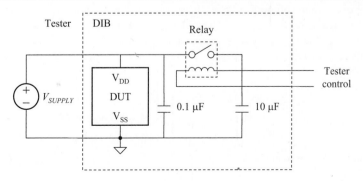

Since the DIB usually includes bypass capacitors for the DUT, each capacitor must be allowed to charge until the average current into or out of the capacitor is stable.

The charging process can take hundreds of milliseconds if the current must stabilize within microamps. Some types of bypass capacitors may even exhibit leakage current greater than the current to be measured. A typical solution to this problem is to connect only a small bypass capacitor (say 0.1 µF) directly to the DUT and then connect a larger capacitor (say 10 µF) through a relay as shown in Figure 3.6. The large bypass capacitor can be disconnected temporarily while the power-down current is measured.

3.4 DC REFERENCES AND REGULATORS

3.4.1 Voltage Regulators

A voltage regulator is one of the most basic analog circuits. The function of a voltage regulator is to provide a well-specified and constant output voltage level from a poorly specified and sometimes fluctuating input voltage. The output of the voltage regulator would then be used as the supply voltage for other circuits in the system. Figure 3.7 illustrates the conversion of a 6- to 12-V ranging power supply to a fixed 5-V output level.

Voltage regulators can be tested using a fairly small number of DC tests. Some of the important parameters for a regulator are output no-load voltage, output voltage or load regulation, input or line regulation, input or ripple rejection, and dropout voltage.

Output no-load voltage is measured by simply connecting a voltmeter to the regulator output with no load current and measuring the output voltage V_O.

Load regulation measures the ability of the regulator to maintain the specified output voltage V_O under different load current conditions I_L. As the output voltage changes with increasing load current, one defines the output voltage regulation as the percentage change in the output voltage (relative to the ideal output voltage, $V_{O,NOM}$) for a specified change in the load current. Load regulation is measured under minimum input voltage conditions

$$\text{load regulation} \equiv 100\% \times \left. \frac{\Delta V_O}{V_{O,\,NOM}} \right|_{\max\,\{\Delta I_L\},\,\text{minimum }V_I} \tag{3.1}$$

The largest load current change, max (ΔI_L), is created by varying the load current from the minimum rated load current (typically 0 mA) to the maximum rated load current.

Load regulation is sometimes specified as the absolute change in voltage, ΔV_O, rather than as a percentage change in V_O. The test definition will be obvious from the specification units (i.e. volts or percentage).

Line regulation or *input regulation* measures the ability of the regulator to maintain a steady output voltage over a range of input voltages. Line regulation is specified as the percentage change in the output voltage as the input line voltage changes over its largest allowable range. Like the load regulation test, line regulation is sometimes specified as an absolute voltage change rather than a percentage. Line regulation is measured under maximum load conditions:

$$\text{line regulation} \equiv 100\% \times \left. \frac{\Delta V_O}{V_{O,\,NOM}} \right|_{\max\{\Delta V_I\},\,\text{maximum }I_L} \tag{3.2}$$

For the regulator shown in Figure 3.7, with the appropriate load connected to the regulator output, the line regulation would be computed by first setting the input voltage to 6 V, measuring

Figure 3.7. 5-V DC voltage regulator.

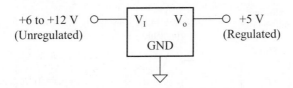

the output voltage, then readjusting the input voltage to 12 V, and again measuring the output voltage to calculate ΔV_O. The line regulation would then be computed using Eq. (3.2).

Input rejection or *ripple rejection* is the ratio of the maximum input voltage variation to the output voltage swing, measured at a particular frequency (commonly 120 Hz) or a range of frequencies. It is a measure of the circuit's ability to reject periodic fluctuations of rectified AC voltage signals applied to the input of the regulator. Input rejection can also be measured at DC using the input voltage range and output voltage swing measured during the line regulation test.

Dropout voltage is the lowest voltage that can be applied between the input and output pins without causing the output to drop below its specified minimum output voltage level. Dropout voltage is tested under maximum current loading conditions. It is possible to search for the exact dropout voltage by adjusting the input voltage until the output reaches its minimum acceptable voltage, but this is a time-consuming test method. In production testing, the input can simply be set to the specified dropout voltage plus the minimum acceptable output voltage. The output is then measured to guarantee that it is equal to or above the minimum acceptable output voltage.

3.4.2 Voltage References

Voltage regulators are commonly used to supply a steady voltage while also supplying a relatively large amount of current. However, many of the DC voltages used in a mixed-signal device do not draw a large amount of current. For example, a 1-V DAC reference does not need to supply 500 mA of current. For this reason, low-power voltage references are often incorporated into mixed-signal devices rather than high-power voltage regulators.

The output of on-chip voltage references may or may not be accessible from the external pins of a DUT. It is common for the test engineer to request a set of test modes so that reference voltages can be measured during production testing. This allows the test program to evaluate the quality of the DC references even if they have no explicit specifications in the data sheet. The design and test engineers can then determine whether failures in the more complicated AC tests may be due to a simple DC voltage error in the reference circuits. DC reference test modes also allow the test program to trim the internal DC references for more precise device operation.

3.4.3 Trimmable References

Many high-performance mixed-signal devices require reference voltages that are trimmed to very exact levels by the ATE tester. DC voltage trimming can be accomplished in a variety of ways. The most common way is to use a programmable reference circuit that can be permanently adjusted to the desired level. One such arrangement is shown in Figure 3.8. The desired level is programmed using fuses, or a nonvolatile digital control mechanism such as EEPROM or flash memory bits. Fuses are blown by forcing a controlled current across each fuse that causes it to vaporize. Fuses can be constructed from either metal or polysilicon. If EEPROM or flash memory is added to a mixed-signal device, then this technology may offer a superior alternative to blown fuses, as EEPROM bits can be rewritten if necessary.

Figure 3.8. Trimmable reference circuit.

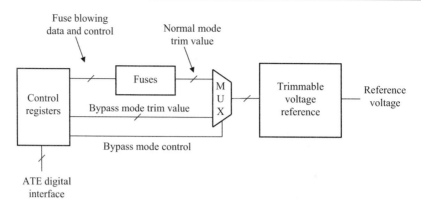

There are various algorithms for finding the digital value that minimizes reference voltage error. In the more advanced trimming architectures such as the one in Figure 3.8, the reference can be experimentally adjusted using a bypass trim value rather than permanently blowing the fuses. In this example, the bypass trim value is enabled using a special test mode control signal, bypass mode control. Once the best trim value has been determined by experimental trials, the fuses are permanently blown to set the desired trim value. Then, during normal operation, the bypass trim value is disabled and the programmed fuses are used to control the voltage reference.

Trimming can also be accomplished using a laser trimming technique. In this technique, a laser is used to cut through a portion of an on-chip resistor to increase its resistance to the desired value. The resistance value, in turn, adjusts the DC level of the voltage reference. The laser trimming technique can also be used to trim gains and offsets of analog circuits. Laser trimming is more complex than trimming with fuses or nonvolatile memory. It requires special production equipment linked to the ATE tester.

Laser trimming must be performed while the silicon wafer is still exposed to open air during the probing process. Since metal fuses can produce a conductive sputter when they vaporize, they too are usually trimmed during the wafer probing process. By contrast, polysilicon fuses and EEPROM bits can be blown either before or after the device is packaged.

There is an important advantage to trimming DC levels after the device has been packaged. When plastic is injected around the silicon die, it can place slight mechanical forces on the die. This, in turn, introduces DC offsets. Because of these DC shifts, a device that was correctly trimmed during the wafer probing process may not remain correctly trimmed after it has been encapsulated in plastic. Another potential DC shift problem relates to the photoelectric effect. Since light shining on a bare die introduces photoelectric DC offsets, a bare die must be trimmed in total darkness. Of course, wafer probers are designed with this requirement in mind. They include a black hood or other mechanism to shield bare die from light sources.

3.5 IMPEDANCE MEASUREMENTS

3.5.1 Input Impedance

Input impedance (Z_{IN}), also referred to as *input resistance*, is a common specification for analog inputs. In general, impedance refers to the behavior of both resistive and reactive (capacitive or inductive) components in the circuit. As the discussion in this chapter is restricted to DC, we are assuming that inductors and capacitors do not participate in circuit operation (act as either a short circuit in the case of an inductor, or an open circuit in the case of a capacitor). Hence, impedance and resistance refer to the same quantity at DC.

Exercises

3.1. The output of a 5-V voltage regulator varies from 5.10 V under no-load condition to 4.85 V under a 5 mA maximum rated load current. What is its load regulation?

ANS. 5%.

3.2. The output of a 5-V voltage regulator varies from 5.05 to 4.95 V when the input voltage is changed from 14 to 6 V under a maximum load condition of 10 mA. What is its line regulation?

ANS. 2%.

3.3. A 9-V voltage regulator is rated to have a load regulation of 3% for a maximum load current of 15 mA. Assuming a no-load output voltage of 9 V, what is the worst-case output voltage at the maximum load current?

ANS. 8.73 V.

Input impedance is a fairly simple measurement to make. If the input voltage is a linear function of the input current (i.e., if it behaves according to Ohm's law), then one simply forces a voltage V and measures a current I, or vice versa, and computes the input impedance according to

$$Z_{IN} = \frac{V}{I} \tag{3.3}$$

Figure 3.9a illustrates the input i–v relationship of a device satisfying Ohm's law. Here we see that the i–v characteristic is a straight line passing through the origin with a slope equal to $1/Z_{IN}$. In many instances, the i–v characteristic of an input pin is a straight line but does not pass through the origin as shown in Figure 3.9b. Such situations typically arise from biasing considerations where the input terminal of a device is biased by a constant current source such as that shown in Figure 3.10 or has in series with it an unknown voltage source to ground, or in series between two components comprising the input series impedance.

In cases such as these, one cannot use Eq. (3.3) to compute the input impedance, as it will not lead correctly to the slope of the i–v characteristic. Instead, one measures the change in the input

Figure 3.9. Input i–v characteristic curves resulting in an impedance function with (a) equal DC and AC operation and (b) unequal DC and AC operation.

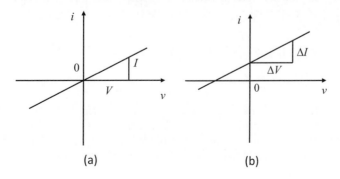

(a)   (b)

current (ΔI) that results from a change in the input voltage (ΔV) and computes the input impedance using

$$Z_{IN} = \frac{\Delta V}{\Delta I}$$

(3.4)

If the input impedance is so low that it would cause excessive currents to flow into the pin, another approach is needed. The alternative method is to force two controlled currents and measure the resulting voltage difference. This is often referred to as a *force-current/measure-voltage* method. Input impedance is again calculated using Eq. (3.4).

EXAMPLE 3.1

In the input impedance test setup shown in Figure 3.10, voltage source SRC1 is set to 2 V and current flowing into the pin is measured to be 0.055 mA. Then SRC1 is set to 1 V and the input current is measured again to be 0.021 mA. What is the input impedance?

Solution:
Input impedance, Z_{IN}, which is a combination of R_{IN} and the input impedance of the block labeled "DUT Circuit," is calculated using Eq. (3.4) as follows

$$Z_{IN} = \frac{2\,V - 1\,V}{0.055\ mA - 0.021\ mA} = 29.41\ k\Omega$$

Note that the impedance could also have been measured by forcing 0.050 and 0.020 mA and measuring the voltage difference. However, the unpredictable value of I_{BIAS} could cause the input voltage to swing beyond the DUT's supply rails. For this reason, the forced-current measurement technique is reserved for low values of resistance.

Figure 3.10. Input impedance test setup.

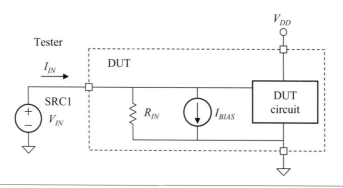

In Example 3.1, the values of the excitation consisting of 2 and 1 V are somewhat irrelevant. We could just as easily have used 2.25 and 1.75 V. However, the larger the difference in voltage, the easier it is to make an accurate measurement of current change. This is true throughout many types of tests. Large changes in voltages and currents are easier to measure than small ones. The test engineer should beware of saturating the input of the device with excessive voltages, though. Saturation could lead to extra input current resulting in an inaccurate impedance measurement. The device data sheet should list the acceptable range of input voltages.

3.5.2 Output Impedance

Output impedance (Z_{OUT}) is measured in the same way as input impedance. It is typically much lower than input impedance; so it is usually measured using a force-current/measure-voltage technique. However, in cases where the output impedance is very high, it may be measured using the force-voltage/measure-current method instead.

EXAMPLE 3.2

In the output impedance test setup shown in Figure 3.11, current source SRC1 is set to 10 mA and the voltage at the pin is measured, yielding 1.61 V. Then SRC1 is set to –10 mA and the output voltage is measured at 1.42 V. What is the total output impedance (R_{OUT} plus the amplifier's output impedance)?

Solution:
Using Eq. (3.4) with Z_{IN} replaced by Z_{OUT}, we write

$$Z_{OUT} = \frac{1.61\ V - 1.42\ V}{10\ mA - (-10\ mA)} = 9.5\ \Omega$$

Figure 3.11. Output impedance test setup.

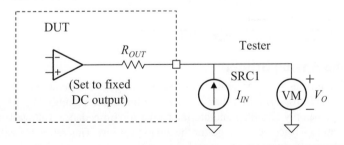

3.5.3 Differential Impedance Measurements

Differential impedance is measured by forcing two differential voltages and measuring the differential current change or by forcing two differential currents and measuring the differential voltage change. Example 3.3 illustrates this approach using forcing currents and measuring the resulting differential voltage. Differential input impedance would be measured in a similar manner.

EXAMPLE 3.3

In the differential output impedance test setup shown in Figure 3.12 current source SRC1 is set to 20 mA, SRC2 is set to –20 mA and the differential voltage at the pins is measured at 201 mV. Then SRC1 is set to –20 mA, SRC2 is set to 20 mA, and the output voltage is measured at –199 mV. What is the differential output impedance?

Solution:
The output impedance is found using Eq. (3.4) to be

$$Z_{OUT} = \frac{201 \text{ mV} - (-199 \text{ mV})}{20 \text{ mA} - (-20 \text{ mA})} = 10 \ \Omega$$

Figure 3.12. Differential output impedance test setup.

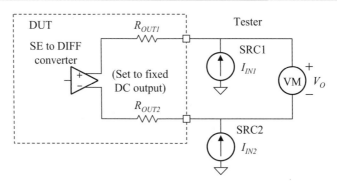

3.6 DC OFFSET MEASUREMENTS

3.6.1 V_{MID} and Analog Ground

Many analog and mixed-signal integrated circuits are designed to operate on a single power supply voltage (V_{DD} and ground) rather than a more familiar bipolar supply (V_{DD}, V_{SS}, and ground). Often these single-supply circuits generate their own low-impedance voltage between V_{DD} and ground that serves as a reference voltage for the analog circuits. This reference voltage, which we will refer to as V_{MID}, may be placed halfway between V_{DD} and ground or it may be placed at some other fixed voltage such as 1.35 V. In some cases, V_{MID} may be generated off-chip and supplied as an input voltage to the DUT.

To simplify the task of circuit analysis, we can define any circuit node to be 0 V and measure all other voltages relative to this node. Therefore, in a single-supply circuit having a V_{DD} of 3 V, a V_{SS} connected to ground, and an internally generated V_{MID} of 1.5 V, we can redefine all voltages relative to the V_{MID} node. Using this definition of 0 V, we can translate our single-supply circuit into a more familiar bipolar configuration with $V_{DD} = +1.5$ V, $V_{MID} = 0$ V, and $V_{SS} = -1.5$ V (Figure 3.13). In order for this approach to be valid, it is assumed that no hidden impedance lies between V_{SS} and ground—a reasonable assumption at low to moderate frequencies, less so at very high frequencies.

Figure 3.13. Redefining V_{MID} as 0 V to simplify circuit analysis.

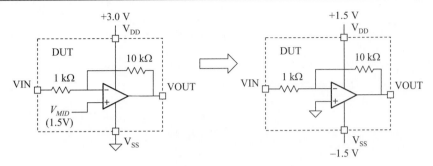

Several integrated circuit design textbooks refer to this type of V_{MID} reference voltage as analog ground, since it serves as the ground reference in single-supply analog circuits. This is an unfortunate choice of terminology from a test engineering standpoint. Analog ground is a term used in the test and measurement industry to refer to a high-quality ground that is separated from the noisy ground connected to the DUT's digital circuits. In fact, the term "ground" has a definite meaning when working with measurement equipment since it is actually tied to earth ground for safety reasons. In this textbook, we will use the term analog ground to refer to a quiet 0 V voltage for use by analog circuits and use the term V_{MID} to refer to an analog reference voltage (typically generated on-chip) that serves as the IC's analog "ground."

3.6.2 DC Transfer Characteristics (Gain and Offset)

The input–output DC transfer characteristic for an ideal amplifier is shown in Figure 3.14. The input–output variables of interest are voltage, but they could just as easily be replaced by current signals. As the real world is rarely accommodating to IC and system design engineers, the actual transfer characteristic for the amplifier would deviate somewhat from the ideal or expected curve. To illustrate the point, we superimpose another curve on the plot in Figure 3.14 and label it "Typical."

In order to maintain correct system operation, design engineers require some assurance that the amplifier transfer characteristic is within acceptable tolerance limits. Of particular interest to the test engineer are the gain and offset voltages shown in the figure. In this section we shall describe the method to measure offset voltages (which is equally applicable to current signals as well), and the next section will describe several methods used to obtain amplifier gain.

3.6.3 Output Offset Voltage ($V_{0,OS}$)

The output offset ($V_{O,OS}$) of a circuit is simply the difference between its ideal DC output and its actual DC output when the input is set to some fixed reference value, normally analog ground or V_{MID}. Output offset is depicted in Figure 3.14 for an input reference value of 0 V. As long as the output is not noisy and there are no AC signal components riding on the DC level, output offset is a trivial test. If the signal is excessively noisy, the noise component must be removed from the DC level in one of two ways. First, the DC signal can be filtered using a low-pass filter. The output of the filter is measured using a DC voltmeter. ATE testers usually have a low-pass filter built into their DC meter for such applications. The low-pass filter can be bypassed during less demanding measurements in order to minimize the overall settling time. The second method of reducing the effects of noise is to collect multiple readings from the DC meter and then mathematically average the results. This is equivalent to a software low-pass filter.

Figure 3.14. Amplifier input–output transfer characteristics in its linear region.

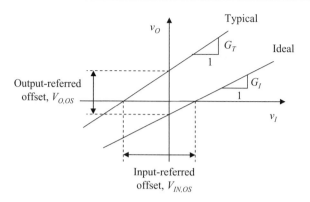

Sometimes sensitive DUT outputs can be affected by the ATE tester's parasitic loading. Some op amps will become unstable and break into oscillations if their outputs are loaded with the stray capacitance of the tester's meter and its connections to the DUT. An ATE meter may add as much as 200 pF of loading on the output of the DUT, depending on the connection scheme chosen by the test engineer. The design engineer and test engineer should evaluate the possible effects of the tester's stray capacitance on each DUT output. It may be necessary to add a buffer amplifier to the DIB to provide isolation between the DUT output and the tester's instruments.

The input impedance of the tester can also shift DC levels when very high-impedance circuit nodes are tested. Consider the circuit in Figure 3.15 where the DUT is assumed to have an output impedance R_{OUT} of 100 kΩ. The DC meter in this example has an input impedance R_{IN} of 1 MΩ. According to the voltage divider principle with two resistors in series, the voltage that appears across the meter V_{MEAS} with respect to the output $V_{O,OS}$ of the DUT is

$$V_{MEAS} = \frac{R_{IN}}{R_{IN} + R_{OUT}} V_{O,OS} = \frac{1 \text{ M}\Omega}{1 \text{ M}\Omega + 100 \text{ k}\Omega} V_{O,OS}$$
$$= 0.909 V_{O,OS}$$

It is readily apparent that a relative error of

$$\text{relative error} = \frac{V_{O,OS} - V_{MEAS}}{V_{O,OS}} = \frac{(1 - 0.909)}{1} = 0.091$$

or 9.1% is introduced into this measurement. A unity gain buffer amplifier may be necessary to provide better isolation between the DUT and tester instrument.

3.6.4 Single-Ended, Differential, and Common-Mode Offsets

Single-ended output offsets are measured relative to some ideal or expected voltage level when the input is set to some specified reference level. Usually these two quantities are the same and are specified on the data sheet. Differential offset is the difference between two outputs of a differential circuit when the input is set to a stated reference level. For simplicity's sake, we shall use $V_{O,OS}$ to denote the output offset for both the single-ended and differential case. It should be clear from the context which offset is being referred to. The output common-mode voltage $V_{O,CM}$ is defined as

Figure 3.15. Meter impedance loading.

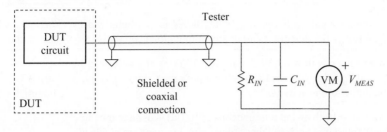

the average voltage level at the two outputs of a differential circuit. Common-mode offset $V_{O,CM,OS}$ is the difference between the output common-mode voltage and the ideal value under specified input conditions.

EXAMPLE 3.4

Consider the single-ended to differential converter shown in Figure 3.16. The two outputs of the circuit are labeled OUTP and OUTN. A 1.5-V reference voltage V_{MID} is applied to the input of the circuit and, ideally, the outputs should both produce V_{MID}. The voltages at OUTP and OUTN, denoted V_P and V_N, respectively, are measured with a meter, producing the following two readings:

$$V_P = 1.507 \text{ V} \quad \text{and} \quad V_N = 1.497 \text{ V}$$

Figure 3.16. Differential output offset test setup.

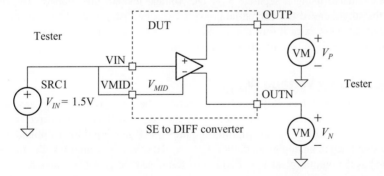

With an expected output reference level of $V_{MID} = 1.50$ V, compute the output differential and common-mode offsets.

Solution:
OUTP single-ended offset voltage, $V_{O,P,OS} = V_P - V_{MID} = +7$ mV
OUTN single-ended offset voltage, $V_{O,N,OS} = V_N - V_{MID} = -3$ mV
   differential offset, $V_{O,D,OS} = V_P - V_N = +10$ mV
Output common-mode voltage, $V_{O,CM} = (V_P + V_N) / 2 = 1.502$ V
   common-mode offset, $V_{O,CM,OS} = V_{O,CM} - V_{MID} = 2$ mV

Exercises	

3.4. An amplifier with a nominal gain of 10 V/V is characterized by $V_{OUT} = 10V_{IN} + 5$. What are its input and output offset voltages?

ANS. +0.5 V (input), 5 V (output).

3.5. An amplifier with a nominal gain of 10 V/V is characterized by $V_{OUT} = 10V_{IN} - V_{IN}^2 + 5$ over a −5 to +5-V input range. What is its input and output offset voltages?

ANS. +0.477 V (input), 5 V (output).

3.6. A voltmeter with an input impedance of 100 kΩ is used to measure the DC output of an amplifier with an output impedance of 500 kΩ. What is the expected relative error made by this measurement?

ANS. 83.3%.

3.7. A differential amplifier has an output OUTP of 3.3 V and an output OUTN of 2.8 V with its input set to a V_{MID} reference level of 3 V. What are the single-ended and differential offsets? The common-mode offset?

ANS. 0.3 V and −0.2V (SE), 0.5 V (DIFF), 50 mV (CM).

3.8. A perfectly linear amplifier has a measured gain of 5.1 V/V and an output offset of −3.2 V. What is the input offset voltage?

ANS. −0.627 V.

In the preceding example, V_{MID} is provided to the device from a highly accurate external voltage source. But what happens when the V_{MID} reference is generated from an on-chip reference circuit which itself has a DC offset? Typically there is a separate specification for the V_{MID} voltage in such cases; the input of the DUT should be connected to the V_{MID} voltage, if it is possible to do so, and the output offsets are then specified relative to the V_{MID} voltage rather than the ideal value.

Thus the inputs and outputs are treated as if V_{MID} was exactly correct. Any errors in the V_{MID} voltage are evaluated using a separate V_{MID} DC voltage test. In this manner, DC offset errors caused by the single-ended to differential converter can be distinguished from errors in the V_{MID} reference voltage. This extra information may prove to be very useful to design engineers who must decide what needs to be corrected in the design.

3.6.5 Input Offset Voltage ($V_{IN,OS}$)

Input or input-referred offset voltage ($V_{IN,OS}$) refers to the negative of the voltage that must be applied to the input of a circuit in order to restore the output voltage to a desired reference level, that is, analog ground or V_{MID}. If an amplifier requires a +10-mV input to be applied to its input to force the output level to analog ground, then $V_{IN,OS} = -10$ mV. It is common in the literature to find $V_{IN,OS}$ defined as the output offset $V_{O,OS}$ divided by the measured gain G of the circuit:

$$V_{IN,OS} = \frac{V_{O,OS}}{G}$$

(3.5)

If an amplifier has a gain of 10 V/V and its output has an output offset of 100 mV, then its input offset voltage is 10 mV. This will always be true, provided that the values used in Eq. (3.5) are derived from the circuit in its linear region of operation. In high-gain circuits, such as an open-loop op amp, it is not uncommon to find the amplifer in a saturated state when measuring the output offset voltage. As such, Eq. (3.5) is not applicable and one must resort to a different technique, as will be explained shortly.

3.7 DC GAIN MEASUREMENTS

3.7.1 Closed-Loop Gain

Closed-loop DC gain is one of the simplest measurements to make, because the input–output signals are roughly comparable in level. Closed-loop gain, denoted G, is defined as the slope of the amplifier input-output transfer characteristic, as illustrated in Figure 3.14. We refer to this gain as closed-loop because it typically contrived from a set of electronic devices configured in a negative feedback loop. It is computed by simply dividing the change in output level of the amplifier or circuit by the change in its input

$$G = \frac{\Delta V_O}{\Delta V_I} \tag{3.6}$$

DC gain is measured using two DC input levels that fall inside the linear region of the amplifier. This latter point is particularly important, because false gain values are often obtained when the amplifier is unknowingly driven into saturation by poorly chosen input levels. The range of linear operation should be included in the test plan.

Gain can also be expressed in decibels (dB). The conversion from volt-per-volt to decibels is simply

$$G\big|_{dB} = 20\log_{10}|G| \tag{3.7}$$

The logarithm function in Eq. (3.7) is a base-10 log as opposed to a natural log.

EXAMPLE 3.5

An amplifier with an expected gain of −10 V/V is shown in Figure 3.17. Both the input and output levels are referenced to an internally generated voltage V_{MID} of 1.5 V. SRC1 is set to 1.4 V and an output voltage of 2.51 V is measured with a voltmeter. Then SRC1 is set to 1.6 V and an output voltage of 0.47 V is measured. What is the DC gain of this amplifier in V/V? What is the gain in decibels?

Figure 3.17. A ×10 amplifier gain test setup.

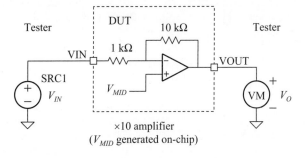

Solution:

The gain of the amplifier is computed using Eq. (3.5) as

$$G = \frac{2.51 \text{ V} - 0.47 \text{ V}}{1.4 \text{ V} - 1.6 \text{ V}} = -10.2 \text{ V/V}$$

or, in terms of decibels

$$G = 20\log_{10}|-10.2| = 20.172 \text{ dB}$$

Gain may also be specified for circuits with differential inputs and/or outputs. The measurement is basically the same.

EXAMPLE 3.6

A fully differential amplifier with an expected gain of +10 V/V is shown in Figure 3.18. SRC1 is set to 1.6 V and SRC2 is set to 1.4 V. This results in a differential input of 200 mV. An output voltage of 2.53 V is measured at OUTP, and an output voltage of 0.48 V is measured at OUTN. This results in a differential output of 2.05 V. Then SRC1 is set to 1.4 V and SRC2 is set to 1.6 V. This results in a differential input level of −200 mV. An output voltage of 0.49 V is measured at OUTP, and an output voltage of 2.52 V is measured at OUTN. The differential output voltage is thus −2.03 V. Using the measured data provided, compute the differential gain of this circuit.

Solution:

The differential gain is found using Eq. (3.5) to be

$$G = \frac{2.05 \text{ V} - 2.03 \text{ V}}{200 \text{ mV} - (-200 \text{ mV})} = +10.2 \text{ V/V}$$

Figure 3.18. Differential ×10 amplifier gain test setup.

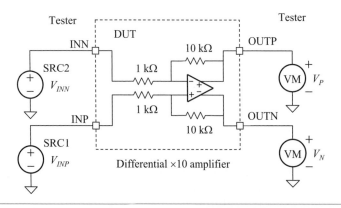

Differential measurements can be made by measuring each of the two output voltages individually and then computing the difference mathematically. Alternatively, a differential voltmeter can be used to directly measure differential voltages. Obviously the differential voltmeter approach will work faster than making two separate measurements. Therefore, the use of a differential voltmeter is the preferred technique in production test programs. Sometimes the differential voltage is very small compared to the DC offset of the two DUT outputs. A differential voltmeter can often give more accurate readings in these cases.

In cases requiring extreme accuracy, it may be necessary to measure the input voltages as well as the output voltages. The DC voltage sources in most ATE testers are well calibrated and stable enough to provide a voltage error no greater than 1 mV in most cases. If this level of error is unacceptable, then it may be necessary to use the tester's high-accuracy voltmeter to measure the exact input voltage levels rather than trusting the sources to produce the desired values. The gain equation in the previous example would then be

$$G = \frac{2.05\ V - 2.03\ V}{V_1 - V_2}$$

where V_1 and V_2 are the actual input voltages measured using a differential voltmeter.

Exercises

3.9. Voltages of 0.8 and 4.1 V appear at the output of a single-ended amplifier when an input of 1.4 and 1.6 V is applied, respectively. What is the gain of the amplifier in V/V? What is the gain in decibels?

ANS. +16.5 V/V, 24.35 dB.

3.10. An amplifier is characterized by $V_{OUT} = 2.5\ V_{IN} + 1$ over an output voltage range of 0 to 10 V. What is the amplifier output for a 2-V input? Similarly for a 3-V input? What is the corresponding gain of this amplifier in V/V for a 1-V input change? What is the gain in decibels?

ANS. 6 V, 8.5 V, +2.5 V/V, 7.96 dB.

3.11. An amplifier is characterized by $V_{OUT} = 2.5\ V_{IN} + 0.25\ V_{IN}^2 + 1$ over an output voltage range of 0 to 12 V. What is the amplifier output for a 2-V input? Similarly for a 3-V input? What is the corresponding gain of this amplifier in V/V for a 1-V input change? What is the gain in decibels? Would a 4-V input represent a valid test point?

ANS. 7 V, 10.75 V, +3.75 V/V, 11.48 dB, No—the output would exceed 12 V.

The astute reader may have noticed that the gain and impedance measurements are fairly similar, in that they both involve calculating a slope from a DC transfer characteristic pertaining to the DUT. Moreover, they do not depend on any value for the offsets, only that the appropriate slope is obtained from the linear region of the transfer characteristic.

3.7.2 Open-Loop Gain

Open-loop gain (abbreviated G_{ol}) is a basic parameter of op amps. It is defined as the gain of the amplifier with no feedback path from output to input. Since many op amps have G_{ol} values of 10,000 V/V or more, it is difficult to measure open-loop gain with the straightforward techniques of the previous examples. It is difficult to apply a voltage directly to the input of an open-loop op amp without causing it to saturate, forcing the output to one power supply rail or the other. For example, if the maximum output level from an op amp is ±5 V and its open-loop gain is equal to 10,000 V/V, then an input-referred offset of only 500 μV will cause the amplifier output to saturate. Since many op amps have input-referred offsets ranging over several millivolts, we cannot predict what input voltage range will result in unsaturated output levels.

We can overcome this problem using a second op amp connected in a feedback path as shown in Figure 3.19. The second amplifier is known as a *nulling amplifier*. The nulling amplifier forces its differential input voltage to zero through a negative feedback loop formed by resistor string R_2 and R_1, together with the DUT op amp. This loop is also known as a servo loop.[2] By doing so, the output of the op amp under test can be forced to a desired output level according to

$$V_{O,DUT} = 2V_{MID} - V_{SRC1} \tag{3.8}$$

where V_{MID} is a DC reference point (grounded in the case of dual-supply op amps, non-grounded for single-supply op amps) and V_{SRC1} is the programmed DC voltage from SRC1. The nulling amplifier and its feedback loop compensate for the input-referred offset of the DUT amplifier. This ensures that the DUT output does not saturate due to its own input-referred offset.

The two matched resistors, R_3, are normally chosen to be around 100 kΩ as a compromise between source loading and op-amp bias-induced offsets. Since the gain around the loop is extremely large, feedback capacitor C is necessary to stabilize the loop. A capacitance value of 1 to 10 nF is usually sufficient. R_{LOAD} provides the specified load resistance for the G_{ol} test.

Under steady-state conditions, the signal that is fed back to the input of the DUT amplifier denoted $V_{IN,DUT}$ is directly related to the nulling amplifier output V_{O-NULL} according to

$$V_{IN,DUT} = V_{IN,DUT}^{+} - V_{IN,DUT}^{-} = \frac{R_1}{R_1 + R_2}\left(V_{O,NULL} - V_{MID}\right) \tag{3.9}$$

where $V_{IN,DUT}^{+}$ and $V_{IN,DUT}^{-}$ are the positive and negative inputs to the DUT amplifier, respectively. Subsequently, the open-loop voltage gain of the DUT amplifier is found from Eqs. (3.6), (3.8), and (3.9) to be given by

$$G_{ol} = \frac{\Delta V_{O,DUT}}{\Delta V_{IN,DUT}} = -\left(\frac{R_1 + R_2}{R_1}\right)\frac{\Delta V_{SRC1}}{\Delta V_{O,NULL}} \tag{3.10}$$

Figure 3.19. Open-loop gain test setup using a nulling amplifier.

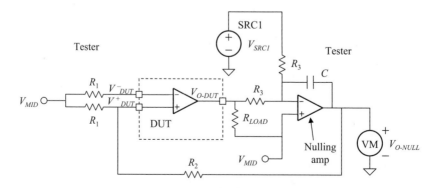

The nulling loop method allows the test engineer to force two desired outputs and then indirectly measure the tiny inputs that caused those two outputs. In this manner, very large gains can be measured without measuring tiny voltages. Of course the accuracy of this approach depends on accurately knowing the values of R_1 and R_2, as well as on by selecting the two resistors labeled as R_3 in Figure 3.19 very nearly identical values (i.e., matched conditions).

In order to maximize the signal handling capability of the test setup shown in Figure 3.19, as well as avoid saturating the nulling amplifer, it is a good idea to set the voltage divider ratio to a value approximately equal to the inverse of the expected open-loop gain of the DUT op amp

$$\frac{R_1}{R_1 + R_2} \approx \frac{1}{G_{ol}} \tag{3.11}$$

from which we can write $R_2 \approx G_{ol} R_1$.

EXAMPLE 3.7

For the nulling amplifier setup shown in Figure 3.19 with $R_1 = 100\ \Omega$, $R_2 = 100\ k\Omega$, and $R_3 = 100\ k\Omega$, together with V_{MID} set to a value midway between the two power supply levels (its actual value is not important because all signals will be referenced to it), SRC1 is set to $V_{MID} + 1$ V and a voltage of $V_{MID} + 2.005$ V is measured at the nulling amplifier output. SRC1 is set to $V_{MID} - 1$ V and a voltage of $V_{MID} + 4.020$ V is measured at the nulling amplifier output. What is the open-loop gain of the amplifier?

Solution:
Open-loop gain is calculated using the following procedure. First the change or swing in the nulling amplifier output $\Delta V_{O,NULL}$ is computed

$$\Delta V_{O,NULL} = 2.005\ V - 4.020\ V = -2.015\ V$$

Then, using Eq. (3.9), the voltage swing at the input of the DUT amplifier, $\Delta V_{IN,DUT}$, is calculated

$$\Delta V_{IN,DUT} = \frac{R_1}{R_1 + R_2} \Delta V_{O,NULL}$$

$$= \frac{100}{100 + 100k}(-2.015\ V)$$

$$= -2.013\ mV$$

Making use of the fact that ΔV_{SRC1} is 2 V, which forces $\Delta V_{O,DUT} = -2$ V, the open-loop gain of the amplifier is found to be

$$G_{ol} = \frac{\Delta V_{O,DUT}}{\Delta V_{IN,DUT}} = \frac{-2\ V}{-2.013\ mV} = 993.5\ V/V$$

If the op amp in the preceding example had an open-loop gain closer to 100 V/V instead of 1000 V/V, then the output of the nulling amplifier would have produced a voltage swing of 20 V instead of 2 V. The nulling amplifier would have been dangerously close to clipping against its output voltage rails (assuming ±15-V power supplies). In fact, if a 5-V op amp were used as the nulling amplifier, it would obviously not be able to produce the 20-V swing.

In the example, the nulling amplifier should have produced two voltages centered around V_{MID}. Instead, it had an average or common-mode offset level of approximately 3 V from this value. A detailed circuit analysis reveals that this offset is caused exclusively by the input-referred offset of the DUT. Hence, the offset that appears at the output of the nulling amplifier, denoted $V_{O,NULL,OS}$, can be used to compute the input-referred offset of the DUT, $V_{IN,DUT,OS}$. Input-referred offset would then be calculated using

$$V_{IN,DUT,OS} = \frac{R_1}{R_1 + R_2} V_{O,NULL,OS} \tag{3.12}$$

Exercises

3.12. For the nulling amplifier setup shown in Figure 3.19 with R_1 = 100 Ω, R_2 = 100 kΩ, and R_3 = 100 kΩ, an SRC1 voltage swing of 1 V results in a 2.3-V swing at the output of the nulling amplifier. What is the open-loop gain in V/V of the DUT amplifier? What is the gain in decibels?   ANS. 435.2 V/V, 52.77 dB.

3.13. For the nulling amplifier setup shown in Figure 3.19 with R_1 = 1 kΩ, R_2 = 100 kΩ, and R_3 = 100 kΩ, an offset of 2.175 V + V_{MID} appears at the output of the nulling op amp when the SRC1 voltage is set to V_{MID}. What is the input offset of the DUT amplifier?   ANS. 21.5 mV.

3.14. For the nulling amplifier setup shown in Figure 3.19 with R_1 = 100 Ω, R_2 = 500 kΩ, and R_3=100 kΩ, and the DUT op amp having an open-loop gain of 4000 V/V, what is the output swing of the nulling amplifier when the SRC1 voltage swings by 1 V?   ANS. 1.25 V.

Because this method involves the same measured data used to compute the open-loop gain, it is a commonly used method to determine the op amp input-referred offset. For the parameters and measurement values described in Example 3.7, the input-referred offset voltage for the DUT is

$$V_{IN,DUT,OS} = \frac{100\ \Omega}{100\ \Omega + 100\ \text{k}\Omega} \left(\frac{4.020\ \text{V} + 2.005\ \text{V}}{2} \right)$$
$$= 3.0\ \text{mV}$$

3.8 DC POWER SUPPLY REJECTION RATIO

3.8.1 DC Power Supply Sensitivity

Power supply sensitivity (PSS) is a measure of the circuit's dependence on a constant supply voltage. Normally it is specified separately with respect to the positive or negative power supply voltages and denoted PSS+ and PSS−. PSS is defined as the change in the output over the change in either power supply voltage with the input held constant

$$PSS^+ \equiv \frac{\Delta V_O}{\Delta V_{PS^+}}\bigg|_{V_{in}\ \text{constant}} \quad \text{and} \quad PSS^- \equiv \frac{\Delta V_O}{\Delta V_{PS^-}}\bigg|_{V_{in}\ \text{constant}} \tag{3.13}$$

In effect, PSS is a type of gain test in which the input is one of the power supply levels.

EXAMPLE 3.8

The input of the ×10 amplifier in Figure 3.20 is connected to its own V_{MID} source forcing 1.5 V. The power supply is set to 3.1 V and a voltage of 1.5011 V is measured at the output of the amplifier. The power supply voltage is then changed to 2.9 V and the output measurement changes to 1.4993 V. What is the PSS of the amplifier in V/V? What is the PSS in decibels?

Solution:
As the positive power supply (V_{DD}) is being changed by SRC1, the positive power supply sensitivity is

$$PSS^+ = \frac{\Delta V_O}{\Delta V_{SRC1}} = \frac{1.5011\ \text{V} - 1.4993\ \text{V}}{3.1\ \text{V} - 2.9\ \text{V}} = 9\ \text{mV/V} = -40.92\ \text{dB}$$

Figure 3.20. Power supply sensitivity test setup.

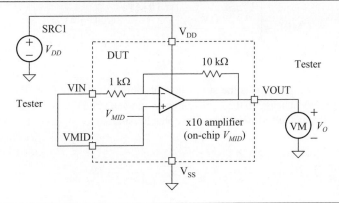

3.8.2 DC Power Supply Rejection Ratio

Power supply rejection ratio (PSRR) is defined as the power supply sensitivity of a circuit divided by the magnitude of the closed-loop gain of the circuit in its normal mode of operation. Normally it is specified separately with respect to each power supply voltage. Mathematically, we write

$$PSRR^+ \equiv \frac{PSS^+}{|G|} \quad \text{and} \quad PSRR^- \equiv \frac{PSS^-}{|G|} \tag{3.14}$$

In Example 3.8, we found PSS$^+$ = 0.009 V/V. In Example 3.5, the DC gain of this same circuit was found to be –10.2 V/V. Hence the PSRR$^+$ would be

$$PSRR^+ = \frac{PSS^+}{|G|} = \frac{0.009 \text{ V/V}}{10.2 \text{ V/V}} = 882 \ \mu V/V$$

Power supply rejection ratio is often converted into decibel units

$$PSRR^+\big|_{dB} = 20\log_{10}\left(882 \ \mu V/V\right) = -61.09 \text{ dB}$$

3.9 DC COMMON-MODE REJECTION RATIO

3.9.1 CMRR of Op Amps

Common-mode rejection ratio (CMRR) is a measurement of a differential circuit's ability to reject a common-mode signal $V_{IN,CM}$ at its inputs. It is defined as the magnitude of the common-mode gain G_{CM} divided by the differential gain G_D, given by

$$CMRR \equiv \left|\frac{G_{CM}}{G_D}\right| \tag{3.15}$$

This expression can be further simplified by substituting for the common-mode gain $G_{CM} = \Delta V_{O,CM}/\Delta V_{IN,CM}$, together with the definition for input-referred offset voltage defined in Eq. (3.5), as follows:

$$CMRR = \left|\frac{\Delta V_{O,CM}\Big/\Delta V_{IN,CM}}{G_D}\right| = \left|\frac{\Delta V_{O,CM}\Big/G_D}{\Delta V_{IN,CM}}\right| = \left|\frac{\Delta V_{IN,OS}}{\Delta V_{IN,CM}}\right| \tag{3.16}$$

The rightmost expression suggests the simplest procedure to measure CMRR; one simply measures $\Delta V_{IN,OS}$ subject to a change in the input common-mode level $\Delta V_{IN,CM}$. One can measure $\Delta V_{IN,OS}$ directly or indirectly, as the following two examples illustrate.

EXAMPLE 3.9

Figure 3.21 shows a simple CMRR test fixture for an op amp. The test circuit is basically a difference-amplifier configuration with the two inputs tied together. V_{MID} is set to 1.5 V and an input common-mode voltage of 2.5 V is applied using SRC1. An output voltage of 1.501 V is measured at the output of the op amp. Then SRC1 is changed to 0.5 V and the output changes to 1.498 V. What is the CMRR of the op amp?

Solution:
As the measurement was made at the output of the circuit, we need to infer from these results the $\Delta V_{IN,OS}$ for the op amp. This requires a few steps: The first is to find the influence of the op amp

input-referred offset voltage $V_{IN,OS}$ on the test circuit output. As in Section 3.7.2, detailed circuit analysis reveals

$$V_0 = \frac{R_I + R_F}{R_I} V_{IN,OS}$$

With all resistors equal and perfectly matched, $V_0 = 2\ V_{IN,OS}$. Hence, $\Delta V_0 = 2\ \Delta V_{IN,OS}$, or, when re-arranged, $\Delta V_{IN,OS} = 0.5\ \Delta V_0$. Subsequently, substituting measured values $\Delta V_0 = 1.501\ V - 1.498\ V = 3$ mV, we find $\Delta V_{IN,OS} = 1.5$ mV. This result can now be substituted into Eq. (3.16), together with $\Delta V_{IN,CM} = \Delta V_{SRC1} = 2.5\ V - 0.5\ V = 2.0\ V$, leading to a CMRR = 750 μV/V or −62.5 dB.

Figure 3.21. Op-amp CMRR test setup.

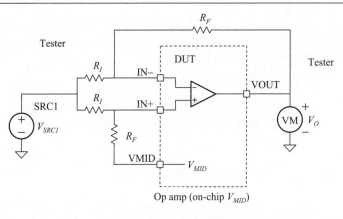

Op amp (on-chip V_{MID})

There is one major problem with this technique for measuring op amp CMRR: The resistors must be known precisely and carefully matched. A CMRR value of −100 dB would require resistor matching to 0.0001%, an impractical value to achieve in practice. A better test circuit setup is the nulling amplifier configuration shown in Figure 3.22. This configuration is very similar to the one used previously to measure the open-loop gain and input offsets of Section 3.7. The basic circuit arrangement is identical, only the excitation and the position of the voltmeter are changed. With this test setup, one can vary the common-mode input to the DUT and measure the differential voltage between the input SRC1 and the nulling amplifier output, which we shall denote as $V_{O,NULL}$. This, in turn, can then be used to deduce the input-referred offset for the DUT amplifier according to

$$V_{IN,DUT,OS} = \frac{R_1}{R_1 + R_2} V_{O,NULL} \tag{3.17}$$

Subsequently, the CMRR of the op amp is given by

$$CMRR = \frac{R_1}{R_1 + R_2} \left| \frac{\Delta V_{O,NULL}}{\Delta V_{SRC1}} \right| \tag{3.18}$$

EXAMPLE 3.10

For nulling amplifier setup shown in Figure 3.22 with R_1 = 100 Ω, R_2 = 100 k Ω, and R_3 = 100 k Ω, together with V_{MID} set to a value midway between the two power supply levels, SRC1 is set to +2.5 V and a differential voltage of 10 mV is measured between SRC1 and the output of the nulling amplifier. Then SRC1 is set to 0.5 V and the measured voltage changes to –12 mV. What is the CMRR of the op amp?

Solution:
Using Eq. (3.17), we deduce

$$\Delta V_{IN,DUT,OS} = \frac{R_1}{R_1 + R_2} \Delta V_{O,NULL}$$

$$= \frac{100}{100 + 100k}\left[10 \text{ mV} - (-12 \text{ mV})\right]$$

$$= 22 \ \mu V$$

for a corresponding ΔV_{SRC1} = 2.5 V – 0.5 V, or 2.0 V. Thus the CMRR is

$$CMRR = \frac{22 \ \mu V}{2.0 \text{ V}} = 11\frac{\mu V}{V} = -99.17 \text{ dB}$$

Figure 3.22. Op amp CMRR test setup using nulling amplifier.

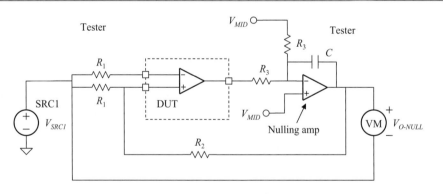

3.9.2 CMRR of Differential Gain Stages

Integrated circuits often use op amps as part of a larger circuit such as a differential input amplifier. In these cases, the CMRR of the op amp is not as important as the CMRR of the circuit as a whole. For example, a differential amplifier configuration such as the one in Figure 3.21 may have terrible CMRR if the resistors are poorly matched, even if the op amp itself has a CMRR of –100 dB. The differential input amplifier CMRR specifications include not only the effects of the op amp, but also the effects of on-chip resistor mismatch. As such, we determine the CMRR using the original definition given in Eq. (3.15). Our next example will illustrate this.

EXAMPLE 3.11

Figure 3.23 illustrates the test setup to measure the CMRR of a differential amplifier having a nominal gain of 10. No assumption about resistor matching is made. Both inputs are connected to a common voltage source SRC1 whose output is set to 2.5 V. A voltage of 1.501 V is measured at the output of the DUT. Then SRC1 is set to 0.5 V and a second voltage of 1.498 V is measured at the DUT output. Next the differential gain of the DUT circuit is measured using the technique described in Section 3.7.1. The gain was found to be 10.2 V/V. What is the CMRR?

Solution:

Since $\Delta V_0 = 1.501\ V - 1.498\ V = 3\ mV$ corresponding to a $\Delta V_{IN,CM} = \Delta V_{SRC1} = 2.0\ V$, the common-mode gain G_{CM} is calculated to be equal to 0.0015 V/V. In addition, we are told that the differential gain G_D is 10.2 V/V; thus we find the CMRR from the following:

$$CMRR = \left| \frac{G_{CM}}{G_D} \right| = \left| \frac{0.0015\,V/V}{10.2\,V/V} \right| = 0.000147 = -76.65\ dB$$

Figure 3.23. A ×10 differential amplifier CMRR test setup.

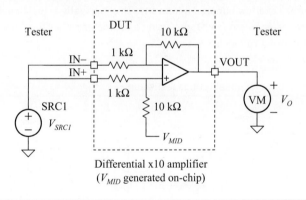

Differential x10 amplifier
(V_{MID} generated on-chip)

Exercises

3.15. An amplifier has an expected CMRR of −100 dB. For a 1-V change in the input common-mode level, what is the expected change in the input offset voltage of this amplifier?   ANS. 10 µV.

3.16. For the nulling amplifier CMRR setup in Figure 3.22 with $R_1 = 100\ \Omega$, $R_2 = 500\ k\Omega$, and $R_3 = 100\ k\Omega$, SRC1 is set to +3.5 V and a differential voltage of 210 mV is measured between SRC1 and the output of the nulling amplifier. Then SRC1 is set to 0.5 V and the measured voltage changes to −120 mV. What is the CMRR of the op amp in decibels?   ANS. 21.99 µV/V, −93.15 dB.

3.10 COMPARATOR DC TESTS

3.10.1 Input Offset Voltage

Input offset voltage for a comparator is defined as the differential input voltage that causes the comparator to switch from one output logic state to the other. The differential input voltage can be ramped from one voltage to another to find the point at which the comparator changes state. This switching point is, however, dependent on the input common-mode level. One usually tests for the input offset voltage under worst-case conditions as outlined in the device test plan.

EXAMPLE 3.12

The comparator in Figure 3.24 has a worst-case input offset voltage of ±50 mV and a midsupply voltage of 1.5V. Describe a test setup and procedure with which to obtain its input offset voltage.

Solution:
The comparator in Figure 3.24 is connected to two voltage sources, SRC1 and SRC2. SRC2 is set to 1.5 V and SRC1 is ramped upward from 1.45 to 1.55 V, as the switching point is expected to lie within this range. When the output changes from logic LO to logic HI, the differential input voltage V_{IN} is measured, resulting in an input offset voltage reading of +5 mV. The V_{IN} voltage could be deduced by simply subtracting 1.5 V from the SRC1 voltage, assuming that the DC sources force voltages to an accuracy of a few hundred microvolts. This is usually a questionable assumption, however. It is best to measure small voltages using a voltmeter rather than assume the tester's DC sources are set to exact voltages.

Figure 3.24. Comparator input offset voltage test setup.

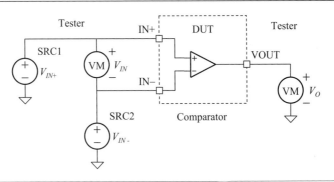

3.10.2 Threshold Voltage

Sometimes a fixed reference voltage is supplied to one input of a comparator, forming a circuit known as a *slicer*. The input offset voltage specification is typically replaced by a single-ended specification, called *threshold voltage*.

The slicer in Figure 3.25 is tested in a manner similar to that of the comparator circuit in the previous example. Assuming that the threshold voltage is expected to fall between 1.45 and 1.55 V, the input voltage from SRC1 is ramped upward from 1.45 to 1.55 V. The output switches states when the input is equal to the slicer's threshold voltage.

Figure 3.25. Slicer threshold voltage test setup.

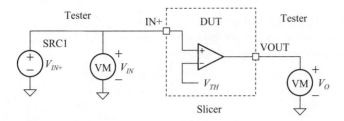

Notice that threshold voltage will be affected by the accuracy of the on-chip voltage reference, V_{TH}. In theory, the threshold voltage should be equal to the sum of the slicer's reference voltage V_{TH} plus the input offset voltage of the comparator. Threshold voltage error is defined as the difference between the actual and ideal threshold voltages.

3.10.3 Hysteresis

In the comparator input offset voltage example, the output changed when the input voltage reached 5 mV. This occurred on a rising input voltage. On a falling input voltage, the threshold may change to a lower voltage. This characteristic is called *hysteresis*, and it may or may not be an intentional design feature. Hysteresis is defined as the difference in threshold voltage between a rising input test condition and a falling input condition.

EXAMPLE 3.13

The comparator in Figure 3.24 is connected to two voltage sources, SRC1 and SRC2. SRC2 is set to 1.5 V and SRC1 is ramped upward from 1.45 to 1.55 V in 1-mV steps. When the output changes from logic LO to logic HI, the differential input voltage is measured, resulting in an input offset voltage reading of +5 mV relative to SRC1. Then the input is ramped downward from 1.55 to 1.45 V and the output switches when the input voltage reaches –3 mV. What is the hysteresis of this comparator?

Solution:
The hysteresis is equal to the difference of the two input offset voltages

$$5 \text{ mV} - (-3 \text{ mV}) = 8 \text{ mV}$$

It should be noted that input offset voltage and hysteresis may change with different common-mode input voltages. Worst-case test conditions should be determined during the characterization process.

3.17. A comparator has an input offset voltage of 50 mV and its positive terminal is connected to a 1-V level, at what voltage on the negative terminal does the comparator change state?   ANS. 0.950 V.

3.18. A slicer circuit is connected to a 1.65 V reference V_{TH} and has a comparator input offset voltage of 11 mV. At what voltage level will the slicer change state?   ANS. 1.661 V.

3.19. A comparator has a measured hysteresis of 9 mV and switches state on a rising input at 2.100 V. At what voltage does the comparator change to a low state on a falling input?   ANS. 2.091 V.

3.11 VOLTAGE SEARCH TECHNIQUES

3.11.1 Binary Searches Versus Step Searches

The technique of ramping input voltage levels until an output condition is met is called a *ramp search*, or *step search*. Step searches are time-consuming and not well suited for production testing. Instead, binary or linear search methods are often used.

To gain a better understanding of these methods, let us consider the general search process. In mathematical terms, let us we denote the input–output behavior of some device under test with some mathematical function, say $y = f(x)$, where x is the input and y the output. Subsequently, to establish the output at some arbitrary level, say $y = D$, we need to find the value of x that satisfies $f(x)-D = 0$. If the inverse of f is known, then we can immediately solve for input as $x = f^{-1}(D)$. Generally, $f(x)$ is not known, since it is specific to each and every device under test. However, through a source-measurement process, the behavior of $y = f(x)$ is encapsulated in the form of a look-up table or by the direct action of a measurement. Consequently, through some search process, we can identify the value of x that satisfies $f(x)-D = 0$. As x is a root of the equation $f(x)-D$, the procedures used to identify the root are known as root-finding algorithms. There are numerous root-finding algorithms, such as bisection (binary), secant (linear), false-position, and Newton-Raphson methods, to name just a few. Any one of these can be adapted to test. Let us begin by describing the binary search method.

To determine an input value such that the output equals a desired target value to within some tolerance, start with two input values. One value is selected such that the output is greater than some desired target value, while the other is selected to obtain an output less than this target value. Let us denote these two output conditions as *output @ input₁* and *output @ input₂*. Subsequently, the binary search process can be described using the following pseudo code:

```
TARGET = desired value
Measure output @ input₁
Measure output @ input₂
output @ input₃ =1000 # initialization
Do WHILE | output @ input₃ - TARGET | ≥ tolerance value
    Set input₃=(input₁+input₂)/2
    Measure output @ input₃
    IF output @ input₃ –TARGET is of opposite sign to output @ input₁ – TARGET Do:
```

> Set input$_1$=input$_1$ and input$_2$=input$_3$
> ELSE
>   Set input$_1$=input$_3$ and input$_2$=input$_2$
> ENDIF
> END Do

A binary search can be applied to the comparator input offset voltage test described in the previous section. Instead of ramping the input voltage from 1.45 to 1.55 V, the comparator input is set half-way between to 1.5 V and the output is observed. If the output is high, then the input is increased by one-quarter of the 100-mV search range (25 mV) to try to make the output go low. If, on the other hand, the output is low, then the input is reduced by 25 mV to try to force the output high. Then the output is observed again. This time, the input is adjusted by one-eighth of the search range (12.5 mV). This process is repeated until the desired input adjustment resolution is reached.

The problem with the binary search technique is that it does not work well in the presence of hysteresis. The binary search algorithm assumes that the input offset voltage is the same whether the input voltage is increased or decreased. If the comparator exhibits hysteresis, then there are two different threshold voltages to be measured. To get around this problem without reverting to the time-consuming ramp search technique, a hybrid approach can be used. A binary search can be used to find the approximate threshold voltage quickly. Then a step search can be used with a much smaller search voltage range.

Another solution to the hysteresis problem is to use a modified binary search algorithm in which the output state of the comparator is returned to a known logic state between binary search approximations. This is achieved by forcing the input either well above or well below the threshold voltage. In this way, steps are always taken in one direction, avoiding hysteresis effects. To measure hysteresis, a binary search is used once with the output state forced high between approximations. Then the input offset is measured again with the output state forced low between approximations. The difference in input offset readings is equal to the hysteresis of the comparator.

3.11.2 Linear Searches

Linear circuits can make use of an even faster search technique called a *linear search*. A linear search is similar to the binary search, except that the input approximations are based on a linear interpolation of input–output relationships. For example, if a 0-mV input to a buffer amplifier results in a 10-mV output and a 1-mV input results in a 20-mV output, then a –1-mV input will probably result in a 0-mV output. The linear search algorithm keeps refining its guesses using a simple straight-line approximation $V_{OUT} = M \times V_{IN} + B$ algorithm until the desired accuracy is reached. The following example will help illustrate this method.

EXAMPLE 3.14

Using a linear search algorithm, find the input offset voltage V_{os} for a ×10 amplifier when the output referred offset voltage is 120 mV.

Solution:
The input to a ×10 amplifier is set to 0 V and the output is measured, yielding a reading of 120 mV. The gain **M** is known to be approximately 10, since this is supposed to be a ×10 amplifier.

The value of offset B can be approximately determined using the $V_{OUT} = M \times V_{IN} + B$ linear equation, that is,

$$120 \text{ mV} = M \times 0 \text{ mV} + B = 10 \times 0 \text{ mV} + B$$
$$\Rightarrow B = 120 \text{ mV (first-pass guess)}$$

Since 0 mV is the desired output, the next estimate for V_{os} can be calculated using the linear equation again

$$0 \text{ mV (desired } V_{OUT}) = M \times V_{IN} + B = 10 \times V_{IN} + 120 \text{ mV}$$

Rewriting this equation to solve for V_{IN}, we get

$$V_{IN} = \frac{(0 \text{ mV } - 120 \text{ mV})}{10} = -12 \text{ mV}$$

Applying the best guess of –12 mV to the input, another output measurement is made, resulting in a reading of 8 mV. Now we have two equations in two unknowns

$$120 \text{ mV} = M \times 0 \text{ mV} + B$$
$$8 \text{ mV} = M (-12 \text{ mV}) + B$$

from which a more accurate estimate of M and B can be made. Solving for the two unknowns

$$M = \frac{120 \text{ mV } - 8 \text{ mV}}{0 \text{ mV} - (-12 \text{ mV})} = 9.333 \text{ V/V}$$
$$B = 10 \text{ mV} - \left[M (-12 \text{ mV}) \right] = 122 \text{ mV}$$

The next input approximation should be close enough to the input offset voltage to produce an output of 0 mV, that is

$$V_{IN} = \frac{(0 \text{ mV} - B)}{M} = \frac{(0 \text{ mV} - 122 \text{ mV})}{9.333} = -13.1 \text{ mV}$$

The input offset voltage of the ×10 amplifier is therefore +13.1 mV, assuming that the circuit is linear. The sign of the input-referred offset voltage is opposite to that found from the search process on account of the definition of offset voltage. In cases where the input–output relationship is not linear, the linear search technique will still work, but will require more iterations of the above process. For each iteration, two linear interpolations are performed using the two most recent input–output data points. This continues until the process converges to within the desired measurement resolution.

The procedure outline above is an example of the secant root finding method. We can easily generalize the principles described above to one involving the following set of iteration equations with pseudo code:

```
TARGET = desired value
k=1 # initialize iteration
Measure output @ input_{k-1}
Measure output @ input_k
output @ input_{k+1} =1000 # initialization
Do WHILE | output @ input_{k+1} – TARGET | ≥ tolerance value
```

$$Set \; input_{k+1} = input_k - \left(output @ input_k - TARGET\right) \times \left(\frac{input_k - input_{k-1}}{output @ input_k - output @ input_{k-1}}\right)$$

```
Measure output @ input_{k+1}
k = k + 1
END Do
```

The reader is encouraged to investigate other root-finding procedures, such as the Newton–Raphson method, to see how well they can be adapted to the search process described above.

EXAMPLE 3.15

For an amplifier characterized by $V_{OUT} = 0.1 + 2.5V_{IN} + 0.01V_{IN}^2 + 0.001V_{IN}^3$ over a 1.0 – 4.0 V output voltage range, determine the input voltage that will establish the output voltage level at 3.00 V using a linear search process. How many search iterations are required for a maximum error of 1 mV? List the input values and corresponding output values as a function of each iteration.

Solution:
We begin by declaring the excitation-measurement process of the DUT as described by the mathematical equation:

$$f = 0.1 + 2.5V_{IN} + 0.01V_{IN}^2 + 0.001V_{IN}^3$$

Next, using a TARGET value of 3.0 V, we perform a linear search on the DUT to find the input voltage level V_{IN} such that the output equals the TARGET value. Beginning with $k = 1$, we declare

$$input_0 = 1.0, f(1.0) = 0.1 + 2.5(1.0) + 0.01(1.0)^2 + 0.001(1.0)^2 = 2.611$$

$$input_1 = 4.0, f(4.0) = 0.1 + 2.5(4.0) + 0.01(4.0)^2 + 0.001(4.0)^2 = 10.324$$

Next, we find the first update to the input level as follows:

$$input_2 = input_1 - \left[f\left(input_1\right) - TARGET \right] \times \left(\frac{input_1 - input_0}{f\left(input_1\right) - f\left(input_0\right)} \right)$$

$$= 4.0 - (10.324 - 3.0) \times \left(\frac{4.0 - 1.0}{10.324 - 2.611} \right) = 1.1513$$

and obtain the DUT output at this input level,

$$f\left(1.1513\right) = 0.1 + 2.5\left(1.1513\right) + 0.01\left(1.1513\right)^2 + 0.001\left(1.1513\right)^2 = 2.9930$$

As we are about 7 mV away from the target, but need to be less than 1 mV, we shall iterate again and find the next input level from

$$input_3 = input_2 - \left[f\left(input_2\right) - TARGET \right] \times \left(\frac{input_2 - input_1}{f\left(input_2\right) - f\left(input_1\right)} \right)$$

$$= 1.1513 - (2.9930 - 3.0) \times \left(\frac{1.1513 - 4.0}{2.9930 - 10.324} \right) = 1.1540$$

and

$$f\left(1.1540\right) = 0.1 + 2.5\left(1.1540\right) + 0.01\left(1.1540\right)^2 + 0.001\left(1.1540\right)^2 = 2.9998$$

Here we are less than 1 mV away from the targeted value of 3.00 V. Hence we stop the iteration at 2 with an input value of 1.1540 V.

Exercises

3.20. For an amplifier characterized by $V_{OUT} = 10V_{IN} - V_{IN}^2 + 5$ over a ±5 V output voltage range, determine the input offset voltage using a binary search process. The input offset voltage is known to fall between -464 and -496 mV. How many search iterations are required for a maximum error of 1 mV? List the input values and corresponding outputs.

ANS. A 32-mV search range with 2-mV resolution is required, requiring four binary iterations: (1) –480 mV, –30.4 mV; (2) –472 mV, +57 mV; (3) –476 mV, +13.4 mV; (4) –478 mV, –8.5 mV. The final estimate is thus –477 mV ($V_{IN,OS}$ = +477 mV; true answer is +477.2 mV).

3.21. Repeat Exercise 3.20 using a linear search process starting with two points at V_{IN} = –250 mV and –750 mV. How many iterations are required for < 1 mV error in $V_{IN,OS}$?

ANS. Two iterations produce estimates of $V_{IN,OS}$ = +471.6 mV and $V_{IN,OS}$ = +477.1 mV.

3.12 DC TESTS FOR DIGITAL CIRCUITS

3.12.1 I_{IH}/I_{IL}

The data sheet for a mixed-signal device usually lists several DC specifications for digital inputs and outputs. Input leakage currents (I_{IH} and I_{IL}) were discussed in Section 3.2.2. Input leakage is also specified for digital output pins that can be set to a high-impedance state.

3.12.2 V_{IH}/V_{IL}

The input high voltage (V_{IH}) and input low voltage (V_{IL}) specify the threshold voltage for digital inputs. It is possible to search for these voltages using a binary search or step search, but it is more common to simply set the tester to force these levels into the device as a go/no-go test. If the device does not have adequate V_{IH} and V_{IL} thresholds, then the test program will fail one of the digital pattern tests that are used to verify the DUT's digital functionality. To allow a distinction between pattern failures caused by V_{IH}/V_{IL} settings and patterns failing for other reasons, the test engineer may add a second identical pattern test that uses more forgiving levels for V_{IH}/V_{IL}. If the digital pattern test fails with the specified V_{IH}/V_{IL} levels and passes with the less demanding settings, then V_{IH}/V_{IL} thresholds are the likely failure mode.

3.12.3 V_{OH}/V_{OL}

V_{OH} and V_{OL} are the output equivalent of V_{IH} and V_{IL}. V_{OH} is the minimum guaranteed voltage for an output when it is in the high state. V_{OL} is the maximum guaranteed voltage when the output is in the low state. These voltages are usually tested in two ways. First, they are measured at DC with the output pin set to static high/low levels. Sometimes a pin cannot be set to a static output level due poor design for test considerations, so only a dynamic test can be performed. Dynamic V_{OH}/V_{OL} testing is performed by setting the tester to expect high voltages above V_{OH} and low voltages below V_{OL}. The tester's digital electronics are able to verify these voltage levels as the outputs toggle during the digital pattern tests. Dynamic V_{OH}/V_{OL} testing is another go/no-go test approach, since the actual V_{OH}/V_{OL} voltages are verified but not measured.

3.12.4 I_{OH}/I_{OL}

V_{OH} and V_{OL} levels are guaranteed while the outputs are loaded with specified load currents, I_{OH} and I_{OL}. The tester must pull current out of the DUT pin when the output is high. This load current is called I_{OH}. Likewise, the tester forces the I_{OL} current into the pin when the pin is low. These currents are intended to force the digital outputs closer to their V_{OH}/V_{OL} specifications, making the V_{OH}/V_{OL} tests more difficult for the DUT to pass. I_{OH} and I_{OL} are forced using a diode bridge circuit in the tester's digital pin card electronics. The diode bridge circuit is discussed in more detail in Chapter 2, "Tester Hardware."

3.12.5 I_{OSH} and I_{OSL} Short-Circuit Current

Digital outputs often include a current-limiting feature that protects the output pins from damage during short-circuit conditions. If the output pin is shorted directly to ground or to a power supply pin, the protection circuits limit the amount of current flowing into or out of the pin. Short-circuit current is measured by setting the output to a low state and forcing a high voltage (usually V_{DD}) into the pin. The current flowing into the pin (I_{OSL}) is measured with one of the tester's current meters. Then the output is set to a high state and 0 V is forced at the pin. The current flowing out of the pin (I_{OSH}) is again measured with a current meter.

3.13 SUMMARY

This chapter has presented only a few of the many DC tests and techniques that the mixed-signal test engineer will encounter. Several chapters or perhaps even a whole book could be devoted to highly accurate DC test techniques. However, this book is intended to address mixed-signal testing. Hopefully, the limited examples given in this chapter will serve as a solid foundation from which the test engineer can build a more diversified DC measurement skill set.

DC measurements are trivial to define and understand, but they can sometimes be excruciatingly difficult to implement. A DC offset of 100 mV is very easy to measure if the required accuracy is ±10 mV. On the other hand, if 1-µV accuracy is required, the test engineer may find this to be one of the more daunting test challenges in the entire project. The accuracy and repeatability requirements of seemingly simple tests like DC offset can present a far more challenging test problem than much more complicated AC tests.

Accuracy and repeatability of measurements is the subject of Chapter 5, following an introductory chapter on data analysis and probability in Chapter 4. This topic pertains to a wide variety of analog and mixed-signal tests. Much of a test engineer's time is consumed by accuracy and repeatability problems. These problems can be one of the most aggravating aspects of mixed-signal testing. The successful resolution of a perplexing accuracy problem can also be one of the most satisfying parts of the test engineer's day.

PROBLEMS

3.1. The output of a 10-V voltage regulator varies from 9.95 V under no-load condition to 9.34 V under a 10-mA maximum rated load current. What is its load regulation?

3.2. The output of a 5-V voltage regulator varies from 4.86 to 4.32 V when the input voltage is changed from 14 to 6 V under a maximum load condition of 10 mA. What is its line regulation?

3.3. A 9-V voltage regulator is rated to have a load regulation of 0.150 V for a maximum load current of 15 mA. Assuming a no-load output voltage of 8.9 V, what is the expected output voltage at the maximum load current?

3.4. A 6-V voltage regulator is rated to have a load regulation of 2% for a maximum load current of 20 mA. Assuming a no-load output voltage of 5.9 V, what is the worst-case output voltage at the maximum load current?

3.5. A voltage of 1.2 V is dropped across an input pin when a 100-µA current is forced into the pin. Subsequently, a 1.254-V level occurs when the current is increased to 200 µA. What is the input resistance?

3.6. The input pin of a device is characterized by the $i - v$ relationship: $i = 0.001 \, v + 100$. What is the resistance seen looking into this pin?

3.7. Voltages of 1.2 and 3.3 V appear at the output of an amplfier when currents of -10 and $+10$ mA, respectively, are forced into its output. What is the output resistance?

3.8. The no-load output voltage of an amplifier is 4 V. When a 600-Ω load is attached to the output, the voltage drops to 3 V. What is the amplifier's output resistance?

3.9. For a ×10 amplifier characterized by $V_{OUT} = 10V_{IN} - V_{IN}^2 + 5$ over a ±5-V range, what are its input and output offset voltages?

3.10. A voltmeter introduces a measurement error of -5% while measuring a 1-V offset from an amplifier. What is the actual reading captured by the voltmeter?

3.11. A voltmeter with an input impedance of 500 kΩ is used to measure the DC output of an amplifier with an output impedance of 500 kΩ. What is the expected relative error made by this measurement?

3.12. A differential amplifier has outputs of 2.4 V (OUTP) and 2.7 V (OUTN) with its input set to a V_{MID} reference level of 2.5 V. What are the single-ended and differential offsets? The common-mode offset? (All offsets are to be measured with respect to V_{MID}.)

3.13. A perfectly linear amplifier has a measured gain of 9.8 V/V and an output offset of 1.2 V. What is the input offset voltage?

3.14. Voltages of 1.3 V and 10.3 V appear at the output of a single-ended amplifier when inputs of 110 mV and 1.3 V are applied, respectively. What is the gain of the amplifier in V/V? What is the gain in decibels?

3.15. An amplifier is characterized by $V_{OUT} = 3.5V_{IN} + 1$ over the input voltage range 0 to 5 V. What is the amplifier output for a 2-V input? Similarly for a 3-V input? What is the corresponding gain of this amplifier in V/V to a the 1-V swing centered around 2.5-V? What is the gain in decibels?

3.16. An amplifier is characterized by $V_{OUT} = 1.5V_{IN} + 0.35V_{IN}^2 + 1$ over the input voltage range 0 to 5 V. What is the amplifier output for a 1-V input? Similarly for a 3-V input? What is the corresponding gain of this amplifier in V/V to a signal with a 1-V swing centered at 1-V? What is the gain in decibels?

3.17. For the nulling amplifier setup shown in Figure 3.19 with $R_1 = 100\ \Omega$, $R_2 = 200\ k\Omega$, and $R_3 = 50\ k\Omega$, an SRC1 input swing of 1 V results in a 130-mV swing at the output of the nulling amplifier. What is the open-loop gain of the DUT amplifier in V/V? What is the gain in decibels?

3.18. For the nulling amplifier setup shown in Figure 3.19 with $R_1 = 200\ \Omega$, $R_2 = 100\ k\Omega$, and $R_3 = 100\ k\Omega$, and a V_{MID} of 2.5 V, an offset of 3.175 V (relative to ground) appears at the output of the nulling op amp when the input is set to V_{MID}. What is the input offset of the DUT amplifier?

3.19. For the nulling amplifier setup shown in Figure 3.19 with $R_1 = 100\ \Omega$, $R_2 = 300\ k\Omega$, and $R_3 = 100\ k\Omega$, and the DUT op amp having an open-loop gain of 1000 V/V, what is the output swing of the nulling amplifier when the input swings by 1 V?

3.20. The input of a ×10 amplifier is connected to a voltage source forcing 1.75 V. The power supply is set to 4.9 V and a voltage of 1.700 V is measured at the output of the amplifier. The power supply voltage is then changed to 5.1 V and the output measurement changes to 1.708 V. What is the PSS? What is the PSRR if the measured gain is 9.8 V/V?

3.21. For nulling amplifier CMRR setup shown in Figure 3.22 with $R_1 = 100\ \Omega$, $R_2 = 300\ k\Omega$, and $R_3 = 100\ k\ \Omega$, SRC1 is set to +3.5 V and a differential voltage of 130 mV is measured between SRC1 and the output of the nulling amplifier. Then SRC1 is set to 1.0 V and the measured voltage changes to –260 mV. What is the CMRR of the op amp in decibels?

3.22. An amplifier has an expected CMRR of –85 dB. For a 1-V change in the input common-mode level, what is the expected change in the input offset voltage of this amplifier?

3.23. A comparator has an input offset voltage of 6 mV and its negative terminal is connected to a 2.5-V level, at what voltage on the positive terminal does the comparator change state?

3.24. A slicer circuit is connected to a 2-V reference and has a threshold voltage error of 20 mV. At what voltage level will the slicer change state?

3.25. If a slicer's 2.5-V reference has an error of +100 mV and the comparator has an input offset of –10 mV, what threshold voltage should we expect?

3.26. A comparator has a measured hysteresis of 10 mV and switches state on a rising input at 2.5 V. At what voltage does the comparator change to a low state on a falling input?

3.27. For an amplifier characterized by $V_{OUT} = 6V_{IN} + 0.5V_{IN}^2 - 2$ over a ±1-V input voltage range, determine the input offset voltage using a linear search process, starting with two points at ±1 V. After how many iterations did the answer change by less than 1 mV? How many iterations would have been required using a binary search from –1 to +1 V?

3.28. For an amplifier characterized by $V_{OUT} = 0.1 + 2.5V_{IN} - 0.8V_{IN}^2$ over a 0.0 to 3.0-V output voltage range, determine the input voltage that will establish the output voltage level at 2.00 V using a linear search process. How many search iterations are required for a maximum error of 1 mV? List the input values and corresponding output as a function of each iteration.

REFERENCES

1. S. Chakravarty and P. J. Thadikaran, *Introduction to IDDQ Testing*, May 1997, Kluwer Academic Publishers, Boston, ISBN 0792399455.
2. Analog Devices application note,* *How to Test Basic Operational Amplifier Parameters*, Analog Devices, Inc., Norwood, MA, July 1982.

---

*The nulling amplifier/servo loop methods presented in this chapter were adapted from the referenced application note to allow compatiblity with single-supply op amps having a V_{MID} reference voltage. The technique has been presented with permission from Analog Devices, Inc.

CHAPTER 4

Data Analysis and Probability Theory

Data analysis is the process by which we examine test results and draw conclusions from them. Using data analysis, we can evaluate DUT design weaknesses, identify DIB and tester repeatability and correlation problems, improve test efficiency, and expose test program bugs. As mentioned in Chapter 1, debugging is one of the main activities associated with mixed-signal test and product engineering. Debugging activities account for about 20% of the average workweek. Consequently, data analysis plays a very large part in the overall test and product-engineering task.

Many types of data visualization tools have been developed to help us make sense of the reams of test data that are generated by a mixed-signal test program. In this chapter we will review several common data analysis tools, such as the datalog, histogram, lot summary, shmoo plot, and the wafer map. We shall also review some statistical and probability concepts as they apply to analog and mixed-signal test. Of particular interest to the engineer is the ability to model large quantity of data with simple mathematical models. Often, these are based on an assumption of the structure of the underlying random behavior and several quantities obtained from the measured data, such as mean and standard deviation. Once a model of the randomness of the data is found, questions about the likelihood an event will occur can be quantified. This helps the test engineer identify with confidence the meaning of a measurement and, ultimately, the conditions under which to perform a measurement in minimum time.

4.1 DATA VISUALIZATION TOOLS

4.1.1 Datalogs (Data Lists)

A datalog, or data list, is a concise listing of test results generated by a test program. Datalogs are the primary means by which test engineers evaluate the quality of a tested device. The format of a datalog typically includes a test category, test description, minimum and maximum test limits, and a measured result. The exact format of datalogs varies from one tester type to another, but datalogs all convey similar information.

Figure 4.1. Example datalog from a Teradyne Catalyst tester.

Sequencer: S_continuity							
1000 Neg PPMU Cont	Failing Pins: 0						
Sequencer: S_VDAC_SNR							
5000 DAC Gain Error	T_VDAC_SNR	-1.00 dB	<	-0.13	dB	< 1.00	dB
5001 DAC S/2nd	T_VDAC_SNR	60.0 dB	<=	63.4	dB		
5002 DAC S/3rd	T_VDAC_SNR	60.0 dB	<=	63.6	dB		
5003 DAC S/THD	T_VDAC_SNR	60.00 dB	<=	60.48	dB		
5004 DAC S/N	T_VDAC_SNR	55.0 dB	<=	70.8	dB		
5005 DAC S/N+THD	T_VDAC_SNR	55.0 dB	<=	60.1	dB		
Sequencer: S_UDAC_SNR							
6000 DAC Gain Error	T_UDAC_SNR	-1.00 dB	<	-0.10	dB	< 1.00	dB
6001 DAC S/2nd	T_UDAC_SNR	60.0 dB	<=	86.2	dB		
6002 DAC S/3rd	T_UDAC_SNR	60.0 dB	<=	63.5	dB		
6003 DAC S/THD	T_UDAC_SNR	60.00 dB	<=	63.43	dB		
6004 DAC S/N	T_UDAC_SNR	55.0 dB	<=	61.3	dB		
6005 DAC S/N+THD	T_UDAC_SNR	55.0 dB	<=	59.2	dB		
Sequencer: S_UDAC_Linearity							
7000 DAC POS ERR	T_UDAC_Lin	-100.0 mV	<	7.2	mV	< 100.0	mV
7001 DAC NEG ERR	T_UDAC_Lin	-100.0 mV	<	3.4	mV	< 100.0	mV
7002 DAC POS INL	T_UDAC_Lin	-0.90 lsb	<	0.84	lsb	< 0.90	lsb
7003 DAC NEG INL	T_UDAC_Lin	-0.90 lsb	<	-0.84	lsb	< 0.90	lsb
7004 DAC POS DNL	T_UDAC_Lin	-0.90 lsb	<	1.23	lsb (F)	< 0.90	lsb
7005 DAC NEG DNL	T_UDAC_Lin	-0.90 lsb	<	-0.83	lsb	< 0.90	lsb
7006 DAC LSB SIZE	T_UDAC_Lin	0.00 mV	<	1.95	mV	< 100.00	mV
7007 DAC Offset V	T_UDAC_Lin	-100.0 mV	<	0.0	mV	< 100.0	mV
7008 Max Code Width	T_UDAC_Lin	0.00 lsb	<	1.23	lsb	< 1.50	lsb
7009 Min Code Width	T_UDAC_Lin	0.00 lsb	<	0.17	lsb	< 1.50	lsb
Bin: 10							

A short datalog from a Teradyne Catalyst tester is shown in Figure 4.1. Each line of the datalog contains a shorthand description of the test. For example, "DAC Gain Error" is the name of test number 5000. The gain error test is part of the S_VDAC_SNR test group and is executed during the T_VDAC_SNR test routine. The minimum and maximum limits for the test are also listed. Using test number 5000 as an example, the lower limit of DAC gain error is –1.00 dB, the upper limit is +1.00 dB, and the measured value for this DUT is –0.13 dB.

The datalog displays an easily recognizable fail flag beside each value that falls outside the test limits. For instance, test 7004 in Figure 4.1 shows a failure in which the measured value is 1.23 LSBs. Since the upper limit is 0.9 LSBs, this test fails. In this particular example, the failure is flagged with an (F) symbol. Hardware and software alarms from the tester also result in a datalog alarm flag, such as (A). Alarms can occur for a variety of reasons, including mathematical divisions by zero and power supply currents that exceed programmed limits. When alarms are generated, the test program halts (unless instructed by the test engineer to ignore alarms). The tester assumes that the DUT is defective and treats the alarm as a failure.

Because the device in Figure 4.1 fails test 7004, it is categorized into bin 10 as displayed at the bottom of the datalog. Bin 1 usually represents a good device, while other bins usually represent various categories of failures and alarms. Sometimes there are multiple grades of shippable devices, which are separated into different passing bins. For example, a certain percentage of 2-GHz microprocessors may fail at 2 GHz, but may operate perfectly well at 1.8 GHz. The 1.8-GHz processors might be sorted into bin 2 and shipped at a lower cost, while the higher-grade 2-GHz processors are sorted into bin 1 to be sold at full price.

4.1.2 Lot Summaries

Lot summaries are generated after all devices in a given production lot have been tested. A lot summary lists a variety of information about the production lot, including the lot number, product number, operator number, and so on. It also lists the yield loss and cumulative yield associated with each of the specified test bins. The overall lot yield is defined as the ratio of the total number of good devices divided by the total number of devices tested:

$$\text{lot yield} = \frac{\text{total good devices}}{\text{total devices tested}} \times 100\% \tag{4.1}$$

Figure 4.2 shows a simplified lot summary, including yields for a variety of test categories. The lot yield is listed in the lot summary, but it does not tell us everything we need to know. If a particular lot exhibits a poor yield, we want to know *why* its yield was low. We want to know what category or categories of tests dominated the failures so we can look into the problem to determine its cause. A lot summary can help us identify which failures are most common to a particular type of DUT. This allows us to focus our attention on the areas of the design, the process, and the test program that might be causing the most failures in production. For this reason, lot summaries also list test categories and what percentage of devices failed each category. Specifically, two metrics are used called test category yield loss and cumulative yield. Yield loss per category is defined as

$$\text{test category yield loss} = \frac{\text{\# failed devices per specific test}}{\text{total devices tested}} \times 100\% \tag{4.2}$$

and cumulative test yield is defined as

$$\text{cumulative yield} = \left(1 - \frac{\text{total failed devices at end of a specific test}}{\text{total devices tested}}\right) \times 100\% \tag{4.3}$$

The lot summary in Figure 4.2 shows that our highest yield loss is due to the RECV channel AC tests. We might think that our XMIT channel has no problems, because it causes only a 0.30% yield loss. However, we have to be careful in making such judgments based on data collected during production. We have to remember that once a DUT fails any test, the tester immediately rejects it and moves on to the next device. After all, there is no point in continuing to test a DUT once it has been disqualified for shipment to the customer.

Since the test program halts after the first DUT failure, the earlier tests will tend to cause more yield loss than later ones, simply because fewer DUTs proceed to the later tests. The earlier failures mask any failures that would have occurred in later tests. For example, any or all the devices that failed the RECV channel tests in Figure 4.2 might also have failed the XMIT channel tests if given the chance. Therefore, during the device characterization phase we may want to instruct the tester to collect data from all tests whether the DUT passes or not. Of course, the extra testing leads to a longer average test time; thus we do not want to perform continue-on-fail testing in production unless necessary.

We can sometimes improve our overall production throughput by moving the more commonly failed tests toward the beginning of the test program. Average test time is reduced by the rearrangement because we do not waste time performing tests that seldom fail only to lose yield to subsequent tests that often fail. Once again, we have to remember that the order of tests may affect the lot summary output. Therefore, whenever we wish to reorder our test program based on yield loss, we would prefer to use lot summaries in which the tester does not halt on the first failure. That way, we know which tests are truly the ones that catch the largest number of defective DUTs.

Figure 4.2. Simplified lot summary.

Lot Number: 122336	
Device Number: TLC1701FN	
Operator Number: 42	
Test Program: F779302.load	

Devices Tested: 10233
Passing Devices: 9392
Test Yield: 91.78%

Bin#	Test Category	Devices Tested	Failures	Yield Loss	Cum. Yield
7	Continuity	10233	176	1.72%	98.28%
2	Supply Currents	10057	82	0.80%	97.48%
3	Digital Patterns	9975	107	1.05%	96.43%
4	RECV Channel AC	9868	445	4.35%	92.08%
5	XMIT Channel AC	9423	31	0.30%	91.78%

When rearranging test programs based on yield loss, we also have to consider the test time that each test consumes. For example, the RECV channel tests in Figure 4.2 may take 800 ms, while the digital pattern tests only takes 50 ms. The digital pattern test is more efficient at identifying failing DUTs since it takes so little test time. Therefore, it might not make sense to move the longer RECV test to the beginning of the program, even though it catches more defective DUTs than does the digital pattern test. Clearly, test program reordering is not a simple matter of moving the tests having the highest yield loss to the beginning of the test program.

4.1.3 Wafer Maps

A wafer map (Figure 4.3) displays the location of failing die on each probed wafer in a production lot. Unlike lot summaries, which only show the number of devices that fail each test category, wafer maps show the physical distribution of each failure category. This information can be very useful in locating areas of the wafer where a particular problem is most prevalent. For example, the continuity failures are most severe at the upper edge of the wafer illustrated in Figure 4.3. Therefore, we might examine the bond pad quality along the upper edge of the wafer to see if we can find out why the continuity test fails most often in this area. Also, the RECV channel failures are most severe near the center of the wafer. This kind of ring-like pattern often indicates a processing problem, such as uneven diffusion of dopants into the semiconductor surface or photomask misalignment. If all steps of the fabrication process are within allowable tolerances, consistent patterns such as this may indicate that the device is simply too sensitive to normal process variations and therefore needs to be redesigned. Wafer maps are a powerful data analysis tool, allowing yield enhancement through a cooperative effort between the design, test, product, and process engineers.

Naturally, it is dangerous to draw too many conclusions from a single wafer map. We need to examine many wafer maps to find patterns of consistent failure distribution. For this reason, some of the more sophisticated wafer mapping tools allows us to overlay multiple wafer maps on top of one another, revealing consistency in failure distributions. From these composite failure maps, we can draw more meaningful conclusions about consistent processing problems, design weaknesses, and test hardware problems.

Figure 4.3. Wafer map.

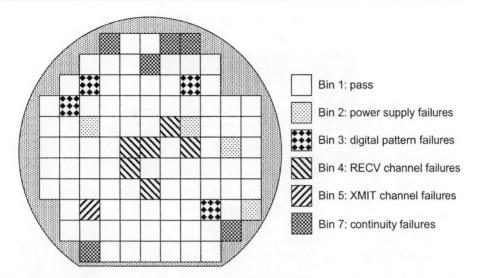

Bin 1: pass

Bin 2: power supply failures

Bin 3: digital pattern failures

Bin 4: RECV channel failures

Bin 5: XMIT channel failures

Bin 7: continuity failures

4.1.4 Shmoo Plots

Shmoo plots were among the earliest computer-generated graphic displays used in semiconductor manufacturing.[1] A shmoo plot is a graph of test results as a function of test conditions. For example, some of the earliest shmoo plots displayed pass/fail test results for PMOS memory ICs as a function of V_{DD} and V_{SS}. The origins of the name "shmoo plot" are not known for certain. According to legend, some of the early plots reminded the engineers of a shmoo, a squash-shaped cartoon character from Al Capp's comic strip "Li'l Abner." Although few shmoo plots are actually shmoo-shaped, the name has remained with us.

The graph in Figure 4.4 is called a *functional shmoo plot*, since it only shows which test conditions produce a passing (functional) or failing (nonfunctional) test result. This type of plot is commonly used to characterize purely digital devices, since digital test programs primarily produce functional pass/fail results from the digital pattern tests. Measured values such as supply current and distortion cannot be displayed using a functional shmoo plot.

Analog and mixed-signal measurements often require a different type of graph, called a *parametric shmoo plot*. Analog and mixed-signal test programs produce many parametric values, such as gain error and signal-to-distortion ratio. Parametric shmoo plots, such as that shown in Figure 4.5, can be used to display analog measurement results at each combination of test conditions rather than merely displaying a simple pass/fail result. Naturally, we always have the option of comparing the analog measurements against test limits, producing pass/fail test results compatible with a simple functional shmoo plot.

Parametric shmoo plots give the test engineer a more complete picture of the performance of mixed-signal DUTs under the specified range of test conditions. This information can tell the engineering team where the device is most susceptible to failure. Assume, for example, that the DUT of Figure 4.5 needs to pass a minimum S/THD specification of 75 dB. The shmoo plot in Figure 4.5 tells us that this DUT is close to failure at about 40°C and it is somewhat marginal at 30°C if our V_{DD} supply voltage is near either end of the allowable range. Once the device weaknesses are understood, then the device design, fabrication process, and test program can be improved to maximize production yield. Also, shmoo plots can help us identify worst-case test conditions. For example, we may choose to perform the S/THD test at both low and high V_{DD} based on the worst-case test conditions indicated by the shmoo plot in Figure 4.5.

Figure 4.4. Functional shmoo plot.

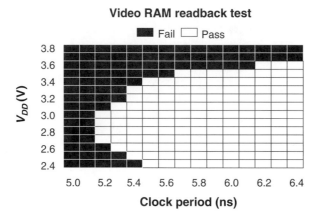

Figure 4.5. Two-dimensional parametric shmoo plot.

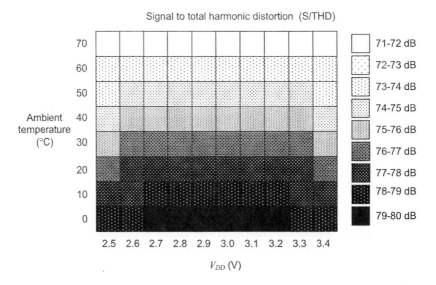

Shmoo plots can be generated, at least in principle, using data collected through manual adjustment of test conditions. However, such a process would be extremely tedious. For example, if we wanted to plot pass/fail results for 10 values of V_{DD} combined with 10 values of master clock period, then we would have to run the test 100 times under 100 different test conditions, adjusting the test conditions by hand each time. Clearly, software automation in the tester is required if the shmoo data collection process is to be a practical one. For this reason, modern ATE tester operating systems often include built-in shmoo plotting tools. These tools not only display the shmoo plots themselves, but they also provide automated adjustment of test conditions and automated collection of test results under each permutation of test conditions.

The shmoo plots illustrated in Figures 4.4 and 4.5 only represent a few of the many types of shmoo plots that can potentially be created. For example, we could certainly imagine a 3D shmoo plot showing pass/fail results for combinations of three test conditions instead of two. It is important to note that any of the many factors affecting DUT performance can be used as shmoo plot test conditions.

Common examples of shmoo test conditions include power supply voltage, master clock frequency (or period), setup and hold times, ambient temperature, I_{OL} or I_{OH} load current, and so on. However, we are free to plot any measured values or pass/fail results as a function of any combination of test conditions. This flexibility makes the shmoo plot a very powerful characterization and diagnostic tool whose usefulness is limited only by the ingenuity and skill of the test or product engineer.

4.1.5 Histograms

When a test program is executed multiple times on a single DUT, it is common to get multiple answers. This is largely a result of the additive noise that is present in the system involving the DUT. For example, a DAC gain error test may show slight repeatability errors if we execute the test program repeatedly, as shown in Figure 4.6. (In this example, only the results from test 5000 have been enabled for display.)

We can view the repeatability of a group of measurements using a visualization tool called a *histogram*. A histogram corresponding to the DAC gain error example is shown in Figure 4.7. It shows a plot of the distribution of measured values as well as a listing of several key statistical values. The plot is divided into a number of vertical histogram cells, each indicating the percentage of values falling within the cell's upper and lower thresholds. For example, approximately 5% of the DAC gain error measurements in this example fell between –0.137 and –0.136 dB. The histogram is a very useful graphical tool that helps us visualize the repeatability of measurements. If the measurement repeatability is good, the distribution should be closely packed, as the example in Figure 4.7 shows. But if repeatability is poor, then the histogram spreads out into a larger range of values. Although histograms are extremely useful for analyzing measurement stability, repeatability studies are not the only use for histograms. They are also used to look at distributions of measurements collected from many DUTs to determine the extent of variability from one device to another. Excessive DUT-to-DUT variability indicates a fabrication process that is out of control or a device design that is too susceptible to normal process variations.

In addition to the numerical results and a plot of the distribution of measured values, the example histogram in Figure 4.7 displays a number of other useful values. For example, the

Figure 4.6. Repeated test executions result in fluctuating measurements.

```
5000 DAC Gain Error      T_VDAC_SNR  -1.000  dB <  -0.127  dB    < 1.000   dB

5000 DAC Gain Error      T_VDAC_SNR  -1.000  dB <  -0.129  dB    < 1.000   dB

5000 DAC Gain Error      T_VDAC_SNR  -1.000  dB <  -0.125  dB    < 1.000   dB

5000 DAC Gain Error      T_VDAC_SNR  -1.000  dB <  -0.131  dB    < 1.000   dB

5000 DAC Gain Error      T_VDAC_SNR  -1.000  dB <  -0.129  dB    < 1.000   dB

5000 DAC Gain Error      T_VDAC_SNR  -1.000  dB <  -0.128  dB    < 1.000   dB

5000 DAC Gain Error      T_VDAC_SNR  -1.000  dB <  -0.132  dB    < 1.000   dB

5000 DAC Gain Error      T_VDAC_SNR  -1.000  dB <  -0.130  dB    < 1.000   dB

5000 DAC Gain Error      T_VDAC_SNR  -1.000  dB <  -0.134  dB    < 1.000   dB

5000 DAC Gain Error      T_VDAC_SNR  -1.000  dB <  -0.131  dB    < 1.000   dB
```

Figure 4.7. Histogram of the DAC gain error test.

Test No	Test Function Name	Test Label	Units
5000	T_VDAC_SNR	DAC Gain Error	db
Lower Test Limit=	-1dB	Upper Test Limit=	1dB
	- DISTRIBUTION	STATISTICS -	
Lower Pop Limit=	-Infinity	Upper Pop Limit=	+Infinity
Total Results=	110	Results Accepted=	110
Underflows=	0	Overflows=	0
Mean=	-0.13003dB	Std Deviation=	0.00292899
Mean – 3 Sigma=	-0.13882dB	Mean + 3 Sigma=	-0.012125dB
Minimum Value=	-0.13594dB	Maximum Value=	-0.12473dB
	- PLOT STATISTICS –		
Lower Plot Limit=	-0.14dB	Upper Plot Limit=	-0.12dB
Cells=	15	Cell Width=	0.0013333dB
Full Scale Percent=	16.36%	Full Scale Count=	18

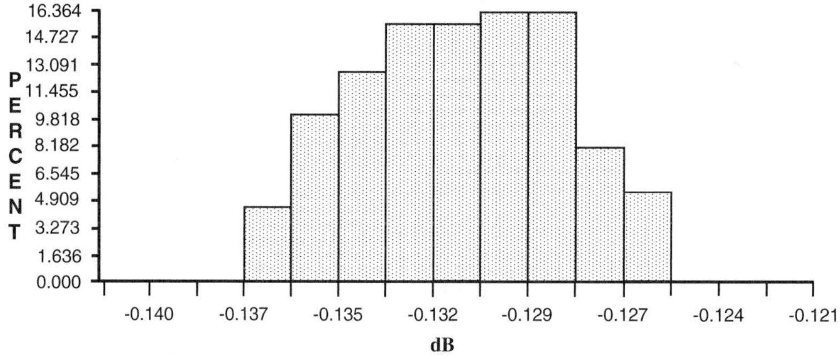

population size is listed beside the heading "Total Results=." It indicates how many times the measurement was repeated. In the case of a DUT-to-DUT variability study, the "Total Results=" value would correspond to the number of DUTs tested rather than the number of measurement repetitions on the same DUT. In either case, the larger the population of results, the trustworthier a histogram becomes. A histogram with fewer than 50 results is statistically questionable because of the limited sample size. Ideally a histogram should contain results from at least 100 devices (or 100 repeated test executions in the case of a single-DUT repeatability study).

While the histogram is useful tool for visualizing the data set, we are often looking for a simpler representation of the data set, especially one that lends itself for comparison to other data sets. To this end, the next section will introduce a statistical description of the data set, where specific metrics of the data are used for comparative purposes. Moreover, we shall introduce some basic concepts of probability theory of random variables to help quantify the mathematical structure present in a data set. It is assumed that the reader has been exposed to probability theory elsewhere, and only the concepts relevant to analog and mixed-signal testing will be covered here. For a more in-depth presentation of statistics, including the derivation of fundamental equations and properties of statistics, the reader should refer to a book on the subject of statistics and probability theory.[2-4]

4.2 STATISTICAL ANALYSIS

4.2.1 Mean (Average) and Standard Deviation (Variance)

The frequency or distribution of data described by a histogram characterizes a sample set in great detail. Often, we look for simpler measures that describe the statistical features of the sample set. The two most important measures are the arithmetic mean μ and the standard deviation σ.

The arithmetic mean or simply the mean μ is a measure of the central tendency, or location, of the data in the sample set. The mean value of a sample set denoted by $x(n)$, $n = 0, 1, 2, \ldots, N-1$, is defined as

$$\mu = \frac{1}{N} \sum_{n=0}^{N-1} x(n) \qquad (4.4)$$

Exercises

4.1. A 5-mV signal is measured with a meter 10 times, resulting in the following sequence of readings: 5 mV, 6 mV, 9 mV, 8 mV, 4 mV, 7 mV, 5 mV, 7 mV, 8 mV, 11 mV. What is the mean value? What is the standard deviation?

ANS. 7 mV, 2.0 mV.

4.2. What are the mean and standard deviation of a set of samples of a coherent sine wave having a DC offset of 5 V and a peak-to-peak amplitude of 1.0 V?

ANS. μ = 5.0 V, σ = 354 mV.

4.3. A 5-mV signal is measured with a meter 10 times, resulting in the following sequence of readings: 7 mV, 6 mV, 9 mV, 8 mV, 4 mV, 7 mV, 5 mV, 7 mV, 8 mV, 11 mV. What is the mean value? What is the standard deviation?

ANS. 7.2 mV, 1.887 mV.

4.4. If 15,000 devices are tested with a yield of 63%, how many devices passed the test?

ANS. 9450 devices.

For the DAC gain error example shown in Figure 4.7 the mean value from 110 measurements is −0.1300 dB. The standard deviation σ, on the other hand, is a measure of the dispersion or uncertainty of the measured quantity about the mean value, μ. If the values tend to be concentrated near the mean, the standard deviation is small. If the values tend to be distributed far from the mean, the standard deviation is large. Standard deviation is defined as

$$\sigma = \sqrt{\frac{1}{N} \sum_{n=0}^{N-1} [x(n) - \mu]^2} \qquad (4.5)$$

Standard deviation and mean are expressed in identical units. In our DAC gain error example, the standard deviation was found to be 0.0029 dB. Another expression for essentially the same quantity is the variance or mean square deviation. It is simply equal to the square of the standard deviation, that is, variance $= \sigma^2$.

Often, the statistics of a sample set are used to estimate the statistics of a larger group or population from which the samples were derived. Provided that the sample size is greater than 30, approximation errors are insignificant. Throughout this chapter, we will make no distinction between the statistics of a sample set and those of the population, because it will be assumed that sample size is much larger than 30.

There is an interesting relationship between a sampled signal's DC offset and RMS voltage and the statistics of its samples. Assuming that all frequency components of the sample set are coherent, the mean of the signal samples is equal to the signal's DC offset. Less obvious is the fact that the standard deviation of the samples is equal to the signal's RMS value, excluding the DC offset. The RMS of a sample set is calculated as the square root of the mean of the squares of the samples

$$\text{RMS} = \sqrt{\frac{1}{N} \sum_{n=0}^{N-1} \left[x\left(n\right) \right]^2} \tag{4.6}$$

If the value of μ in Eq. (4.5) is zero (i.e., if the sample set has no DC component), then Eq. (4.5) becomes identical to the RMS calculation above. Thus we can calculate the standard deviation of the samples of a coherent signal by calculating the RMS of the signal after subtracting the average value of the sample set (i.e., the DC offset).

4.2.2 Probabilities and Probability Density Functions

The histogram in Figure 4.7 exhibits a feature common to many analog and mixed-signal measurements. The distribution of values has a shape similar to a bell. The bell curve (also called a *normal distribution* or *Gaussian distribution*) is a common one in the study of statistics. According to the central limit theorem,[5] the distribution of a set of random variables, each of which is equal to a summation of a large number ($N > 30$) of statistically independent random values, trends toward a Gaussian distribution. As N becomes very large, the distribution of the random variables becomes Gaussian, whether or not the individual random values themselves exhibit a Gaussian distribution. The variations in a typical mixed-signal measurement are caused by a summation of many different random sources of noise and crosstalk in both the device and the tester instruments. As a result, many mixed-signal measurements exhibit the common Gaussian distribution.

Figure 4.8 shows the histogram count from Figure 4.7 superimposed on a plot of the corresponding Gaussian probability density function (PDF). The PDF is a function that defines the probability that a randomly chosen sample X from the statistical population will fall near a particular value. In a Gaussian distribution, the most likely value of X is near the mean value, μ. Thus the PDF has a peak at $x = \mu$.

Notice that the height of the histogram cells only approximates the shape of the true Gaussian curve. If we collect thousands of test results instead of the 110 used in this example, the height of the actual histogram cells should more closely approach the shape of the probability density

Figure 4.8. Continuous normal (Gaussian) distribution for DAC gain example.

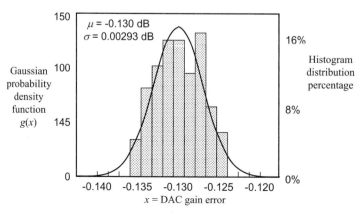

function. This is the nature of statistical concepts. Actual measurements only approach the theoretical ideal when large sample sets are considered.

The bell-shaped probability density function $g(x)$ for any Gaussian distribution having a mean μ and standard deviation σ is given by the equation

$$g(x) = \frac{1}{\sigma\sqrt{2\pi}} e^{\frac{-(x-\mu)^2}{2\sigma^2}} \qquad \textbf{(4.7)}$$

Since the shape of the histogram in Figure 4.8 approximates the shape of the Gaussian pdf, it is easy to assume that we can calculate the expected percentage of histogram counts at a value by simply plugging the value of into Eq. (4.7). However, a PDF represents the probability *density*, rather than the probability itself. We have to perform integration on the PDF to calculate probabilities and expected histogram counts.

The probability that a randomly selected value in a population will fall between the values a and b is equal to the area under the PDF curve bounded by $x = a$ and $x = b$, as shown in Figure 4.9. Stating this more precisely, for any probability density function $f(x)$, the probability P that a randomly selected value X will fall between the values a and b is given by

$$P(a < X < b) = \int_a^b f(x)dx \qquad \textbf{(4.8)}$$

The value of P must fall between 0 (0% probability) and 1 (100% probability). In the case where $a = -\infty$ and $b = \infty$, the value of P must equal 1, since there is a 100% probability that a randomly chosen value will be a number between $-\infty$ and $+\infty$. Consequently, the total area underneath any PDF must always be equal to 1.

As $f(x)$ is assumed continuous, the probability that a random variable X is *exactly* equal to any particular value is zero. In such case we can replace either or both of the signs $<$ in Eq. (4.8) by $\leq$, allowing us to write

$$P(a \leq X \leq b) = \int_a^b f(x)dx \qquad \textbf{(4.9)}$$

Incorporating the equality in the probability expression is therefore left as a matter of choice.

The probability that a Gaussian distributed randomly variable X will fall between the values of a and b can be derived from Eq. (4.8) by substituting Eq. (4.7) to obtain

$$P(a \leq X \leq b) = \int_a^b g(x)\,dx = \int_a^b \frac{1}{\sigma\sqrt{2\pi}} e^{\frac{-(x-\mu)^2}{2\sigma^2}}\,dx \qquad \textbf{(4.10)}$$

Unfortunately, Eq. (4.10) cannot be solved in closed form. However, it can easily be solved using numerical integration methods. For instance, in our Gaussian DAC gain example, let us say we want to predict what percentage of measured results should fall into the seventh histogram cell. From the histogram, we see that there are 15 evenly spaced cells between –0.140 and –0.120 dB. The seventh cell represents all values falling between the values

$$a = -0.140 \text{ dB} + 7\,\frac{\left[-0.120 \text{ dB} - (-0.140 \text{ dB})\right]}{15} = -0.132 \text{ dB}$$

Figure 4.9. The probability over the range *a* to *b* is the area under the PDF, *f(x)*, in that interval.

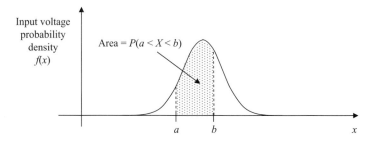

and

$$b = -0.140 \text{ dB} + 8 \frac{\left[-0.120 \text{ dB} - (-0.140 \text{ dB})\right]}{15} = -0.1307 \text{ dB}$$

We can calculate the probability that a randomly selected DAC gain error measurement X will fall between -0.1320 and -0.1307 using the equation

$$P\left(-0.132 < X < -0.1307\right) = \int_{-0.132}^{-0.1307} \frac{1}{\sigma\sqrt{2\pi}} e^{\frac{-(x-\mu)}{2\sigma^2}} dx$$

where $\mu = -0.130$ dB and $\sigma = 0.00293$ dB. Using mathematical analysis software, the value of $P(-0.1320 < X < -0.1307)$ is found to be 0.163, or 16.3%. Theoretically, then, we should see 16.3% of the 110 DAC gain error measurements fall between -0.1307 and -0.1320. Indeed the seventh histogram cell shows that approximately 15.5% of the measurements fall between these values.

This example is fairly typical of applied statistics. Actual distributions never exactly match a true Gaussian distribution. Notice, for example, that the ninth and tenth histogram cells in Figure 4.8 are badly out of line with the ideal Gaussian curve. Also, notice that there are no values outside the fourth and twelfth cells. An ideal Gaussian distribution would extend to infinity in both directions. In other words, if one is willing to wait billions of years, one should eventually see an answer of +200 dB in the DAC gain error example. In reality, of course, the answer in the DAC gain example will never stray more than a few tenths of a decibel away from the average reading of -0.130 dB, since the actual distribution is only near Gaussian.

Nevertheless, statistical analysis predicts actual results well enough to be very useful in analyzing test repeatability and manufacturing process stability. The comparison between ideal results and actual results is close enough to allow some general statements. First, the peak-to-peak variation of the observed data (e.g., maximum observed value – minimum observed value) is roughly equal to six times the standard deviation of a near-Gaussian distribution, that is,

$$\text{peak-to-peak value} \approx 6\sigma \tag{4.11}$$

In the DAC gain distribution example, the standard deviation is 0.00293 dB. Therefore, we would expect to see values ranging from approximately -0.139 dB to -0.121 dB. These values are displayed in the example histogram in Figure 4.7 beside the labels "Mean -3 sigma" and "Mean $+3$ sigma." The actual minimum and maximum values are also listed. They range from -0.136 to -0.125 dB, which agrees fairly well with the ideal values of $\mu \pm 3\sigma$.

At this point we should note a common misuse of statistical analysis. We have used as our example a gain measurement, expressed in decibels. Since the decibel is based on a logarithmic transformation, we should actually use the equivalent V/V measurements to calculate statistical quantities such as mean and standard deviation. For example, the average of three decibel values, 0, −20, and −40 dB, is 20 dB. However, the true average of these values as calculated using V/V is given by

$$\text{average gain} = \frac{1 + \frac{1}{10} + \frac{1}{100}}{3} = 0.37 \ \frac{V}{V} = -8.64 \text{ dB}$$

A similar discrepancy arises in the calculation of standard deviation. Therefore, a Gaussian-distributed sample set converted into decibel form is no longer Gaussian. Nevertheless, we often use the nonlinearized statistical calculations from a histogram to evaluate parameters expressed in decibel units as a time-saving shortcut. The discrepancy between linear and logarithmic calculations of mean and standard deviation become negligible as the range of decibel values decreases. For example, the range of values in the histogram of Figure 4.7 is quite small; so the errors in mean and standard deviation are minor. The reader should be careful when performing statistical analysis of decibel values ranging over several decibels.

4.2.3 The Standard Gaussian Cumulative Distribution Function $\Phi(z)$

Computing probabilities involving Gaussian distributions is complicated by the fact that numerical integration methods must be used to solve the definite integrals involved. Fortunately, a simple change of variable substitution can be used to convert the integral equations into one involving a Gaussian distribution with zero mean and unity standard deviation. This then enables a set of tables or approximations that require numerical evaluation to be used to solve the probabilities associated with an arbitrary Gaussian distribution. Hence, the test engineer can completely avoid the need for numerical integration routines. Let us consider how this is done.

The probability that a randomly selected value X will be less than a particular value x can be calculated directly from Eq. (4.9). We set $a = -\infty$ and $b = x$ and write

$$F(x) = P(X < x) = P(-\infty < X < x) = \int_{-\infty}^{x} f(y) \, dy \qquad \textbf{(4.12)}$$

This integral is central to probability theory and is given a special name called the *cumulative distribution function (CDF)*. Here we view $F(x)$ as an ordinary function of the variable x. The probability that X lies in the range a to b can then be expressed in terms of the difference of $F(x)$ evaluated at $x = a$ and $x = b$ according to

$$P(a < X < b) = F(b) - F(a) \qquad \textbf{(4.13)}$$

In the case of a Gaussian distribution, $F(x)$ is equal to

$$F(x) = \int_{-\infty}^{x} \frac{1}{\sigma\sqrt{2\pi}} e^{\frac{-(y-\mu)^2}{2\sigma^2}} \, dy \qquad \textbf{(4.14)}$$

If we consider the simple change of variable $z = (y - \mu)/\sigma$, Eq. (4.14) can be rewritten as

$$F(x) = \int_{-\infty}^{\left(\frac{x-\mu}{\sigma}\right)} \frac{1}{\sqrt{2\pi}} e^{\frac{-z^2}{2}} \, dz \qquad (4.15)$$

Except for the presence of μ and σ in the upper integration limit, the integration kernel no longer depends on these two values. Alternatively, one can view $F(x)$ in Eq. (4.15) as the CDF of a Gaussian distribution having zero mean and unity standard deviation. By tabulating a single function, say

$$\Phi(z) = \frac{1}{\sqrt{2\pi}} \int_{-\infty}^{z} e^{\frac{-u^2}{2}} \, du \qquad (4.16)$$

we can write $F(x)$ as

$$F(x) = \Phi\left(\frac{x - \mu}{\sigma}\right) \qquad (4.17)$$

In other words, to determine the value of a particular CDF involving a Gaussian random variable with mean μ and standard deviation σ at a particular point, say, x, we simply normalized x by subtracting the mean value followed by a division by σ, that is, $z = (y - \mu)/\sigma$, and compute $\Phi(z)$.

The function $\Phi(z)$ is known as the *standard* Gaussian CDF. The variable z is known as the *standardized point of reference*. Traditionally, $\Phi(z)$ has been evaluated by looking up tables that list $\Phi(z)$ for different values of z. A short tabulation of $\Phi(z)$ is provided in Appendix A. Rows of $\Phi(z)$ are interleaved between rows of z. A plot of $\Phi(z)$ versus z is also provided in Figure 4.10. For reference, we see from this plot that $\Phi(-\infty) = 0$ and $\Phi(\infty) = 1$. Also evident is the antisymmetry about the point $(0, 0.5)$, giving rise to the relation $\Phi(-z) = 1 - \Phi(z)$.

More recently, the following expression[6] has been found to give reasonably good accuracy (less than 0.1%) for $\Phi(z)$ with $\alpha = 1/\pi$ and $\beta = 2\pi$:

$$\Phi(z) \approx \begin{cases} \left[1 - \left(\dfrac{1}{(1-\alpha)z + \alpha\sqrt{z^2 + \beta}} \right) \dfrac{1}{\sqrt{2\pi}} e^{\frac{-z^2}{2}} \right] & 0 \le z \le \infty \\[6mm] \left[\left(\dfrac{1}{(\alpha-1)z + \alpha\sqrt{z^2 + \beta}} \right) \dfrac{1}{\sqrt{2\pi}} e^{\frac{-z^2}{2}} \right] & -\infty < z < 0 \end{cases} \qquad (4.18)$$

Equation (4.18) is generally very useful when we require a standardized value that is not contained in the table of Appendix A or any other Gaussian CDF table for that matter. To illustrate the application of the standard cdf, the probability that a Gaussian distributed random variable X with mean μ and standard deviation σ lies in the range a to b is written as

$$P(a < X < b) = \Phi\left(\frac{b - \mu}{\sigma}\right) - \Phi\left(\frac{a - \mu}{\sigma}\right) \qquad (4.19)$$

Figure 4.10. Standard Gaussian cumulative distribution function.

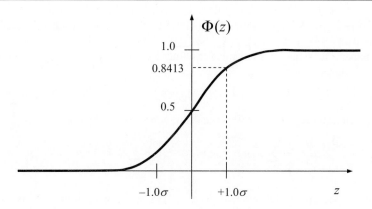

Equation (4.19) is a direct result of substituting Eq. (4.17) into Eq. (4.13). Equation (4.19) can be used to generate certain rules of thumb when dealing with Gaussian random variables. The probability that a random variable will fall within:

1σ of its mean is

$$P\left(\mu-\sigma<X<\mu+\sigma\right)=\Phi\left(\frac{\mu+\sigma-\mu}{\sigma}\right)-\Phi\left(\frac{\mu-\sigma-\mu}{\sigma}\right)=\Phi(1)-\Phi(-1)=0.6826$$

2σ of its mean is

$$P\left(\mu-2\sigma<X<\mu+2\sigma\right)=\Phi\left(\frac{\mu+2\sigma-\mu}{\sigma}\right)-\Phi\left(\frac{\mu-2\sigma-\mu}{\sigma}\right)=\Phi(2)-\Phi(-2)=0.9544$$

3σ of its mean is

$$P\left(\mu-3\sigma<X<\mu+3\sigma\right)=\Phi\left(\frac{\mu+3\sigma-\mu}{\sigma}\right)-\Phi\left(\frac{\mu-3\sigma-\mu}{\sigma}\right)=\Phi(3)-\Phi(-3)=0.9974$$

It is also instructive to look at several limiting cases associated with Eq. (4.19); specifically, when $a=-\infty$ and b is an arbitrary value, we find

$$P\left(X<b\right)=P\left(-\infty<X<b\right)=\Phi\left(\frac{b-\mu}{\sigma}\right)-\Phi\left(\frac{-\infty-\mu}{\sigma}\right)=\Phi\left(\frac{b-\mu}{\sigma}\right) \qquad \textbf{(4.20)}$$

Conversely, with a arbitrary and $b=\infty$, we get

$$P\left(a<X\right)=P\left(a<X<\infty\right)=\Phi\left(\frac{\infty-\mu}{\sigma}\right)-\Phi\left(\frac{a-\mu}{\sigma}\right)=1-\Phi\left(\frac{a-\mu}{\sigma}\right) \qquad \textbf{(4.21)}$$

Of course, as previously mentioned, with $a = -\infty$ and $b = \infty$, we obtain $P\left(-\infty < X < \infty\right) = 1$.

Notice that the probability that X will fall outside the range $\mu \pm 3\sigma$ is extremely small, that is, $P\left(X < \mu - 3\sigma\right) + P\left(\mu + 3\sigma < X\right) = \Phi(-3) + 1 - \Phi(3) = 0.0026$. As a result, one would not expect many measurement results to fall very far beyond $\mu \pm 3.0\sigma$ in a relatively small-size test set.

To summarize, the steps involved in calculating a probability involving a Gaussian random variable are as follows:

1. Estimate the mean μ and standard deviation σ of the random variable from the sample set using

$$\mu = \frac{1}{N}\sum_{n=0}^{N-1} x(n) \quad \text{and} \quad \sigma = \sqrt{\frac{1}{N}\sum_{n=0}^{N-1}\left[x(n) - \mu\right]^2}$$

2. Determine the probability interval limits, a and b, and write a probability expression in terms of the standard Gaussian cumulative distribution function $\Phi(z)$ according to

$$P\left(a < X < b\right) = \Phi\left(\frac{b - \mu}{\sigma}\right) - \Phi\left(\frac{a - \mu}{\sigma}\right)$$

3. Evaluate $\Phi(z)$ through a computer program, or a look-up table (Appendix A) or by using the numerical approximation given in Eq. (4.18).

The following examples will help to illustrate this procedure.

EXAMPLE 4.1

A DC offset measurement is repeated many times, resulting in a series of values having an average of 257 mV. The measurements exhibit a standard deviation of 27 mV. What is the probability that any single measurement will return a value larger than 245 mV?

Solution:
If X is used to denote the Gaussian random variable, then we want to know $P(245\text{ mV} < X)$. Comparing the probability limits with the expression listed in Eq. (4.21), we can state $a = 245$ mV and $b = \infty$. Furthermore, since $\mu = 257$ mV and $\sigma = 27$ mV, we can write

$$P\left(245\text{ mV} < X\right) = 1 - \Phi\left(\frac{245\text{ mV} - 257\text{ mV}}{27\text{ mV}}\right) = 1 - \Phi(-0.44)$$

Referring to Appendix A we see that the value $\Phi(-0.44)$ is not listed. However, it lies somewhere between 0.3085 and 0.3446; thus either we can make a crude midpoint interpolation of 0.33 or, alternatively, we can use the approximation for $\Phi(z)$ given in Eq. (4.18) and write

$$\Phi(-0.44) \approx \left[\frac{1}{\left(\frac{1}{\pi} - 1\right)(-0.44) + \left(\frac{1}{\pi}\right)\sqrt{(-0.44)^2 + 2\pi}}\right]\frac{1}{\sqrt{2\pi}}e^{\frac{-(-0.44)^2}{2}} = 0.3262$$

We shall select the latter value of 0.3262 and state that the probability that the measurement will be **greater** than 245 mV is equal to 1 − 0.3262, or 0.6738. Consequently, there is a 67.38% chance that any individual measurement will exceed 245 mV.

EXAMPLE 4.2

A sample of a DC signal in the presence of noise is to be compared against a reference voltage level. If the sample is greater than the 1-V threshold, the comparator output goes to logic 1. If 100 samples were taken, but only five ones were detected, what is the value of the signal, assuming that the noise is zero mean and has a standard deviation of 10 mV?

Solution:
Modeling the PDF of the signal plus noise at the receiver input, we can write

$$f(v) = \frac{1}{10^{-2}\sqrt{2\pi}} e^{\frac{-(v-\mu)^2}{2(10^{-2})^2}}$$

where the mean value, μ, is equal to the unknown DC signal level. Next, we are told that there is a 5% chance of the sample exceeding the 1-V threshold. This is equivalent to the mathematical statement $p\,(1.0 \le v < \infty) = 5/100$. Because the noise is Gaussian, we can write the following from Eq. (4.19):

$$P\,(1.0 \le V < \infty) = 0.05 = 1 - \Phi\left(\frac{1.0 - \mu}{10^{-2}}\right)$$

or

$$\Phi\left(\frac{1.0 - \mu}{10^{-2}}\right) = 0.95$$

which allows us to write

$$\frac{1.0 - \mu}{10^{-2}} = \Phi^{-1}(0.95) \Rightarrow \mu = 1.0 - 10^{-2} \times \Phi^{-1}(0.95)$$

We can estimate the value of $\Phi^{-1}(0.95)$ from Appendix A, to be somewhere between 1.6 and 1.7. or we can make use of a computer program and find $\Phi^{-1}(0.95) = 1.645$. This leads to

$$\mu = 1.0 - 10^{-2} \times 1.645 = 0.98$$

Therefore the signal level is 0.98 V.

Exercises

4.5. A 5-mV signal is measured with a meter ten times resulting in the following sequence of readings: 5 mV, 6 mV, 9 mV, 8 mV, 4 mV, 7 mV, 5 mV, 7 mV, 8 mV, 11 mV. Write an expression for the PDF for this measurement set assuming the distribution is Gaussian.

ANS. $g(v) = \dfrac{1}{(2 \times 10^{-3})\sqrt{2\pi}} e^{-\frac{(v - 7 \times 10^{-3})^2}{2(2 \times 10^{-3})^2}}$;

$\mu = 7.0 \text{ mV}; \sigma = 2.0 \text{ mV}.$

4.6. A set of measured data is modeled as a Gaussian distribution with 1.0-V mean value and a standard deviation of 0.1 V. What is the probability a single measurement will fall between 0.8 and 0.9 V.

ANS. $P(0.8 \le V \le 0.9) = 0.18.$

4.2.4 Verifying Gaussian Behavior: The Kurtosis and Normal Probability Plot

It is fairly common to encounter distributions that are non-Gaussian. Two common deviations from the familiar bell shape are bimodal distributions, such as that shown in Figure 4.11, and distributions containing outliers as shown in Figure 4.12. When evaluating measurement repeatability on a single DUT, these distributions are a warning sign that the test results are not sufficiently repeatable. When evaluating process stability (i.e., consistency from DUT to DUT), these plots may indicate a weak design or a process that needs to be improved. In general, looking at the histogram plot to see if the distribution accurately follows a Gaussian distribution is difficult to do by inspection. Instead, one prefers to work with some form of mathematical measure of normality.

A commonly used metric to see how closely a given data set follows a Gaussian distribution is the kurtosis. The kurtosis is defined as

$$\text{kurtosis} = \frac{1}{N} \sum_{n=1}^{N} \left[\frac{x(n) - \mu}{\sigma} \right]^4 \qquad (4.22)$$

where μ is the mean and σ is the standard deviation of a random sequence x. A Gaussian distribution regardless of its mean value or standard deviation has a kurtosis of three. Any significant deviation from this value will imply that the distribution is non-Gaussian. However, one generally

Figure 4.11. Bimodal distribution.

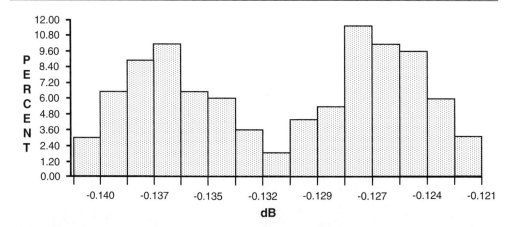

Figure 4.12. Distribution with statistical outliers.

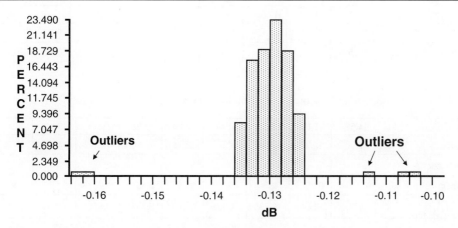

must be cautious of making decisions from a point metric like kurtosis alone. Instead, a normal probability plot[7] is a better approach for observing whether or not a data set is approximately normally distributed. The data are plotted against a theoretical normal distribution in such a way that the points should form an approximate straight line. Departures from this straight line indicate departures from normality.

The steps for constructing a normal probability plot are best illustrated by way of an example. Let us assume that we have the following 10 data points with a kurtosis of 2.5 (generalizing these steps to larger data sets should be straightforward):

Data	12.04	14.57	12.00	12.02	13.48	12.70	12.86	15.13	10.10	14.14

Step 1: Sort the data from the smallest to the largest.

Data	12.04	14.57	12.00	12.02	13.48	12.70	12.86	15.13	10.10	14.14
Data Sort	10.10	12.00	12.02	12.04	12.70	12.86	13.48	14.14	14.57	15.13

Step 2: Assign each data point a position number, k, from 1 to N (length of data vector, 10).

Data	12.04	14.57	12.00	12.02	13.48	12.70	12.86	15.13	10.10	14.14
Data Sort	10.10	12.00	12.02	12.04	12.70	12.86	13.48	14.14	14.57	15.13
Position k	1	2	3	4	5	6	7	8	9	10

Step 3: Use the position values to calculate the corresponding uniformly distributed probability, where the individual probabilities are found from the expression $p_k = (k - 0.5)/N$, where in this case $N = 10$.

Data	12.04	14.57	12.00	12.02	13.48	12.70	12.86	15.13	10.10	14.14
Data Sort	10.10	12.00	12.02	12.04	12.70	12.86	13.48	14.14	14.57	15.13
Position k	1	2	3	4	5	6	7	8	9	10
Prob. p_k	0.05	0.15	0.25	0.35	0.45	0.55	0.65	0.75	0.85	0.95

Step 4: Compute the standard z scores corresponding to each probability, that is, $\Phi^{-1}(p_k)$.

Data	12.04	14.57	12.00	12.02	13.48	12.70	12.86	15.13	10.10	14.14
Data Sort	10.10	12.00	12.02	12.04	12.70	12.86	13.48	14.14	14.57	15.13
Position k	1	2	3	4	5	6	7	8	9	10
Prob. p_k	0.05	0.15	0.25	0.35	0.45	0.55	0.65	0.75	0.85	0.95
$\Phi^{-1}(p_k)$	-1.645	-1.036	-0.674	-0.385	-0.126	0.126	0.385	0.674	1.036	1.645

Step 5: Plot the sorted measured data versus the standard normal values, $\Phi^{-1}(p_k)$ as shown in Figure 4.13.

Figure 4.13. A normal probability plot used to visually validate that a data set follows Gaussian behavior ($N = 10$).

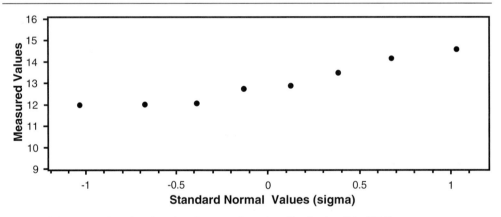

Figure 4.14. A normal probability plot of a near-Gaussian distribution ($N = 1000$).

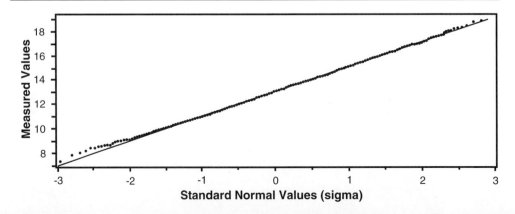

As is evident from Figure 4.13, the data appear to roughly follow a straight line. There is no reason to think that the data are severely non-Gaussian even though the kurtosis does not equal to 3. Keep in mind that our confidence in our decision either way is questionable since we are working with only 10 samples. If the sample set is increased to 1000, the straight-line behavior of the probability plot becomes quite obvious as shown in Figure 4.14. We would then say with great confidence that the data behave Gaussian over a range of $\pm 2\sigma$ but less so beyond these values. It is also interesting to note that the kurtosis is very close to 3 at 2.97. This example serves to illustrate the limitation of a single-point metric in judging the normality of the data.

4.3 NON-GAUSSIAN DISTRIBUTIONS FOUND IN MIXED-SIGNAL TEST

A few examples of non-Gaussian probability distributions functions that one encounters in analog and mixed-signal test problems will be described in this section. Both continuous and discrete probability distributions will be described. The reader will encounter these distributions later on in this textbook.

4.3.1 The Uniform Probability Distribution

A random variable X that is equally likely to take on values within a given range [A,B] is said to be uniformly distributed. The random variable may be discrete or continuous. The PDF of a continuous uniform distribution is a rectangular function, as illustrated in Figure 4.15. Note that the height of this PDF must be chosen so that the area enclosed by the rectangular function is unity. Mathematically, the PDF of a uniform distribution is given by

$$f(x)=\begin{cases} \dfrac{1}{B-A}, & A \leq x \leq B \\ \\ 0, & \text{elsewhere} \end{cases} \tag{4.23}$$

The probability that a uniformly distributed random variable X will fall in the interval a to b, where $-A \leq a < b \leq B$ is obtained by substituting Eq. (4.23) into Eq. (4.9) to get

$$P(a \leq X \leq b)=\int_a^b f(x)\,dx = \int_a^b \frac{1}{B-A}\,dx = \frac{b-a}{B-A} \tag{4.24}$$

The mean and standard deviation of an ideal uniform distribution are given by

$$\mu = \frac{A+B}{2} \tag{4.25}$$

and

$$\sigma = \frac{(B-A)}{\sqrt{12}} \tag{4.26}$$

In the discrete case, the PDF (or what is also referred to as the probability mass function) of a uniformly distributed random variable within the interval [A,B] is a series of equally spaced, equally weighted impulse functions described as

$$f(x) = \frac{1}{B-A+1} \sum_{k=A}^{B} \delta(x-k) \tag{4.27}$$

In much the same way as the continuous case, the probability that a uniformly distributed random integer X will fall in the interval bounded by integers a and b, where $A \le a < b \le B$ is obtained by substituting Eq. (4.27) into Eq. (4.9) to get

$$P(a \le X \le b) = \int_{a}^{b} f(x)\,dx = \frac{b-a}{B-A+1} \tag{4.28}$$

The mean and standard deviation of the discrete uniform distribution are given by

$$\mu = \frac{A+B}{2} \quad \text{and} \quad \sigma = \sqrt{\frac{(B-A+1)^2 - 1}{12}} \tag{4.29}$$

Uniform distributions occur in at least two instances in mixed-signal test engineering. The first instance is random number generators found in various programming languages, such as the *rand* function in MATLAB and the *random* function in C. This type of function returns a randomly chosen number between a minimum and maximum value (typically 0 and 1, respectively). The numbers are supposed to be uniformly distributed between the minimum and maximum values. There should be an equal probability of choosing any particular number, and therefore a histogram of the resulting population should be perfectly flat. One measure of the quality of a random number generator is the degree to which it can produce a perfectly uniform distribution of values.

The second instance in which we commonly encounter a uniform distribution is the errors associated with the quantization process of an analog-to-digital converter (ADC). It is often assumed that statistical nature of these errors is uniformly distributed between –1/2 LSB (least significant bit) and +1/2 LSB. This condition is typically met in practice with an input signal that is sufficiently random. From Eq. (4.25), we see that the average error is equal to zero. Using Eq. (4.26) and, assuming that the standard deviation and RMS value are equivalent, we expect that the ADC will generate $1/\sqrt{12}$ LSB of RMS noise when it quantizes a signal. In the case of a

Figure 4.15. Uniform distribution pdf.

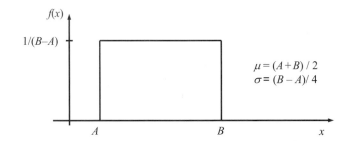

full-scale sinusoidal input having a peak of 2^{N-1} LSBs (or an RMS of $2^{N-1}/\sqrt{2}$), the signal-to-noise ratio (SNR) at the output of the ADC is

$$SNR = 20 \, \log\left(\frac{\text{signal RMS}}{\text{noise RMS}}\right) = 20 \, \log\left(\frac{2^{N-1}/\sqrt{2} \text{ LSB}}{1/\sqrt{12} \text{ LSB}}\right) \tag{4.30}$$

leading to

$$SNR = 20 \, \log_2\left(\sqrt{6} \times 2^{N-1}\right) \tag{4.31}$$

Simplifying further, we see that the SNR depends linearly on the number of bits, N, according to

$$SNR = \frac{20 \log_2\left(\sqrt{6} \times 2^{N-1}\right)}{\log_2\left(10\right)} = 1.761 \text{ dB} + 6.02 \text{ dB} \times N \tag{4.32}$$

In the situation where we know the SNR, we can deduce from Eq. (4.32) that the equivalent number of bits for the ADC, that is, ENOB = N, is

$$ENOB = \frac{SNR - 1.761 \text{ dB}}{6.02 \text{ dB}} \tag{4.33}$$

Exercises

4.7. A random number generator produces the digits: N = 0, 1, 2, ..., 9 uniformly. What is the cumulative distribution function of this data set? What is the mean and standard deviation of this pdf?

ANS. $P\left(0 \leq X \leq x\right) = F\left(x\right) = 0.1 + 0.1x$,
$x \in \{0,1,2,...,9\}; \mu = 4.5; \sigma = 2.87$.

4.8. A triangular PDF has the form $f(x) = \begin{cases} 4x, & 0 \leq x < 0.5 \\ 4 - 4x, & 0.5 \leq x \leq 1 \end{cases}$.

What is the corresponding cumulative distribution function? What is the mean and standard deviation?

ANS. $F(x) = \begin{cases} 0, & x < 0 \\ 2x^2, & 0 \leq x < 0.5 \\ 1 - 2(1-x)^2, & 0.5 \leq x \leq 1 \\ 1, & 1 < x \end{cases}$;

$\mu = \frac{1}{2}; \sigma = \frac{1}{\sqrt{24}}$

4.3.2 The Sinusoidal Probability Distribution

Another non-Gaussian distribution that is commonly found in analog and mixed-signal test engineering problems is the sinusoidal distribution shown in Figure 4.16. The PDF of a sinusoidal signal described by $A \sin\left(2\pi f \cdot t\right) + B$ is given by

$$f\left(x\right) = \begin{cases} \dfrac{1}{\pi\sqrt{A^2 - \left(x - B\right)^2}}, & -A + B \leq x \leq A + B \\ \\ 0, & \text{elsewhere} \end{cases} \tag{4.34}$$

The probability that a sinusoidal distributed random variable X will fall in the interval a to b, where $-A + B \leq a < b \leq A + B$ is obtained by substituting Eq. (4.34) into Eq. (4.9) to get

$$P\left(a < X < b\right) = \int_a^b \frac{1}{\pi\sqrt{A^2 - \left(x - B\right)^2}}\, dx = \frac{1}{\pi}\left[\sin^{-1}\left(\frac{b - B}{A}\right) - \sin^{-1}\left(\frac{a - B}{A}\right)\right] \qquad \textbf{(4.35)}$$

The mean and standard deviation of an ideal sinusoidal distribution with offset B are given by

$$\mu = B \qquad\qquad \textbf{(4.36)}$$

and

$$\sigma = \frac{A}{\sqrt{2}} \qquad\qquad \textbf{(4.37)}$$

Sinusoidal distributions occur in at least two instances in mixed-signal test engineering. The first instance is in ADC testing where the ADC transfer characteristic is derived from the code distribution that results from sampling a full-scale sine-wave input described by $A \sin\left(2\pi f \cdot t\right) + B$. Because an ideal D-bit ADC has $2^D - 1$ code edges located at $-A + B + V_{LSB} \times n$, $n = 0, 1, \dots 2^D - 1$, where V_{LSB} is the distance between adjacent code edges, the probability of a random sample X falling in the region bounded by consecutive code edges can be derived from Eq. (4.35) as

$$P\left[-A + B + n \times V_{LSB} < X < -A + B + \left(n + 1\right) \times V_{LSB}\right]$$
$$= \frac{1}{\pi}\sin^{-1}\left[\frac{\left(n + 1\right) \times V_{LSB} - A}{A}\right] - \frac{1}{\pi}\sin^{-1}\left[\frac{n \times V_{LSB} - A}{A}\right] \qquad \textbf{(4.38)}$$

Figure 4.16. The PDF of a sinusoidal signal described by $A\sin\left(2\pi f \cdot t\right) + B$.

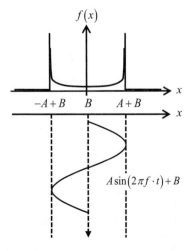

If N samples of the ADC are collected, we can then predict (based on the above probability argument) that the number of times we should see a code between n and $n + 1$, denoted by $H(n)$, is

$$H(n)=\frac{N}{\pi}\left\{\sin^{-1}\left[\frac{(n+1)\times V_{LSB}}{A}-1\right]-\sin^{-1}\left[\frac{n\times V_{LSB}}{A}-1\right]\right\} \qquad (4.39)$$

Another situation where sinusoidal distributions are used is in serial I/O channel testing. A sinusoidal phase modulated signal is created and used to excite the clock recovery circuit of a digital receiver. The amplitude of the sine wave is then varied until a specific level of probability of error (also referred to as bit error rate) is achieved. In this situation, we are looking for the probability that the sinusoidal distribution with zero mean lies in the region between some threshold value, say t_{TH}, and the upper limit of the sine wave, A. Using Eq. (4.35), we can state this as

$$P\left(t_{TH}<X<A\right)=\frac{1}{\pi}\left[\frac{\pi}{2}-\sin^{-1}\left(\frac{t_{TH}}{A}\right)\right]=\frac{1}{\pi}\cos^{-1}\left(\frac{t_{TH}}{A}\right) \qquad (4.40)$$

EXAMPLE 4.3

A 3-bit ADC is excited by a 1024-point full-scale sine wave described by $1.0\sin\left(2\pi\frac{1}{1024}\cdot n\right)+2.0$, $n = 0,1,...,1023$. Arrange the voltage axis between 1.00 and 3.00 V into 2^3 equal regions and count the number of samples from the sine-wave that fall into each region. Compare with the counts given by Eq. (4.39).

Solution:
Subdividing the voltage range of 1.00 to 3.00 V into 8 regions suggests that the distance between each code is given by

$$V_{LSB}=\frac{3.00-1.00}{2^3}=0.25\ V$$

Next, organizing all 1024 samples into eight regions we obtain the table of code distribution shown below. In addition, in the rightmost column we list the corresponding code counts as predicted by Eq. (4.39). As is clearly evident, the results compare.

Index, n	Code Region (V)	Samples Falling in Code Region	Predicted Count, H
0	1.00–1.25	235	235.57
1	1.25–1.50	106	105.75
2	1.50–1.75	88	88.31
3	1.75–2.00	82	82.36
4	2.00–2.25	84	82.36
5	2.25–2.50	88	88.31
6	2.50–2.75	106	105.75
7	2.75–3.00	235	235.57

A plot of the above two data sets is shown below for comparative purposes. Clearly, the two data sets are in close agreement.

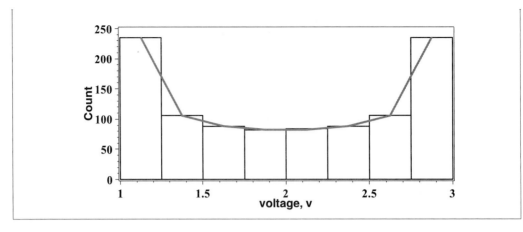

4.3.3 The Binomial Probability Distribution

Consider an experiment having only two possible outcomes, say **A** and **B**, which are mutually exclusive. If the probability of **A** occurring is p, then the probability of **B** will be $q = 1 - p$. If the experiment is repeated N times, then the probability of **A** occurring k times is given by

$$P[X = k] = \frac{N!}{k!(N-k)!} p^k (1-p)^{N-k} \tag{4.41}$$

Exercises	

4.9. An 8-bit ADC converts an analog input voltage level into 2^8 code levels ranging from 0 to $2^8 - 1$. If the ADC is excited by a 1024-point full-scale sine wave described by $0.5\sin(2\pi\, 1/1024 \cdot n) + 0.5$, $n = 0,1,...,1023$, what is the expect number of codes appearing at the output corresponding to code 0? What is the code count for code 2^8-1? What is the distance between adjacent code edges?

ANS. $H(0) = H(2^8 - 1)$ = 40.77; *LSB* = 3.9 mV.

4.10. What is the probability that a 1-V amplitude sine wave lies between 0.6 and 0.8 V? How about lying between −0.1 and 0.1 V? Again for 0.9 V and 1.0 V?

ANS. $P(0.6 \leq V \leq 0.8)$ = 0.09; $P(-0.1 \leq V \leq 0.1)$ = 0.06; $P(0.9 \leq V \leq 1.0)$ = 0.14

The binomial probability density function can then be written as

$$f(x) = \sum_{k=0}^{N} \frac{N!}{k!(N-k)!} p^k (1-p)^{N-k} \delta(x-k) \tag{4.42}$$

The probability that a binomial distributed random integer X will fall in the interval bounded by integers a and b, where $0 \leq a < b \leq N$, is obtained by substituting Eq. (4.42) into Eq. (4.9) to get

$$P\left(a\le X\le b\right)=\int_{a}^{b} f\left(x\right) dx =\int_{a}^{b}\sum_{k=0}^{N}\frac{N!}{k!\left(N-k\right)!}p^{k}\left(1-p\right)^{N-k}\delta\left(x-k\right) dx \qquad (4.43)$$

or, when simplified,

$$P\left(a\le X\le b\right)=\sum_{k=a}^{b}\frac{N!}{k!\left(N-k\right)!}p^{k}\left(1-p\right)^{N-k} \qquad (4.44)$$

The mean and standard deviation of the binomial PDF is

$$\mu = N\cdot p \quad \text{and} \quad \sigma =\sqrt{N\cdot p\cdot\left(1-p\right)} \qquad (4.45)$$

The binomial distribution arises in applications where there are two types of objects or outcomes (e.g., heads/tails, correct/erroneous bits, good/defective items) and we are interested in the number of times a specific object type occurs in a randomly selected batch of size N. We shall see the application of the binomial distribution to model the bit error rate in data communications.

EXAMPLE 4.4

A lot of 100 devices is tested during production, where it is assumed that a single bad device has a probability of 0.1 of failing. What is the probability that three bad dies will show up during this test?

Solution:
Each die from the lot of 100 has equal chance of failing with a probability $p = 0.1$. Consequently, the probability that three dies out of 100 will fail is determined using Eq. (4.41) as follows:

$$P\left[X = 3\right]=\frac{3!}{3!\left(100-3\right)!}\left(0.1\right)^{3}\left(1-0.1\right)^{100-3} = 0.0059$$

Exercises

4.11. A lot of 500 devices is tested during production, where it is assumed that a single device has a probability of 0.99 of passing. What is the probability that two bad dies will show up during this test?

ANS. $P[X = 2] = 0.0836$

4.12. A sample of 2 dies are selected from a large lot and tested. If the probability of passing is 5/6, what is the PDF of two bad dies appearing during this test?

ANS. $f\left(x\right)=\frac{25}{36}\delta\left(x\right)$
$+\frac{10}{36}\delta\left(x-1\right)+\frac{1}{36}\delta\left(x-2\right)$

4.13. Data are transmitted over a channel with a single-bit error probability of 10^{-12}. If 10^{12} bits are transmitted, what is the probability of one bit error. What is the average number of errors expected?

ANS. $P[X = 1] = 0.3678$; $\mu = 1$

4.4 MODELING THE STRUCTURE OF RANDOMNESS

At the heart of any measurement problem is the need to quantify the structure of the underlying randomness of the data set. For data that are modeled after a single probability density function, this is generally a straightforward procedure. For instance, the measured data shown in Figure 4.17a was found to have a mean value of 1 and standard deviation of 3. If we assume that the randomness of this data set is Gaussian, then we can describe the PDF of this sequence as

$$g_1(x) = \frac{1}{3\sqrt{2\pi}} e^{\frac{-(x-1)^2}{18}}$$

(4.46)

where x represents any value of the data sequence.

Conversely, if we assume that the randomness of the data set is to be modeled as a uniformly distributed random process, then according to the formulas for mean and standard deviation for the uniform distribution given in Eq. (4.25) and Eq. (4.26), we write

$$1 = \frac{A+B}{2} \quad \text{and} \quad 3 = \frac{(B-A)}{\sqrt{12}}$$

(4.47)

leading to

$$A = 1 - 3\sqrt{3} \quad \text{and} \quad B = 1 + 3\sqrt{3}$$

(4.48)

Figure 4.17. (a) A random set of measured data and (b) its corresponding histogram with various PDFs superimposed.

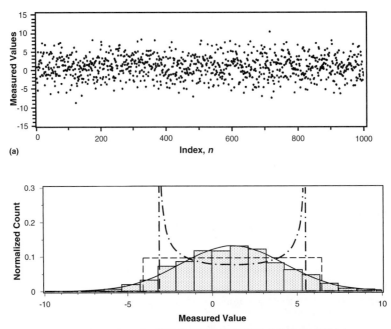

(a)

(b)

Legend: Gaussian PDF (-), Sinusoidal PDF (- .), Uniform PDF (--)

from which we write

$$g_2(x) = \begin{cases} \dfrac{1}{6\sqrt{3}}, & 1-3\sqrt{3} \le x \le 1-3\sqrt{3} \\ \\ 0, & \text{elsewhere} \end{cases} \tag{4.49}$$

Likewise, if we assume that the randomness is to be modeled as a sinusoidal distribution, then we can immediately see

$$B = 1 \quad \text{and} \quad A = 3\sqrt{2} \tag{4.50}$$

from Eqs (4.36) and (4.37), allowing us to write

$$g_3(x) = \begin{cases} \dfrac{1}{\pi\sqrt{18-(x-1)^2}}, & 1-3\sqrt{2} \le x \le 1+3\sqrt{2} \\ \\ 0, & \text{elsewhere} \end{cases} \tag{4.51}$$

Validating one's assumptions of the underlying randomness is, of course, essential to the modeling questions. One method is to look at a plot of the histogram of the measured data and superimpose the PDF over this plot. This is demonstrated in Figure 4.17b for the Gaussian, uniform, and sinusoidal pdfs. As is evident form this plot, the Gaussian PDF models the randomness the best in this case.

4.4.1 Modeling a Gaussian Mixture Using the Expectation-Maximization Algorithm

In some measurement situations, randomness can come from multiple sources, and one that cannot be modeled by a single Gaussian or similar pdf. We saw such a situation in Figure 4.11 and Figure 4.12 where the histogram either was bimodal or contained outliers. To model either situation, we require a different approach than that just described. It is based on the expectation–maximization (EM) algorithm.[8] The algorithm begins by writing the general form of the PDF as

$$f(x) = \alpha_1 f_1(x, \mu_1, \sigma_1) + \cdots + \alpha_i f_i(x, \mu_i, \sigma_i) + \cdots + \alpha_G f_G(x, \mu_G, \sigma_G) \tag{4.52}$$

where $f_i(x, \mu_i, \sigma_i)$ is the ith component PDF and α_i is its corresponding weighting factor. The sum of all weighting factors must total to one, that is, $\sum_{k=1}^{G} \alpha_k = 1$. To ensure convergence, $f_i(x, \mu_i, \sigma_i)$ should be continuous in x over the range of interest. In this textbook, we shall limit our discussion to individual PDFs having Gaussian form, that is,

$$f_i(x, \mu_i, \sigma_i) = \frac{1}{\sigma_i\sqrt{2\pi}} e^{-\frac{(x-\mu_i)^2}{2\sigma_i^2}}$$

It is the objective of the EM algorithm to find the unknown parameters $\alpha_1, \alpha_2, \ldots, \alpha_G$, $\mu_1, \mu_2, \ldots, \mu_G$ and $\sigma_1, \sigma_2, \ldots, \sigma_G$ based on a set of data samples. The EM algorithm begins by first defining the data weighting function Φ_i for each PDF component (i from 1 to G) as follows

$$\Phi_i(x, \mu, \sigma, \alpha) = \frac{\alpha_i f_i(x, \mu_i, \sigma_i)}{\alpha_1 f_1(x, \mu_1, \sigma_1) + \cdots + \alpha_i f_i(x, \mu_i, \sigma_i) + \cdots + \alpha_G f_G(x, \mu_G, \sigma_G)} \quad (4.53)$$

where the model parameters are defined in vector form as

$$\begin{aligned} \mu &= [\mu_1, \quad \ldots, \quad \mu_i, \quad \ldots, \quad \mu_G] \\ \sigma &= [\sigma_1, \quad \ldots, \quad \sigma_i, \quad \ldots, \quad \sigma_G] \\ \alpha &= [\alpha_1, \quad \ldots, \quad \alpha_i, \quad \ldots, \quad \alpha_G] \end{aligned} \quad (4.54)$$

According to the development of the EM algorithm, the following set of compact recursive equations for each model parameter can be written:

$$\begin{aligned}
\alpha_1^{(n+1)} &= \frac{1}{N} \sum_{i=1}^{N} \Phi_1\left(x_i, \mu^{(n)}, \sigma^{(n)}, \alpha^{(n)}\right), & \mu_1^{(n+1)} &= \frac{1}{N\alpha_1^{(n+1)}} \sum_{i=1}^{N} x_i \Phi_1\left(x_i, \mu^{(n)}, \sigma^{(n)}, \alpha^{(n)}\right), \\
\alpha_2^{(n+1)} &= \frac{1}{N} \sum_{i=1}^{N} \Phi_2\left(x_i, \mu^{(n)}, \sigma^{(n)}, \alpha^{(n)}\right), \ldots, & \mu_2^{(n+1)} &= \frac{1}{N\alpha_2^{(n+1)}} \sum_{i=1}^{N} x_i \Phi_2\left(x_i, \mu^{(n)}, \sigma^{(n)}, \alpha^{(n)}\right), \ldots, \\
\alpha_G^{(n+1)} &= \frac{1}{N} \sum_{i=1}^{N} \Phi_G\left(x_i, \mu^{(n)}, \sigma^{(n)}, \alpha^{(n)}\right) & \mu_G^{(n+1)} &= \frac{1}{N\alpha_G^{(n+1)}} \sum_{i=1}^{N} x_i \Phi_G\left(x_i, \mu^{(n)}, \sigma^{(n)}, \alpha^{(n)}\right)
\end{aligned}$$

$$\sigma_1^{(n+1)} = \sqrt{\frac{1}{N\alpha_1^{(n+1)}} \sum_{i=1}^{N} \left(x_i - \mu_1^{(n+1)}\right)^2 \Phi_1\left(x_i, \mu^{(n)}, \sigma^{(n)}, \alpha^{(n)}\right)}, \quad (4.55)$$

$$\sigma_2^{(n+1)} = \sqrt{\frac{1}{N\alpha_2^{(n+1)}} \sum_{i=1}^{N} \left(x_i - \mu_2^{(n+1)}\right)^2 \Phi_2\left(x_i, \mu^{(n)}, \sigma^{(n)}, \alpha^{(n)}\right)}, \ldots,$$

$$\sigma_G^{(n+1)} = \sqrt{\frac{1}{N\alpha_G^{(n+1)}} \sum_{i=1}^{N} \left(x_i - \mu_G^{(n+1)}\right)^2 \Phi_G\left(x_i, \mu^{(n)}, \sigma^{(n)}, \alpha^{(n)}\right)}$$

The above model parameters are evaluated at each data point, together with an initial guess of the model parameters. The process is repeated until the change in any one step is less than some desired root-mean-square error tolerance, defined as

$$\text{error}^{(n+1)} < \text{TOLERANCE} \quad (4.56)$$

where

$$\text{error}^{(n+1)} = \left[\begin{aligned} &\left(\mu_1^{(n+1)} - \mu_1^{(n)}\right)^2 + \cdots + \left(\mu_G^{(n+1)} - \mu_G^{(n)}\right)^2 + \left(\sigma_1^{(n+1)} - \sigma_1^{(n)}\right)^2 + \cdots \\ &+ \left(\sigma_G^{(n+1)} - \sigma_G^{(n)}\right)^2 + \left(\alpha_1^{(n+1)} - \alpha_1^{(n)}\right)^2 + \cdots + \left(\alpha_G^{(n+1)} - \alpha_G^{(n)}\right)^2 \end{aligned} \right]^{1/2}$$

To help clarify the situation, let us assume we want to fit two independent Gaussian distributions to a bimodal distribution such as that shown Figure 4.11. In this case, we shall assume that the PDF for the overall distribution is as follows

$$f(x) = \alpha_1 \left[\frac{1}{\sigma_1 \sqrt{2\pi}} e^{\frac{-(x-\mu_1)^2}{2\sigma_1^2}} \right] + \alpha_2 \left[\frac{1}{\sigma_2 \sqrt{2\pi}} e^{\frac{-(x-\mu_2)^2}{2\sigma_2^2}} \right] \tag{4.57}$$

where the unknown model parameters are $\mu_1, \sigma_1, \mu_2, \sigma_2, \alpha_1$ and α_2. Next, we write the data weighting functions for this particular modeling situation as

$$\Phi_1(x, \mu, \sigma, \alpha) = \frac{\dfrac{\alpha_1}{\sigma_1 \sqrt{2\pi}} e^{\frac{-(x-\mu_1)^2}{2\sigma_1^2}}}{\dfrac{\alpha_1}{\sigma_1 \sqrt{2\pi}} e^{\frac{-(x-\mu_1)^2}{2\sigma_1^2}} + \dfrac{\alpha_2}{\sigma_2 \sqrt{2\pi}} e^{\frac{-(x-\mu_2)^2}{2\sigma_2^2}}}$$

$$\Phi_2(x, \mu, \sigma, \alpha) = \frac{\dfrac{\alpha_2}{\sigma_2 \sqrt{2\pi}} e^{\frac{-(x-\mu_2)^2}{2\sigma_2^2}}}{\dfrac{\alpha_1}{\sigma_1 \sqrt{2\pi}} e^{\frac{-(x-\mu_1)^2}{2\sigma_1^2}} + \dfrac{\alpha_2}{\sigma_2 \sqrt{2\pi}} e^{\frac{-(x-\mu_2)^2}{2\sigma_2^2}}} \tag{4.58}$$

From Eq. (4.55), we write the update equations for each model parameter as

$$\alpha_1^{(n+1)} = \frac{1}{N} \sum_{i=1}^{N} \Phi_1\left(x_i, \mu^{(n)}, \sigma^{(n)}, \alpha^{(n)}\right), \qquad \mu_1^{(n+1)} = \frac{1}{N \alpha_1^{(n+1)}} \sum_{i=1}^{N} x_i \Phi_1\left(x_i, \mu^{(n)}, \sigma^{(n)}, \alpha^{(n)}\right)$$

$$\alpha_2^{(n+1)} = \frac{1}{N} \sum_{i=1}^{N} \Phi_2\left(x_i, \mu^{(n)}, \sigma^{(n)}, \alpha^{(n)}\right), \qquad \mu_2^{(n+1)} = \frac{1}{N \alpha_2^{(n+1)}} \sum_{i=1}^{N} x_i \Phi_2\left(x_i, \mu^{(n)}, \sigma^{(n)}, \alpha^{(n)}\right)$$

$$\sigma_1^{(n+1)} = \sqrt{\frac{1}{N \alpha_1^{(n+1)}} \sum_{i=1}^{N} \left(x_i - \mu_1^{(n+1)}\right)^2 \Phi_1\left(x_i, \mu^{(n)}, \sigma^{(n)}, \alpha^{(n)}\right)} \tag{4.59}$$

$$\sigma_2^{(n+1)} = \sqrt{\frac{1}{N \alpha_2^{(n+1)}} \sum_{i=1}^{N} \left(x_i - \mu_2^{(n+1)}\right)^2 \Phi_2\left(x_i, \mu^{(n)}, \sigma^{(n)}, \alpha^{(n)}\right)}$$

The following two examples will help to illustrate the EM approach more clearly.

EXAMPLE 4.5

A set of DC measurements was made on an amplifier where the normalized histogram of the data set revealed a bimodal distribution as shown below:

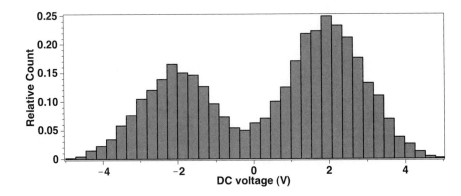

Assuming that the data consist of 10,000 samples and are stored in a vector, model the distribution of these data using a two-Gaussian mixture having the following general form:

$$f(x) = \alpha_1 \left\{ \frac{1}{\sigma_1\sqrt{2\pi}} e^{\frac{-(x-\mu_1)^2}{2\sigma_1^2}} \right\} + \alpha_2 \left\{ \frac{1}{\sigma_2\sqrt{2\pi}} e^{\frac{-(x-\mu_2)^2}{2\sigma_2^2}} \right\}$$

Solution:

To begin, we recognize from the histogram plot that two peaks occur at $V = -2$ and $V = 2.2$. We could use these values as our initial guess for the means of the two-Gaussian mixture. However, lets use some values quite a distance from these values, say $V = -1.0$ and $V = 1.0$. Next, recognizing that 67% of the area under one Gaussian represents ±1 sigma about the mean value, the standard deviation is about 1 V each. To stress the algorithm, we will start with a standard deviations of 0.5 V each. For lack of any further insight, we shall assume that each Gaussian is equally likely to occur resulting in the two alpha coefficients starting off with a value of 0.5 each. Using the data weighting functions for a two-Gaussian mixture found from Eq. (4.58), together with the update equations of Eq. (4.59), we wrote a computer program to iterate through these equations to find the following model parameters after 20 iterations:

$$\alpha_1^{[21]} = 0.3959, \quad \mu_1^{[21]} = -2.0235, \quad \sigma_1^{[21]} = 0.98805$$
$$\alpha_2^{[21]} = 0.6041, \quad \mu_2^{[21]} = 1.9695, \quad \sigma_2^{[21]} = 1.0259$$

Based on Eq. (4.56), we have an RMS error of 0.97 mV. Superimposing the PDF predicted by the two-Gaussian mixture model onto the histogram data above (solid line) reveals a very good fit as illustrated below. Also, superimposed on the plot is the PDF curve corresponding to the initial guess (dash–dot line).

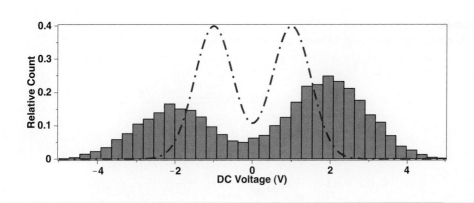

EXAMPLE 4.6

Ten thousand zero crossings were collected using an oscilloscope. A histogram of these zero-crossings is plotted below:

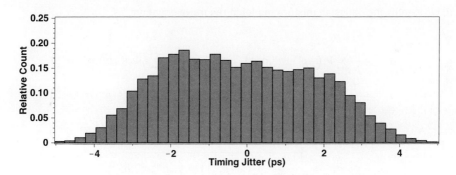

It is assumed that this data set consists of three independent Gaussian distributions described by

$$f(t) = \alpha_1 \left\{ \frac{1}{\sigma_1 \sqrt{2\pi}} e^{\frac{-(t-\mu_1)^2}{2\sigma_1^2}} \right\} + \alpha_2 \left\{ \frac{1}{\sigma_2 \sqrt{2\pi}} e^{\frac{-(t-\mu_2)^2}{2\sigma_2^2}} \right\} + \alpha_3 \left\{ \frac{1}{\sigma_3 \sqrt{2\pi}} e^{\frac{-(t-\mu_3)^2}{2\sigma_3^2}} \right\}$$

Using the EM algorithm, compute the model parameters of this PDF?

Solution:

We begin by first identifying the three data weighting functions as follows:

$$\Phi_1\left(t,\mu,\sigma,\alpha\right)=\cfrac{\dfrac{\alpha_1}{\sigma_1\sqrt{2\pi}}e^{\frac{-(t-\mu_1)^2}{2\sigma_1^2}}}{\dfrac{\alpha_1}{\sigma_1\sqrt{2\pi}}e^{\frac{-(t-\mu_1)^2}{2\sigma_1^2}}+\dfrac{\alpha_2}{\sigma_2\sqrt{2\pi}}e^{\frac{-(t-\mu_2)^2}{2\sigma_2^2}}+\dfrac{\alpha_3}{\sigma_3\sqrt{2\pi}}e^{\frac{-(t-\mu_3)^2}{2\sigma_3^2}}}$$

$$\Phi_2\left(t,\mu,\sigma,\alpha\right)=\cfrac{\dfrac{\alpha_2}{\sigma_2\sqrt{2\pi}}e^{\frac{-(t-\mu_2)^2}{2\sigma_2^2}}}{\dfrac{\alpha_1}{\sigma_1\sqrt{2\pi}}e^{\frac{-(t-\mu_1)^2}{2\sigma_1^2}}+\dfrac{\alpha_2}{\sigma_2\sqrt{2\pi}}e^{\frac{-(t-\mu_2)^2}{2\sigma_2^2}}+\dfrac{\alpha_3}{\sigma_3\sqrt{2\pi}}e^{\frac{-(t-\mu_3)^2}{2\sigma_3^2}}}$$

$$\Phi_3\left(t,\mu,\sigma,\alpha\right)=\cfrac{\dfrac{\alpha_3}{\sigma_3\sqrt{2\pi}}e^{\frac{-(t-\mu_3)^2}{2\sigma_3^2}}}{\dfrac{\alpha_1}{\sigma_1\sqrt{2\pi}}e^{\frac{-(t-\mu_1)^2}{2\sigma_1^2}}+\dfrac{\alpha_2}{\sigma_2\sqrt{2\pi}}e^{\frac{-(t-\mu_2)^2}{2\sigma_2^2}}+\dfrac{\alpha_3}{\sigma_3\sqrt{2\pi}}e^{\frac{-(t-\mu_3)^2}{2\sigma_3^2}}}$$

Next, we identify the Gaussian mixture model fit recursive equations as

$$\alpha_1^{(n+1)}=\frac{1}{N}\sum_{i=1}^{N}\Phi_1\left(t_i,\mu^{(n)},\sigma^{(n)},\alpha^{(n)}\right),\quad \mu_1^{(n+1)}=\frac{1}{N\alpha_1^{(n+1)}}\sum_{i=1}^{N}t_i\Phi_1\left(t_i,\mu^{(n)},\sigma^{(n)},\alpha^{(n)}\right)$$

$$\alpha_2^{(n+1)}=\frac{1}{N}\sum_{i=1}^{N}\Phi_2\left(t_i,\mu^{(n)},\sigma^{(n)},\alpha^{(n)}\right),\quad \mu_2^{(n+1)}=\frac{1}{N\alpha_2^{(n+1)}}\sum_{i=1}^{N}t_i\Phi_2\left(t_i,\mu^{(n)},\sigma^{(n)},\alpha^{(n)}\right)$$

$$\alpha_3^{(n+1)}=\frac{1}{N}\sum_{i=1}^{N}\Phi_3\left(t_i,\mu^{(n)},\sigma^{(n)},\alpha^{(n)}\right),\quad \mu_3^{(n+1)}=\frac{1}{N\alpha_3^{(n+1)}}\sum_{i=1}^{N}t_i\Phi_3\left(t_i,\mu^{(n)},\sigma^{(n)},\alpha^{(n)}\right)$$

$$\sigma_1^{(n+1)}=\sqrt{\frac{1}{N\alpha_1^{(n+1)}}\sum_{i=1}^{N}\left(t_i-\mu_1^{(n+1)}\right)^2\Phi_1\left(t_i,\mu^{(n)},\sigma^{(n)},\alpha^{(n)}\right)}$$

$$\sigma_2^{(n+1)}=\sqrt{\frac{1}{N\alpha_2^{(n+1)}}\sum_{i=1}^{N}\left(t_i-\mu_2^{(n+1)}\right)^2\Phi_2\left(t_i,\mu^{(n)},\sigma^{(n)},\alpha^{(n)}\right)}$$

$$\sigma_3^{(n+1)}=\sqrt{\frac{1}{N\alpha_3^{(n+1)}}\sum_{i=1}^{N}\left(t_i-\mu_3^{(n+1)}\right)^2\Phi_3\left(t_i,\mu^{(n)},\sigma^{(n)},\alpha^{(n)}\right)}$$

Using the captured data, together with our initial guess of the model parameters,

$$\alpha_1^{(1)}=0.3,\quad \mu_1^{(1)}=-1.0,\quad \sigma_1^{(1)}=0.5$$
$$\alpha_2^{(1)}=0.3,\quad \mu_2^{(1)}=0,\quad \sigma_2^{(1)}=0.5$$
$$\alpha_3^{(1)}=0.4,\quad \mu_3^{(1)}=1.0,\quad \sigma_3^{(1)}=0.5$$

we iterate through the above equations 100 times and find the following set of parameters:

$$\alpha_1^{(101)} = 0.4782, \quad \mu_1^{(101)} = -1.8150, \quad \sigma_1^{(101)} = 1.0879$$
$$\alpha_2^{(101)} = 0.2644, \quad \mu_2^{(101)} = 0.4486, \quad \sigma_2^{(101)} = 0.9107$$
$$\alpha_3^{(101)} = 0.2572, \quad \mu_3^{(101)} = 2.1848, \quad \sigma_3^{(101)} = 0.9234$$

Double-checking the curve fit, we superimpose the final PDF model on the histogram of the original data below, together with intermediate results after 5, 10, and 20 iterations. We also illustrate our first guess of the PDF. As is clearly evident from below, we have a very good fit, and this occurs after about 20 iterations.

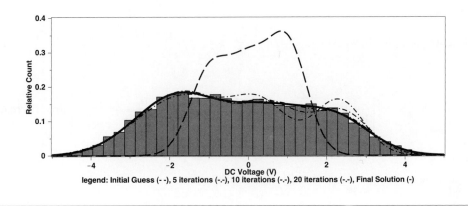

legend: Initial Guess (- -), 5 iterations (-.-), 10 iterations (-.-), 20 iterations (-.-), Final Solution (-)

The EM algorithm is a well-known statistical tool for solving maximum likelihood problems. It is interesting to note that a survey of the literature via the Internet reveals that this algorithm is used in a multitude of applications from radar to medical imaging to pattern recognition. The algorithm provides an iterative formula for the estimation of the unknown parameters of the Gaussian mixture and can be proven to monotonically increase the likelihood that the estimate is correct in each step of the iteration. Nonetheless, the EM algorithm is not without some drawbacks. An initial guess is required to start the process; the closer to the final solution, the better, because the algorithm will converge much more quickly. And secondly, the algorithm cannot alter the assumption about the number of Gaussians in the mixture and hence can lead to sub-optimum results. Thirdly, local maxima or saddle points can prematurely stop the search process on account of the nonlinearities associated between the densities and their corresponding parameters μ and σ. Nonetheless, the EM algorithm provides an excellent means to model jitter distribution problems.[8] In fact, due to the underlying importance of the EM algorithm to statistics in general, advances are being made to the EM algorithm regularly, such as that described in Vlassis and Likas.[9]

4.4.2 Probabilities Associated with a Gaussian Mixture Model

The ultimate goal of any pdf-modeling problem is to answer questions related to the probability that a specific type of event can occur. We encounter this question previously in Subsections

4.2.2–4.2.3 for a single Gaussian pdf. Here we shall extend this concept to an arbitrary Gaussian mixture of the general form,

$$f(x) = \alpha_1 \left\{ \frac{1}{\sigma_1 \sqrt{2\pi}} e^{\frac{-(x-\mu_1)^2}{2\sigma_1^2}} \right\} + \alpha_2 \left\{ \frac{1}{\sigma_2 \sqrt{2\pi}} e^{\frac{-(x-\mu_2)^2}{2\sigma_2^2}} \right\} + \cdots + \alpha_G \left\{ \frac{1}{\sigma_G \sqrt{2\pi}} e^{\frac{-(x-\mu_G)^2}{2\sigma_G^2}} \right\} \qquad (4.60)$$

The probability that a random event X lies in the interval bounded by a and b is given by

$$P(a \le X \le b) = \int_a^b f(x)\,dx \qquad (4.61)$$

Substituting Eq. (4-60), we write

$$P(a \le X \le b) = \int_a^b \left[\frac{\alpha_1}{\sigma_1 \sqrt{2\pi}} e^{\frac{-(x-\mu_1)^2}{2\sigma_1^2}} + \frac{\alpha_2}{\sigma_2 \sqrt{2\pi}} e^{\frac{-(x-\mu_2)^2}{2\sigma_2^2}} + \cdots + \frac{\alpha_G}{\sigma_G \sqrt{2\pi}} e^{\frac{-(x-\mu_G)^2}{2\sigma_G^2}} \right] dx \qquad (4.62)$$

Distributing the integral across each term, we further write

$$P(a \le X \le b) = \alpha_1 \int_a^b \frac{1}{\sigma_1 \sqrt{2\pi}} e^{\frac{-(x-\mu_1)^2}{2\sigma_1^2}} dx + \alpha_2 \int_a^b \frac{1}{\sigma_2 \sqrt{2\pi}} e^{\frac{-(x-\mu_2)^2}{2\sigma_2^2}} dx + \cdots + \alpha_G \int_a^b \frac{1}{\sigma_G \sqrt{2\pi}} e^{\frac{-(x-\mu_G)^2}{2\sigma_G^2}} dx \qquad (4.63)$$

Finally, using the standard Gaussian CDF notation, we write Eq. (4.63) as

$$P(a \le X \le b) = \alpha_1 \left[\Phi\left(\frac{b-\mu_1}{\sigma_1} \right) - \Phi\left(\frac{a-\mu_1}{\sigma_1} \right) \right] + \alpha_2 \left[\Phi\left(\frac{b-\mu_1}{\sigma_2} \right) - \Phi\left(\frac{a-\mu_2}{\sigma_2} \right) \right] + \cdots$$
$$+ \alpha_G \left[\Phi\left(\frac{b-\mu_G}{\sigma_G} \right) - \Phi\left(\frac{a-\mu_G}{\sigma_G} \right) \right] \qquad (4.64)$$

It should be evident from the above result that the probabilities associated with a Gaussian mixture model is a straightforward extension of a single Gaussian PDF and offers very little additional difficulty.

EXAMPLE 4.7

A bimodal distribution is described by the following pdf:

$$f(x) = \frac{\alpha_1}{\sigma_1\sqrt{2\pi}} e^{\frac{-(x-\mu_1)^2}{2\sigma_1^2}} + \frac{\alpha_2}{\sigma_2\sqrt{2\pi}} e^{\frac{-(x-\mu_2)^2}{2\sigma_2^2}}$$

with coefficients,

$$\alpha_1 = 0.3959, \quad \mu_1 = -2.0235, \quad \sigma_1 = 0.98805$$
$$\alpha_2 = 0.6041, \quad \mu_2 = 1.9695, \quad \sigma_2 = 1.0259$$

A plot of this PDF is provided in Example 4.5 above. What is the probability that some single random event X is greater than 0? What about greater than 10?

Solution:
According to Eq. (4.64), we can write

$$P(0 \le X < \infty) = \alpha_1\left[\Phi\left(\frac{\infty - \mu_1}{\sigma_1}\right) - \Phi\left(\frac{0 - \mu_1}{\sigma_1}\right)\right] + \alpha_2\left[\Phi\left(\frac{\infty - \mu_1}{\sigma_2}\right) - \Phi\left(\frac{0 - \mu_2}{\sigma_2}\right)\right]$$

which further simplifies to

$$P(0 \le X) = 2 - \alpha_1\Phi\left(\frac{-\mu_1}{\sigma_1}\right) - \alpha_2\Phi\left(\frac{-\mu_2}{\sigma_2}\right)$$

Substituting the appropriate Gaussian mixture model parameters, we write

$$P(0 \le X) = 2 - 0.3959 \times \Phi\left(\frac{2.0235}{0.98805}\right) - 0.6041 \times \Phi\left(\frac{-1.9695}{1.0259}\right)$$

Solving each standard CDF term, we write

$$P(0 \le X) = 2 - 0.3959 \times 0.9797 - 0.6041 \times 0.02744$$

which finally reduces to

$$P(0 \le X) = 0.5955 .$$

Likewise, for $P(10 \le X)$ we write

$$P(10 \le X) = 2 - 0.3959 \times \Phi\left(\frac{10 + 2.0235}{0.98805}\right) - 0.6041 \times \Phi\left(\frac{10 - 1.9695}{1.0259}\right)$$

which, when simplified, leads to

$$P\left(10 \leq X\right) \approx 1 - 0.6041 \times \Phi\left(\frac{10 - 1.9695}{1.0259}\right) = 1.5 \times 10^{-15}$$

Exercises

4.14. A measurement set is found to consist of two Gaussian distributions having the following model parameters: $\alpha_1 = 0.7$, $\mu_1 = -2.$, $\sigma_1 = 1$, $\alpha_2 = 0.3$, $\mu_2 = 2$, and $\sigma_2 = 0.5$. What is the probability that some single random measurement X is greater than 0? What about greater than 2? What about less than –3?

ANS. $P[0 \leq X]$
$= 0.3159$; $P[2 \leq X]$
$= 0.1500$; $P[X \leq -3]$
$= 0.1110$.

4.5 SUMS AND DIFFERENCES OF RANDOM VARIABLES

Up to this point in the chapter we have described various mathematical functions that can be used to represent the behavior of a random variable, such as the bimodal shape of a histogram of data samples. In this section we will go one step further and attempt to describe the general shape of a probability distribution as a sum of several random variables. This perspective provides greater insight into the measurement process and provides an understanding of the root cause of measurement variability. The test engineer will often make use of these principles to identify the root cause of device repeatability problems or product variability issues.

Consider a sum of two independent random variables X_1 and X_2 as follows:

$$Y = X_1 + X_2 \tag{4.65}$$

If the PDF of X_1 and X_2 are described by $f_1(x)$ and $f_2(x)$, then it has been shown[2] that the PDF of the random variable Y, denoted by $f_y(x)$, is given by the convolution integral as

$$f_y(x) = \int_{-\infty}^{\infty} f_1(x - \tau) f_2(\tau) d\tau \tag{4.66}$$

It is common practice to write this convolution expression using a shorthand notation as follows

$$f_y(x) = f_1(x) \otimes f_2(x) \tag{4.67}$$

The above theory extends to the summation of multiple independent random variables. For example,

$$Y = X_1 + X_2 + X_3 + \cdots + X_R \tag{4.68}$$

where the PDF of the total summation is expressed as

$$f_y(x) = f_1(x) \otimes f_2(x) \otimes f_3(x) \otimes \cdots \otimes f_R(x) \tag{4.69}$$

The above theory can also be extended to sums or differences of scaled random variables according to

$$Y = \alpha_1 X_1 + \alpha_2 X_2 + \alpha_3 X_3 + \cdots + \alpha_R X_R \tag{4.70}$$

where the PDF of the sum is given by

$$f_y(x) = \frac{1}{|\alpha_1|} f_1\left(\frac{x}{\alpha_1}\right) \otimes \frac{1}{|\alpha_2|} f_2\left(\frac{x}{\alpha_2}\right) \otimes \frac{1}{|\alpha_3|} f_3\left(\frac{x}{\alpha_3}\right) \otimes \cdots \otimes \frac{1}{|\alpha_R|} f_R\left(\frac{x}{\alpha_R}\right) \tag{4.71}$$

The following three examples will help to illustrate the application of this theory to various measurement situations.

EXAMPLE 4.8

A DUT with zero input produces an output offset of 10 mV in the presence of a zero mean Gaussian noise source with a standard deviation of 10 mV, as illustrated in the figure accompanying. Write a PDF description of the voltage that the voltmeter reads.

Solution:
Let us designate the offset voltage V_{OFF} as a random variable and the noise produced by the DUT (V_{noise}) as the other random variable. As the offset voltage is constant for all time, we can model the PDF of this voltage term using a unit impulse centered around a 10-mV level as $f_1(v) = \delta(v - 10 \times 10^{-3})$. The PDF of the noise source is simply $f_2(v) = \dfrac{1}{(10 \times 10^{-3})\sqrt{2\pi}} e^{-v^2 / 2(10 \times 10^{-3})^2}$. As the noise

signal is additive with the offset level, we find the PDF of the voltage seen by the voltmeter as the convolution of the two individual pdfs, leading to

$$f_{VM}(v) = f_1(v) \otimes f_2(v) = \int_{-\infty}^{\infty} \delta(v - \tau - 10 \times 10^{-3}) \frac{1}{(10 \times 10^{-3})\sqrt{2\pi}} e^{-\tau^2 / 2(10 \times 10^{-3})^2} d\tau$$

Next, using the sifting property of the impulse function, we write the voltmeter PDF as

$$f_{VM}(x) = \frac{1}{(10 \times 10^{-3})\sqrt{2\pi}} e^{-(v - 10 \times 10^{-3})^2 / 2(10 \times 10^{-3})^2}$$

A plot of this PDF is shown below:

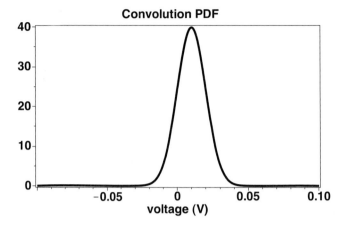

The voltage seen by the voltmeter is a Gaussian distrubed random variable with a mean of 10 mV and a standard deviation of 10 mV.

While the PDF in the above example could have been written directly from the problem description, it serves to illustrate the relationship of the observed data to the physics of the problem. Let us repeat the above example, but this time let us assume that the offset voltage can take on two different levels.

EXAMPLE 4.9

A DUT with zero input produces an output offset that shifts equally between two different levels of 10 mV and −20 mV, depending on the DUT temperature. The noise associated with the DUT is a zero mean Gaussian noise source with a standard deviation of 10 mV. Write a PDF description of the voltage that the voltmeter reads.

Solution:
As the offset voltage can take on two different voltage values, the PDF of the offset voltage becomes $f_1(v) = 1/2 \cdot \delta(v + 20 \times 10^{-3}) + 1/2 \cdot \delta(v - 10 \times 10^{-3})$. The PDF of the noise signal is described

as $f_2(v) = \dfrac{1}{(10 \times 10^{-3})\sqrt{2\pi}} e^{-v^2 / 2(10 \times 10^{-3})^2}$, leading to the PDF of the voltage seen by the voltmeter as

$$f_{VM}(v) = \int_{-\infty}^{\infty} \left[\frac{1}{2} \cdot \delta(v - \tau + 20 \times 10^{-3}) + \frac{1}{2} \cdot \delta(v - \tau - 10 \times 10^{-3}) \right] \frac{1}{(10 \times 10^{-3})\sqrt{2\pi}} e^{-\tau^2 / 2(10 \times 10^{-3})^2} d\tau$$

which reduces to

$$f_{VM}(x) = \frac{1}{2}\frac{1}{(10\times10^{-3})\sqrt{2\pi}}e^{-(x+20\times10^{-3})^2 / 2(10\times10^{-3})^2} + \frac{1}{2}\frac{1}{(10\times10^{-3})\sqrt{2\pi}}e^{-(x-10\times10^{-3})^2 / 2(10\times10^{-3})^2}$$

A plot of this PDF is shown below:

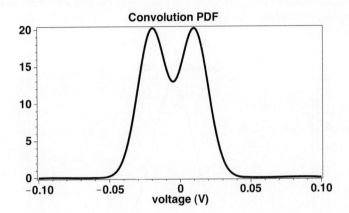

Here we see that the voltage seen by the voltmeter is a bimodal Gaussian distrubed random variable with centers of −20 mV and 10 mV.

Let us again repeat the above example, but this time let us assume that the offset voltage can take on a range of different levels.

EXAMPLE 4.10

A DUT with zero input produces an output offset that is uniformly distributed between −20 mV and 10 mV. The noise associated with the DUT is a zero mean Gaussian noise source with a standard deviation of 10 mV. Write a PDF description of the voltage that the voltmeter reads.

Solution:
As the offset voltage can take on a range of voltage values between −20 mV and 10 mV, the PDF of the offset voltage becomes $f_1(v) = \dfrac{1}{30\times10^{-3}}$ for $-20\times10^{-3} \leq v \leq 10\times10^{-3}$. The PDF of the

noise signal is described as $f_z(v) = \dfrac{1}{(10 \times 10^{-3})\sqrt{2\pi}} e^{-v^2 / 2(10 \times 10^{-3})^2}$, leading to the PDF of the voltage seen by the voltmeter as

$$f_{VM}(v) = \int_{-20 \times 10^{-3}}^{10 \times 10^{-3}} \frac{1}{30 \times 10^{-3}} \frac{1}{(10 \times 10^{-3})\sqrt{2\pi}} e^{-(v-\tau)^2 / 2(10 \times 10^{-3})^2} d\tau$$

Unlike the previous two examples, there is no obvious answer to this intergral equation. Instead, we resort to solving this integral using a numerical integration method. The solution is shown in the plot below.

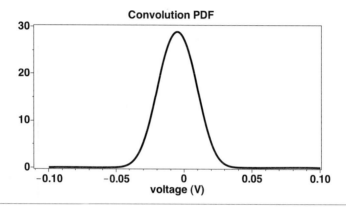

4.5.1 The Central Limit Theorem

As mentioned previously, the central limit theorem states that if a large number of random variables are independent, then the density of their sum Eq. (4.68) tends to a Gaussian distribution with mean μ and standard deviation σ as $R \to \infty$ described by

$$f_y(x) = \frac{1}{\sigma\sqrt{2\pi}} e^{-(x-\mu)^2 / 2\sigma^2} \qquad (4.72)$$

This fact can be viewed as a property of repeat application of the convolution operation on positive functions. For instance, let us consider the sum of three random variables, each with a uniform PDF between 0 and 1, that is, for i from 1 to 3, we obtain

$$f_i(x) = \begin{cases} 1, & 0 \le x \le 1 \\ 0, & \text{otherwise} \end{cases} \qquad (4.73)$$

Applying the convolution on the first two terms of the summation, we write

$$f_{X_1+X_2}(x) = \int_{-\infty}^{\infty} \left[\begin{cases} 1, & 0 \le x-\tau \le 1 \\ 0, & \text{otherwise} \end{cases}\right]\left[\begin{cases} 1, & 0 \le \tau \le 1 \\ 0, & \text{otherwise} \end{cases}\right] d\tau = \begin{cases} 0, & x < 0 \\ x, & x < 1 \\ 2-x, & x < 2 \\ 0, & 2 \le x \end{cases} \quad \textbf{(4.74)}$$

Next we repeat this integration with the above result and write

$$f_{X_1+X_2+X_3}(x) = \int_{-\infty}^{\infty} \left[\begin{cases} 1, & 0 \le x-\tau \le 1 \\ 0, & \text{otherwise} \end{cases}\right]\left[\begin{cases} 0, & \tau < 0 \\ \tau, & \tau < 1 \\ 2-\tau, & \tau < 2 \\ 0, & 2 \le \tau \end{cases}\right] d\tau = \begin{cases} 0, & x < 0 \\ x^2/2, & x < 1 \\ -1-(x-1)^2/2+2x-x^2/2, & x < 2 \\ 4-2x+(x-1)^2/2, & x < 3 \\ 0, & 3 \le x \end{cases} \quad \textbf{(4.75)}$$

Figure 4.18 illustrates how the uniform distribution described above tends toward a Gaussian function by repeated application of the convolution operation.

The importance of the central limit theorem for analog and mixed-signal test is that repeated measurements of some statistic, such as a mean or standard deviation, generally follows a Gaussian distribution, regardless of the underlying random behavior from which the samples were derived. This fact will figure prominently in the concept of measurement repeatability and accuracy of the next chapter.

Figure 4.18. Illustrating repeated application of the convolution operation (a) PDF of X_1, (b) PDF of $X_1 + X_2$, (c) PDF of $X_1 + X_2 + X_3$.

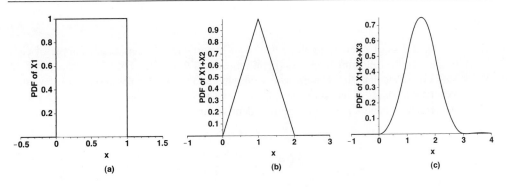

4.15. Two random variables are combined as the difference according to $y = X_1 - X_2$, where the PDF for X_1 is

given by $f_1(x) = \dfrac{1}{\sigma_1\sqrt{2\pi}} e^{-(x-\mu_1)^2/2\sigma_1^2}$ and the PDF for X_2 as

$f_2(x) = \dfrac{1}{\sigma_2\sqrt{2\pi}} e^{-(x-\mu_2)^2/2\sigma_2^2}$. What is the PDF of Y?

ANS. $f_Y(x) = \dfrac{1}{\sqrt{\sigma_1^2 + \sigma_2^2}\sqrt{2\pi}} e^{-[x-(\mu_1-\mu_2)]^2/2(\sigma_1^2+\sigma_2^2)}$

4.6 SUMMARY

There are literally hundreds, if not thousands, of ways to view and process data gathered during the production testing process. In this chapter, we have examined only a few of the more common data displays, such as the datalog, wafer map, and histogram. We reviewed several statistical methods in which to model large quantities of data with simple mathematical models. These models capture the underlying structure of randomness, and not the randomness itself. For the most part, we focused on models based on Gaussian or normal distributions. Probability theory was then introduced to answer questions related to the likelihood that an event will occur. This idea is central to most, if not all, production measurement approaches, because obtaining just the right amount of information with the least effort is the objective of any production engineer. We also looked at several other types of distributions that are commonly found in analog and mixed-signal test applications, such as the uniform, binomial and sinusoidal distribution. Near the end of the chapter the idea of modeling an arbitrary distribution as mixture of multiple Gaussian distributions was introduced. A numerical technique based on the expectation-maximization algorithm was described. This idea will figure prominently in Chapter 14 where we quantify clock and data jitter associated with high-speed digital I/O channels. Finally, we looked at the operation of adding several random variables together from a probability perspective. The theory related to this operation can often help the test engineer identify sources of variability. Finally, we concluded this chapter with a discussion on the central limit theorem, which essentially states that any operation that adds multiple samples together such as a mean or standard deviation metric, will behave as a Gaussian random variable. This fact will figure significantly in the next chapter on measurement repeatability and accuracy.

PROBLEMS

4.1. The thickness of printed circuit boards is an important characteristic. A sample of eight boards had the following thickness (in millimeters): 1.60, 1.55, 1.65, 1.57, 1.55, 1.62, 1.52, and 1.67. Calculate the sample mean, sample variance, and sample standard deviation. What are the units of measurement for each statistic?

4.2. A random variable X has the probability density function

$$f(x) = \begin{cases} ce^{-3x}, & x > 0 \\ 0, & x \le 0 \end{cases}$$

(a) Find the value of the constant c.
(b) Find $P(1 < X < 2)$

(c) Find $P(X \le 1)$

(d) Find $P(X < 1)$

(e) Find the cumulative distribution function $F(x)$.

4.3. Compare the tabulated results for $\Phi(z)$ listed in the table of Appendix A with those generated by Eq. (4.18). Provide a plot of the two curves. What is the worst-case error?

4.4. For the following specified values of z, estimate the value of $\Phi(z)$ using Eq. (4.18) and compare the results with those obtained from the table of Appendix A. Use linear interpolation where necessary.

(a) $z = -3.0$, (b) $z = -1.9$, (c) $z = -0.56$, (d) $z = -0.24$, (e) $z = 0.0$,

(f) $z = -0.09$, (g) $z = +0.17$, (h) $z = +3.0$, (i) $z = +5.0$, (j) $z = -5.0$.

4.5. Using first principles, show that the following relationships are true.

$P(X < b) = 1 - P(X > b)$

$P(a < X < b) = 1 - P(X < a) - P(X > b)$

$P(|X| > c) = 1 - P(-c < X < c)$

4.6. Calculate the following probabilities associated with a Gaussian-distributed random variable having the following mean and standard deviation values:

(a) $P(0 < X < 30 \text{ mV})$ when $\mu = 0$, $\sigma = 10$ mV

(b) $P(-30 \text{ mV} < X < 30 \text{ mV})$ when $\mu = 1$ V, $\sigma = 10$ mV

(c) $P(-1.5 \text{ V} < X < 1.4 \text{ V})$ when $\mu = 0$, $\sigma = 1$ V

(d) $P(-300 \text{ mV} < X < -100 \text{ mV})$ when $\mu = -250$ mV, $\sigma = 50$ mV

(e) $P(X < 250 \text{ mV})$ when $\mu = 100$ mV, $\sigma = 100$ mV

(f) $P(-200 \text{ mV} > X)$ when $\mu = -75$ mV, $\sigma = 150$ mV

(g) $P(|X| < 30 \text{ mV})$ when $\mu = 0$, $\sigma = 10$ mV

(h) $P(|X| > 30 \text{ mV})$ when $\mu = 0$, $\sigma = 10$ mV

4.7. In each of the following equations, find the value of z that makes the probability statement true. Assume a Gaussian distributed random variable with zero mean and unity standard deviation.

(a) $\Phi(z) = 0.9452$

(b) $P(Z < z) = 0.7881$

(c) $P(Z < z) = 0.2119$

(d) $P(Z > z) = 0.2119$

(e) $P(|Z| < z) = 0.5762$

4.8. In each of the following equations, find the value of x that makes the probability statement true. Assume a Gaussian distributed random variable with $\mu = -1$ V and $\sigma = 100$ mV.

(a) $P(X < x) = 0.9$

(b) $P(X < x) = 0.3$

(c) $P(X > x) = 0.3$

(d) $P(|X| < x) = 0.4$

4.9. It has been observed that a certain measurement is a Gaussian-distributed random variable, of which 25% are less than 20 mV and 10% are greater than 70 mV. What are the mean and standard deviation of the measurements?

4.10. If X is a uniformly distributed random variable over the interval (0, 100), what is the probability that the number lies between 23 and 33? What are the mean and standard deviation associated with this random variable?

4.11. It has been observed that a certain measurement is a uniformly distributed random variable, of which 25% are less than 20 mV and 10% are greater than 70 mV. What are the mean and standard deviation of the measurements?

4.12. A DC offset measurement is repeated many times, resulting in a series of values having an average of -110 mV. The measurements exhibit a standard deviation of 51 mV. What is the probability that any single measurement will return a positive value? What is the probability that any single measurement will return a value less than -200 mV? Provide sketches of the pdf, label critical points, and highlight the areas under the PDF that corresponds to the probabilities of interest.

4.13. A series of AC gain measurements are found to have an average value of 10.3 V/V and a variance of 0.1 (V/V)². What is the probability that any single measurement will return a value less than 9.8 V/V? What is the probability that any single measurement will lie between 10.0 V/V and 10.5 V/V? Provide sketches of the pdf, label critical points, and highlight the areas under the PDF that correspond to the probabilities of interest.

4.14. A noise measurement is repeated many times, resulting in a series of values having an average RMS value of 105 µV. The measurements exhibit a standard deviation of 21 µV RMS. What is the probability that any single noise measurement will return an RMS value larger than 140 µV? What is the probability that any single measurement will return an RMS value less than 70 µV? What is the probability that any single measurement will return an RMS value between 70 and 140 µV? Provide sketches of the pdf, label critical points, and highlight the areas under the PDF that correspond to the probabilities of interest.

4.15. The gain of a DUT is measured with a meter 10 times, resulting in the following sequence of readings: 0.987 V/V, 0.966 V/V, 0.988 V/V, 0.955 V/V, 1.00 V/V, 0.978 V/V, 0.981 V/V, 0.979 V/V, 0.978 V/V, 0.958 V/V. Write an expression for the PDF for this measurement set assuming the distribution is Gaussian.

4.16. The total harmonic distortion (THD) of a DUT is measured 10 times, resulting in the following sequence of readings expressed in percent: 0.112, 0.0993, 0.0961, 0.0153, 0.121, 0.00911, 0.219, 0.101, 0.0945, 0.0767. Write an expression for the PDF for this measurement set assuming the distribution is Gaussian.

4.17. The total harmonic distortion (THD) of a DUT is measured 10 times, resulting in the following sequence of readings expressed in percent: 0.112, 0.0993, 0.0961, 0.0153, 0.121, 0.00911, 0.219, 0.101, 0.0945, 0.0767. Write an expression for the PDF for this measurement set, assuming that the distribution is uniformily distributed.

4.18. A 6-bit ADC is excited by a 1024-point full-scale sine wave described by $1.0\sin\left(2\pi\dfrac{1}{1024}\cdot n\right)+2.0,\ \ n=0,1,...,1023$. Arrange the voltage axis between 1.00 and 3.00 V into 2^6 equal regions and count the number of samples from the sine-wave that fall into each region. Compare with the actual counts with that predicted by theory.

4.19. An 8-bit ADC operates over a range of $0-10$ V. Using a 5-V 1024-point input sine wave centered around a 5-V level, determine the expected code distribution. Verify your results using a histogram of captured data.

4.20. A 6-bit ADC operates over a range of $0-10$ V. Using a 3-V 1024-point input sine wave centered around a 5-V level, how many times does the code $n=50$ appear at the output?

4.21. A 6-bit ADC operates over a range of $0-10$ V. Using a 3-V 1024-point input sine wave centered around a 6-V level, how many times does the code $n=50$ appear at the output?

4.22. A random number generator produces the digits $N=$ -10, -9, -8,...., 9, 10 uniformly. What is the cumulative distribution function of this data set? What is the mean and standard deviation of this pdf?

4.23. A random number generator produces the digits: $N=$ 100, 101,...., 999,1000 uniformly. What is the cumulative distribution function of this data set? What is the mean and standard deviation of this pdf?

4.24. A PDF has the form $f(x) = \begin{cases} 4\left(x-x^3\right), & 0 \le x \le 1 \\ 0, & otherwise \end{cases}$. What is the corresponding cumulative distribution function? What is the mean and standard deviation of this pdf?

4.25. A lot of 1000 devices is tested during production, where it is assumed that a single device has a probability of 0.99 of passing. What is the probability that five bad dies will show up during this test?

4.26. A lot of 1000 devices is tested during production. The device is categorized into five bins, where bins 1 to 4 each have a probability of 0.1 of occurring and bin 5 with a probability of 0.6. What is the probability that 75 dies will appear in bin 1 or in bin 2?

4.27. A sample of 4 dies are selected from a large lot and tested. If the probability of passing is 5/6, what is the PDF of four bad dies appearing during this test?

4.28. Data is transmitted over a channel with a single-bit error probability of 10^{-12}. If 10^{13} bits are transmitted, what is the probability of one bit error? What is the average number of errors expected?

4.29. Data are transmitted over a channel with a single-bit error probability of 10^{-10}. If 10^9 bits are transmitted, what is the probability of one bit error. What is the average number of errors expected?

4.30. A measurement set is found to consist of two Gaussian distributions having the following model parameters: $\alpha_1 = 0.5$, $\mu_1 = 2.5$, $\sigma_1 = 0.5$, $\alpha_2 = 0.5$, $\mu_2 = 3.0$, and $\sigma_2 = 0.2$. What is the probability that some single random measurement X is greater than 0? What about greater than 2? What about less than 3?

4.31. A measurement set is found to consist of three Gaussian distributions having the following model parameters: $\alpha_1 = 0.3$, $\mu_1 = 1.0$, $\sigma_1 = 1$, $\alpha_2 = 0.4$, $\mu_2 = -1.0$, $\sigma_2 = 1.5$, $\alpha_3 = 0.3$, $\mu_3 = 0$, and $\sigma_3 = 0.5$. What is the probability that some single random measurement X is greater than 0? What about greater than 2? What about less than -3?

4.32. A DUT with zero input produces an output offset of -5 mV in the presence of a zero mean Gaussian noise source with a standard deviation of 3 mV. Write a PDF description of the voltage that the voltmeter reads.

4.33. A DUT with zero input produces an output offset that shifts equally between two different levels of 2 mV and −2 mV, depending on the DUT temperature. The noise associated with the DUT is a zero mean Gaussian noise source with a standard deviation of 5 mV. Write a PDF description of the voltage that the voltmeter reads.

4.34. A DUT with zero input produces an output offset that is uniformly distributed between -5 mV and 7 mV. The noise associated with the DUT is a zero mean Gaussian noise source with a standard deviation of 5 mV. Write a PDF description of the voltage that the voltmeter reads.

4.35. Two random variables are combined as the difference according to $y = 1/2\, X_1 + X_2$ where the PDF for X_1 is given by $f_1(x) = \dfrac{1}{\sigma_1\sqrt{2\pi}} e^{-(x-\mu_1)^2/2\sigma_1^2}$ and the PDF for X_2 as $f_2(x) = \dfrac{1}{\sigma_2\sqrt{2\pi}} e^{-(x-\mu_2)^2/2\sigma_2^2}$. What is the PDF of Y?

4.36. Two random variables are combined as the difference according to $y = 1/2\, X_1 + 3X_2$, where the PDF for X_1 is Gaussian with a mean of 1 and a standard deviation of 1. The PDF for X_2 is Gaussian with a mean of -1 and a standard deviation of 2. What is the PDF of Y?

4.37. Two random variables are combined as the difference according to $y = X_1 - 1/2\, X_2$, where the PDF for X_1 is Gaussian with a mean of 2 and a standard deviation of 2. The PDF for X_2 is Gaussian with a mean of −1 and a standard deviation of 1. What is the PDF of Y?

REFERENCES

1. K. Baker and J. van Beers, Shmoo plotting: The black art of IC testing, in *Proceedings International Test Conference*, 1996, pp. 932, 933.

2. A. Papoulis, *Probability, Random Variables, and Stochastic Processes*, 3rd edition, McGraw-Hill, New York, December 1991, ISBN 0070484775.

3. J. S. Bendat and A. G. Piersol, *Random Data: Analysis and Measurement Procedures*, John Wiley & Sons, New York, April 1986, ISBN 0471040002.

4. G. R. Cooper and C. D. McGillem, *Probabilistic Methods of Signal and System Analysis*, 3rd edition, Oxford University Press, New York, 1999, ISBN 0195123549.

5. M. J. Kiemele, S. R. Schmidt and R. J. Berdine, *Basic Statistics, Tools for Continuous Improvement*, 4thedition, Air Academy Press, Colorado Springs, CO, 1997, ISBN 1880156067, pp. 9–71.

6. P.O. Börjesson and C. E. W. Sundberg, Simple approximations of the error function $Q(x)$ for communication applications, *IEEE Transactions On Communications*, pp. 639–43, March 1979.

7. J. Chambers, W. Cleveland, B. Kleiner, and P. Tukey, *Graphical Methods for Data Analysis*, Wadsworth, Belmont, CA, 1983.

8. A. P. Dempster, N. M. Laird, and D. B. Rubin, Maximum likelihood from in-complete data via the EM algorithm. *Journal of the Royal Statistical Society: Series B*, **39**(1), pp. 1–38, November 1977.

9. N. Vlassis and A. Likas, A kurtosis-based dynamic approach to Gaussian mixture modeling, *IEEE Transactions on Systems, Man, and Cybernetics—Part A: Systems and Humans*, **29**(4), pp. 393–399, July 1999.

CHAPTER 5

Yield, Measurement Accuracy, and Test Time

Testing is an important and essential phase in the manufacturing process of integrated circuits. In fact, the only way that manufacturers can deliver high-quality ICs in a reasonable time is through clever testing procedures. The IC manufacturing process involves three major steps: fabrication, testing, and packaging. Today, manufacturing costs associated with mixed-signal ICs is being dominated by the test phase (i.e., separating bad dies from good ones), although packaging costs are becoming quite significant in some large ICs. In order to create clever test procedures, one needs to have a clear understanding of the tradeoffs involved. In particular, the test engineer needs to understand the needs of their business (making ICs for toys or the automotive industry), the cost of test, and the quality of the product produced. It is the intent of this chapter to outline these tradeoffs, beginning with a discussion on manufacturing yield, followed by a discussion on measurement accuracy, and then moving to a discussion of test time. As the needs of a business is highly variable, we will make comments throughout this chapter where appropriate.

5.1 YIELD

The primary goal of a semiconductor manufacturer is to produce large quantities of ICs for sale to various electronic markets—that is, cell phones, ipods, HDTVs, and so on. Semiconductor factories are highly automated, capable of producing millions of ICs over a 24-hour period, every day of the week. For the most part, these ICs are quite similar in behavior, although some will be quite different from one another. A well-defined means to observe the behavior of a set of large elements, such as ICs, is to categorize their individual behavior in the form of a histogram, as shown in Figure 5.1. Here we illustrate a histogram of the offset voltage associated with a lot of devices. We see that 15% of devices produced in this lot had an offset voltage between −0.129 V and −0.128 V. We can conjecture that the probability of another lot producing devices with an offset voltage in this same range is 15%. Of course, how confident we are with our conjecture is the basis of all things statistical; we need to capture more data to support our claim. This we will address shortly; for now, let us consider the "goodness" of what we produced.

In general, the component data sheet defines the "goodness" of an analog or mixed-signal device. As a data sheet forms the basis of any contract between a supplier and a buyer, we avoid any subjective argument of why one measure is better or worse than another; it is simply a matter of data sheet definition. Generally, the goodness of an analog and mixed-signal device is defined by a range of acceptability, bounded by a lower specification limit (LSL) and an upper specification limit (USL), as further illustrated in Figure 5.1. These limits would be found on the device data sheet. Any device whose behavior falls outside this range would be considered as a bad device. This particular example considers a device with a two-sided limit. Similarly, the same argument applies to a one-sided limit; just a different diagram is used.

Testing is the process of separating good devices from the bad ones. The yield of a given lot of material is defined as the ratio of the total good devices divided by the total devices tested:

$$\text{yield} = \frac{\text{total good devices}}{\text{total devices tested}} \times 100\% \tag{5.1}$$

If 10,000 parts are tested and only 7000 devices pass all tests, then the yield on that lot of 10,000 devices is 70%. Because testing is not a perfect process, mistakes are made, largely on account of the measurement limitations of the tester, noise picked up at the test interface, and noise produced by the DUT itself. The most critical error that can be made is one where a bad device is declared good, because this has a direct impact on the operations of a buyer. This error is known as an escape. As a general rule, the impact that an escape has on a manufacturing process goes up exponentially as it moves from one assembly level to another. Hence, the cost of an escape can be many orders of magnitude greater than the cost of a single part. Manufacturers make use of test metrics to gauge the goodness of the component screening process. One measure is the defect level (DL) and it is defined as

$$\text{DL} = \frac{\text{total bad devices declared good}}{\text{total devices declared good}} \times 100\% \tag{5.2}$$

Figure 5.1. Histogram showing specification limits and regions of acceptance and rejection.

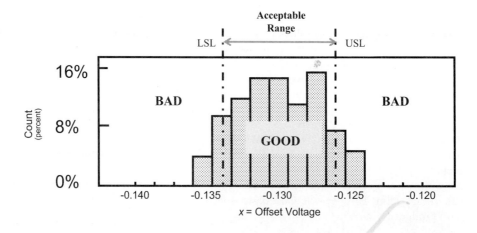

or, when written in terms of escapes, we write

$$DL = \frac{\text{total escapes}}{\text{total devices declared good}} \times 100\% \qquad (5.3)$$

Often DL is expressed in the number of defects-per-million or what is more commonly stated as parts per million (ppm).

It is important to note that a measure of defect level is a theoretical concept based on a probability argument and one that has no empirical basis, because if we knew which devices were escapes, then we would be able to identify them as bad and remove them from the set of good devices. Nonetheless, companies do estimate their defect levels from various measures based on their field returns, or by analyzing the test data during the test process using a secondary screening procedure.

Exercises

5.1. If 15,000 devices are tested with a yield of 63%, how many devices passed the test?   ANS. 9450 devices.

5.2. A new product was launched with 100,000 sales over a one-year time span. During this time, seven devices were returned to the manufacturer even though an extensive test screening procedure was in place. What is the defect level associated with this testing procedure in parts per million?   ANS. 70 ppm.

5.2 MEASUREMENT TERMINOLOGY

The cause of escapes in an analog or mixed-signal environment is largely one that is related to the measurement process itself. A value presented to the instrument by the DUT will introduce errors, largely on account of the electronic circuits that make up the instrument. These errors manifest themselves in various forms. Below we shall outline these errors first in a qualitative way, and then we will move onto a quantitative description in the next section on how these errors interrelate and limit the measurement process.

5.2.1 Accuracy and Precision

In conversational English, the terms *accuracy* and *precision* are virtually identical in meaning. *Roget's Thesaurus*[1] lists these words as synonyms, and *Webster's Dictionary*[2] gives almost identical definitions for them. However, these terms are defined very differently in engineering textbooks.[3-5] Combining the definitions from these and other sources gives us an idea of the accepted technical meaning of the words:

Accuracy: The difference between the average of measurements and a standard sample for which the "true" value is known. The degree of conformance of a test instrument to absolute standards, usually expressed as a percentage of reading or a percentage of measurement range (full scale).

Precision: The variation of a measurement system obtained by repeating measurements on the same sample back-to-back using the same measurement conditions.

According to these definitions, precision refers only to the repeatability of a series of measurements. It does not refer to consistent errors in the measurements. A series of measurements can be incorrect by 2 V, but as long as they are consistently wrong by the same amount, then the measurements are considered to be precise.

This definition of precision is somewhat counterintuitive to most people, since the words *precision* and *accuracy* are so often used synonymously. Few of us would be impressed by a "precision" voltmeter exhibiting a consistent 2-V error! Fortunately, the word *repeatability* is far more commonly used in the test-engineering field than the word *precision*. This textbook will use the term *accuracy* to refer to the overall closeness of an averaged measurement to the true value and *repeatability* to refer to the consistency with which that measurement can be made. The word *precision* will be avoided.

Unfortunately, the definition of accuracy is also somewhat ambiguous. Many sources of error can affect the accuracy of a given measurement. The accuracy of a measurement should probably refer to all possible sources of error. However, the accuracy of an instrument (as distinguished from the accuracy of a measurement) is often specified in the absence of repeatability fluctuations and instrument resolution limitations. Rather than trying to decide which of the various error sources are included in the definition of accuracy, it is probably more useful to discuss some of the common error components that contribute to measurement inaccuracy. It is incumbent upon the test engineer to make sure all components of error have been accounted for in a given specification of accuracy.

5.2.2 Systematic or Bias Errors

Systematic or bias errors are those that show up consistently from measurement to measurement. For example, assume that an amplifier's output exhibits an offset of 100 mV from the ideal value of 0 V. Using a digital voltmeter (DVM), we could take multiple readings of the offset over time and record each measurement. A typical measurement series might look like this:

     101 mV, 103 mV, 102 mV, 101 mV, 102 mV, 103 mV, 103 mV, 101 mV, 102 mV …

This measurement series shows an average error of about 2 mV from the true value of 100 mV. Errors like this are caused by consistent errors in the measurement instruments. The errors can result from a combination of many things, including DC offsets, gain errors, and nonideal linearity in the DVM's measurement circuits. Systematic errors can often be reduced through a process called *calibration*. Various types of calibration will be discussed in more detail in Section 5.4.

5.2.3 Random Errors

In the preceding example, notice that the measurements are not repeatable. The DVM gives readings from 101 to 103 mV. Such variations do not surprise most engineers because DVMs are relatively inexpensive. On the other hand, when a two million dollar piece of ATE equipment cannot produce the same answer twice in a row, eyebrows may be raised.

Inexperienced test engineers are sometimes surprised to learn that an expensive tester cannot give perfectly repeatable answers. They may be inclined to believe that the tester software is defective when it fails to produce the same result every time the program is executed. However, experienced test engineers recognize that a certain amount of random error is to be expected in analog and mixed-signal measurements.

Random errors are usually caused by thermal noise or other noise sources in either the DUT or the tester hardware. One of the biggest challenges in mixed-signal testing is determining whether the random errors are caused by bad DIB design, by bad DUT design, or by the tester itself. If the source of error is found and cannot be corrected by a design change, then averaging or filtering of measurements may be required. Averaging and filtering are discussed in more detail in Section 5.6.

5.2.4 Resolution (Quantization Error)

In the 100-mV measurement list, notice that the measurements are always rounded off to the nearest millivolt. The measurement may have been rounded off by the person taking the measurements, or perhaps the DVM was only capable of displaying three digits. ATE measurement instruments have similar limitations in measurement resolution. Limited resolution results from the fact that continuous analog signals must first be converted into a digital format before the ATE computer can evaluate the test results. The tester converts analog signals into digital form using analog-to-digital converters (ADCs).

ADCs by nature exhibit a feature called *quantization error*. Quantization error is a result of the conversion from an infinitely variable input voltage (or current) to a finite set of possible digital output results from the ADC. Figure 5.2 shows the relationship between input voltages and output codes for an ideal 3-bit ADC. Notice that an input voltage of 1.2 V results in the same ADC output code as an input voltage of 1.3 V. In fact, any voltage from 1.0 to 1.5 V will produce an output code of 2.

If this ADC were part of a crude DC voltmeter, the meter would produce an output reading of 1.25 V any time the input voltage falls between 1.0 and 1.5 V. This inherent error in ADCs and measurement instruments is caused by quantization error. The resolution of a DC meter is often limited by the quantization error of its ADC circuits.

If a meter has 12 bits of resolution, it means that it can resolve a voltage to one part in $2^{12}-1$ (one part in 4095). If the meter's full-scale range is set to ±2 V, then a resolution of approximately 1 mV can be achieved (4 V/4095 levels). This does not automatically mean that the meter is accurate to 1 mV, it simply means the meter cannot resolve variations in input voltage smaller than 1 mV. An instrument's resolution can far exceed its accuracy. For example, a 23-bit voltmeter might be able to produce a measurement with a 1-μV resolution, but it may have a systematic error of 2 mV.

Figure 5.2. Output codes versus input voltages for an ideal 3-bit ADC.

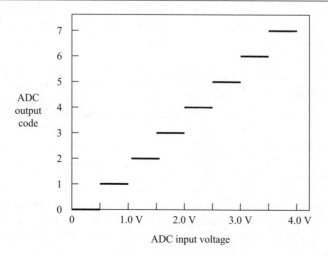

ADC input voltage

5.2.5 Repeatability

Nonrepeatable answers are a fact of life for mixed-signal test engineers. A large portion of the time required to debug a mixed-signal test program can be spent tracking down the various sources of poor repeatability. Since all electrical circuits generate a certain amount of random noise, measurements such as those in the 100-mV offset example are fairly common. In fact, if a test engineer gets the same answer 10 times in a row, it is time to start looking for a problem. Most likely, the tester instrument's full-scale voltage range has been set too high, resulting in a measurement resolution problem. For example, if we configured a meter to a range having a 10-mV resolution, then our measurements from the prior example would be very repeatable (100 mV, 100 mV, 100 mV, 100 mV, etc.). A novice test engineer might think that this is a terrific result, but the meter is just rounding off the answer to the nearest 10-mV increment due to an input ranging problem. Unfortunately, a voltage of 104 mV would also have resulted in this same series of perfectly repeatable, perfectly incorrect measurement results. Repeatability is desirable, but it does not in itself guarantee accuracy.

5.2.6 Stability

A measurement instrument's performance may drift with time, temperature, and humidity. The degree to which a series of supposedly identical measurements remains constant over time, temperature, humidity, and all other time-varying factors is referred to as *stability*. Stability is an essential requirement for accurate instrumentation.

Shifts in the electrical performance of measurement circuits can lead to errors in the tested results. Most shifts in performance are caused by temperature variations. Testers are usually equipped with temperature sensors that can automatically determine when a temperature shift has occurred. The tester must be recalibrated anytime the ambient temperature has shifted by a few degrees. The calibration process brings the tester instruments back into alignment with known electrical standards so that measurement accuracy can be maintained at all times.

After the tester is powered up, the tester's circuits must be allowed to stabilize to a constant temperature before calibrations can occur. Otherwise, the measurements will drift over time as the tester heats up. When the tester chassis is opened for maintenance or when the test head is opened up or powered down for an extended period, the temperature of the measurement electronics will typically drop. Calibrations then have to be rerun once the tester recovers to a stable temperature.

Shifts in performance can also be caused by aging electrical components. These changes are typically much slower than shifts due to temperature. The same calibration processes used to account for temperature shifts can easily accommodate shifts of components caused by aging. Shifts caused by humidity are less common, but can also be compensated for by periodic calibrations.

5.2.7 Correlation

Correlation is another activity that consumes a great deal of mixed-signal test program debug time. Correlation is the ability to get the same answer using different pieces of hardware or software. It can be extremely frustrating to try to get the same answer on two different pieces of equipment using two different test programs. It can be even more frustrating when two supposedly identical pieces of test equipment running the same program give two different answers.

Of course correlation is seldom perfect, but how good is good enough? In general, it is a good idea to make sure that the correlation errors are less than one-tenth of the full range between the minimum test limit and the maximum test limit. However, this is just a rule of thumb. The exact requirements will differ from one test to the next. Whatever correlation errors exist, they must be considered part of the measurement uncertainty, along with nonrepeatability and systematic errors.

The test engineer must consider several categories of correlation. Test results from a mixed-signal test program cannot be fully trusted until the various types of correlation have been verified. The more common types of correlation include tester-to-bench, tester-to-tester, program-to-program, DIB-to-DIB, and day-to-day correlation.

Tester-to-Bench Correlation

Often, a customer will construct a test fixture using bench instruments to evaluate the quality of the device under test. Bench equipment such as oscilloscopes and spectrum analyzers can help validate the accuracy of the ATE tester's measurements. Bench correlation is a good idea, since ATE testers and test programs often produce incorrect results in the early stages of debug. In addition, IC design engineers often build their own evaluation test setups to allow quick debug of device problems. Each of these test setups must correlate to the answers given by the ATE tester. Often the tester is correct and the bench is not. Other times, test program problems are uncovered when the ATE results do not agree with a bench setup. The test engineer will often need to help debug the bench setup to get to the bottom of correlation errors between the tester and the bench.

Tester-to-Tester Correlation

Sometimes a test program will work on one tester, but not on another presumably identical tester. The differences between testers may be catastrophically different, or they may be very subtle. The test engineer should compare all the test results on one tester to the test results obtained using other testers. Only after all the testers agree on all tests is the test program and test hardware debugged and ready for production.

Similar correlation problems arise when an existing test program is ported from one tester type to another. Often, the testers are neither software compatible nor hardware compatible with one another. In fact, the two testers may not even be manufactured by the same ATE vendor. A myriad of correlation problems can arise because of the vast differences in DIB layout and tester software between different tester types. To some extent, the architecture of each tester will determine the best test methodology for a particular measurement. A given test may have to be executed in a very different manner on one tester versus another. Any difference in the way a measurement is taken can affect the results. For this reason, correlation between two different test approaches can be very difficult to achieve. Conversion of a test program from one type of tester to another can be one of the most daunting tasks a mixed-signal test engineer faces.

Program-to-Program Correlation

When a test program is streamlined to reduce test time, the faster program must be correlated to the original program to make sure no significant shifts in measurement results have occurred. Often, the test reduction techniques cause measurement errors because of reduced DUT settling time and other timing-related issues. These correlation errors must be resolved before the faster program can be released into production.

DIB-to-DIB Correlation

No two DIBs are identical, and sometimes the differences cause correlation errors. The test engineer should always check to make sure that the answers obtained on multiple DIB boards agree. DIB correlation errors can often be corrected by *focused calibration* software written by the test engineer (this will be discussed further in Section 5.4).

5.3. A 5-mV signal is measured with a meter 10 times, resulting in the following sequence of readings: 5 mV, 6 mV, 9 mV, 8 mV, 4 mV, 7 mV, 5 mV, 7 mV, 8 mV, 11 mV. What is the average measured value? What is the systematic error?
ANS. 7 mV, 2 mV.

5.4. A meter is rated at 8-bits and has a full-scale range of ±5 V. What is the measurement uncertainty of this meter, assuming only quantization errors from an ideal meter ADC?
ANS. ±19.6 mV.

5.5. A signal is to be measured with a maximum uncertainty of ±0.5 μV. How many bits of resolution are required by an ideal meter having a ±1 V full-scale range?
ANS. 21 bits.

Day-to-Day Correlation

Correlation of the same DIB and tester over a period of time is also important. If the tester and DIB have been properly calibrated, there should be no drift in the answers from one day to the next. Subtle errors in software and hardware often remain hidden until day-to-day correlation is performed. The usual solution to this type of correlation problem is to improve the focused calibration process.

5.2.8 Reproducibility

The term *reproducibility* is often used interchangeably with *repeatability*, but this is not a correct usage of the term. The difference between reproducibility and repeatability relates to the effects of correlation and stability on a series of supposedly identical measurements. Repeatability is most often used to describe the ability of a single tester and DIB board to get the same answer multiple times as the test program is repetitively executed.

Reproducibility, by contrast, is the ability to achieve the same measurement result on a given DUT using any combination of equipment and personnel at any given time. It is defined as the statistical deviation of a series of supposedly identical measurements taken over a period of time. These measurements are taken using various combinations of test conditions that ideally should not change the measurement result. For example, the choice of equipment operator, tester, DIB board, and so on, should not affect any measurement result.

Consider the case in which a measurement is highly repeatable, but not reproducible. In such a case, the test program may consistently pass a particular DUT on a given day and yet consistently fail the same DUT on another day or on another tester. Clearly, measurements must be both repeatable and reproducible to be production-worthy.

5.3 A MATHEMATICAL LOOK AT REPEATABILITY, BIAS, AND ACCURACY

To gain a better understanding of the meaning of accuracy and its impact on a measurement,[6] consider the circuit diagram of a voltage–reference–voltmeter arrangement shown in Figure 5.3. Here we model the reference level with a DC voltage source with value V_{REF}. The voltmeter is modeled with an ideal meter with reading $V_{MEASURED}$ and two voltage sources in series with the reference. One voltage source represents the offset introduced by the voltmeter (V_{OFF}), and the other represents the noise generated by the voltmeter (V_{noise}). By KVL, we can write the voltmeter value as

$$V_{MEASURED} = V_{REF} + V_{OFF} + V_{noise} \qquad (5.4)$$

If we repeat a sequence of measurements involving the same reference, we would obtain a set of values that would in general be all different on account of the noise that is present. To eliminate the effects of this noise, one could instead take the average value of a large number of samples as the measurement. For instance, if we take the expected or average value of each side of Eq. (5.4), we write

$$E\left[V_{MEASURED}\right] = E\left[V_{REF} + V_{OFF} + V_{noise}\right] \qquad (5.5)$$

Recognizing that the expectation operation distributes across addition, we can write

$$E\left[V_{MEASURED}\right] = E\left[V_{REF}\right] + E\left[V_{OFF}\right] + E\left[V_{noise}\right] \qquad (5.6)$$

Assuming that the noise process is normal with zero mean, together with the fact that V_{REF} and V_{OFF} are constants, we find that the expected measured value becomes

$$E\left[V_{MEASURED}\right] = V_{REF} + V_{OFF} \qquad (5.7)$$

As long as the sample set is large, then averaging will eliminate the effects of noise. However, if the sample size is small, a situation that we often find in practice, then our measured value will vary from one sample set to another. See, for example, the illustration in Figure 5.4a involving two sets of samples. Here we see the mean values $\mu_{M,1}$ and $\mu_{M,2}$ are different. If we increase the total number of samples collected to, say, N, we would find that the mean values of the two distributions approach one another in a statistical sense. In fact, the mean of the means will converge to $V_{REF} + V_{OFF}$ with a standard deviation of $\sigma_M / \sqrt{N}$ as illustrated by the dashed line distribution shown in Figure 5.4b. We should also note that the distribution of means is indeed Gaussian as required by the central limit theorem.

Metrology (the science of measurement) is interested in quantifying the level of uncertainty present in a measurement. Three terms from metrology are used in test to describe this uncertainty: repeatability, accuracy, and bias.

Figure 5.3. Modeling a measurement made by a voltmeter with offset and noise component.

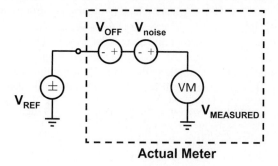

Actual Meter

Assume that N samples are taken during some measurement process and that these samples are assigned to vector x. The mean value of the measurement is quantified as

$$\mu_M = \frac{1}{N} \sum_{k=1}^{N} x[k] \tag{5.8}$$

The repeatability of a measurement refers to the standard deviation associated with the measurement set, that is,

$$\sigma_M = \sqrt{\frac{1}{N} \sum_{k=1}^{N} \left(x[k] - \mu_M \right)^2} \tag{5.9}$$

For the example shown in Figure 5.4b, repeatability refers to the spread of the measurement samples about the sample mean value μ_M. The larger the spread, the less repeatable the measurement will be. We can now define repeatability as the variation (quantified by σ_M) of a measurement system obtained by repeating measurements on the same sample back-to-back using the same measurement conditions.

Bias error or systematic error is the difference between the reference value and the average of a large number of measured values. Bias error can be mathematically described as

$$\beta = V_{REF} - E\left[V_{MEASURED}\right] \tag{5.10}$$

where $E\left[V_{MEASURED}\right]$ is derived through a separate measurement process involving a (very) large number of samples, that is,

$$E\left[V_{MEASURED}\right] = \frac{1}{N} \sum_{k=1}^{N} x[k] \bigg|_{N \text{ LARGE}} \tag{5.11}$$

This step is usually conducted during the characterization phase of the product rather than during a production run to save time. In essence, $E\left[V_{MEASURED}\right]$ converges to $V_{REF} + V_{OFF}$ (the noiseless value) and β equals the negative of the instrument offset, that is,

$$\beta = -V_{OFF} \tag{5.12}$$

Finally, we come to the term *accuracy*. Since test time is of critical importance during a production test, the role of the test engineer is to make a measurement with just the right amount of uncertainty—no more, no less. This suggests selecting the test conditions so that the accuracy of the measurement is just right.

Like bias error, accuracy is defined in much the same way, that being the difference between the known reference value and the expected value of the measurement process. However, accuracy accounts for the error that is introduced due to the repeatability of the measurement that is caused by the small sample set. Let us define the difference between the reference level V_{REF} and an estimate of the mean value, given by $\widehat{V_{MEASURED}} = \frac{1}{N} \sum_{k=1}^{N} x[k] \bigg|_{N \text{ SMALL}}$, as the measurement error:

$$E = V_{REF} - \widehat{V}_{MEASURED} \tag{5.13}$$

Absolute accuracy of a measurement is then defined in terms of a range of possible errors as

$$E_{MIN} \leq \text{accuracy} \leq E_{MAX} \tag{5.14}$$

Figure 5.4. (a) Small sets of different measurements will have different mean values. (b) The mean value of a large sample set will converge to V_{REF} with an offset V_{OFF}. (c) Distribution of measurement errors.

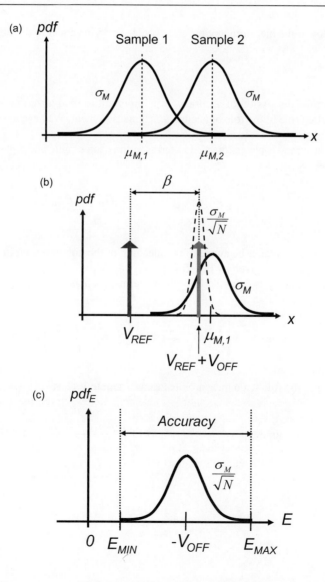

where

$$E_{MIN} = \min \left\{ V_{REF} - \min \left[\widehat{V}_{MEASURED} \right], V_{REF} - \max \left[\widehat{V}_{MEASURED} \right] \right\}$$

$$E_{MAX} = \max \left\{ V_{REF} - \min \left[\widehat{V}_{MEASURED} \right], V_{REF} - \max \left[\widehat{V}_{MEASURED} \right] \right\}$$

It is common practice to refer to the range bounded by E_{MAX} and E_{MIN} as the uncertainty range of the measurement process, or simply just accuracy, as shown in Figure 5.4c. Clearly, some measure of the distribution of measurement errors must be made to quantify accuracy. We have more to say about this in a moment.

As a reference check, if the measurement process is noiseless, then we have

$$\max \left[\widehat{V}_{MEASURED} \right] = \min \left[\widehat{V}_{MEASURED} \right] = V_{REF} + V_{OFF} \tag{5.15}$$

and the accuracy of the measurement process would simply be equal to the offset term, that is,

$$accuracy = -V_{OFF} \tag{5.16}$$

At this point in the discussion of accuracy, it is important to recognize that measurement offset plays an important role in the role of accuracy in a measurement. We have a lot more to say about this in a moment.

Absolute accuracy is also expressed in terms of the center value and a plus–minus difference measure defined as

$$accuracy = \frac{E_{MAX} + E_{MIN}}{2} \pm \frac{|E_{MAX} - E_{MIN}|}{2} \tag{5.17}$$

It is important that the reader be aware of the meaning of the notation used in Eq. (5.17) and how it maps to the error bound given by Eq. (5.14).

Sometimes accuracy is expressed in relative terms, such as a percentage of reference value, that is,

$$accuracy_{\%} = \left[\frac{1}{2} \frac{E_{MAX} + E_{MIN}}{V_{REF}} \pm \frac{1}{2} \frac{|E_{MAX} - E_{MIN}|}{V_{REF}} \right] \times 100\% \tag{5.18}$$

or a percentage of the full-scale measurement range (denote V_{FS}) as

$$accuracy_{\%FS} = \left[\frac{1}{2} \frac{E_{MAX} + E_{MIN}}{V_{FS}} \pm \frac{1}{2} \frac{|E_{MAX} - E_{MIN}|}{V_{FS}} \right] \times 100\% \tag{5.19}$$

EXAMPLE 5.1

A measurement of a 1-V reference level is made 100 times, where the minimum reading is 0.95 V and the maximum reading is 1.14 V. What is the absolute accuracy of these measurements? What is the relative accuracy with respect to a full-scale value of 5 V?

Solution:
According to Eq. (5.14), the smallest and largest errors are

$$E_{MIN} = \min\{1-0.95, 1-1.14\} = -0.14$$
$$E_{MAX} = \max\{1-0.95, 1-1.14\} = 0.05$$

leading to an absolute accuracy bound of

$$-0.14 \leq \text{accuracy} \leq 0.05$$

or expressed with respect to the center as

$$\text{accuracy} = \frac{-0.14+0.05}{2} \pm \frac{|-0.14-0.05|}{2} = -0.045 \pm 0.095 \text{ V}$$

In terms of a relative accuracy referenced to the 5-V full-scale level, we obtain

$$\text{accuracy}_{\%FS} = \frac{-0.045 \pm 0.095 \text{ V}}{5 \text{ V}} \times 100\% = -0.9 \pm 1.9\%$$

It is interesting to observe the statistics of this estimator, $\widehat{V}_{MEASURED}$, when noise is present. Regardless of the nature of the noise, according to the central limit theorem, the estimator $\widehat{V}_{MEASURED}$ will follow Gaussian statistics. This is because the estimator is a sum of several random variables as described in the previously chapter. The implications of this is that if one were to repeat a measurement a number of times and create a histogram of resulting estimator values, one would find that it has a Gaussian shape. Moreover, it would have a standard deviation of approximately $\sigma_M/\sqrt{N}$. As a first guess, we could place the center of the Gaussian distribution at the value of the estimator $\widehat{V}_{MEASURED} = \frac{1}{N}\sum_{k=1}^{N} x[k]$. Hence, the pdf of the estimator values would appear as

$$g(v) = \frac{\sqrt{N}}{\sigma_M \sqrt{2\pi}} e^{\frac{-\left(v - \widehat{V}_{MEASURED}\right)^2}{2\left(\sigma_M/\sqrt{N}\right)^2}} \tag{5.20}$$

Consequently, we can claim with a 99.7% probability that the true mean of the measurement will lie between $\widehat{V}_{MEASURED} - 3\dfrac{\sigma_M}{\sqrt{N}}$ and $\widehat{V}_{MEASURED} + 3\dfrac{\sigma_M}{\sqrt{N}}$. One can introduce an α term to the previous range term and generalize the result to a set of probabilities—for example,

$$P\left(\widehat{V}_{MEASURED} - \alpha\frac{\sigma_M}{\sqrt{N}} \leq E\left[V_{MEASURED}\right] \leq \widehat{V}_{MEASURED} + \alpha\frac{\sigma_M}{\sqrt{N}}\right) = \begin{cases} 0.667, & \alpha = 1 \\ 0.950, & \alpha = 2 \\ 0.997, & \alpha = 3 \end{cases} \qquad (5.21)$$

One can refer to the α term as a confidence parameter, because the larger its value, the greater our confidence (probability) that the noiseless measured value lies within the range defined by

$$\widehat{V}_{MEASURED} - \alpha\frac{\sigma_M}{\sqrt{N}} \leq E\left[V_{MEASURED}\right] \leq \widehat{V}_{MEASURED} + \alpha\frac{\sigma_M}{\sqrt{N}} \qquad (5.22)$$

In the statistical literature this range is known as the confidence interval (CI). The extremes of measurement estimator can then be identified as

$$\max\left[\widehat{V}_{MEASURED}\right] = \widehat{V}_{MEASURED} + \alpha\frac{\sigma_M}{\sqrt{N}}$$

$$\min\left[\widehat{V}_{MEASURED}\right] = \widehat{V}_{MEASURED} - \alpha\frac{\sigma_M}{\sqrt{N}} \qquad (5.23)$$

where $\widehat{V}_{MEASURED}$ is any one estimate of the mean of the measurement and σ_M is the standard deviation of the measurement process usually identified during a characterization phase. Substituting Eq. (5.23) into Eq. (5.17), together with definitions given in Eq. (5.14), we write the accuracy expression as

$$\text{accuracy} = V_{REF} - \widehat{V}_{MEASURED} \pm \alpha\frac{\sigma_M}{\sqrt{N}} \qquad (5.24)$$

As a first-order approximation, let us assume

$$V_{REF} - \widehat{V}_{MEASURED} \approx \beta \qquad (5.25)$$

Then Eq. (5.24) becomes

$$\text{accuracy} = \beta \pm \alpha\frac{\sigma_M}{\sqrt{N}} \qquad (5.26)$$

This is the **fundamental** equation for **measurement accuracy**. It illustrates the dependency of accuracy on the bias error, repeatability, and the number of samples. It also suggest several ways in which to improve measurement accuracy:

1. Remove the bias error β by calibrating to a known reference value.
2. Decrease the intrinsic amount of noise σ_M associated with a measurement by purchasing more expensive instruments with a lower noise floor or by improving the device interface board (DIB) design and test interface.
3. Increase the size of the sample set N to reduce the influence of measurement repeatability; increase the time of test; or alter the algorithm that is used to extract the mean value.

The next few sections will address these three points in greater detail.

EXAMPLE 5.2

A DC offset measurement is repeated 100 times, resulting in a series of values having an average of 257 mV and a standard deviation of 27 mV. In what range does the noiseless measured value lie for a 99.7% confidence? What is the accuracy of this measurement assuming the systematic offset is zero?

Solution:

Using Eq. (5.22) with $\alpha = 3$, we can bound the noiseless measured value to lie in the range defined by

$$257 \text{ mV} - 3 \times \frac{27 \text{ mV}}{\sqrt{100}} \leq E\left[V_{MEASURED}\right] \leq 257 \text{ mV} + 3 \times \frac{27 \text{ mV}}{\sqrt{100}}$$

or

$$248.9 \text{ mV} \leq E\left[V_{MEASURED}\right] \leq 265.1 \text{ mV}$$

The accuracy of this measurement (where a measurement is the average of 100 voltage samples) would then be ±8.1 mV with a 99.7% confidence. Alternatively, if we repeat this measurement 1000 times, we can expect that 997 measured values (i.e., each measured value corresponding to 100 samples) will lie between 248.9 mV and 265.1 mV.

Exercises

5.6. A series of 100 measurements is made on the output of an op-amp circuit whereby the distribution was found to be Gaussian with mean value of 12.5 mV and a standard deviation of 10 mV. Write an expression for the pdf of these measurements?

ANS. $g(v) = \dfrac{1}{(10^{-2})\sqrt{2\pi}} e^{-\frac{(v-12.5\times10^{-3})^2}{2(10^{-2})^2}}$;

$\mu = 12.5$ mV; $\sigma = 10.0$ mV.

5.7. A series of 100 measurements is made on the output of an op-amp circuit whereby the distribution was found to be Gaussian with mean value of 12.5 mV and a standard deviation of 1 mV. If this experiment is repeated, write an expression for the pdf of the mean values of each of these experiments?

ANS. $f(v) = \dfrac{1}{(10^{-3})\sqrt{2\pi}} e^{-\frac{(v-12.5\times10^{-3})^2}{2(10^{-3})^2}}$;

$\mu = 12.5$ mV; $\sigma = 1$ mV.

Exercises

5.8. A series of 100 measurements is made on the output of an op-amp circuit whereby the distribution was found to be Gaussian with mean value of 12.5 mV and a standard deviation of 1 mV. If this experiment is repeated and the mean value is compared to a reference level of 10 mV, what is the mean and standard deviation of the error distribution that results? Write an expression for the pdf of these errors?

ANS. $h(v) = \dfrac{1}{(10^{-3})\sqrt{2\pi}} e^{-\frac{(v-2.5\times10^{-3})^2}{2(10^{-3})^2}}$;

$\mu = 2.5$ mV; $\sigma = 1$ mV.

5.9. A series of 100 measurements is made on the output of an op amp circuit whereby the distribution was found to be Gaussian with mean value of 12.5 mV and a standard deviation of 1 mV. If this experiment and the mean value is compared to a reference value of 10 mV, in what range will the expected value of the error lie for a 99.7% conference interval.

ANS. -0.5 mV $\leq E[\text{error}]$ ≤ 5.5 mV

5.4 CALIBRATIONS AND CHECKERS

Measurement accuracy can be improved by eliminating the bias error β associated with a measurement process. This section will look at the several ways in which to remove this error so that the tester is performing with maximum accuracy at all times during its operation.

5.4.1 Traceability to Standards

Every tester and bench instrument must ultimately correlate to standards maintained by a central authority, such as the National Institute of Standards and Technology (NIST). In the United States, this government agency is responsible for maintaining the standards for pounds, gallons, inches, and electrical units such as volts, amperes, and ohms. The chain of correlation between the NIST and the tester's measurements involves a series of calibration steps that transfers the "golden" standards of the NIST to the tester's measurement instruments.

Many testers have a centralized standards reference, which is a thermally stabilized instrument in the tester mainframe. The standards reference is periodically replaced by a freshly calibrated reference source. The old one is sent back to a certified calibration laboratory, which recalibrates the reference so that it agrees with NIST standards. Similarly, bench instruments are periodically recalibrated so that they too are traceable to the NIST standards. By periodically refreshing the tester's traceability link to the NIST, all testers and bench instruments can be made to agree with one another.

5.4.2 Hardware Calibration

Hardware calibration is a process of physical "knob tweaking" that brings a piece of measurement instrumentation back into agreement with calibration standards. For instance, oscilloscope probes often include a small screw that can be used to nullify the overshoot in rapidly rising digital edges. This is one common example of hardware calibration.

One major problem with hardware calibration is that it is not a convenient process. It generally requires a manual adjustment of a screw or knob. Robotic screwdrivers might be employed to allow partial automation of the hardware calibration process. However, the use of robotics is an elaborate solution to the calibration problem. Full automation can be achieved using a simpler procedure known as *software calibration*.

Figure 5.5. (a) Modeling a voltmeter with an ideal voltmeter and a nonideal component in cascade. (b) Calibrating the nonideal effects using a software routine.

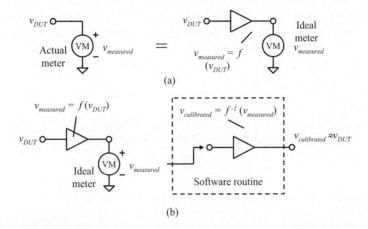

(a)

(b)

5.4.3 Software Calibration

Using software calibration, ATE testers are able to correct hardware errors without adjusting any physical knobs. The basic idea behind software calibration is to separate the instrument's ideal operation from its nonidealities. Then a *model* of the instrument's nonideal operation can be constructed, followed by a *correction* of the nonideal behavior using a mathematical routine written in software. Figure 5.5 illustrates this idea for a voltmeter.

In Figure 5.5a a "real" voltmeter is modeled as a cascade of two parts: (1) an ideal voltmeter and (2) a black box that relates the voltage across its input terminals v_{DUT} to the voltage that is measured by the ideal voltmeter, $v_{measured}$. This relationship can be expressed in more mathematical terms as

$$v_{measured} = f\left(v_{DUT}\right) \tag{5.27}$$

where $f(\cdot)$ indicates the functional relationship between $v_{measured}$ and v_{DUT}.

The true functional behavior $f(\cdot)$ is seldom known; thus one assumes a particular behavior or model, such as a first-order model given by

$$v_{measured} = Gv_{DUT} + \text{offset} \tag{5.28}$$

where G and *offset* are the gain and offset of the voltmeter, respectively. These values must be determined from measured data. Subsequently, a mathematical procedure is written in software that performs the inverse mathematical operation

$$v_{calibrated} = f^{-1}\left(v_{measured}\right) \tag{5.29}$$

where $v_{calibrated}$ replaces v_{DUT} as an estimate of the true voltage that appears across the terminals of the voltmeter as depicted in Figure 5.5b. If $f(\cdot)$ is known precisely, then $v_{calibrated} = v_{DUT}$.

In order to establish an accurate model of an instrument, precise reference levels are necessary. The number of reference levels required to characterize the model fully will depend on its order—that is, the number of parameters used to describe the model. For the linear or firstorder model described, it has two parameters, G and *offset*. Hence, two reference levels will be required.

    To avoid conflict with the meter's normal operation, relays are used to switch in these reference levels during the calibration phase. For example, the voltmeter in Figure 5.6 includes a pair of calibration relays, which can connect the input to two separate reference levels, V_{ref1} and V_{ref2}. During a system level calibration, the tester closes one relay and connects the voltmeter to V_{ref1} and measures the voltage, which we shall denote as $v_{measured1}$. Subsequently, this process is repeated for the second reference level V_{ref2} and the voltmeter provides a second reading, $v_{measured2}$.

    Based on the assumed linear model for the voltmeter, we can write two equations in terms of two unknowns

$$v_{measured1} = GV_{ref1} + \text{offset}$$
$$v_{measured2} = GV_{ref2} + \text{offset} \tag{5.30}$$

Using linear algebra, the two model parameters can then be solved to be

$$G = \frac{v_{measured2} - v_{measured1}}{V_{ref2} - V_{ref1}} \tag{5.31}$$

and

$$\text{offset} = \frac{v_{measured1}V_{ref2} - v_{measured2}V_{ref1}}{V_{ref2} - V_{ref1}} \tag{5.32}$$

The parameters of the model, G and offset, are also known as *calibration factors*, or cal factors for short.

    When subsequent DC measurements are performed, they are corrected using the stored calibration factors according to

$$v_{calibrated} = \frac{v_{measured} - \text{offset}}{G} \tag{5.33}$$

This expression is found by isolating v_{DUT} on one side of the expression in Eq. (5.28) and replacing it by $v_{calibrated}$.

Figure 5.6. DC voltmeter gain and offset calibration paths.

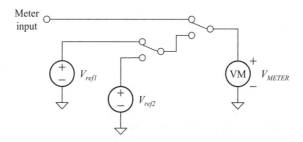

Of course, this example is only for purposes of illustration. Most testers use much more elaborate calibration schemes to account for linearity errors and other nonideal behavior in the meter's ADC and associated circuits.. Also, the meter's input stage can be configured many ways, and each of these possible configurations needs a separate set of calibration factors. For example, if the input stage has 10 different input ranges, then each range setting requires a separate set of calibration factors. Fortunately for the test engineer, most instrument calibrations happen behind the scenes. The calibration factors are measured and stored automatically during the tester's periodic system calibration and checker process.

5.4.4 System Calibrations and Checkers

Testers are calibrated on a regular basis to maintain traceability of each instrument to the tester's calibration reference source. In addition to calibrations, software is also executed to verify the functionality of hardware and make sure it is production worthy. This software is called a *checker program*, or *checker* for short. Often calibrations and checkers are executed in the same program. If a checker fails, the repair and maintenance (R&M) staff replaces the failing tester module with a good one. After replacement, the new module must be completely recalibrated.

There are several types of calibrations and checkers. These include calibration reference source replacement, performance verification (PV), periodic system calibrations and checkers, instrument calibrations at load time, and focused calibrations. Calibration reference source replacement and recalibration was discussed in Section 5.4.1. A common replacement cycle time for calibration sources is once every six months.

Exercises

5.10. A meter reads 0.5 mV and 1.1 V when connected to two precision reference levels of 0 and 1 V, respectively. What are the offset and gain of this meter? Write the calibration equation for this meter.

ANS. 0.5 mV, 1.0995 V/V, $V_{calibrated} = (V_{measured} - 0.5$ mV$)/1.0995$.

5.11. A meter is assumed characterized by a second-order equation of the form: $V_{measured} = offset + G_1 V_{calibrated} + G_2 V_{calibrated}^2$. How many precision DC reference levels are required to obtain the parameters of this second-order expression?

ANS. Three.

5.12. A meter is assumed characterized by a second-order equation of the form $V_{measured} = offset + G_1 v_{IN} + G_2 v_{IN}^2$. Write the calibration equation for this meter in terms of the unknown calibration factors.

ANS.

$$V_{calibrated} = \frac{-G_1 + \sqrt{G_1^2 + 4G_2 \left(V_{measured} - offset\right)}}{2G_2} \quad \text{or}$$

$$V_{calibrated} = \frac{-G_1 - \sqrt{G_1^2 + 4G_2 \left(V_{measured} - offset\right)}}{2G_2}$$

depending on the data conditions.

To verify that the tester is in compliance with all its published specifications, a more extensive process called *performance verification* may be performed. Although full performance verification is typically performed at the tester vendor's production floor, it is seldom performed on the production floor. By contrast, periodic system calibrations and checkers are performed on a regular basis in a production environment. These software calibration and checker programs verify that all the system hardware is production worthy.

Since tester instrumentation may drift slightly between system calibrations, the tester may also perform a series of fine-tuning calibrations each time a new test program is loaded. The extra calibrations can be limited to the subset of instruments used in a particular test program. This helps to minimize program load time. To maintain accuracy throughout the day, these calibrations may be repeated on a periodic basis after the program has been loaded. They may also be executed automatically if the tester temperature drifts by more than a few degrees.

Finally, focused calibrations are often required to achieve maximum accuracy and to compensate for nonidealities of DIB board components such as buffer amplifiers and filters. Unlike the ATE tester's built-in system calibrations, focused calibration and checker software is the responsibility of the test engineer. Focused calibrations fall into two categories: (1) focused instrument calibrations and (2) focused DIB calibrations and checkers.

5.4.5 Focused Instrument Calibrations

Testers typically contain a combination of slow, accurate instruments and fast instruments that may be less accurate. The accuracy of the faster instruments can be improved by periodically referencing them back to the slower more accurate instruments through a process called *focused calibration*. Focused calibration is not always necessary. However, it may be required if the test engineer needs higher accuracy than the instrument is able to provide using the built-in calibrations of the tester's operating system.

A simple example of focused instrument calibration is a DC source calibration. The DC sources in a tester are generally quite accurate, but occasionally they need to be set with minimal DC level error. A calibration routine that determines the error in a DC source's output level can be added to the first run of the test program. A high-accuracy DC voltmeter can be used to measure the actual output of the DC source. If the source is in error by 1 mV, for instance, then the requested voltage is reduced by 1 mV and the output is retested. It may take several iterations to achieve the desired value with an acceptable level of accuracy.

A similar approach can be extended to the generation of a sinusoidal signal requiring an accurate RMS value from an arbitrary waveform generator (AWG). A high-accuracy AC voltmeter is used to measure the RMS value from the AWG. The discrepancy between the measured value and the desired value is then used to adjust the programmed AWG signal level. The AWG output level will thus converge toward the desired RMS level as each iteration is executed.

EXAMPLE 5.3

A 2.500-V signal is required from a DC source as shown in Figure 5.7. Describe a calibration procedure that can be used to ensure that 2.500 V ± 500 μV does indeed appear at the output of the DC source.

Solution:
The source is set to 2.500 V and a high-accuracy voltmeter is connected to the output of the source using a calibration path internal to the tester. Calibration path connections are made through one or more relays such as the ones in Figure 5.6. Assume the high-accuracy voltmeter reads 2.510 V from the source. The source is then reprogrammed to 2.500 V − 10 mV and the output is remeasured. If the second meter reading is 2.499 V, then the source is reprogrammed

to 2.500 V – 10 mV + 1 mV and measured again. This process is repeated until the meter reads 2.500 V (plus or minus 500 µV). Once the exact programmed level is established, it is stored as a calibration factor (e.g., calibration factor = 2.500 V – 10 mV + 1 mV = 2.491 V). When the 2.500-V DC level is required during subsequent program executions, the 2.491 V calibration factor is used as the programmed level rather than 2.500 V. Test time is not wasted searching for the ideal level after the first calibration is performed. However, calibration factors may need to be regenerated every few hours to account for slow drifts in the DC source. This recalibration interval is dependent on the type of tester used.

Figure 5.7. DC source focused calibration.

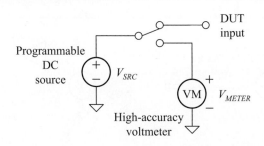

Another application of focused instrument calibration is spectral leveling of the output of an AWG. An important application of AWGs is to provide a composite signal consisting of *N* sine waves or *tones* all having equal amplitude at various frequencies and arbitrary phase. Such waveforms are in a class of signals commonly referred to as *multitone* signals. Mathematically, a multitone signal *y(t)* can be written as

$$y(t) = A_0 + A_1 \sin\left(2\pi f_1 t + \phi_1\right) + \cdots + A_N \sin\left(2\pi f_N t + \phi_N\right) = A_0 + \sum_{k=1}^{N} A_k \sin\left(2\pi f_k t + \phi_k\right) \quad (5.34)$$

where A_k, f_k, and ϕ_k denote the amplitude, frequency, and phase, respectively, of the *k*th tone. A multitone signal can be viewed in either the time domain or in the frequency domain. Timedomain views are analogous to oscilloscope traces, while frequency-domain views are analogous to spectrum analyzer plots. The frequency-domain graph of a multitone signal contains a series of vertical lines corresponding to each tone frequency and whose length* represents the root-mean-square (RMS) amplitude of the corresponding tone. Each line is referred to as a *spectral line*. Figure 5.8 illustrates the time and frequency plots of a composite signal consisting of three tones of frequencies 1, 2.5, and 4.1 kHz, all having an RMS amplitude of 2 V. Of course, the peak amplitude of each sinusoid in the multitone is simply $\sqrt{2} \times 2$ or 2.82 V, so we could just as easily plot these values as peak amplitudes rather than RMS. This book will consistently display frequency-domain plots using RMS amplitudes.

---

*Spectral density plots are commonly defined in engineering textbooks with the length of the spectral line representing one-half the amplitude of a tone. In most test engineering work, including spectrum analyzer displays, it is more common to find this length defined as an RMS quantity.

Figure 5.8. Time-domain and frequency-domain views of a three-tone multitone.

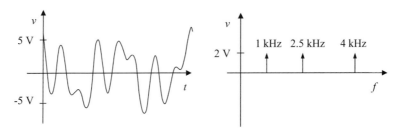

Figure 5.9. Modeling an AWG as a cascaded combination of an ideal source and frequency-dependent gain block.

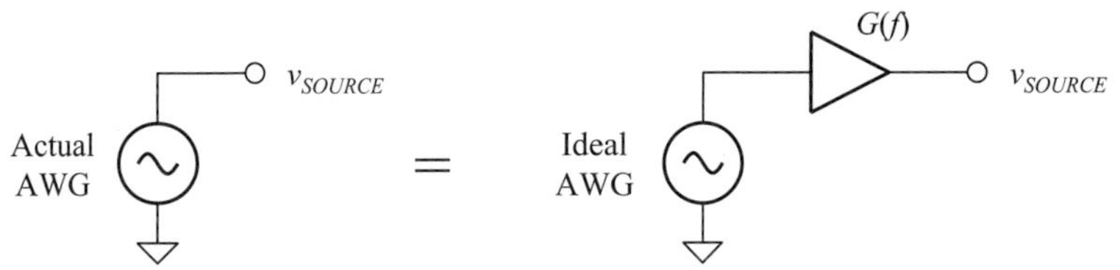

The AWG produces its output signal by passing the output of a DAC through a low-pass antiimaging filter. Due to its frequency behavior, the filter will not have a perfectly flat magnitude response. The DAC may also introduce frequency-dependent errors. Thus the amplitudes of the individual tones may be offset from their desired levels. We can therefore model this AWG multitone situation as illustrated in Figure 5.9. The model consists of an ideal source connected in cascade with a linear block whose gain or magnitude response is described by $G(f)$, where f is the frequency expressed in hertz. To correct for the gain change with frequency, the amplitude of each tone from the AWG is measured individually using a high-accuracy AC voltmeter. The ratio between the actual output and the requested output corresponds to $G(f)$ at that frequency. This gain can then be stored as a calibration factor that can subsequently be retrieved to correct the amplitude error at that frequency. The calibration process is repeated for each tone in the multitone signal. The composite signal can then be generated with corrected amplitudes by dividing the previous requested amplitude at each frequency by the corresponding AWG gain calibration factor. Because the calibration process equalizes the amplitudes of each tone, the process is called multitone leveling.

As testers continue to evolve and improve, it may become increasingly unnecessary for the test engineer to perform focused calibrations of the tester instruments. Focused calibrations were once necessary on almost all tests in a test program. Today, they can sometimes be omitted with little degradation in accuracy. Nevertheless, the test engineer must evaluate the need for focused calibrations on each test. Even if calibrations become unnecessary in the future, the test engineer should still understand the methodology so that test programs on older equipment can be comprehended.

Calibration of circuits on the DIB, on the other hand, will probably always be required. The tester vendor has no way to predict what kind of buffer amplifiers and other circuits will be placed on the DIB board. The tester operating system will never be able to provide automatic calibration of these circuits. The test engineer is fully responsible for understanding the calibration requirements of all DIB circuits.

5.13. A DC source is assumed characterized by a third-order equation of the form $V_{MEASURED} = 0.005 + V_{PROGRAMMED} - 0.003\, V_{PROGRAMMED}^3$ and is required to generate a DC level of 2.6 V. However, when programmed to produce this level, only 2.552 V is measured. Using iteration, determine a value of the programmed source voltage that will establish a measured voltage of 2.6 V to within a ± 1-mV accuracy.

ANS. 2.651 V.

5.14. An AWG has a gain response described by $\left(\sqrt{1+\left(\frac{f}{10^3}\right)^2}\right)^{-1}$ and is to generate three tones at frequencies of 1, 2, and 3 kHz. What are the calibration factors?

ANS. 0.707, 0.447, and 0.316.

5.4.6 Focused DIB Circuit Calibrations

Often circuits are added to a DIB board to improve the accuracy of a particular test or to buffer the weak output of a device before sending it to the tester electronics. As the signal-conditioning DIB circuitry is added in cascade with the test instrument, a model of the test setup is identical to that given in Figure 5.5a. The only difference is that functional block $v_{measured} = f(v_{DUT})$ includes both the meter and the DIB's behavior. As a result, the focused instrument calibrations of Section 5.4.3 can be used with no modifications. Conversely, the meter may already have been calibrated so that the functional block $f(\cdot)$ covers the DIB circuitry only. One must keep track of the extent of the calibration to avoid any double counting.

EXAMPLE 5.4

The op-amp circuit in Figure 5.10 has been added to a DIB board to buffer the output of a DUT. The buffer will be used to condition the DC signal from the DUT before sending it to a calibrated DC voltmeter resident in the tester. If the output is not buffered, then we may find that the DUT breaks into oscillations as a result of the stray capacitance arising along the lengthy signal path leading from the DUT to the tester. The buffer prevents these oscillations by substantially reducing stray capacitance at the DUT output. In order to perform an accurate measurement, the behavior of the buffer must be accounted for. Outline the steps to perform a focused DC calibration on the op-amp buffer stage.

Solution:
To perform a DC calibration of the output buffer amplifier, it is necessary to assume a model for the op-amp buffer stage. It is reasonable to assume that the buffer is fairly linear over a wide range of signal levels, so that the following linear model can be used:

$$v_{measured} = Gv_{DUT} + \text{offset}$$

Subsequently, following the same procedure as outlined in Section 5.4.3, a pair of known voltages are applied to the input of the buffer from source SRC1 via the relay connection and the output of the buffer is measured with a voltmeter. This temporary connection is called a calibration path. As an example, let SRC1 force 2 V and assume that an output voltage of 2.023 V is measured using the voltmeter. Next the input is dropped to 1 V, resulting in an output voltage of 1.012 V. Using Eq. (5.31), we find the buffer has gain given by

$$G = \frac{2.023\ V - 1.012\ V}{2\ V - 1\ V} = 1.011\ V/V$$

and the offset is found from Eq. (5.32) to be

$$offset = \frac{1.012\ V \cdot 2\ V - 2.023\ V \times 1\ V}{2\ V - 1\ V} = 1\ mV$$

Hence, the DUT output v_{DUT} and the voltmeter value $v_{measured}$ are related according to

$$v_{measured} = 1.011\ V/V \times v_{DUT} + 0.001\ V$$

The goal of the focused DC calibration procedure is to find an expression that relates the DUT output in terms of the measured value. Hence, by rearranging the expression and replacing $v_{calibrated}$ for v_{DUT}, we obtain

$$v_{calibrated} = \frac{v_{measured} - 0.001\ V}{1.011\ V/V}$$

For example, if the voltmeter reads 1.732 V, the actual voltage appearing at its terminals is actually

$$v_{calibrated} = \frac{1.732\ V - 0.001\ V}{1.011\ V/V} = 1.712\ V$$

If the original uncalibrated answer had been used, there would have been a 20-mV error! This example shows why focused DUT calibrations are so important to accurate measurements.

Figure 5.10. DC calibration for op-amp buffer circuit.

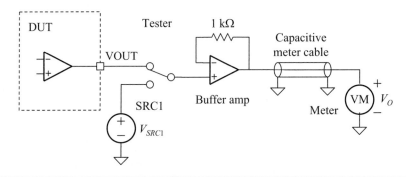

When buffer amplifiers are used to assist the measurement of AC signals, a similar calibration process must be performed on each frequency that is to be measured. Like the AWG calibration example, the buffer amplifier also has a nonideal frequency response and will affect the reading of the meter. Its gain variation, together with the meter's frequency response, must be measured at each frequency used in the test during a calibration run of the test program. Assuming that the meter has already been calibrated, the frequency response behavior of the DIB circuitry must be correctly accounted for. This is achieved by measuring the gain in the DIB's signal path at each specific test frequency. Once found, it is stored as a calibration factor. If additional circuits such as filters, ADCs, and so on, are added on the DIB board and used under multiple configurations, then each unique signal path must be individually calibrated.

5.5 TESTER SPECIFICATIONS

The test engineer should exercise diligence when evaluating tester instrument specifications. It can be difficult to determine whether or not a particular tester instrument is capable of making a particular measurement with an acceptable level of accuracy. The tester specifications usually do not include the effects of uncertainty caused by instrument repeatability limitations. All the specification conditions must be examined carefully. Consider the following DC meter example.

A DC meter consisting of an analog-to-digital converter and a programmable gain amplifier (PGA) is shown in Figure 5.11. The programmable gain stage is used to set the range of the meter so that it can measure small signals as well as large ones. Small signals are measured with the highest gain setting of the PGA, while large signals are measured with the lowest gain setting. This ranging process effectively changes the resolution of the ADC so that its quantization error is minimized.

Calibration software in the tester compensates for the different PGA gain settings so that the digital output of the meter's ADC can be converted into an accurate voltage reading. The calibration software also compensates for linearity errors in the ADC and offsets in the PGA and ADC. Fortunately, the test engineer does not have to worry about these calibrations because they happen automatically.

Table 5.1 shows an example of a specification for a fictitious DC meter, the DVM100. This meter has five different input ranges, which can be programmed in software. The different ranges allow small voltages to be measured with greater accuracy than large voltages. The accuracy is specified as a percentage of the measured value, but there is an accuracy limit of 1 mV for the lower ranges and 2.5 mV for the higher ranges.

Figure 5.11. Simplified DC voltmeter with input ranging amplifier.

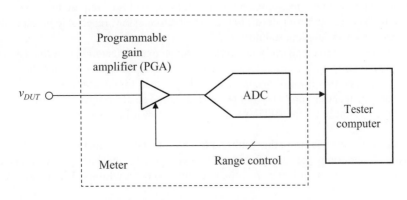

Table 5.1. DVM100 DC Voltmeter Specifications

Range	Resolution	Accuracy (% of Measurement)
±0.5 V	15.25 µV	±0.05 % or 1 mV (whichever is greater)
±1 V	30.5 µV	±0.05 % or 1 mV
±2 V	61.0 µV	±0.05 % or 1 mV
±5 V	152.5 µV	±0.10 % or 2.5 mV
±10 V	305.2 mV	±0.10 % or 2.5 mV

Note: All specs apply with the measurement filter enabled.

This accuracy specification probably assumes that the measurement is made 100 or more times and averaged. For a single nonaveraged measurement, there may also be a repeatability error to consider. It is not clear from the table above what assumptions are made about averaging. The test engineer should make sure that all assumptions are understood before relying on the accuracy numbers.

EXAMPLE 5.5

A DUT output is expected to be 100 mV. Our fictitious DC voltmeter, the DVM100, is set to the 0.5-V range to achieve the optimum resolution and accuracy. The reading from the meter (with the meter's input filter enabled) is 102.3 mV. Calculate the accuracy of this reading (excluding possible repeatability errors). What range of outputs could actually exist at the DUT output with this reading?

Solution:
The measurement error would be equal to ±0.05% of 100 mV, or 50 µV, but the specification has a lower limit of 1 mV. The accuracy is therefore ±1 mV. Based on the single reading of 102.3 mV, the actual voltage at the DUT output could be anywhere between 101.3 and 103.3 mV.

In addition to the ranging hardware, the meter also has a low-pass filter in series with its input. The filter can be bypassed or enabled, depending on the measurement requirements. Repeatability is enhanced when the low-pass filter is enabled, since the filter reduces electrical noise in the input signal. Without this filter the accuracy would be degraded by nonrepeatability. The filter undoubtedly adds settling time to the measurement, since all low-pass filters require time to stabilize to a final DC value. The test engineer must often choose between slow, repeatable measurements and fast measurements with less repeatability.

It may be possible to empirically determine through experimentation that this DC voltmeter has adequate resolution and accuracy to make a DC offset measurement with less than 100 µV of error. However, since this level of accuracy is far better than the instrument's ±1-mV specifications, the instrument should probably not be trusted to make such a measurement in production. The accuracy might hold up for 100 days and then drift toward the specification limits of 1 mV on day 101.

Another possible scenario is that multiple testers may be used that do not all have 100-µV performance. Tester companies are often conservative in their published specifications, meaning that the instruments are often better than their specified accuracy limits. This is not a license to use the instruments to more demanding specifications. It is much safer to use the specifications as printed, since the vendor will not take any responsibility for use of instruments beyond their official specifications.

Sometimes the engineer may have to design front-end circuitry such as PGAs and filters onto the DIB board itself. The DIB circuits might be needed if the front-end circuitry of the meter is inadequate for a high-accuracy measurement. Front-end circuits may also be added if the signal from the DUT cannot be delivered cleanly through the signal paths to the tester instruments. Very high-impedance DUT signals might be susceptible to externally coupled noise, for example. Such signals might benefit from local buffering and amplification before passing to the tester instrument. The test engineer must calibrate any such buffering or filtering circuits using a focused DIB calibration.

Exercises

5.15. A voltmeter is specified to have an accuracy of ±1% of programmed range. If a DC level is measured on a ±1 V range and appears on the meter as 0.5 V, what are the minimum and maximum DC levels that might have been present at the meter's input during this measurement?

ANS. 0.5 V ±10 mV (i.e., the input could lie anywhere between 490 and 510 mV).

5.6 REDUCING MEASUREMENT ERROR WITH GREATER MEASUREMENT TIME

In this section we shall look at several commonly used filtering techniques for extracting the mean value of a measurement process. As we shall see, filtering will improve measurement accuracy at the expense of longer test time.

5.6.1 Analog Filtering

Analog filters are often used in tester hardware to remove unwanted signal components before measurement. A DC voltmeter may include a low-pass filter as part of its front end. The purpose of the filter is to remove all but the lowest-frequency components. It acts as a hardware averaging circuit to improve the repeatability of the measurement. More effective filtering is achieved using a filter with a low cutoff frequency, since a lower cutoff frequency excludes more electrical noise. Consequently, a lower frequency cutoff corresponds to better repeatability in the final measurement. For instance, if the low-pass filter has cutoff frequency ω_b and the variation of a measurement has a bandwidth of ω_{DUT} (assumed to be much less than ω_b) with an RMS value of V_{DUT}, then we can compute the value of the RMS noise that passes through the filter according to

$$V_{n_o} = V_{DUT} \sqrt{\frac{\omega_b}{\omega_{DUT}}} \quad \text{V} \tag{5.35}$$

The above expression illustrates the noise reduction gained by filtering the output. The smaller the ratio ω_b / ω_{DUT}, the greater the noise reduction. Other types of filtering circuits can be placed on the DIB board when needed. For example, a very narrow bandpass filter may be placed on the DIB board to clean up noise components in a sine wave generated by the tester. The filter allows a much more ideal sine wave to the input of the DUT than the tester would otherwise be able to produce.

An important drawback to filtering a signal prior to measurement is the additional time required for the filter to settle to its steady-state output. The settling time is inversely proportional to the filter cutoff frequency. Thus, there is an inherent tradeoff between repeatability and test time. The following example will quantify this tradeoff for a first-order system.

EXAMPLE 5.6

The simple RC low-pass circuit shown in Figure 5.12 is used to filter the output of a DUT containing a noisy DC signal. For a particular measurement, the signal component is assumed to change from 0 to 1 V, instantaneously. How long does it take the filter to settle to within 1% of its final value? By what factor does the settling time increase when the filter's 3-dB bandwidth is decreased by a factor of 10?

Figure 5.12. RC low-pass filter.

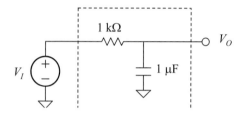

Solution:
From the theory of first-order networks, the step response of the circuit starting from rest (i.e., $v_I = 0$) is

$$v_o(t) = S\left(1 - e^{-t/\tau}\right) \tag{5.36}$$

where $S = 1$ V is the magnitude of the step and $\tau = RC = 10^{-3}$ s. Moreover, the 3-dB bandwidth ω_b (expressed in rad/s) of a first-order network is $1/RC$, so we can rewrite the above expression as

$$v_o(t) = S\left(1 - e^{-\omega_b t}\right) \tag{5.37}$$

Clearly, the time $t = t_s$ the output reaches an arbitrary output level of V_0 is then

$$t_s = -\frac{\ln\left(\dfrac{S - V_0}{S}\right)}{\omega_b} \tag{5.38}$$

Furthermore, we recognize that $(S - V_o)/S$ is the settling error ε or the accuracy of the measurement, so we can rewrite Eq. (5.38) as

$$t_S = -\frac{\ln(\varepsilon)}{\omega_b} \qquad\qquad (5.39)$$

Hence, the time it takes to reach within 1% of 1 V, or 0.99 V, is 4.6 ms. Since settling time and 3-dB bandwidth are inversely related according to Eq. (5.39), a tenfold decrease in bandwidth leads to a tenfold increase in settling time. Specifically, the settling time becomes 46 ms.

Exercises

5.16. What is the 3-dB bandwidth of the RC circuit of Figure 5.12 expressed in hertz, when R = 1 kΩ and C = 2.2 nF?   ANS. 72.34 kHz.

5.17. How long does it take a first-order RC low-pass circuit with R = 1 kΩ and C = 2.2 nF to settle to 5% of its final value?   ANS. 6.6 µs.

5.18. By what factor should the bandwidth of an RC low-pass filter be decreased in order to reduce the variation in a DC measurement from 250 µV RMS to 100 µV RMS. By what factor does the settling time increase.   ANS. The bandwidth should be decreased by 6.25 (= 2.5^2). Settling time increases by 6.25.

5.19. The variation in the output signal of a DUT is 1 mV RMS. Assume that the DUT's output follows a first-order frequency response and has a 3-dB bandwidth of 100 Hz. Estimate the output noise voltage spectral density.   ANS. 6.37×10^{-9} V²/Hz.

5.20. The variation in the output RMS signal of a DUT is 1 mV, but it needs to be reduced to a level closer to 500 µV. What filter bandwidth is required to achieve this level of repeatability? Assume that the DUT's output follows a first-order frequency response and has a 3-dB bandwidth of 1000 Hz.   ANS. 250 Hz.

5.21. The output of a DUT has an uncertainty of 10 mV. How many samples should be combined in order to reduce the uncertainty to 100 µV?   ANS. 10,000.

5.6.2 Averaging

Averaging defined by the expression $\frac{1}{N}\sum_{k=1}^{N} x[k]$ is a specific form of discrete-time filtering. Averaging can be used to improve the repeatability of a measurement. For example, we can average the following series of nine voltage measurements and obtain an average of 102 mV.

101 mV, 103 mV, 102 mV, 101 mV, 102 mV, 103 mV, 103 mV, 101 mV, 102 mV

There is a good chance that a second series of nine unique measurements will again result in something close to 102 mV. If the length of the series is increased, the answer will become more repeatable and reliable. But there is a point of diminishing returns. To reduce the effect of noise on the voltage measurement by a factor of two, one has to take four times as many readings and average them. At some point, it becomes prohibitively expensive (i.e., from the point of view of test time) to improve repeatability. In general, if the RMS variation in a measurement is again denoted V_{DUT}, then after averaging the measurement N time, the RMS value of the resulting averaged value will be

$$V_{n_o} = \frac{V_{DUT}}{\sqrt{N}} \quad \text{V} \tag{5.40}$$

Here we see the output noise voltage reduces the input noise before averaging by the factor $\sqrt{N}$. Hence, to reduce the noise RMS voltage by a factor of two requires an increase in the sequence length, N, by a factor of four.

AC measurements can also be averaged to improve repeatability. A series of sine wave signal level measurements can be averaged to achieve better repeatability. However, one should not try to average readings in decibels. If a series of measurements is expressed in decibels, they should first be converted to linear form using the equation $V = 10^{dB/20}$ before applying averaging. Normally, the voltage or gain measurements are available before they are converted to decibels in the first place; thus the conversion from decibels to linear units or ratios is not necessary. Once the average voltage level is calculated, it can be converted to decibels using the equation $dB = 20 \log_{10}(V)$. To understand why we should not perform averaging on decibels, consider the sequence 0, −20, -40 dBV. The average of these values is −20 dBV. However, the actual voltages are 1 V, 100 mV, and 10 mV. Thus the correct average value is (1 V + 0.1 V + 0.01 V) / 3 = 37 mV, or −8.64 dBV.

5.7 GUARDBANDS

Guardbanding is an important technique for dealing with the uncertainty of each measurement. If a particular measurement is known to be accurate and repeatable with a worst-case uncertainty of $\pm\varepsilon$, then the final test limits should be tightened from the data sheet specification limits by ε to make sure that no bad devices are shipped to the customer. In other words,

$$\begin{aligned}
\text{guardbanded upper test limit} &= \text{upper specification limit} - \varepsilon \\
\text{guardbanded lower test limit} &= \text{lower specification limit} + \varepsilon
\end{aligned} \tag{5.41}$$

So, for example, if the data sheet limit for the offset on a buffer output is −100 mV minimum and 100 mV maximum, and an uncertainty of ±10 mV exists in the measurement, the test program limits should be set to −90 mV minimum and 90 mV maximum. This way, if the device output is 101 mV and the error in its measurement is −10 mV, the resulting reading of 91 mV will cause a failure as required. Of course, a reading of 91 mV may also represent a device with an 81-mV output and a +10-mV measurement error.

In such cases, guardbanding has the unfortunate effect of disqualifying good devices. Ideally, we would like all guardbands to be set to 0 so that no good devices will be discarded. To minimize the guardbands, we must improve the repeatability and accuracy of each test, but this typically requires longer test times. There is a balance to be struck between repeatability and the number of good devices rejected. At some point, the added test time cost of a more repeatable measurement

Table 5.2. DUT Output and Measured Values

DUT Output (mV)	Measured Value (mV)
105	101
101	107
98	102
96	95
86	92
72	78

outweighs the cost of discarding a few good devices. This tradeoff is illustrated in Figure 5.13 on the histogram of some arbitrary offset voltage test data for two different-sized guardbands. With larger guardbands, the region of acceptability is reduced; hence fewer good devices will be shipped.

EXAMPLE 5.7

Table 5.2 lists a set of output values from a DUT together with their measured values. It is assumed that the upper specification limit is 100 mV and the measurement uncertainty is ±6 mV. How many good devices are rejected because of the measurement error? How many good devices are rejected if the measurement uncertainty is increased to ±10 mV?

Solution:
From the DUT output column on the left, four devices are below the upper specification limit of 100 mV and should be accepted. The other two should be rejected. Now with a measurement uncertainty of ±6 mV, according to Eq. (5.41) the guardbanded upper test limit is 94 mV. With the revised test limit, only two devices are acceptable. The others are all rejected. Hence, two otherwise good devices are disqualified.

If the measurement uncertainty increases to ±10 mV, then the guardbanded upper test limit becomes 90 mV. Five devices are rejected and only one is accepted. Consequently, three otherwise good devices are disqualified.

In practice, we need to set ε equal to 3 to 6 times the standard deviation of the measurement to account for measurement variability. A diagram illustrating the impact of shifting the test limits away from the specification limits on the probability density is provided in Figure 5.14. This diagram shows a marginal device with an average (true) reading equal to the upper specification limit. The upper and lower specification limits (USL and LSL, respectively) have each been tightened by $\varepsilon = 3\sigma$. The tightened upper and lower test limits (UTL and LTL, respectively) reject marginal devices such as this, regardless of the magnitude of the measurement error. A more stringent guardband value of $\varepsilon = 6\sigma$ gives us an extremely low probability of passing a defective device, but this is sometimes too large a guardband to allow a manufacturable yield.

Figure 5.13. (a) Guardbanding the specification limits. (b) Illustrating the implications of large guardbands on the region of acceptability.

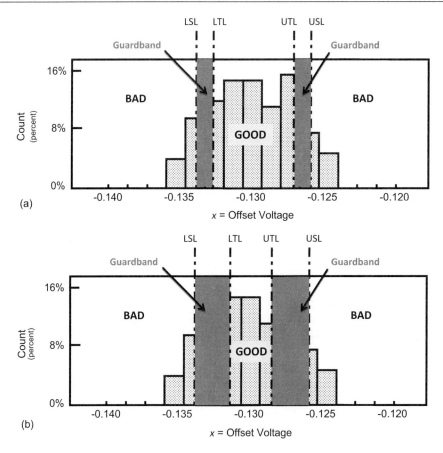

(a)

(b)

EXAMPLE 5.8

A DC offset measurement is repeated many times, resulting in a series of values having an average of 257 mV. The measurements exhibit a standard deviation of 27 mV. If our specification limits are 250 ±50 mV, where would we have to set our 6σ guardbanded upper and lower test limits?

Solution:

The value of σ is equal to 27 mV; thus the width of the 6σ guardbands would have to be equal to 162 mV. The upper test limit would be 300 mV − 162 mV, and the lower test limit would be 200 mV + 162 mV. Clearly, there is a problem with the repeatability of this test, since the lower guardbanded test limit is higher than the upper guardbanded test limit! Averaging would have to be used to reduce the standard deviation.

Figure 5.14. Guardbanded measurement with Gaussian distribution.

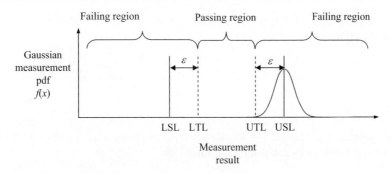

If a device is well-designed and a particular measurement is sufficiently repeatable, then there will be few failures resulting from that measurement. But if the distribution of measurements from a production lot is skewed so that the average measurement is close to one of the test limits, then production yields are likely to fall. In other words, more good devices will fall within the guardband region and be disqualified. Obviously, a measurement with poor accuracy or poor repeatability will just exacerbate the problem.

The only way the test engineer can minimize the required guardbands is to improve the repeatability and accuracy of the test, but this requires longer test times. At some point, the test time cost of a more repeatable measurement outweighs the cost of throwing away a few good devices. Thus there are inherent tradeoffs between repeatability, test time, guardbands, and production yield.

The standard deviation of a test result calculated as the average of N values from a statistical population is given by

$$\sigma_{ave} = \frac{\sigma}{\sqrt{N}} \tag{5.42}$$

So, for example, if we want to reduce the value of a measurement's standard deviation σ by a factor of two, we have to average a measurement four times. This gives rise to an unfortunate exponential tradeoff between test time and repeatability.

We can use Gaussian statistical analysis to predict the effects of nonrepeatability on yield. This allows us to make our measurements repeatable enough to give acceptable yield without wasting time making measurements that are *too* repeatable. It also allows us to recognize the situations where the average device performance or tester performance is simply too close to failure for economical production.

EXAMPLE 5.9

How many times would we have to average the DC measurement in Example 5.8 to achieve 6σ guardbands of 10 mV? If each measurement takes 5 ms, what would be the total test time for the averaged measurement?

Solution:

The value of σ_{ave} must be equal to 10 mV divided by 6 to achieve 6σ guardbands. Rearranging Eq. (5.42), we see that N must be equal to

$$N = \left(\frac{\sigma}{\sigma_{ave}}\right)^2 = \left(\frac{27 \text{ mV}}{10 \text{ mV}/6}\right)^2 = 262 \text{ measurements}$$

The total test time would be equal to 262 times 5 ms, or 1.31 s. This is clearly unacceptable for production testing of a DC offset. The 27-mV standard deviation must be reduced through an improvement in the DIB hardware or the DUT design.

Above we stated that the guardbands should be selected to be between 3 and 6 standard deviations of the measurement. Here we recast this statement in terms of the desired defect level. Consider the situation depicted in Figure 5.14 for a marginal device. The probability that a bad part will have a measured value that is less than UTL is given by

$$P\left(X < UTL\right) = \Phi\left(\frac{\varepsilon}{\sigma_n}\right) \tag{5.43}$$

If N devices are produced, the defect level in ppm as defined by Eq. (5.3) can be written as

$$DL\left[ppm\right] = \frac{\#escapes}{N} \times 10^6 = \Phi\left(\frac{\varepsilon}{\sigma_n}\right) \times 10^6 \tag{5.44}$$

Rearranging Eq. (5.44) and isolating the guardband term we find

$$\varepsilon = \sigma_n \times \left|\Phi^{-1}\left(\frac{DL}{10^6}\right)\right| \tag{5.45}$$

The upper and lower test limits can then be found Eq. (5.41) above.

EXAMPLE 5.10

A DC offset test is performed on a DUT with lower and upper specification limits of –5 mV and 5 mV, respectively. The expected RMS level of the noise present during the test is 1 mV. If a defect level of less than 200 ppm is required, what should be the test limits?

Solution:
According to Eq. (5.45), the guardband is

$$\varepsilon = 10^{-3} \times \left| \Phi^{-1}\left(\frac{200}{10^6} \right) \right| = 35.40 \text{ mV}$$

then from Eq. (5-41), we find

$$\text{LTL} = -5 + \varepsilon = -1.45 \text{ mV} \qquad \text{and} \qquad \text{UTL} = 5 - \varepsilon = 1.45 \text{ mV}$$

Exercises

5.22. A series of AC RMS measurements reveal an average value of 1.25 V and a standard deviation of 35 mV. If our specification limits were 1.2 V plus or minus 150 mV, where would we have to set our 3σ guardbanded upper and lower test limits? If 6σ guardbands are desired, how many times would we have to average the measurement to achieve guardbands of 40 mV?

ANS. 3σ guardbanded test limits are: 1.15 and 1.245 V. N = 28.

5.23. A device is expected to exhibit a worst-case offset voltage of ±50 mV and is to be measured using a voltmeter having an accuracy of only ±5 mV. Where should the guardbanded test limits be set?

ANS. ±45 mV.

5.24. The guardband of a particular measurement is 10 mV and the test limit is set to ±25 mV. What are the original device specification limits?

ANS. ±35 mV.

5.25. The following lists a set of output voltage values from a group of DUTs together with their measured values: {(2.3, 2.1), (2.1, 1.6), (2.2, 2.1), (1.9, 1.6), (1.8, 1.7), (1.7, 2.1), (1.5, 2.0)}. If the upper specification limit is 2.0 V and the measurement uncertainty is ±0.5 V, how many good devices are rejected due to the measurement error?

ANS. Four devices (all good devices are rejected by the 1.5-V guardbanded upper test limit).

5.8 EFFECTS OF MEASUREMENT VARIABILITY ON TEST YIELD

Consider the case of a measurement result having measurement variability caused by additive Gaussian noise. This test has a lower test limit (LTL) and an upper test limit (UTL). If the true measurement result is exactly between the two test limits, and the repeatability error never exceeds ±1/2 (UTL − LTL), then the test will always produce a passing result. The repeatability error never gets large enough to push the total measurement result across either of the test limits. This situation is depicted in Figure 5.15 where the pdf plot is shown.

On the other hand, if the average measurement is exactly equal to either the LTL or the UTL, then the test results will be unstable. Even a tiny amount of repeatability error will cause the test to randomly toggle between a passing and failing result when the test program is repeatedly executed. Assuming the statistical distribution of the repeatability errors is symmetrical, as in the

Figure 5.15. Probability density plot for measurement result between two test limits.

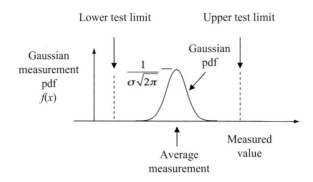

case of the Gaussian pdf, the test will produce an equal number of failures and passing results. This is illustrated by the pdf diagram shown in Figure 5.16. The area under the pdf is equally split between the passing region and the failing region; so we would expect 50% of the test results to pass and 50% to fail.

For measurements whose average value is close to but not equal to either test limit, the analysis gets a little more complicated. Consider an average measurement μ that is δ_1 units below the upper test limit as shown in Figure 5.17.

Any time the repeatability error exceeds δ_1 the test will fail. In effect, the measurement noise causes an erroneous failure. The probability that the measurement error will *not* exceed δ_1 and cause a failure is equal to the area underneath the portion of the pdf that is less than the UTL. This area is equal to the integral of the pdf from minus infinity to the UTL of the measurement results. In other words, the probability that a measurement will not fail the upper test limit as adopted from Eq. 5.19 is

$$P\left(X < \mathrm{UTL}\right) = \Phi\left(\frac{\mathrm{UTL} - \mu}{\sigma}\right) \tag{5.46}$$

Conversely, the probability of a failing result due to the upper test limit is

$$P\left(\mathrm{UTL} < X\right) = 1 - \Phi\left(\frac{\mathrm{UTL} - \mu}{\sigma}\right) \tag{5.47}$$

Similar equations can be written for the lower test limit.

If the distribution of measurement values becomes very large relative to the test limits, then we have to consider the area in both failing regions as shown in Figure 5.18. Clearly, if the true measurement result μ is near either test limit, or if the standard deviation σ is large, the test program has a much higher chance of rejecting a good DUT.

Considering both UTL failures and LTL failures, the probability of a passing result given this type of measurement repeatability according to Eq. (5.19) is

$$P\left(\mathrm{LTL} < X < \mathrm{UTL}\right) = \Phi\left(\frac{\mathrm{UTL} - \mu}{\sigma}\right) - \Phi\left(\frac{\mathrm{LTL} - \mu}{\sigma}\right) \tag{5.48}$$

Figure 5.16. Probability density plot for nonrepeatable measurement centered at the UTL.

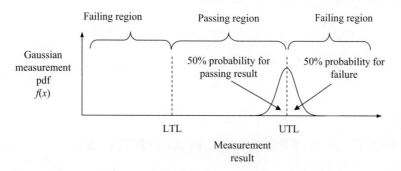

Figure 5.17. Probability density plot for average reading, μ, slightly below UTL by δ_1.

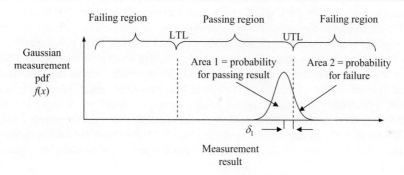

The probability of a failing result due to measurement variability is

$$P\left(X < \text{LTL or UTL} < X\right) = P\left(X < \text{LTL}\right) + P\left(\text{UTL} < X\right) = 1 + \Phi\left(\frac{\text{LTL} - \mu}{\sigma}\right) - \Phi\left(\frac{\text{UTL} - \mu}{\sigma}\right) \quad \textbf{(5.49)}$$

EXAMPLE 5.11

A DC offset measurement is repeated many times, resulting in a series of values having an average of 257 mV. The measurements exhibit a standard deviation of 27 mV. What is the probability that a nonaveraged offset measurement will fail on any given test program execution? Assume an upper test limit of 300 mV and a lower test limit of 200 mV.

Solution:
The probability that the test will lie outside the test limits of 200 and 300 mV is obtained by substituting the test limits into Eq. (5.49),

$$P(X < 200 \text{ mV}) + P(300 \text{ mV} < X) = 1 + \Phi\left(\frac{200 \text{ mV} - 257 \text{ mV}}{27 \text{ mV}}\right) - \Phi\left(\frac{300 \text{ mV} - 257 \text{ mV}}{27 \text{ mV}}\right)$$

$$= 1 + \Phi(-2.11) - \Phi(1.59)$$

Using Table 5.1, we estimate the cdf values as

$$P(X < 200 \text{ mV}) + P(300 \text{ mV} < X) \cong 1 + 0.0179 - 0.9452 = 0.0727$$

Here we see that there is a 7.27% chance of failure, even though the true DC offset value is known to be within acceptable limits.

5.9 EFFECTS OF REPRODUCIBILTY AND PROCESS VARIATION ON YIELD

Measured DUT parameters vary for a number of reasons. The factors affecting DUT parameter variation include measurement repeatability, measurement reproducibility, and the stability of the process used to manufacture the DUT. So far we have examined only the effects of measurement repeatability on yield, but the equations in the previous sections describing yield loss due to measurement variability are equally applicable to the total variability of DUT parameters.

Inaccuracies due to poor tester-to-tester correlation, day-to-day correlation, or DIB-to-DIB correlation appear as reproducibility errors. Reproducibility errors add to the yield loss caused by repeatability errors. To accurately predict yield loss caused by tester inaccuracy, we have to include both repeatability errors and reproducibility errors. If we collect averaged measurements using multiple testers, multiple DIBs, and repeat the measurements over multiple days, we can calculate the mean and standard deviation of the reproducibility errors for each test. We can then combine the standard deviations due to repeatability and reproducibility using the equation

$$\sigma_{tester} = \sqrt{\left(\sigma_{repeatability}\right)^2 + \left(\sigma_{reproducibility}\right)^2} \tag{5.50}$$

Yield loss due to total tester variability can then be calculated using the equations from the previous sections, substituting the value of σ_{tester} in place of σ.

The variability of the actual DUT performance from DUT to DUT and from lot to lot also contributes to yield loss. Thus the overall variability can be described using an overall standard deviation, calculated using an equation similar to Eq. (5.50), that is,

Figure 5.18. Probability density with large standard deviation.

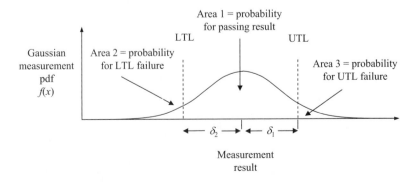

$$\sigma_{total} = \sqrt{\left(\sigma_{repeatability}\right)^2 + \left(\sigma_{reproducibility}\right)^2 + \left(\sigma_{process}\right)^2} \tag{5.51}$$

Since σ_{total} ultimately determines our overall production yield, it should be made as small as possible to minimize yield loss. The test engineer must try to minimize the first two standard deviations. The design engineer and process engineer should try to reduce the third.

EXAMPLE 5.12

A six-month yield study finds that the total standard deviation of a particular DC offset measurement is 37 mV across multiple lots, multiple testers, multiple DIB boards, and so on. The standard deviation of the measurement repeatability is found to be 15 mV, while the standard deviation of the reproducibility is found to be 7 mV. What is the standard deviation of the actual DUT-to-DUT offset variability, excluding tester repeatability errors and reproducibility errors? If we could test this device using perfectly accurate, repeatable test equipment, what would be the total yield loss due to this parameter, assuming an average value of 2.430 V and test limits of 2.5 V ± 100 mV?

Solution:
Rearranging Eq. (5.51), we write

$$
\begin{aligned}
\sigma_{process} &= \sqrt{\left(\sigma_{total}\right)^2 - \left(\sigma_{repeatability}\right)^2 - \left(\sigma_{reproducibility}\right)^2} \\
&= \sqrt{(37 \text{ mV})^2 - (15 \text{ mV})^2 - (7 \text{ mV})^2} \\
&= 33 \text{ mV}
\end{aligned}
$$

Thus, even if we could test every device with perfect accuracy and no repeatability errors, we would see a DUT-to-DUT variability of $\sigma = 33$ mV. The value of μ is equal to 2.430 V; thus our overall yield loss for this measurement is found by substituting the above values into Eq. (5.49) as

$$
\begin{aligned}
P(X < 2.4 \text{ V}) + P(2.6 \text{ V} < X) &= 1 + \Phi\left(\frac{2.4 \text{ V} - 2.43 \text{ V}}{33 \text{ mV}}\right) - \Phi\left(\frac{2.6 \text{ V} - 2.43 \text{ V}}{33 \text{ mV}}\right) \\
&= 1 + \Phi(-0.91) - \Phi(5.15)
\end{aligned}
$$

From Table 5.1, $\Phi(-0.91) \cong \Phi(-0.9) = 0.1841$, and we estimate $\Phi(5.15) \cong 1$; hence

$$P(X < 2.4 \text{ V}) + P(2.6 \text{ V} < X) \cong 1 + 0.1841 - 1 = 0.1841$$

We would therefore expect an 18% yield loss due to this one parameter, due to the fact that the DUT-to-DUT variability is too high to tolerate an average value that is only 30 mV from the lower test limit. Repeatability and reproducibility errors would only worsen the yield loss; so this device would probably not be economically viable. The design or process would have to be modified to achieve an average DC offset value closer to 2.5 V.

The probability that a particular device will pass all tests in a test program is equal to the product of the passing probabilities of each individual test. In other words, if the values P_1, P_2, $P_3, \ldots, P_n$ represent the probabilities that a particular DUT will pass each of the n individual tests in a test program, then the probability that the DUT will pass all tests is equal to

$$P\left(\text{DUT passes all tests}\right) = P_1 \times P_2 \times P_3 \times \cdots \times P_n \tag{5.52}$$

Equation (5.52) is of particular significance, because it dictates that each of the individual tests must have a very high yield if the overall production yield is to be high. For example, if each of the 200 tests has a 2% chance of failure, then each test has only a 98% chance of passing. The yield will therefore be $(0.98)^{200}$, or 1.7%! Clearly, a 1.7% yield is completely unacceptable. The problem in this simple example is not that the yield of any one test is low, but that so many tests combined will produce a large amount of yield loss.

EXAMPLE 5.13

A particular test program performs 857 tests, most of which cause little or no yield loss. Five measurements account for most of the yield loss. Using a lot summary and a continue-on-fail test process, the yield loss due to each measurement is found to be:

Test #1: 1%, Test #2: 5%, Test #3: 2.3%, Test #4: 7%, Test #5: 1.5%
All other tests combined 0.5%

What is the overall yield of this lot of material?

Solution:
The probability of passing each test is equal to 1 minus the yield loss produced by that test. The values of $P_1, P_2, P_3, \ldots, P_5$ are therefore

$$P_1 = 99\%, P_2 = 95\%, P_3 = 97.7\%, P_4 = 93\%, P_5 = 98.5\%$$

If we consider all other tests to be a sixth test having a yield loss of 0.5%, we get a sixth probability

$$P_6 = 99.5\%$$

Using Eq. (5.52) we write

$$P(\text{DUT passes all tests}) = 0.99 \times 0.95 \times 0.977 \times 0.93 \times 0.985 \times 0.995 = 0.8375$$

Thus we expect an overall test yield of 83.75%.

Because the yield of each individual test must be very high, a methodology called *statistical process control* (SPC) has been adopted by many companies. The goal of SPC is to minimize the total variability (i.e., to try to make $\sigma_{total} = 0$) and to center the average test result between the upper and lower test limits [i.e. to try to make $\mu = $ (UTL+LTL)/2]. Centering and narrowing the measurement distribution leads to higher production yield, since it minimizes the area of the Gaussian pdfs that extend into the failing regions as depicted in Figure 5.19. In the next section, we will briefly

Figure 5.19. DUT-to-DUT mean and standard deviation determine yield.

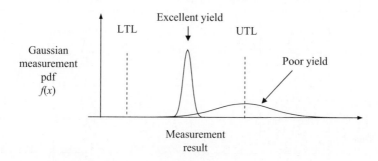

examine the SPC methodology to see how it can help improve the quality of the manufacturing process, the quality of the test equipment and software, and most important the quality of the devices shipped to the customer.

5.10 STATISTICAL PROCESS CONTROL

5.10.1 Goals of SPC

Statistical process control (SPC) is a structured methodology for continuous process improvement. SPC is a subset of total quality control (TQC), a methodology promoted by the renowned quality expert, Joseph Juran.[7,8] SPC can be applied to the semiconductor manufacturing process to monitor the consistency and quality of integrated circuits.

Exercises

5.26. An AC gain measurement is repeated many times, resulting in a series of values having an average of 2.3 V/V. The measurements exhibit a standard deviation of 0.15 V/V. What is the probability that the gain measurement will fail on any given test program execution? Assume an upper test limit of 2.4 V/V and a lower test limit of 2.2 V/V.

ANS. 0.50.

5.27. A particular test program performs 600 tests, most of which cause little or no yield loss. Four measurements account for most of the yield loss. The yield loss due to each measurement is found to be: Test #1: 1.5%, Test #2: 4%, Test #3: 5.3%, Test #4: 2%. All other tests combined 5%. What is the overall yield loss of this lot of material?

ANS. Yield loss = 16.63%.

SPC provides a means of identifying device parameters that exhibit excessive variations over time. It does not identify the root cause of the variations, but it tells us when to look for problems. Once an unstable parameter has been identified using SPC, the engineering and manufacturing team searches for the root cause of the instability. Hopefully, the excessive variations can be reduced or eliminated through a design modification or through an improvement in one of the

Figure 5.20. Process stability conclusions.

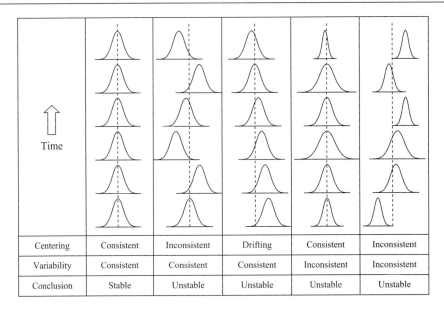

Centering	Consistent	Inconsistent	Drifting	Consistent	Inconsistent
Variability	Consistent	Consistent	Consistent	Inconsistent	Inconsistent
Conclusion	Stable	Unstable	Unstable	Unstable	Unstable

many manufacturing steps. By improving the stability of each tested parameter, the manufacturing process is brought under control, enhancing the inherent quality of the product.

A higher level of inherent quality leads to higher yields and less demanding test requirements. If we can verify that a parameter almost never fails, then we may be able to stop testing that parameter on a DUT-by-DUT basis. Instead, we can monitor the parameter periodically to verify that its statistical distribution remains tightly packed and centered between the test limits. We also need to verify that the mean and standard deviation of the parameter do not fluctuate wildly from lot to lot as shown in the four rightmost columns of Figure 5.20.[*] Once the stability of the distributions has been verified, the parameter might only be measured for every tenth device or every hundredth device in production. If the mean and standard deviation of the limited sample set stays within tolerable limits, then we can be confident that the manufacturing process itself is stable. SPC thus allows statistical sampling of highly stable parameters, dramatically reducing testing costs.

5.10.2 Six-Sigma Quality

If successful, the SPC process results in an extremely small percentage of parametric test failures. The ultimate goal of SPC is to achieve six-sigma quality standards for each specified device parameter. A parameter is said to meet six-sigma quality standards if its standard deviation is no greater than 1/12 of the difference between the upper and lower specification limits *and* the center of its statistical distribution is no more than 1.5σ away from the center of the upper and lower test limits. These criteria are illustrated in Figure 5.21.

---

[*]The authors acknowledge the efforts of the Texas Instruments SPC Guidelines Steering Team, whose document "Statistical Process Control Guidelines, The Commitment of Texas Instruments to Continuous Improvement Through SPC" served as a guide for several of the diagrams in this section.

Figure 5.21. Six-sigma quality standards lead to low defect rates (< 3.4 defective parts per million).

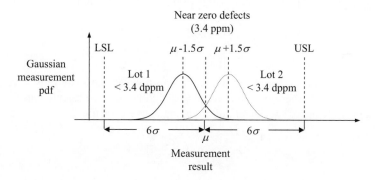

Six-sigma quality standards result in a failure rate of less than 3.4 defective parts per million (dppm). Therefore, the chance of an untested device failing a six-sigma parameter is extremely low. This is the reason we can often eliminate DUT-by-DUT testing of six-sigma parameters.

5.10.3 Process Capability: C_p, and C_{pk}

Process capability is the inherent variation of the process used to manufacture a product. Process capability is defined as the $\pm 3\sigma$ variation of a parameter around its mean value. For example, if a given parameter exhibits a 10-mV standard deviation from DUT to DUT over a period of time, then the process capability for this parameter is defined as 60 mV.

The centering and variation of a parameter are defined using two process stability metrics, C_p and C_{pk}. The process potential index, C_p, is the ratio between the range of passing values and the process capability

$$C_p = \frac{USL - LSL}{6\sigma} \tag{5.53}$$

C_p indicates how tightly the statistical distribution of measurements is packed, relative to the range of passing values. A very large C_p value indicates a process that is stable enough to give high yield and high quality, while a C_p less than 2 indicates a process stability problem. It is impossible to achieve six-sigma quality with a C_p less than 2, even if the parameter is perfectly centered. For this reason, six-sigma quality standards dictate that all measured parameters must maintain a C_p of 2 or greater in production.

The process capability index, C_{pk}, measures the process capability with respect to centering between specification limits

$$C_{pk} = C_p \left(1 - k\right) \tag{5.54}$$

where

$$k = \frac{\left|T - \mu\right|}{0.5\left(USL - LSL\right)} \tag{5.55}$$

Here T is the specification target (ideal measured value) and μ is the average measured value. The target value T is generally placed in the middle of the specification limits, defined as

$$T = \frac{\text{USL} + \text{LSL}}{2} \qquad (5.56)$$

For one-sided specifications, such as a signal-to-distortion ratio test, we only have an upper or lower specification limit. Therefore, we have to use slightly different calculations for C_p and C_{pk}. In the case of only the upper specification limit being defined, we use

$$C_{pk} = C_p = \frac{\text{USL} - \mu}{3\sigma} \qquad (5.57)$$

Alternatively, with only the lower specification limit defined, we use

$$C_{pk} = C_p = \frac{\mu - \text{LSL}}{3\sigma} \qquad (5.58)$$

The value of C_{pk} must be 1.5 or greater to achieve six-sigma quality standards as shown in Figure 5.21.

EXAMPLE 5.14

The values of an AC gain measurement are collected from a large sample of the DUTs in a production lot. The average reading is 0.991 V/V and the upper and lower specification limits are 1.050 and 0.950 V/V, respectively. The standard deviation is found to be 0.0023 V/V. What is the process capability and the values of C_p and C_{pk} for this lot? Does this lot meet six-sigma quality standards?

Solution:
The process capability is equal to 6 sigma, or 0.0138 V/V. The values of C_p and C_{pk} are given by Eqs. (5.53)–(5.55):

$$C_p = \frac{\text{USL} - \text{LSL}}{6\sigma} = \frac{1.050 - 0.950}{0.0138} = 7.245$$

$$k = \frac{|T - \mu|}{0.5(\text{USL} - \text{LSL})} = \frac{|1 - 0.991|}{0.5(1.050 - 0.950)} = 0.18$$

$$C_{pk} = C_p(1 - k) = 5.94$$

This parameter meets six-sigma quality requirements, since the values of C_p is greater than 2 and C_{pk} is greater than 1.5.

5.10.4 Gauge Repeatability and Reproducibility

As mentioned previously in this chapter, a measured parameter's variation is partially due to variations in the materials and the process used to fabricate the device and partially due to the tester's repeatability errors and reproducibility errors. In the language of SPC, the tester is known as a

Table 5.3. %GRR Acceptance Criteria

Measurement Cp	%GRR	Rating
1	100	Unacceptable
3	33	Unacceptable
5	20	Marginal
10	10	Acceptable
50	2	Good
100	1	Excellent

gauge. Before we can apply SPC to a manufacturing process, we first need to verify the accuracy, repeatability, and reproducibility of the gauge. Once the quality of the testing process has been established, the test data collected during production can be continuously monitored to verify a stable manufacturing process.

Gauge repeatability and reproducibility, denoted GRR, is evaluated using a metric called *measurement C_p*. We collect repeatability data from a single DUT using multiple testers and different DIBs over a period of days or weeks. The composite sample set represents the combination of tester repeatabilty errors and reproducibility errors [as described by Eq. (5.50)]. Using the composite mean and standard deviation, we calculate the measurement C_p using Eq. (5.53). The gauge repeatability and reproducibility percentage (precision-to-tolerance ratio) is defined as

$$\%\mathrm{GRR} = \frac{100}{\mathrm{measurement}\ C_p} \qquad (5.59)$$

The general criteria for acceptance of gauge repeatability and reproducibility are listed in Table 5.3.

5.11 SUMMARY

In this chapter we have introduced the concept of accuracy and repeatability and shown how these concepts affect device quality and production test economics. We have examined many contributing factors leading to inaccuracy and nonrepeatability. Using software calibrations, we can eliminate or at least reduce many of the effects leading to measurement inaccuracy. Measurement repeatability can be enhanced through averaging and filtering, at the expense of added test time. The constant balancing act between adequate repeatability and minimum test time represents a large portion of the test engineer's workload. One of fundamental skills that separates good test engineers from average test engineers is the ability to quickly identity and correct problems with measurement accuracy and repeatability. Doing so while maintaining low test times and high yields is the mark of a great test engineer.

Statistical process control not only allows us to evaluate the quality of the process, including the test and measurement equipment, but also tells us when the manufacturing process is not stable. We can then work to fix or improve the manufacturing process to bring it back under control. We have really only scratched the surface of SPC and TQC in this chapter. Although every test engineer may not necessarily get involved in SPC directly, it is important to understand the basic concepts. The limited coverage of this topic is only intended as an introduction to the subject rather than a complete tutorial. For a comprehensive treatment of these subjects, the reader is encouraged to refer to books devoted to TQC and Six Sigma.[7–10]

PROBLEMS

5.1. If 20,000 devices are tested with a yield of 98%, how many devices failed the test?

5.2. A new product was launched with 20,000 sales over a one-year time span. During this time, three devices were returned to the manufacturer even though an extensive test screening procedure was in place. What is the defect level associated with this testing procedure in parts per million?

5.3. A 55-mV signal is measured with a meter 10 times, resulting in the following sequence of readings: 57 mV, 60 mV, 49 mV, 58 mV, 54 mV, 57 mV, 55 mV, 57 mV, 48 mV, 61 mV. What is the average measured value? What is the systematic error?

5.4. A DC voltmeter is rated at 14 bits of resolution and has a full-scale input range of ±5 V. Assuming the meter's ADC is ideal, what is the maximum quantization error that we can expect from the meter? What is the error as a percentage of the meter's full-scale range?

5.5. A 100-mV signal is to measured with a worst-case error of ±10 μV. A DC voltmeter is set to a full-scale range of ±1 V. Assuming that quantization error is the only source of inaccuracy in this meter, how many bits of resolution would this meter need to have to make the required measurement? If the meter in our tester only has 14 bits of resolution but has binary-weighted range settings (i.e., ±1 V, ±500 mV, ±250 mV, etc.), how would we make this measurement?

5.6. A voltmeter is specified to have an accuracy error of ±0.1% of full-scale range on a ±1-V scale. If the meter produces a reading of 0.323 V DC, what is the minimum and maximum DC levels that might have been present at the meter's input during this measurement?

5.7. A series of 100 gain measurements was made on a DUT whereby the distribution was found to be Gaussian with mean value of 10.1 V/V and a standard deviation of 0.006 V/V. Write an expression for the pdf of these measurements.

5.8. A series of 100 gain measurements was made on a DUT whereby the distribution was found to be Gaussian with mean value of 10.1 V/V and a standard deviation of 0.006 V/V. If this experiment is repeated, write an expression for the pdf of the mean values of each of these experiments?

5.9. A series of 100 gain measurements was made on a DUT whereby the distribution was found to be Gaussian with mean value of 10.1 V/V and a standard deviation of 0.006 V/V. If this experiment is repeated and the mean value is compared to a reference gain value of 10 V/V, what is the mean and standard deviation of the error distribution that results? Write an expression for the pdf of these errors.

5.10. A series of 100 gain measurements was made on a DUT whereby the distribution was found to be Gaussian with mean value of 10.1 V/V and a standard deviation of 0.006 V/V. If this experiment and the mean value is compared to a reference value of 10 V/V, in what range will the expected value of the error lie for a 99.7% conference interval.

5.11. A meter reads −1.039 V and 1.121 V when connected to two highly accurate reference levels of −1 V and 1 V, respectively. What is the offset and gain of this meter? Write the calibration equation for this meter.

5.12. A DC source is assumed characterized by a third-order equation of the form: $V_{SOURCED} = 0.004 + V_{PROGRAMMED} + 0.001V_{PROGRAMMED}^{2} - 0.007V_{PROGRAMMED}^{3}$ and is required to generate a DC level of 1.25 V. However, when programmed to produce this level, 1.242 V is measured. Using iteration, determine a value of the programmed source voltage that will establish a measured voltage of 1.25 V to within a ± 0.5 mV accuracy.

5.13. An AWG has a gain response described by $G(f) = \sqrt{\dfrac{1}{1 + \left(\dfrac{f}{4000}\right)^2}}$ and is to generate three

tones at frequencies of 1, 2, and 3 kHz. What are the gain calibration factors? What voltage levels would we request if we wanted an output level of 500 mV RMS at each frequency?

5.14. Several DC measurements are made on a signal path that contains a filter and a buffer amplifier. At input levels of 1 and 3 V, the output was found to be 1.02 and 3.33 V, respectively. Assuming linear behavior, what is the gain and offset of this filter-buffer stage?

5.15. Using the setup and results of Problem 5.14, what is the calibrated level when a 2.13 V level is measured at the filter-buffer output? What is the size of the uncalibrated error?

5.16. A simple RC low-pass circuit is constructed using a 1-kΩ resistor and a 10-μF capacitor. This RC circuit is used to filter the output of a DUT containing a noisy DC signal. If the DUT's noise voltage has a constant spectral density of 100 nV/$\sqrt{\text{Hz}}$, what is the RMS noise voltage that appears at the output of the RC filter? If the we decrease the capacitor value to 2.2 μF, what is the RMS noise voltage at the RC filter output?

5.17. Assume that we want to allow the RC filter in Problem 5.16 to settle to within 0.2% of its final value before making a DC measurement. How much settling time does the first RC filter in Problem 5.16 require? Is the settling time of the second RC filter greater or less than that of the first filter?

5.18. A DC meter collects a series of repeated offset measurements at the output of a DUT. A first-order low-pass filter such as the first one described in Problem 5.16 is connected between the DUT output and the meter input. A histogram is produced from the repeated measurements. The histogram shows a Gaussian distribution with a 50-mV difference between the maximum value and minimum value. It can be shown that the standard deviation, σ, of the histogram of a repeated series of identical DC measurements on one DUT is proportional to the RMS noise at the meter's input. Assume that the difference between the maximum and minimum measured values is roughly equal to 6σ. How much would we need to reduce the cutoff frequency of the low-pass filter to reduce the nonrepeatability of the measurements from 50 to 10 mV? What would this do to our test time, assuming that the test time is dominated by the settling time of the low-pass filter?

5.19. The DUT in Problem 5.16 can be sold for $1.25, assuming that it passes all tests. If it does not pass all tests, it cannot be sold at all. Assume that the more repeatable DC offset measurement in Problem 5.16 results in a narrower guardband requirement, causing the production yield to rise from 92% to 98%. Also assume that the cost of testing is known to be 3.5 cents per second and that the more repeatable measurement adds 250 ms to the test time. Does the extra yield obtained with the lower filter cutoff frequency justify the extra cost of testing resulting from the filter's longer settling time?

5.20. A series of DC offset measurements reveal an average value of 10 mV and a standard deviation of 11 mV. If our specification limits were 0 ± 50 mV, where would we have to set our 3σ guardbanded upper and lower test limits? If 6σ guardbands are desired, how many times would we have to average the measurement to achieve guardbands of 20 mV?

5.21. A DC offset measurement is repeated many times, resulting in a series of values having an average of −100 mV. The measurements exhibit a standard deviation of 38 mV. What is the probability that the offset measurement will fail on any given test program execution? Assume an upper test limit of 0 mV and a lower test limit of −150 mV. Provide a sketch of the pdf, label critical points, and highlight the area under the pdf that corresponds to the probability of interest.

5.22. A gain measurement is repeated many times, resulting in a series of values having an average of 6.5 V/V. The measurements exhibit a standard deviation of 0.05 V/V. If our specification limits are 6.0 ± 0.5 V/V, where would we have to set our 3 σ guardbanded upper and lower test limits? If 6σ guardbands are desired, how many times would we have to average the measurement to achieve guardbands of 0.1 V/V?

5.23. A DC offset test is performed on a DUT with lower and upper specification limits of −12 mV and 12 mV, respectively. The expected RMS level of the noise present during the test is 1.5 mV. If a defect level of less than 200 ppm is required, what should be the test limits?

5.24. A device is expected to exhibit a worst-case offset voltage of ±10 mV and is to be measured using a voltmeter having an accuracy of only ±500 μV. Where should the guardbanded test limits be set?

5.25. The guardband of a particular measurement is 0.2 V/V and the test limits are set to 6.1 V/V and 6.2 V/V. What are the original device specification limits?

5.26. A series of DC measurements reveal the following list of values:

$$\{0 \text{ mV}, -10 \text{ mV}, 1.5 \text{ mV}, 9.5 \text{ mV}, -8.5 \text{ mV}, 13.2 \text{ mV},$$
$$18.5 \text{ mV}, -17.2 \text{ mV}, 5.3 \text{ mV}, \text{ and } 6.2 \text{ mV}\}$$

If our specification limits were 0 ± 50 mV, where would we have to set our 3 σ guardband upper and lower test limits? Provide a sketch to illustrate the probability density function and show the test limits. If 6σ guardbands are desired, how many times would we have to average the measurement to achieve guardbands of 24 mV?

5.27. The following contains a list of output voltage values from a DUT together with their actual measured values (i.e., sets of (true value, measured value)):

$$\{(1.9, 1.81), (2.1, 1.75), (2.1, 1.77), (1.8, 1.79), (1.9, 1.71), (2.1, 1.95),$$
$$(2.2, 2.11), (1.7, 1.89), (1.5, 1.7)\}$$

If the upper specification limit is 2 V and the guardbanded upper test limit is set to 1.8 V, answer the following questions:
(a) How many good devices are rejected on account of measurement error?
(b) How many devices escape the test?
(c) If the upper test limit is reduced to 1.74 V, how many devices escape on account of measurement error?

5.28. An AC gain measurement is repeated many times, resulting in a series of values having an average of 0.99 V/V. The measurements exhibit a standard deviation of 0.2 V/V. What is the probability that the gain measurement will fail on any given test program execution? Assume an upper test limit of 1.2 V/V and a lower test limit of 0.98 V/V. Provide a sketch of the pdf, label critical points, and highlight the area under the pdf that corresponds to the probability of interest.

5.29. The standard deviation of a measurement repeatability is found to be 12 mV, while the standard deviation of the reproducibility is found to be 8 mV. Determine the standard deviation of the tester's variability. If process variation contributes an additional 10 mV of uncertainity to the measurement, what is the total standard deviation of the overall measurement?

5.30. An extensive study of yield finds that the total standard deviation of a particular DC offset measurement is 25 mV across multiple lots, multiple testers, multiple DIB boards, and so on. The standard deviation of the measurement repeatability is found to be 19 mV, while the standard deviation of the reproducibility is found to be 11 mV. What is the standard deviation of the actual DUT-to-DUT offset variability, excluding tester repeatability errors and reproducibility errors? If we could test this device using perfectly accurate, repeatable test equipment, what would be the total yield loss due to this parameter, assuming an average value of 2.235 V and test limits of 2.25 V ± 40 mV.

5.31. A particular test program performs 1000 tests, most of which cause little or no yield loss. Seven measurements account for most of the yield loss. The yield loss due to each measurement is found to be: Test #1: 1.1%; Test #2: 6%; Test #3: 3.3%; Test #4: 8%; Test #5: 2%; Test #6: 2%; Test #7: 3%; all other tests: 1%. What is the overall yield of this lot of material?

5.32. The values of an AC noise measurement are collected from a large sample of the DUTs in a production lot. The average RMS reading is 0.12 mV and the upper and lower RMS specification limits are 0.15 and 0.10 mV, respectively. The standard deviation is found to be 0.015 mV. What is the process capability and the values of C_p and C_{pk} for this lot? Does this lot meet six-sigma quality standards?

REFERENCES

1. A. H. Moorehead, et.al., *The New American Roget's College Thesaurus*, New American Library, New York, 1985, p. 6.

2. *Webster's New World Dictionary*, Simon and Schuster, New York, August 1995, pp. 5, 463.

3. B. Metzler, *Audio Measurement Handbook*, Audio Precision, Inc., Beaverton, OR, August 1993, p. 147.

4. W. D. Cooper, *Electronic Instrumentation and Measurement Techniques*, 2nd edition, Prentice Hall, Englewood Cliffs, NJ, 1978, ISBN 0132517108, pp. 1, 2.

5. R. F. Graf, *Modern Dictionary of Electronics*, Newnes Press, Boston, July 1999, ISBN 0750698667, pp. 5, 6, 584.

6. G. W. Roberts and S. Aouini, An overview of mixed-signal production test: Past, present and future, *IEEE Design & Test of Computers*, **26**. (5), pp. 48–62, September/October 2009.

7. J. M. Juran (editor) and A. Blanford Godfrey, *Juran's Quality Handbook*, 5th edition, January 1999, McGraw-Hill, New York, ISBN 007034003X.

8. M. J. Kiemele, S. R. Schmidt, and R. J. Berdine, *Basic Statistics, Tools for Continuous Improvement*, 4th edition, Air Academy Press, Colorado Springs, CO, 1997, ISBN 1880156067, pp. 9–71.

9. Thomas Pyzdek, *The Complete Guide to Six Sigma*, Quality Publishing, Tucson, AZ, 1999, ISBN 0385494378.

10. Forrest W. Breyfogle, *Implementing Six Sigma: Smarter Solutions Using Statistical Methods*, 2nd edition, June 7, 1999, John Wiley & Sons, New York, ISBN 0471296597

CHAPTER 6

DAC Testing

Data converters (digital-to-analog and analog-to-digital) are used in all aspects of system and circuit design, from audio and video players to cellular telephones to ATE test hardware. When used in conjunction with computers and microprocessors, low-cost mixed-signal systems and circuits have been created that have high noise immunity and an ability to store, retrieve, and transmitted analog information in digital format. Such systems have fueled the growth in the use of the Internet, and this growth continues to push data converter technology to higher operating frequencies and larger bandwidths, along with higher conversion resolution and accuracy.

In this chapter, we will focus on testing the *intrinsic parameters* of a digital-to-analog converter (DAC). The next chapter will look at testing the intrinsic parameters of an analog-to-digital converter (ADC). Intrinsic parameters are those parameters that are intrinsic to the circuit itself and whose parameters are not dependent on the nature of the stimulus. This includes such measurements as absolute error, integral nonlinearity (INL), and differential nonlinearity (DNL). For the most part, intrinsic measurements are related to the DC behavior of the device. In contrast, the AC or transmission parameters, such as gain, gain tracking, signal-to-noise ratio, and signal to harmonic distortion, are strongly dependent on the nature of the stimulus signal. For instance, the amplitude and frequency of the sine wave used in a signal-to-distortion test will often affect the measured result. We defer a discussion of data converter transmission parameters until Chapter 11, after the mathematical details of AC signaling and measurement are described.

When testing a DAC or ADC, it is common to measure both intrinsic parameters and transmission parameters for characterization. However, it is often unnecessary to perform the full suite of transmission tests and intrinsic tests in production. The production testing strategy is often determined by the end use of the DAC or ADC. For example, if a DAC is to be used as a programmable DC voltage reference, then we probably do not care about its signal-to-distortion ratio at 1 kHz. We care more about its worst-case absolute voltage error. On the other hand, if that same DAC is used in a voice-band codec to reconstruct voice signals, then we have a different set of concerns. We do not care as much about the DAC's absolute errors as we care about their end effect on the transmission parameters of the composite audio channel, comprising the DAC, low-pass filter, output buffer amplifiers, and so on.

This example highlights one of the differences between digital testing and specification-oriented mixed-signal testing. Unlike digital circuits, which can be tested based on what they *are* (NAND gate, flip-flop, counter, etc.), mixed-signal circuits are often tested based on what they *do* in the system-level application (precision voltage reference, audio signal reconstruction circuit, video signal generator, etc.). Therefore, a particular analog or mixed-signal subcircuit may be copied from one design to another without change, but it may require a totally different suite of tests depending on its intended functionality in the system-level application.

6.1 BASICS OF DATA CONVERTERS

6.1.1 Principles of DAC and ADC Conversion

Digital-to-analog converters, denoted as DACs, can be considered as decoding devices that accept some input value in the form of an integer number and produces as output a corresponding analog quantity in the form of a voltage, current or other physical quantity. In most engineering applications such analog quantities are conceived as approximation to real numbers. Adapting this view, we can model the DAC decoding process involving a voltage output with an equation of the form

$$v_{OUT} = G_{DAC} D_{IN} \tag{6.1}$$

where D_{IN} is some integer value, v_{OUT} is a real-valued output value, and G_{DAC} is some real-valued proportionality constant. Because the input D_{IN} is typically taken from a digital system, it may come in the form of a D-bit-wide base-2 unsigned integer number expressed as

$$D_{IN} = b_0 + b_1 2^1 + b_2 2^2 + \cdots + b_{D-1} 2^{D-1} \tag{6.2}$$

where the coefficients $b_{D-1}, b_{D-2}, ..., b_2, b_1, b_0$ have either a 0 or 1 value. A commonly used symbol for a DAC is that shown in Figure 6.1. Coefficient b_{D-1} is regarded as the most significant bit (MSB), because it has the largest effect on the number and the coefficient b_0 is known as the least significant bit (LSB), as it has the smallest effect on the number.

For a single LSB change at the input (i.e., $\Delta D_{IN} = 1$ LSB), we see from Eq. (6.1) that the smallest voltage change at the output is $\Delta v_{OUT} = G_{DAC} \times 1$ L.. Because this quantity is called upon frequently, it is designated as V_{LSB} and is referred to as the *least significant bit step size*. The transfer characteristic of a 4-bit DAC with decoding equation

$$v_{OUT} = \frac{1}{10} D_{IN}, \qquad D_{IN} \in \{0, 1, ..., 15\} \tag{6.3}$$

is shown in Figure 6.2a. Here the DAC output ranging from 0 to 1.5 V is plotted as a function of the digital input. For each input digital word, a single analog voltage level is produced, reflecting the one-to-one input–output mapping of the DAC. Moreover, the LSB step size, V_{LSB}, is equal to 0.1 V.

Figure 6.1. A D-bit digital-to-analog converter.

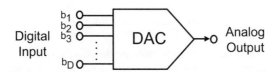

Figure 6.2. (a) DAC code-to-voltage transfer curve (b) ADC voltage-to-code transfer curve.

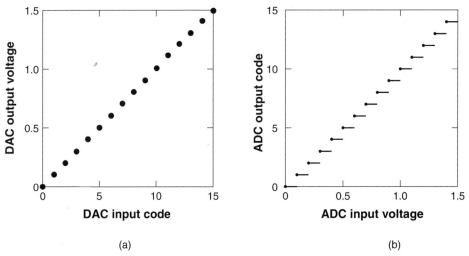

(a)   (b)

Alternatively, we can speak about the gain of the DAC (G_{DAC}) as the ratio of the range of output values to the range of input values as follows

$$G_{DAC} = \frac{v_{OUT,\max} - v_{OUT,\min}}{D_{IN,\max} - D_{IN,\min}} \tag{6.4}$$

If we denote the full-scale output range as $V_{FSR} = v_{OUT,\max} - v_{OUT,\min}$ and the input integer range as $2D-1$ (for above example, $15-0 = 2^4-1$), then the DAC gain becomes

$$G_{DAC} = \frac{V_{FSR}}{2^D - 1} \tag{6.5}$$

expressed in terms of volts per bit. Consequently, the LSB step size for the ideal DAC in volts can be written as

$$V_{LSB} = \frac{V_{FSR}}{2^D - 1} \tag{6.6}$$

Interesting enough, if the terms $v_{OUT,\min}$ and $v_{OUT,\max}$ covers some arbitrary voltage range corresponding to an arbitrary range of digital inputs, then the DAC input-output behavior can be described in identical terms to the ideal DAC described above except that an offset term is added as follows

$$v_{OUT} = G_{DAC}D_{IN} + \text{offset} \tag{6.7}$$

Analog-to-digital converters, denoted as ADCs, can be considered as encoding devices that map an input analog level into a digital word of fixed bit length, as captured by the symbol shown in Figure 6.3. Mathematically, we can represent the input–output encoding process in general terms with the equation

$$D_{OUT} = G_{ADC}v_{IN} + \text{offset} \tag{6.8}$$

Returning to our number line analogy, we recognize the ADC process is one that maps an input analog value that is represented on a real number line to a value that lies on an integer number line.

Figure 6.3. A D-bit analog-to-digital converter.

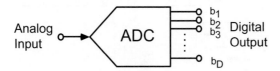

However, not all numbers on a real number line map directly to a value on the integer number line. Herein lies the challenge with the encoding process. One solution to this problem is to divide the analog input full-scale range (V_{FSR}) into 2^D equal-sized intervals according to

$$V_{LSB} = \frac{V_{FSR}}{2^D} \tag{6.9}$$

and assign each interval a code number. Mathematically, we can write this in the form of a set of inequalities as follows:

$$D_{OUT} = \begin{cases} 0, & V_{FS-} \leq v_{IN} < V_{LSB} \\ 1, & V_{LSB} \leq v_{IN} < 2V_{LSB} \\ \vdots & \vdots \\ 2^D - 2, & \left(2^D - 2\right)V_{LSB} \leq v_{IN} < \left(2^D - 1\right)V_{LSB} \\ 2^D - 1, & \left(2^D - 1\right)V_{LSB} \leq v_{IN} \leq V_{FS+} \end{cases} \tag{6.10}$$

where V_{FS-} and V_{FS+} defined the ADC full-scale range of operation, that is, $V_{FSR} = V_{FS+} - V_{FS-}$. The transfer characteristic for a 4-bit ADC is shown in Figure 6.2b for a full-scale input range between 0 and 1.5 V and a LSB step size V_{LSB} of 0.09375 V.

The transfer characteristic of an ADC is not the same across all ADCs, unlike the situation that one finds for DACs. The reason for this comes back to the many-to-one mapping issue described above. Two common approaches used by ADC designers to define ADC operation are based on the mathematical principle of rounding or truncating fractional real numbers. The transfer characteristics of these two types of ADCs are shown in Figure 6.4 for a 4-bit example with $V_{FS-} = 0$ V and $V_{FS+} = 1.5$ V. If the ADC is based on the rounding principle, then the ADC transfer characteristics can be described in general terms as

$$D_{OUT} = \begin{cases} 0, & V_{FS-} \leq v_{IN} < \frac{1}{2}V_{LSB} \\ 1, & \frac{1}{2}V_{LSB} \leq v_{IN} < \frac{3}{2}V_{LSB} \\ \vdots & \vdots \\ 2^D - 2, & \left[\left(2^D - 2\right) - \frac{1}{2}\right]V_{LSB} \leq v_{IN} < \left[\left(2^D - 1\right) - \frac{1}{2}\right]V_{LSI} \\ 2^D - 1, & \left[\left(2^D - 1\right) - \frac{1}{2}\right]V_{LSB} \leq v_{IN} \leq V_{FS+} \end{cases} \tag{6.11}$$

Figure 6.4. Alternative definitions of the ADC transfer characteristic. (a) ADC operation based on the rounding operation; (b) ADC operation based on the truncation operation.

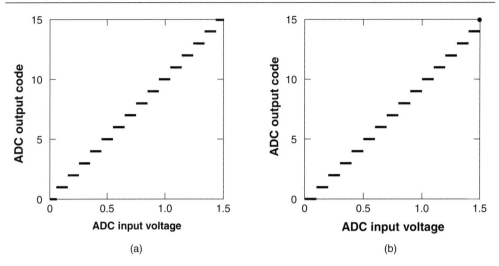

(a)   (b)

or for the truncating principle as follows

$$D_{OUT} = \begin{cases} 0, & V_{FS-} \leq v_{IN} < V_{LSB} \\ 1, & V_{LSB} \leq v_{IN} < 2V_{LSB} \\ \vdots & \vdots \\ 2^D - 2, & \left(2^D - 2\right)V_{LSB} \leq v_{IN} < \left(2^D - 1\right)V_{LSB} \\ 2^D - 1, & \left(2^D - 1\right)V_{LSB} = v_{IN} = V_{FS+} \end{cases} \tag{6.12}$$

In both of these two cases, the full-scale range is no longer divided into equal segments. These new definitions lead to a different value for the LSB step size as that described earlier. For these two cases, it is given by

$$V_{LSB} = \frac{V_{FSR}}{2^D - 1} \tag{6.13}$$

Finally, to complete our discussion on ideal ADCs, we like to point out that the proportionality constant G_{ADC} in Eq. (6.8) can be expressed in terms of bits per volt as

$$G_{ADC} = \frac{D_{OUT,\max} - D_{OUT,\min}}{v_{IN,\max} - v_{IN,\min}} = \frac{2^D - 1}{V_{FSR}} \tag{6.14}$$

For both the truncating- and rounding-based ADC, its gain is equal to the reciprocal of the LSB step size as is evident when Eqs. (6.13) and (6.14) are compared.

Exercises

6.1. A 4-bit DAC has a full-scale voltage range of 0 to 5 V. What is the gain of this DAC?

ANS. $G_{DAC} = 0.33 \ \text{V/bit.}$

6.2. A 10-bit DAC has a full-scale voltage range of -5.0 to 5.0 V. What is the LSB step size?

ANS. $V_{LSB} = 9.77 \ \text{mV.}$

6.3. A 3-bit ADC has a full-scale voltage range of 0 to 5.0 V. What is the gain of this ADC if its internal operation is based on rounding?

ANS. $G_{DAC} = 1.4 \ \text{bits/V.}$

6.4. A 10-bit ADC has a full-scale voltage range of -5.0 to 5.0 V. What is the LSB step size if its internal operation is based on truncation?

ANS. $V_{LSB} = 9.77 \ \text{mV.}$

6.1.2 Data Formats

There are several different encoding formats for ADCs and DACs including unsigned binary, sign/magnitude, two's complement, one's complement, mu-law, and A-law. One common omission in device data sheets is DAC or ADC data format. The test engineer should always make sure the data format has been clearly defined in the data sheet before writing test code.

The most straightforward data format is unsigned binary written as $b_{D-1}b_{D-2}...b_2b_1b_0$ whose equivalent base-10 integer value is found from

$$D_{IN} = b_{N-1}2^{N-1} + b_{N-2}2^{N-2} + \quad + b_1 2^1 + b_0 2^0 \tag{6.15}$$

Unsigned binary format places the lowest voltage at code 0 and the highest voltage at the code with all 1's. For example, an 8-bit DAC with a full-scale voltage range of 1.0 to 3.0 V would have the code-to-voltage relationship shown in Table 6.1.

One LSB step size is equal to the full-scale voltage range, $V_{FS+} - V_{FS-}$, divided by the number of DAC codes (e.g., 2D) minus one

$$V_{LSB} = \frac{V_{FS+} - V_{FS-}}{\# \ \text{DAC codes} \ - 1} \tag{6.16}$$

Table 6.1. Unsigned Binary Format for an 8-Bit DAC

Code	Voltage
00000000 (integer 0)	1.0 V
00000001 (integer 1)	1.0 V + 1 VLSB = 1.007843 V
...	
01111111 (integer 127)	1.0 V + 127 VLSBs = 1.996078 V
10000000 (integer 128)	1.0 V + 128 VLSBs = 2.003922 V
...	
11111111 (integer 255)	1.0 V + 255 VLSBs = 3.0 V

In this example, the voltage corresponding to one LSB step size is equal to (3.0 V – 1.0 V)/255 = 7.843 mV. Sometimes the full-scale voltage is defined with one an additional imaginary code above the maximum code (i.e., code 256 in our 8-bit example). If so, then the LSB size would be (3.0 V – 1.0 V)/256 = 7.8125 mV. This source of ambiguity should be clarified in the data sheet.

Another common data format is two's complement; written exactly the same as an unsigned binary number, for example, $b_{D-1}b_{D-2}\cdots b_2 b_1 b_0$. A two's complement binary representation is converted to its equivalent base-10 integer value using the equation

$$D_{IN} = -b_{D-1}2^{D-1} + b_{D-2}2^{D-2} + b_{D-3}2^{D-3} + \cdots + b_1 2^1 + b_0 2^0 \qquad (6.17)$$

A two's complement binary formatted number can be used to express both positive and negative integer values. Positive numbers are encoded the same as an unsigned binary in two's complement, except that the most significant bit must always be zero. When the most significant bit is one, the number is negative. To multiply a two's complement number by –1, all bits are inverted and one is added to the result. The two's complement encoding scheme for an 8-bit DAC is shown in Table 6.2. As is evident from the table, all outputs are made relative to the DAC's midscale value of 2.0 V. This level corresponds to input digital code 0. Also evident from this table is the LSB is equal to 5 mV. The midscale (MS) value is computed from either of the following two expressions using knowledge of the lower and upper limits of the DAC's full-scale range, denoted V_{FS-} and V_{FS+}, respectively, together with the LSB step size obtained from Eq. (6.16), as follows:

$$V_{MS} = V_{FS-} + \frac{\#\,DAC\,codes}{2} V_{LSB} \qquad (6.18)$$

or

$$V_{MS} = V_{FS+} - \left(\frac{\#\,DAC\,codes}{2} - 1\right) V_{LSB} \qquad (6.19)$$

Note that the two's complement encoding scheme is slightly asymmetrical since there are more negative codes than positive ones.

One's complement format is similar to two's complement, except that it eliminates the asymmetry by defining 11111111 as minus zero instead of minus one, thereby making 11111111 a redundant code. One's complement format is not commonly used in data converters because it is not quite as compatible with mathematical computations as two's complement or unsigned binary formats.

Table 6.2. Two's Complement Format for an 8-Bit DAC

Code	Voltage
10000000 (integer –128)	2.0 V – 128 VLSBs = 1.360 V
10000001 (integer –127)	2.0 V – 127 VLSBs = 1.365 V
...	
11111111 (integer –1)	2.0 V – 1 VLSB = 1.995 V
00000000 (integer 0)	2.0 V (midscale voltage)
00000001 (integer –1)	2.0 V + 1 VLSB = 2.005 V
...	
01111110 (integer 126)	2.0 V + 126 VLSBs = 2.630 V
01111111 (integer 127)	2.0 V + 127 VLSBs = 2.635 V

Sign/magnitude format is occasionally used in data converters. In sign/magnitude format, the most significant bit is zero for positive values and one for negative values. A sign/magnitude formatted binary number expressed in the form $b_{D-1}b_{D-2}\cdots b_2b_1b_0$ has the following base-10 equivalent integer value:

$$D_{IN} = (-1)^{b_{N-1}} \times \left(b_{N-2}2^{N-2} + b_{N-3}2^{N-3} + \cdots + b_1 2^1 + b_0 2^0 \right) \tag{6.20}$$

Like one's complement, sign/magnitude format also has a redundant negative zero value. Table 6.3 shows sign/magnitude format for the 8-bit DAC example. The midscale level corresponding to input code 0 for this type of converter is

$$V_{MS} = V_{FS-} + \left(\frac{\# \text{ DAC codes}}{2} - 1 \right) V_{LSB} = V_{FS+} - \left(\frac{\# \text{ DAC codes}}{2} - 1 \right) V_{LSB} \tag{6.21}$$

where V_{LSB} is given by

$$V_{LSB} = \frac{V_{FS+} - V_{FS-}}{\# \text{ DAC codes } - 2} \tag{6.22}$$

Two other data formats, mu-law and A-law, were developed in the early days of digital telephone equipment. Mu-law is used in North American and related telephone systems, while A-law is used in European telephone systems. Today the mu-law and A-law data formats are sometimes found not only in telecommunications equipment but also in digital audio applications, such as PC sound cards. These two data formats are examples of companded encoding schemes.

Companding is the process of compressing and expanding a signal as it is digitized and reconstructed. The idea behind companding is to digitize or reconstruct large amplitude signals with coarse converter resolution while digitizing or reconstructing small amplitude signals with finer resolution. The companding process results in a signal with a fairly constant signal to quantization noise ratio, regardless of the signal strength.

Compared with a traditional linear converter having the same number of bits, a companding converter has worse signal-to-noise ratio when signal levels are near full scale, but better signal-to-noise ratios when signal levels are small. This tradeoff is desirable for telephone conversations, since it limits the number of bits required for transmission of digitized voice. Companding is therefore a simple form of lossy data compression.

Table 6.3. Sign/Magnitude Format for an 8-Bit DAC

Code	Voltage
11111111 (integer −127)	2.0 V−127 VLSBs = 1.365 V
11111110 (integer −126)	2.0 V−126 VLSBs = 1.370 V
...	
10000001 (integer −1)	2.0 V−1 VLSB = 1.995 V
10000000 (integer −0)	2.0 V
00000000 (integer 0)	2.0 V
00000001 (integer 1)	2.0 V + 1 VLSB = 2.005 V
...	
01111110 (integer 126)	2.0 V + 126 VLSBs = 2.630 V
01111111 (integer 127)	2.0 V + 127 VLSBs = 2.635 V

Figure 6.5. Comparison of linear and companding 4-bit ADC-to-DAC transfer curves.

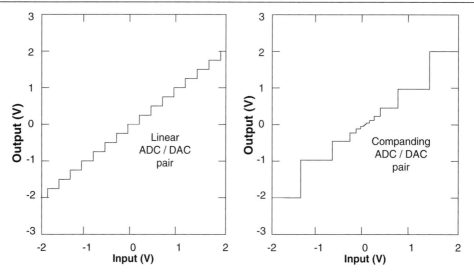

Figure 6.5 shows the transfer curve of a simple 4-bit companded ADC followed by a 4-bit DAC. In a true logarithmic companding process such as the one in Figure 6.5, the analog signal is passed through a linear-to-logarithmic conversion before it is digitized. The logarithmic process compresses the signal so that small signals and large signals appear closer in magnitude. Then the compressed signal may be digitized and reconstructed using an ADC and DAC. The reconstructed signal is then passed through a logarithmic-to-linear conversion to recover a companded version of the original signal.

Exercises

6.5. A 4-bit DAC has a full-scale voltage range of 0 to 5 V. The input is formatted using an unsigned binary number representation. List all possible ideal output levels. What output level corresponds to the DAC input code 0? What is the VLSB?

ANS. Code 0 to 15: 0, 0.333, 0.666, 0.999, 1.33, 1.66, 2.00, 2.33, 2.66, 3.00, 3.33, 3.66, 4.00, 4.33, 4.66, 5.00 V; Code 0 = 0 V; VLSB: = 0.333 V.

6.6. A 4-bit ADC has a full-scale voltage range of −5 to 5 V. The internal operation of the ADC is based on truncation and the input digital signal is formatted using a two's complement binary number representation. List all possible ideal output levels. What is the LSB step size?

ANS. Code −8 to 7: −5, −4.33, −3.67, −3.00, −2.33, −1.66, −1.00, −0.33, .34, 1.00, 1.67, 2.34, 3.00, 3.67, 4.34, 5.0 V; Code 0 = 0.34 V; VLSB: = 0.666 V.

6.7. What is the ideal gain of a 10-bit DAC with a full-scale voltage range of 0 to 5 V?

ANS. 4.888 mV/bit.

6.8. A 7-bit DAC has a full-scale voltage range of 0.5 to 2.5 V. The input is formatted using a 2's complement number representation. What is the midscale voltage level? What is the expected output voltage level if the input digital code is 1001101?

ANS. 1.5079 V, 0.7047 V.

The mu-law and A-law encoding and decoding rules are a sign/magnitude format with a piecewise linear approximation of a true logarithmic encoding scheme. They define a varying LSB size that is small near 0 and larger as the voltage approaches plus or minus full scale. Each of the piecewise linear sections is called a *chord*. The steps in each chord are of a constant size. The piecewise approximation was much easier to implement in the early days of digital telecommunications than a true logarithmic companding scheme, since the piecewise linear sections could be implemented with traditional binary weighted ADCs and DACs. Today, the A-law and mu-law encoding and decoding process is often performed using lookup tables combined with linear sigma-delta ADCs and DACs having at least 13 bits of resolution. A more complete discussion of A-law and mu-law codec testing can be found in Matthew Mahoney's book, *DSP-based Testing of Analog and Mixed-Signal Circuits*.[1]

6.1.3 Comparison of DACs and ADCs

Although this chapter is devoted to DAC testing, many of the concepts presented are closely tied to ADC testing. For instance, the code-to-voltage transfer characteristics for a DAC are similar to the voltage-to-code characteristics of an ADC. However, it is very important to note that a DAC represents a one-to-one mapping function whereas an ADC represents a many-to-one mapping. This distinction is illustrated in Figure 6.2(a) and (b). For each digital input code, a DAC produces only one output voltage.

An ADC, by contrast, produces the same output code for many different input voltages. In fact, because an ADC's circuits generate random noise and because any input signal will include a certain amount of noise, the ADC decision levels represent *probable* locations of transitions from one code to the next. We will discuss the probabilistic nature of ADC decision levels and their effect on ADC testing in Chapter 7. While DACs also generate random noise, this noise can be removed through averaging to produce a single, unambiguous voltage level for each DAC code. Therefore, the DAC transfer characteristic is truly a one-to-one mapping of codes to voltages.

The difference between DAC and ADC transfer characteristics prevents us from using complementary testing techniques on DACs and ADCs. For example, a DAC is often tested by measuring the output voltage corresponding to each digital input code. The test engineer might be tempted to test an ADC using the complementary approach, applying the ideal voltage levels for each code and then comparing the actual output code against the expected code. Unfortunately, this approach is completely inappropriate in most ADC testing cases, since it does not characterize the location of each ADC decision level. Furthermore, this crude testing approach will often fail perfectly good ADCs simply because of gain and offset variations that are within acceptable limits.

Although there are many differences in the testing of DACs and ADCs, there are enough similarities that we have to treat the two topics as one. In Chapter 7 we will see how ADC testing is similar to DAC testing and also how it differs. In this chapter, however, we will concentrate mainly on DAC testing. Also, we will concentrate mostly on voltage output DACs. Current output DACs are tested using the same techniques, using either a current mode DC voltmeter or a calibrated current-to-voltage translation circuit on the device interface board (DIB).

6.1.4 DAC Failure Mechanisms

The novice test engineer may be inclined to think that all N-bit DACs are created equal and are therefore tested using the same techniques. As we will see, this is not the case. There are many different types of DACs, including binary-weighted architectures, resistive divider architectures, pulse-width-modulated (PWM) architectures, and pulse-density-modulated (PDM) architectures (commonly known as *sigma-delta* DACs). Furthermore, there are hybrids of these architectures, such as the multibit sigma-delta DAC and segmented resistive divider DACs. Each of these DAC architectures has a unique set of strengths and weaknesses. Each architecture's weaknesses determines its likely failure mechanisms, and these in turn drive the testing methodology. As previously noted, the requirements of the DAC's system-level application also determine the testing methodology.

Before we discuss testing methodologies for each type of DAC, we first need to outline the DC and dynamic tests commonly performed on DACs. The DC tests include the usual specifications like gain, offset, power supply sensitivity, and so on. They also include converter-specific tests such as absolute error, monotonicity, integral nonlinearity (INL), and differential nonlinearity (DNL), which measure the overall quality of the DAC's code-to-voltage transfer curve. The dynamic tests are not always performed on DACs, especially those whose purpose is to provide DC or low-frequency signals. However, dynamic tests are common in applications such as video DACs, where fast settling times and other high-frequency characteristics are key specifications.

6.2 BASIC DC TESTS

6.2.1 Code-Specific Parameters

DAC specifications sometimes call for specific voltage levels corresponding to specific digital codes. For instance, an 8-bit two's complement DAC may specify a voltage level of 1.360 V ± 10 mV at digital code −128 and a voltage level of 2.635 V ± 10 mV at digital code +127. (See Section 6.1.2 for a description of converter data formats such as unsigned binary and two's complement.) Alternatively, DAC code errors can be specified as a percentage of the DAC's full-scale range rather than an absolute error. In this case, the DAC's full-scale range must first be measured to determine the appropriate test limits. Common code-specific parameters include the maximum full-scale (V_{FS+}) voltage, minimum full-scale (V_{FS-}) voltage, and midscale (V_{MS}) voltage. The midscale voltage typically corresponds to 0 V in bipolar DACs or a center voltage such as $V_{DD}/2$ in unipolar (single power supply) DACs. It is important to note that although the minimum full-scale voltage is often designated with the V_{FS-} notation, it is not necessarily a negative voltage.

6.2.2 Full-Scale Range

Full-scale range (V_{FSR}) is defined as the voltage difference between the maximum voltage and minimum voltage that can be produced by a DAC. This is typically measured by simply measuring the DAC's positive full-scale voltage, V_{FS+}, and then measuring the DAC's negative full-scale voltage, V_{FS-}, and subtracting

$$V_{FSR} = V_{FS+} - V_{FS-} \qquad\qquad (6.23)$$

6.2.3 DC Gain, Gain Error, Offset, and Offset Error

It is tempting to say that the DAC's offset is equal to the measured midscale voltage, V_{MS}. It is also tempting to define the gain of a DAC as the full-scale range divided by the number of spaces, or steps, between codes. These definitions of offset and gain are approximately correct, and in fact they are sometimes found in data sheets specified exactly this way. They are quite valid in a perfectly linear DAC. However, in an imperfect DAC, these definitions are inferior because they are very sensitive to variations in the V_{FS-}, V_{MS}, and V_{FS+} voltage outputs while being completely insensitive to variations in all other voltage outputs.

Figure 6.6 shows a simulated DAC transfer curve for a rather bad 4-bit DAC. Notice that code 0 does not produce 0 V, as it should. However, the overall curve has an offset near 0 V. Also, notice that the gain, if defined as the full-scale range divided by the number of spaces between codes, does not match the general slope of the curve. The problem is that the V_{FS+}, V_{FS-}, and V_{MS} voltages are not in line with the general shape of the transfer curve.

A less ambiguous definition of gain and offset can be found by computing the best-fit line for these points and then computing the gain and offset of this line. For high-resolution DACs with

Figure 6.6. Endpoint/midpoint-referenced gain and offset for a 4-bit DAC.

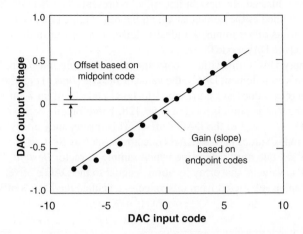

reasonable linearity, the errors between these two techniques become very small. Nevertheless, the best-fit line approach is independent of DAC resolution; thus it is the preferred technique.

A best-fit line is commonly defined as the line having the minimum squared errors between its ideal, evenly spaced samples and the actual DAC output samples. For a sample set $S(i)$, where i ranges from 0 to $N-1$ and N is the number of samples in the sample set, the best-fit line is defined by its slope (DAC gain) and offset using a standard linear equation having the form

$$Best\_fit\_line = \text{gain} \times i + \text{offset} \quad \text{for } i = 0, 1, \ldots, N-1 \tag{6.24}$$

The equations for slope and offset can be derived using various techniques. One technique minimizes the partial derivatives with respect to slope and offset of the squared errors between the sample set S and the best-fit line. Another technique is based on linear regression.[2] The equations derived from the partial derivative technique are

$$\text{gain} = \frac{N\,K_4 - K_1\,K_2}{N\,K_3 - K_1^2}, \qquad \text{offset} = \frac{K_2}{N} - \text{gain}\,\frac{K_1}{N} \tag{6.25}$$

where

$$K_1 = \sum_{i=0}^{N-1} i, \quad K_2 = \sum_{i=0}^{N-1} S(i), \quad K_3 = \sum_{i=0}^{N-1} i^2, \quad K_4 = \sum_{i=0}^{N-1} i\,S(i)$$

The derivation details are left as an exercise in the problem set found at the end of this chapter. These equations translate very easily into a computer program.

The values in the array *Best_fit_line* represent samples falling on the least-squared-error line. The program variable *Gain* represents the gain of the DAC, in volts per bit. This gain value is the average gain across all DAC samples. Unlike the gain calculated from the full-scale range divided by the number of code transitions, the slope of the best-fit line represents the true gain of the DAC. It is based on all samples in the DAC transfer curve and therefore is not especially sensitive to any one code's location. Gain error, ΔG, expressed as a percent, is defined as

$$\Delta G = \left(\frac{G_{ACTUAL}}{G_{IDEAL}} - 1 \right) \times 100\% \tag{6.26}$$

Likewise, the best-fit line's calculated offset is not dependent on a single code as it is in the midscale code method. Instead, the best-fit line offset represents the offset of the total sample set. The DAC's offset is defined as the voltage at which the best-fit line crosses the *y* axis. The DAC's offset error is equal to its offset minus the ideal voltage at this point in the DAC transfer curve. The *y* axis corresponds to DAC code 0.

In unsigned binary DACs, this voltage corresponds to *Best_fit_line(1)* in the MATLAB routine. However, in two's complement DACs, the value of *Best_fit_line(1)* corresponds to the DAC's V_{FS-} voltage, and therefore does not correspond to DAC code 0. In an 8-bit two's complement DAC, for example, the 0 code point is located at $i = 128$. Therefore, the value of the program variable *Offset* does not correspond to the DAC's offset. This discrepancy arises simply because we cannot use negative index values in MATLAB code arrays such as *Best_fit_line(–128)*. Therefore, to find the DAC's offset, one must determine which sample in vector *Best_fit_line* corresponds to the DAC's 0 code. The value at this array location is equal to the DAC's offset. The ideal voltage at the DAC 0 code can be subtracted from this value to calculate the DAC's offset error.

EXAMPLE 6.1

A 4-bit two's complement DAC produces the following set of voltage levels, starting from code –8 and progressing through code +7:

$$-780 \text{ mV}, -705 \text{ mV}, -530 \text{ mV}, -455 \text{ mV}, -400 \text{ mV}, -325 \text{ mV}, -150 \text{ mV}, -75 \text{ mV},$$
$$120 \text{ mV}, 195 \text{ mV}, 370 \text{ mV}, 445 \text{ mV}, 500 \text{ mV}, 575 \text{ mV}, 750 \text{ mV}, 825 \text{ mV}$$

These code levels are shown in Figure 6.7. The ideal DAC output at code 0 is 0 V. The ideal gain is equal to 100 mV/bit. Calculate the DAC's gain (volts per bit), gain error, offset, and offset error using a best-fit line as reference.

Figure 6.7. A 4-bit DAC transfer curve and best-fit line.

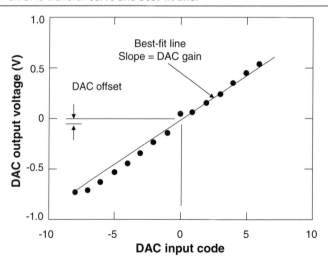

Solution:
We calculate gain and offset using the previous MATLAB routine, resulting in a gain value of 109.35 mV/bit and an offset value of –797.64 mV. The gain error is found from Eq. (6.26) to be

$$\Delta G = \left(\frac{109.35 \text{ mV}}{100 \text{ mV}} - 1 \right) \times 100\% = 9.35\%$$

Because this DAC uses a two's complement encoding scheme, this offset value is the offset of the best-fit line, not the offset of the DAC at code –8.

The DAC's offset is found by calculating the best-fit line's value at DAC code 0, which corresponds to i = 8

$$DAC \ offset = gain \times 8 + offset$$
$$= 109.35 \ mV \times 8 - 797.64 \ mV$$
$$= 77.16 \ mV$$

$$DAC \ offset \ error = DAC \ offset - ideal \ offset$$
$$= 77.16 \ mV - 0 \ V = 77.16 \ mV$$

Clearly, when the ideal offset is 0 V, the DAC offset and offset error are identical. Many DACs have an ideal offset of $V_{DD}/2$ or some other nonzero value. These DACs are commonly used in applications requiring a single power supply. In such a case, the offset should be nonzero, but the offset error should always be zero.

6.2.4 LSB Step Size

The least significant bit (LSB) step size is defined as the average step size of the DAC transfer curve. It is equal to the gain of the DAC, in volts per bit. Although it is possible to measure the approximate LSB size by simply dividing the full-scale range by the number of code transitions, it is more accurate to measure the gain of the best-fit line to calculate the average LSB size. Using the results from the previous example, the 4-bit DAC's LSB step size is equal to 109.35 mV.

6.2.5 DC PSS

DAC DC power supply sensitivity (PSS) is easily measured by applying a fixed code to the DAC's input and measuring the DC gain from one of its power supply pins to its output. PSS for a DAC is therefore identical to the measurement of PSS in any other circuit, as described in Section 3.8.1. The only difference is that a DAC may have different PSS performance depending on the applied digital code. Usually, a DAC will exhibit the worst PSS performance at its full-scale and/or minus full-scale settings because these settings tie the DAC output directly to a voltage derived from the power supply. Worst-case conditions should be used once they have been determined through characterization of the DAC.

Exercises

6.9. A 4-bit unsigned binary DAC produces the following set of voltage levels, starting from code 0 and progressing through to code 15:

1.0091, 1.2030, 1.3363, 1.5617, 1.6925, 1.9453, 2.0871, 2.3206, 2.4522, 2.6529, 2.8491, 2.9965, 3.1453, 3.3357, 3.4834, 3.6218

The ideal DAC output at code 0 is 1 V and the ideal gain is equal to 200 mV/bit. The data sheet for this DAC specifies offset and offset using a best-fit line, evaluated at code 0. Gain is also specified using a best-fit line. Calculate the DAC's gain (volts per bit), gain error, offset, and offset error.

ANS. G = 177.3 mV/bit; ΔG = −11.3%; offset = 1.026 V; offset error = 26.1 mV.

6.10. Estimate the LSB step size of the DAC described in Exercise 6.7 using its measured full-scale range (i.e. using the endpoint method). What are the gain error and offset error?

ANS. LSB = 174.2 mV; ΔG = −12.9%; offset error = 9.1 mV.

6.3 TRANSFER CURVE TESTS

6.3.1 Absolute Error

The ideal DAC transfer characteristic or transfer curve is one in which the step size between each output voltage and the next is exactly equal to the desired LSB step size. Also, the offset error of the transfer curve should be zero. Of course, physical DACs do not behave in an ideal manner; so we have to define figures of merit for their actual transfer curves.

One of the simplest, least ambiguous figures of merit is the DAC's maximum and minimum absolute error. An absolute error curve is calculated by subtracting the ideal DAC output curve from the actual measured DAC curve. The values on the absolute error curve can be converted to LSBs by dividing each voltage by the ideal LSB size, V_{LSB}. The conversion from volts to LSBs is a process called *normalization*.

Mathematically, if we denote the ith value on the ideal and actual transfer curves as $S_{IDEAL}(i)$ and $S(i)$, respectively, then we can write the normalized absolute error transfer curve $\Delta S(i)$ as

$$\Delta S(i) = \frac{S(i) - S_{IDEAL}(i)}{V_{LSB}} \tag{6.27}$$

EXAMPLE 6.2

Assuming an ideal gain of 100 mV per LSB and an ideal offset of 0 V at code 0, calculate the absolute error curve for the 4-bit DAC of the previous example. Express the results in terms of LSBs.

Solution:
The ideal DAC levels are –800, –700, ..., +700 mV. Subtracting these ideal values from the actual values, we can calculate the absolute voltage errors $\Delta S(i)$ as:

+20 mV, 25 mV, +70 mV, +45 mV, 0 mV, 225 mV, +50 mV, +25 mV, +120 mV,
+95 mV, +170 mV, +145 mV, +100 mV, +75 mV, +150 mV, +125 mV

The maximum absolute error is +170 mV and the minimum absolute error is –25 mV. Dividing each value by the ideal LSB size (100 mV), we get the normalized error curve shown in Figure 6.8. This curve shows that this DAC's maximum and minimum absolute errors are +1.7 and –0.25 LSBs, respectively. In a simple 4-bit DAC, this would be considered very bad performance, but this is an imaginary DAC designed for instructional purposes. In high-resolution DACs, on the other hand, absolute errors of several LSBs are common. The larger normalized absolute error in high-resolution DACs is a result of the smaller LSB size. Therefore, absolute error testing is often replaced by gain, offset, and linearity testing in high-resolution DACs.

Figure 6.8. Normalized DAC absolute error curve.

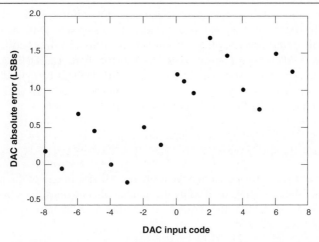

6.3.2 Monotonicity

A monotonic DAC is one in which each voltage in the transfer curve is larger than the previous voltage, assuming a rising voltage ramp for increasing codes. (If the voltage ramp is expected to decrease with increasing code values, we simply have to make sure that each voltage is less than the previous one.) While the 4-bit DAC in the previous examples has a terrible set of absolute errors, it is nevertheless monotonic. Monotonicity testing requires that we take the discrete first derivative of the transfer curve, denoted here as $S'(i)$, according to

$$S'(i) = S(i + 1) - S(i) \tag{6.28}$$

If the derivatives are all positive for a rising ramp input or negative for a falling ramp input, then the DAC is said to be monotonic.

EXAMPLE 6.3

Verify monotonicity in the previous DAC example.

Solution:

The first derivative of the DAC transfer curve is calculated, yielding the following values

75 mV, 175 mV, 75 mV, 55 mV, 75 mV, 175 mV, 75 mV, 195 mV, 75 mV,
175mV, 75 mV, 55 mV, 75 mV, 175 mV, 75 mV

Notice that there are only 15 first derivative values, even though there are 16 codes in a 4-bit DAC. This is the nature of the discrete derivative, since there are one fewer *changes* in voltage than there are voltages. Since each value in this example has the same sign (positive), the DAC is monotonic.

6.3.3 Differential Nonlinearity

Notice that in the monotonicity example the step sizes are not uniform. In a perfect DAC, each step would be exactly 100 mV corresponding to the ideal LSB step size. Differential nonlinearity (DNL) is a figure of merit that describes the uniformity of the LSB step sizes between DAC codes. DNL is also known as *differential linearity error* or DLE for short. The DNL curve represents the error in each step size, expressed in fractions of an LSB. DNL is computed by calculating the discrete first derivative of the DACs transfer curve, subtracting one LSB (i.e., V_{LSB}) from the derivative result, and then normalizing the result to one LSB

$$\text{DNL}(i) = \frac{S(i+1) - S(i) - V_{LSB}}{V_{LSB}} \text{ LSB} \tag{6.29}$$

As previously mentioned, we can define the average LSB size in one of three ways. We can define it as the actual full-scale range divided by the number of code transitions (number of codes minus 1) or we can define the LSB as the slope of the best-fit line. Alternatively, we can define the LSB size as the ideal DAC step size.

Exercises

6.11. Assuming an ideal gain of 200 mV/bit and an ideal offset of 1 V at code 0, calculate the absolute error transfer curve for the 4-bit DAC of Exercise 6.7. Normalize the result to a single LSB step size.

ANS. 0.0455, 0.0150, -0.3185, −0.1915, −0.5375, −0.2735, −0.5645, −0.3970, −0.7390 −0.7355 −0.7545 −1.0175 −1.2735 −1.3215 −1.5830 −1.8910

6.12. Compute the discrete first derivative of the DAC transfer curve given in Exercise 6.7. Is the DAC output monotonic?

ANS. 0.1939, 0.1333, 0.2254, 0.1308, 0.2528, 0.1418, 0.2335, 0.1316, 0.2007, 0.1962, 0.1474, 0.1488, 0.1904, 0.1477, 0.1384
The DAC is monotonic since there are no negative values in the discrete derivative.

The choice of LSB calculations depends on what type of DNL calculation we want to perform. There are four basic types of DNL calculation method: best-fit, endpoint, absolute, and best-straight-line. Best-fit DNL uses the best-fit line's slope to calculate the average LSB size. This is probably the best technique, since it accommodates gain errors in the DAC without relying on the values of a few individual voltages. Endpoint DNL is calculated by dividing the full-scale range by the number of transitions. This technique depends on the actual values for the maximum full-scale (V_{FS+}) and minimum full-scale (V_{FS-}) levels. As such it is highly sensitive to errors in these two values and is therefore less ideal than the best-fit technique. The absolute DNL technique uses the ideal LSB size derived from the ideal maximum and minimum full-scale values. This technique is less commonly used, since it assumes the DAC's gain is ideal.

The best-straight-line method is similar to the best-fit line method. The difference is that the best-straight-line method is based on the line that gives the best answer for integral nonlinearity (INL) rather than the line that gives the least squared errors. Integral nonlinearity will be discussed later in this chapter. Since the best-straight-line method is designed to yield the best possible answer, it is the most relaxed specification method of the four. It is used only in cases where the DAC or ADC linearity performance is not critical. Thus the order of methods from most relaxed to most demanding is best-straight line, best-fit, endpoint, and absolute.

The choice of technique is not terribly important in DNL calculations. Any of the three techniques will result in nearly identical results, as long as the DAC does not exhibit grotesque gain or linearity errors. DNL values of ±1/2 LSB are usually specified, with typical DAC performance of ±1/4 LSB for reasonably good DAC designs. A 1% error in the measurement of the LSB size would result in only a 0.01 LSB error in the DNL results, which is tolerable in most cases. The choice of technique is actually more important in the integral nonlinearity calculation, which we will discuss in the next section.

EXAMPLE 6.4

Calculate the DNL curve for the 4-bit DAC of the previous examples. Use the best-fit line to define the average LSB size. Does this DAC pass a ±1/2 LSB specification for DNL? Use the endpoint method to calculate the average LSB size. Is this result significantly different from the best-fit calculation?

Solution:
The first derivative of the transfer curve was calculated in the previous monotonicity example. The first derivative values are

75 mV, 175 mV, 75 mV, 55 mV, 75 mV, 175 mV, 75 mV, 195 mV, 75 mV, 175 mV,
75 mV, 55 mV, 75 mV, 175 mV, 75 mV

The average LSB size, 109.35 mV, was calculated in Example 6.1 using the best-fit line calculation. Dividing each step size by the average LSB size yields the following normalized derivative values (in LSBs)

0.686, 1.6, 0.686, 0.503, 0.686, 1.6, 0.686, 1.783, 0.686, 1.6, 0.686, 0.503, 0.686, 1.6, 0.686

Subtracting one LSB from each of these values gives us the DNL values for each code transition of this DAC expressed as a fraction of an LSB

−0.314, 0.6, −0.314, −0.497, −0.314, 0.6, −0.314, 0.783, −0.314, 0.6, −0.314, −0.497, −0.314, 0.6, −0.314

Note that there is one fewer DNL value than there are DAC codes.

   Figure 6.9a shows the DNL curve for this DAC. The maximum DNL value is +0.783 LSB, while the minimum DNL value is –0.497. The minimum value is within the –1/2 LSB test limit, but the

Figure 6.9. 4-bit DAC DNL curve (a) best-fit method (b) endpoint method.

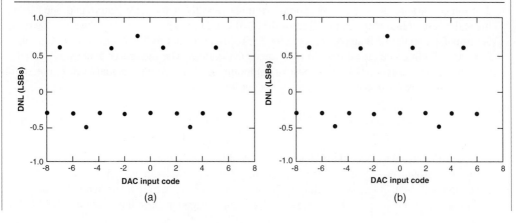

(a)   (b)

maximum DNL value exceeds the +1/2 LSB limit. Therefore, this DAC fails the DNL specification of ±1/2 LSB.

The average LSB step size calculated using the endpoint method is given by

$$V_{LSB} = \frac{V_{FS+} - V_{FS-}}{\text{number of codes} - 1} = \frac{825 \text{ mV} - (-780 \text{ mV})}{16 - 1} = 107 \text{ mV}$$

The DNL curve calculated using the endpoint method gives the following values, which have been normalized to an LSB size of 107 mV:

−0.299, 0.636, −0.299, −0.486, −0.299, 0.636, −0.299, 0.822, −0.299,

0.636, −0.299, −0.486, −0.299, 0.636, −0.299

The corresponding DNL curve is shown in Figure 6.9b. Using the endpoint calculation, we get slightly different results. Instead of a maximum DNL result of +0.783 LSB and a minimum DNL of −0.497 LSB, we get +0.822 and −0.486 LSB, respectively. This might be enough of a difference compared to the best-fit technique to warrant concern. Unless the endpoint method is explicitly called for in the data sheet, the best-fit method should be used since it is the least sensitive to abnormalities in any one DAC voltage.

Exercises

6.13. Calculate the DNL curve for the 4-bit DAC of Exercise 6.7. Use the best-fit line to define the average LSB size. Does this DAC pass a ±1/2 LSB specification for DNL?

ANS. 0.0937, −0.2481, 0.2714, −0.2622, 0.4259, −0.2002, 0.3170, −0.2577, 0.1320, 0.1067 −0.1686, −0.1607, 0.0739, −0.1669, −0.2194; pass

6.14. Calculate the DNL curve for the 4-bit DAC of Exercise 6.7. Use the endpoint method to calculate the average LSB size. Does this DAC pass a ±1/2 LSB specification for DNL?

ANS. 0.1132 −0.2347, 0.2941 −0.2491, 0.4514 −0.1859, 0.3406 −0.2445, 0.1523, 0.1264 −0.1537 −0.1457, 0.0931 −0.1520 −0.2054; pass

6.3.4 Integral Nonlinearity

The integral nonlinearity curve is a comparison between the actual DAC curve and one of three lines: the best-fit line, the endpoint line, or the ideal DAC line. The INL curve, like the DNL curve, is normalized to the LSB step size. As in the DNL case, the best-fit line is the preferred reference line, since it eliminates sensitivity to individual DAC values. The INL curve can be calculated by subtracting the reference DAC line (best-fit, endpoint, or ideal) from the actual DAC curve, dividing the results by the average LSB step size according to

$$INL(i) = \frac{S(i) - S_{REF}(i)}{V_{LSB}} \tag{6.30}$$

Note that using the ideal DAC line is equivalent to calculating the absolute error curve. Since a separate absolute error test is often specified, the ideal line is seldom used in INL testing. Instead, the endpoint or best-fit line is generally used. As in DNL testing, we are interested in the maximum and minimum value in the INL curve, which we compare against a test limit such as ±1/2 LSB.

EXAMPLE 6.5

Calculate the INL curve for the 4-bit DAC in the previous examples. First use an endpoint calculation, then use a best-fit calculation. Does either result pass a specification of ±1/2 LSB? Do the
two methods produce a significant difference in results?

Solution:

Using an endpoint calculation method, the INL curve for the 4-bit DAC of the previous examples
is calculated by subtracting a straight line between the V_{FS-} voltage and the V_{FS+} voltage from the
DAC output curve. The difference at each point in the DAC curve is divided by the average LSB
size, which in this case is calculated using an endpoint method. As in the endpoint DNL example,
the average LSB size is equal to 107 mV. The results of the INL calculations are (again, these
values are expressed in LSBs)

$$0.0, -0.299, 0.336, 0.037, -0.449, -0.748, -0.112, -0.411,$$
$$0.411, 0.112, 0.748, 0.449, -0.037, -0.336, 0.299, 0.0$$

Figure 6.10a shows this endpoint INL curve. The maximum INL value is +0.748 LSB, and the
minimum INL value is –0.748. This DAC does not pass an INL specification of ±1/2 LSB.

Using a best-fit calculation method, the INL curve for the 4-bit DAC of the previous examples is
calculated by subtracting the best-fit line from the DAC output curve. Each point in the difference
curve is divided by the average LSB size, which in this case is calculated using the best-fit line
method. As in the best-fit DNL example, the average LSB size is equal to 109.35 mV. The results
of the INL calculations are

$$0.161, -0.153, 0.448, 0.133, -0.364, -0.678, -0.077, -0.392, 0.392,$$
$$0.077, 0.678, 0.364, -0.133, -0.448, 0.153, -0.161$$

The maximum value is +0.678, and the minimum value is –0.678. These INL results are better
than the endpoint INL values, but still do not pass a ±1/2 LSB test limit. The best-fit INL curve
is shown in Figure 6.10b for comparison with the endpoint INL curve. The two INL curves are
somewhat similar in shape, but the individual INL values are quite different. Remember that the
DNL curves for endpoint and best-fit calculations were nearly identical. So, as previously stated,

Figure 6.10. 4-bit DAC INL curve (a) endpoint method (b) best-fit method.

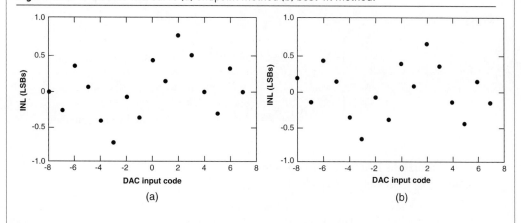

the choice of calculation technique is much more important for INL curves than for DNL curves. Notice also that while an endpoint INL curve always begins and ends at zero, the best-fit curve does not necessarily behave this way. A best-fit curve will usually give better INL results than an endpoint INL calculation. This is especially true if the DAC curve exhibits a bowed shape in either the upward or downward direction. The improvement in the INL measurement is another strong argument for using a best-fit approach rather than an absolute or endpoint method, since the best-fit approach tends to increase yield.

The INL curve is the integral of the DNL curve, thus the term "integral nonlinearity"; DNL is a measurement of how consistent the step sizes are from one code to the next. INL is therefore a measure of accumulated errors in the step sizes. Thus, if the DNL values are consistently larger than zero for many codes in a row (step sizes are larger than 1 LSB), the INL curve will exhibit an upward bias. Likewise, if the DNL is less than zero for many codes in a row (step sizes are less than 1 LSB), the INL curve will have a downward bias. Ideally, the positive error in one code's DNL will be balanced by negative errors in surrounding codes and vice versa. If this is true, then the INL curve will tend to remain near zero. If not, the INL curve may exhibit large upward or downward bends, causing INL failures.

The INL integration can be implemented using a running sum of the elements of the DNL. The ith element of the INL curve is equal to the sum of the first $i-1$ elements of the DNL curve plus a constant of integration. When using the best-fit method, the constant of integration is equal to the difference between the first DAC output voltage and the corresponding point on the best-fit curve, all normalized to one LSB. When using the endpoint method, the constant of integration is equal to zero. When using the absolute method, the constant is set to the normalized difference between the first DAC output and the ideal output. In any running sum calculation it is important to use high-precision mathematical operations to avoid accumulated math error in the running sum. Mathematically, we can express this process as

$$INL(i) = \sum_{k=0}^{i-1} DNL(k) + C \tag{6.31}$$

where

$$C = \begin{cases} \dfrac{S(0) - Best\_fit\_line(0)}{V_{LSB}} & \text{for best-fit linearity method} \\ 0 & \text{for endpoint linearity method} \\ \dfrac{S(0) - S_{IDEAL}(0)}{V_{LSB}} & \text{for absolute linearity method} \end{cases}$$

and $i = 0$ indicates the DAC level corresponding to V_{FS-}.

Conversely, DNL can be calculated by taking the first derivative of the INL curve

$$DNL(i) = INL'(i) = INL(i+1) - INL(i) \tag{6.32}$$

This is usually the easiest way to calculate DNL. The first derivative technique works well in DAC testing, but we will see in the next chapter that the DNL curve for an ADC is easier to capture than the INL curve. In ADC testing it is more common to calculate the DNL curve first and then integrate it to calculate the INL curve. In either case, whether we integrate DNL to get INL or differentiate INL to get DNL, the results are mathematically identical.

Integral nonlinearity and differential nonlinearity are sometimes referred to by the names integral linearity error (ILE) and differential linearity error (DLE). However, the terms INL and DNL seem to be more prevalent in data sheets and other literature. We will use the terms INL and DNL throughout this text.

Exercises

6.15. Calculate the INL curve for a 4-bit unsigned binary DAC whose DNL curve is described by the following values (in LSBs)

0.0937, −0.2481, 0.2714, −0.2622, 0.4259, −0.2002, 0.3170, −0.2577, 0.1320, 0.1067, −0.1686, −0.1607, 0.0739, −0.1669, −0.2194

The DAC output for code 0 is 1.0091 V. Assume that the best-fit line has a gain of 177.3 mV/bit and an offset of 1.026 V.

ANS. −0.0959, −0.0022, −0.2503, 0.0210, −0.2412, 0.1847, −0.0155, 0.3016, 0.0438, 0.1759, 0.2825, 0.1139, −0.0467, 0.0272, -0.1397, −0.3591

6.3.5 Partial Transfer Curves

A customer or systems engineer may specify that only a portion of a DAC or ADC transfer curve must meet certain specifications. For example, a DAC may be designed so that its V_{FS-} code corresponds to 0 V. However, due to analog circuit clipping as the DAC output signal approaches ground, the DAC may clip to a voltage of 100 mV. If the DAC is designed to perform a specific function that never requires voltages below 100 mV, then the customer may not care about this clipping. In such a case, the DAC codes below 100 mV are excluded from the offset, gain, INL, DNL, and so on. specifications. The test engineer may then treat these codes as if they do not exist. This type of partial DAC and ADC testing is becoming more common as more DACs and ADCs are designed into custom applications with very specific requirements. General-purpose DACs are unlikely to be specified using partial curves, since the customer's application needs are unknown.

6.3.6 Major Carrier Testing

The techniques discussed thus far for measuring INL and DNL are based on a testing approach called *all-codes testing*. In all-codes testing, all valid codes in the transfer curve are measured to determine the INL and DNL values. Unfortunately, all-codes testing can be a very time-consuming process.

Depending on the architecture of the DAC, it may be possible to determine the location of each voltage in the transfer curve without measuring each one explicitly. We will refer to this as selected-code testing. Selected-code testing can result in significant test-time savings, which of course represents substantial savings in test cost. There are several selected-code testing techniques, the simplest of which is called the *major carrier* method.

Many DACs are designed using an architecture in which a series of binary-weighted resistors or capacitors are used to convert the individual bits of the converter code into binary-weighted currents or voltages. These currents or voltages are summed together to produce the DAC output. For instance, a binary-weighted unsigned binary D-bit DAC's output can be described as a sum of binary-weighted voltage or current values, $W_0, W_1, \ldots, W_n$, multiplied by the individual bits of the DAC's input code, $b_{D-1}, b_{D-2}, ..., b_2, b_1$. The DAC's output value is therefore equal to

$$\text{DAC Output} = b_0 W_0 + b_1 W_1 + \cdots + b_{D-2} W_{D-2} + b_{D-1} W_{D-1} + \text{DC Base} \qquad (6.33)$$

where

- DC base is the DAC output value with a V_{FS-} input code.
- DAC code bits $b_{D-1}, b_{D-2}, ..., b_2, b_1, b_0$; take on values of 1 or 0.

If this idealized model of the DAC is sufficiently accurate, then we only need to make $D+1$ measurements of DAC behavior and solve for the unknown model parameters: $W_0, W_1, ..., W_{D-1}$ and the DC Base term. Subsequently, we can cycle through all binary values using Eq. (6.33) and compute the entire DAC transfer curve. This DAC testing method is called the major carrier technique. The major carrier approach can be used for ADCs as well as DACs. The assumption of sufficient DAC or ADC model accuracy is only valid if the actual superposition errors of the DAC or ADC are low. This may or may not be the case. The superposition assumption can only be determined through characterization, comparing the all-codes DAC output levels with the ones generated by the major carrier method.

The most straightforward way to obtain each model parameter W_n is to set code bit b_n to 1 and all other to zero. This is then repeated for each code bit for n from 0 to $D-1$. However, the resulting output levels are widely different in magnitude. This makes them difficult to measure accurately with a voltmeter, since the voltmeter's range must be adjusted for each measurement. A better approach that alleviates the accuracy problem is to measure the step size of the major carrier transitions in the DAC curve, which are all approximately 1 LSB in magnitude. A major carrier transition is defined as the voltage (or current) transition between the DAC codes 2^n-1 and 2^n. For example, the transition between binary 00111111 and 01000000 is a major carrier transition for $n = 6$. Major carrier transitions can be measured using a voltmeter's sample-and-difference mode, giving highly accurate measurements of the major carrier transition step sizes.

Once the step sizes are known, we can use a series of inductive calculations to find the values of $W_0, W_1, ..., W_{D-1}$. We start by realizing that we have actually measured the following values:

$$\text{DC base} = \text{measured DAC output with minus full-scale code}$$
$$V_0 = W_0$$
$$V_1 = W_1 - W_0$$
$$V_2 = W_2 - (W_1 + W_0)$$
$$V_3 = W_3 - (W_2 + W_1 + W_0)$$
$$\cdots$$
$$V_n = W_n - (W_{n-1} + W_{n-2} + W_{n-3} + \cdots + W_0)$$

The value of the first major transition, V_0, is a direct measurement of the value of W_0 (the step size of the least significant bit). The value of W_1 can be calculated by rearranging the second equation: $W_1 = V_1 + W_0$. Once the values of W_0 and W_1 are known, the value of W_2 is calculated by rearranging the third equation: $W_2 = V_2 + W_1 + W_0$, and so forth. Once the values of $W_0 - W_n$ are known, the complete DAC curve can be reconstructed for each possible combination of input bits $b_0 - b_n$ using the original model of the DAC described by Eq. (6.33).

The major carrier technique can also be used on signed binary and two's complement converters, although the codes corresponding to the major carrier transitions must be chosen to match the converter encoding scheme. For example, the last major transition for our two's complement 4-bit DAC example happens between code 1111 (decimal −1) and 0000 (decimal 0). Aside from these minor modifications in code selection, the major carrier technique is the same as the simple unsigned binary approach.

EXAMPLE 6.6

Using the major carrier technique on the 4-bit DAC example, we measure a DC base of -780 mV setting the DAC to V_{FS-} (binary 1000, or -8). Then we measure the step size between 1000 (-8) and 1001 (-7). The step size is found to be 75 mV. Next we measure the step size between 1001 (-7) and 1010 (-6). This step size is 175 mV. The step size between 1011 (-5) and 1100 (-4) is 55 mV and the step size between 1111 (-1) and 0000 (0) is 195 mV. Determine the values of W_0, W_1, W_2, and W_3. Reconstruct the voltages on the ramp from DAC code -8 to DAC code $+7$.

Solution:

Rearranging the set of equations $V_n = W_n - (W_{n-1} + W_{n-2} + W_{n-3} + \ldots + W_0)$ to solve for W_n, we obtain

$$\text{DC baseline} = \text{measured DAC output with } V_{FS-} \text{ code } = -780 \text{ mV}$$
$$W_0 = V_0 = 75 \text{ mV}$$
$$W_1 = V_1 + W_0 = 175 \text{ mV} + 75 \text{ mV} = 250 \text{ mV}$$
$$W_2 = V_2 + W_1 + W_0 = 380 \text{ mV}$$
$$W_3 = V_3 + W_2 + W_1 + W_0 = 900 \text{ mV}$$

For a two's complement DAC, we have to realize that the most significant bit is inverted in polarity compared to an unsigned binary DAC. Therefore, the DAC model for our 4-bit DAC is given by

$$\text{DAC output} = b_0 W_0 + b_1 W_1 + b_2 W_2 + \overline{b_3} W_3 + \text{DC base} \tag{6.34}$$

Using this two's complement version of the DAC model, the 16 voltage values of the DAC curve are reconstructed as shown in Table 6.4. Notice that these values are exactly equal to the all-codes results in Figure 6.8. The example DAC was created using a binary-weighted model with perfect superposition; so it is no surprise the major carrier technique works for this imaginary DAC. Real DACs and ADCs often have superposition errors that make the major carrier technique unusable.

Table 6.4. DAC Transfer Curve Calculated Using the Major Carrier Technique

DAC Code	Calculation	Output Voltage (mV)
1000	DC Base	-780
1001	W_0+DC Base	-705
1010	W_1+DC Base	-530
1011	W_1+W_0+DC Base	-455
1100	W_2+DC Base	-400
1101	W_2+W_0+DC Base	-325
1110	W_2+W_1+DC Base	-150
1111	W_2+W_1+W0+DC Base	-75
0000	W_3+DC Base	120
0001	W_3+W_0+DC Base	195
0010	W_3+W_1+DC Base	370
0011	W_3+W_1+W_0+DC Base	445
0100	W_3+W_2+DC Base	500
0101	W_3+W_2+W_0+DC Base	575
0110	W_3+W_2+W_1+DC Base	750
0111	W_3+W_2+W_1+W_0+DC Base	825

6.3.7 Other Selected-Code Techniques

Besides the major carrier method, other selected-code techniques have been developed to reduce the test time associated with all-codes testing. The simplest of these is the segmented method. This method only works for certain types of DAC and ADC architectures, such as the 12-bit segmented DAC shown in Figure 6.11. Although most segmented DACs are actually constructed using a different architecture than that in Figure 6.11, this simple architecture is representative of how segmented DACs can be tested.

The example DAC uses a simple unsigned binary encoding scheme with 12 data bits, D11-D0. It consists of two portions, a 6-bit coarse resolution DAC and a 6-bit fine resolution DAC. The LSB step size of the coarse DAC is equal to the full-scale range of the fine DAC plus one fine DAC LSB. In other words, if the combined 12-bit DAC has an LSB size of V_{LSB}, then the fine DAC also has a step size of V_{LSB}, while the coarse DAC has a step size of $2^6 \times V_{LSB}$. The output of these two 6-bit DACs can therefore be summed together to produce a 12-bit DAC

$$\text{DAC output} = \text{coarse DAC contribution} + \text{fine DAC contribution} \qquad (6.35)$$

Both the fine DAC and the coarse DAC are designed using a resistive divider architecture rather than a binary-weighted architecture. Since major carrier testing can only be performed on binary-weighted architectures, an all-codes testing approach must be used to verify the performance of each of the two 6-bit resistive divider DACs. However, we would like to avoid testing each of the 212, or 4096 codes of the composite 12-bit DAC. Using superposition, we will test each of the two 6-bit DACs using an all-codes test. This requires only 2 × 26, or 128 measurements. We will then combine the results mathematically into a 4096-point all-codes curve using a linear model of the composite DAC.

Let us assume that through characterization, it has been determined that this example DAC has excellent superposition. In other words, the step sizes of each DAC are independent of the setting of the other DAC. Also, the summation circuit has been shown to be highly linear. In a case such as this, we can measure the all-codes output curve of the coarse DAC while the fine DAC is set to 0 (i.e., D5-D0 = 000000). We store these values into an array $V_{DAC\text{-}COARSE}(n)$, where n takes on the values 0 to 63, corresponding to data bits D11-D6. Then we can measure the all-codes output curve for the fine DAC while the coarse DAC is set to 0 (i.e., D11-D6 = 000000). These voltages are stored in the array $V_{DAC\text{-}FINE}(n)$, where n corresponds to data bits D5-D0.

Figure 6.11. Segmented DAC conceptual block diagram.

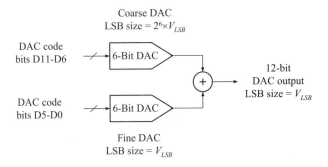

Although we have only measured a total of 128 levels, superposition allows us to recreate the full 4096-point DAC output curve by a simple summation. Each DAC output value $V_{DAC}(i)$ is equal to the contribution of the coarse DAC plus the contribution of the fine DAC

$$V_{DAC}(i) = V_{DAC-FINE}(i \text{ AND } 000000111111) + V_{DAC-COARSE}\left(\frac{i \text{ AND } 111111000000}{64}\right) \quad (6.36)$$

where i ranges from 0 to 4095.

Thus a full 4096-point DAC curve can be mathematically reconstructed from only 128 measurements by evaluating this equation at each value of i from 0 to 4095. Of course, this technique is totally dependent on the architecture of the DAC. It would be inappropriate to use this technique on a nonsegmented DAC or a segmented DAC with large superposition errors.

A more advanced selected-codes technique was developed at the National Institute of Standards and Technology (NIST). This technique is useful for all types of DACs and ADCs. It does not make any assumptions about superposition errors or converter architecture. Instead, it uses linear algebra and data collected from production lots to create an empirical model of the DAC or ADC. The empirical model only requires a few selected codes to recreate the entire DAC or ADC transfer curve. Although the details of this technique are beyond the scope of this book, the original NIST paper is listed in the references at the end of this chapter.[3]

Another similar technique uses wavelet transforms to predict the overall performance of converters based on a limited number of measurements.[4] Again, this topic is beyond the scope of this book.

Exercises

6.16. The step sizes between the major carries of a 4-bit unsigned binary DAC were measured to be as follows

code 0→1: 0.2010 V; code 1→2: 0.1987 V; code 3→4: 0.1877 V; code 7→8: 0.1998 V Determine the values of W_0, W_1, W_2, and W_3.

ANS. $W_0 = 0.2010$, $W_1 = 0.3997$, $W_2 = 0.7884$, $W_3 = 1.5889$

6.4 DYNAMIC DAC TESTS

6.4.1 Conversion Time (Settling Time)

So far we have discussed only low-frequency DAC performance. The DAC DC tests and transfer curve tests measure the DAC's static characteristics, requiring the DAC to stabilize to a stable voltage or current level before each output level measurement is performed. If the DAC's output stabilizes in a few microseconds, then we might step through each output state at a high frequency, but we are still performing static measurements for all intents and purposes.

A DAC's performance is also determined by its dynamic characteristics. One of the most common dynamic tests is settling time, commonly referred to as *conversion time*. Conversion time is defined as the amount of time it takes for a DAC to stabilize to its final static level *within a specified error band* after a DAC code has been applied. For instance, a DAC's settling time may be defined as 1 μs to ±1/2 LSB. This means that the DAC output must stabilize to its final value plus or minus a 1/2 LSB error band no more than 1 μs after the DAC code has been applied.

This test definition has one ambiguity. Which DAC codes do we choose to produce the initial and final output levels? The answer is that the DAC must settle from any output level to any other level within the specified time. Of course, to test every possibility, we might have to measure

Figure 6.12. DAC settling time measurement (a) referenced to a digital signal; (b) referenced to the DAC output 50% point.

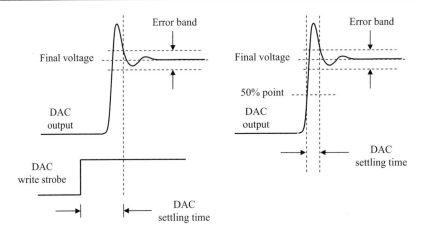

millions of transitions on a typical DAC. As with any other test, we have to determine what codes represent the worst-case transitions. Typically settling time will be measured as the DAC transitions from minus full-scale (V_{FS-}) to plus full-scale (V_{FS+}) and vice versa, since these two tests represent the largest voltage swing.

The 1/2 LSB example uses an error band specification that is referenced to the LSB size. Other commonly used definitions require the DAC output to settle within a certain percentage of the full-scale range, a percentage of the final voltage, or a fixed voltage range. So we might see any of the following specifications:

> settling time = 1 μs to ± 1% of full-scale range
> settling time = 1 μs to ± 1% of final value
> settling time = 1 μs to ±1 mV

The test technique for all these error-band definitions is the same; we just have to convert the error-band limits to absolute voltage limits before calculating the settling time. The straightforward approach to testing settling time is to digitize the DAC's output as it transitions from one code to another and then use the known time period between digitizer samples to calculate the settling time. We measure the final settled voltage, calculate the settled voltage limits (i.e., ±1/2 LSB), and then calculate the time between the digital signal transition that initiates a DAC code change and the point at which the DAC first stays within the error band limits, as shown in Figure 6.12a.

In extremely high frequency DACs it is common to define the settling time not from the DAC code change signal's transition but from the time the DAC passes the 50% point to the time it settles to the specified limits as shown in Figure 6.12b. This is easier to calculate, since it only requires us to look at the DAC output, not at the DAC output relative to the digital code.

6.4.2 Overshoot and Undershoot

Overshoot and undershoot can also be calculated from the samples collected during the DAC settling time test. These are defined as a percentage of the voltage swing or as an absolute voltage. Figure 6.13 shows a DAC output with 10% overshoot and 2% undershoot on a V_{FS-} to V_{FS+} transition.

Figure 6.13. DAC overshoot and undershoot measurements.

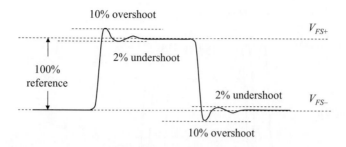

Figure 6.14. DAC rise and fall time measurements.

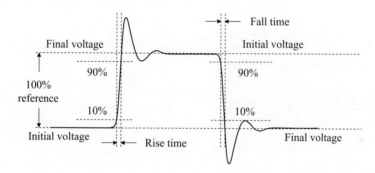

6.4.3 Rise Time and Fall Time

Rise and fall time can also be measured from the digitized waveform collected during a settling time test. Rise and fall times are typically defined as the time between two markers, one of which is 10% of the way between the initial value and the final value and the other of which is 90% of the way between these values as depicted in Figure 6.14. Other common marker definitions are 20% to 80% and 30% to 70%.

6.4.4 DAC-to-DAC Skew

Some types of DACs are designed for use in matched groups. For example, a color palette RAM DAC is a device that is used to produce colors on video monitors. A RAM DAC uses a random access memory (RAM) lookup table to turn a single color value into a set of three DAC output values, representing the red, green, and blue intensity of each pixel. These DAC outputs must change almost simultaneously to produce a clean change from one pixel color to the next. The degree of timing mismatch between the three DAC outputs is called *DAC-to-DAC skew*. It is measured by digitizing each DAC output and comparing the timing of the 50% point of each output to the 50% point of the other outputs. There are three skew values (R-G, G-B, and B-R), as illustrated in Figure 6.15. Skew is typically specified as an absolute time value, rather than a signed value.

6.4.5 Glitch Energy (Glitch Impulse)

Glitch energy, or glitch impulse, is another specification common to high-frequency DACs. It is defined as the total area under the voltage-time curve of the glitches in a DAC's output as it

Figure 6.15. DAC-to-DAC skew measurements.

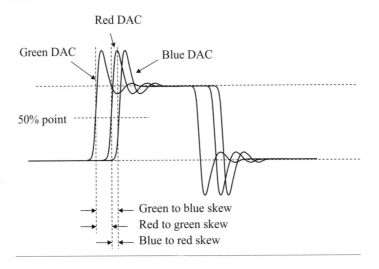

Figure 6.16. Glitch energy measurements.

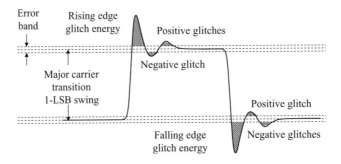

switches across the largest major transition (i.e., 01111111 to 10000000 in an 8-bit DAC) and back again. As shown in Figure 6.16, the glitch area is defined as the area that falls *outside* the rated error band. These glitches are caused by a combination of capacitive/inductive ringing in the DAC output and skew between the timing of the digital bits feeding the binary-weighted DAC circuits. The parameter is commonly expressed in picosecond-volts (ps-V) or equivalently, picovolt-seconds (pV-s). (These are not actually units of energy, despite the term *glitch energy*.) The area under the negative glitches is considered positive area, and should be added to the area under the positive glitches. Both the rising-edge glitch energy and the falling-edge glitch energy should be tested.

6.4.6 Clock and Data Feedthrough

Clock and data feedthrough is another common dynamic DAC specification. It measures the cross-talk from the various clocks and data lines in a mixed-signal circuit that couple into a DAC output. There are many ways to define this parameter; so it is difficult to list a specific test technique.

However, clock and data feedthrough can be measured using a technique similar to all the other tests in this section. The output of the DAC is digitized with a high-bandwidth digitizer. Then the various types of digital signal feedthrough are analyzed to make sure they are below the defined test limits. The exact test conditions and definition of clock and data feedthrough should be provided in the data sheet. This measurement may require time-domain analysis, frequency-domain analysis, or both.

6.5 TESTS FOR COMMON DAC APPLICATIONS

6.5.1 DC References

As previously mentioned, the test list for a given DAC often depends on its intended functionality in the system-level application. Many DACs are used as simple DC references. An example of this type of DAC usage is the power level control in a cellular telephone. As the cellular telephone user moves closer or farther away from a cellular base station (the radio antenna tower), the transmitted signal level from the cellular telephone must be adjusted. The transmitted level may be adjusted using a transmit level DAC so that the signal is just strong enough to be received by the base station without draining the cellular telephone's battery unnecessarily.

If a DAC is only used as a DC (or slow-moving) voltage or current reference, then its AC transmission parameters are probably unimportant. It would probably be unnecessary to measure the 1-kHz signal to total harmonic distortion ratio of a DAC whose purpose is to set the level of a cellular telephone's transmitted signal. However, the INL and DNL of this DAC would be extremely important, as would its absolute errors, monotonicity, full-scale range, and output drive capabilities (output impedance).

DACs used as DC references are usually measured using the intrinsic parameters listed in this chapter, rather than the transmission parameters outlined in Chapter 11. Notable exceptions are signal-to-noise ratio and idle channel noise (ICN). These may be of importance if the DC level must exhibit low noise. For example, the cellular telephone's transmitted signal might be corrupted by noise on the output of the transmit level control DAC, and therefore we might need to measure the DAC's noise level.

Dynamic tests are not typically performed on DC reference DACs, with the exception of settling time. The settling time of typical DACs is often many times faster than that required in DC reference applications; so even this parameter is frequently guaranteed by design rather than being tested in production.

6.5.2 Audio Reconstruction

Audio reconstruction DACs are those used to reproduce digitized sound. Examples include the voice-band DAC in a cellular telephone and the audio DAC in a PC sound card. These DACs are more likely to be tested using the transmission parameters of Chapter 11, since their purpose is to reproduce arbitrary audio signals with minimum noise and distortion.

The intrinsic parameters (i.e., INL and DNL) of audio reconstruction DACs are typically measured only during device characterization. Linearity tests can help track down any transmission parameter failures caused by the DAC. It is often possible to eliminate the intrinsic parameter tests once the device is in production, keeping only the transmission tests.

Dynamic tests are not typically specified or performed on audio DACs. Any failures in settling time, glitch energy, and so on, will usually manifest themselves as failures in transmission parameters such as signal-to-noise, signal-to-distortion, and idle channel noise.

Exercises

6.17. The step response of a DAC obtained from an oscilloscope is as follows:

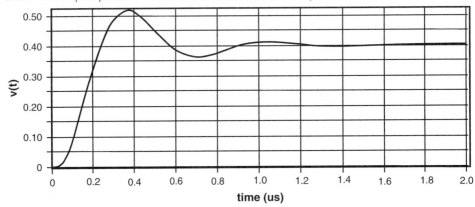

The data sheet states that the settling time is 1 μs (error band = ± 20 mV). Does this DAC settle fast enough to meet the settling time specification? Also, determine the overshoot of this signal and its rise time. Estimate the total glitch energy during the positive-going transition.

ANS. Actual settling time = 0.82 μs; yes; overshoot = 30%; rise time = 0.2 μs. Glitch energy = 0.5(0.3)(0.13) + 0.5(0.5)(-0.033) + 0.5(0.6)(0.01) = 14 ns-V (triangle approximation).

6.5.3 Data Modulation

Data modulation is another purpose to which DACs are often applied. The cellular telephone again provides an example of this type of DAC application. The IF section of a cellular telephone base-band modulator converts digital data into an analog signal suitable for transmission, similar to those used in modems (see Section 11.1.2). Like the audio reconstruction DACs, these DACs are typically tested using sine wave or multitone transmission parameter tests.

Again, the intrinsic tests like INL and DNL may be added to a characterization test program to help debug the design. However, the intrinsic tests are often removed after the transmission parameters have been verified. Dynamic tests such as settling time may or may not be necessary for data modulation applications.

Data modulation DACs also have very specific parameters such as error vector magnitude (EVM) or phase trajectory error (PTE). Parameters such as these are very application-specific. They are usually defined in standards documents published by the IEEE, NIST, or other government or industry organization. The data sheet should provide references to documents defining application-specific tests such as these. The test engineer is responsible for translating the measurement requirements into ATE-compatible tests that can be performed on a production tester. ATE vendors are often a good source of expertise and assistance in developing these application-specific tests.

6.5.4 Video Signal Generators

As discussed earlier, DACs can be used to control the intensity and color of pixels in video cathode ray tube (CRT) displays. However, the type of testing required for video DACs depends on the nature of their output. There are two basic types of video DAC application, RGB and NTSC. An RGB (red-green-blue) output is controlled by three separate DACs. Each DAC controls the intensity of an electron beam, which in turn controls the intensity of one of the three primary colors of each pixel as the beam is swept across the CRT. In this application, each DAC's output voltage or current directly control the intensity of the beam. RGB DACs are typically used in computer monitors.

The NTSC format is used in transmission of standard (i.e., non-HDTV) analog television signals. It requires only a single DAC, rather than a separate DAC for each color. The picture intensity, color, and saturation information is contained in the time-varying offset, amplitude, and phase of a 3.54-MHz sinusoidal waveform produced by the DAC. Clearly this is a totally different DAC application than the RGB DAC application. These two seemingly similar video applications require totally different testing approaches.

RGB DACs are tested using the standard intrinsic tests like INL and DNL, as well as the dynamic tests like settling time and DAC-to-DAC skew. These parameters are important because the DAC outputs directly control the rapidly changing beam intensities of the red, green, and blue electron beams as they sweep across the computer monitor. Any settling time, rise time, fall time, undershoot, or overshoot problems show up directly on the monitor as color or intensity distortions, vertical lines, ghost images, and so on.

The quality of the NTSC video DAC, by contrast, is determined by its ability to produce accurate amplitude and phase shifts in a 3.54-MHz sine wave while changing its offset. This type of DAC is tested with transmission parameters like gain, signal-to-noise, differential gain, and differential phase (topics of Chapters 10 and 11).

6.6 SUMMARY

DAC testing is far less straightforward than one might at first assume. Although DACs all perform the same basic function (digital-to-analog conversion), the architecture of the DAC and its intended application determine its testing requirements and methodologies. A large variety of standard tests have been defined for DACs, including transmission parameters, DC intrinsic parameters, and dynamic parameters. We have to select DAC test requirements carefully to guarantee the necessary quality of the DAC without wasting time with irrelevant or ineffective tests.

ADC testing is very closely related to DAC testing. Many of the DC and intrinsic tests defined in this chapter are very similar to those performed on ADCs. However, due to the many-to-one transfer characteristics of ADCs, the measurement of the ADC input level corresponding to each output code is much more difficult than the measurement of the DAC output level corresponding to each input code. Chapter 7, "ADC Testing," explains the various ways the ADC transfer curve can be measured, as well as the many types of ADC architectures and applications the test engineer will likely encounter.

PROBLEMS

6.1. Given a set of N points denoted by $S(i)$, derive the parameters of a straight line described by

$$\text{Best\_fit\_line}(i) = \text{gain} \times i + \text{offset} \quad \text{for } i = 0, 1, \ldots, N-1$$

that minimizes the following mean-square error criteria

$$\overline{e^2} = \sum_{i=0}^{N-1} \left[S(i) - \text{Best\_fit\_line}(i) \right]^2 = \sum_{i=0}^{N-1} \left[S(i) - \text{gain} \times i + \text{offset} \right]^2$$

Hint: Find partial derivatives $\partial \overline{e^2} / \partial \, gain$ and $\partial \overline{e^2} / \partial \, offset$, set them both to zero, and solve for the two unknowns, gain and offset, from the system of two equations.

6.2. The output levels of a 4-bit DAC produces the following set of voltage levels, starting from code 0 and progressing through to code 15:

0.0465, 0.3255, 0.7166, 1.0422, 1.5298, 1.8236, 2.1693, 2.5637,

2.8727, 3.3443, 3.6416, 4.0480, 4.3929, 4.7059, 5.0968, 5.5050

What is the full-scale range of this DAC?

6.3. A 4-bit DAC has a full-scale voltage range of 0 to 1.0 V. The input is formatted using an unsigned binary number representation. List all possible ideal output levels. What output level corresponds to the DAC input code 0?

6.4. A 5-bit DAC has a full-scale voltage range of 2.0 to 4.0 V. The input is formatted using a 2's complement number representation. List all possible ideal output levels. What output level corresponds to the DAC input code 0?

6.5. A 4-bit ADC has a full-scale voltage range of −10 to 10 V. The internal operation of the ADC is based on truncation and the input digital signal is formatted using a two's complement binary number representation. List all possible ideal output levels. What is the LSB step size?

6.6. A 5-bit ADC has a full-scale voltage range of 0 to 5 V. The internal operation of the ADC is such that all code widths are equal. Also, the input digital signal is formatted using a unsigned binary number representation. List all possible ideal output levels. What is the LSB step size?

6.7. A 4-bit unsigned binary DAC produces the following set of voltage levels, starting from code 0 and progressing through to code 15

0.0465, 0.3255, 0.7166, 1.0422, 1.5298, 1.8236, 2.1693, 2.5637,
2.8727, 3.3443, 3.6416, 4.0480, 4.3929, 4.7059, 5.0968, 5.5050

The ideal DAC output at code 0 is 0 V and the ideal gain is equal to 400 mV/bit. Answer the following questions assuming a best-fit line is used as a reference.
(a) Calculate the DAC's gain (volts per bit), gain error, offset and offset error.
(b) What is the LSB step size?
(c) Calculate the absolute error transfer curve for this DAC. Normalize the result to one LSB.
(d) Is the DAC output monotonic?
(e) Compute the DNL curve for this DAC. Does this DAC pass a ±1/2 LSB specification for DNL?

6.8. A 4-bit unsigned binary DAC produces the following set of voltage levels, starting from code 0 and progressing through to code 15

0.0465, 0.3255, 0.7166, 1.0422, 1.5298, 1.8236, 2.1693, 2.5637,
2.8727, 3.3443, 3.6416, 4.0480, 4.3929, 4.7059, 5.0968, 5.5050

The ideal DAC output at code 0 is 0 V and the ideal gain is equal to 400 mV/bit. Answer the following questions assuming an endpoint-to-endpoint line is used as a reference.
(a) Calculate the DAC's gain (volts per bit), gain error, offset and offset error.
(b) What is the LSB step size?
(c) Calculate the absolute error transfer curve for this DAC. Normalize the result to one LSB.
(d) Is the DAC output monotonic?
(e) Compute the DNL curve for this DAC. Does this DAC pass a ±1/2 LSB specification for DNL?

6.9. A 4-bit two's complement DAC produces the following set of voltage levels, starting from code -8 and progressing through to code +7

−0.9738, −0.8806, −0.6878, −0.6515, −0.3942, −0.3914, −0.2497, −0.1208,
−0.0576, 0.1512, 0.2290, 0.4460, 0.4335, 0.5999, 0.6743, 0.8102

The ideal DAC output at code 0 is 0 V and the ideal gain is equal to 133.3 mV/bit. Answer the following questions assuming a best-fit line is used as a reference.
(a) Calculate the DAC's gain (volts per bit), gain error, offset and offset error.

(b) Estimate the LSB step size using its measured full-scale range. What is the gain error and offset error?

(c) Calculate the absolute error transfer curve for this DAC. Normalize the result to one LSB.

(d) Is the DAC output monotonic?

(e) Compute the DNL curve for this DAC. Does this DAC pass a ±1/2 LSB specification for DNL?

6.10. A 4-bit two's complement DAC produces the following set of voltage levels, starting from code −8 and progressing through to code +7

−0.9738, −0.8806, −0.6878, −0.6515, −0.3942, −0.3914, −0.2497, −0.1208,
−0.0576, 0.1512, 0.2290, 0.4460, 0.4335, 0.5999, 0.6743, 0.8102

The ideal DAC output at code 0 is 0 V and the ideal gain is equal to 133.3 mV/bit. Answer the following questions assuming a endpoint-to-endpoint line is used as a reference.

(a) Calculate the DAC's gain (volts per bit), gain error, offset and offset error.

(b) Estimate the LSB step size using its measured full-scale range. What is the gain error and offset error?

(c) Calculate the absolute error transfer curve for this DAC. Normalize the result to one LSB.

(d) Is the DAC output monotonic?

(e) Compute the DNL curve for this DAC. Does this DAC pass a ±1/2 LSB specification for DNL?

6.11. Calculate the INL curve for a 4-bit unsigned binary DAC whose DNL curve is described by the following values

−0.0815, −0.1356, −0.1133, 0.0057, 0.0218, 0.1308, −0.0361, −0.0950,
0.1136, −0.1633, 0.2101, 0.0512, 0.0119, −0.0706, −0.0919

The DAC output for code 0 is −0.4919 V. Assume that the best-fit line has a gain of 63.1 mV/bit and an offset of −0.5045 V. Does this DAC pass a ±1/2 LSB specification for INL?

6.12. Calculate the DNL curve for a 4-bit DAC whose INL curve is described by the following values

0.1994, 0.1180, −0.0177, −0.1310, −0.1253, −0.1036, 0.0272, −0.0089,
−0.1039, 0.0096, −0.1537, 0.0565, 0.1077, 0.1196,0.0490, −0.0429

Does this DAC pass a ±1/2 LSB specification for DNL?

6.13. The step sizes between the major carries of a 5-bit unsigned binary DAC were measured to be as follows

code 0 → 1: 0.1939 V, code 1 → 2: 0.1333 V, code 3 → 4: 0.1308 V, code 7 → 8: 0.1316 V, code 15 → 16: 0.1345 V

Determine the values of W_0, W_1, W_2, W_3, and W_4. Reconstruct the voltages on the ramp from DAC code 0 to DAC code 31 if the DC base value is 100 mV.

6.14. The step sizes between the major carries of a 4-bit two's complement DAC were measured to be as follows:

code −8 → −7: 0.1049 V, code −7 → −6: 0.1033 V, code −5 → −4: 0.0998 V, code −1 → 0: 0.1016 V

Determine the values of W_0, W_1, W_2, and W_3. Reconstruct the voltages on the ramp from DAC code −8 to DAC code +7 if the DC base value is 500 mV.

6.15. Can a major carrier test technique be used to describe a 4-bit unsigned DAC if the output levels beginning with code 0 were found to be the following

0.0064, 0.0616, 0.1271, 0.1812, 0.2467, 0.3206, 0.3856, 0.4406,

0.5021, 0.5716, 0.6364, 0.6880, 0.7662, 0.8262, 0.8871, 0.9480

What if the DAC output levels were described by the following

0.0064, 0.0616, 0.1271, 0.1823, 0.2478, 0.3030, 0.3684, 0.4236, 0.4851, 0.5403, 0.6058, 0.6610, 0.7264, 0.7816, 0.8471, 0.9023

Explain your reasoning.

6.16. The step response of a DAC obtained from an oscilloscope is as follows:

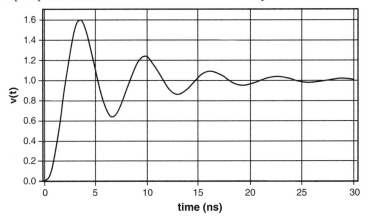

The data sheet states that the settling time is 10 ns (error band = ±50 mV). Does this DAC settle fast enough to meet this specification? Also, determine the overshoot of this signal and its rise time. Estimate the total glitch energy during the positive-going transition.

6.17. Using MATLAB or equivalent software, evaluate the following expression for the step response of a circuit using a time step of no larger than 20 ns

$$v(t) = 1 + \frac{e^{-\zeta \omega_n t}}{\sqrt{1-\zeta^2}} \sin\left(\omega_n t \sqrt{1-\zeta^2} - \cos^{-1}\zeta\right)$$

where $\omega_n = 2\pi \times 100$ MHz and $\zeta = 0.3$. Determine the time for circuit to settle to within 1% of its final value. Determine the rise time.

REFERENCES

1. M. Mahoney, *Tutorial DSP-Based Testing of Analog and Mixed-Signal Circuits*, The Computer Society of the IEEE, 1730 Massachusetts Avenue N.W., Washington D.C. 200361903, 1987, ISBN: 0818607858.

2. G. W. Snedecor and W. G. Cochran, *Statistical Methods*, Eighth Edition, Iowa State University Press, 1989, ISBN: 0813815614, pp. 149–176.

3. G. N. Stenbakken, T. M. Souders, *Linear Error Modeling of Analog and Mixed-Signal Devices*, Proc. International Test Conference, 1991.

4. T. Yamaguchi, M. Soma, *Dynamic Testing of ADCs Using Wavelet Transforms*, Proc. International Test Conference, 1997, pp. 379–88.

CHAPTER 7

ADC Testing

As mentioned in Chapter 6, "DAC Testing," there are many similarities between DAC testing and ADC testing. However, there are also a few notable differences. In this chapter, we will examine their differences as they relate to the intrinsic parameters of an ADC such as DC offset, INL, and DNL. A discussion will then follow about testing the dynamic operation of ADCs.

7.1 ADC TESTING VERSUS DAC TESTING

7.1.1 Comparison of DACs and ADCs

The primary difference between DAC and ADC testing relates to the fundamental difference in their transfer curves. As discussed in Chapter 6, the DAC transfer curve is a one-to-one mapping function, while the ADC transfer curve is a many-to-one mapping functionas captured by the 4-bit DAC and ADC example shown in Figure 7.1. In this chapter, we will see that the ADC curve in Figure 7.1b is one that never occurs in practice. The output codes generated by a real-world ADC are affected by noise from the input circuits. As a result, an ADC curve is statistical in nature rather than deterministic. In other words, for a given input voltage, it may not be possible to predict exactly what output code will be produced. Before we can study testing methods for ADCs, we should first examine the statistical nature of a true ADC transfer curve.

7.1.2 Statistical Behavior of ADCs

To understand the statistical nature of ADCs, we have to model the ADC as a combination of a perfect ADC and a noise source with no DC offset. The noise source represents the combination of the noise portion of the real-world input signal plus the self-generated noise of the ADCs input circuits. Figure 7.2 shows this noisy ADC model.

Applying a DC level to the noisy ADC, we can begin to understand the statistical nature of ADC decision levels. A noise-free ADC might be described by a simple output/input relationship such as

$$\text{output code} = Quantize \text{ (input voltage)} \qquad \textbf{(7.1)}$$

Figure 7.1. Comparing transfer curves. (a) DAC and (b) ADC.

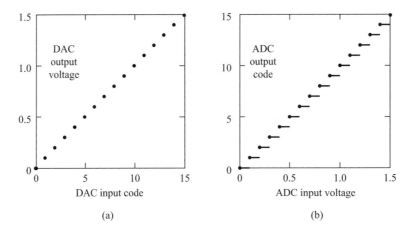

(a)

(b)

Figure 7.2. ADC model including input noise.

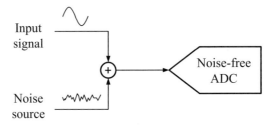

where the function *Quantize* ( ) represents the noise-free ADC's quantization process. The noisy ADC can be described using a similar equation

$$\text{output code} = Quantize \text{ (input voltage + noise voltage)} \qquad (7.2)$$

Now consider the case of a noisy ADC with a DC input voltage. If the DC input level lies exactly between two ADC decision levels, and the noise voltage rarely exceed $\pm\frac{1}{2}\,V_{LSB}$, then the ADC will for the most part produce the same output code. The noise voltage for all practical purposes never gets large enough to push the total voltage across either of the adjacent decision levels. We depict this situation in Figure 7.3 where we described the input signal v_{IN} with a probability density function given by

$$f\left(v_{IN}\right)=\frac{1}{\sigma_n\sqrt{2\pi}}e^{\frac{-\left(v_{IN}-V_{DC}\right)^2}{2\sigma_n^{\,2}}} \qquad (7.3)$$

Here we assume that the noise present at the ADC input is modeled as a Gaussian-distributed random variable with zero mean and a standard deviation of σ_n (i.e., the RMS noise voltage).

On the other hand, if the input DC voltage is exactly equal to a decision level (i.e., $v_{IN} = V_{TH}$) as depicted in Figure 7.4, then even a tiny amount of noise voltage will cause the quantization process to randomly dither between the two codes on each side of the decision level. While we

Figure 7.3. Probability density plot for DC input between two decision levels.

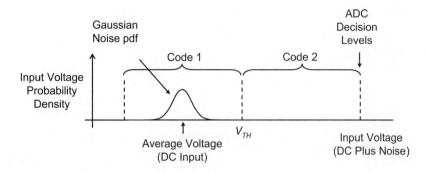

Figure 7.4. Probability density plot for DC input equal to a decision level.

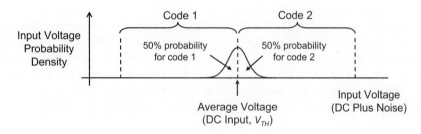

are assuming that the noise is Gaussian distributed, this conclusion will be same regardless of the nature of the noise as long as its pdf is symmetrical about its mean value.[1] Since the area under the pdf is equally split between code 1 and code 2, we would expect 50% of the ADC conversions to produce code 1 and 50% of the conversions to produce code 2.

For input voltages that are close but not equal to the decision levels, the process is little more complicated but tractable using the probability theory from Chapter 4. Consider the DC input V_{DC} as being some value less than the decision level V_{TH} that separates code 1 and code 2, such as situation depicted in Figure 7.5. The probability that the input v_{IN} will trip the ADC quantizer to the next code value is given by

$$P\left(V_{TH} < v_{IN}\right) = \int_{V_{TH}}^{\infty} \frac{1}{\sigma_n \sqrt{2\pi}} e^{\frac{-\left(v_{IN} - V_{DC}\right)^2}{2\sigma_n^2}} dv_{IN} = 1 - \Phi\left(\frac{V_{TH} - V_{DC}}{\sigma_n}\right) \qquad (7.4)$$

where $\Phi(z)$ is the standard Gaussian cumulative distribution function. Likewise, the probability that the input signal will not trip the ADC decision level is

$$P\left(v_{IN} < V_{TH}\right) = \Phi\left(\frac{V_{TH} - V_{DC}}{\sigma_n}\right) \qquad (7.5)$$

We can therefore conclude that N samples of the ADC output will contain $N \times \Phi\left(\Delta V / \sigma_n\right)$ code 1 codes and $N \times \left[1 - \Phi\left(\Delta V / \sigma_n\right)\right]$ code 2 codes where $\Delta V = V_{TH} - V_{DC}$. Of course, this assumes that the noise level is sufficiently small that other codes are not tripped.

Figure 7.5. Probability density plot for DC input input less than the decision level V_{TH}.

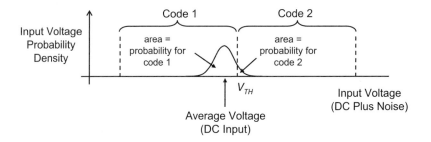

EXAMPLE 7.1

An ADC input is set to 2.453 V DC. The noise of the ADC and DC signal source is characterized to be 10 mV RMS and is assumed to be perfectly Gaussian. The transition between code 134 and 135 occurs at 2.461 V DC for this particular ADC. Therefore, the value 134 is the expected output from the ADC. What is the probability that the ADC will produce code 135 instead of 134? If we collected 200 samples from the output of the ADC, how many would we expect to be 134 and how many would be 135? How might we determine that the transition between code 134 and 135 occurs at 2.461 V DC? How might we characterize the effective RMS input noise?

Solution:
With an input of 2.453 V DC, the ADC's input noise would have to exceed 2.461 V – 2.453 V = 8 mV to cause the ADC to trip to code 135. This value is equal to +0.8σ, since σ = 10 mV. From Appendix A, the Gaussian cdf of +0.8σ is equal to 0.7881. Therefore, there is a 78.81% probability that the noise will *not* be sufficient to trip the ADC to code 135. Thus we can expect 78.81% of the conversions to produce code 134 and 21.19% of the conversions to produce code 135. If we collect 200 samples from the ADC, we would expect 78.81% of the 200 conversions (approximately 158 conversions) to produce code 134. We would expect the remaining 21.19% of the conversions (42 samples) to produce code 135.

To determine the transition voltage, we simply have to adjust the input voltage up or down until 50% of the samples are equal to 134% and 50% are equal to 135. To determine the value of σ, we can adjust the input voltage until we get 84.13% of the conversions to produce code 134. The difference between this voltage and the transition voltage is equal to 1.0σ, which is equal to the effective RMS input noise of the ADC.

Because the circuits of an ADC generate random noise, the ADC decision levels represent *probable* locations of transitions from one code to the next. In the previous example, we saw that an input noise level of 10 mV would cause a 2.453-V DC input voltage to produce code 134 only 79% of the time and code 135 21% of the time. Therefore, with an input voltage of 2.453V, we will get an *average* output code of 134 × 079 + 135 × 0.21 = 134.21. Of course, the ADC cannot produce code 134.21. This value only represents the average output code we can expect if we collect many samples.

If we plot the average output code from a typical ADC versus DC input levels, we will see the true transfer characteristics of the ADC. Figure 7.6 shows a true ADC transfer curve compared

Figure 7.6. ADC probable output code transfer curve.

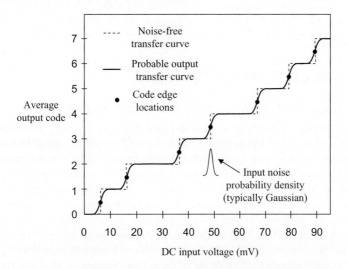

to the idealized, noise-free transfer curve. The center of the transition from one code to the next (i.e., the decision level) is often called a *code edge*. The wider the distribution of the Gaussian input noise, the more rounded the transitions from one code to the next will be. In fact, the true ADC transfer characteristic is equal to the convolution of the Gaussian noise probability density function with the noise-free transfer curve.

Code edge measurement is one of the primary differences between ADC and DAC testing. DAC voltages can simply be measured one at a time using a DC voltmeter or digitizer. By contrast, ADC code edges can only be measured using an iterative process in which the input voltage is adjusted until the output samples dither equally between two codes. Because of the statistical nature of the ADC's transfer curve, each iteration of the search requires 100 or more conversions to achieve a repeatable average value. Since this brute-force approach would lead to very long test times in production, a number of faster methodologies have been developed to locate code edges. Unfortunately, these production techniques generally result in somewhat less exact measurements of code edge voltages.

Exercises

7.1. If V is normally distributed with zero mean and a standard deviation of 2 mV, find $P(V < 4$ mV$)$. Repeat for $P(V > -1$ mV$)$. Repeat for $P(-1$ mV $< V < 4$ mV$)$.

ANS. $P(V < 4$ mV$) = 0.9772$; $P(V > -1$ mV$) = 0.6915$; $P(-1$ mV $< V < 4$ mV$) = 0.6687$

7.2. If V is normally distributed with zero mean and a standard deviation of 100 mV, what is the value of V_{IN} such that $P(V < V_{IN}) = 0.9641$.

ANS. $V_{IN} = 180$ mV.

7.3. An ADC input is set to 1.4 V DC. The noise of the ADC and DC signal source is characterized to be 15 mV RMS and is assumed to be perfectly Gaussian. The transition between code 90 and 91 occurs at 1.4255 V DC. If 500 samples of the ADC output are collected, how many do we expect to be code 90 and how many would be code 91?

ANS. # of code 90 = 95.54% or ~ 478 and # of code 91 = 4.46% (~22).

In the next section, we will examine the various ways in which the code edges of an ADC can be measured, both for characterization and production. Once the code edges have been located, we can apply all the same tests to ADCs that we applied to DACs. Tests such as INL, DNL, DC gain, and DC offset are commonly performed using the code edge information.

7.2 ADC CODE EDGE MEASUREMENTS

7.2.1 Edge Code Testing Versus Center Code Testing

To measure ADC intrinsic parameters such as INL and DNL, we first have to convert the many-to-one transfer curve of the ADC into a one-to-one mapping function similar to that of a DAC. Then we simply apply the same testing methods and criteria from Chapter 6 to the one-to-one transfer curve of the ADC. There are two ways to convert the many-to-one transfer curve of an ADC into a one-to-one curve. These two methods are known as *center code testing* and *edge code testing*. Figure 7.7 illustrates the difference between edge code testing and center code testing. Code centers are defined as the midpoint between the code edges. For example, consider a case in which the decision level between code 57 and code 58 corresponds to an input voltage of 100 mV and the decision level between codes 58 and 59 corresponds an input of 114 mV. In this example, the center of code 58 corresponds to the average of these two voltages, (114 mV + 100 mV)/2 = 107 mV.

Figure 7.7 highlights the problem with center code testing. Notice that the code centers fall very nearly on a straight line, while the code edges show much less linear behavior. The averaging process in the definition of code centers produces an artificially low DNL result compared to edge code testing. Because the code widths in Figure 7.7 alternate between wide and narrow codes, the averaging process effectively smooth's these variations out, leaving a transfer characteristic that looks like it has fairly evenly spaced steps. Because center code testing produces an artificially low DNL value, this technique should be avoided. The edge code method is a more discerning test, and is therefore the preferred means of translating the transfer curve of an ADC to the one-to-one mapping needed for INL and DNL measurements.

Figure 7.7. Code edges and code centers.

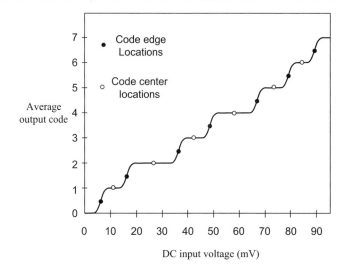

We can search for code edges in one of several different ways. Three common techniques are the step search or binary search method, the hardware servo method, and the histogram method. In the next section, we will see how each of these techniques is applied, and we will examine the strengths and weaknesses of each method. Since all the various ADC edge measurement techniques are slower than simply measuring an output voltage, ADC testing is generally much slower than DAC testing.

7.2.2 Step Search and Binary Search Methods

The most obvious method to find the edge between two ADC codes is to simply adjust the input voltage of the ADC up or down until the output codes are evenly divided between the first code and the second. To achieve repeatable results, we need to collect about 50 to 100 samples from the ADC so that we have a statistically significant number of conversions. The input voltage adjustment could be performed using a simple step search, but a faster method is to use a binary search to quickly find the input voltage corresponding to the ADC code edge. (Step searches and binary searches were discussed in Section 3.11.)

Binary searches are an acceptable production test method for comparators and slicer circuits, which are effectively one-bit ADCs. However, if we try to apply a binary search technique to multibit ADCs in production, we run into a major problem. If we use a binary search with, say, five iterations, we have to collect 100 samples for each iteration. This would result in a total of 500 collected samples *per code edge*. A D-bit ADC has $2^D - 1$ code edges. Therefore, the test time for most ADCs would be far too high. For example, a 10-bit ADC operating at a sampling rate of 100 kHz would require a total data collection time of 500 codes times $2^{10} - 1$ edges times the sample period (1/100 kHz). Thus the total collection time would be $500 \times 1023 \times 10\ \mu s = 5.115\ s$! Clearly, this is not a production-worthy solution.

7.2.3 Servo Method

A much better method for measuring code edges in production is the use of a servo circuit. Figure 7.8 shows a simplified block diagram of an ADC servo measurement setup. The output codes from the ADC are compared against a value programmed into the search value register. If the ADC output is greater than or equal to the expected value, the integrator ramps downward. If it is less than the expected value, the integrator ramps upward.

Figure 7.8 ADC servo test setup.

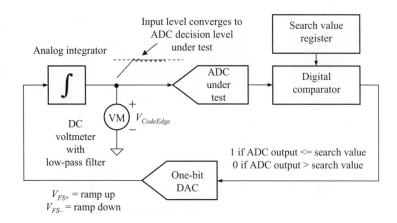

Eventually, the integrator finds the desired code edge and fluctuates back and forth across its transition level. The average voltage at the ADC input, $V_{CodeEdge}$, represents the lower edge of the code under test. This voltage can easily be measured using a DC voltmeter with a low-pass filtered input. The servo search process is repeated for each code edge in the ADC transfer curve.

The servo method is actually a fast hardware version of the step search. Unlike the step search or binary search methods, the servo method does not perform averaging before moving from one input voltage to the next. The continuous up/down adjustment of the servo integrator coupled with the averaging process of the filtered voltmeter act together to remove the effects of the ADC's input noise. Because of its speed, the servo technique is generally more production-worthy than the step search or binary search methods.

Although the servo method is faster than the binary search method, it is also fairly slow compared with a more common production testing technique, the histogram method. Histogram testing requires an input signal with a known voltage distribution. There are two commonly used histogram methods: the linear ramp method and the sinusoidal method.

7.2.4 Linear Ramp Histogram Method

The simplest way to perform a histogram test is to apply a rising or falling linear ramp to the input of the ADC and collect samples from the ADC at a constant sampling rate. The ADC samples are captured as the input ramp slowly moves from one end of the ADC conversion range to the other. The ramp is set to rise or fall slowly enough that each ADC code is "hit" several times, as shown in Figure 7.9. The number of occurrences of each code is directly proportional to the width of the code. In other words, wide codes are hit more often than narrow codes. For example, if the voltage spacing between the upper and lower decision levels for code 2 are twice as wide as the spacing for code 1, then we expect code 2 to occur twice as often as code 1. The reason for this is that it takes the linear ramp input signal twice as long to sweep through code 2 as it takes to sweep through code 1. Of course, this method assumes that the ramp is perfectly linear and that the ADC sampling rate is constant throughout the entire ramp. This condition is easily maintained in mixed-signal ATE testers.

Figure 7.9. ADC samples from linear ramp histogram test.

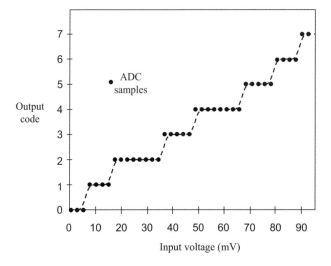

The number of occurrences of each code is plotted as a histogram, as illustrated in Figure 7.10. Ideally, each code should be hit the same number of times, but this would only be true for a perfectly linear ADC. The histogram shows us which codes are hit more often, indicating that they are wider codes. For example, we can see from the histogram in Figure 7.10 that codes 2 and 4 are twice as wide as codes 1 and 6.

Let us denote the number of hits that occur for the ith code word of a D-bit ADC as $H(i)$ for $i = 0,1,\ldots,2^D - 1$. Next, let us define the average number of hits for each code word, excluding the number of hits included in the two end codes, as

$$H_{Average} = \frac{1}{2^D - 2} \sum_{i=1}^{2^D - 2} H(i) \tag{7.6}$$

Dividing $H(i)$ by $H_{Average}$, we obtain the width of each code word in units of LSBs as

$$\text{code width}(i) = \frac{H(i)}{H_{Average}}, \quad i = 1,2,\ldots,2^D - 2 \tag{7.7}$$

Excluding the highest and lowest code count is necessary, because these two codes do not have a defined code width. In effect, the end codes are infinitely wide. For example, code 0 in an unsigned binary ADC has no lower decision level, since there is no code corresponding to −1. In many practical situations, the input ramp signal extends beyond the upper and lower ranges of the

Figure 7.10. LSB normalization translates ADC code histogram into LSB code widths.

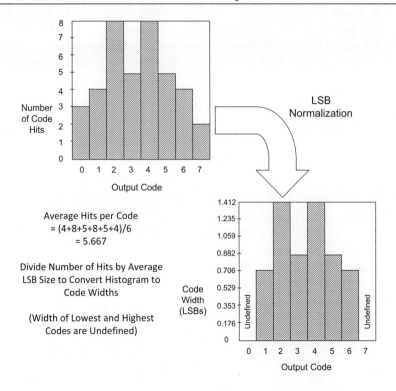

ADC resulting in an increase code count for these two code words. These meaningless hits should be ignored in the linear ramp histogram analysis.

7.2.5 Conversion from Histograms to Code Edge Transfer Curves

To calculate absolute or best-fit INL and DNL curves, we have to determine the absolute voltage for each decision level. Unfortunately, an LSB code width plot such as the one in Figure 7.10 tells us the width of each code in LSBs rather than volts. To convert the code width plot into voltage units, we need to measure the average LSB size of the ADC, in volts. This can be done using a binary search or servo method to find the upper and lower code edge voltages, V_{UE} and V_{LE}. In a D-bit ADC, there are $2^D - 2$ LSBs between these two code edges. Therefore, the average LSB size can be calculated as follows:

$$V_{AveCodeWidth} = V_{LSB} = \frac{V_{UE} - V_{LE}}{2^D - 2} \tag{7.8}$$

The code width plot can then be converted to volts by multiplying each value by the average code width, in volts

$$V_{CodeWidth}(i) = V_{LSB} \times \text{LSB code width}(i) \tag{7.9}$$

The following example will illustrate this approach.

EXAMPLE 7.2

A binary search method is used to find the transition between code 0 and code 1 of the ADC in Figure 7.9. The code edge is found to be 53 mV. A second binary search determines the code edge between codes 6 and 7 to be 2.77 V. What is the average LSB step size for this 3-bit ADC? Based on the data contained in the histogram of Figure 7.10, what is the width of each of the 8 codes, in volts?

Solution:
The average LSB size is equal to

$$V_{LSB} = \frac{2.77 \text{ V} - 0.053 \text{V}}{2^3 - 2} = 452.8 \text{ mV}$$

Therefore, the code width for each code is:

    Code 0: Undefined (infinite width)
    Code 1: 0.706 LSBs × 452.8 mV = 319.68 mV
    Code 2: 1.412 LSBs × 452.8 mV = 639.35 mV
    Code 3: 0.882 LSBs × 452.8 mV = 399.37 mV
    Code 4: 1.412 LSBs × 452.8 mV = 639.35 mV
    Code 5: 0.882 LSBs × 452.8 mV = 399.37 mV
    Code 6: 0.706 LSBs × 452.8 mV = 319.68 mV
    Code 7: Undefined (infinite width)

If we wish to calculate the absolute voltage level of each code edge, we simply perform a running sum on the code widths expressed in volts, starting with the voltage V_{LE}, as follows

$$V_{CodeEdge}(i) = \begin{cases} V_{LE}, & i=0 \\ V_{LE} + \sum_{k=1}^{i} V_{CodeWidth}(k), & i=1,2,\ldots,2^D-2 \end{cases}$$
(7.10)

Alternatively, we can write a recursive equation for the code edges as follows

$$V_{CodeEdge}(i) = V_{CodeEdge}(i-1) + V_{LSB} \times V_{CodeWidth}(i), \quad i=1,2,\ldots,2^D-2$$
(7.11)

where we begin with $V_{CodeEdge}(0) = V_{LE}$. The resulting code edge transfer curve is equivalent to a DAC output transfer curve, except that it will only have $2^D - 1$ values rather than 2D values.

EXAMPLE 7.3

Using the results of Example 7.2, reconstruct the 3-bit ADC transfer curve for each decision level.

Solution:
The transition from code 0 to code 1 was measured using a binary search. It was 53 mV. The other codes edges can be calculated using a running sum:

Code 0 to Code 1: 53 mV
Code 1 to Code 2: 53 mV + 319.68 mV = 372.68 mV
Code 2 to Code 3: 372.68 mV + 639.35 mV = 1011.9 mV
Code 3 to Code 4: 1011.9 mV + 399.37 mV = 1411.5 mV
Code 4 to Code 5: 1411.5 mV + 639.35 mV = 2050.8 mV
Code 5 to Code 6: 2050.8 mV + 399.37 mV = 2450.4 mV
Code 6 to Code 7: 2450.4 mV + 319.68 mV = 2770.0 mV

7.2.6 Accuracy Limitations of Histogram Testing

The accuracy of any code width or edge is inversely proportional to the average number of hits per code. Consider an input ramp to an ADC that extends over the ADC input range, V_{UE}–V_{LE}, with ramp duration T_R. If N_1 samples are collected from the ADC at a sampling rate of F_S over this time duration, then we can write

$$T_R = \frac{N_1}{F_S}$$
(7.12)

Furthermore, if N_1 samples are collected over the ADC input range then each sample represents the response to a voltage change ΔV given by

$$\Delta V = \frac{V_{UE} - V_{LE}}{N_1}$$
(7.13)

7.4. A linear histogram test was performed on an unsigned 4-bit ADC, resulting in the following distribution of code hits beginning with code 0

4, 5, 5, 7, 8, 4, 2, 4, 4, 3, 6, 3, 4, 6, 5 , 9

A binary search was performed on the first transition between codes 0 and 1 and found the code edge to be at 125 mV. A second binary search was performed and found the code edge between codes 14 and 15 to be 3.542 V. What is the average LSB size for this 4-bit ADC? Determine the width of each code, in volts.

ANS. LSB = 224.1 mV; code 0: undefined, code 1: 258.9 mV; code 2: 258.9 mV; code 3: 362.4 mV; code 4: 414.2 mV, code 5: 207.1 mV; code 6: 103.5 mV; code 7: 207.1 mV; code 8: 207.1 mV; code 9: 155.3 mV; code 10: 310.6 mV; code 11: 155.3 mV; code 12: 207.1 mV; code 13: 310.6 mV; code 14: 258.9 mV; code 15: undefined.

7.5. For the distribution of code hits obtained for the 4-bit ADC listed in Exercise 7.4, determine the location of the code edges.

ANS. Beginning with code 0–1 transition: 0.1250 V, 0.3839 V, 0.6427 V, 1.0051 V, 1.4193 V, 1.6264 V, 1.7300 V, 1.9370 V, 2.1441 V, 2.2995 V, 2.6101 V, 2.7654 V, 2.9725 V, 3.2831 V, 3.5420 V.

Conversely, we can also express this voltage step or voltage resolution in terms of the average number of code hits $H_{Average}$ and the LSB step size by combining Eqs. (7.6) and (7.8) with (7.13) to arrive at

$$\Delta V = \frac{V_{LSB}}{H_{Average}} \tag{7.14}$$

By dividing each side of this equation by V_{LSB}, we obtain voltage resolution expressed in LSBs as

$$\frac{\Delta V}{V_{LSB}} = \pm \frac{1}{2} \frac{1}{H_{Average}} \text{ [LSB]} \tag{7.15}$$

For example, if we measure an average of 5 hits per code, then the code width or code edge would, on average, have one-fifth of an LSB of resolution. If one LSB step size is equivalent to 452.8 mV, as in the last example, then the code width and edge would have a possible error of ±45.28 mV. To improve the accuracy of the histogram test, the average number of hits per code must be increased.

To understand how the average number of hits per code can be increased, consider combining Eqs. (7.12), (7.13), and (7.14), together with Eq. (7.8), to arrive at

$$H_{Average} = \frac{F_S}{2^D - 2} \times T_R \tag{7.16}$$

Clearly, a higher average number of hits per code is achieveable by using a longer ramp duration, a higher ADC sampling frequency, or a smaller ADC resolution. The latter two parameters are generally set by the DUT, so the test engineer really has only one option: Run the ramp very slowly.

Figure 7.11. Uncertainty caused by random noise.

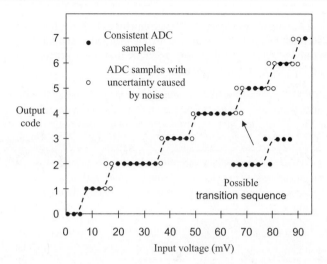

This, in turn, drives up the time of the test. Nonetheless, for characterization this is an acceptable solution. Typically, code hits on the order of several hundreds is selected. The larger sample set also helps to improve the repeatibility of the test.

In production testing, however, we can only afford to collect a relatively small number of samples from each code, typically 16 or 32. Otherwise the test time becomes excessive. Therefore, even a perfect ADC will not produce a flat histogram in production testing because the limited number of samples collected gives rise to a limited code width resolution and repeatability. We can see that the samples in Figure 7.9 are spread too far apart to resolve small fractions of an LSB.

In addition to the accuracy limitation caused by limited resolution, we also face a repeatability limitation.[2] If we look carefully at Figure 7.9, we notice that several of the codes occur so close to a decision level that the ADC noise will cause the results to vary from one test execution to the next. This variability will happen *even if our input signal is exactly the same during each test execution.* Figure 7.11 illustrates the uncertainty in output codes caused by noise in the ADC circuits.

In many cases, we find that the raw data sequence from the ADC may zigzag up and down as the output codes near a transition from one code to the next. In Figure 7.11, for instance, we see that it is possible to achieve an ADC output sequence 4, 4, 4, 4, 4, 5, 4, 5, 5, 5 rather than the ideal sequence 4, 4, 4, 4, 4, 4, 5, 5, 5, 5. Unfortunately, this is the nature of histogram testing of ADCs. The results will be variable and somewhat unrepeatable unless we collect many samples per code. In histogram testing, as in many other tests, there is an inherent tradeoff between good repeatability and low test time. It is the test engineer's responsibility to balance the need for low test time with the need for acceptable accuracy and repeatability.

7.2.7 Rising Ramps Versus Falling Ramps

Most ADC architectures include one or more analog comparators in their design. Since comparators may be subject to hysteresis, we occasionally find a discrepancy between code edges measured using a rising ramp and code edges measured using a falling ramp. The most complete way to test an ADC is to test parameters such as INL and DNL using both a rising ramp and a falling

Exercises

7.6. A 10-bit 5-V ADC operates with a sampling frequency of 1 MHz. If a linear histogram test is to be conducted on this ADC, what should be the minimum duration of the ramp signal so that the average code count is at least 10 hits? 100 hits?   ANS. $T_R >$ 10.22 ms; $T_R >$ 102.2 ms.

7.7. A 10-bit 5-V ADC operates with a sampling frequency of 1 MHz. If a linear histogram test is to be conducted on this ADC with a ramp signal of 15-ms duration, estimate the voltage resolution of this test.   ANS. $\Delta V =$ 332.6 µV.

7.8. A 10-bit 5-V ADC operates with a sampling frequency of 1 MHz. If a linear histogram test is to be conducted on this ADC with a ramp signal of 15-ms duration, how many samples need to be collected for processing?   ANS. $N =$ 15,000.

Figure 7.12. Types of linear histogram inputs.

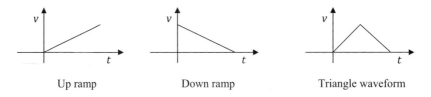

Up ramp   Down ramp   Triangle waveform

ramp. Both methods must produce a passing result before the ADC is considered good. However, the extra test doubles the test time; thus we prefer to use only one ramp. If characterization shows that we have a good match between the rising ramp and falling ramp, then we can drop back to a single test for production. Alternatively, if characterization shows that either the rising ramp or falling ramp always produces the worst-case results, then we can use only the worst-case test condition to save test time.

A compromise solution is to ramp the signal up at twice the normal rate and then ramp it down again (Figure 7.12). This triangle waveform approach tests both the falling and rising edge locations, averaging their results. It takes no longer than a single ramp technique, but it cancels the effects of hysteresis. A separate test could then be performed to verify that the ADC's hysteresis errors are within acceptable limits. The hysteresis test could be performed at only a few codes, saving test time compared to the two-pass ramp solution.

7.2.8 Sinusoidal Histogram Method

Sinusoidal histogram tests were originally used to compensate for the relatively poor linearity of early AWG instruments. Since it is easier to produce a pure sinusoidal waveform than to produce a perfectly linear ramp, early testers often relied on sinusoidal histogram testing for high-resolution ADCs. A second, more common reason to use the sinusoidal histogram method is that it allows better characterization of the dynamic performance of the ADC. The linear histogram technique is basically a static performance test. Because the input voltage is ramped slowly, the input level only changes by a fraction of an LSB from one ADC sample to the next. Sometimes we need to test the ADC transition levels in a more dynamic, real-world situation. To do this, we can use a high-frequency sinusoidal input signal. Our goal is to make the ADC respond to the rapidly changing inputs of a sinusoid rather than the slowly varying voltages of a ramp. In theory, we could use a high-frequency triangle wave to achieve this result, but high-frequency linear triangles are much more difficult to produce than high-frequency sinusoids.

Ramp inputs have an even distribution of voltages over the entire ADC input range. Sinusoids, on the other hand, have an uneven distribution of voltages. A sine wave spends much more time near the upper and lower peak than at the center. As a result, we would expect to get more code hits at the upper and lower codes than at the center of the ADC's transfer curve, even when testing a perfect ADC. Fortunately, the distribution of voltage levels in a pure sinusoid is well defined; thus we can compensate for the uneven distribution of voltages inherent to sinusoidal waveforms.

Figure 7.13 shows a sinusoidal waveform that is quantized by a 4-bit ADC. Notice that there are only 15 decision levels in a 4-bit ADC and that the sine wave is programmed to exceed the upper and lower decision levels by a fairly wide margin. The reason we program the sine wave to exceed the ADC's full-scale range is that we have to make sure that the sine wave passes through all the codes if we want to get a histogram of all code widths. If we expand the time scale to view a quarter period of the waveform, we can see how the distribution of output codes is nonuniform due to the sinusoidal distribution of voltages, as shown in Figure 7.14. Clearly we get more code hits near the peaks of the sine wave than at the center, even for this simple example.

In order to understand the details of this test setup more clearly, consider the illustration shown in Figure 7.15. The diagram consists of three parts. At the center is the transfer characteristic of a 4-bit ADC, that is, 16 output code levels expressed as a function of the ADC input voltage. Below the ADC input voltage axis is a rotated graph of the ADC input voltage as a function of time. Here the input signal is a sine wave of amplitude peak and with DC offset, Offset, all

Figure 7.13. Sinusoidal input for 4-bit ADC.

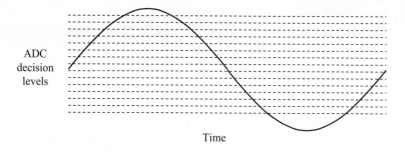

Figure 7.14. Close-up view of 4-bit ADC sinusoidal input.

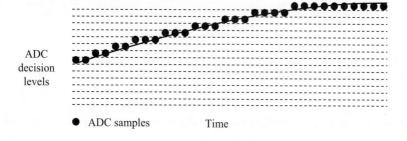

Figure 7.15. Sinusoidal histogram for an ideal ADC.

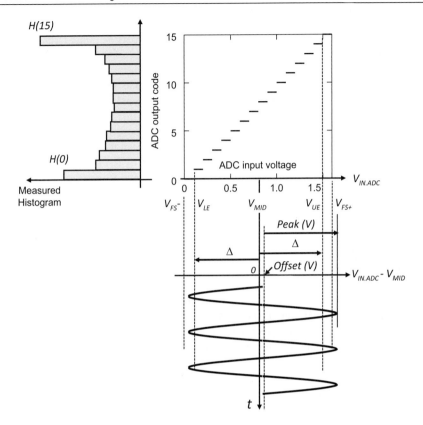

referenced to the middle level of the ADC input as defined by the distance between the upper and lower decision levels (V_{UE} and V_{LE}), that is,

$$V_{MID} = V_{LE} + \Delta \tag{7.17}$$

where

$$\Delta = \frac{V_{UE} - V_{LE}}{2} \tag{7.18}$$

Looking back at the expression for the LSB step size V_{LSB} in Eq. (7.8), we recognize that we can also write Δ as

$$\Delta = \left(2^{D-1} - 1\right)V_{LSB} \tag{7.19}$$

On the left-hand side of the ADC transfer curve, we have plot of a histogram of the ADC code levels (albeit rotated by 90 degrees). Here we see that the histogram exhibits a "bathtub"-like shape. If we try to use this histogram result the same way we use the linear ramp histogram results, the upper and lower codes would appear to be much wider than the middle codes. Clearly, we need to normalize our histogram to remove the effects of the sinusoidal waveform's nonuniform voltage distribution.

The normalization process is slightly complicated because we do not really know what the gain and offset of the ADC will be a priori. Additionally, we may not know the exact offset and amplitude of the sinusoidal input waveform. Fortunately, we have a piece of information at our disposal that tells us the level and offset of the signal as the ADC sees it.

The number of hits at the upper and lower codes in our histogram can be used to calculate the input signal's offset and amplitude. For example, in Figure 7.13, we can see that we will get more hits at the lower code than at the upper code. The lower codes will be hit more often because the sinusoid has a negative offset. The mismatch between these two numbers tells us the offset, while the number of total hits tells us the amplitude. Consider the pdf for the input voltage seen by the ADC is

$$f(v) = \begin{cases} \dfrac{1}{\pi\sqrt{\text{peak}^2 - (v - \text{offset})^2}}, & -\text{peak} + \text{offset} \le v \le \text{peak} + \text{offset} \\ 0, & \text{otherwise} \end{cases} \tag{7.20}$$

The probability that the input signal is less than the lowest code decision level now defined by $-\Delta$(i.e., relative to the ADC mid-level) is given by

$$P(V_{IN,ADC} < -\Delta) = \int_{-\text{Peak}+\text{Offset}}^{-\Delta} \frac{1}{\pi\sqrt{\text{peak}^2 - (v - \text{offset})^2}} \, dv \tag{7.21}$$

which, after some calculus and algebra, reduces to

$$P(V_{IN,ADC} < -\Delta) = \frac{1}{\pi}\left[\sin^{-1}\left(\frac{-\Delta - \text{offset}}{\text{peak}}\right) + \frac{\pi}{2}\right] \tag{7.22}$$

Likewise, the probability that the input signal is larger than the highest code decision level $+\Delta$ is found in a similar manner as

$$P(\Delta < V_{IN,ADC}) = \int_{\Delta}^{\text{peak}+\text{offset}} \frac{1}{\pi\sqrt{\text{peak}^2 - (v - \text{offset})^2}} \, dv = \frac{1}{\pi}\left[\frac{\pi}{2} - \sin^{-1}\left(\frac{\Delta - \text{offset}}{\text{peak}}\right)\right] \tag{7.23}$$

If N samples are collected from the ADC output (including end code counts), then the expected number of code hits for code 0 and code $2^D - 1$ is simply given by the

$$H(0) = N \times P(V_{IN,ADC} < -\Delta) = \frac{N}{\pi}\left[\sin^{-1}\left(\frac{-\Delta - \text{offset}}{\text{peak}}\right) + \frac{\pi}{2}\right]$$

$$H(2^D - 1) = N \times P(\Delta < V_{IN,ADC}) = \frac{N}{\pi}\left[\frac{\pi}{2} - \sin^{-1}\left(\frac{\Delta - \text{offset}}{\text{peak}}\right)\right] \tag{7.24}$$

Here we see we have two equations and two unknowns, which leads to the following solution:

$$\text{offset} = \left(\frac{C_2 - C_1}{C_2 + C_1}\right)\Delta = \left(\frac{C_2 - C_1}{C_2 + C_1}\right)(2^{D-1} - 1)V_{LSB} \tag{7.25}$$

and

$$\text{peak} = \frac{2}{C_2 + C_1}\Delta = \frac{2}{C_2 + C_1}\left(2^{D-1} - 1\right)V_{LSB} \tag{7.26}$$

where

$$C_1 = \cos\left(\pi \frac{H\left(2^D - 1\right)}{N}\right) \quad \text{and} \quad C_2 = \cos\left(\pi \frac{H(0)}{N}\right)$$

We should note that N should be large enough that each ADC code is hit at least 16 times. The common rule of thumb is to collect at least 32 samples for each code in the ADC's transfer curve. For example, an 8-bit converter would require $2^8 \times 32 = 8192$ samples. Of course, some codes will be hit more often than 32 times and some will be hit less often than 32 times due to the curved nature of the sinusoidal input.

Once we know the values of peak and offset, we can calculate the ideal sine wave distribution of code hits, denoted $H_{sinewave}$, that we would expect from a perfectly linear ADC excited by a sinusoid. The equation for the ith code count, once again, excluding the upper and lower code counts, is

$$H_{sinewave}(i) = \frac{N}{\pi}\left[\sin^{-1}\left(\frac{\left(i + 1 - 2^{D-1}\right)V_{LSB} - \text{offset}}{\text{peak}}\right) - \sin^{-1}\left(\frac{\left(i - 2^{D-1}\right)V_{LSB} - \text{offset}}{\text{peak}}\right)\right],$$
$$i = 1, 2, \ldots, 2^D - 2 \tag{7.27}$$

The term $H_{sinewave}(i)$ represent probable numbers of hits per code and are therefore not necessarily integers.[3] The ideal hit counts for each ADC code should therefore be calculated using floating-point calculations.

We obtain the width of the ith code word in units of LSBs by dividing the actual ith code count by $H_{sinewave}(i)$

$$\text{LSB code width}(i) = \frac{H(i)}{H_{sinewave}(i)}, \quad i = 1, 2, \ldots, 2^D - 2 \tag{7.28}$$

Figure 7.16 illustrates the sinusoidal histogram normalization process for an idealized 4-bit ADC. Once we have calculated the normalized histogram, we are ready to convert the code widths into a code edge plot, using the same steps as we used for the linear ramp histogram method.

This example is based on an ideal ADC with equal code widths. Even with this idealized simulation, the normalized histogram does not result in equal code width measurements. This simulated example was based on a sample size of 32 samples per ADC code (16 ADC codes × 32 samples per code = 512 collected samples). As we can see in Figure 7.16, many of the codes were hit fewer than 20 times in this simulation. Like the linear ramp histogram method, the number of hits per code limits the measurement resolution of a sinusoidal histogram. If we had collected hundreds of samples for each code in this 4-bit ADC example, the results would have been much closer to a flat histogram. Also, the repeatability of code width measurements will improve with a larger sample size. Unfortunately, a larger sample size requires a longer test time. Again, we are faced with a tradeoff between low test time and high accuracy. We'll explore this in greater detail shortly. Let us first look at an example.

Figure 7.16. Sinusoidal histogram for an ideal 4-bit ADC (simulated).

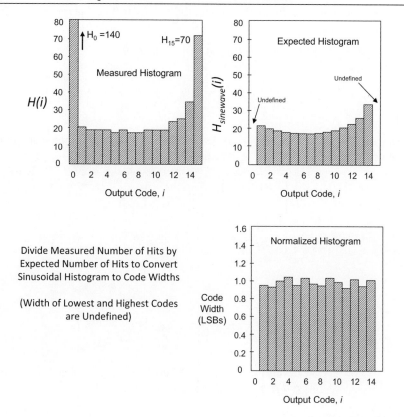

EXAMPLE 7.4

The distribution of code hits for an two's complement 4-bit ADC excited by a sinusoidal signal beginning with code −8 is as follows

$$170, 61, 55, 48, 44, 42, 40, 39, 39, 40, 41, 42, 45, 50, 72, 196$$

A binary search was performed on the first transition between codes −8 and −7 and found the code edge to be at 330 mV. A second binary search was performed and found the code edge between codes 6 and 7 to be 4.561 V. What is the average LSB size for this 4-bit ADC? What is the mid-level of the ADC input? What is the offset and amplitude of the sinusoidal signal seen by the ADC relative to its mid-level? What is expected or ideal sinusoidal distribution of code hits corresponding to this input signal? Determine the width of each code, as well as the code edges (all in volts). Plot the transfer characteristics of this ADC.

Solution:
According to Eq. (7.8) we find the average LSB step size is

$$V_{LSB} = \frac{4.561\,V - 0.330V}{2^4 - 2} = 302.2 \text{ mV}$$

Next, we compute the sinusoidal signal seen by the ADC relative to the input mid-level according to the following

$$V_{MID} = V_{LE} + \Delta = 0.330 + 2.115 = 2.445 \text{ V}$$

where

$$\Delta = \frac{V_{UE} - V_{LE}}{2} = \frac{4.561 - 0.330}{2} = 2.115 \text{ V}$$

The parameters of the sine wave seen by the ADC relative to mid-level is then found from the code hits data such that

$$C_1 = \cos\left(\pi \frac{H(2^D - 1)}{N}\right) = \cos\left(\frac{196}{1024}\pi\right) = 0.8245,$$

$$C_2 = \cos\left(\pi \frac{H(0)}{N}\right) = \cos\left(\frac{170}{1024}\pi\right) = 0.8670$$

leading to

$$peak = \frac{2}{C_2 + C_1}\Delta = \frac{2}{0.8670 + 0.8245} \times 2.115 = 2.5011 \text{ V}$$

$$offset = \left(\frac{C_2 - C_1}{C_2 + C_1}\right)\Delta = \left(\frac{0.8670 - 0.8245}{0.8670 + 0.8245}\right) \times 2.115 = 0.05309 \text{ V}$$

Next, the expected number of code hits for an ideal ADC is found from Eq. (7.27), resulting in the following list of code hits beginning code –7 and ending with code 6:

$$H_{sinewave} = 67.42, 54.33, 47.78, 44.02, 41.67, 40.23, 39.56, 39.44, 39.88, 41.09,$$
$$43.10, 46.25, 51.56, 61.50$$

Subsequently, the width of each code (-7 to 6) expressed in LSBs is found from Eq. (7.28) to be

code width = 0.9048, 1.012, 1.005, 0.9995, 1.008, 0.9943, 0.9858, 0.9888, 1.003, 0.9978, 0.9745, 0.9730, 0.9697, 1.171

The code width of each code (-7 to 6) is scaled by V_{LSB} to obtain the code width in volts:

code width = 0.2734, 0.3058, 0.3037, 0.3020, 0.3046, 0.3005, 0.2979, 0.2988, 0.3031, 0.3015, 0.2945, 0.2940, 0.2930, 0.3539

Finally, the code edges in are found through application of Eq. (7.11) using the above code widths in volts:

code edges = 0.330, 0.6034, 0.9092, 1.213, 1.515, 1.820, 2.120, 2.418, 2.717, 3.020, 3.322, 3.616, 3.910, 4.203, 4.557

A plot of transfer characteristics corresponding to this ADC is shown below:

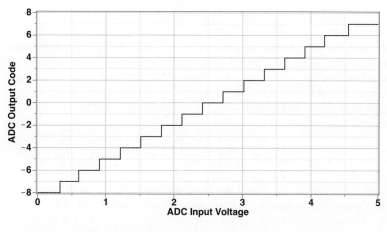

Exercises

7.9. The distribution of code hits for an unsigned 4-bit ADC excited by a sinusoidal signal beginning with code 0 is as follows

$$137, 80, 52, 60, 40, 51, 36, 48, 36, 48, 37, 52,$$
$$42, 64, 80, 160$$

A binary search was performed on the first transition between codes 0 and 1 and found the code edge to be at −4.921 V. A second binary search was performed and found the code edge between codes 14 and 15 to be 4.952 V. What is the offset and amplitude of the input sinusoidal signal? What is expected or ideal sinusoidal distribution of code hits? Finally, what is the distribution of code widths (in volts) for this ADC?

ANS. Offset = 0.0849 V; peak = 5.500 V.

Ideal sinusoidal distribution (code 1 to 14): 80.75, 60.56, 51.95, 47.19, 44.36, 42.70, 41.90, 41.81, 42.43, 43.86, 46.37, 50.55, 57.90, 73.60.

Code widths (code 1 to 14): 0.7003, 0.6060, 0.8159, 0.5980, 0.8103, 0.5948, 0.8082, 0.6070, 0.7983, 0.5949, 0.7912, 0.5864, 0.7800, 0.7694.

To gain a better idea of the distribution of codes and its impact on the performance of the sinusoidal histogram test, consider the minimum and maximum number of code hits corresponding to an ideal D-bit ADC excited by a peak-to-peak sine wave signal equal to the full-scale range of the ADC with frequency F_T as follows:

$$\text{maximum number of code hits} = \frac{1}{\pi}\left(\frac{F_S}{F_T}\right)\left[\frac{\pi}{2} - \sin^{-1}\left(1 - \frac{1}{2^{D-1}}\right)\right]$$

$$\text{minimum number of code hits} = \frac{1}{2\pi}\left(\frac{F_S}{F_T}\right)\sin^{-1}\left(\frac{1}{2^{D-1}}\right)$$

(7.29)

While not explicit in the above equation, one can show that the total number of samples collected from the ADC output when excited by a single cycle of the sine wave input is given by

$$N = \frac{F_S}{F_T} \tag{7.30}$$

Like before, both F_S and D are parameters of the DUT, leaving F_T as the only parameter that the test engineer can use to optimize the test (e.g., minimize test time). The lower the test frequency, F_T, the greater the number of minimum code hits and, in turn, the longer the test time. The resolution of the measurement is bounded by the largest step size that takes place during the sampling process. Each step change in the input signal level can be described by

$$\Delta V[n] = \frac{V_{FSR}}{2}\left\{ \sin\left[2\pi \frac{F_T}{F_S}(n+1) \right] - \sin\left[2\pi \frac{F_T}{F_S} n \right] \right\} \tag{7.31}$$

The largest step change occurs around the zero crossing point of the sine wave, resulting in the sinusoidal historgam method having a worst-case voltage resolution of

$$\max\{\Delta V\} = \frac{V_{FSR}}{2}\sin\left(2\pi \frac{F_T}{F_S} \right) \tag{7.32}$$

If we assume the full-scale range is equivalent to $2DV_{LSB}$, Eq. (7.32) can be rewrtitten as

$$\max\{\Delta V\} = 2^{D-1}V_{LSB}\sin\left(2\pi \frac{F_T}{F_S} \right) \tag{7.33}$$

Furthermore, because F_T is typically much smaller than F_S, we can approximate the worst-case voltage resolution of the sinusoidal histogram test as

$$\max\{\Delta V\} = \pm 2^{D-1}\pi V_{LSB}\frac{F_T}{F_S} \tag{7.34}$$

Clearly, the lower the test frequency, the higher the resolution but longer the test time.

Exercises

7.10. A 10-bit 5-V ADC operates with a sampling frequency of 1 MHz. If a sinusoidal histogram test is to be conducted on this ADC, what should be the maximum frequency of the input signal so that the average minimum code count is at least 10 hits? 100 hits?

ANS. F_T = 1865.1 Hz; F_T = 186.51 Hz.

7.11. A 10-bit 5-V ADC operates with a sampling frequency of 1 MHz. If a sinusoidal histogram test is to be conducted on this ADC with a test time of no more than 15 ms, estimate the worst-case voltage resolution for this test.

ANS. ΔV = ±1.047 mV; ΔV = ±0.1072 LSB.

7.12. A 10-bit 5-V ADC operates with a sampling frequency of 1 MHz. If a sinusoidal histogram test is to be conducted on this ADC with an input frequency of 100 Hz, how many samples need to be collected for processing? How long will it take to collect these samples?

ANS. N = 10,000; test time = 10 ms.

7.3 DC TESTS AND TRANSFER CURVE TESTS

7.31. DC Gain and Offset

Once we have produced a code edge transfer curve for an ADC, we can test the ADC much as we would test a DAC. Since a code edge transfer curve is a one-to-one mapping function, we can apply all the same DC and transfer curve tests outlined in Chapter 6, "DAC Testing." There are a few minor differences to consider. For example, a D-bit ADC has one fewer code edge than an D-bit DAC has outputs. A more important difference is that the ideal ADC transfer curve may be ambiguously defined. The test engineer should realize that there are several ways to define the ideal performance of an ADC.

Figure 7.17 shows two alternate definitions of an 8-bit ADC's ideal performance. The first alternative is to define the ideal location of the first code edge, V_{LE}, at a voltage corresponding to $+\frac{1}{2}$ LSB above the V_{FS-} level. The second alternative is to define the ideal location of the first code edge at a voltage corresponding to $+1$ LSB above V_{FS-}.

When measuring DC offsets and other absolute voltage levels, it is very important that we understand exactly what the ideal transfer curve is supposed to be. Otherwise, we may introduce errors of $\pm\frac{1}{2}$ LSB. Unfortunately, there is little consistency from one ADC data sheet to the next as to the intended ideal performance. This is another issue that the test engineer must clarify before writing the test program.

Once the ideal curve has been established, DC gain and offset can be measured in a manner similar to DAC DC gain and offset. The gain and offset are measured by calculating the slope and offset of the best-fit line. If the converter is defined using the definition illustrated in Figure 7.17a, we have to remember that the ideal line would have an offset of $+\frac{1}{2}$ LSB.

Unfortunately, there are many other ways to define gain and offset. In some data sheets, the offset is defined simply as the offset of the first code edge from its ideal position and the gain is defined as the ratio of the actual voltage range divided by the ideal voltage range from V_{FS-} to V_{FS+}. Other definitions abound; thus, the test engineer is responsible for determining the correct methodology for each ADC to be tested. Of course, ambiguities in the data sheet should be clarified to prevent correlation headaches caused by misunderstandings in data sheet definitions.

Figure 7.17. Alternate definitions of ADC transfer curves.

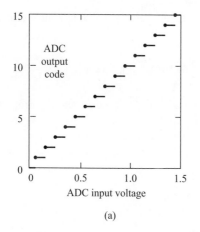

(a)

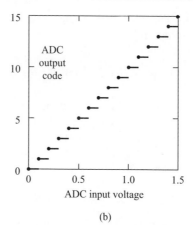

(b)

7.3.2 INL and DNL

Except for the fact that an ADC code edge transfer curve has one fewer value than an equivalent DAC curve, we can calculate ADC INL and DNL exactly the same way as DAC INL and DNL. If we use the histogram method, we can take a shortcut in measuring INL and DNL. Specifically, once the code widths are known, the endpoint DNL expressed in units of LSBs can be determined by subtracting one LSB from each code width as follows

$$\text{endpoint } DNL\,(i) = \text{LSB code width}\,(i) - 1, \quad i = 1, 2, \ldots, 2^D - 2 \tag{7.35}$$

Subsequently, as described in Chapter 6, the DNL curve can then be integrated using a running sum to calculate the endpoint INL curve in units of LSBs according to the following

$$\text{endpoint } INL\,(i) = \begin{cases} 0 & i = 1 \\ \sum\limits_{k=1}^{i-1} \text{Endpoint } DNL\,(k), & i = 2, 3, \ldots, 2^D - 2 \\ 0 & i = 2^D - 1 \end{cases} \tag{7.36}$$

Using this shortcut method, we never even have to compute the absolute voltage level for each code edge, unless we need that information for a separate test, such as gain or offset.

As with DAC INL and DNL testing, a best-fit approach is the preferred method for calculating ADC INL and DNL. As discussed in Chapter 6, "DAC Testing," best-fit INL and DNL testing results in a more meaningful, repeatable reference line than endpoint testing, since the best-fit reference line is less dependent on any individual code's edge location. We can convert an endpoint INL curve to a best-fit INL curve by first calculating the best-fit line for the endpoint INL curve, for example,

$$\text{best-fit endpoint } INL[i] = \text{gain} \times i + \text{offset}, \quad i = 1, \ldots, 2^D - 1$$

Subtracting the best-fit line from the endpoint INL curve yields the best-fit INL curve, that is,

$$\text{best-fit } INL[i] = \text{endpoint } INL[i] - \text{best-fit endpoint } INL[i], \quad i = 1, \ldots, 2^D - 1 \tag{7.37}$$

The best-fit DNL curve is then calculated by taking the discrete time first derivative of the best-fit INL curve according to

$$\text{best-fit } DNL\,(i) = \text{best-fit } INL\,(i) - \text{best-fit } INL\,(i-1), \quad i = 1, 2, \ldots, 2^D - 2 \tag{7.38}$$

Notice that the histogram method captures an endpoint DNL curve and then integrates the DNL curve to calculate endpoint INL. This is unlike the DAC methodology and the ADC servo/search methodologies, which start with a measurement of absolute voltage levels to measure INL and then calculate the DNL through discrete time first derivatives. The following example will illustrate this method.

EXAMPLE 7.5

A linear histogram test was performed on an unsigned 4-bit ADC resulting in the following distribution of code hits beginning with code 0:

$$4, 5, 5, 7, 8, 4, 2, 4, 4, 3, 6, 3, 4, 6, 5, 9$$

Determine the best-fit DNL and INL characteristicsof this ADC.

Solution:

We begin by first finding the endpoint DNL characteristics for this ADC. As the average code hit is 4.714, we find that the code width (in LSBs) beginning with code 1 and ending with code 14 is:

Code Widths:

[0, undefined], [1, 1.061], [2, 1.061], [3, 1.485], [4, 1.697], [5, 0.8485], [6, 0.4243], [7, 0.8485], [8, 0.8485], [9, 0.6364], [10, 1.273], [11, 0.6364], [12, 0.8485], [13, 1.273], [14, 1.061], [15, undefined]

Subsequently, using Eq. (7.35), we find that the endpoint DNL characteristics beginning with the 0 to 1 code transition and ending with the 13th to 14th code transition is:

Endpoint DNL:

[1, 0.061], [2, 0.061], [3, 0.485], [4, 0.697], [5, −0.1515], [6, −0.5757], [7, −0.1515], [8, −0.1515], [9, −0.3636], [10, 0.273], [11, −0.3636], [12, −0.1515], [13, 0.273], [14, 0.061]

Integrating the DNL function according to Eq. (7.36), we find the endpoint INL characteristics beginning with code 1 and ending with code 15 as follows:

Endpoint INL:

[1, 0], [2, 0.061], [3, 0.122], [4, 0.607], [5, 1.304], [6, 1.152], [7, 0.5763], [8, 0.4248], [9, 0.2733], [10, −0.0903], [11, 0.1827], [12, −0.1809], [13, −0.3324], [14, −0.0594], [15, 0]

Using the regression analysis equations of Chapter 6, we find that the gain and offset of the best-fit line parameters associated with the endpoint INL curve is −0.04393 and 0.5769, respectively. The best-fit line corresponding to the endpoint INL is then given by the expression

$$\text{best-fit endpoint } INL[i] = -0.04393 \times i + 0.5769, \quad i = 1,...,15$$

Evaluating this function for n from 1 to 15, and it subtracting from the endpoint INL data set, that is,

$$\text{best-fit } INL[i] = \text{endpoint } INL[i] - \text{best-fit endpoint } INL[i], \quad i = 1,...,15$$

we obtain the following set of best-fit INL points:

Best-Fit INL:

[1, −0.5330], [2, −0.4280], [3, −0.3231], [4, 0.2058], [5, 0.9467], [6, 0.8387], [7, 0.3069], [8, 0.1993], [9, 0.0918], [10, −0.2279], [11, 0.0890], [12, −0.2306], [13, −0.3382], [14, −0.0213], [15, 0.0821]

Below is a plot of three sets of data corresponding to the INL characteristics of the ADC: endpoint INL, the regression line for the endpoint INL, and the corresponding best-fit INL. As is clearly evident, the endpoint INL and best-fit INL are different.

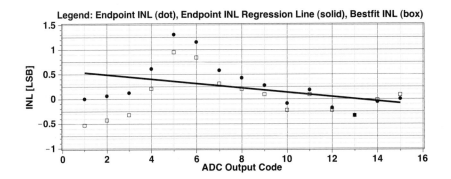

Finally, we compute the best-fit DNL characteristics of the ADC by differentiating the best-fit INL curve using the first-order difference operation given in Eq. (7.38):

Best-Fit DNL:
    [1, 0.1050], [2, 0.1049], [3, 0.5289], [4, 0.7409], [5, –0.1080], [6, –0.5318], [7, –0.1076],
    [8, –0.1075], [9, –0.3197], [10, 0.3169], [11, –0.3196], [12, –0.1076], [13, 0.3169], [14, 0.1034]

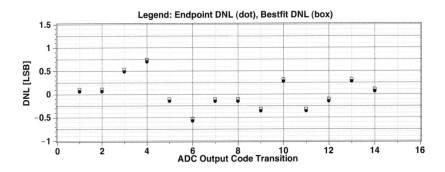

Above, we see from above that the DNL is very similar for both the endpoint and best-fit reference lines.

7.3.3 Monotonicity and Missing Codes

One final difference between ADC testing and DAC testing relates to differences in their weaknesses. For example, a DAC may be nonmonotonic, while an ADC will usually be monotonic *if it is tested statically.* For an ADC to be nonmonotonic, one or more of its code widths has to be

Exercises

7.13. A linear histogram test was performed on an unsigned binary 4-bit ADC resulting in the following distribution of code hits beginning with code 0:

$$4, 5, 5, 7, 8, 7, 9, 5, 6, 3, 6, 7, 9, 6, 5, 9$$

Determine the endpoint DNL curve for this ADC.

ANS. DNL for code transitions from 1 to 14: −0.2045, −0.2045, 0.1136, 0.2727, 0.1136, 0.4318, −0.2045, −0.0455, −0.5227, −0.0455, 0.1136, 0.4318, −0.0455, −0.2045.

7.14. For the code distribution described in Exercise 7.13, determine the endpoint INL curve for this 4-bit ADC.

ANS. INL atcode edge 1 to 15:0, −0.2045, −0.4091, −0.2955, −0.0227, 0.0909, 0.5227, 0.3182, 0.2727, −0.2500, −0.2955, −0.1818, 0.2500, 0.2045, 0.

7.15. Compute the best-fit INL characteristic of the ADC described in Exercise 7.13.

ANS. INL atcode edge 1 to 15: 0.09521, −0.1254, −0.3460, −0.2480, 0.00905, 0.1071, 0.5231, 0.3025, 0.2410, −0.2977, −0.3592, −0.2612, 0.1548, 0.0933, −0.1286.

7.16. Compute the best-fit DNL characteristic of the ADC described in Exercise 7.13.

ANS. DNL for code transitions from 1 to 14: −0.2206, −0.2206, 0.0980, 0.2570, 0.09805, 0.4160, −0.2206, −0.0615, −0.5387, −0.0615, 0.0980, 0.4160, −0.0615, −0.2219.

negative. (One example of this is an ADC whose DC reference voltage is somehow drastically perturbed as the input voltage varies. However, this failure mechanism is quite rare.) Nevertheless, an ADC can *appear* to be nonmonotonic when its input is changing rapidly.[3]

For this reason, we do not typically test ADCs for monotonicity when we use slowly changing inputs (as in search or linear ramp INL and DNL tests). However, when testing ADCs with rapidly changing inputs, the ADC may behave as if it were nonmonotonic due to slew rate limitations in its comparator(s). These monotonicity errors show up as signal-to-noise ratio failures in some ADCs and as sparkling in others. (Sparkling is a dynamic failure mode discussed in Section 7.4.3.)

Unlike DACs, ADCs are often tested for missing codes. A missing code is one whose voltage width is zero. This means that the missing code can never be hit, regardless of the ADC's input voltage. A missing code appears as a missing step on an ADC transfer curve, as illustrated in Figure 7.18. Since DACs always produce a voltage for each input code, DACs cannot have missing codes. Although a true missing code is one that has zero width, missing codes are often defined as any code having a code width smaller than some specified value, such as 1/10 LSB. Technically, a code having a width of 1/10 LSB is not missing, but the chances of it being hit are low enough that it is considered to be missing from the ADC transfer curve.

Figure 7.18. (a) Monotonicity errors in DACs and (b) missing codes in ADCs.

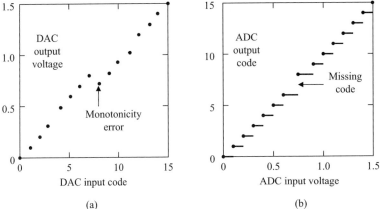

(a)   (b)

7.4 DYNAMIC ADC TESTS

7.4.1 Conversion Time, Recovery Time, and Sampling Frequency

DACs have many dynamic tests such as settling time, rise and fall time, overshoot, and under-shoot. ADCs do not exhibit these same features, since they do not have an analog output. Instead, an ADC may have any or all the following timing specifications: maximum sampling frequency, maximum conversion time, and minimum recovery time. There are many ways to design ADCs and ADC digital interfaces. Let us look at a few of the common interfacing strategies.

One common interface scheme is shown in Figure 7.19. The ADC begins a conversion cycle when the CONVERT signal is asserted high. After the conversion cycle is completed, the ADC asserts a DATA_READY signal that indicates the conversion is complete. Then the data are read from the ADC using a READ signal.

Maximum conversion time is the maximum amount of time it takes an ADC to produce a digital output after the CONVERT signal is asserted. The ADC is guaranteed to produce a valid output within the maximum conversion time. It is tempting to say that an ADC's maximum sampling frequency is simply the inverse of the maximum conversion time. In many cases this is true. Some ADCs require a minimum recovery time, which is the minimum amount of time the system must wait before asserting the next CONVERT signal. The maximum sampling frequency is therefore given by the equation

$$F_{max} = \frac{1}{T_{convert} + T_{recovery}} \tag{7.39}$$

We typically test $T_{convert}$ by measuring the period of time from the CONVERT signal's active edge to the DATA_READY signal's active edge. We have to verify that the $T_{convert}$ time is less than or equal to the maximum conversion time specification. For this measurement, we can use a time measurement system (TMS) instrument, or we can sometimes use the tester's digital pattern compare function if we can tolerate a less accurate pass/fail test. We can verify the F_{max} specification (and thus the $T_{recovery}$ specification) by simply operating the converter at its maximum sampling rate, F_{max}, and verifying that it passes all its dynamic performance specifications at this frequency.

Figure 7.19. ADC sample timing.

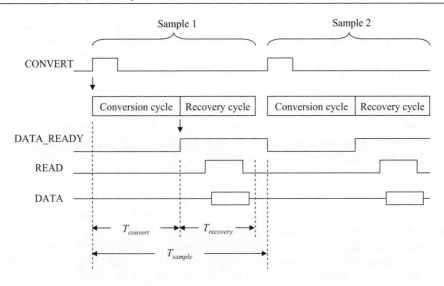

Figure 7.20. ADC conversion cycles with internally generated CONVERT signal.

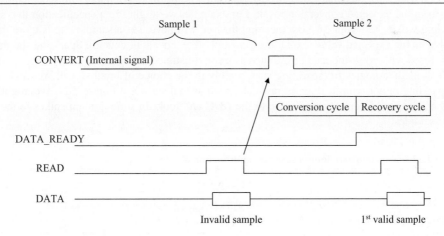

In many ADC designs, the CONVERT signal is generated automatically after the ADC output data is read, as shown in Figure 7.20. This type of converter requires no externally supplied CONVERT signal. The first sample read from the ADC must therefore be discarded, since no conversion is performed until after the first READ pulse initiates the first conversion cycle.

Sometimes ADCs simply perform continuous conversions at a constant sampling rate. The CONVERT signal is generated at a fixed frequency derived from the device master clock. This architecture is very common in ADC channels such as those in a cellular telephone voice band interface or multimedia audio device. The continuous conversions can usually be disabled by a register bit or other control mechanism to minimize power consumption when conversions are not needed. These devices sometimes generate a DATA_READY signal that must be used to synchronize the tester with an asynchronous data stream. DUT-defined timing can be a difficult situation to deal with, since ATE testers are not designed to operate in a slave mode with the DUT driving digital timing.

Clearly, there are many ways to design ADCs. The test engineer has to deal with many different permutations of interfacing possibilities, each with its own testing requirements.

7.4.2 Aperture Jitter

Noise associated with sampling clock can introduce error into the data conversion process. Typically, clock noise or what is also referred to as aperture jitter is guaranteed by acceptable signal-to-noise ratio (SNR) performance. It may or may not be tested in production, depending on the required sampling rate of the ADC. Very high-frequency ADCs typically must be tested for aperture jitter in production.

7.4.3 Sparkling

Sparkling is a phenomenon that happens most often in high-speed flash converters, such as those described ahead in Section 7.5.3, due to digital timing race conditions. It is the tendency for an ADC to occasionally produce a conversion that has a larger than expected offset from the expected value. We can think of a sparkle sample as one that is a statistical outlier from the Gaussian distribution in Figure 7.6. Sparkling shows up in a time-domain plot as sudden variations from the expected values. It got its name from early flash ADC applications, in which the sample outliers produced white sparkles on a video display. Sparkling is specified as a maximum acceptable deviation from the expected conversion result. For example, we might see a specification that states sparkling will be less than 2 LSBs, meaning that we will never see a sample that is more than 2 LSBs from the expected value (excluding gain and offset errors, of course). Sparkling should not be confused with noise-induced errors such as those illustrated in Figure 7.11.

Test methodologies for sparkling vary, mainly in the choice of input signal. We might look for sparkling in our ramp histogram raw data, such as that shown in Figure 7.21. We might also apply a very high-frequency sine wave to the ADC and look for time-domain spikes in the collected samples.

Figure 7.21. Sparkling in a linear ramp histogram sample set.

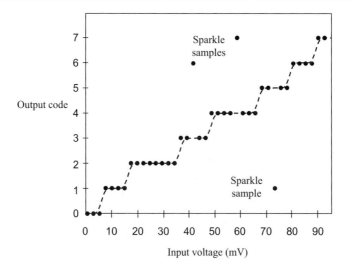

Since it is a random digital failure process, sparkling often produces intermittent test results. Sparkling is generally caused by a weakness in the ADC design that must be eliminated through good design margin rather than being screened out by exhaustive testing. Nevertheless, ADC sparkling tests are often added to a test program as a quick sanity check, making use of samples collected for one of the required parametric tests.

7.5 TESTS FOR COMMON ADC APPLICATIONS

7.5.1 DC Measurements

Like DACs, ADCs can be used for a variety of purposes. The ADC's application often determines its required parameters. For example, an ADC may be used to measure absolute voltage levels, as in a DC voltmeter or battery monitor. In this type of application, we do not usually care about transmission parameters like signal-to-noise ratio. We will typically only need to know the DC gain, DC offset, INL, DNL, and worst-case absolute voltage errors in decision levels, relative to the ideal decision levels. Idle channel noise will sometimes be specified, to ensure that results obtained from the ADC are not unrepeatable due to excessive noise.

Successive approximation ADCs and integrating ADCs are the most common converter type used for DC measurements. Sigma-delta designs[4,5] are seldom used due to their inherent tendency to produce self-tones with certain DC inputs.

7.5.2 Audio Digitization

Audio digitization is a very common application for ADCs, especially high-resolution ADCs. When the resolution exceeds 12 or 13 bits, it becomes very expensive to perform transfer curve tests such as INL and DNL because of the large number of code edges that must be measured. Fortunately, transmission parameters such as frequency response, signal to distortion ratio, idle channel noise, and so on, are more meaningful measures of audio digitizer performance. These sampled channel tests are much less time-consuming to measure than INL and DNL, especially when testing ADCs with 16 or more bits of resolution. Sigma–delta ADCs[5] have become the most common architecture for audio digitization application.

As previously mentioned, self-tones are a potential source of trouble when sigma-delta ADCs are used in audio digitization applications. Because of the way the human mind processes sound, very low-amplitude self-tones are much easier to hear than white noise at equivalent signal levels. It is impractical to test self-tones at every possible DC input level. Self-tones should at least be tested with the analog input tied to ground or $V_{DD}/2$ (or whatever voltage represents the converter's midscale input level). When characterization indicates that a particular ADC design is not prone to self-tone generation, then this test is often eliminated in production.

7.5.3 Data Transmission

Data transmission applications differ from audio applications mainly in terms of the sampling rates and the frequency range of the transmitted signals. Data transmission ADCs, such as those found in modems, hard disk drive read channels, and cellular telephone intermediate frequency (IF) sections, often digitize signals that are well above the audio band. These applications typically require lower-resolution ADCs, but may require much higher sampling rates. Aperture jitter is often a prime concern for these applications, especially if the signal frequency band extends past a few tens of megahertz. Excessive aperture jitter can introduce apparent noise in the digitized signal, ruining the performance of the ADC.

Signal-to-noise ratio, group delay distortion, and other transmission parameters are often specified in data transmission applications. Also, data transmission specifications such as error vector magnitude (EVM), phase trajectory error (PTE), and bit error rate (BER) may also need to be tested. Some of these parameters are so numerous that we cannot possibly cover them in this book. The test engineer will have to learn about these and other application-specific testing requirements by studying the relevant standards documents. ATE vendors can also be a tremendous source of expertise when learning about new testing requirements and methodologies.

Most ADC architectures are well suited for low-frequency data transmission applications (with the exception of integrating converters). High-frequency applications may require fast successive approximation ADCs, semiflash ADCs, or even full-flash ADCs, depending on the required sampling rates.

7.5.4 Video Digitization

NTSC video signal digitization is another key application for high-speed ADCs. These applications require the faster ADC types (flash, semiflash, or pipelined successive approximation ADCs). The test list for these types of converters usually includes transmission parameters as well as differential gain and differential phase measurements. Like other high-speed applications, aperture jitter is a key performance specification for video digitization applications. Sparkling is particularly noticeable in video applications, thus, this potential weakness should be thoroughly characterized and/or tested in production.

7.6 SUMMARY

ADC testing is very closely related to DAC testing. Many of the DC and intrinsic tests defined in this chapter are very similar to those performed on DACs. The most important difference is that the ADC code edge transfer curve is harder and much more time-consuming to measure than the DAC transfer curve. However, once the many-to-one statistical mapping of an ADC has been converted to a one-to-one code edge transfer curve, the DC and transfer curve tests are very similar in nature to those encountered in DAC testing. This chapter by no means represents an exhaustive list of all possible ADC types and testing methodologies. There is a seemingly endless variety of ADC architectures and methods for defining their performance. Hopefully, this chapter will provide a solid starting point for the beginning test engineer.

PROBLEMS

7.1. If V is normally distributed with zero mean and a standard deviation of 50 mV, find $P(V < 40$ mV$)$. Repeat for $P(V > 10$ mV$)$. Repeat for $P(-10$ mV $< V < 40$ mV$)$.

7.2. If V is normally distributed with mean 10 mV and standard deviation 50 mV, find $P(V < 40$ mV$)$. Repeat for $P(V > 10$ mV$)$. Repeat for $P(-10$ mV $< V < 40$ mV$)$.

7.3. If V is normally distributed with zero mean and standard deviation 200 mV, what is the value of ΔV such that $P(V < \Delta V) = 0.6$?

7.4. An ADC input is set to 3.340 V DC. The noise of the ADC and DC signal source is characterized to be 15 mV RMS and is assumed to be perfectly Gaussian. The transition between code 324 and 325 occurs at 3.350 V DC for this particular ADC; therefore the value 324 is the expected output from the ADC. What is the probability that the ADC will produce code 325 instead of 324? If we collected 400 samples from the output of the ADC, how many would we expect to be code 324 and how many would be code 325?

7.5. An ADC input is set to 1.000 V DC. The transition between code 65 and 66 occurs at 1.025 V DC for this particular ADC. If 200 samples of the ADC output are collected and 176 of them are code 65 and the remaining code 66, what is the RMS value of the noise at the input of this particular ADC?

7.6. An ADC input is set to 2.000 V DC. The noise of the ADC and DC signal source is characterized to be 10 mV RMS and is assumed to be perfectly Gaussian. The transition between code 115 and 116 occurs at 1.990 V DC and the transition between code 116 and 117 occurs at 2.005 V DC for this particular ADC. If 500 samples of the ADC output are collected, how many do we expect to be code 115, code 116, and code 117?

7.7. A linear histogram test was performed on an unsigned binary 3-bit ADC, resulting in the following distribution of code hits beginning with code 0:

$$5, 6, 4, 6, 7, 7, 5, 6$$

A binary search was performed on the first transition between codes 0 and 1 and found the code edge to be at 10 mV. A second binary search was performed and found the code edge between codes 6 and 7 to be 1.25 V. What is the average LSB size for this 3-bit ADC? Determine the width of each code, in volts. Also, determine the location of the code edges. Plot the transfer curve for this ADC.

7.8. A linear histogram test was performed on a two's complementary 4-bit ADC resulting in the following distribution of code hits beginning with code −8:

$$12, 15, 13, 12, 10, 12, 12, 14, 14, 13, 15, 19, 16, 14, 20, 19$$

A binary search was performed on the first transition between codes −8 and −7 and found the code edge to be at 75 mV. A second binary search was performed and found the code edge between codes 6 and 7 to be 4.56 V. What is the average LSB size for this 4-bit ADC? Determine the width of each code, in volts. Also, determine the location of the code edges. Plot the transfer curve for this ADC.

7.9. A linear histogram test was performed on an unsigned binary 3-bit ADC, resulting in the following distribution of code hits beginning with code 0:

$$6, 6, 5, 6, 4, 6, 5, 6$$

A binary search was performed on the first transition between codes 0 and 1 and found the code edge to be at 32 mV. A second binary search was performed and found the code edge between codes 6 and 7 to be 3.125 V. What is the average LSB size for this 3-bit ADC? What is the measurement accuracy of this test, in volts?

7.10. A 12-bit ADC operates with a sampling frequency of 25 MHz. If a linear histogram test is to be conducted on this ADC, what should be the minimum duration of the ramp signal so that the average code count is at least 6 hits? What about for 100 hits?

7.11. A 10-bit ADC with a 10-V full-scale range operates with a sampling frequency of 60 MHz. If a linear histogram test is to be conducted on this ADC with a ramp signal of 100-μs duration, estimate the voltage resolution of this test. How many samples need to be collected for this test?

7.12. A sinusoidal histogram test was performed on an unsigned binary 4-bit ADC, resulting in the following distribution of code hits beginning with code 0:

$$137, 81, 60, 52, 47, 44, 43, 42, 42, 42, 44, 46, 50, 57, 72, 166$$

A binary search was performed on the first transition between codes 0 and 1 and found the code edge to be at 14 mV. A second binary search was performed and found the code edge between codes 14 and 15 to be 0.95 V. What is the average LSB size for this 4-bit ADC? Determine the width of each code, in volts. Also, determine the location of the code edges. Plot the transfer curve for this ADC.

7.13. A sinusoidal histogram test was performed on an two's complementary binary 4-bit ADC resulting in the following distribution of code hits beginning with code −8:

$$251, 163, 104, 118, 80, 99, 71, 93, 70, 94, 72, 101, 82, 124, 163, 315$$

A binary search was performed on the first transition between codes −8 and −7 and found the code edge to be at 20 mV. A second binary search was performed and found the code edge between codes 6 and 7 to be 9.94 V. What is the average LSB size for this 4-bit ADC? Determine the width of each code, in volts. Also, determine the location of the code edges. Plot the transfer curve for this ADC.

7.14. A sinusoidal histogram test was performed on an unsigned binary 4-bit ADC, resulting in the following distribution of code hits beginning with code 0:

$$137, 81, 60, 52, 47, 44, 43, 42, 42, 42, 44, 46, 50, 57, 72, 166$$

A binary search was performed on the first transition between codes 0 and 1 and found the code edge to be at 14 mV. A second binary search was performed and found the code edge between codes 14 and 15 to be 0.95 V. What was the input DC bias level relative to $V_{SS} = 0$ V used to perform this test? What is the minimum and maximum input signal level applied to the ADC relative to VSS?

7.15. A sinusoidal histogram test was performed on an two's complementary binary 4-bit ADC resulting in the following distribution of code hits beginning with code −8:

$$251, 163, 104, 118, 80, 99, 71, 93, 70, 94, 72, 101, 82, 124, 163, 315$$

A binary search was performed on the first transition between codes −8 and −7 and found the code edge to be at 20 mV. A second binary search was performed and found the code edge between codes 6 and 7 to be 9.94 V. What was the input DC bias level relative to $V_{SS} = 0$ V used to perform this test? What is the minimum and maximum input signal level applied to the ADC relative to V_{SS}?

7.16. A 12-bit ADC operates with a sampling frequency of 25 MHz. If a sinusoidal histogram test is to be conducted on this ADC, what should be the maximum frequency of the input signal so that the average minimum code count is at least 10 hits? 100 hits? How many samples will be collected? How long will it take to collect these samples?

7.17. A 10-bit 5-V ADC operates with a sampling frequency of 25 MHz. If a sinusoidal histogram test is to be conducted on this ADC with a test time of no more than 15 ms, estimate the worst-case voltage resolution for this test. How many samples need to be collected for this test?

7.18. A linear histogram test was performed on a two's complementary 4-bit ADC, resulting in the following distribution of code hits beginning with code −8:

$$20, 15, 14, 12, 11, 12, 12, 14, 14, 13, 15, 16, 16, 14, 20, 23$$

Determine the endpoint DNL and INL curves for this ADC. Compare these results to those obtained with a best-fit reference line.

7.19. Determine the endpoint DNL and INL curves for the histogram data provided in Problem 7.8. Compare these results to those obtained with a best-fit reference line.

7.20. Determine the endpoint DNL and INL curves for the histogram data provided in Problem 7.9. Compare these results to those obtained with a best-fit reference line.

REFERENCES

1. M. J. Kiemele, S. R. Schmidt, and R. J. Berdine, *Basic Statistics, Tools for Continuous Improvement*, 4th edition, Air Academy Press, Colorado Springs, CO, pp. 9–71, 1997, ISBN 1880156067.

2. M. Mahoney, *Tutorial DSP-Based Testing of Analog and Mixed-Signal Circuits*, The Computer Society of the IEEE, Washington, D.C., p. 137, 1987, ISBN 0818607858.

3. Reference 2, pp. 147–154.

4. J. C. Candy and G. C. Temes, *Oversampling Delta-Sigma Data Converters: Theory, Design, and Simulation*, IEEE Press, New York, 1992, ISBN 0879422858.

5. S. R. Norsworthy, R. Schreier and G. C. Temes, *Delta-Sigma Data Converters: Theory, Design, and Simulation*, IEEE Press, New York, 1996, ISBN 0780310454.

CHAPTER 8

Sampling Theory

U p to this point in the textbook, we have been discussing DC and other very low-frequency measurements of electronic devices. It is at this juncture that we will turn our attention to the testing of electronic devices with AC signals using digital signal processing (DSP) techniques. Such testing techniques have profoundly changed the design and use of automatic test equipment (ATE) found in most, if not all, production facilities. At the heart of these methods is the principle of sampling. While sampling has been known since the late 1920s in its present form, its widespread use in the design community came into being in the early 1970s with advancements in monolithic circuit integration techniques. Today, most electronic equipment, such as cell phones, personal handheld devices, computers, and so on, make use of these same design principles owing to its flexibility, programmability, testability, and cost.

In this chapter we describe the theory of sampling and a related concept of reconstruction. Much of the mathematical notation that we use for the remainder of this textbook will be defined here. We shall discuss the sampling theorem and the effects of aliasing. We shall also describe the impact of quantization effects and clock jitter (noise) on the overall conversion process. We will then turn our attention to the creation of repetitive analog signals to be used as the test stimulus and the need for a coherent sample set. Signal coherence is one of the most important system conditions for a fast and accurate DSP-based testing scheme. The chapter closes with a short discussion of clock synchronization across different sampling systems.

8.1 ANALOG MEASUREMENTS USING DSP

8.1.1 Traditional Versus DSP-Based Testing of AC Parameters

AC measurements such as gain and frequency response can be measured with relatively simple analog instrumentation, as mentioned in Section 2.4. To measure gain, an AC continuous sine wave generator can be programmed to source a single tone at a desired voltage level, V_{in}, and at a desired frequency. A true RMS voltmeter can then measure the output response from the DUT, V_{out}. Then gain can be calculated using a simple formula: gain $= V_{out}/V_{in}$.

The pure analog approach to AC testing suffers from a few problems, however. First, it is relatively slow when AC parameters must be tested at multiple frequencies. For example, each frequency in a frequency response test must be measured separately, leading to a lengthy testing process. Second, traditional analog instrumentation is unable to measure distortion in the presence of the fundamental tone. Thus the fundamental tone must be removed with a notch filter, adding to test hardware complexity. Third, analog testing measures RMS noise along with RMS signal, making results unrepeatable unless we apply averaging or bandpass filtering.

In the early 1980s, a new approach to production testing of AC parameters was widely adopted in the ATE industry. The new approach became known as *DSP-based testing*.[1] Digital signal processing (DSP) is a powerful methodology that allows faster, more accurate, more repeatable measurements than traditional AC measurements using an RMS voltmeter. A mixed-signal test engineer will never be fully competent without a strong background in signal processing theory. Unfortunately, a full treatment of sampling theory and DSP is well beyond the scope of this book. Other texts have covered the subject of signal processing in much more detail.[2–4]

The reader is assumed to already have a strong theoretical background in DSP, although this book will undoubtedly fall into the hands of the DSP novice as well. We will review the basics of sampling theory and DSP as they apply to mixed-signal testing, without giving the subject an in-depth treatment. Hopefully, this introductory coverage will both refresh the experienced reader's memory of DSP and allow the novice to understand the fundamentals of DSP-based testing.

Before we can discuss DSP-based testing, we must first understand sampling theory for both analog-to-digital converters and digital-to-analog converters. In this chapter, we will examine the basics of sampling theory before proceeding to a more detailed study of DSP-based testing in Chapter 9.

8.2 SAMPLING AND RECONSTRUCTION

8.2.1 Use of Sampling and Reconstruction in Mixed-Signal Testing

Sampling and reconstruction are the processes by which signals are converted from the continuous (i.e., analog) signal domain to the discrete (i.e., digital) signal domain and back again. Both sampling and reconstruction are used extensively in mixed-signal testing. The ATE tester samples and reconstructs signals to stimulate the DUT and measure its response. The DUT may also sample and reconstruct signals as part of its normal operation. Both mathematical and physical sampling and reconstruction occur as the DUT is tested. Figure 8.1 illustrates the various types of sampling and reconstruction that occur when the voice-band interface circuit of Figure 1.2 is tested.

In a purely mathematical world, a continuous waveform can be sampled and then reconstructed without loss of signal quality, as long as a few constraints are met. Unfortunately, a number of imperfections are introduced in the physical world that makes the conversion between continuous time and discrete time fall short of the mathematical theory. Many of these imperfections will be discussed in this section.

8.2.2 Sampling: Continuous-Time and Discrete-Time Representation

Many signals in the physical world around us are continuous (i.e., analog) in nature. Familiar examples of real-world analog signals include sound waves, light intensity, temperature, and pressure. Many modern electronic systems, such as the cellular telephone example in Chapter 1, must convert the continuous signals in the physical world into discrete digital representations

Figure 8.1. Various test signals associated with a voice-band interface circuit.

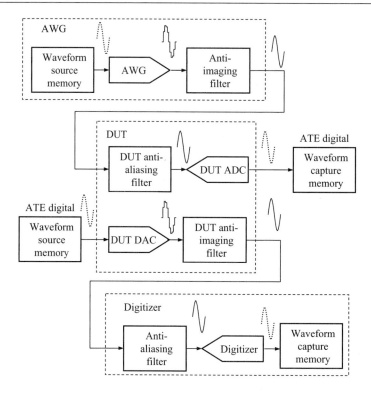

compatible with digital storage, digital transmission, and mathematical processing. Continuous signals are often described by mathematical equations, such as

$$v(t) = A \, \sin\left(2\pi f_o t + \phi\right)$$

(8.1)

where $v(t)$ is a continuous function of time t, whose value in this particular case changes in a sinusoidal manner with amplitude A, frequency f_o, and phase shift ϕ.

Sampling is a process in which a continuous-time signal is converted into a sequence of discrete samples uniformly spaced at intervals of T_s seconds, often written as

$$v[n] = v(t)\big|_{t = nT_s}$$

(8.2)

where $v[n]$ defines the values of $v(t)$ at the sampling instants defined at $t = nT_s$. Such a process is depicted in Figure 8.2. We refer to T_s as the *sampling period* and its reciprocal $F_s = 1/T_s$, as the *sampling frequency* or *sampling rate,* and n as an arbitrary integer. To simplify our notation, it is common practice to drop the T_s term in the argument of Eq. (8.2) because it is assumed to be constant for all time. The continuous waveform $v(t)$ is said to exist in continuous time, while the sampled waveform $v[n]$ is said to exist in discrete time. r example, substituting Eq. (8.1) into (8.2), we can write

$$v[n] = A \, \sin\left(2\pi f_o nT_s + \phi\right) = A \, \sin\left(2\pi \frac{f_o}{F_s} n + \phi\right)$$

(8.3)

Figure 8.2. Continuous-time signal and its sampled equivalent.

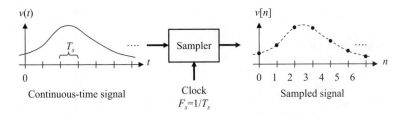

For reasons that will become clear later in this chapter, we often impose the condition that the ratio f_o/F_s be a rational fraction, $f_o/F_s = M/N$, where M and N are integers, allowing one to write

$$v[n] = A \sin\left(2\pi \frac{M}{N} n + \phi\right) \tag{8.4}$$

Discrete signals such as this can then be stored in computer arrays and processed using DSP functions.

Up to this point we have defined a sampled waveform in the discrete-time domain as a sequence of numbers defined by $v[n]$. We can also define a sampled waveform as a continuous function of time. The use of this alternative notation is important in the next section where the samples are converted back into the original continuous-time signal. To enable such a description, we must make use of the concept of impulse functions. Mathematically, an impulse function, denoted by $\delta(t)$, is defined as having zero amplitude everywhere except at $t = 0$, where it is infinitely large in such a way that it contains unit area under its curve, as depicted by the following two rules

$$\delta(t) = 0, \; t \neq 0 \tag{8.5}$$

and

$$\int_{-\infty}^{\infty} \delta(t) \, dt = 1 \tag{8.6}$$

It is important to realize that no function in the ordinary sense can satisfy these two rules. However, we can imagine a sequence of pulse like functions that have progressively taller and thinner peaks, with the area under the curve remaining equal to unity as illustrated inFigure 8.3a. If we take this argument to the limit, letting the pulse width go to zero while the pulse height goes to infinity, then we have what we refer to as an impulse function. It should be obvious from this description that we are going to encounter some difficulty in graphing the impulse function. Hence, an impulse is graphically represented by an arrow whose height is equal to the area (voltage × time) under the impulse, as shown in Figure 8.3b.

An important property of impulse functions is the so-called sifting property, defined by

$$\int_{-\infty}^{\infty} v(t) \, \delta(t - t_0) \, dt = v(t_0) \tag{8.7}$$

Here the impulse function selects or sifts out a particular value of the function $v(t)$, namely, the value at $t = t_0$, in the integration process. If $v_a(t)$ denotes a signal that has been uniformly sampled every T_s seconds, then we can make use of the sifting property and write the following mathematical representation for $v_a(t)$ in terms of a series of evenly spaced, equally sized

Figure 8.3. Impulse definition.

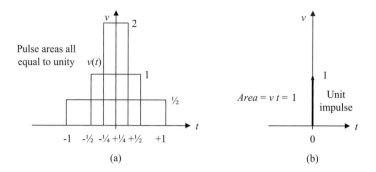

(a)

(b)

Figure 8.4. Continuous-time representation of a sampled signal as a series of impulses created by multiplying the original continuous-time signal by a unit impulse train.

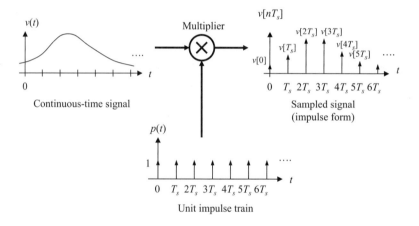

impulse functions, commonly referred to as a *unit impulse train*:

$$v_a(t) = \sum_{n=-\infty}^{\infty} v(t)\, \delta(t - nT_s) \tag{8.8}$$

Figure 8.4 illustrates the impulse representation of a sequence of samples from a continuous-time signal. Mathematically, the impulses are equal to the multiplication of the continuous-time signal times a unit impulse train.

Equivalently, through direct application of the sifting property of the impulse function, we can write Eq. (8.8) as

$$v_a(t) = \sum_{n=-\infty}^{\infty} v(nT_s)\, \delta(t - nT_s) \tag{8.9}$$

Note that $v_a(t)$ is not defined at the sampling instants because $\delta(t-nT_s)$ is not defined at $t = nT_s$. However, one must keep in mind that the values of $v_a(t)$ at the sampling instants are embedded in the area carried by each impulse function. It should now be clear that $v_a(t)$ and $v[n]$ are different but equivalent models of the sampling process in the continuous-time and discrete-time domains, respectively. In order to keep track of which domain we are working in (i.e., continuous or discrete), we shall make use of parentheses to encompass the argument of a continuous-time signal, $v(nT_s)$, and square brackets, $v[n]$, to denote a discrete-time signal.

8.2.3 Reconstruction

The inverse operation of sampling is reconstruction. Reconstruction is a process in which a sampled waveform (impulse form) is converted into a continuous waveform by a circuit such as a digital-to-analog converter (DAC) and an anti-imaging filter. In effect, reconstruction is the operation that fills in the missing waveform that appears between samples. In essence, the combined effect of the DAC and filter can be modeled as a single reconstruction operation denoted with impulse response $p(t)$ as shown in Figure 8.5. Mathematically speaking, the reconstruction operation performs interpolation between sampled values. The relationship between the input and output of the reconstruction process can be described by a convolution integral given by

$$v_R(t) = \int_{-\infty}^{\infty} v_a(t-\tau) p(\tau) d\tau \tag{8.10}$$

Substituting the impulse representation for the input signal $v_a(t)$ from Eq. (8.9), we can write the reconstructed signal several ways as

$$v_R(t) = \int_{-\infty}^{\infty} \sum_{n=-\infty}^{\infty} v(nT_s) \delta(t-nT_s-\tau) p(\tau) d\tau = \sum_{n=-\infty}^{\infty} \int_{-\infty}^{\infty} v(nT_s) \delta(t-nT_s-\tau) p(\tau) d\tau \tag{8.11}$$

Through the direct application of the sifting property of the impulse function, the output signal from the reconstruction process can then be described as

$$v_R(t) = \sum_{n=-\infty}^{\infty} v(nT_s) p(t-nT_s) \tag{8.12}$$

The shape of the impulse response defines the shape of the waveform between adjacent samples. Thus $p(t)$ is commonly referred to as the *characteristic pulse shape* of the reconstruction operation. Equation (8.12) states that each sample is multiplied by a delayed version of $p(t)$ and the resulting waveforms are added together to form $v_R(t)$. In other words, at each sample time $t = nT_s$, a pulse $p(t-nT_s)$ is generated with an amplitude proportional to the sample value $v(nT_s)$. Collectively, all the pulses are summed to form the output continuous signal $v_R(t)$. The general form of Eq. (8.12) appears often in the study of linear, time-invariant continuous-time systems. It is a special case of *convolution* and we say that the output is obtained by *convolving* the continuous-time equivalent signal of $v(nT_s)$ with $p(t)$. The following example will help to illustrate this concept.

Most DACs make use of a square characteristic pulse, as it is easiest to realize in practice. The sum of all shifted and scaled square pulses will result in a "staircase" continuous-time waveform, as shown in Figure 8.7. It is also evident that the staircase waveform is a rather poor

Figure 8.5. Reconstructing a continuous-time signal from a data sequence.

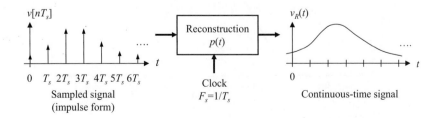

EXAMPLE 8.1

An input sequence $v[n]$ derived from a sinusoid has the following sampled values {0, 0.50, 0.87, 1.00, 0.87, 0.50, 0} corresponding to $n = 0, \ldots, 6$. Everywhere else the sequence is assumed to be zero. Using a triangular reconstruction pulse shape $p(t)$ defined as follows

$$p(t) = \begin{cases} 1 - |t - 1|, & 0 \le t < 2 \\ 0 & \text{elsewhere} \end{cases} \qquad (8.13)$$

plot the output waveform $v_R(t)$. Assume a sampling period, T_s, of 1 s.

Solution:
To begin, a plot of the characteristic pulse $p(t)$ is shown in Figure 8.6a. As is evident, $p(t)$ is a triangular waveform with a pulse duration that lasts for 2 s and has a peak value of 1. Following Eq. (8.12), with the limits of summation changed from 0 to 6 (as all other sample values are assumed equal to zero), we can write the reconstructed waveform as

$$v_R(t) = \sum_{n=0}^{6} v(nT_s) p(t - nT_s)$$

or when expanded as

$$v_R(t) = v(0)p(t) + v(T_s)p(t - T_s) + v(2T_s)p(t - 2T_s) + \cdots + v(6T_s)p(t - 6T_s)$$

Now we can substitute an expression for each shifted $p(t)$ according to Eq. (8.13). However, it is more instructive to demonstrate this by superimposing all the pulses weighted by the sampled value on one time axis as shown in Figure 8.6b. At any particular time point, we can add up the contribution from each pulse, and form a single point on the reconstructed waveform. This is shown in the figure for $t = 3.6$ s. This same operation can be repeated for all the remaining time points. The result is a straight-line interpolation between adjacent sampled values.

Figure 8.6. Convolving a triangular pulse with a sequence of sampled values. (a) Triangular impulse system response. (b) System output response.

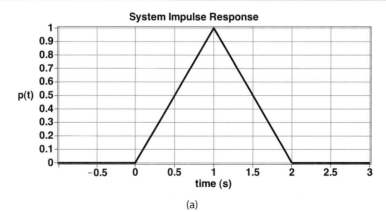

(a)

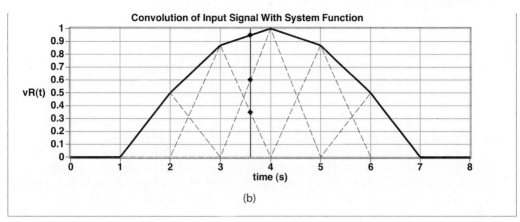

Figure 8.7. Illustrating the reconstruction operation with a DAC and an anti-imaging filter circuit.

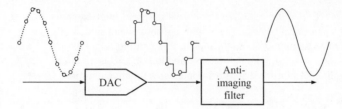

approximation of the original waveform. A better approximation would certainly be obtained by increasing the number of steps per period used to reconstruct the waveform. However, the upper frequency range of the DAC limits this approach. It is also clear from Figure 8.7 that the reconstructed waveform $v_R(t)$ contains a large amount of undesirable high-frequency energy, as the reconstructed signal is made up of various sized pulses. To eliminate this high-frequency energy, the DAC is usually followed by a post-filtering circuit, typically one with a low-pass characteristic having a cutoff frequency of at most one-half F_s. Such a filter is known under different names as a smoothing or anti-imaging filter. Collectively, the DAC and the anti-imaging filter are called a *reconstruction filter*.

Cascading a filter after the DAC effectively alters the characteristic pulse $p(t)$ of the reconstruction process and provides a much better approximation to the original waveform. In fact, perfect reconstruction can be obtained if the characteristic pulse of the overall reconstruction process has the following form:

$$p(t) = \begin{cases} 1, & t = 0 \\ \dfrac{\sin\left(\pi t / T_s\right)}{\left(\pi t / T_s\right)}, & \text{otherwise} \end{cases} \qquad (8.14)$$

This is a very long pulse, and its infinite length implies that to reconstruct a signal at time t exactly requires all the samples, not just those around that time. Substituting Eq. (8.14) into (8.12) allows us to write an exact interpolation formula for recovering the continuous-time information from the sampled values as

$$v_R(t) = \begin{cases} v(nT_s), & t = nT_s \\ \displaystyle\sum_{n=-\infty}^{\infty} v(nT_s) \dfrac{\sin\left[\dfrac{\pi}{T_s}(t - nT_s)\right]}{\dfrac{\pi}{T_s}(t - nT_s)}, & \text{otherwise} \end{cases} \qquad (8.15)$$

It is interesting to note that $v_R(t)$ is equal to $v(nT_s)$ at all the sampling instants.

In practice, a perfect reconstruction operation can only be approached, not actually realized. Consequently, some imperfections are introduced in the reconstruction process. There are two main sources of errors: (1) aperture effect due to the characteristic pulse shape and (2) magnitude and phase errors related to the anti-imaging filter. Both types of errors lead to

Exercises

8.1. Using MATLAB or an equivalent software program, plot 64 samples of a sine wave having unity amplitude, zero phase shift, and a period of 16 samples. We shall refer to this plotting range as the observation interval. (The stem command in MATLAB is an effective method for plotting discrete samples as a function of time.)

ANS. Setting $A = 1$, $\phi = 0$, $N = 16$, and $M = 1$, we get:

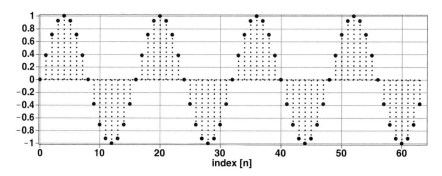

8.2. Using MATLAB or an equivalent software program, plot 16 samples of a sine wave having amplitude of 2, $\pi/4$ phase shift, and a period of 16/3 samples.

ANS. Setting $A = 2$, $\phi = \pi/4$, $N = 16$, and $M = 3$, we get:

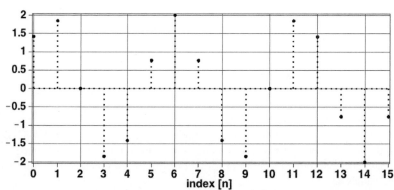

frequency-dependent magnitude and phase errors. If either error is an important parameter of a particular test, then they would need to be measured and corrected using a focused calibration procedure.

8.2.4 The Sampling Theorem and Aliasing

The sampling examples of the previous subsections are all performed in accordance with the *sampling theorem*. Shannon introduced the idea back in 1949 for application in communication systems. For this reason, it is sometimes referred to as the *Shannon sampling theorem*. However, interest and knowledge of the sampling theorem in engineering applications can be traced back to Nyquist in 1928 and as far back as 1915 in the literature of mathematicians. For a historical account of the sampling theorem, interested readers can refer to Jerri[6] for a detailed account. Specifically, the sampling theorem for band-limited signals can be stated in two separate but equivalent ways:

EXAMPLE 8.2

Reconstruct the sample set that results from a single sine wave with parameters: $A = 1$, $\phi = 0$, $N = 32$, and $M = 13$ using (a) zero-order hold, (b) first-order hold, and (c) perfect reconstruction.

Solution:
A plot of the 32 samples from the sine wave is shown below:

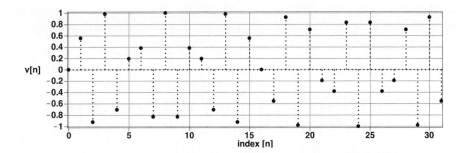

As is evident from this plot, no apparent appearance of a sine wave is present in the data set. This happens when the frequency of the sine wave approaches the frequency $F_s/2$, or in terms of M and N, as M approaches $N/2$. This does not mean that the data set is in error; it is just difficult to see the sine wave shape. If the data set is convolved with a zero-order hold operation, then we obtain the continuous-time waveform shown below. Superimposed on the plot are the original data points. Clearly, the data samples lie on the interpolated waveform. Looking at the interpolated waveform, it would be a stretch to say that this waveform is sinusoidal in shape. However, the information related to the sine wave is contained in the interpolated waveform. We just have to do more post-processing (e.g., linear filtering). We shall discuss this further in the next subsection.

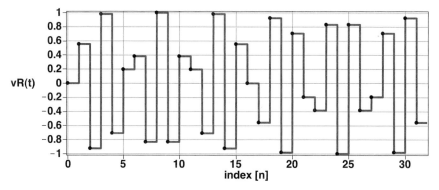

If the data set is convolved with a first-order hold operation, then we obtain the continuous-time waveform shown below:

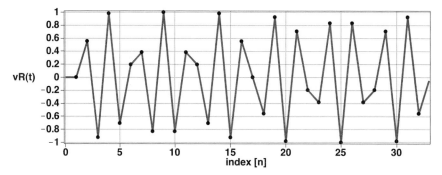

This interpolated waveform is quite different than that corresponding to the zero-order hold operation, but it too does not exhibit a sinusoidal shape. Finally, we perform the perfect reconstruction operation using the interpolation function described by Eq. (8.14) and obtain the waveform shown below:

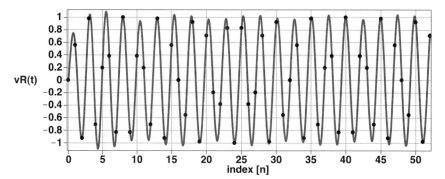

Here the reconstructed waveform has a very distinct sinusoidal shape completing 13 full cycles of the sine wave over the 32 sample points. We note that the waveform has a slight transient component at the start of the waveform but disappears after a few cycles. This waveform is almost identical to the unity amplitude sine wave with zero phase with parameters $M = 13$ and $N = 32$. We therefore conclude that zero-order and first-order hold interpolation is not sufficient to recover the original analog waveform from a set of samples. Additional filtering or post processing is necessary.

Exercises

8.3. Reconstruct the sampled signal displayed in Exercise 8.2 using

the rectangular pulse described by $p(t)=\begin{cases}1, & 0\leq t<1 \\ 0, & \text{elsewhere}\end{cases}$

ANS.

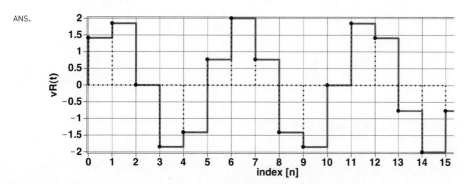

The Sampling Theorem

1. A continuous-time signal with frequencies no higher than F_{max} is completely described by specifying the values of the signal at instants of time separated by $1/(2F_{max})$ seconds.

2. A continuous-time signal with frequencies no higher than F_{max} may be completely recovered from knowledge of its samples taken at the rate of $2F_{max}$ per second.

The sampling rate $2F_{max}$ is called the *Nyquist rate*, and its reciprocal is called the *Nyquist interval*. The Nyquist rate is the minimum sampling rate allowable by the sampling theorem. Although somewhat confusing at times, the *Nyquist frequency* refers to F_{max}.

The first part of the sampling theorem is exploited by ATE digitizers. Part 2 of the theorem is exploited by waveform generators. For example, a 10-kHz sine wave appearing at the output of a DUT can theoretically be sampled by the digitizer at 20.1 kHz with no loss of signal information. However, if it is sampled at a slightly lower frequency of 19.9 kHz, specific information about its characteristics are lost. To better understand this, consider the setup shown in Figure 8.8 consisting of a sampler driven by a sine wave, followed by a perfect reconstruction operation.

Ideally, if Shannon's theorem is satisfied, the output of this arrangement should correspond exactly with the input signal (i.e., have exactly the same amplitude, phase, and frequency). In the case shown here, less than two samples per period are taken from the input sine wave; hence it violates the sampling theorem. Such a waveform is said to be undersampled. Subsequently, the signal reconstructed from these samples, shown on the right-hand side in Figure 8.8, has the same amplitude as the input signal but has a much lower frequency (as an estimate of the reconstruction operation consider joining adjacent samples with straight lines). The sampling and reconstruction process has distorted the input signal. The phenomenon of a higher frequency sinusoid acquiring the identity of a lower-frequency sinusoid after sampling is called *aliasing*.

Figure 8.8. Undersampled sine wave and its reconstructed image.

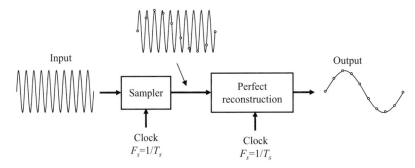

To avoid aliasing in practice, it is important to limit the bandwidth of the signals that appear at the input to the digitizer to less than the one-half the Nyquist rate. In general, practical signals are not limited to a fixed range of frequencies, but have a frequency spectrum that decay to zero as the frequency approaches infinity. As a result, it is not always clear how to satisfy the sampling theorem. To avoid this ambiguity a low-pass *anti-aliasing* filter is placed before the digitizer to attenuate the high-frequency components in the signal so that their aliases become insignificant.

While aliasing is generally an effect that is to be avoided, the process of undersampling has been used to an advantage in several applications. As we will see in Section 11.3, undersampling is used to extend the measurement capabilities of an arbitrary waveform digitizer. Aliasing may also be advantageously utilized by a DUT as part of its normal operation. The cellular telephone base-band interface is one such example that might use undersampling to convert high-frequency inputs to lower-frequency signals to be digitized by a slow ADC.

Finally, we would like to address a commonly asked question: What happens if we sample a sinusoidal at exactly twice its frequency? The answer is that information may be lost. To understand this, consider that an arbitrary sinusoidal (i.e., one with arbitrary amplitude and phase) can always be represented as the sum of a sine and cosine signal operating at the same frequency:

$$C \sin\left(2\pi f_o t + \phi\right) = A \cos\left(2\pi f_o t\right) + B \sin\left(2\pi f_o t\right) \tag{8.16}$$

Therefore, analyzing the effect of sampling a sine and cosine signal allows us to generalize the result for a signal having an arbitrary phase, ϕ. Figure 8.9 illustrates the samples derived from a sine and cosine signal sampled at twice their frequency. As is evident, all the samples from the sine wave are zero, whereas those from the cosine signal are not. Clearly, any information contained in the sine wave such as its amplitude would be lost and unobtainable from the samples. We can therefore conclude that one should not attempt to sample at exactly twice the Nyquist rate.

8.2.5 Quantization Effects

Mathematical sampling can be achieved with no loss of signal quality. A computer can come very close to mathematical perfection. For example, the following MATLAB routine can be used to create 64 samples of a sine wave with unity amplitude and zero phase shift:

```
pi = 3.14159265359;
for k = 1:64,
    v(k) = 1*sin(2*pi/64*k);
end;
```

Figure 8.9. Sine and cosine waves sampled at twice the signal frequency.

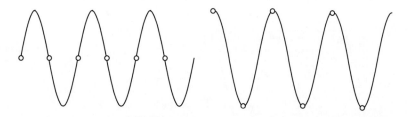

As the time index k is incremented in unit steps, the sampling period is by default equal to unity, resulting in a unity sampling frequency. Therefore, the frequency of the sampled sinusoid is 1/64 Hz, as $M = 1$ and $N = 64$. The quality of the sine wave is limited only by the tiny amounts of mathematical error in the computation process. This sampling process would result in a nearly perfect sampled representation of the sine wave. It would have almost no distortion and very little noise. The ADC included in a digitizer, on the other hand, will always introduce some amount of noise and distortion. The noise introduced by an ADC can be classified as (1) quantization noise and (2) circuit related noise such as thermal and shot noise. Distortion, on the other hand, is a result of nonlinear circuit behavior and component mismatches.

In a perfectly designed and manufactured ADC, the majority of the noise will be caused by the quantization error of the conversion process. Figure 8.10 shows a set of samples obtained from a sine wave that has been digitized by a 4-bit ADC. For example, the quantized waveform in Figure 8.10 could be stored in a computer memory as the sample set {7,11,14,14,11,7,4, 1,1,4, 7,11,14,14,11,7,4,1,1,4,7,11,14,14,11,7,4,1,1,4}. Also shown in Figure 8.10 is the original analog waveform superimposed on a regular spaced set of grid lines, together with an expanded view of a single sample shown on the right-hand side.

The vertical grid lines correspond to the sampling instances, with time increasing from left to right. The horizontal grid lines correspond to the ADC input decision levels with corresponding output code level identified. The distance between adjacent horizontal grid lines is the LSB step size (V_{LSB}). A single LSB step size sets the largest distance that the ADC output will be from a sample obtained directly from the original waveform (see the expanded view on the right of Figure 8.10). In general, an ideal D-bit ADC with a full-scale analog input range of V_{FSR} whose quantization operation is based on rounding or truncation has an LSB step size of

$$V_{LSB} = \frac{V_{FSR}}{2^D - 1} \tag{8.17}$$

If the ADC quantization process is one where all ideal code widths are of equal length, then the LSB step size will be given by

$$V_{LSB} = \frac{V_{FSR}}{2^D} \tag{8.18}$$

See Section 6.1 of DAC Testing for further details about the various ADC transfer characteristics.

The quantization errors in Figure 8.10 do not look especially severe at first glance. However, if we were to reconstruct a continuous-time waveform from these samples, the analog waveform would contain a significant noise component as illustrated in Figure 8.11. If we separate the errors from the quantized samples, we can see how much noise the quantization process has introduced.

Figure 8.10. Quantized sine wave samples.

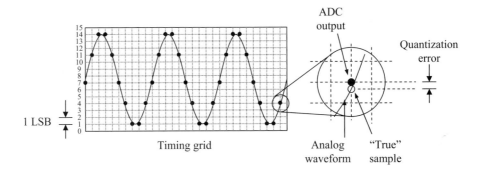

Exercises

8.4. To illustrate the effects of aliasing, compare 24 samples of a sinusoid with unity amplitude, zero phase shift, and a period of 12 s derived using a sampling rate of 1 Hz and a sampling rate of 1/8 Hz. Use MATLAB or an equivalent software program for your analysis.

ANS.   Setting A=1, φ=0, f_o=1/12, and F_s=1, we get:

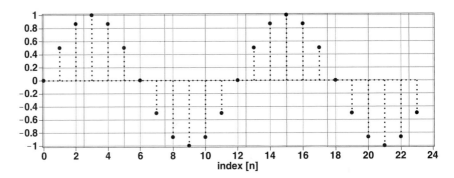

SETTING A = 1, ϕ = 0, f_o = 1/12, and F_s = 1/8, we get

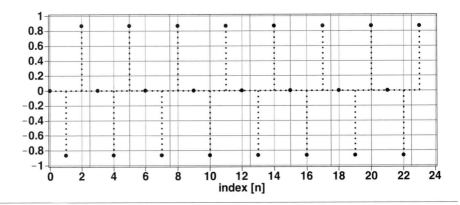

Figure 8.11. Illustrating the noise component that is associated with a quantization operation.

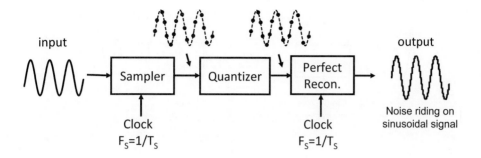

Figure 8.12. Representing the quantized waveform as a sum of the original sampled signal and a quantization error signal.

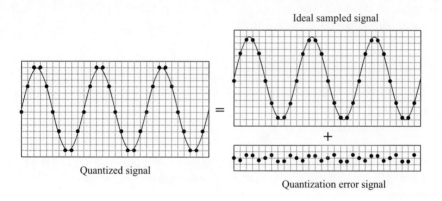

In Figure 8.12, the quantized waveform is equal to the sum of the ideal sampled waveform and an error waveform. The error waveform $v_q[n]$ is the quantization error or noise added by ADC quantization process based on rounding. Statistically speaking, the quantization errors of a random input signal exhibit a uniform probability distribution[5] from $-\frac{1}{2} V_{LSB}$ to $+\frac{1}{2} V_{LSB}$, assuming a perfect ADC. Moreover, the ideal quantization error sequence $v_q[n]$ resembles a random sequence having an average and RMS value given by

$$v_{q-AVE} = 0 \quad \text{and} \quad v_{q-RMS} = \frac{V_{LSB}}{\sqrt{12}} \tag{8.19}$$

A truncating ADC has the exact same RMS error value but the average value will be $V_{LSB}/2$. In a statistical sense, we can model the behavior of the quantizer as a summation element as illustrated in Figure 8.13 where one input to the quantizer is the sampled signal and the other the error sequence with average and RMS values given by Eq. (8.19). The corresponding ADC output signal will contain a noise component with an RMS value of $V_{LSB}/\sqrt{12}$. This same level of error will also appear with the signal after perfect reconstruction.

Obviously, quantization error can be reduced by using an ADC with more bits of resolution (consider combining Eqs. (8.17) and (8.19)). Higher resolution would provide more

Figure 8.13. Modeling the quantization operation with a summation element with one input coming from the sampler, the other input corresponding to the error sequence $v_q[n]$.

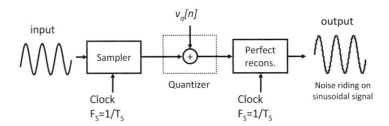

vertical graticules on the plots in Figure 8.10, reducing the size of each LSB step. Adding an extra bit of ADC resolution reduces the size of each LSB step by one-half, thereby reducing the RMS value of the quantization noise by a factor of two, or 6 decibels (6 dB). A 16-bit ADC is theoretically capable of a 97.76-dB signal-to-noise ratio (SNR) with a full-scale sine wave input. A 15-bit ADC would therefore be capable of 91.76 dB SNR, and so on. (See Chapter 10, "Analog Channel Testing" for an explanation of the decibel unit and SNR measurements.)

EXAMPLE 8.3

Compute the quantization error sequence that results from exciting a 3-bit ADC with a full-scale amplitude sinusoidal signal of unity amplitude, zero mean, zero phase, $M = 1$, and $N = 64$. Also, compute the RMS value of the quantization error and compare this result with its theoretical predicted value.

Solution:
To aid us in this investigation we shall make use of the following MATLAB routine for an ideal 3 bit quantizer performing a rounding operation typical of an ADC having a full-scale input range between −1 and +0.75:

```
% 3-Bit Quantizer (-1 <= X <= +0.75)
D = 3;                        % # of bits of resolution
VFSR = 1.75;                  % Full scale voltage range
VLSB = FS/(2^D-1);            % Least significant bit step size
Y = round(X/VLSB)*VLSB;       % rounds to nearest integer level, then scale by VLSB
```

A quantizer is the element of the ADC that limits the continuous input signal, say X, to discrete values denoted by Y–n this case, values of −1, −0.75, −0.5, −0.25, 0, 0.25, 0.5, and 0.75. The ADC would then interpret these levels and provide an output digital representation, for example in a two's complement form. The transfer characteristic, Y vs. X, for this quantizer is shown in Figure 8.14.

Figure 8.14. Ideal quantizer transfer characteristic.

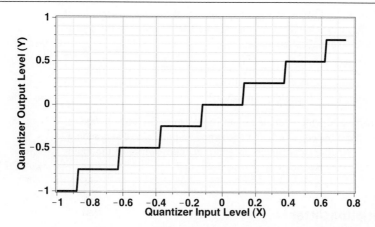

Now passing a near full-scale sinusoid having the parameters, $A = 0.75$, $\phi = 0$, $N = 64$, and $M = 1$ through the 3-bit quantizer, we get the error sequence shown in Figure 8.15.

Figure 8.15. Quantization error sequence.

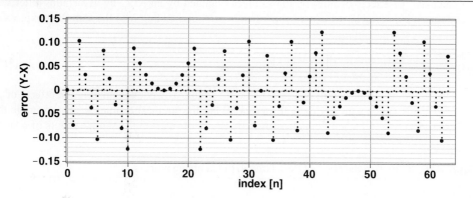

Here we see that the error sequence has symmetrical response bounded between −0.125 or +0.125 and has a mean value of −1.0842e−18 or nearly zero. The RMS value of the error is computed to be 0.0670. According to the quantization theory presented earlier, the error sequence should have an average value of 0 and an RMS value of $0.25/\sqrt{12} = 0.0722$ based on an LSB step size of 0.25 V. For all intents and purposes, the results of this simulation agree reasonably well. The discrepancy is largely a result of the quantizer's low resolution of 3 bits. If we increased its resolution, we would discover a much closer correspondence between experiment and theory.

Exercises

8.5. What is the LSB of an ideal 8-bit ADC that has a full-scale input range of 0–1 V? What is the expected RMS value of the corresponding quantization noise?

<small>ANS.</small> 3.9 mV, 1.13 mV.

8.6. If an ideal 7-bit ADC has an RMS quantization noise component of 1.4 mV, what is the quantization noise for a 5-bit ADC having an identical full-scale input range?

<small>ANS.</small> 2%.

8.7. A 4-bit ADC with an analog input range from −1.5 to +1.5 V gives an output of code of 4 for a code range beginning at 0 and ending at 15. What are the minimum and maximum values of the input voltage corresponding to this output code?

<small>ANS.</small> −0.7 V, −0.5 V.

8.2.6 Sampling Jitter

Another source of signal quality degradation is sampling jitter. Jitter is the error in the placement of each clock edge controlling the timing of each ADC or DAC sample. Figure 8.16 illustrates the effect of jitter on the sampling process of an ADC. Here we make use of the same regular spaced grid as that used in Section 8.2.5 except that this time we added an additional set of vertical dotted lines to indicate the actual clock edge subject to random clock jitter.

As is evident in this situation, the actual sample can differ quite significantly from the ideal sample and the size of this error is proportional to the magnitude of the jitter. Mathematically, we can calculate the effects of jitter on the samples obtained by an ADC by associating jitter with a random timing variable, which we shall denote as t_j, and adding it to the sampling expression given in Eq. (8.2) according to

$$v[n] = v(t)\big|_{t = nT_s + t_j} \tag{8.20}$$

Due to the nature of t_j, $v[n]$ is now a random variable as well. Calculating the effects of jitter can become mathematically complicated in all but the simplest examples. One example that allows us to draw some useful conclusions is the study of jitter on the sample points of a single

Figure 8.16. Illustrating the effect of clock jitter on the sampling process.

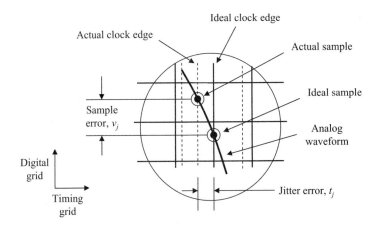

sinusoid with peak amplitude A_o and frequency f_o. The phase shift is assumed equal to zero without loss of generality. Without jitter, the sample points are

$$v[n] = v(t)\big|_{t=nT_s} = A_o \sin(2\pi f_o nT_s) \tag{8.21}$$

With jitter present, according to Eq. (8.20), the samples become

$$v[n] = v(t)\big|_{t=nT_s+t_j} = A_o \sin(2\pi f_o(nT_s+t_j)) \tag{8.22}$$

We can separate this expression into two parts, one that includes the deterministic component and the other due to jitter. To see this, consider using the trigonometric identity $\sin(A+B) = \sin(A)\cos(B) + \cos(A)\sin(B)$ so that we can rewrite Eq. (8.22) as

$$v[n] = A_o \sin(2\pi f_o nT_s)\cos(2\pi f_o t_j) + A_o \cos(2\pi f_o nT_s)\sin(2\pi f_o t_j) \tag{8.23}$$

Since the magnitude of the jitter t_j is assumed to be small compared to the sampling period T_s, we can approximate Eq. (8.23) as

$$v[n] \approx A_o \sin(2\pi f_o nT_s) + A_o 2\pi f_o t_j \cos(2\pi f_o nT_s) \tag{8.24}$$

Here we made use of the fact that when x is small, $\cos(x) \approx 1$ and $\sin(x) \approx x$. Now we have the jitter term separated from the deterministic term, allowing us to claim that the error in the sample due to jitter, denoted as v_j, is

$$v_j[n] \approx A_o 2\pi f_o t_j \cos(2\pi f_o nT_s) \tag{8.25}$$

Recognizing that the derivative of a sine wave is a cosine wave further allows us to write the jitter-induced error in terms of the magnitude of the jitter and the slope of the signal at the sample point

$$v_j[n] \approx \left[\frac{dv(t)}{dt}\bigg|_{t=nT_s} \right] \cdot t_j \tag{8.26}$$

This result should be readily apparent from Figure 8.16. It suggests that a timing error will induce a larger sample error at the rapidly rising or falling points of a sine wave than at its peak or trough.

Assuming that the jitter t_j has an RMS value of t_{j-RMS} and is independent of $v(t)$, we can approximate the RMS value of the error sequence $v_j[n]$ as the product of the RMS value of t_j and the RMS value of the derivative of $v(t)$ at each sampling instant. For a sampled sinusoidal signal with peak amplitude A_o and frequency f_o, the RMS value of the jitter-induced error is

$$v_{j-RMS} \approx \frac{2\pi A_o f_o}{\sqrt{2}} t_{j-RMS} \tag{8.27}$$

Assuming the jitter-induced error is independent of the quantization-induced error, we can model each effect separately as shown in Figure 8.17. The total combined RMS error can then be combined in a statistical sense as

$$v_{ERROR-RMS} = \sqrt{\left(\frac{2\pi \cdot A \cdot f_o}{\sqrt{2}} t_{j-RMS} \right)^2 + \left(\frac{V_{LSB}}{\sqrt{12}} \right)^2} \tag{8.28}$$

Figure 8.17. Modeling the sampling and quantization process as a linear combination of two independent error sequences.

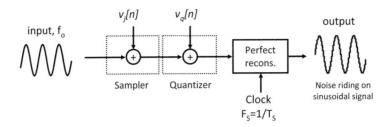

Consider the following ADC example.

EXAMPLE 8.4

A 12-bit ADC operating at a sampling rate of 20 MHz is used to sample a signal with a bandwidth of 500 kHz. The ADC operates off a single supply of 5 V. What is the worst-case RMS value of the error associated with the sampled signal when the clock signal has an RMS noise component of 500 ps?

Solution:
A 12-bit ADC assumed to have a full-scale range of 5 V has an LSB step size given by

$$V_{LSB} = \frac{5\ V}{2^{12} - 1} = 1.221\ mV$$

The RMS error associated with the ADC conversion process accounting for both quantization error and clock jitter with a worst-case amplitude signal given by 5 V peak-to-peak is therefore

$$V_{ERROR-RMS} = \sqrt{\left(\frac{2\pi \cdot {}^{5\ V}\!/_{2} \cdot 500\ kHz \cdot 500\ ps}{\sqrt{2}}\right)^2 + \left(\frac{1.221\ mV}{\sqrt{12}}\right)^2} = 2.79\ mV$$

It is interesting to note that if the above error was induced by quantization error alone, then the effective number of bits this ADC would achieve is found from

$$\sqrt{12} \times V_{ERROR-RMS} = \frac{V_{FSR}}{2^{D_{effective}} - 1} \Rightarrow D_{effective} = \log_2\left[\frac{V_{FSR}}{\sqrt{12} \times V_{ERROR-RMS}} + 1\right] \tag{8.29}$$

Substituting the appropriate parameters, we find $D_{effective}$ = 9 bits. Hence, clock jitter has induced a loss of 3 bits from ideal ADC operation.

At this point in our discussion we can use this result to set a limit on the maximum tolerable jitter allowable based on the ADC's speed and resolution. We first have to define the amount of jitter-induced noise that we are willing to tolerate. Let us define an LSB step size upper limit on the tolerable amount of jitter-induced noise, that is,

$$v_{j-RMS} < V_{LSB} = \frac{V_{FSR}}{2^D - 1} \tag{8.30}$$

Substituting Eq. (8.27) and rearranging allows us to bound the jitter according to

$$t_{j-RMS} < \frac{V_{FSR}}{\sqrt{2}\pi A_o f_o \left(2^D - 1\right)}$$ (8.31)

Further more, if we assume a full-scale input sinusoid, $V_{FSR} = 2A_o$, then we can find a lower limit on the maximum allowable jitter given by

$$t_{j-RMS} < \frac{\sqrt{2}}{\pi f_o \left(2^D - 1\right)}$$ (8.32)

Conversely, for a D-bit ADC having an RMS sampling jitter t_{j-RMS}, the maximum sampling frequency that can be used (i.e., $F_{s-MAX} = 2f_{o-MAX}$) is

$$F_{s-MAX} < \frac{2\sqrt{2}}{\pi t_{j-RMS} \left(2^D - 1\right)}$$ (8.33)

or we can conclude that the maximum conversion resolution (expressed in number of bits) available with a maximum sampling frequency F_{s-MAX} and RMS sampling jitter t_{j-RMS} is

$$D_{MAX} < \log_2 \left(\frac{2\sqrt{2}}{\pi t_{j-RMS} F_{s-MAX}} + 1 \right)$$ (8.34)

Figure 8.18 illustrates the trade-off between ADC effective resolution and RMS clock jitter for a range of sampling frequencies under worst-case conditions of full-range input signal swing and an input frequency set at the Nyquist frequency of $F_s/2$. As is quite evident form this plot, the effective ADC resolution decreases dramatically with increasing clock jitter. Furthermore, to achieve high ADC resolution at high sampling rates requires very low level of clock jitter.

The effect of sampling jitter on the operation of a DAC can be described by similar mathematical expressions derived for the ADC. Consider that the effect of clock jitter on the output of

Figure 8.18. Worst-case tradeoff of the effective ADC resolution and the RMS clock jitter for a range of ADC sampling frequencies.

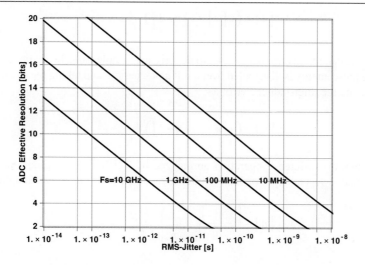

Figure 8.19. The effect of clock jitter on the actual DAC output can be separated into an ideal output and a jitter-induced error signal.

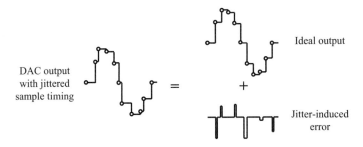

Figure 8.20. Modeling the jitter-induced error that occurs during the reconstruction process.

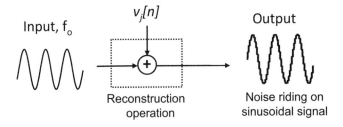

a DAC can be separated from its ideal operation as shown in Figure 8.19. Here the actual output waveform is separated into an ideal waveform and one that contains the jitter induced noise. Mathematically, the jitter-induced error can be described as

$$v_j(t) = \left[v(nT_s) - v\big((n-1)T_s\big) \right] \left[u\big(t - nT_s - t_j\big) - u\big(t - nT_s\big) \right] \tag{8.35}$$

where $u(t)$ is a unit step function. With an error pulse occurring on average once every clock period, we can consider that the effective energy contributed by each pulse at the sampling instant is

$$e_p[n] = \big(v[n] - v[n-1]\big)^2 \frac{t_j}{T_s} \tag{8.36}$$

Furthermore, we can relate this energy back to the original sample value by dividing Eq. (8.36) by T_s; that is, the pulse energy is distributed over a full clock period, and taking the square-root value, then we can write the jitter-induced error as

$$v_j[n] = \big(v[n] - v[n-1]\big)\sqrt{\frac{t_j}{T_s}} \tag{8.37}$$

Recognizing that the difference operation normalized by T_s is a discrete-time representation of differentiation allows us to approximate the jitter-induced error (for high oversampling ratios) as

$$v_j[n] \approx \left[\frac{dv(t)}{dt}\bigg|_{t=nT_s} \right] \sqrt{t_j T_s} \tag{8.38}$$

This expression is similar to that given for the jitter-induced error of the ADC, except that t_j is replaced by $\sqrt{t_j T_s}$. Hence we can make use of Eqs. (8.27)–(8.34) with the appropriate change of variable. For instance, the RMS value of the jitter-induced error voltage error would be written as

$$v_{j-RMS} = \frac{2\pi \cdot A \cdot f_o}{\sqrt{2}} \cdot \sqrt{t_{j-RMS} T_s} \tag{8.39}$$

In an identical manner as for the ADC conversion process, the DAC reconstruction process can be modeled with as a linear combination of the digital signal and an error sequence whose RMS value is given by Eq. (8.39).

This example serves to illustrate the sensitivity of the DAC to clock jitter and how bad the measurement accuracy can become if clock jitter is not properly accounted for.

If the jitter-induced voltage error is less than the LSB step size of the DAC, then the clock jitter will have little effect on the operation of the DAC. This condition is met when

$$t_{j-RMS} \le F_s \left[\frac{\sqrt{2}}{2\pi} \frac{V_{FSR}}{(2^D - 1)} \frac{1}{A \cdot f_o} \right]^2 \tag{8.40}$$

This was obtained by combining (8.30) with (8.39) and rearranging to obtain Eq. (8.40). Under worst-case conditions (i.e., input amplitude of $V_{FSR}/2$ and $f_o = F_s/2$), one can show that the maximum conversion resolution achievable for a DAC with an RMS clock jitter of t_{j-RMS} is given by

$$D_{MAX} < \log_2 \left(\frac{2}{\pi} \sqrt{\frac{2}{t_{j-RMS} F_{s-MAX}}} + 1 \right) \tag{8.41}$$

Plotting this expression for a range of sampling frequencies we obtain the graph shown in Figure 8.22. In comparison to the equivalent ADC worst-case effective number of bits shown in Figure 8.18, we see that the DAC is much more sensitive to clock jitter. This was also observed in Example 8.5.

Figure 8.22. Worst-case tradeoff of the effective DAC resolution and the RMS clock jitter for a range of DAC sampling frequencies.

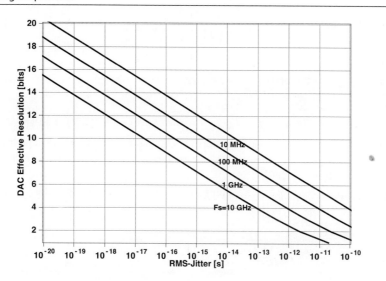

EXAMPLE 8.5

An AWG and DIG in a particular ATE is constructed from an 18-bit DAC and a 16-bit ADC with a full-scale value of 10 V. A 1 V peak sinusoidal signal is to be sourced at 4.5 MHz and applied to a DUT with an expected gain of unity. The AWG and DIG are clocked at 32 MHz as shown in the test setup of Figure 8.21. Measurements have shown that the clock source has a jitter noise component of 200 ps RMS. What is the expected RMS error in the samples collected by the digitizer? What is the effective resolution of the AWG-Digitizer combination?

Figure 8.21. An AWG-Digitizer setup for measuring the AC gain of a DUT.

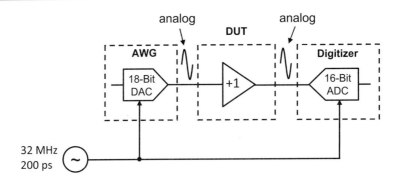

Solution:
We begin by determining the jitter—induced error associated with the DAC of the AWG. Using Eq. (8.39), we compute the AWG. DAC induced RMS error as

$$V_{AWG-RMS} = \frac{2\pi \cdot A \cdot f_o}{\sqrt{2}} \cdot \sqrt{\frac{t_{j-RMS}}{F_s}} = \frac{2\pi \cdot 1\,V \cdot 4.5\,MHz}{\sqrt{2}} \cdot \sqrt{\frac{200\,ps}{32\,MHz}} = 49.9\,mV$$

Next, we compute the corresponding jitter induced RMS error associated with the digitizer-ADC as

$$V_{DIG-RMS} = \sqrt{\left(\frac{2\pi \cdot A \cdot f_o}{\sqrt{2}} t_{j-RMS}\right)^2 + \left(\frac{V_{LSB}}{\sqrt{12}}\right)^2}$$

$$= \sqrt{\left(\frac{2\pi \cdot 1\,V \cdot 4.5\,MHz \cdot 200\,ps}{\sqrt{2}}\right)^2 + \left(\frac{1}{\sqrt{12}}\frac{10}{2^{16}-1}\right)^2} = 3.99\,mV$$

Finally, assuming that the errors are independent, we find that the total RMS error associated with the measurement setup is

$$V_{DIG-RMS} = \sqrt{\left(V_{AWG-RMS}\right)^2 + \left(V_{DIG-RMS}\right)^2} = \sqrt{\left(49.9\,mV\right)^2 + \left(3.99\,mV\right)^2} = 50.1\,mV$$

Any sequence of measurements will therefore contain about a 50 mV RMS error. It is obvious that this error will largely be from the AWG reconstruction process (i.e., 50-mV error from the AWG compared to 4 mV from the DIG).

If the AWG-digitization process is assumed to be ideal and is only limited by the quantization error associated with the ADC, then the effective number of bits can be found from

$$D_{effective} = \log_2\left[\frac{V_{FSR}}{\sqrt{12} \times V_{ERROR-RMS}} + 1\right] = \log_2\left[\frac{10\,V}{\sqrt{12} \times 50\,mV} + 1\right] = 5.9\,bits$$

In either the DAC or ADC case, according to Eq. (8.27) doubling the timing jitter doubles the noise level. Also, doubling the signal amplitude or signal frequency doubles the jitter-induced noise. Testers often have particular sampling frequencies or other conditions that produce minimum sampling jitter. For instance, a particular tester may produce minimum jitter if the digital pattern is exercised at the tester's master clock frequency divided by 2N, where N is any integer. As another example, a particular digitizer may operate with minimum jitter when its phase-locked loop phase discriminator input is near 16 kHz. If extremely low noise measurements are to be performed, the test engineer should understand which sampling rates provide the least jitter in each of the tester's instruments and subsystems.

8.3 REPETITIVE SAMPLE SETS

8.3.1 Finite and Infinite Sample Sets

In many mixed-signal systems such as a cellular telephone, the waveforms sampled by the system's ADC sub-blocks are nonrepetitive. In the cellular telephone example, the caller's voice is a random signal that seldom, if ever, repeats. The cellular telephone digitizes the caller's voice and processes the samples in real time in a continuous process. For all intents and purposes, we can consider the cellular telephone sample sets to be infinite in length.

In the DSP-based testing environment, on the other hand, signals are often created and measured using a finite sample set of a few hundred or a few thousand samples. If desired, the finite sample sets in mixed-signal testers can be repeated endlessly, allowing easier debugging with spectrum analyzers and oscilloscopes. During production testing, however, the sample sets are only allowed to repeat long enough to collect the necessary measurement information. The use of repetitive, finite sample sets drives a number of ATE-specific limitations which the test engineer must understand. For example, Figure 8.23 shows a short sequence of 16 samples that repeats endlessly. Notice how sample 16 feeds smoothly into sample 1 at the end of each sequence. This smooth wraparound results from a property known as *coherence*. Coherence is one of the most important enabling factors for fast and accurate DSP-based test.

Exercises

8.8. What is the RMS value of the error induced by an ADC having an RMS sampling jitter of 100 ps while measuring a 1-V amplitude sinusoid with a frequency of 100 kHz?   ANS. *44.4 μV.*

8.9. What is the maximum sampling jitter that a 6-bit ADC can tolerate when it has a full-scale input range of 0–3 V and is converting a 100-kHz, 1-V peak sinusoid?   ANS. *0.107 μs.*

8.10. What is the maximum sampling jitter that a 5-bit DAC can tolerate when it has a maximum sampling rate of 10 MHz?   ANS. *84.3 ps.*

8.11. If a 6-bit DAC has a sampling jitter of 500 ps RMS, what is its maximum sampling rate?   ANS. *408.5 kHz.*

8.12. If an ADC is controlled by a clock circuit with a minimum clock period of 1 μs and RMS jitter of 2.5 ns, what is the maximum conversion resolution possible with the ADC?   ANS. *8.5 bits.*

Figure 8.23. Finite sample set, repeated indefinitely.

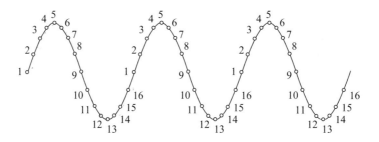

8.3.2 Coherent Signals and Noncoherent Signals

In the example waveform of Figure 8.23, the last sample of the first iteration wraps smoothly into the first sample of the second iteration because there is exactly one sine wave cycle represented by the 16 samples. If we reconstruct this sample set at a sampling frequency F_s, then the sine wave would have a frequency of $F_s/16$. This frequency is known as the *fundamental frequency* or *primitive frequency*, F_f. In general, the fundamental frequency F_f of N samples collected at a sampling rate of F_s is

$$F_f = \frac{F_s}{N} \tag{8.42}$$

The period of the fundamental frequency is called the *primitive period* or *unit test period* (UTP)

$$UTP = \frac{1}{F_f} \tag{8.43}$$

The amount of time required to collect a set of N samples at a rate of F_s is also equal to one UTP

$$UTP = \frac{N}{F_s} \tag{8.44}$$

In practice, it usually takes an extra fraction of a UTP to allow the DUT and ATE hardware to settle to a stable state before a sample set is collected.

The fundamental frequency is often called the *frequency resolution*. The reason for this alternate terminology is that the only coherent frequencies that can be produced with a repeating sample set are those frequencies that are integer multiples of the fundamental frequency. Hence, in terms of N and the sampling frequency, the conerent frequencies F_c are

$$F_c = M\frac{F_s}{N} \tag{8.45}$$

where M is an integer 0, 1, 2,..., $N/2$. The astute reader will recognize that we first made use of coherent frequencies in Section 8.2.1 in the development of Eq. (8.4), where $f_o = F_c$.

As an example, if we source the samples in Figure 8.23 at a rate of 16 kHz, then the fundamental frequency would be 16 kHz/16 = 1 kHz. The sine wave in Figure 8.23 would appear at 1 kHz. The next-highest frequency we could produce with 16 samples at this sampling rate is 2 kHz. If we wanted to produce a 1.5-kHz sine wave, then we would have a noncoherent sample set as shown in Figure 8.24.

If we wanted to produce a 1.5-kHz sine wave using a coherent sample set, then we would have to choose a sampling system with a fundamental frequency equal to 1.5 kHz/N, where N is

Figure 8.24. Noncoherent sample sets cannot be looped properly.

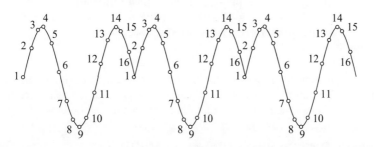

any integer. We might choose $F_f = 500$ Hz, for example, and then use the third multiple of the fundamental frequency to produce the 1.5-kHz sine wave. A fundamental frequency of 500 Hz could be achieved using 32 samples instead of 16 (16 kHz/32 = 500 Hz). We would then calculate a sine wave with three cycles in 32 samples according to

```
pi = 3.14159265359;
for k = 1:32,
    sinewave(k) = sin(2*pi*3/32*(k-1));
end
```

Since the fundamental frequency determines the frequency resolution of a measurement, it might seem that minimizing the fundamental frequency would be a great idea. In the absence of test time constraints, a fundamental frequency of 1 Hz would provide good flexibility in test frequency choice. Remember, though, that the UTP drives the test time. Since one UTP is equal to $1/F_f$, a 1-Hz frequency resolution would require 1 s of data collection time. For most production tests, this would be unacceptable.

Many test situations call for the application of a coherent multitone signal to excite a device. Such a signal is created by simply adding together a set of P unique sine waves (i.e., having different coherent frequencies) according to the following formula

$$v[n] = \sum_{i=1}^{P} A_i \sin\left(2\pi \frac{M_i}{N} n + \phi_i \right) \tag{8.46}$$

Here each sine wave is assigned a unique amplitude A_i, phase shift ϕ_i, and frequency designated by $(M_i/N)F_s$. The integers represented by M_i are commonly referred to as the *Fourier spectral bins*.

Any signal made up of a sum of coherent signals is also coherent. If one or more of the frequency components are noncoherent, though, the entire waveform will be noncoherent. Although noncoherent sample sets cannot be used to generate continuous signals through a looping process, they can be analyzed with DSP operations using a preprocessing operation called *windowing*. However, windowing is an inferior production measurement technique compared to coherent, nonwindowed testing. Windowing will be discussed in Chapter 9, "DSP-Based Testing."

Returning to the 16-kHz sampling example, we could create a multitone signal with frequencies at 1.5, 2.5, and 3.5 kHz using an expanded calculation given by the following MATLAB routine

```
pi = 3.14159265359;
phase1 = 0, phase2 = 0, phase3 = 0;
```

```
for k = 1:32,
    multitone(k) = sin(2*pi*3/32*(k-1) + phase1*pi/180)...
        + sin(2*pi*5/32*(k-1) + phase2*pi/180)...
        + sin(2*pi*7/32*(k-1) + phase3*pi/180;
end
```

The endpoints of this waveform would wrap smoothly from end to beginning because the wave-form is coherent. The multitone signal calculated would be described as a three-tone multitone waveform with equal amplitudes at the third, fifth, and seventh spectral bins.

8.3.3 Peak-to-RMS Control in Coherent Multitones

Notice that in the multitone example in Section 8.3.2, all the frequency components are created at the same phase (0 degrees). The problem with this type of waveform is that it may have an extremely large peak-to-RMS ratio, especially as the number of tones increases. Consider the 7-tone multitone signal in Figure 8.25. The first waveform consists entirely of sine waves, while the second waveform consists entirely of cosine waves. These waveforms exhibit a spiked shape that is unacceptable for most testing purposes since it tends to cause signal clipping in the DUT's circuits.

The peak-to-RMS ratio of a multitone can be adjusted by shifting the phase of each tone to a randomly chosen value. The waveform in Figure 8.26 shows how randomly selected phases for

Figure 8.25. Pure sine and pure cosine seven-tone multitones.

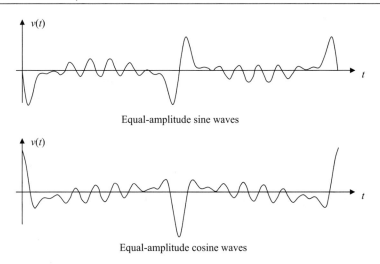

Equal-amplitude sine waves

Equal-amplitude cosine waves

Figure 8.26. Seven-tone multitone created with random phases.

the seven tones of Figure 8.25 produces a much less "spikey" waveform. Unfortunately, there is no equation to calculate phases to give a desired peak-to-RMS ratio. In many test programs, phases are chosen using a pseudorandom number generator with a uniform probability distribution between 0 and 2π radians. If the desired peak-to-RMS ratio is not achieved with one set of pseudorandom phases, then the program tries again until the desired ratio is found. These phase values can be generated each time the program loads, or they can be hard-coded into the test program once they have been determined through trial and error. This second approach prevents a pseudorandom algorithm from choosing one set of phases on a given day and another set of phases at a later date. For example, this might happen when an upgraded tester operating system includes a change to the pseudorandom algorithm. Theoretically, a different set of phases should not cause any shift in measurement results, but the use of hard-coded phases removes one more unknown factor from the measurement correlation effort.

What is the ideal peak-to-RMS ratio for a multitone signal? At first it might seem that it would be best to let the pseudorandom process search for a minimum peak-to-RMS ratio. This would provide the largest RMS voltage for a given peak-to-peak operating range. Larger RMS signals provide better noise immunity and improved repeatability. But this kind of signal is susceptible to large shifts in peak-to-RMS ratio if any of the filters in the ATE tester or DUT cause frequency-dependent phase shifts. A change in peak-to-RMS ratio could lead to a clipped signal, which would ruin the measurement accuracy.

In many end applications, the DUT will usually see a peak-to-RMS ratio of about 10 to 11 decibels (a ratio of about 3.35:1). Although the 10- to 11-dB range appears in many data sheets without explanation, it is based on the approximate peak-to-RMS levels encountered in typical analog signals. This range is roughly equal to the peak-to-RMS ratio of broadband signals having near-Gaussian-distributed amplitudes and random phases. As it happens, this ratio also tends to produce a multitone whose peak-to-RMS ratio is least sensitive to phase shifts from filters. For this reason, the pseudorandom phase selection process should be set to search for a peak-to-RMS ratio of between 10 and 11 decibels. A multitone signal must contain at least six tones to hit a peak-to-RMS ratio of 3.35:1. For signals having fewer tones, the target ratio is not terribly important, but it is still a good idea to use pseudorandom phase shifts for the tones rather than adding pure sine or cosine waveforms.

8.3.4 Spectral Bin Selection

One of the common mistakes a novice test engineer makes is to choose spectral bins by simply calculating the nearest integer multiple of the fundamental frequency. For example, if the test engineer wanted a 2-kHz sine wave using a 16-kHz sampling rate and 32 samples, then according to Eq. (8.45) the nearest Fourier spectral bin is

$$M = \frac{2 \text{ kHz}}{\left(16 \text{ kHz}/32\right)} = 4$$

which corresponds exactly with spectral bin 4. The problem with bin 4 is that it is not mutually prime with the number of samples, 32. (Mutually prime numbers are ones containing no common factors.) The number $4 = 2^2$ is not mutually prime with the number $32 = 2^5$, so this choice of bin, sampling frequency, and number of points is a poor one.

One of the problems with non-mutually-prime spectral bins is that they may cause the quantization noise of a coherent signal to contain periodic errors instead of errors that are randomly distributed over the UTP. Consider the sine example with 4 cycles in 32 samples. If we look at a quantized version of this signal from a 4-bit ADC in Figure 8.27, we see that the quantization errors repeat four times in the sample set. The same problem occurs with DAC converters as well. Furthermore, a nonprime spectral bin hits fewer code levels on the DAC and ADC; Thus we are

Figure 8.27. Non-mutually prime spectral bin selection leads to periodic errors.

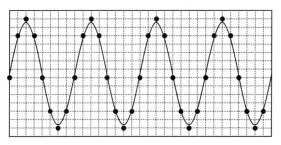

Non-mutually-prime spectral bin

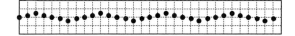

Periodic quantization errors
appear as gain error and
harmonic distortion

just testing the same points repeatedly. Repetitively exercising the same code levels results in less robust fault coverage in the DAC and ADC circuits.

Another problem with non-mutually-prime bins is that they tend to lead to overlaps between test tones, harmonic distortion components, and intermodulation distortion components. The use of mutually prime bins does not necessarily prevent intermodulation distortion overlaps, but it makes them less likely. Whether mutually prime bins are chosen or not, one should verify that all distortion components fall into spectral bins that do not coincide with bins containing important signal information.

Consider the example of a three-tone multitone at 1, 2, and 3 kHz. The problem with this multitone is that there is a great deal of distortion overlap. The second and third harmonic distortion of the 1-kHz tone falls on top of the 2- and 3-kHz test tones, respectively. Also, the second order intermodulation distortion between the 2-kHz tone and the 3-kHz tone appears at 1 and 4 kHz, corrupting the 1-kHz test tone. All these overlaps would cause errors in any measurement involving the 1-, 2-, or 3-kHz tones (gain, frequency response, distortion, etc.). A better approach is to use test frequencies close to the desired frequencies, but located at spectral bins that do not cause any intermodulation or harmonic distortion overlaps.

EXAMPLE 8.6

Select the spectral bins for a three-tone signal at 1, 2, and 3 kHz with no more than ±50-Hz error in the signal frequencies. The signal should take no more than 50 ms to repeat. Use a 16-kHz sampling rate.

Solution:
With a maximum UTP of 50 ms and a sampling rate of 16 kHz, the number of sample points is found from Eq. (8.44) to be

$$N \leq 50 \text{ ms} \times 16 \text{ kHz} = 800$$

An important constraint on the number of sample points used in most test systems is that N must be a power of two (i.e., 2P, where P is an integer). The reason for this will be explained in more detail in Chapter 9, but it is because we will ultimately use the Fast Fourier Transform (FFT) algorithm to measure the response of the DUT to the three-tone signal. Therefore, we shall select N equal to 512. We want the highest possible N in order to achieve the greatest frequency resolution or the smallest fundamental frequency. Working with 512 samples, the fundamental frequency becomes

$$F_f = \frac{16 \text{ kHz}}{512} = 31.25 \text{ Hz}$$

Subsequently, the closest spectral bin numbers that correspond to the 1-, 2-, and 3-kHz signals are found using Eq. (8.45), together with Eq. (8.42), to be

$$M_1 = \frac{1 \text{ kHz}}{31.25 \text{ Hz}} = 32 \text{ (non-mutually-prime, shift to 31)}$$

$$M_2 = \frac{2 \text{ kHz}}{31.25 \text{ Hz}} = 64 \text{ (non-mutually-prime, shift to 63)}$$

$$M_3 = \frac{3 \text{ kHz}}{31.25 \text{ Hz}} = 96 \text{ (non-mutually-prime, shift to 97)}$$

In all three cases, the computed spectral bin values were all even numbers sharing a factor of 2 with the number of samples, 512. Shifting the result by one in either a positive or negative direction eliminates their dependence on the common factor of 2. The resulting test frequencies, f_1, f_2, and f_3, are then

$$f_1 = 31 \times 31.25 \text{ Hz} = 968.75 \text{ Hz}$$
$$f_2 = 63 \times 31.25 \text{ Hz} = 1968.75 \text{ Hz}$$
$$f_3 = 97 \times 31.25 \text{ Hz} = 3031.25 \text{ Hz}$$

In all three cases, the chosen test frequencies are within the desired ±50-Hz error margin and are therefore acceptable.

We now have to verify that there are no distortion overlaps using spectral bins 31, 63, and 97. First we list the harmonics of the three test tone bins (stopping at the Nyquist bin, which is located at 8 kHz, or bin 256). Harmonics are defined as all the frequencies at an integer multiple of the test tone and computed according to the following table:

Harmonic Distortion Terms

Harmonic/Test Tone	M_A	$2M_A$	$3M_A$	$4M_A$	$5M_A$	$6M_A$	$7M_A$	$8M_A$
M_1	31	62	93	124	155	186	217	248
M_2	63	126	189	256	—	—	—	—
M_3	97	194	—	—	—	—	—	—

None of these harmonics overlaps with the other harmonics or with the test tones; Thus the harmonic distortion o verlap criterion is met. Next we look for intermodulation components. Intermodulation components appear at the sum and difference of any two tones, that is, $2F_1 - F_2$, second- and third-order distortions. (Second-order distortions are those in which the magnitude of the integers in front of F_1 and F_2 add up to 2; third-order distortions are those in which the magnitude of the integers add up to 3; etc.) The following is a series of tables listing the intermodulation interaction between the three test tones:

Second-Order IMD Sum Terms ($M_A + M_B$)

Test Tone Bin	$M_1 = 31$	$M_2 = 63$	$M_3 = 97$
$M_1 = 31$	62	94	128
$M_2 = 63$	94	126	160
$M_3 = 97$	128	160	194

Second-Order IMD Difference Terms ($M_A - M_B$)

Test Tone Bin	$M_1 = 31$	$M_2 = 63$	$M_3 = 97$
$M_1 = 31$	0	32	66
$M_2 = 63$	32	0	34
$M_3 = 97$	66	34	0

Third-Order IMD Sum Terms ($2M_A + M_B$)

Test Tone Bin	$M_1 = 31$	$M_2 = 63$	$M_3 = 97$
$M_1 = 31$	93	125	159
$M_2 = 63$	157	189	223
$M_3 = 97$	225	—	—

Third-Order IMD Difference Terms ($2M_A - M_B$)

Test Tone Bin	$M_1 = 31$	$M_2 = 63$	$M_3 = 97$
$M_1 = 31$	31	1	35
$M_2 = 63$	95	63	29
$M_3 = 97$	163	131	97

Here the letter *A* and *B* represent the row and column spectral bin, respectively. By cycling through all possible combinations of the sums and difference terms, a complete listing of the intermodualtion products can be easily identified. As is evident from this table, none of the intermodulation components falls on top of a test tone located in bins 31, 63, and 97 (except possibly along the diagonal of the matrix corresponding to the third order intermodulation distortion products where these can be ignored). Thus, the chosen spectral bins are good ones.

To avoid overlaps between harmonic distortion components and signal components, we should guarantee not only that the tones are mutually prime with the number of samples, but that they are also mutually prime with one another. For example, bins 3 and 9 are both mutually prime with 512 samples, but they are not mutually prime with one another. Consequently, the third harmonic of spectral bin 3 coincides with the tone at bin 9, resulting in an overlap.

The lack of overlap between harmonic distortion components and test tones in this example is guaranteed by a choice of mutually prime bins. In addition, none of the harmonics interferes with any of the intermodulation components. The choice of mutually prime bins does not guarantee a lack of overlap between intermodulation components and test tones or distortion components, but it does reduce the likelihood of such overlaps. Since there are no overlaps in this example, we can measure gain, frequency response, harmonic distortion, intermodulation distortion, and signal-to-noise ratio with the same set of collected samples. The ability to measure multiple parameters using a single data collection cycle is an advantage of multitone testing. This technique saves a tremendous amount of test time compared with single-tone testing approaches.

As we have seen, multitone DSP-based testing only provides accurate measurements if the test engineer is careful with the selection of test tones. Careless selection of spectral bins will lead to

Exercises

8.13. What is the fundamental frequency of 512 samples collected at a rate of 1 MHz? What is the corresponding UTP?

ANS. 1/512 MHz, 512 µs.

8.14. How many cycles of a 2.1375-kHz sine wave are completed in a 7.953216-ms UTP?

ANS. 17 cycles.

8.15. What is the nearest coherent frequency to 20 kHz when 512 samples are collected at a rate of 44 kHz? How many cycles are completed in one UTP?

ANS. 20.023 kHz, 233 cycles.

8.16. Using a hand analysis, compute the peak-to-RMS ratio of a two-tone multitone described by $A\sin(\omega_1 t) + B\sin(\omega_2 t)$.

ANS. $\sqrt{2}\,\dfrac{A+B}{\sqrt{A^2+B^2}}$

8.17. Select the spectral bins of a two-tone signal at 15 and 30 kHz such that minimum distortion overlap occurs. Assume that the sampling rate is 44.8 kHz and that the UTP must be less than 100 ms.

ANS. 1501, 2999.

answers that may be slightly incorrect. If we had chosen bin 62 for the 2-kHz tone, for example, then the second harmonic distortion from the 1-kHz tone would have affected the measured level of the 2-kHz tone by a small but significant amount.

In most cases, we choose a sample set consisting of an even number of samples. Thus the mutually prime rule prevents us from using even-numbered spectral bins. In the previous example, we chose bin 63 instead of 62 because it was mutually prime with the number of samples. There is a second reason that we chose 63 and not 62. Combinations of odd

Figure 8.28. Even spectral bins lead to asymmetrical peaks relative to the signal's DC offset.

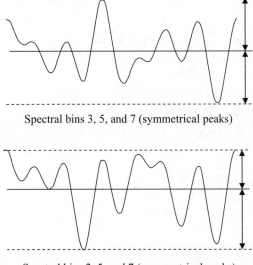

Spectral bins 3, 5, and 7 (symmetrical peaks)

Spectral bins 2, 5, and 7 (asymmetrical peaks)

harmonics and even harmonics in a multitone signal result in a signal with asymmetrical positive and negative peaks relative to the DC offset of the signal. The DC offset of such an asymmetrical multitone is not centered between the maximum and minimum voltages. This gives poor fault coverage for the circuit under test because it exercises one side of the signal range more than the other.

Figure 8.28 shows the difference between an all-odd multitone and a mixed odd-even multitone. Of course a single-tone signal with an even harmonic does not have the asymmetry problem, but it may lead to the kind of quantization noise modulation illustrated in Figure 8.27. The bottom line is that odd-numbered, mutually prime spectral bins should always be used whenever possible. If the situation is truly desperate, non-mutually-prime or even-numbered bins can be used as a last resort.

8.4 SYNCHRONIZATION OF SAMPLING SYSTEMS

8.4.1 Simultaneous Testing of Multiple Sampling Systems

Many DUTs contain both ADC and DAC channels, as in the case of the voice-band interface of Figure 8.1. These channels are often tested simultaneously in an ATE test program to minimize test time. Simultaneous testing requires a digital pattern loop containing the appropriate samples to excite the DAC channel. At the same time, an AWG converts another set of samples into analog form to excite the ADC channel. The response of the ADC is collected directly into ATE memory for later processing. The DAC response must be digitized before being stored into ATE memory. The response of each channel would then be analyzed through a post-processing frequency-domain operation and judged suitable or not. For example, we might test the gain of the ADC channel and the DAC channel simultaneously using this approach.

Unfortunately, crosstalk between the ADC and DAC channels can lead to small gain errors if we use the same test tones in both channels. Often, the gain errors are small enough that we can live with them. However, if the DAC and ADC have channel-to-channel crosstalk specifications, we can save some test time by measuring the crosstalk during the gain test. All we have to do is select slightly different test tones on the DAC side from those used on the ADC side. Then the feedthrough from DAC to ADC will show up in different bins from the ADC signals and vice versa. This is made possible by operating the various components of the ATE and DUT at different test frequencies but ensuring that they have the same UTP. In turn, this also implies that the fundamental frequency F_f is the same for all components and will guarantee coherent sampling sets in both signal paths. Let us consider the following example.

The preceding example demonstrates one of the reasons we prefer to use the same fundamental frequency for both ADC and DAC. It allows us to make coherent crosstalk measurements between two supposedly isolated signal paths. The other reason is that the UTP for the ADC and DAC is identical by definition, assuming we drive the ADC and DAC from the same digital pattern loop. For instance, if it takes 30 ms to collect the DAC channel samples, then it also takes 30 ms to collect the ADC channel samples. Identical UTPs drive identical fundamental frequencies, since the UTP is the inverse of the fundamental frequency. Sometimes, though, the ADC and DAC are not designed to sample at the same frequency. Fortunately, the sampling frequencies are often related by a simple integer multiple (i.e., 16 and 32 kHz). In these cases, we can simply collect more samples on one channel than on the other to achieve identical fundamental frequencies.

Matching all the fundamental frequencies in a particular test would be easy if we could simply request any arbitrary sampling rate from the ATE instruments. Unfortunately, many ATE testers have a limited choice of sampling frequencies. Henry Ford once said that you could purchase a Model T automobile in any color you wanted, as long as you wanted black. Sometimes particular tester architecture gives us a similar choice of sampling rates. For example, we might

EXAMPLE 8.6

A DUT's DAC and ADC both operate at a 32-kHz sampling rate. Find a sampling system that tests the gain of a DAC channel and an ADC channel simultaneously, as shown in Figure 8.1. Use three tones at 1, 2, and 3 kHz, with a maximum test tone error of 100 Hz.

Solution:

Let us start by setting the number of samples in the waveform exciting the DUT's DAC and ADC at $N = 1024$. Subsequently, the fundamental frequency F_f will be

$$F_f = \frac{32 \text{ kHz}}{1024} = 31.25 \text{ Hz}$$

The desired three tones will then be located in spectral bins of 32, 64, and 96, resulting in the desired test frequencies of 1, 2 and 3 kHz.

Beginning with the ADC channel test, we shall select the sampling rate of the AWG to be 16 kHz and impose the constraint that it has a fundamental frequency of 31.25 Hz. This in turn requires that the sample set consist of 512 samples. Using the sampling rate and spectral bins from the prior example, we will create an AWG waveform with 512 samples, having test tones at bins 31, 63, and 97. We will source this signal from the AWG at a 16-kHz sampling rate. To achieve the same fundamental frequency as the AWG signal, the ADC must collect 1024 samples (32 kHz sampling rate/1024 points = 16 kHz sampling rate/512 points). Using this sampling system we know that the ADC samples will form a coherent sample set.

Next let us consider the DAC channel test. By feeding 1024 samples to the DAC at 32 kHz, we would create a sampling system with the same fundamental frequency as the ADC. If we then set the digitizer sampling rate to 16 kHz and collect 512 samples from the DAC output, we would again achieve a fundamental frequency of 31.25 Hz, guaranteeing coherence. To allow simultaneous testing of the ADC and DAC gain, we need to select different spectral bins than those used to test the ADC. We can choose bins 33, 67, and 95 to meet all our testing criteria.

have a choice of any sampling frequency we want as long we want a multiple of 4 Hz. In the remaining sections we shall examine some of the ATE clocking architectures that the test engineer might encounter.

8.4.2 ATE Clock Sources

Mixed-signal testers use a variety of different approaches to clock generation. The most common clock generation schemes involve phase-locked loops, frequency synthesizers, or flying adders. Each of these has strong points and weak points that the test engineer will have to deal with. Ultimately, though, all the clocks in a mixed-signal tester should be referenced to a single master clock so that all instrumentation can be synchronized to achieve coherent sampling systems during each test.

The phase-locked loop (PLL) frequency generator is a circuit that produces an output clock equal to a reference clock times M over N, where M and N are integers. An example ATE PLL-based clocking architecture is shown in Figure 8.29. It consists of several counter stages and a voltage-controlled oscillator (VCO). This PLL is used to generate the sampling clock for a digitizer. It can use either a fixed 10-MHz internal frequency reference or an externally supplied reference frequency.

Figure 8.29. ATE PLL-based digitizer clocking source.

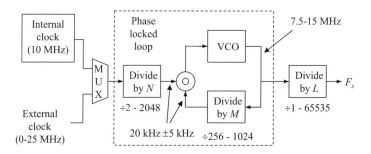

The external reference is required if a DAC output is to be digitized, since a reference clock would have to come from the same digital pattern generator feeding the DAC its samples, frame syncs, and other digital signals. The PLL shown in Figure 8.29 operates by first dividing the reference frequency F_{REF} by N, then by multiplying the result by M through the divide-by-M counter in the negative feedback loop around the VCO. Finally, the frequency of the VCO output can be further divided by another counter stage, which divides the output by integer L resulting in the output sampling frequency F_s given by

$$F_s = \frac{M}{NL} F_{REF} \qquad\qquad (8.47)$$

This particular example imposes a number of restrictions on the test engineer. First, the externally supplied reference clock must be between 0 Hz and 20 MHz. Next, the value of N must be between 2 and 2048. The output of the divide-by-N stage should be as close to 20 kHz as possible for maximum stability of the PLL. Other frequencies will work, but will introduce additional jitter into the clock. The VCO output must be between 5 and 10 MHz. The value of M must be between 256 and 1024. Finally, the value of L must be between 1 and 65535. Every time the PLL is reconfigured, it must be allowed to settle to a stable state, adding a bit of wait time between tests. Clearly, this clocking architecture is very inflexible and puts a large burden on the test engineer.

More modern testers allow the test engineer to select a wider range of frequencies using a frequency synthesizer. Frequency synthesizers work by taking a reference clock (10 MHz, for example) and passing it through a series of dividers and frequency mixers to produce a very stable output frequency with very little jitter. These synthesizers also take significant time to stabilize (25−50 ms), but that is the price paid for low jitter. Synthesizers are not entirely flexible either. For instance, a particular synthesizer may only be able to produce integer multiples of 4 Hz.

Flying adders can allow an even more flexible clocking source with little settling time, but they may introduce a little more jitter than a frequency synthesizer. A flying adder works by using a

---

Exercises

8.18. An ATE PLL-based clock source such as the one in Figure 8.29 is set with the divide-by-N counter equal to 1024, the divide-by-M equal to 512, and the divide-by-L equal to 64. With a reference frequency of 25 MHz, what is the output sampling frequency? Are all frequency constraints met in this configuration?

ANS. 195.3125 kHz.
Yes, intermediate frequencies are 24.414 kHz and 12.5 MHz.

Figure 8.30. Clock generation using flying adder delays.

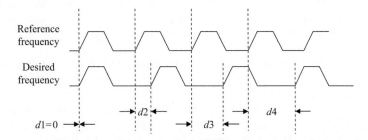

high-frequency reference clock and calculating the difference between the desired clock edges and the clock edges produced by the reference clock. Each desired clock edge is generated by delaying each reference clock edge by a carefully calculated amount of time as shown in Figure 8.30.

A new delay time has to be calculated for each delayed clock edge. The calculation is performed on the fly by an adder circuit, thus the terminology "flying adder." Because the edges are generated by programmable delay circuits, flying adders are sometimes more prone to jitter than frequency synthesizers. However, the frequency stabilization time for a flying adder clock circuit is nearly instantaneous.

8.5 SUMMARY

In this chapter, we have presented an introduction to sampling theory and coherent sampling systems as they are applied in mixed-signal ATE testing. While the treatment of this material is not as thorough as one might encounter in a signal processing textbook, this level of coverage should be adequate for beginning mixed-signal test engineers. Thus far, we have only seen how coherent multitone sample sets are created, sourced, and captured. In the next chapter, we will explore the use of digital signal processing algorithms, such as the fast Fourier transform (FFT), in the analysis of the samples collected during a coherent mixed-signal test. As we will see, digital signal processing allows a combination of low test time and high accuracy not possible with conventional, purely analog instrumentation.

PROBLEMS

8.1. For the following parameters of a sampled sine wave (A, ϕ, N, and M), calculate the frequency of the resulting sine wave assuming a sampling rate of 1 Hz. What are the UTP and the fundamental frequency related to this collection of samples?

    (a) $A = 2$, $\phi = 0$, $N = 32$, and $M = 1$ (d) $A = 1$, $\phi = 0$, $N = 64$, and $M = 33$

    (b) $A = 1$, $\phi = \pi/2$, $N = 64$, and $M = 13$ (e) $A = 1$, $\phi = 0$, $N = 128$, and $M = 65$

    (c) $A = 2$, $\phi = \pi/8$, $N = 64$, and $M = 5$ (f) $A = 0.5$, $\phi = 5\pi/8$, $N = 32$, and $M = 2.5$

8.2. Using MATLAB or an equivalent software program, plot the samples of the signals described in Problem 8.1 over its UTP.

8.3. Repeat Problems 8.1 and 8.2 with a sampling rate of 8 kHz.

8.4. Using a square characteristic pulse, reconstruct the samples obtained in Problem 8.1 and determine the frequency of the resulting sine wave. Identify any situation where aliasing occurs.

8.5. Using perfect interpolation reconstruct the samples generated from Problem 8.1. How does the reconstructed waveform compare to a zero-order hold function (square pulse) and a first-order hold function (triangular pulse)?

8.6. If a digital sinusoidal signal described by $A = 1$, $\phi = 0$, $N = 32$, and $M = 5$ is played through a DAC and speaker arrangement whose sampling rate is 8 kHz, what analog frequency will be heard? Repeat for $A = 1$, $\phi = 0$, $N = 32$, and $M = 25$. *Hint:* Reconstruct the discrete signal using the perfect interpolation formula in MATLAB and observe the frequency of the reconstructed signal.

8.7. What is the LSB step size of an ideal 12-bit rounding ADC that has a full-scale input range of 0–3 V? What is the expected RMS value of the corresponding quantization noise?

8.8. If an ideal 8-bit rounding ADC has a 400-μV RMS quantization noise component, what would be the noise component for a 5-bit ADC having the same input range?

8.9. A 6-bit truncating ADC with an analog input range from −1.5 to +1.5 V gives an output of code of 37 for a code range beginning at 0 and ending at 63. What are the minimum and maximum values of the input voltage corresponding to this output code?

8.10. A 6-bit rounding ADC with an analog input range from −1.5 to +1.5 V gives an output of code of 37 for a code range beginning at 0 and ending at 63. What are the minimum and maximum values of the input voltage corresponding to this output code?

8.11. What is the RMS value of the error induced by an ADC having an RMS sampling jitter of 250 ps while measuring a 1-V amplitude sinusoid with a frequency of 20 kHz?

8.12. What is the maximum allowable sampling jitter that a 10-bit ADC can tolerate when it has a full-scale input range of 3 V and converting a 1-V amplitude sinusoid with a frequency of 20 kHz? Assume a 1-LSB step size maximum allowable voltage error.

8.13. What is the maximum allowable sampling jitter that an 8-bit DAC can tolerate when it has a maximum sampling rate of 10 MHz? Assume a 1-LSB step size maximum allowable voltage error.

8.14. If a 6-bit DAC has an RMS sampling jitter of 100 ps, what is the maximum sampling rate? Assume a 1-LSB step size maximum allowable voltage error.

8.15. If an ADC or DAC is controlled by a clock circuit with a minimum clock period of 10 ns and RMS jitter of 250 ps, what is the maximum conversion resolution possible with either the ADC or DAC? Assume a 1-LSB step size maximum allowable voltage error.

8.16. What is the RMS value of the uncertainty associated with a signal sampled by a 10-bit ADC having an RMS clock jitter of 100 ps. The ADC has a full-scale range of 5 V and Nyquist frequency of 1 MHz. *Hint:* Independent errors add in a mean-squared sense, that is, $v^2_{total-RMS} = v^2_{q-RMS} + v^2_{j-RMS}$.

8.17. What is the RMS value of the uncertainty associated with a signal sampled by a 6-bit ADC having an RMS clock jitter of 1 ns. The ADC has a full-scale range of 5 V and Nyquist frequency of 1 MHz.

8.18. Plot the quantization error sequence that results after exciting a 6-bit rounding ADC with a full-scale amplitude sine wave. Use the MATLAB routine given in Example 8.3 for the quantizer. Compute the mean and RMS value of the quantization noise sequence. Repeat for an 8-bit quantizer. How does the mean and RMS value compare with theory in the two cases?

8.19. What is the fundamental frequency of 1024 samples collected at a rate of 20 kHz? What is the corresponding UTP?

8.20. How many cycles of a 20-kHz sine wave are completed in a 0.8-ms UTP? How many cycles of a 20-kHz sine wave are completed in a 0.79-ms UTP? If the signals are repeated indefinitely, which ones are coherent?

8.21. What are the coherent frequencies associated with 16 samples collected at a rate of 1 kHz?

8.22. Using MATLAB or an equivalent software program, plot a multitone signal consisting of three unity amplitude sine waves with frequencies of 1, 5, and 11 kHz over a UTP of 1 ms. Assume that the phase shifts are all zero.

8.23. Using the random selection method described in Section 8.3.3, search for the phases of fifty 1-mV amplitude tones such that the peak-to-RMS value is in a ratio of approximately 3.35:1.

Provide a plot to illustrate your result. Investigate the sensitivity of the peak-to-RMS value by changing the phase of each tone by 1% and computing the change in the peak-to-RMS value.

8.24. The Newman phase selection criterion[7] relies on selecting the phase of the kth tone of a multitone signal according the quadratic expression given by $\phi_k = (\pi/N)(k-1)^2$. Using this equation, determine the phases of the 50-tone multitone signal of Problem 8.23 and compare its peak-to-RMS value to the ideal value of 3.35. Investigate the sensitivity of the peak-to-RMS value by changing the phase of each tone by 1% and computing the change in the peak-to-RMS value.

8.25. Using MATLAB or an equivalent software program, plot a multitone signal consisting of 100 unity-amplitude sine waves distributed across a frequency range of 100 kHz using a UTP of 1 ms. Determine the phases of each tone such that the peak-to-RMS value approaches the ideal value of 3.35.

8.26. Select the spectral bins of a four tone signal at 1, 2, 3, and 4 kHz such that minimum distortion overlap occurs. Assume that the sampling rate is 44.8 kHz and that the UTP must be less than 100 ms. The accuracy of the test frequencies should be less than ±100 Hz. Justify your answer by providing the appropriate distortion tables.

8.27. Create the appropirate distortion tables for the following signalling situations:

(a) $N = 1024, M_1 = 15, M_2 = 23, M_3 = 27$.
(b) $N = 1024, M_1 = 15, M_2 = 23, M_3 = 27, M_4 = 33$.
(c) $N = 1024, M_1 = 15, M_2 = 23, M_3 = 27, M_4 = 33, M_5 = 39, M_6 = 49$.

8.28. An ATE PLL-based clock source is set with the divide-by-N counter equal to 4096, the divide-by-M equal to 512, and the divide-by-L equal to 128. With a reference frequency of 200 MHz, what is the output sampling frequency?

8.29. An ATE PLL-based clock source has a divide-by-N counter with a range from 1 to 1024, a divide-by-M counter with a range from 1 to 256, and a divide-by-L with a range of 1 to 65535. With a reference frequency of 200 MHz, what is the range of the output sampling frequency?

REFERENCES

1. M. Mahoney, *Tutorial DSP-Based Testing of Analog and Mixed-Signal Circuits*, The Computer Society of the IEEE, Washington D.C., 1987, ISBN 0818607858.

2. A. V. Oppenheim and R. W. Schafer, *Discrete-Time Signal Processing*, Prentice Hall, Englewood Cliffs, NJ, 1989, ISBN 013216292X.

3. A. V. Oppenheim et al., *Signals and Systems*, Prentice Hall, Englewood Cliffs, NJ, 1997, ISBN 0138147574.

4. W. McC. Siebert, *Circuits, Signals, and Systems*, The MIT Press, Cambridge, MA, 1985, ISBN 0262192292.

5. P. G. Hoel, S. C. Port, and C. J. Stone, *Introduction to Probability Theory*, Houghton Mifflin Company, Boston, 1971, ISBN 039504636X, p. 118.

6. A. J. Jerri, The Shannon sampling theorem—Its various extensions and applications: A tutorial review, *Proceedings of the IEEE*, **65**, pp. 1565–1596, 1997.

7. E. V. Ouderaa, J. Schoukens, and J. Renneboog, Peak factor minimization of the input and output signals of linear systems, *IEEE Transactions on Instrumentation and Measurement*, **37**(2), pp. 207–212, June 1988.

CHAPTER 9

DSP-Based Testing

In Chapter 8, we briefly touched on the advantages of DSP-based testing before beginning our review of sampling theory. In this chapter, we will take a more detailed look at digital signal processing and the power it gives us in testing mixed-signal devices. Although a full study of DSP is beyond the scope of this book, many good texts have been written on the subject.[1–3] In this chapter, we will review the basics of DSP, limiting our discussion to discrete (i.e., sampled) waveforms of finite length. More specifically, this chapter will review the concept of a Fourier series and its application to continuous-time and discrete-time periodic signals. As much of the work related to DSP-based testing involve discrete waveforms, we shall turn our attention to the discrete-time Fourier series (DTFS), discrete Fourier transform (DFT), and their implementation using the fast-Fourier transform (FFT) algorithm. The interplay between time and frequency will also be described. In some test situations, coherency is difficult to achieve, in such situation windowing is used. A detail look at windowing will be provided, from a description of the terminology used to its impact on spectral frequency leakage. Finally, we shall end the chapter with a look at the inverse-FFT and its application to spectral filtering, such as for noise reduction.

9.1 ADVANTAGES OF DSP-BASED TESTING

9.1.1 Reduced Test Time

Coherent DSP-based testing gives us several advantages over traditional measurement techniques. The first advantage of DSP-based testing is reduced test time. Since we can create and measure signals with multiple frequencies at the same time, we can perform many parametric measurements in parallel. If we need to test the frequency response and phase response of a filter, for example, we can perform a series of gain and phase measurements at a dozen or so frequencies simultaneously.

DSP-based testing allows us to send all the filter test frequencies through the device under test (DUT) at the same time. Once we have collected the DUT's output response using a digitizer

286

or capture memory, DSP allows us to separate each test tone in the output waveform from all the other test tones. We can then calculate a separate gain and phase measurement at each frequency without running many separate tests. We can also measure noise and distortion at the same time that we measure gain and phase shift, further reducing test time.

9.1.2 Separation of Signal Components

Separation of signal components from one another gives us a second huge advantage over non-DSP-based measurements. We can isolate noise and distortion components from one another and from the test tones. This allows for much more accurate and repeatable measurements of the primary test tones and their distortion components.

Using coherent test tones, we are always guaranteed that all the distortion components will fall neatly into separate Fourier spectral bins rather than being smeared across many bins (as is the case with noncoherent signal components). DSP-based testing is a major advantage in the elimination of errors and poor repeatability.

9.1.3 Advanced Signal Manipulations

In this chapter we will see how DSP-based testing allows us to manipulate digitized output waveforms to achieve a variety of results. We can perform interpolations between the samples to achieve better time resolution. We can apply mathematical filters to emphasize or diminish certain frequency components. We can remove noise from signals to achieve better accuracy. All these techniques are made possible by the application of digital signal processing to sampled DUT outputs.

9.2 DIGITAL SIGNAL PROCESSING

9.2.1 DSP and Array Processing

Before we embark on a review of digital signal processing, let us take some time to define exactly what a digital signal is. Then we shall look at the different ways we can process digital signals. There is a slight semantic difference between digital signal processing and array processing. An array, or vector, is a set of similar numbers, such as a record of all the heights of the students in a class. An example of array processing would be the calculation of the average height of the students. A digital signal is also a set of numbers (i.e., voltage samples), but the samples are ordered in time. Digital signal processing is thus a subset of array processing, since it is limited to mathematical operations on *time-ordered* samples. However, since most arrays processed on an ATE tester are in fact digital signals, most automated test equipment (ATE) languages categorize all array processing operations under the umbrella of DSP. So much for semantics!

ATE digital signal processing is often accomplished on a specialized computer called an *array processor*. However, tester computers have become faster over the years, making a separate array processor unnecessary in some of the newer testers. Depending on the sophistication of the tester's operating system, the presence or absence of a separate array processor may not even be apparent to the user.

There are many array processing functions that prove useful in mixed-signal test engineering. One simple example is the averaging function. The average of a series of samples is equivalent to the DC offset of the signal. If we want to measure the RMS noise of a digitized waveform, we must first remove the DC offset. Otherwise the DC offset would add to the RMS calculation, resulting in an erroneous noise measurement. We can compute the DC offset of the digitized waveform using an averaging function. Subtracting the offset from each sample in the waveform eliminates the DC offset. In MATLAB, we might accomplish the DC removal using the following simple routine:

```
% DC Removal Routine
  % Calculate the DC offset (average) of a waveform, x
    average = sum(x)/length(x);
  % Subtract offset from each waveform sample
    x = x-average;
```

This DC removal routine can be considered a digital signal processing operation, although a dedicated array processor is not needed to perform the calculations. Fortunately, we are able to take advantage of many built-in array processing operations in mixed-signal testers rather than writing them from scratch. These built-in operations are streamlined by the ATE vendor to allow the fastest possible processing time on the available computation hardware. For example, some of the computations may take place in parallel using multiple array processors, thus saving test time. Although tester languages vary widely, the following pseudocode is representative of typical ATE array processing operations:

```
float offset, waveform [256];
offset = average (waveform [1 to 256]);
waveform = waveform – offset;
```

Digital signal processing operations are somewhat more complicated than the simple array processing functions listed. Operations such as the discrete Fourier transform (DFT), fast Fourier transform (FFT), inverse FFT, and filtering operations will require a little more explanation.

9.2.2 Fourier Analysis of Periodic Signals

The tremendous advantage of DSP-based testing is the ability to measure many different frequency components simultaneously, minimizing test time. For example, we can apply a seven-tone multitone signal such as the one in Figure 8.26 to a low-pass filter. The filter will amplify or attenuate each frequency component by a different amount according to the filter's transfer function. It may also shift the phase of each frequency component.

It is easy to see how we can apply a multitone signal to the input of a filter. We simply compute the sample set in Figure 8.26 and apply it through an arbitrary waveform generator (AWG) to the input of the filter. A digitizer can then be used to collect samples from the output of the filter. But how do we then extract the amplitude of the individual frequency components from the composite signal at the filter's output?

The answer is a Fourier analysis. It is a mathematical method that gives us the power to split a composite signal into its individual frequency components. It is based on the fact that we can describe any signal in either the time domain or the frequency domain. Fourier analysis allows us to convert time-domain signal information into a description of a signal as a function of frequency. Fourier analysis allows us to convert a signal from the time domain to the frequency domain and back again without losing any information about the signal in either domain. This powerful capability is at the heart of mixed-signal testing.

Jean Baptiste Joseph Fourier was a clever French mathematician who developed Fourier analysis for the study of heat transfer in solid bodies. His technique was published in 1822. Almost 200 years later, the importance of Fourier's work in today's networked world is astounding. Applications of his method extend to cellular telephones, disk drives, speech recognition systems, radar systems, and mixed-signal testing, to name just a few.

9.2.3 The Trigonometric Fourier Series

Fourier's initial work showed how a mathematical series of sine and cosine terms could be used to analyze heat conduction problems. This became known as the *trigonometric Fourier series* and was probably the first systematic application of a trigonometric series to a problem. At the time of his death in 1830, he had extended his methods to include the Fourier integral, leading to the concept of a *Fourier transform*. The Fourier transform is largely applied to the analysis of aperiodic signals. Mixed-signal test engineering is primarily concerned with the discrete form of the Fourier series and, specifically, coherent sample sets. Therefore, we shall limit our discussion mainly to the Fourier series.

Let $x(t)$ denote a periodic signal with period T such that it satisfies

$$x(t) = x(t+T) \tag{9.1}$$

for all values of $-\infty < t < \infty$. Using a trigonometric Fourier series expansion of this signal, we are able to resolve the signal into an infinite sum of cosine and sine terms according to

$$x(t) = a_0 + \sum_{k=1}^{\infty} \left[a_k \cos(k \times 2\pi f_o t) + b_k \sin(k \times 2\pi f_o t) \right] \tag{9.2}$$

The first term in the series a_0 represents the DC or average component of $x(t)$. The coefficients a_k and b_k represents the amplitudes of the cosine and sine terms, respectively. They are commonly referred to as the spectral or Fourier coefficients and are determined from the following integral equations:

$$a_0 = \frac{1}{T} \int_0^T x(t)\, dt$$

$$a_k = \frac{2}{T} \int_0^T x(t) \cos\left(k \frac{2\pi}{T} t \right) dt, \quad k \geq 1 \tag{9.3}$$

$$b_k = \frac{2}{T} \int_0^T x(t) \sin\left(k \frac{2\pi}{T} t \right) dt, \quad k \geq 1$$

The frequency of each cosine and sine term is an integer multiple of the fundamental frequency $f_o = 1/T$, referred to as a harmonic. For instance, the quantity kf_o represents the kth harmonic of the fundamental frequency.

A more compact form of the trigonometric Fourier series is

$$x(t) = \sum_{k=0}^{\infty} c_k \cos(k \times 2\pi f_o t - \phi_k) \tag{9.4}$$

where

$$c_k = \sqrt{a_k^2 + b_k^2} \tag{9.5}$$

and

$$\phi_k = \begin{cases} \phi_o & \text{if } a_k \geq 0, b_k \geq 0 \\ -\phi_o & \text{if } a_k \geq 0, b_k < 0 \\ \pi - \phi_o & \text{if } a_k < 0, b_k \geq 0 \\ \phi_o - \pi & \text{if } a_k < 0, b_k < 0 \end{cases} \qquad \text{where} \quad \phi_o = \tan^{-1}\left(\frac{|b_k|}{|a_k|} \right) \tag{9.6}$$

This result follows from the trigonometric identity: $\cos(A+B)=\cos(A)\cos(B)-\sin(A)\sin(B)$. We should also note that the phase term is subtracted from the argument of the cosine function. This is known as a *phase lag* representation rather than a *phase lead*. Phase lag is defined as the phase angle corresponding to a zero crossing of some signal minus the zero-crossing phase angle of a reference signal, that is, output phase angle minus reference phase angle. Also, to avoid any numerical difficulty associated with the arctangent function, we define the phase term as $-\pi \le \phi \le \pi$ using a four-quadrant definition.

We prefer the magnitude and phase form representation of the Fourier series as it lends itself more directly to graphical form. Specifically, vertical lines can be drawn at discrete frequency points corresponding to $0, f_o, 2f_o, 3f_o$, and so on, with their heights proportional to the amplitudes of the corresponding frequency components, that is, c_k versus kf_o. Such a plot conveys the *magnitude spectrum*

EXAMPLE 9.1

Determine the Fourier series representation of the 5-V, 10-kHz clock signal shown in Figure 9.1 and plot the corresponding magnitude and phase spectrum.

Solution:
The spectral coefficients are determined according to Eq. (9-3) as follows

$$a_0 = \frac{1}{10^{-4}}\left(\int_0^{0.5\times10^{-4}} 0 \, dt + \int_{0.5\times10^{-4}}^{10^{-4}} 5 \, dt\right) = \frac{1}{10^{-4}} \, 5 \left(10^{-4} - 0.5\times10^{-4}\right) = 2.5$$

$$a_k = \frac{2}{10^{-4}}\left(\int_0^{0.5\times10^{-4}} 0 \cos\left(k\cdot10^4\cdot2\pi t\right)dt + \int_{0.5\times10^{-4}}^{10^{-4}} 5 \cos\left(k\cdot10^4\cdot2\pi t\right)dt\right)$$

$$= \frac{5}{k\pi} \, \sin\left(k \, 10^{-4}\cdot2\pi \, t\right)\Big|_{0.5\times10^{-4}}^{10^{-4}} = 0$$

$$b_k = \frac{2}{10^{-4}}\left(\int_0^{0.5\times10^{-4}} 0 \sin\left(k \, 10^4\cdot2\pi \, t\right)dt + \int_{0.5\times10^{-4}}^{10^{-4}} 5 \sin\left(k\cdot10^4\cdot2\pi \, t\right)dt\right)$$

$$= -\frac{5}{k\pi} \, \cos\left(k \, 10^{-4}\cdot2\pi \, t\right)\Big|_{0.5\times10^{-4}}^{10^{-4}} = \frac{5}{k\pi}\left[\cos(k\pi)-\cos(k\cdot2\pi)\right] = \frac{5}{k\pi}\left[(-1)^k - 1\right]$$

Figure 9.1. 10-kHz clock signal.

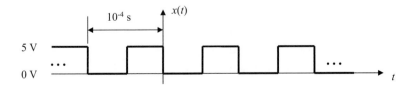

For even values of k, the term $[(-1)^k - 1] = 0$, hence the clock signal consists of only odd harmonics. Therefore,

$$b_k = \begin{cases} 0 & k \text{ even} \\ -\dfrac{10}{k\pi} & k \text{ odd} \end{cases}$$

The Fourier series representation then becomes

$$x(t) = 2.5 - \sum_{k=1,\, k \text{ odd}}^{\infty} \frac{10}{k\pi} \sin\left(k \cdot 2\pi \cdot 10^4 \cdot t\right)$$

or, when written in the form of Eq. (9–4), becomes

$$x(t) = 2.5 + \sum_{k=1,\, k \text{ odd}}^{\infty} \frac{10}{k\pi} \cos\left(k \cdot 2\pi \cdot 10^4 \cdot t + \frac{\pi}{2}\right)$$

as $\cos(x + \pi/2) = -\sin(x)$ and $\cos(x - \pi/2) = \sin(x)$. The corresponding magnitude and phase spectrum plots are then as shown in Figure 9.2.

Figure 9.2. Magnitude and phase spectrum plots of 10-kHz clock signal.

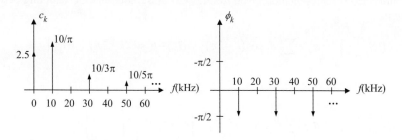

of $x(t)$. Likewise, a *phase spectrum* can be drawn in the exact same manner except that the heights of the vertical lines are proportional to the phases ϕ_k of the corresponding frequency components. The following example illustrates these two plots for a 5-V, 10-kHz clock signal.

It is instructive to view the behavior of the Fourier series representation for the clock signal of the previous example consisting of 10 and 50 terms. This we show in Figure 9.3 superimposed on one period of the clock signal. Clearly, as the number of terms in the series is increased, the Fourier series representation more closely resembles the clock signal. Some large amplitude oscillatory behavior occurs at the jump discontinuity. This is known as *Gibb's phenomenon* and is a result of truncating terms in the Fourier series representation.

Figure 9.3. Observing the Fourier series representation of a clock signal for: (a) 10 terms, and (b) 50 terms. Superimposed on the plot is the original clock signal.

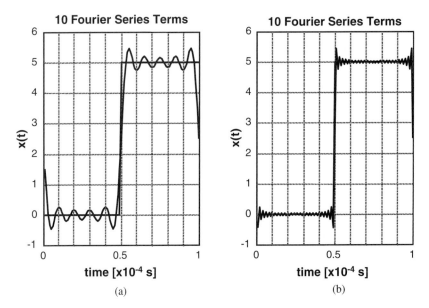

(a)   (b)

9.2.4 The Discrete-Time Fourier Series

As mentioned previously, sampling is an important step in mixed-signal testing. To understand the effects of sampling on an arbitrary periodic signal, consider sampling the general form of the Fourier series representation given in Eq. (9.2). Assuming that the sampling period is T_s, we can write

$$x(t)\big|_{t=nT_s} = a_0 + \sum_{k=1}^{\infty}\left[a_k \cos\left(k \cdot 2\pi f_o \cdot nT_s\right) + b_k \sin\left(k \cdot 2\pi f_o \cdot nT_s\right)\right] \tag{9.7}$$

As $F_s = 1/T_s$, we can rewrite Eq. (9.7) as

$$x(t)\big|_{t=nT_s} = a_0 + \sum_{k=1}^{\infty}\left\{a_k \cos\left[k\left(\frac{2\pi f_o}{F_s}\right)n\right] + b_k \sin\left[k\left(\frac{2\pi f_o}{F_s}\right)n\right]\right\} \tag{9.8}$$

In this form we see that the frequency of the fundamental is a fraction f_o/F_s of 2π. Furthermore, this term no longer has units of radians per second but rather just radians. To distinguish this from our regular notion of frequency, it is commonly referred to as a *normalized frequency*, because it has been normalized by the sampling frequency F_s. Except for the time reference T_s on the left-hand side of the equation, the information about the time scale is lost. This is further complicated by the fact that one usually uses the shorthand notation

$$x[n] = x(t)\big|_{t=nT_s} \tag{9.9}$$

and eliminates the time reference altogether. The discrete-time signal $x[n]$ is simply a sequence of numbers with no reference to the underlying time scale. Hence, the original samples cannot be reconstructed without knowledge of the original sampling frequency. Therefore, a sampling period or frequency must always be associated with a discrete-time signal, $x[n]$.

For much of the work in this textbook, we are concerned with coherent sampling sets, that is, $UTP = NT_s$, or, equivalently, $f_o = 1/UTP = F_f = F_s/N$. Equation (9.8) can then be reduced to

$$x[n] = a_0 + \sum_{k=1}^{\infty} \left\{ a_k \cos\left[k\left(\frac{2\pi}{N} \right) n \right] + b_k \sin\left[k\left(\frac{2\pi}{N} \right) n \right] \right\} \tag{9.10}$$

where the frequency of the fundamental reduces to $1/N$, albeit normalized by F_s.

As the original continuous-time signal $x(t)$ is periodic and sampled coherently, $x[n]$ will be periodic with respect to the sample index n, according to

$$x[n] = x[n + N] \tag{9.11}$$

Because the roles of n and k are interchangeable in the arguments of the sine and cosine terms of Eq. (9.10), it suggests that $x[n]$ will also repeat with index k over the period N. Through a detailed trigonometric development, we can rewrite the infinite series given by Eq. (9.10) as the sum of $N/2$ trigonometric terms as follows (here it is assumed N is even, as is often the case in testing applications):

$$x[n] = \tilde{a}_0 + \sum_{k=1}^{N/2-1} \left\{ \tilde{a}_k \cos\left[k\left(\frac{2\pi}{N} \right) n \right] + \tilde{b}_k \sin\left[k\left(\frac{2\pi}{N} \right) n \right] \right\} + \tilde{a}_{N/2} \cos\left(\pi n \right) \tag{9.12}$$

where

$$\tilde{a}_0 = \sum_{m=0}^{\infty} a_{0+mN}, \qquad \tilde{a}_k = \sum_{m=0}^{\infty} \left(a_{k+mN} + a_{N-k+mN} \right), \qquad \tilde{a}_{N/2} = \sum_{m=0}^{\infty} a_{N/2+mN}$$

$$\tilde{b}_k = \sum_{m=0}^{\infty} \left(b_{k+mN} - b_{N-k+mN} \right) \tag{9.13}$$

Here $\tilde{a}_k$ represents the amplitude of the cosine component located at the kth Fourier spectral bin. Likewise, $\tilde{b}_k$ is the amplitude of the sine component located at the kth Fourier spectral bin. Of course, $\tilde{a}_0$ represents the DC or average value of the sample set. The equations of Eq. (9.13) represent the sum of all aliases terms that arise during the sampling process. Since we are assuming that the original continuous-time signal is frequency band-limited, the sums in Eq. (9.13) will converge to finite values.

As before, Eq. (9.12) can be written in a more compact form using magnitude and phase notation as

$$x[n] = \sum_{k=0}^{N/2} \tilde{c}_k \cos\left[k\left(\frac{2\pi}{N} \right) n - \tilde{\phi}_k \right] \tag{9.14}$$

where

$$\tilde{c}_k = \sqrt{\tilde{a}_k^2 + b_k^2} \tag{9.15}$$

and

$$
\tilde{\phi}_k =
\begin{cases}
\tilde{\phi}_o, & \text{if } \tilde{a}_k \geq 0, \tilde{b}_k \geq 0 \\
-\tilde{\phi}_o, & \text{if } \tilde{a}_k \geq 0, \tilde{b}_k < 0 \\
\pi - \tilde{\phi}_o, & \text{if } \tilde{a}_k < 0, \tilde{b}_k \geq 0 \\
\tilde{\phi}_o - \pi, & \text{if } \tilde{a}_k < 0, \tilde{b}_k < 0
\end{cases}
\qquad \text{where} \qquad
\tilde{\phi}_o = \tan^{-1}\left(\frac{|\tilde{b}_k|}{|\tilde{a}_k|}\right)
\qquad (9.16)
$$

Equation (9.14) can then be graphically displayed using a magnitude and phase spectrum plot.

The importance of the above expressions cannot be understated, because it relates the spectrum of a sampled signal to the original continuous-time signal. Furthermore, it suggests that a discrete time signal has a spectrum that consists of, at most, $N/2$ unique frequencies. Moreover, these frequencies are all harmonically related to the primitive or fundamental frequency of $1/N$ radians. This representation is known as a *discrete-time Fourier series (DTFS)* representation of $x[n]$ written in trigonometric form. It will form the basis for all the computer analysis in this text.

Exercises

9.1. Find the trigonometric Fourier series representation of a square-wave $x(t)$ having a period of 2 s and whose behavior is described by

$$
x(t) =
\begin{cases}
+1 & \text{if } 0 < t < 1 \\
-1 & \text{if } 1 < t < 2
\end{cases}
$$

ANS. $x(t) = \dfrac{1}{2} - \dfrac{2}{\pi}\left(\cos\dfrac{\pi}{2}t - \dfrac{1}{3}\cos\dfrac{3\pi}{2}t + \dfrac{1}{5}\cos\dfrac{5\pi}{2}t - + \cdots\right)$

9.2. Express the Fourier series in Exercise 9.1 using cosine terms only.

ANS. $x(t) = \dfrac{4}{\pi}\left[\cos\left(\pi t - \dfrac{\pi}{2}\right) + \dfrac{1}{3}\cos\left(3\pi t - \dfrac{\pi}{2}\right) + \dfrac{1}{5}\cos\left(5\pi t - \dfrac{\pi}{2}\right) + \cdots\right]$

9.3. Find the Fourier series representation of a square-wave $x(t)$ having a period of 4 s and whose behavior is described by

$$
x(t) =
\begin{cases}
1 & \text{if } -1 < t < 1 \\
0 & \text{if } 1 < t < 3
\end{cases}
$$

ANS. $x(t) = \dfrac{1}{2} - \dfrac{2}{\pi}\left(\cos\dfrac{\pi}{2}t - \dfrac{1}{3}\cos\dfrac{3\pi}{2}t + \dfrac{1}{5}\cos\dfrac{5\pi}{2}t - + \cdots\right)$

EXAMPLE 9.2

Calculate the DTFS representation of the 10-kHz clock signal of Example 9.1 when sampled at a 100-kHz sampling rate.

Solution:
With a 100-kHz sampling rate, 10 points per period will be collected in one period of the 10-kHz clock signal (i.e., $N = 10$). Using the equations for the spectral coefficients in (9.13), together with the Fourier series result of Example 9.1, we find that the $\tilde{a}_k$ coefficients are as follows:

$$\tilde{a}_0 = \sum_{m=0}^{\infty} a_{0+10m} = a_0 = 2.5$$

$$\tilde{a}_k = \sum_{m=0}^{\infty} \left(a_{k+10m} + a_{10-k+10m} \right) = 0, \qquad k \in \{1,2,3,4\}$$

$$\tilde{a}_5 = \sum_{m=0}^{\infty} a_{5+10m} = 0$$

Subsequently, the $\tilde{b}_k$ coefficients for $k = 1, 2, \ldots, 4$ (by definition, $\tilde{b}_0 = 0$ and $\tilde{b}_5 = 0$) are found as follows:

$$\tilde{b}_k = \sum_{m=0}^{\infty} \left(b_{k+10m} - b_{10-k+10m} \right) = \begin{cases} 0, & k \text{ even} \\ -\dfrac{10}{\pi} \displaystyle\sum_{m=0}^{\infty} \left(\dfrac{1}{(k+10m)} - \dfrac{1}{(10-k+10m)} \right), & k \text{ odd} \end{cases}$$

Here the summation involves the difference between two harmonic progressions where no closed-form summation formulas are known to exist. Subsequently, a numerical routine was written that summed the first 100 terms of this series. The result is

$$\tilde{b}_1 = -3.07516, \qquad \tilde{b}_2 = 0, \qquad \tilde{b}_3 = -0.72528, \qquad \tilde{b}_4 = 0$$

According to Eq. (9.12), the discrete-time Fourier series representation for the clock signal then becomes

$$x[n] = 2.5 - 3.07516 \, \sin\left[1\left(\frac{2\pi}{10} \right) n \right] - 0.72528 \, \sin\left[3\left(\frac{2\pi}{10} \right) n \right]$$

or, using magnitude and phase notation, we write

$$x[n] = 2.5 + 3.07516 \, \cos\left[1\left(\frac{2\pi}{10} \right) n + \frac{\pi}{2} \right] + 0.72528 \, \cos\left[3\left(\frac{2\pi}{10} \right) n + \frac{\pi}{2} \right]$$

It is important for the reader to verify the samples produced by the above discrete-time Fourier series. Evaluating $x[n]$ for one complete period (i.e., $n = \{0, 1, 2, \ldots, 9\}$), we obtain $x = \{2.5, 0.0027, 0.0017, 0.0017, 0.0027, 2.5, 4.9973, 4.9983, 4.9983, 4.9973\}$. As expected, all the samples correspond quite closely with samples from the original signal, as shown in Figure 9.4. The small difference can be contributed to the error that results from including only 100 terms in the summation of the $\tilde{b}_k$ coefficients. Of particular interest is the value that the discrete-time Fourier series assigns to the waveform at the jump discontinuity. In general, the sample value at a jump discontinuity is ambiguous and undefined. Fourier analysis resolves this problem by assigning the sample value of the discontinuity as the midway point of the jump. In this particular case, the midpoint of each jump discontinuity is $(5-0)/2 = 2.5$.

Figure 9.4. Comparing the samples of a DTFS representation and the original clock signal. Also shown is the DTFS as a continuous function of n.

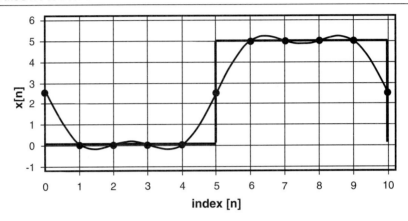

Also shown in the plot of Figure 9.4 is a graph of the DTFS representation as a continuous function of sample index, n. It is interesting to note that the samples are the intersection of this continuous-time function with the original clock signal.

When analyzing a waveform collected from a DUT we want to know the spectral coefficients $\tilde{a}_k$ and $\tilde{b}_k$ (or $\tilde{c}_k$ and $\tilde{\phi}_k$) so that we can see how much signal power is present at each frequency and deduce the DUT's overall performance. Thus far, we have only considered how to compute the spectral coefficients from a Fourier series expansion of a continuous-time waveform. As we shall see, there is a more direct way to compute the spectral coefficients of a discrete-time Fourier series from N samples of the continuous-time waveform. In fact, we shall outline two methods: one that highlights the nature of the problem in algebraic terms and the other involving a closed-form expression for the coefficients in terms of the sampled values.

To begin, let us consider that we have N samples, denoted $x[n]$ for $n = 0, 1, 2, \ldots, N-1$, Consider that each one of these samples must satisfy the discrete-time Fourier series expansion of Eq. (9.12). This is rather unlike the Fourier series expansion for a continuous-time signal that consists of an infinite number of trigonometric terms.

In practice, the summation must be limited to a finite number of terms, resulting in an approximation error. The discrete-time Fourier series, on the other hand, consists of only N trigonometric terms and, hence, there is no error in its representation. Therefore, we can write directly from Eq. (9.12) at sampling instant $n = 0$.

$$x[0] = \tilde{a}_0 + \sum_{k=1}^{N/2-1} \left\{ \tilde{a}_k \cos[0] + \tilde{b}_k \sin[0] \right\} + \tilde{a}_{N/2} \cos[0] = \tilde{a}_0 + \tilde{a}_1 + \cdots + \tilde{a}_{N/2-1} + \tilde{a}_{N/2} \quad \textbf{(9.17)}$$

Next, at $n = 1$, we write

$$x[1] = \tilde{a}_0 + \sum_{k=1}^{N/2-1} \left\{ \tilde{a}_k \cos\left[k\left(\frac{2\pi}{N}\right) \right] + \tilde{b}_k \sin\left[k\left(\frac{2\pi}{N}\right) \right] \right\} + \tilde{a}_{N/2} \cos[\pi]$$

$$= \tilde{a}_0 + \tilde{a}_1 \cos\left[\left(\frac{2\pi}{N}\right) \right] + \tilde{b}_1 \sin\left[\left(\frac{2\pi}{N}\right) \right] + \tilde{a}_2 \cos\left[2\left(\frac{2\pi}{N}\right) \right] + \tilde{b}_2 \sin\left[2\left(\frac{2\pi}{N}\right) \right] + \cdots \quad \textbf{(9.18)}$$

$$+ \tilde{a}_{N/2-1} \cos\left[\left(\frac{N}{2}-1\right)\left(\frac{2\pi}{N}\right) \right] + \tilde{b}_{N/2-1} \sin\left[\left(\frac{N}{2}-1\right)\left(\frac{2\pi}{N}\right) \right] + \tilde{a}_{N/2} \cos[\pi]$$

Similarly, for $n = 2$, we write

$$x[2] = \tilde{a}_0 + \sum_{k=1}^{N/2-1} \left\{ \tilde{a}_k \cos\left[2k\left(\frac{2\pi}{N}\right) \right] + \tilde{b}_k \sin\left[2k\left(\frac{2\pi}{N}\right) \right] \right\} + \tilde{a}_{N/2} \cos(2\pi)$$

$$= \tilde{a}_0 + \tilde{a}_1 \cos\left[2\left(\frac{2\pi}{N}\right) \right] + \tilde{b}_1 \sin\left[2\left(\frac{2\pi}{N}\right) \right] + \tilde{a}_2 \cos\left[4\left(\frac{2\pi}{N}\right) \right] + \tilde{b}_2 \sin\left[4\left(\frac{2\pi}{N}\right) \right] + \cdots \quad \textbf{(9.19)}$$

$$+ \tilde{a}_{N/2-1} \cos\left[2\left(\frac{N}{2}-1\right)\left(\frac{2\pi}{N}\right) \right] + \tilde{b}_{N/2-1} \sin\left[2\left(\frac{N}{2}-1\right)\left(\frac{2\pi}{N}\right) \right] + \tilde{a}_{N/2} \cos(2\pi)$$

Continuing for all remaining sampling instants, up to $n = N-1$, we can write the Nth equation as

$$x[N-1] = \tilde{a}_0 + \sum_{k=1}^{N/2-1} \left\{ \tilde{a}_k \cos\left[(N-1)k\left(\frac{2\pi}{N}\right) \right] + \tilde{b}_k \sin\left[(N-1)k\left(\frac{2\pi}{N}\right) \right] \right\} + \tilde{a}_{N/2} \cos[(N-1)\pi]$$

$$= \tilde{a}_0 + \tilde{a}_1 \cos\left[(N-1)\left(\frac{2\pi}{N}\right) \right] + \tilde{b}_1 \sin\left[(N-1)\left(\frac{2\pi}{N}\right) \right] + \tilde{a}_2 \cos\left[2(N-1)\left(\frac{2\pi}{N}\right) \right]$$

$$+ \tilde{b}_2 \sin\left[2(N-1)\left(\frac{2\pi}{N}\right) \right] + \cdots + \tilde{a}_{N/2-1} \cos\left[\left(\frac{N}{2}-1\right)(N-1)\left(\frac{2\pi}{N}\right) \right] \quad \textbf{(9.20)}$$

$$+ \tilde{b}_{N/2-1} \sin\left[\left(\frac{N}{2}-1\right)(N-1)\left(\frac{2\pi}{N}\right) \right] + \tilde{a}_{N/2} \cos[(N-1)\pi]$$

Finally, on observing the behavior of each one of these equations, we find that all the trigonometric terms have numerical values and the only unknown terms are the spectral coefficients. In essence, we have a system of N linear equations in N unknowns. Straightforward linear algebra can then be used to compute the spectral coefficients. For instance, if we define vectors

$$C = \begin{bmatrix} a_0 & a_1 & b_1 & a_2 & b_2 & \cdots & a_{N-1} \end{bmatrix}$$

and

$$X = \begin{bmatrix} x[0] & x[1] & x[2] & \cdots & x[N-1] \end{bmatrix}$$

together with matrix

$$W = \begin{bmatrix} 1 & \cos\left[(1)(0)\left(\frac{2\pi}{N}\right)\right] & \sin\left[(1)(0)\left(\frac{2\pi}{N}\right)\right] & \cdots & \sin\left[\left(\frac{N}{2}-1\right)(0)\left(\frac{2\pi}{N}\right)\right] & \cos\left[\left(\frac{N}{2}\right)(0)\left(\frac{2\pi}{N}\right)\right] \\ 1 & \cos\left[(1)(1)\left(\frac{2\pi}{N}\right)\right] & \sin\left[(1)(1)\left(\frac{2\pi}{N}\right)\right] & \cdots & \sin\left[\left(\frac{N}{2}-1\right)(1)\left(\frac{2\pi}{N}\right)\right] & \cos\left[\left(\frac{N}{2}\right)(1)\left(\frac{2\pi}{N}\right)\right] \\ 1 & \cos\left[(1)(2)\left(\frac{2\pi}{N}\right)\right] & \sin\left[(1)(2)\left(\frac{2\pi}{N}\right)\right] & \cdots & \sin\left[\left(\frac{N}{2}-1\right)(2)\left(\frac{2\pi}{N}\right)\right] & \cos\left[\left(\frac{N}{2}\right)(2)\left(\frac{2\pi}{N}\right)\right] \\ \vdots & \vdots & \vdots & & \vdots & \vdots \\ 1 & \cos\left[(1)(N-1)\left(\frac{2\pi}{N}\right)\right] & \sin\left[(1)(N-1)\left(\frac{2\pi}{N}\right)\right] & \cdots & \sin\left[\left(\frac{N}{2}-1\right)(N-1)\left(\frac{2\pi}{N}\right)\right] & \cos\left[\left(\frac{N}{2}\right)(N-1)\left(\frac{2\pi}{N}\right)\right] \end{bmatrix}$$

we can write

$$X = WC \qquad\qquad (9.21)$$

Multiplying both sides by W^{-1}, the spectral coefficients are found as follows

$$C = W^{-1}X \qquad\qquad (9.22)$$

The next example will illustrate this method on the clock signal samples from Example 9.2.

EXAMPLE 9.3

Consider from Example 9.2 that the clock signal samples are $x = \{2.5, 0.0, 0.0, 0.0, 0.0, 2.5, 5.0, 5.0, 5.0, 5.0\}$. Compute the spectral coefficients of the DTFS using linear algebra.

Solution:
Using the procedure described, we can write the following system of linear equations in matrix form:

$$\begin{bmatrix} 2.5 \\ 0 \\ 0 \\ 0 \\ 0 \\ 2.5 \\ 5 \\ 5 \\ 5 \\ 5 \end{bmatrix} = \begin{bmatrix} 1 & 1 & 0 & 1 & 0 & 1 & 0 & 1 & 0 & 1 \\ 1 & 0.809 & 0.588 & 0.309 & 0.951 & -0.309 & 0.951 & -0.809 & 0.588 & -1 \\ 1 & 0.309 & 0.951 & -0.809 & 0.588 & -0.809 & -0.588 & 0.309 & -0.951 & 1 \\ 1 & -0.309 & 0.951 & -0.809 & -0.588 & 0.809 & -0.588 & 0.309 & 0.951 & -1 \\ 1 & -0.809 & 0.588 & 0.309 & -0.951 & 0.309 & 0.951 & -0.809 & -0.588 & 1 \\ 1 & -1.000 & 0.000 & 1.000 & 0.000 & -1.000 & 0.000 & 1.000 & 0.000 & -1 \\ 1 & -0.809 & -0.588 & 0.309 & 0.951 & 0.309 & -0.951 & -0.809 & 0.588 & 1 \\ 1 & -0.309 & -0.951 & -0.809 & 0.588 & 0.809 & 0.588 & 0.309 & -0.951 & -1 \\ 1 & 0.309 & -0.951 & -0.809 & -0.588 & -0.809 & 0.588 & 0.309 & 0.951 & 1 \\ 1 & 0.809 & -0.588 & 0.309 & -0.951 & -0.309 & -0.951 & -0.809 & -0.588 & -1 \end{bmatrix} \begin{bmatrix} a_0 \\ a_1 \\ b_1 \\ a_2 \\ b_2 \\ a_3 \\ b_3 \\ a_4 \\ b_4 \\ a_5 \end{bmatrix}$$

Using the matrix routines available in MATLAB, we obtain the following spectral coefficients:

$$
\begin{bmatrix}
\tilde{a}_0 \\
\tilde{a}_1 \\
\tilde{b}_1 \\
\tilde{a}_2 \\
\tilde{b}_2 \\
\tilde{a}_3 \\
\tilde{b}_3 \\
\tilde{a}_4 \\
\tilde{b}_4 \\
\tilde{a}_5
\end{bmatrix}
=
\begin{bmatrix}
2.500 \\
0 \\
-3.078 \\
0 \\
0 \\
0 \\
-0.7265 \\
0 \\
0 \\
0
\end{bmatrix}
\Rightarrow
\begin{bmatrix}
\tilde{c}_0 \\
\tilde{\phi}_0 \\
\tilde{c}_1 \\
\tilde{\phi}_1 \\
\tilde{c}_2 \\
\tilde{\phi}_2 \\
\tilde{c}_3 \\
\tilde{\phi}_3 \\
\tilde{c}_4 \\
\tilde{\phi}_4 \\
\tilde{c}_5 \\
\tilde{\phi}_5
\end{bmatrix}
=
\begin{bmatrix}
2.500 \\
0 \\
3.078 \\
-\pi/2 \\
0 \\
0 \\
0.7265 \\
-\pi/2 \\
0 \\
0 \\
0 \\
0
\end{bmatrix}
$$

Exercises

9.4. Using the summation formulae in Eqs. (9.12) and (9.13), determine the DTFS representation of the Fourier series representation of $x(t)$ given in exercise 9.1. Use 10 points per period and limit the series to 100 terms. Also, express the result in magnitude and phase form.

ANS. $1.2301 \sin\left[\left(\dfrac{2\pi}{10}\right)n\right] + 0.2901 \sin\left[3\left(\dfrac{2\pi}{10}\right)n\right]$

$= 1.2301 \cos\left[\left(\dfrac{2\pi}{10}\right)n - \dfrac{\pi}{2}\right] + 0.2901 \cos\left[3\left(\dfrac{2\pi}{10}\right)n - \dfrac{\pi}{2}\right]$

9.5. A sampled signal consists of the following four samples: {0.7071, 0.7071, −0.7071, −0.7071}. Using linear algebra, determine the DTFS representation for these samples. Also, express the result in magnitude and phase form.

ANS. $0.7071 \cos\left[\left(\dfrac{2\pi}{4}\right)n\right] + 0.7071 \sin\left[\left(\dfrac{2\pi}{4}\right)n\right] = 1.0 \cos\left[\left(\dfrac{2\pi}{4}\right)n - \dfrac{\pi}{4}\right]$

9.6. A sampled signal consists of the following set of samples {0, 1, 2, 3, 2, 1}. Determine the DTFS representation for this sample set using the orthogonal basis method. Also, express the result in magnitude and phase form.

ANS. $1.5 - 1.333 \cos\left[\left(\dfrac{2\pi}{6}\right)n\right] - 0.1667 \cos\left[3\left(\dfrac{2\pi}{6}\right)n\right]$

$= 1.5 + 1.333 \cos\left[\left(\dfrac{2\pi}{6}\right)n - \pi\right] + 0.1667 \cos\left[3\left(\dfrac{2\pi}{6}\right)n - \pi\right]$

On comparison with the results in Example 9.2, we see that they agree reasonably well (the small difference is attributed to the series truncation as explained in the last example). The latter set of magnitude and phase coefficients was derived using Eq. (9.15).

The preceding example highlights the fact that the spectral coefficients of the discrete-time Fourier series are determined by straightforward linear algebraic methods. Another method exists for finding the spectral coefficients and one that is much more insightful as it provides a closed-form solution for each spectral coefficient. To arrive at this solution, we need to consider the orthogonal property of cosines and sines. Specifically, consider the following set of orthogonal basis functions

$$\sum_{n=0}^{N-1} \cos\left[p\left(\frac{2\pi}{N}\right)n\right]\cos\left[k\left(\frac{2\pi}{N}\right)n\right] = \begin{cases} 0 & \text{for } p \neq k \\ N/2 & \text{for } p = k \neq 0 \\ N & \text{for } p = k = 0 \end{cases}$$

$$\sum_{n=0}^{N-1} \sin\left[p\left(\frac{2\pi}{N}\right)n\right]\sin\left[k\left(\frac{2\pi}{N}\right)n\right] = \begin{cases} 0 & \text{for } p \neq k \\ N/2 & \text{for } p = k \neq 0 \end{cases} \qquad (9.23)$$

$$\sum_{n=0}^{N-1} \sin\left[p\left(\frac{2\pi}{N}\right)n\right]\cos\left[k\left(\frac{2\pi}{N}\right)n\right] = 0 \quad \text{for all } p$$

Armed with these identities, we multiply Eq. (9.12) with $\cos\left[k\left(\frac{2\pi}{N}\right)n\right]$ and sum n from 0 to N-1 on both sides to obtain[*]

$$\tilde{a}_k = \begin{cases} \dfrac{1}{N}\sum_{n=0}^{N-1} x[n]\cos\left[k\left(\frac{2\pi}{N}\right)n\right], & k = 0, N/2 \\[4mm] \dfrac{2}{N}\sum_{n=0}^{N-1} x[n]\cos\left[k\left(\frac{2\pi}{N}\right)n\right], & k = 1,2,\ldots,N/2-1 \end{cases} \qquad (9.24)$$

Likewise, we multiply Eq. (9.12) with $\sin\left[k\left(\frac{2\pi}{N}\right)n\right]$ and sum n from 0 to N-1 on both sides to obtain

$$\tilde{b}_k = \frac{2}{N}\sum_{n=0}^{N-1} x[n]\sin\left[k\left(\frac{2\pi}{N}\right)n\right], \quad k = 1,2,\ldots,N/2-1 \qquad (9.25)$$

The details are left as an exercise for the reader in problem 9.7.

---

[*]In many textbooks, a single expression for a_k is usually written as was done for b_k. This is achieved by writing the discrete-time Fourier series as

$$x[n] = \frac{\tilde{a}_0}{2} + \sum_{k=1}^{N/2-1}\left\{\tilde{a}_k \cos\left[k\left(\frac{2\pi}{N}\right)n\right] + \tilde{b}_k \sin\left[k\left(\frac{2\pi}{N}\right)n\right]\right\} + \frac{\tilde{a}_{N/2}}{2}\cos(\pi n)$$

where the end terms in the series are scaled by a factor of 2.

EXAMPLE 9.4

Repeat Example 9.3 but compute the spectral coefficients of the DTFS using the orthogonal basis method.

Solution:

With x = {2.5, 0.0, 0.0, 0.0, 0.0, 2.5, 5.0, 5.0, 5.0, 5.0}, we can compute from Eq. (9.24) the following coefficients:

$$\tilde{a}_0 = \frac{1}{10}[2.5+0+0+0+0+2.5+5+5+5+5] = 2.5$$

$$\tilde{a}_1 = \frac{2}{10}\left\{ 2.5\cos\left[(1)\left(\frac{2\pi}{10}\right)(0)\right] + 2.5\cos\left[(1)\left(\frac{2\pi}{10}\right)(5)\right] + 5\cos\left[(1)\left(\frac{2\pi}{10}\right)(6)\right] \right.$$
$$\left. + 5\cos\left[(1)\left(\frac{2\pi}{10}\right)(7)\right] + 5\cos\left[(1)\left(\frac{2\pi}{10}\right)(8)\right] + 5\cos\left[(1)\left(\frac{2\pi}{10}\right)(9)\right] \right\} = 0$$

Continuing, we find $\tilde{a}_2 = 0$, $\tilde{a}_3 = 0$, $\tilde{a}_4 = 0$, and $\tilde{a}_5 = 0$. Likewise, from Eq. (9.25) we have

$$\tilde{b}_1 = \frac{2}{10}\left\{ 2.5\sin\left[(1)\left(\frac{2\pi}{10}\right)(0)\right] + 2.5\sin\left[(1)\left(\frac{2\pi}{10}\right)(5)\right] + 5\sin\left[(1)\left(\frac{2\pi}{10}\right)(6)\right] \right.$$
$$\left. + 5\sin\left[(1)\left(\frac{2\pi}{10}\right)(7)\right] + 5\sin\left[(1)\left(\frac{2\pi}{10}\right)(8)\right] + 5\sin\left[(1)\left(\frac{2\pi}{10}\right)(9)\right] \right\} = -3.077$$

$$\tilde{b}_2 = \frac{2}{10}\left\{ 2.5\sin\left[(2)\left(\frac{2\pi}{10}\right)(0)\right] + 2.5\sin\left[(2)\left(\frac{2\pi}{10}\right)(5)\right] + 5\sin\left[(2)\left(\frac{2\pi}{10}\right)(6)\right] \right.$$
$$\left. + 5\sin\left[(2)\left(\frac{2\pi}{10}\right)(7)\right] + 5\sin\left[(2)\left(\frac{2\pi}{10}\right)(8)\right] + 5\sin\left[(2)\left(\frac{2\pi}{10}\right)(9)\right] \right\} = 0$$

$$\tilde{b}_3 = \frac{2}{10}\left\{ 2.5\sin\left[(3)\left(\frac{2\pi}{10}\right)(0)\right] + 2.5\sin\left[(3)\left(\frac{2\pi}{10}\right)(5)\right] + 5\sin\left[(3)\left(\frac{2\pi}{10}\right)(6)\right] \right.$$
$$\left. + 5\sin\left[(3)\left(\frac{2\pi}{10}\right)(7)\right] + 5\sin\left[(3)\left(\frac{2\pi}{10}\right)(8)\right] + 5\sin\left[(3)\left(\frac{2\pi}{10}\right)(9)\right] \right\} = -0.7265$$

and $\tilde{b}_4 = 0$ and $\tilde{b}_5 = 0$.

Not surprising, the results are identical to those found in Example 9.3.

9.2.5 Complete Frequency Spectrum

One of the most important insights that can be obtained from the closed-form expression for the spectral coefficients is how they behave beyond the range $k = 0,\ldots,N/2$. To begin, consider evaluating Eq. (9.24) over the range $k = N/2,\ldots,N$. On doing so, we write

$$\tilde{a}_k = \begin{cases} \dfrac{1}{N}\displaystyle\sum_{n=0}^{N-1} x[n]\cos\left[k\left(\dfrac{2\pi}{N}\right)n\right], & k = N/2, N \\[4mm] \dfrac{2}{N}\displaystyle\sum_{n=0}^{N-1} x[n]\cos\left[k\left(\dfrac{2\pi}{N}\right)n\right], & k = N/2+1,\ldots,N-1 \end{cases} \tag{9.26}$$

Here, only the cosine terms are affected by the index k. Next, if we consider the change of variable substitution, $k \to N - k$, together combined with the trigonometric identity $\cos\left[(N-k)(2\pi/N)n\right] = \cos\left[k(2\pi/N)n\right]$, Eq. (9.26) can be rewritten as

$$\tilde{a}_{N-k} = \begin{cases} \dfrac{1}{N}\displaystyle\sum_{n=0}^{N-1} x[n]\cos\left[k\left(\dfrac{2\pi}{N}\right)n\right], & k = 0, N/2 \\[4mm] \dfrac{2}{N}\displaystyle\sum_{n=0}^{N-1} x[n]\cos\left[k\left(\dfrac{2\pi}{N}\right)n\right], & k = 1,\ldots,N/2-1 \end{cases} \tag{9.27}$$

Recognizing that Eq. (9.26) is equivalent to Eq. (9.27) allows us to conclude over the range of $k = 1,\ldots,N-1$ that

$$\tilde{a}_{N-k} = \tilde{a}_k \tag{9.28}$$

Following the same reasoning as above in Eqs. (9.26)–(9.28), together with the trigonometric identity $\sin\left[(N-k)(2\pi/N)n\right] = -\sin\left[k(2\pi/N)n\right]$, one will find

$$\tilde{b}_{N-k} = -\tilde{b}_k \tag{9.29}$$

Converting this result into magnitude and phase form, we find using Eq. (9.15)

$$\begin{aligned} \tilde{c}_{N-k} &= \tilde{c}_k \\ \tilde{\phi}_{N-k} &= -\tilde{\phi}_k \end{aligned} \qquad \text{for all } k \tag{9.30}$$

Here we see that the magnitude spectrum excluding the DC bin has even symmetry about bin $N/2$. Similarly, the phase spectrum exhibits an odd symmetry about $N/2$. These two situations are highlighted in Figure 9.5 for an arbitrary discrete-time periodic signal.

Next, let us consider the periodicity of the spectrum. Consider replacing k in Eq. (9.24) by $k + N$ so that we write

$$\tilde{a}_{k+N} = \begin{cases} \dfrac{1}{N}\displaystyle\sum_{n=0}^{N-1} x[n]\cos\left[(k+N)\left(\dfrac{2\pi}{N}\right)n\right], & k = 0, N/2 \\[4mm] \dfrac{2}{N}\displaystyle\sum_{n=0}^{N-1} x[n]\cos\left[(k+N)\left(\dfrac{2\pi}{N}\right)n\right], & k = 1,\ldots,N/2-1 \end{cases} \tag{9.31}$$

EXAMPLE 9.5

Plot the magnitude and phase spectrum for the clock signal of Example 9.3 over 10 frequency bins.

Solution:
The magnitude spectrum for the clock signal of Example 9.2 is shown in Figure 9.6a and the corresponding phase spectrum appears in Figure 9.6b.

Due to the periodicity of the cosine function, Eq. (9.31) simplifies directly to

$$
\tilde{a}_{k+N} = \begin{cases} \dfrac{1}{N}\displaystyle\sum_{n=0}^{N-1} x[n]\cos\left[k\left(\dfrac{2\pi}{N}\right)n\right], & k = 0, N/2 \\[4mm] \dfrac{2}{N}\displaystyle\sum_{n=0}^{N-1} x[n]\cos\left[k\left(\dfrac{2\pi}{N}\right)n\right], & k = 1,\ldots,N/2-1 \end{cases}
\tag{9.32}
$$

which is equal to $\tilde{a}_k$. Hence, we write

$$
\tilde{a}_{k+N} = \tilde{a}_k \quad \text{for all } k
\tag{9.33}
$$

Following a similar line of reasoning, one can write

$$
\tilde{b}_{k+N} = \tilde{b}_k \quad \text{for all } k
\tag{9.34}
$$

Through the direct application of Eq. (9.15), we can write

$$
\begin{aligned} \tilde{c}_{k+N} &= \tilde{c}_k \\ \tilde{\phi}_{k+N} &= \tilde{\phi}_k \end{aligned} \quad \text{for all } k
\tag{9.35}
$$

We can therefore conclude from above that the spectrum of a periodic signal $x[n]$ with period N is also periodic with period N. Therefore, combining spectral symmetry, together with its periodicity, the frequency spectrum of a discrete-time periodic signal is defined for all frequencies. Figure 9.7 illustrates the full frequency spectrum of an arbitrary signal. To aid the reader, adjacent periods of the spectrum are indicated with dashed boxes.

At this point in our discussion, we should point out that most test vendors only provide spectral information corresponding to the Nyquist interval, $k = 0,1,\ldots, N/2$. Although less important today, 30 years ago when DSP-based ATE started to appear on the market, memory was expensive. Attempts to minimize memory usage were paramount. This led to the elimination of redundant spectral information.

Figure 9.5. Illustrating the spectral symmetry about $N/2$ for $k = 0, 1, \ldots, N - 1$: (a) magnitude spectrum and (b) phase spectrum.

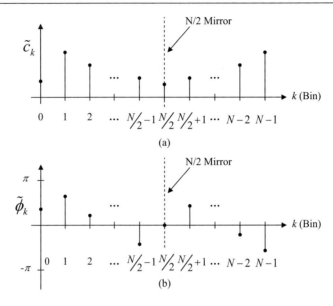

(a)

(b)

Figure 9.6. (a) Magnitude and (b) phase spectrum for the 10-kHz clock signal over 10 frequency bins.

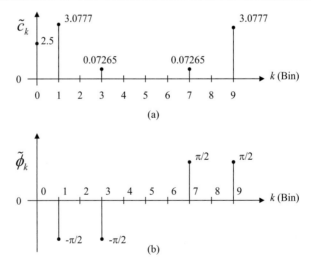

(a)

(b)

Figure 9.7. Illustrating the spectral periodicity: (a) magnitude spectrum and (b) phase spectrum.

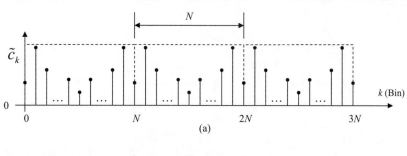

(a)

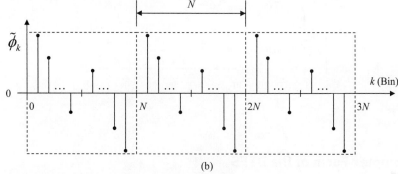

(b)

9.2.6 Time and Frequency Denormalization

The time and frequency scale associated with a data sequence $x[n]$ is described in terms of normalized time and frequency, according to the sample indexes, n and k, respectively. To obtain the actual time and frequency scales associated with the original samples, one must perform the operation of time and frequency *denormalization*. To achieve this, knowledge of the sampling period T_s or sampling frequency F_s is required.

To reconstruct the original time scale, one simply multiplies the sample index, n by T_s, according to the translation

$$n \rightarrow nT_s \tag{9.36}$$

Conversely, the frequency scale is restored when one multiplies the sample index k by F_s/N, according to the translation

$$k \rightarrow k\frac{F_s}{N} \tag{9.37}$$

Figure 9.8 illustrates the frequency denormalization procedure for the magnitude of the spectrum for the clock signal described in Example 9.5. For this particular case, $F_s = 100$ kHz and $N = 10$, resulting in a frequency denormalization scale factor of 10 kHz.

Figure 9.8. Illustrating the procedure of denormalizing a frequency axis.

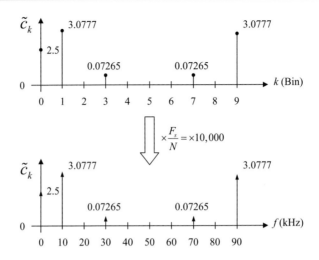

9.2.7 Complex Form of the DTFS

In most DSP textbooks, the DTFS is expressed in complex form using Euler's equation:

$$e^{j\varphi} = \cos(\varphi) + j\sin(\varphi) \tag{9.38}$$

where j is a complex number equal to $\sqrt{-1}$. The main reason for this choice lies with the ease in which the exponential function can be algebraically manipulated in contrast to trigonometric formulas. To convert the DTFS representation in Eq. (9.12) into complex form, consider that the cosine and sine functions can be written as

$$\cos(\varphi) = \frac{e^{j\varphi} + e^{-j\varphi}}{2} \qquad \sin(\varphi) = \frac{e^{j\varphi} - e^{-j\varphi}}{2j} \tag{9.39}$$

When the preceding two formulae are substituted into Eq. (9.12), together with several algebraic manipulations, we get

$$x[n] = \tilde{a}_0 + \sum_{k=1}^{N/2} \left(\frac{\tilde{a}_k - j\tilde{b}_k}{2} \right) e^{jk\left(\frac{2\pi}{N}\right)n} + \sum_{k=N/2}^{N-1} \left(\frac{\tilde{a}_{N-k} + j\tilde{b}_{N-k}}{2} \right) e^{jk\left(\frac{2\pi}{N}\right)n} \tag{9.40}$$

The above expression for the DTFS is more conveniently written in complex form as

$$x[n] = \sum_{k=0}^{N-1} X(k) e^{jk\left(\frac{2\pi}{N}\right)n} \tag{9.41}$$

Exercises

9.7. Plot the magnitude and phase spectrum of the DTFS representation of a sampled signal described by

$$x[n] = 0.6150 \sin\left[\left(\frac{2\pi}{8}\right)n\right] - 0.2749 \sin\left[2\left(\frac{2\pi}{8}\right)n\right]$$
$$+ 0.1451 \sin\left[3\left(\frac{2\pi}{8}\right)n\right] + 0.0649 \cos\left[4\left(\frac{2\pi}{8}\right)n\right]$$

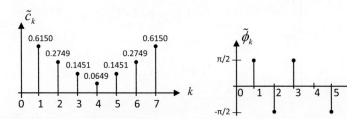
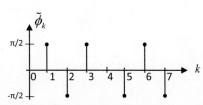

ANS.

9.8. Plot the frequency-denormalized magnitude spectrum of the following DTFS representation of a sampled signal described by

$$x[n] = 0.6150 \sin\left[\left(\frac{2\pi}{8}\right)n\right] - 0.2749 \sin\left[2\left(\frac{2\pi}{8}\right)n\right] + 0.1451 \sin\left[3\left(\frac{2\pi}{8}\right)n\right] + 0.0649 \cos\left[4\left(\frac{2\pi}{8}\right)n\right]$$

assuming a sampling rate of 16kHz.

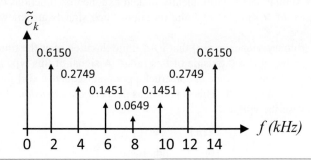

ANS.

where

$$X(k) = \begin{cases} \tilde{a}_0, & k = 0 \\ \dfrac{\tilde{a}_k - j\tilde{b}_k}{2}, & k = 1, 2, \ldots, N/2 - 1 \\ \tilde{a}_{N/2}, & k = N/2 \\ \dfrac{\tilde{a}_{N-k} + j\tilde{b}_{N-k}}{2}, & k = N/2 + 1, \ldots, N - 1 \end{cases} \qquad \textbf{(9.42)}$$

As is evident, the coefficients $X(k)$ in front of each exponential term are, in general, complex numbers. For the most part, the real component of each term relates to one-half the cosine coefficient of the trigonometric series. Conversely, the imaginary part of each term is related to one-half the sine coefficients. The exceptions are the $X(0)$ and $X(N/2)$ terms. These two terms are directly related to the cosine terms with a scale factor of one. The fact that the scale factor is not evenly distributed among each term can be a source of confusion for some.

Alternative forms of Eq. (9.42) can also be written. For instance, we can rewrite Eq. (9.42) in polar form, together with substitutions from Eq. (9.15), as

$$
X(k) = \begin{cases}
\tilde{c}_0 e^{j0}, & k = 0 \\
\dfrac{1}{2} \tilde{c}_k e^{-j\tilde{\phi}_k}, & k = 1, 2, \ldots, N/2 - 1 \\
\tilde{c}_{N/2} e^{j0}, & k = N/2 \\
\dfrac{1}{2} \tilde{c}_{N-k} e^{+j\tilde{\phi}_{N-k}}, & k = N/2 + 1, \ldots, N - 1
\end{cases} \tag{9.43}
$$

9.3 DISCRETE-TIME TRANSFORMS

9.3.1 The Discrete Fourier Transform

In the previous section it was shown how a sequence of N samples repeated indefinitely can be represented *exactly* with a set of $N/2$ harmonically related sinusoidal pairs and a DC component, or in terms of N harmonically related complex exponential functions. In this section, we shall demonstrate that a similar set of harmonically related exponential functions can also be used to represent a sequence of N samples of finite duration. Such signals are known as *discrete-time aperiodic* signals.

Consider an arbitrary sequence $y[n]$ that is of finite duration over the time interval $n = 0$ *to* $N-1$. Next, consider that $y[n] = 0$ outside of this range. A signal of this type is shown in Figure 9.9a. From this aperiodic signal, we can construct a periodic sequence $x[n]$ for which $y[n]$ is one period, as illustrated in Figure 9.9b. This is known as the *periodic extension* of $y[n]$. In mathematical terms, we can describe $y[n]$ as

$$
y[n] = \begin{cases}
x[n], & n = 0, 1, \ldots, N - 1 \\
0, & \text{otherwise}
\end{cases} \tag{9.44}
$$

If we consider the complex form of the DTFS representation for $x[n]$, we can write $y[n]$ as

$$
y[n] = \begin{cases}
\displaystyle\sum_{k=0}^{N-1} X(k) e^{jk\left(\frac{2\pi}{N}\right)n}, & n = 0, 1, \ldots, N - 1 \\
0, & \text{otherwise}
\end{cases} \tag{9.45}
$$

As we choose the period N to be larger, $y[n]$ matches $x[n]$ over a longer interval, and as $N \to \infty$, $y[n] = x[n]$ for any finite value of n. Thus, for very large N, the spectrum of $y[n]$ is identical to $x[n]$. However, we note that as $N \to \infty$, the form of the mathematics in Eq. (9.45)

Figure 9.9. (a) Arbitrary signal of finite duration; (b) periodic extension of infinite duration.

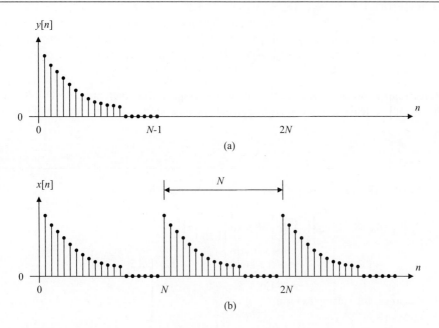

(a)

(b)

changes. The spectral makeup of the discrete-time signal no longer consists of harmonically related discrete frequencies, but rather becomes a continuous function of frequency. Under such conditions, the representation in Eq. (9.45) in the limit becomes known as a *Fourier transform* and is written with an integral operation as follows

$$y[n] = \frac{1}{2\pi} \int_{-\pi}^{\pi} Y\left(e^{j\omega}\right) e^{j\omega n} d\omega \tag{9.46}$$

where

$$Y\left(e^{j\omega}\right) = \sum_{n=-\infty}^{\infty} y[n] e^{-j\omega n} \tag{9.47}$$

As in all computer applications, of which mixed-signal testing is just one example, we must limit our discussion to finite values of N and, preferably (because time is always a major concern), to small values of N. This implies that we really have no other choice but to work directly with Eq. (9.45). It has been shown that the spectral coefficients $X(k)$ associated with Eq. (9.45) are directly related to samples of $Y\left(e^{j\omega}\right)$ uniformly spaced according to

$$X(k) = \left. \frac{Y\left(e^{j\omega}\right)}{N} \right|_{\omega = \frac{2\pi}{N} k} \tag{9.48}$$

Substituting Eq. (9.47) into (9.48), and limiting the summation to the maximum possible nonzero values of $y[n]$, $n = 0,1,\ldots, N-1$, we can write

$$X(k) = \frac{1}{N} \sum_{n=0}^{N-1} y[n] e^{-jk\left(\frac{2\pi}{N}\right)n} \tag{9.49}$$

Due to the importance of the interplay between the spectral coefficients of the DTFS representing the periodic extension of the aperiodic signal and its Fourier transform, the set of coefficients

Exercises

9.9. A DTFS representation for a sequence of data is given by

$$x[n] = 0.25 + 1.0 \cos\left[\left(\frac{2\pi}{10}\right)n\right] + 0.5 \sin\left[\left(\frac{2\pi}{10}\right)n\right]$$

$$+ 0.2 \cos\left[3\left(\frac{2\pi}{10}\right)n\right] - 0.2 \sin\left[3\left(\frac{2\pi}{10}\right)n\right] + 0.2 \cos\left[5\left(\frac{2\pi}{10}\right)n\right]$$

Express $x[n]$ in complex form.

ANS. $$x[n] = 0.25 + (0.5 - j0.25)e^{j\left(\frac{2\pi}{10}\right)n} + (0.1 + j0.1)e^{j3\left(\frac{2\pi}{10}\right)n}$$

$$+ 0.2e^{j5\left(\frac{2\pi}{10}\right)n} + (0.1 - j0.1)e^{j7\left(\frac{2\pi}{10}\right)n} + (0.5 + j0.25)e^{j9\left(\frac{2\pi}{10}\right)n}$$

9.10. A DTFS representation expressed in complex form is given by

$$x[n] = 2 + (1 + j1)e^{j\left(\frac{2\pi}{8}\right)n} + (1 - j1)e^{j3\left(\frac{2\pi}{8}\right)n} + (1 + j1)e^{j5\left(\frac{2\pi}{8}\right)n} + (1 - j1)e^{j7\left(\frac{2\pi}{8}\right)n}$$

Express $x[n]$ in trigonometric form.

ANS. $$x[n] = 2 + 2.0 \cos\left[\left(\frac{2\pi}{8}\right)n\right] - 2.0 \sin\left[\left(\frac{2\pi}{8}\right)n\right] + 2.0 \cos\left[3\left(\frac{2\pi}{8}\right)n\right] + 2.0 \sin\left[3\left(\frac{2\pi}{8}\right)n\right]$$

9.11. The following vector describes the spectral coefficients of a DTFS expressed in complex rectangular form

$$X = \begin{bmatrix} 1 & 0.25 + j0.25 & 4 - j1 & 0 & 0.3 & 0 & 4 + j1 & 0.25 - j0.25 \end{bmatrix}$$

Write the DTFS in trigonometric form using a magnitude and phase notation.

ANS. $$x[n] = 1 + 0.7071 \cos\left[\left(\frac{2\pi}{8}\right)n + \frac{\pi}{4}\right] + 8.2462 \cos\left[2\left(\frac{2\pi}{8}\right)n - 0.0780\pi\right] + 0.3 \cos\left[4\left(\frac{2\pi}{8}\right)n\right]$$

$\{X(0), X(1), \ldots, X(N-1)\}$ in Eq. (9.49) is referred to as the *discrete Fourier transform* (DFT) of $y[n]$. Because the DFT is essentially a special interpretation of a DTFS, the algorithms in Section 9.2.5 can also be used to produce the spectral coefficients of the DFT. However, as explained in the next section, a more efficient algorithm is available to compute the DFT of a discrete-time aperiodic signal or, for that matter, the spectral coefficients of a DTFS. This algorithm is known as the *fast Fourier transform* (FFT) and represents one of the most significant developments in digital signal processing. However, before we move on to this topic, we shall first consider an important degenerate case of the DFT.

Consider the situation where an aperiodic signal has infinite duration or exists with nonzero values over a much longer time than the observation interval of N samples. This is illustrated in Figure 9.10a for an exponentially decaying waveform. Under such conditions it is impossible to represent this signal exactly with a periodic signal having a finite period. Instead, one can only approximate the waveform over the observation interval as shown in Figure 9.10b using a periodic extension of the finite duration signal shown in Figure 9.10c. The error is a form of *time-domain aliasing* and is directly related to the jump discontinuity that occurs at the wraparound point of the periodically extended waveform. The spectral coefficients determined by the DFT would then correspond to samples of the Fourier transform of the signal shown in Figure 9.10c, not Figure 9.10a.

Figure 9.10. (a) Arbitrary signal of infinite duration; (b) periodic extension of infinite duration of short portion of signal; (c) signal approximation of finite duration.

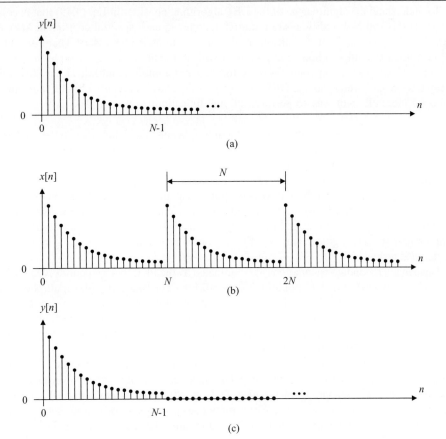

In practice it is important to keep the jump discontinuity to a minimum if the DFT is to reveal the spectral properties of the original waveform. The most common method is to extend the observation interval to large values of N so that the combined energy of the aliased components is made insignificant relative to the energy in the signal. Furthermore, some reduction in the overall observation time can be achieved if the method of windowing is used. Windowing is a mathematical process that alters the shape of the signal over the observation interval and gradually forces it to decay to zero at both ends. This eliminates the discontinuities in the periodically extended waveform at the expense of decreased frequency resolution. The net result is a concentration of the aliasing energy or *frequency leakage* into a few spectral bins instead of having it spread across many different bins.

Because test time is always paramount in a mixed-signal test, one should avoid discontinuities in the periodically extended waveform. In the words of Chapter 8, one should restrict all signals to be coherent rather than noncoherent. We shall delay the introduction of our examples until we first describe the principles of the fast Fourier transform.

9.3.2 The Fast Fourier Transform

In the early 1960s, J. Tukey invented a new algorithm for performing the DFT computations in a much more efficient manner. J. W. Cooley, a programmer at IBM, translated Tukey's algorithm into computer code and the Cooley–Tukey fast Fourier transform (FFT) was born.[4] It is now known that this algorithm actually dates back at least a century. The great German mathematician C. F. Gauss is known to have developed the same algorithm.

To understand the significance of the DFT algorithm, consider in Eq. (9.49) that N complex multiplications and $N-1$ additions are required to compute each spectral coefficient $X(k)$. For N spectral coefficients, another N multiplications and additions are necessary. Therefore, in total, $(N-1)N$ complex multiplications and additions will be required. As an example, to perform a DFT on 1024 samples, a computer has to perform over one million multiplications. To minimize test time, we would prefer that the DFT computation time be as small as possible; thus obviously we need a more efficient way to perform the multiplications in a DFT. This is where the FFT comes in.

The FFT works by partitioning each of the multiplications and additions in the DFT in such a way that there are many redundant calculations. The redundancy is removed by "folding" the redundant calculations on top of one another and performing each calculation only once. The folding operation forms a so-called *butterfly network* because of the butterfly shapes in the calculation flowchart.[5] There are several different ways to split the calculations and fold the redundancies into one another. The butterfly network can be laid out in a decimation-in-frequency configuration or a decimation-in-time configuration. Fortunately, the details of the FFT algorithm itself are largely unimportant to the test engineer, since the FFT operation is built into the operating system of most mixed-signal testers.

Since many redundant calculations are eliminated by the FFT, it only requires $N \log_2(N)$ complex multiplications. For a 1024-point FFT, only 1024×10 or 10240 complex multiplications are required. Compared to the one million complex multiplications required by the complex version of the DFT, this represents a huge reduction in computation time. The difference between the FFT and DFT becomes more extreme with larger sample sets, since the DFT produces an exponential increase in computations as the sample size increases.

Although the FFT produces the same output as an equivalent DFT, the more common FFTs can only operate on a sample size that is equal to 2n, where n is an integer. For instance, it is not possible to perform a standard Cooley–Tukey FFT on 380 samples, although a DFT would have no problem doing so. The limited choice of sample sizes is the major difference between the DFT and the FFT, other than the difference in computation time. Nevertheless, the savings in test time is so huge that test engineers usually have no choice but to use the FFT with its limited sample size flexibility. It is quite possible that improvements in computation speeds will eventually make the FFT obsolete in mixed-signal testers, allowing DFTs instead. Until then, the mixed-signal test engineer should be prepared to work with 2n samples for most tests.

9.3.3 Interpreting the FFT Output

The output format of a mixed-signal tester's FFT depends somewhat on the vendor's operating system. In older testers, the format of the FFT output was arranged as an N-point array with the DC and Nyquist levels in the first two array elements followed by the cosine/sine pairs for each spectral bin. Today, most testers incorporate commercial DSP chips sets that compute the FFT using complex arithmetic and store the complex numbers in an array beginning with the DC bin followed by successive harmonic bins up to the Nyquist bin. The same format is also used for most numerical software packages such as MATLAB.

One has to be careful, though, when interpreting the FFT output. Many ATE versions of the FFT do not produce peak voltage outputs. Some produce voltage squared (power) outputs, some produce voltage outputs multiplied by the number of samples over 2 (i.e., a 1-V input with 1024 points produces an FFT output of 512 units), and so on. This suggests that the test engineer must become familiar with the FFT routine that they intend to use and determine all the necessary scale factors. In addition, it has been the authors' preference to adjust the scale factors so that the FFT produces RMS levels instead of peak levels. For reasons that will become clear in the next chapter, many test metrics call for the combination of the power of several spectral components. Working with RMS values simplifies this approach.

To better understand the steps involved, let us consider the manner in which MATLAB performs the FFT and the corresponding scale factors needed to produce RMS spectral levels. First, the FFT that is performed in MATLAB is given by the equation

$$Y(k) = \sum_{n=0}^{N-1} y[n] e^{-jk\left(\frac{2\pi}{N}\right)n} \tag{9.50}$$

In turn, the complex spectral coefficients $X(k)$ of the DTFS representation of the periodic extension of $y[n]$ are given by

$$X(k) = \frac{Y(k)}{N} \tag{9.51}$$

and the corresponding cosine/sine coefficients are found according to Eq. (9.42) to be

$$\tilde{a}_k = \begin{cases} \text{Re}\left\{X(k)\right\}, & k = 0, N/2 \\ 2\,\text{Re}\left\{X(k)\right\}, & k = 1, 2, \ldots, N/2 - 1 \end{cases} \tag{9.52}$$

and

$$\tilde{b}_k = \begin{cases} 0, & k = 0, N/2 \\ -2\,\text{Im}\left\{X(k)\right\}, & k = 1, 2, \ldots, N/2 - 1 \end{cases} \tag{9.53}$$

where Re{ } and Im{ } denote the real and imaginary parts of a complex number, respectively. Again, we must alert the reader to the different scale factors in front of these terms.

The magnitude and phaserepresentation is determined from Eq. (9.15) to be

$$
\tilde{c}_k = \begin{cases} |X(k)|, & k = 0, N/2 \\ 2|X(k)|, & k = 1, \ldots, N/2 - 1 \end{cases} \tag{9.54}
$$

where $|X(k)| = \sqrt{\text{Re}\{X(k)\}^2 + \text{Im}\{X(k)\}^2}$ and the phase is computed using the four-quadrant formula given below

$$
\tilde{\phi}_k = \begin{cases} -\tilde{\phi}_o & \text{if } \text{Re}[X(k)] \geq 0, \text{Im}[X(k)] \geq 0, \\ \tilde{\phi}_o & \text{if } \text{Re}[X(k)] \geq 0, \text{Im}[X(k)] < 0, \\ \tilde{\phi}_o - \pi & \text{if } \text{Re}[X(k)] < 0, \text{Im}[X(k)] \geq 0, \\ \pi - \tilde{\phi}_o & \text{if } \text{Re}[X(k)] < 0, \text{Im}[X(k)] < 0, \end{cases} \quad \text{where} \quad \tilde{\phi}_o = \tan^{-1}\left\{ \frac{\left|\text{Im}[X(k)]\right|}{\left|\text{Re}[X(k)]\right|} \right\} \tag{9.55}
$$

In many situations we shall find it more convenient to report the spectral coefficients in terms of their RMS values. To do so, we divide the spectral coefficients $\tilde{c}_k$ (except the DC term described $k = 0$) by $\sqrt{2}$ to obtain

$$
\tilde{c}_{k-RMS} = \begin{cases} \tilde{c}_k, & k = 0 \\ \dfrac{\tilde{c}_k}{\sqrt{2}}, & k = 1, \ldots, N/2 \end{cases} \tag{9.56}
$$

On substituting Eq. (9.54) into (9.56), we obtain

$$
\tilde{c}_{k-RMS} = \begin{cases} |X(k)|, & k = 0 \\ \sqrt{2}|X(k)|, & k = 1, \ldots, N/2 - 1 \\ \dfrac{1}{\sqrt{2}}|X(k)|, & k = N/2 \end{cases} \tag{9.57}
$$

The following example will further illustrate this procedure.

At this point in our discussion of the FFT, it would be instructive to consider the spectral properties of a coherent sinusoidal signal and a noncoherent sinusoidal signal having equal amplitudes.

Extending the observation interval in Example 9.8 certainly helped to decrease the amount of frequency leakage, which, in turn, helped to improve the measurement accuracy of the noncoherent

EXAMPLE 9.6

Using the FFT routine in MATLAB (or some other equivalent software), compute the spectral coefficients $\{a_k\}$ and $\{b_k\}$ of a multitone signal having the following eight samples; {0.1414, 1.0, −0.1414, −0.8, −0.1414, 1.0, 0.1414, −1.2}. These samples were derived from a signal with the following DTFS representation

$$x[n] = \cos\left[2\left(\frac{2\pi}{8}\right)n - \frac{\pi}{2}\right] + 0.2\,\cos\left[3\left(\frac{2\pi}{8}\right)n - \frac{\pi}{4}\right]$$

Also report the magnitude (in RMS) and the phase of each spectral coefficient.

Solution:
With the samples of the multitone signal described as

$$x = [0.1414, 1.0, −0.1414, −0.8, −0.1414, 1.0, 0.1414, −1.2],$$

the FFT routine in MATLAB produces the following output, together with the scaled result:

$$Y = FFT(x) = \begin{bmatrix} 0 \\ 0 \\ 0 - j4.0000 \\ 0.5657 - j0.5657 \\ 0 \\ 0.5657 + j0.5657 \\ 0 + j4.0000 \\ 0 \end{bmatrix} \quad \Rightarrow \quad X = \frac{Y}{8} = \begin{bmatrix} 0 \\ 0 \\ 0 - j0.5000 \\ 0.0707 - j0.0707 \\ 0 \\ 0.0707 + j0.0707 \\ 0 + j0.5000 \\ 0 \end{bmatrix}$$

Subsequently, the cosine/sine spectral coefficients are determined from Eqs. (9.52) and (9.53) to be

$$\begin{bmatrix} \tilde{a}_0 \\ \tilde{a}_1 \\ \tilde{a}_2 \\ \tilde{a}_3 \\ \tilde{a}_4 \end{bmatrix} = \begin{bmatrix} 0 \\ 0 \\ 0 \\ 0.1414 \\ 0 \end{bmatrix} \quad \text{and} \quad \begin{bmatrix} \tilde{b}_0 \\ \tilde{b}_1 \\ \tilde{b}_2 \\ \tilde{b}_3 \\ \tilde{b}_4 \end{bmatrix} = \begin{bmatrix} 0 \\ 0 \\ 1.0 \\ 0.1414 \\ 0 \end{bmatrix}$$

Finally, the corresponding magnitude (in RMS) and phase terms (in radians) are as follows:

$$\begin{bmatrix} \tilde{c}_{0-RMS} \\ \tilde{c}_{1-RMS} \\ \tilde{c}_{2-RMS} \\ \tilde{c}_{3-RMS} \\ \tilde{c}_{4-RMS} \end{bmatrix} = \begin{bmatrix} 0 \\ 0 \\ 0.7071 \\ 0.1414 \\ 0 \end{bmatrix} \quad \text{and} \quad \begin{bmatrix} \tilde{\phi}_0 \\ \tilde{\phi}_1 \\ \tilde{\phi}_2 \\ \tilde{\phi}_3 \\ \tilde{\phi}_4 \end{bmatrix} = \begin{bmatrix} 0 \\ 0 \\ \pi/2 \\ \pi/4 \\ 0 \end{bmatrix}$$

EXAMPLE 9.7

Using the FFT routine in MATLAB (or some other equivalent software), compute the spectral coefficients of a coherent and noncoherent sinusoidal signal with parameters $A = 1$, $\phi = 0$, $M = 3$, $N = 64$, and $A = 1$, $\phi = 0$, $M = \pi$ (3.14156), $N = 64$. For each case, plot the RMS magnitude of the spectrum in dB relative to a 1-V RMS reference level.

Solution:

The following MATLAB routine was written to produce 64 samples of the coherent and noncoherent waveforms and to perform the corresponding FFT analysis:

```
% coherent signal definition-x -
N=64; M=3; A=1; P=0;    % signal definition
     for n=1:N,
               x(n)=A*sin(2*pi*M/N*(n-1)+P);
     end;
% noncoherent signal definition-z -
     N=64; M=pi; A=1; P=0;
     for n=1:N,
               z(n)=A*sin(2*pi*M/N*(n-1)+P);
     end;
% perform Fourier analysis
     X=fft(x)/length(x);
     Z=fft(z)/length(z);
```

The results of the FFT analysis for the coherent waveform are shown in Figure 9.11. The time-domain waveform is shown on the left, while the corresponding magnitude of the spectrum is shown on the right. The spectrum is expressed in dB relative to a 1-V RMS reference level. When this definition is used, the decibel units are referred to as dBV. Mathematically, it is written as

$$c_{k-RMS}(dBV) = 20 \log_{10}\left(\frac{c_{k-RMS}}{1-\text{V RMS}}\right) \tag{9.58}$$

Also, it is customary to plot the frequency-domain data as a continuous curve by interpolating between frequency samples instead of using a line spectrum. In some cases, one uses a zero-order interpolation operation to produce a step or bar graph of the spectrum, as we shall use in the next few examples, or a first-order interpolation operation.

The results for the noncoherent waveform were also found and are shown in Figure 9.12. On comparing the magnitude of the two spectra, we clearly see a significant difference. In the case of the coherent sinusoidal waveform, a single spike occurs in bin 3 with over 300 dB of separation distance from all other spectral coefficients. A closer look at the numbers indicates that the tone has an RMS value of 0.707106 (-3.0103 dBV in the plot) or an amplitude of 1. This is exactly the value specified in the code that is used to generate the coherent sinusoid. In the case of the noncoherent sinusoidal waveform, no single spike occurs. Rather, the single-tone waveform appears to consist of many frequency components. If other signal components were present, then they would be corrupted by the power in these leaked components. What is worse, it is extremely difficult to determine the amplitude of the noncoherent sinusoidal signal with its power smeared across many frequency locations.

Figure 9.11. (a) Coherent waveform time-domain plot and (b) corresponding spectrum plot.

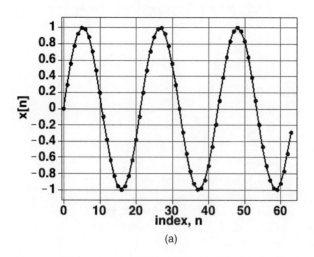

(a)

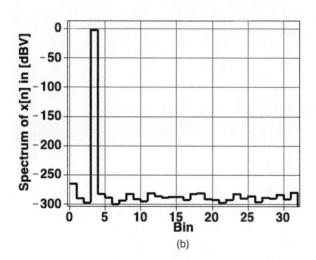

(b)

The most straightforward method to improve the measurement accuracy of a noncoherent waveform is to increase its observation interval. Generally speaking, this approach is used by most benchtop instruments found in one's laboratory, such as spectrum analyzers, multimeters, and digitizing oscilloscopes. Samples taken from the input signal are unrelated to the sampling rate of the instrument, and are therefore noncoherent. Instruments of this type are usually not expected to generate a result in a very short time, such as 25 ms. Rather, they are only required to produce a result every 1 or 2 s, which is usually more than adequate. Consequently, one can construct a less complex, noncoherent measurement system. Our next example will illustrate the effect of a longer observation interval on the spectrum of a noncoherent sinusoidal waveform.

Figure 9.12. (a) Noncoherent waveform time-domain plot and (b) corresponding spectrum plot.

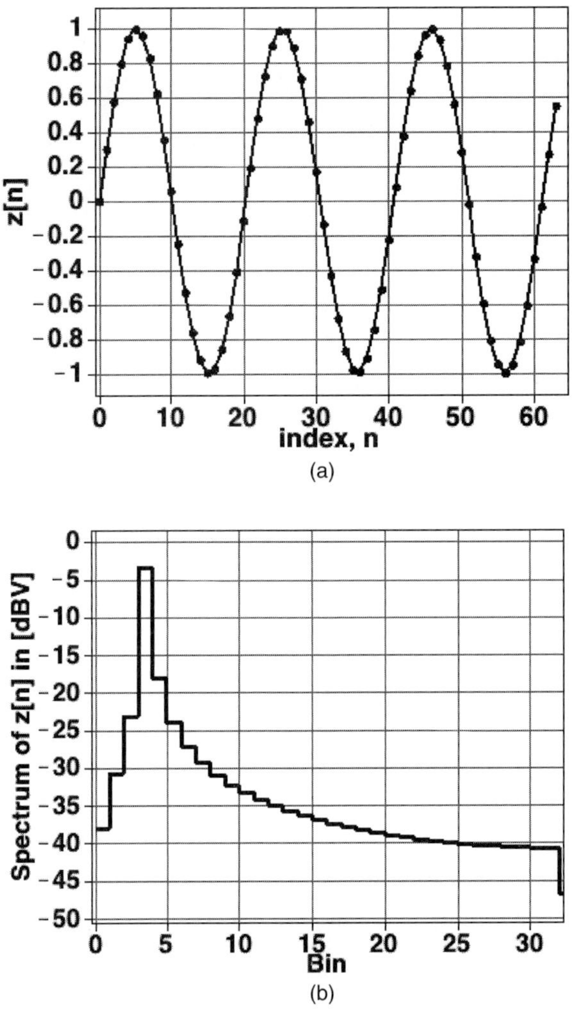

(a)

(b)

EXAMPLE 9.8

Extend the observation interval of the noncoherent waveform of Example 9.7 by collecting 8192 samples instead of 64. Plot the corresponding magnitude of the resulting spectrum. Determine the amplitude of the input signal from its spectrum.

Solution:
The MATLAB code for the noncoherent signal from Example 9.7 was modified as follows:

```
% noncoherent signal definition–z –
        NOI=8192;                          % observation interval
```

```
N=64; M=pi; A=1; P=0;              % signal definition
for n=1:NOI,
    z(n)=A*sin(2*pi*M/N*(n-1)+P);
end;
```

Here we distinguish between N, the number of samples in one UTP, and N_{OI}, the number of samples collected over the entire observation interval. In other words, N_{OI}/N represents the number of UTPs that the signal will complete in the observation interval.

The Fourier analysis was then repeated and the corresponding spectrum was found as shown in Figure 9.13. On the left is the plot of the time-domain waveform over the last 64 samples of the full 8192 samples, as any more samples would fill the graph and mask all detail. On the right is the magnitude of its spectrum.

Figure 9.13. (a) Noncoherent waveform time-domain plot and (b) frequency spectrum with expanded observation interval.

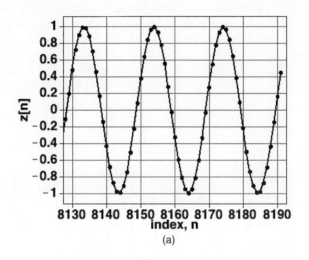

(a)

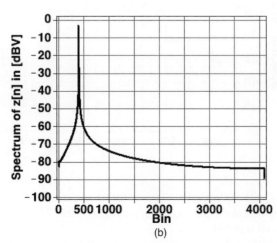

(b)

When we refer back to the spectrum derived in Example 9.7 and compare it to the one derived here, we see that the general shape of the magnitude of the spectrum is much more concentrated around a single frequency and that the frequency leakage components are much smaller. The astute reader may be wondering about the scale of the axis. In this case we are plotting the index or bin of each frequency components from 0 to 8191, whereas in the previous case our frequency index is from 0 to 63. It is important to realize that each bin is equivalent to $Bin\left(F_s/N_{OI}\right)$ Hz. In other words, the frequency range is identical in each case; only the frequency granularity is different.

In order to estimate the amplitude of the input waveform, one cannot rely on the peak value of the spectrum as was done in the coherent waveform case. Rather, we must use several frequency components centered around the peak spectral concentration to estimate the waveform amplitude. To see this more clearly, we provide in Figure 9.14 an expanded view of the spectrum around the spectral peak. Here we see that a peak spectral value of −3.2318 dBV occurs in bin 402 [bin $(M/N)N_{OI}$]. Ideally, the spectral peak value should be −3.0103 dBV [= $20\log_{10}\left(0.707 \text{ V RMS}/1 \text{ V RMS}\right)$]. To improve the estimate, we must take into consideration the power associated with the side tones. As the magnitude of these side tones drop off fairly quickly, let us consider that the power associated with the input signal is mainly associated with the power of the five tones before and after the spectral peak. On doing so, the amplitude estimate becomes −3.0355 dBV. Including more side tones into this calculation will only help to improve the estimate. Generally speaking, side tones less than −60 dB below the spectral peak value will improve the accuracy to within 0.1%.

Figure 9.14. Expanded view of the noncoherent tone around the spectral peak.

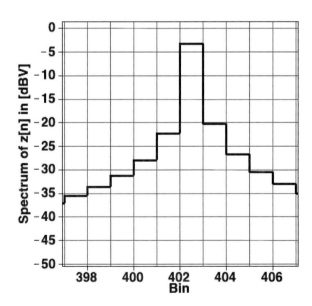

We could also improve the accuracy of the estimate by further increasing the observation interval. For example, increasing the observation interval to 131,072 samples will improve the amplitude estimate to −3.0109 dB.

waveform. However, as in most production test situations, one is always searching for a faster solution. In Section 9.3.1 the method of windowing was suggested as a possibility. We shall investigate this claim further in the next subsection.

Exercises

9.12. Evaluate $x[n] = 1.0 \sin\left[3\,(2\pi/16)n + \pi/8\right]$ for $n = 0,1,\ldots,15$. Using MATLAB, compute the FFT of the 10 samples of $x[n]$ and determine the corresponding $\{a_k\}$ and $\{b_k\}$ spectral coefficients.

ANS. $\{a_3 + jb_3\} = \{0.3827 + j0.9239\}$; all others are zero.

9.13. Evaluate $x[n] = (n-3)^2$ for $n = 0,1,\ldots,15$. Using MATLAB, compute the FFT of the 16 samples of $x[n]$ and determine the corresponding $\{c_k\}$ and $\{\phi_k\}$ spectral coefficients. Express the magnitude coefficients in RMS form.

ANS. $\{c_k\} = \{41.5, 52.8, 24.3, 16.4, 12.8, 10.8, 9.75, 9.17, 4.50\}$; $\{c_{k\text{-}RMS}\} = \{41.5, 37.3, 17.2, 11.6, 9.05, 7.67, 6.89, 6.49, 3.18\}$; $\{\phi_k\} = \{0, -1.25, -1.70, -1.99, -2.24, -2.47, -2.70, -2.92, -3.14\}$.

9.14. A signal has a 0.5-V amplitude. Express its amplitude in units of dBV.

ANS. -9.03 dBV.

9.15. A signal has a period of 1 ms and is sampled at a rate of 128 kHz. If 128 samples are collected, what is the frequency resolution of the resulting FFT? If the number of samples collected increases to 8192 samples, what is the frequency resolution of the resulting FFT?

ANS. 1 kHz, 15.625 Hz.

9.3.4 Windowing

At the heart of a DTFS is the assumption that the signal of interest continues for all time, albeit as a periodic extension of an N-point sample set. If the signal is coherent with the UTP, then the periodic extended waveform will remain coherent over any integer multiple of the UTP. In the limit, as the observation interval ($LxUTP$, L is an integer) goes to infinity, the DTFS representation will have a line spectrum that is identical to that described by the Fourier transform of the infinite-duration sampled sinusoidal signal as illustrated in Figure 9.15a. However, if the sampled signal is not coherent in the UTP, then the resulting DTFS will have a line spectrum that differs significantly from that corresponding to the continuous-running sampled sinusoidal signal. This should not be too surprising given that the periodically extended sampled signal and the continuous-running sampled sinusoidal signals are very different signals, as captured by the illustration on the left-hand side of Figure 9.15b. It should also be of no surprise that the DTFS representation of the noncoherent sampled sinusoidal contains numerous line spectra, as the waveform deviates significantly from an ideal sine wave. Now we learned in Section 9.3.1 that the DTFS representation of a sampled signal represents the sampled spectrum of a finite duration signal, that is, one that lasts for a fixed

Figure 9.15. Comparing the DTFS representation and Fourier transform of various types of sinusoidal signals: (a) coherent with UTP, (b) noncoherent with UTP, and (c) noncoherent vs. aperiodic signal.

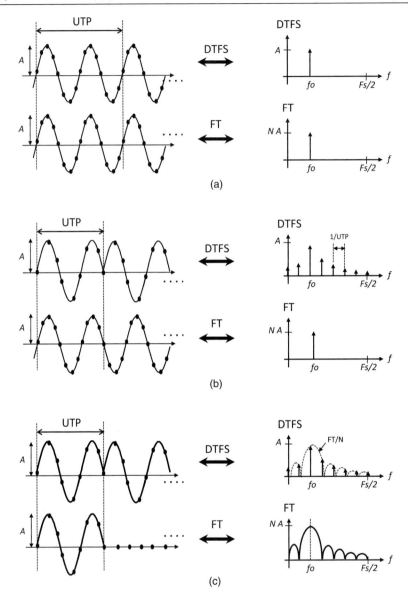

duration of a UTP, as depicted in Figure 9.15c. We see from this illustration that the shape of the line spectrum associated with the noncoherent signal is directly related to the Fourier transform of this aperiodic signal. More importantly, the shape of the line spectrum can be traced back to the rectangular function that converts the infinite-duration sampled sine wave to one of finite duration, leading to the term *windowing*.

Figure 9.16. Illustrating the rectangular windowing effect on a coherent signal with UTP and the resulting single-line DTFS representation.

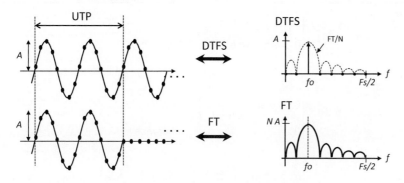

It is interesting to note that the DTFS representation of a coherent sinusoidal signal also represents samples of the Fourier transform of the finite duration signal. However, due to the relationship between the frequency of the signal, f_o, and the sampling frequency F_s, only one point in the entire Nyquist interval ($F_s/2$) is nonzero as shown in Figure 9.16. This, in turn, gives the appearance of a single line spectrum, as well as an appearance identical to that given by its Fourier transform.

Windowing is the operation of converting an infinite duration sampled signal to one of finite duration. Mathematically, if $x[n]$ represents a sampled signal of infinite duration and $w[n]$ as the samples corresponding to the windowing operation, then the widowed sequence can be written as

$$y[n] = w[n]x[n] \tag{9.59}$$

Windowing is not limited to a rectangular shaped function, because many types of windows have been developed for spectral analysis—*largely to control the shape of the spectral decomposition.* Table 9.1 lists some of the more commonly used window functions and their corresponding N-point sequence. This includes the Hanning (or Hann), Blackman, and Kaiser windows. The Hanning window is used in about 95% of all windowing applications. When in doubt, begin with the Hanning window.

Because multiplication in the time domain corresponds to convolution in the frequency domain, this fact becomes the guiding principle in the design of different window functions. The frequency domain description of a window function is derived through the application of the Fourier transform provided in Eq. (9.47). While window functions have no convenient closed form (except for the rectangular function), window attributes have been collected and tabulated for easy access.[6] Figure 9.17 illustrates the general shape of a window function in the frequency domain. In many ways, the shape of this window resembles that of an amplitude-based filter, and many of the attributes associated with a window are filter-like parameters such as DC frequency gain, main-lobe bandwidth (MLBW), first side-lobe attenuation, rate of attenuation decay and equivalent noise bandwidth (ENBW). One can show that the DC gain of the window, $|W(0)|$ is proportional to the sum of the window coefficients as

$$|W(0)| = \frac{1}{N}\sum_{n=0}^{N-1} w[n] \tag{9.60}$$

Table 9.1. Some Commonly Used Window Types and Their Corresponding N-Point Sequence

Window Type	Window Function $w[n]$
Rectangular	$w[n] = \begin{cases} 1, & 0 \le n \le N-1 \\ 0 & \text{otherwise} \end{cases}$
Hanning (Hann)	$w[n] = \begin{cases} 0.5 - 0.5\cos\left(\dfrac{2\pi}{N-1}n\right) & 0 \le n \le N-1 \\ 0 & \text{otherwise} \end{cases}$
Blackman	$w[n] = \begin{cases} 0.42 - 0.5\cos\left(\dfrac{2\pi}{N-1}n\right) + 0.08\cos\left(\dfrac{4\pi}{N-1}n\right) & 0 \le n \le N-1 \\ 0 & \text{otherwise} \end{cases}$
Kaiser	$w[n,\beta] = \begin{cases} \dfrac{I_0\left(\pi\beta\sqrt{1-\left(\dfrac{2n}{N-1}-1\right)^2}\right)}{I_0(\pi\beta)}, & 0 \le n \le N-1 \\ 0 & \text{otherwise} \end{cases}$ where $$I_0(x) = 1 + \sum_{k=1}^{\infty}\left[\left(\frac{1}{k!}\frac{x}{2}\right)^k\right]^2 \approx 1 + \sum_{k=1}^{20}\left[\left(\frac{1}{k!}\frac{x}{2}\right)^k\right]^2$$ (zeroth-order modified Bessel function of the first kind)

Figure 9.17. Some attributes of a window function in the frequency domain.

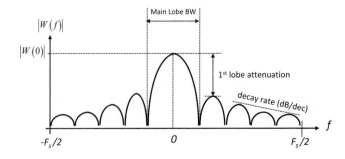

Likewise, the area under the squared-window magnitude response $|W(f)|^2$, denoted by the window shape factor ε^2, can be shown to be equal to

$$\varepsilon^2 = \int_{-F_s/2}^{F_s/2} |W(f)|^2 \, df = \frac{1}{N} \sum_{n=0}^{N-1} w^2[n] \tag{9.61}$$

Correspondingly, the ENBW is then defined as

$$\text{ENBW} = \frac{\int_{-F_s/2}^{F_s/2} |W(f)|^2 \, df}{[W(0)]^2} = N \frac{\sum_{n=0}^{N-1} w^2[n]}{\left[\sum_{n=0}^{N-1} w[n]\right]^2} \tag{9.62}$$

Table 9.1 lists commonly used window types, such as rectangular, Hanning, Blackman, and Kaiser, to name just a few. Table 9.2 summarizes their corresponding window metrics. One important window metric that is not self-evident is the processing gain (PG). We'll describe this next.

Consider a set of samples derived from a complex sine wave with frequency f_o and amplitude A, that is,

$$x[n] = Ae^{+j2\pi f_o n} \tag{9.63}$$

Here we make use of a complex-valued sine wave rather than a real-valued one because it simplifies the presentation. Let us further assume these samples are multiplied by some arbitrary window function $w[n]$. The Fourier transform of the widowed signal is then found using Eq. (9.47) to be

$$Y\left(e^{j\omega}\right) = A \sum_{n=0}^{N-1} w[n] e^{+j2\pi f_o n} \, e^{-j\omega n} \tag{9.64}$$

Table 9.2. Some Window Types and Their Corresponding Performance Metrics

Window Type	$G_{COHERENT}$ $= |W(0)|$	$G_{NONCOHERENT} =$ $\sqrt{\int_{-F_s/2}^{F_s/2} |W(f)|^2 \, df}$	MLBW (bins)	ENBW (bins)	1st Lobe Atten. (dB)	Decay Rate (dB/oct)	PG
Rectangular	1.00	1.00	2	1.00	13	−6	1.00
Hanning (Hann)	0.50	0.612	4	1.50	26	−18	0.666
Blackman	0.42	0.552	6	1.72	50.4	−18	0.579
Kaiser (β = 10)	0.22	0.398	20	3.19	247	−6	0.313

Correspondingly, the FFT samples this frequency domain signal with $\omega = \dfrac{2\pi}{N}k$, $k = 0$, $N\text{-}1$, leading to the widowed DTFS spectral coefficient (signified by the subscript W) as

$$X_W(k) = \left. \frac{Y(e^{j\omega})}{N} \right|_{\omega = \frac{2\pi}{N}k} = \frac{A}{N} \sum_{n=0}^{N-1} w[n] e^{j2\pi\left(f_o - \frac{k}{N}\right)n}, \qquad k = 0,1 \qquad \textbf{(9.65)}$$

If the input frequency is coherent with the mth bin of the FFT, then Eq. (9.65) reduces to

$$\left| X_W(k) \right| = \begin{cases} \dfrac{A}{N} \displaystyle\sum_{n=0}^{N-1} w[n], & k = m \\[2mm] 0, & k = 0,1,...,N-1, k \neq m \end{cases} \qquad \textbf{(9.66)}$$

For the mth bin, we recognize that the magnitude of the spectral coefficient is proportional to the amplitude of the input signal. This allows us to define the *coherent signal gain* of the windowing operation as

$$G_{COHERENT} \triangleq \frac{\left| X_W(m) \right|}{A} = \frac{1}{N} \sum_{n=0}^{N-1} w[n] \qquad \textbf{(9.67)}$$

Interestingly enough, the coherent gain is equal to the DC gain of the window.

If we repeat this analysis but this time consider the input as a set of N samples derived from a Gaussian distribution with zero mean and standard deviation σ_M, one can show that the expected or average noncoherent power of each FFT bin is

$$E\left\{ \left| X_{W-RMS}(k) \right|^2 \right\} = \frac{\sigma_M^2}{N^2} \sum_{n=0}^{N-1} w^2[n], \quad k = 0,1,...,N-1 \qquad \textbf{(9.68)}$$

In the case of the rectangular window, the average power per bin is $\sigma_{\eta,bin}^2 = \sigma_M^2/N$, so we define the *noncoherent power gain* for the mth bin as

$$G_{NONCOHERENT}^2 \triangleq \frac{E\left\{ \left| X_{W-RMS}(m) \right|^2 \right\}}{\sigma_{\eta,bin}^2} = \frac{1}{N} \sum_{n=0}^{N-1} w^2[n] \qquad \textbf{(9.69)}$$

Here noncoherent power gain is equivalent to the window shape factor ε^2 and we notice that it is less than unity except for a rectangular window. On the surface this would imply that the windowing reduces the noise in each bin and that this would be advantageous. However, the signal level reduces at a faster rate due to the coherence gain, hence the apparent reduction in noise power is further offset by the reduction in signal level. To capture this attribute, we speak in terms of the processing gain (PG), where it is defined as the ratio of the output signal-power-to-noise-power-per-bin to that corresponding to the S/N of the rectangular window, that is,

$$PG \triangleq \frac{[S/N]_{OUT}}{[S/N]_{IN}} = \frac{\dfrac{\left| X_W(m) \right|^2}{2} \bigg/ E\left\{ \left| X_{W-RMS}(m) \right|^2 \right\}}{\dfrac{A^2}{2} \bigg/ \sigma_{n,bin}^2} = \frac{1}{N} \frac{\left[\displaystyle\sum_{n=0}^{N-1} w[n] \right]^2}{\displaystyle\sum_{n=0}^{N-1} w^2[n]} \qquad \textbf{(9.70)}$$

For windows other than rectangular, the sum of window coefficients is always less than the corresponding sum of squared coefficients (as positive and negative coefficients can cancel), PG is always less than one. Therefore, we can conclude that windowing reduces the quality of the spectral coefficient estimation. The astute reader will recognize that the PG is equal to the inverse of the ENBW given in Eq. (9.62). This is reasonable, because a larger noise bandwidth will allow greater noise to contribute to each spectral coefficient estimate.

In order to appreciate the coherent and noncoherent properties of the windowing process associated with the FTT/DTFS spectra estimation process, it is common practice to represent this as a bank of filters as shown in Figure 9.18. Here the center frequency of each filter is tuned to a bin of the FFT and whose frequency domain characteristics are those associated with the window function. Coherent sinusoidal signals will pass directly thorough the center of the appropriate filter with gain $G_{COHERENT}$, for example,

$$X_W\left(k\right) = G_{COHERENT} \times X\left(k\right)$$

(9.71)

where $X(k)$ are the DTFS spectral coefficients that result from a rectangular window. Of course, we could just as easily write this in terms of the c_k spectral coefficients through the application of Eq. (9.54).

Figure 9.18. Representing the windowing-FFT extraction process as a bank of filters whose frequency characteristics are those of the window function tuned to each bin of the FFT.

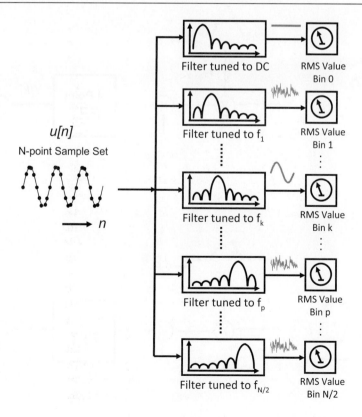

In the case of noncoherent signals, the situation is more complex. If a single noncoherent sine wave signal is processed through a windowing operation within the main-lobe bandwidth of the kth bin, then we describe the spectral coefficient estimate corresponding to bin k as

$$X_W(k) = G_{NONCOHERENT} \times \sqrt{\sum_{l=-MLBW/2}^{MLBW/2} X^2(k+l)} \qquad (9.72)$$

If the signal consists of several noncoherent sine waves that are far enough apart so that their main lobes do not overlap, then Eq. (9.72) can be used for each bin k, $k = 0, ..., N-1$. However, if the tones are too close to one another, Eq. (9.72) will be in error, because the spectra will overlap. So unless a window with a main lobe bandwidth less than the frequency separation is found, windowing alone will not improve the spectral coefficient estimates. Widowing combined with an increase observation interval can help to improve this situation. The next few examples will help to illustrate these tradeoffs.

To close this subsection, we illustrate the set up for using windowing with appropriate scale factors for spectral analysis in Figure 9.19. Figure 9.19a illustrates the coherent case and Figure 9.19b illustrates the appropriate scale factor for the noncoherent case. Scaling by the

Figure 9.19. Highlighting the scale factors associated with windowing when (a) coherent signals are being processed; (b) noncoherent signals are being processed.

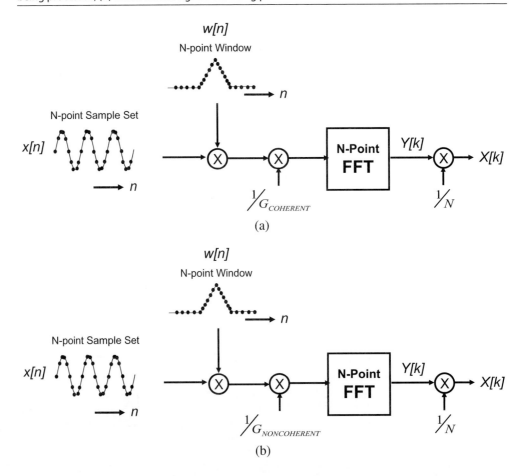

appropriate gain factor right after the widow operation equalizes the power in the window data with that associated with the original waveform. However, as the power across the main lobe of the window is being combined in a mean-square sense, the noise in each bin is also accumulated resulting in an increase in the noise power associated with the spectral estimate, given by

$$\sigma_{X_k} = \sigma_{n,bin}\sqrt{\mathrm{MLBW}}$$

(9.73)

where $\sigma_{n,bin}$ is the average RMS noise level per bin and σ_{X_k} is the noise associated with the kth spectral estimate.

EXAMPLE 9.9

Through the application of a Hanning window, compute the magnitude of the spectra of the noncoherent waveforms described in Examples 9.7 and 9.8 consisting of 64 and 8192 samples, respectively. Compare the spectra with the results from a rectangular window (i.e., the nonwindowed results obtained previously).

Solution:
Let us begin by investigating the spectral properties of the noncoherent waveform of Example 9.7, consisting of 64 samples (Figure 9.12). A MATLAB script was written to perform the windowing operation. The code is

```
% noncoherent signal definition-z -
    NOI=64;                    % observation interval
    N=64; M=pi; A=1; P=0;      % signal definition
    for n=1:NOI,
            z(n)=A*sin(2*pi*M/N*(n-1)+P);
    end;
% Hanning windowing operation
    for n=1:NOI,
            w(n)=0.5*(1-cos(2*pi*(n-1)/(N-1)));
    end;
    Gnoncoherent=sqrt(sum(w.*w)/NOI);
    u=1/Gnoncoherent * z .* w;
```

The results of running the above code is shown in Figure 9.20. Specifically, in Figure 9.20a the noncoherent sine wave z[n] is shown plotted over 64 points. In Figure 9.20b, 64 points associated with the window function w[n] are shown. The corresponding window data sequence, with the appropriate noncoherent gain factor accounted for, is shown in Figure 9.20c. As can be seen from Table 9.2, the noncoherent gain for the Hanning window is 0.612. Here we see that the window effectively squeezes the endpoints of the noncoherent waveform toward zero, forcing the endpoints of the waveform to meet smoothly. An FFT was then performed on the modified waveform from which the magnitude of the spectrum was derived. The windowed spectrum is shown in Figure 9.20d. Superimposed on the plot is the spectrum of the original noncoherent waveform derived in Example 9.6 without windowing (i.e., rectangular windowing). As is evident, the Hanning windowed spectrum is much more concentrated about a single frequency than the rectangular windowed spectrum.

Figure 9.20. Windowing results (a) original noncoherent waveform, (b) Hanning window, (c) windowed data, and (d) spectrum magnitude.

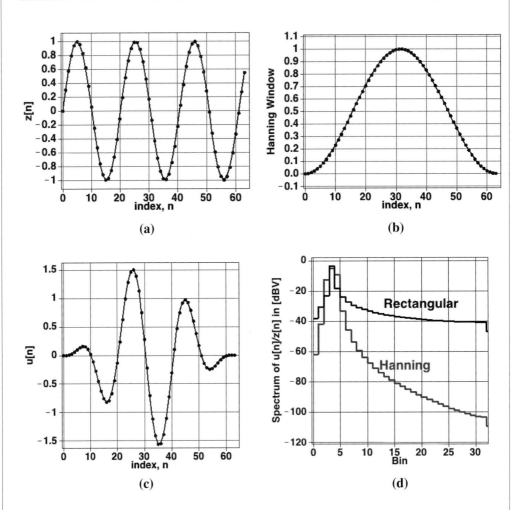

(a)

(b)

(c)

(d)

In order to estimate the amplitude of the signal present in the windowed data, an expanded view of the spectrum about its peak is shown in Figure 9.21. As side tones are present about the peak value of the spectrum, we must consider these tones in the estimate of the signal amplitude. Performing a square-root-of-sum-of-squares calculation involving the RMS value of the signals present in bins 1 to 5 (within the main lobe bandwidth of the window), we obtain a combined RMS value of 0.7070567. This, in turn, implies an amplitude estimate of 0.99993. For all intents and purposes, this is unity and was found with only 64 samples. Unless improved frequency resolution is necessary, preprocessing with the Hanning window is just as effective as the coherent measurement in this example. However, repeatability of measurements is degraded in windowed systems in the presence of random noise, because the processing gain is less than unity.

If we increase the number of samples to 8192 and repeat the windowing operation, we find the spectrum much more closely concentrated about a single frequency. The results are shown

Figure 9.21. Expanded view of the windowed, noncoherent tone around the spectral peak.

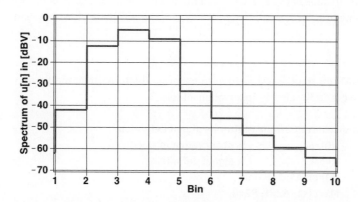

Figure 9.22. Spectral comparison of rectangular and Hanning windows.

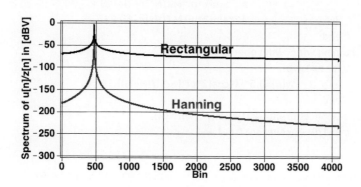

in Figure 9.22, together with the spectrum resulting from a rectangular window. Clearly, the Hanning window has a very narrow spectrum, very much like the coherent case. It would be suitable for making measurements in situations where more than one tone is present in the input signal. The drawback, of course, is that a longer observation interval is necessary and that side tones have to be dealt with.

In the previous example, windowing was used to improve the measurement of a noncoherent sinusoidal signal. In fact, an accurate estimate was obtained without extending the observation interval over and above that of a coherent sinusoidal signal. This example may incorrectly give the reader the impression that windowing can resolve the frequency leakage problem associated with noncoherent signaling with no added time expense. As we shall see in this next example consisting of a noncoherent multitone signal, this is indeed not the case.

EXAMPLE 9.10

Determine the magnitude of the spectrum of a multitone signal consisting of three noncoherent tones with parameters, $A = 1$, $\phi = 0$, $M = \pi$, $N = 64$; $A = 1$, $\phi = 0$, $M = \pi+1$, $N = 64$; and $A = 1$, $\phi = 0$, $M = \pi+2$, $N = 64$.

Solution:

The following MATLAB routine was written to generate the three-tone multitone signal with a Hanning window over a 64-sample observation interval:

```
% noncoherent 3-tone multitone signal definition
NOI=64;              % observation interval
N=64;                % signal definition
M1=pi; A1=1; P1=0;
M2=pi+1; A2=1; P2=0;
M3=pi+2; A3=1; P3=0;
for n=1: NOI,
        z(n)=A1*sin(2*pi*M1/N*(n-1)+P1) +
                A2*sin(2*pi*M2/N*(n-1)+P2) +
                A3*sin(2*pi*M3/N*(n-1)+P3);
end;
% Hanning windowing operation
    for n=1:NOI,
            w(n)=0.5*(1-cos(2*pi*(n-1)/(N-1)));
    end;
    Gnoncoherent=sqrt(sum(w.*w)/NOI);
    u=1/Gnoncoherent * z .* w;
```

Figure 9.23. Windowed three-tone multitone spectra with (a) $NOI = 64$ and (b) 8192.

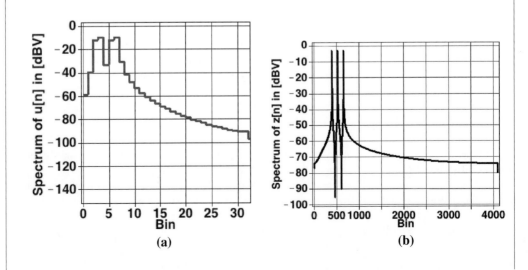

(a)   (b)

The routine was executed and the corresponding magnitude of the frequency spectrum was derived using the frequency-domain conversion routine of Example 9.7. The result is shown in Figure 9.23 on the left-hand side. The plot extends over the Nyquist frequency range of 32 points.

Surprisingly, the spectrum appears to have only two spectral peaks. It is as if only two tones were present in the input signal. This is a direct result of frequency leakage. If the observation interval is increased, say, to 8192 samples, then better frequency resolution is obtained and a clear separation of each frequency component is evident. This is shown on the right-hand side of Figure 9.23. With the side tones around each spectral peak below 100 dB, each tone can be accurately measured with a relative error of no more than 0.001% using the method of the previous example.

9.4 THE INVERSE FFT

9.4.1 Equivalence of Time- and Frequency-Domain Information

A discrete Fourier transform produces a frequency-domain representation of a discrete-time waveform. This was shown to be equivalent to a discrete-time Fourier series representation of a periodically extended waveform. The transformation is lossless, meaning that all information about the original signal is maintained in the transformation. Since no information is lost in the transformation from the time domain to the frequency domain, it seems logical that we should be able to take a frequency-domain signal back into the time domain to reconstruct the original signal. Indeed, this is possible and can be seen directly from Eqs. (9.21) and (9.22). If we substitute the expression for the frequency-domain coefficients given by Eq. (9.22) back into (9.21), we clearly see that we obtain our original information

$$\mathbf{X} = \mathbf{W}\left[\mathbf{W}^{-1}\mathbf{X}\right] = \mathbf{W}\mathbf{W}^{-1}\mathbf{X} = \mathbf{X} \qquad (9.74)$$

In practice, the form of the mathematics used to perform the frequency-to-time operation is very similar to that used to perform the time-to-frequency operation. In fact, the same FFT algorithm can be used, except for some possible array rearrangements and some predictable scale factor changes. When the FFT is used to perform the frequency-to-time transformation, it is referred to as an *inverse-FFT*. The calculations are so similar that some testers perform an inverse FFT using the same syntax as the forward FFT. A flag is set to determine whether the FFT is forward or inverse.

It is worth noting that the magnitude of a spectrum alone cannot be converted back into the time domain. Phase information at each test frequency must be combined with the magnitude information in the form of a complex number using either rectangular or polar notation. The specific format will depend on the vendor's operating system. Our next example will illustrate this procedure using MATLAB.

Exercises

9.16. The results of an FFT analysis of a noncoherent sinusoid indicate the following significant dBV values around the spectral peak: -38.5067, -36.3826, -33.6514, -29.7796, -22.9061, -7.7562, -25.3151. Estimate the amplitude of the corresponding tone.

ANS. 0.596.

9.17. By what factor does the repeatability of a set of spectral measurements made on a coherent signal decrease with the application of a Blackman window with signal power correction?

ANS. Sample set scaled by $1/G_{COHERENT} = 2.38$ followed by a RMS noise change of $G_{NONCOHERENT} = 0.552$ due to window shape, resulting in an increase RMS noise level of $G_{NONCOHERENT}/G_{COHERENT} = 1.31 \left(\sqrt{ENBW}\right)$. Repeatability therefore suffers by a factor of 1.31.

9.18. If a sampled sine wave corresponding to a bin of an FFT has an amplitude of 1 V, what is the expected amplitude observed through a Hanning window without correction?

ANS. $0.5 \ (= G_{COHERENT})$.

9.19. If the average RMS noise voltage per bin associated with an N-point rectangular window FFT is 100 µV, what is the expected RMS noise level per bin if a Hanning window is used without correction assuming the input signal is noncoherent with the UTP? What would this noise level per bin be after correction? What is the noise level associated with a signal within the main lobe of the window?

ANS. 61.2 µV (observed directly); 100 µV (with correction); 200 µV.

EXAMPLE 9.11

A discrete-time signal is described by its DTFS representation as

$$x[n] = 1 + 2\,\cos\left[2\left(\frac{2\pi}{8}\right)n + \frac{\pi}{4}\right] + 0.5\,\cos\left[3\left(\frac{2\pi}{8}\right)n\right]$$

Determine the time-domain samples using MATLAB's inverse FFT routine.

Solution:

The inverse FFT is performed in MATLAB using a vector Y that contains the samples of the Fourier transform $Y(k)$ in complex form. As described in Section 9.3, these are related to the coefficients of the DTFS according to

$$Y(k) = NX(k)$$

To obtain $X(k)$, we first note from $x[n]$ that the spectral coefficients of the DTFS written in trigonometric form are immediately obvious as

$$\begin{bmatrix} \tilde{c}_0 \\ \tilde{c}_1 \\ \tilde{c}_2 \\ \tilde{c}_3 \\ \tilde{c}_4 \end{bmatrix} = \begin{bmatrix} 1 \\ 0 \\ 2 \\ 0.5 \\ 0 \end{bmatrix} \quad \text{and} \quad \begin{bmatrix} \tilde{\phi}_0 \\ \tilde{\phi}_1 \\ \tilde{\phi}_2 \\ \tilde{\phi}_3 \\ \tilde{\phi}_4 \end{bmatrix} = \begin{bmatrix} 0 \\ 0 \\ -\pi/4 \\ 0 \\ 0 \end{bmatrix}$$

Subsequently, $X = [X(0)\ X(1)\dots X(N-1)]$ and $Y = [Y(0)\ Y(1)\dots Y(N-1)]$ with $N = 8$ are determined from Eq. (9.43), and from their interrelationship described, to be

$$X = \begin{bmatrix} 1 \\ 0 \\ 0.7071 + j0.7071 \\ 0.25 \\ 0 \\ 0.25 \\ 0.7071 - j0.7071 \\ 0 \end{bmatrix} \quad \Rightarrow \quad Y = 8 \ X = \begin{bmatrix} 8 \\ 0 \\ 5.6568 + j5.6568 \\ 2 \\ 0 \\ 2 \\ 5.6568 - j5.6568 \\ 0 \end{bmatrix}$$

Submitting the Y vector to MATLAB's inverse FFT routine, we obtain the following time-domain samples:

$$x = IFFT\{Y\} = \begin{bmatrix} 2.9142 \\ -0.7678 \\ -0.4142 \\ 2.7678 \\ 1.9141 \\ -0.0606 \\ -0.4142 \\ 2.0606 \end{bmatrix}$$

Of course, these time-domain samples agree with those obtain by evaluating $x[n]$ directly at the sampling instances.

> **Exercises**

9.20. Using MATLAB, compute the IFFT of the following sequence of complex numbers: {1.875, 0.75–j0.375, 0.625,
0.75+j0.375}.   ANS. {1.0, 0.5, 0.25, 0.125}.

9.21. For the following signal x[n], compute the RMS value of
each frequency component (beginning with DC):

$$x[n] = 1 + 2\ \cos\left[2\left(\frac{2\pi}{8}\right)n + \frac{\pi}{4}\right] + 0.5\ \cos\left[3\left(\frac{2\pi}{8}\right)n\right]$$

ANS. 1, 1.414, 0.3536;

Using Parseval's theorem, compute the RMS value of x[n].   1.7676.

9.4.2 Parseval's Theorem

The RMS value X_{RMS} of a discrete time periodic signal $x[n]$ is defined as the square root of the sum of the individual samples squared, normalized by the number of samples N, according to

$$X_{RMS} = \sqrt{\frac{1}{N} \sum_{n=0}^{N-1} x^2[n]} \tag{9.75}$$

Parseval's theorem for discrete-time periodic signals states that the RMS value X_{RMS} of a periodic signal $x[n]$ described by a DTFS written in trigonometric terms is given by

$$X_{RMS} = \sqrt{c_0^2 + \frac{1}{2} \sum_{k=1}^{N/2} c_k^2} \tag{9.76}$$

When the magnitude of each spectral coefficient is expressed as an RMS value, Eq. (9.76) can be rewritten as a square-root-of-sum-of-squares calculation given by

$$X_{RMS} = \sqrt{\sum_{k=0}^{N/2} c_{k-RMS}^2} \tag{9.77}$$

In this text, we shall make greatest use of this form of Parseval's theorem. It is the easiest form to remember, because there are no extra scale factors to keep track of.

The importance of Parseval's theorem is that it allows the computation of the RMS value associated with all aspects of a signal such as distortion, noise, and so on, to be made directly from the DTFS description. For example, the RMS noise level associated with a signal is the square root of sum of squares of all bins that do not contain signal-related power or DC given by

$$X_n = \sqrt{\sum_{k=1, k \neq signal}^{N/2} c_{k-RMS}^2} \tag{9.78}$$

9.4.3 Frequency-Domain Filtering

A useful feature of DSP-based testing is the ability to apply arbitrary filter functions to a set of collected samples, simulating electronic filters in traditional analog instrumentation. Filtering can be applied either in the time domain or in the frequency domain. Time-domain filtering is accomplished by convolving the sampled waveform by the impulse response of the desired filter. Frequency-domain filtering is performed by multiplying the signal spectrum by the desired filter's frequency response. Frequency-domain filtering is faster to implement than time-domain filtering, because the spectrum of a signal is already available in the computer. As such, time-domain filtering is rarely used in mixed-signal testing applications.

Filter functions are selected to provide a pre-described frequency response such as a low-pass or bandpass behavior. For instance, a low-pass filter is used to suppress the high-frequency content of a signal while allowing the low-frequency energy to pass relatively unchanged. In another application, filters are used to alter the phase of a signal to improve its transient behavior (i.e., reduce ringing). In general, the z-domain transfer function $H(z)$ of an arbitrary discrete-time filter is described by

$$H(z) = \frac{a_0 + a_1 z^1 + \cdots + a_N z^N}{1 + b_1 z^1 + \cdots + b_N z^N} \tag{9.79}$$

From the convolution property of discrete-time Fourier transforms, the transform $Y_{out}\left(e^{j\omega T_s}\right)$ of the filter output is related to the transform $Y_{in}\left(e^{j\omega T_s}\right)$ of the filter input according to

$$Y_{out}\left(e^{j\omega T_s}\right) = H\left(e^{j\omega T_s}\right) Y_{in}\left(e^{j\omega T_s}\right) \tag{9.80}$$

If we limit the input to discrete frequencies

$$\omega = \frac{2\pi}{N} k, \qquad k = 0, 1, \ldots, N - 1 \tag{9.81}$$

Eq. (9.80) can be rewritten as

$$Y_{out}\left(e^{j\frac{2\pi}{N} kT}\right) = H\left(e^{j\frac{2\pi}{N} kT_s}\right) Y_{in}\left(e^{j\frac{2\pi}{N} kT_s}\right) \tag{9.82}$$

or, with the shorthand notation of section 9.3.1, for $k = 0, 1, \ldots, N-1$, as

$$Y_{out}(k) = H(k) Y_{in}(k) \tag{9.83}$$

or in vector form as

$$\mathbf{Y}_{out} = \mathbf{H} \cdot \mathbf{Y}_{in} \tag{9.84}$$

where the dot operation represents an inner product. If we define the DTFS transforms for the input and output signals, that is, $\mathbf{Y}_{in} = \mathrm{FFT}\left[\mathbf{y}_{in}/N\right]$ and $\mathbf{Y}_{out} = \mathrm{FFT}\left[\mathbf{y}_{in}/N\right]$, then Eq. (9.81) can be written as

$$\mathrm{FFT}\left[\frac{\mathbf{y}_{out}}{N}\right] = \mathbf{H} \cdot \mathrm{FFT}\left[\frac{\mathbf{y}_{in}}{N}\right] \tag{9.85}$$

Taking the inverse FFT of each side, we obtain the time-domain samples of the output signal as

$$\mathbf{y}_{out} = N \cdot \text{IFFT}\left\{\mathbf{H} \cdot \text{FFT}\left[\frac{\mathbf{y}_{in}}{N}\right]\right\} \tag{9.86}$$

which is further reduced to

$$\mathbf{y}_{out} = \text{IFFT}\left\{\mathbf{H} \cdot \text{FFT}\left[\mathbf{y}_{in}\right]\right\} \tag{9.87}$$

as $\alpha \cdot FFT\left[x\right] = FFT\left[\alpha \cdot x\right]$. The procedure to filter a signal in the frequency domain is now clear. We first perform an FFT operation on the input signal to bring it into the frequency domain. Next, each frequency component of the input is scaled by the corresponding filter response $H(k)$ to produce the filtered frequency-domain output. Finally, an inverse FFT can be performed on the filtered output to produce the filtered time-domain signal. Often, this last step can be eliminated because we can extract all desired information from the filtered spectrum in the frequency domain.

9.4.4 Noise Weighting

Noise weighting is one common example of DSP-based filtering in mixed-signal test programs. Weighting filters are called out in many telecom and audio specifications because the human ear is more sensitive to noise in some frequency bands than others. The magnitude of the frequency response of the A-weighting filter is shown in Figure 9.24. It is designed to approximate the frequency response of the average human ear. For matters related to hearing, phase variations have little effect on the listener and are therefore ignored in noise-related tests. By weighting the noise from a telephone or audio circuit before measuring its RMS level, we can get a more accurate idea of how good or bad the telephone or audio equipment will sound to the consumer.

In traditional bench instruments, the weighting filter is applied to the analog signal before it is passed to an RMS voltmeter. In DSP-based testing, we can perform the same filtering function mathematically, after the unweighted DUT signal has been sampled. Application of a mathematical filter to a sampled waveform means the ATE tester does not have to include a physical A-weighting filter in its measurement instruments. The resulting reduction in tester complexity reduces tester cost and improves reliability. Application of the A-weighting filter is a simple matter of multiplying its magnitude by the magnitude of the FFT of the signal under test.

Figure 9.24. A-weighting filter magnitude response.

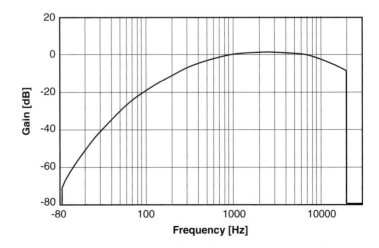

9.22. The gain and phase of a particular system at 1 kHz is 0.8 and -π/4, respectively. Determine the spectral coefficient of the DTFS that corresponds to the output signal at 1 kHz when excited by a signal with a spectral coefficient described by 0.25 – j0.35. Express the result in rectangular and polar form.

ANS.   −0.0566 – j0.3394, 0.3441e+j1.7341.

A very simple form of noise filtering can be used to measure the noise over a particular bandwidth. For example, if a specification calls for a noise level of 10 μV RMS over a band of 100 Hz to 1 kHz, then we can simply apply a brick wall bandpass filter to the FFT results, eliminating all noise components that do not fall within this frequency range. The remaining noise can then be measured by performing an inverse FFT followed by an RMS calculation. The same results can be also achieved by adding up the signal power in all spectral bins from 100 Hz to 1 kHz. To do this we simply square the RMS value of each frequency component from 100 Hz to 1 kHz, add them all together, and then take the square root of the total to obtain the RMS value of the noise according to

$$V_{N-rms} = \sqrt{\sum_{k=B_L}^{B_U} c_{k-RMS}^2} \qquad (9.88)$$

Here B_L and B_U are the spectral bins corresponding to the lower and upper frequencies of the brick wall filter (excluding any DC component). In this particular case, B_L and B_U correspond to the 100 Hz and 1 kHz frequencies, respectively.

9.5 SUMMARY

Coherent DSP-based testing allows the mixed-signal test engineer to perform AC measurements in a few tens of milliseconds. These same measurements might otherwise take hundreds or thousands of milliseconds using traditional analog bench instruments. The AWG, digitizer, source memory, and capture memory of a mixed-signal ATE tester allow us to translate signals between continuous time and sampled time. Digital signal processing operations such as the FFT and inverse FFT allow us to perform operations that are unavailable using traditional non-DSP measurement methodologies. Time-domain interpolations, frequency-domain filtering, and noise reduction functions are just a few of the powerful operations that DSP-based testing makes available to the accomplished mixed-signal test engineer.

We are fortunate in mixed-signal ATE to be able to use coherent sampling systems to bypass mathematical windowing. Bench instruments such as spectrum analyzers and digitizing oscilloscopes must use windowing extensively. The signals entering a spectrum analyzer or oscilloscope are generally noncoherent, since they are not synchronized to the instrument's sampling rate. However, a spectrum analyzer is not usually expected to produce an accurate reading in only 25 ms; thus it can overcome the repeatability problem inherent in windowing by simply averaging results or collecting additional samples. ATE equipment must be fast as well as accurate; thus windowing is normally avoided whenever possible. Fortunately, mixed-signal testers give us control of both the signal source and sampling processes during most tests. Synchronization of input waveforms and sampling processes affords us the tremendous accuracy/cost advantage of coherent DSP-based testing.

Despite its many advantages, DSP-based testing also places a burden of knowledge upon the mixed-signal test engineer. Matthew Mahoney, author of "DSP-Based Testing of Analog and Mixed-Signal Circuits,"[7] once told an amusing story about a frustrated student in one of his DSP-based testing seminars. Exasperated by the complexity of digital signal processing, the distressed student suddenly exclaimed "But this means we must know something!" Indeed, compared to the push-the-button/read-the-answer simplicity of bench equipment, DSP-based testing requires us to know a whole lot of "something."

The next two chapters will explore the use of DSP-based measurements in testing analog and sampled channels. In Chapter 10, "Analog Channel Testing," we will explore the various types of DSP-based tests that are commonly performed on nonsampled channels such as amplifiers and analog filters. Chapter 11, "Sampled Channel Testing," will then extend these DSP-based testing concepts to sampled channels such as ADCs, DACs, and switched capacitor filters.

PROBLEMS

9.1   Find the trigonometric Fourier series representation of the functions displayed in (a)–(d). Assume that the period in all cases is 1 ms.

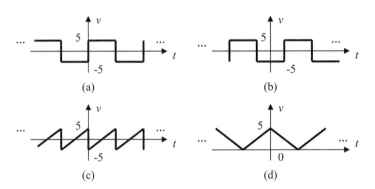

(a)   (b)

(c)   (d)

9.2.   Find the trigonometric Fourier series representation of a sawtooth waveform $x(t)$ having a period of 2 s and whose behavior is described by $x(t)=t$ if $-1<t<1$. Using MATLAB, numerically compare your FS representation with $x(t)$.

9.3.   Express the Fourier series in problem 9.2 using cosine terms only.

9.4.   Using the summation formulae in Eq. (9.13), determine the DTFS representation of the Fourier series representation of $x(t)$ given in Exercise 9.1. Use 10 points per period and limit the series to 100 terms. Also, express the result in magnitude and phase form.

9.5.   A sampled signal consists of the following six samples: $\{0.1, 0.1, 1, -3, 4, 0\}$. Using linear algebra, determine the DTFS representation for these samples. Also, express the result in magnitude and phase form. Using MATLAB, numerically compare the samples from your DTFS representation with the actual samples given here.

9.6.   A sampled signal consists of the following set of samples $\{0, -1, 2.4, 4, -0.125, 3.4\}$. Determine the DTFS representation for this sample set using the orthogonal basis method. Also, express the result in magnitude and phase form. Using MATLAB, numerically compare the samples from your DTFS representation with the actual samples given here.

9.7. Derive Eqs. (9.24) and (9.25) from first principles. Begin by multiplying the DTFS representation given in Eq. (9.12) by $\cos\left[k\left(2\pi/N\right)n\right]$, then sum n on both sides from 0 to N–1. Reduce the expression by using the set of trigonometric identities given in Eq. (9.23). Repeat using $\sin\left[k\left(2\pi/N\right)n\right]$.

9.8. Plot the magnitude and phase spectrum of the following FS representations of $x(t)$:

(a) $x(t)=\dfrac{2}{\pi}\left(\sin \pi t -\tfrac{1}{2}\sin 2\pi t +\tfrac{1}{3}\sin 3\pi t -\tfrac{1}{4}\sin 4\pi t +\cdots\right)$

$x(t)=\dfrac{2}{\pi}\left[\cos\left(\pi t -\dfrac{\pi}{2}\right)+\tfrac{1}{2}\cos\left(2\pi t +\dfrac{\pi}{2}\right)\right.$

(b) $\left.+\tfrac{1}{3}\cos\left(3\pi t -\dfrac{\pi}{2}\right)+\tfrac{1}{4}\cos\left(4\pi t +\dfrac{\pi}{2}\right)+\cdots\right]$

(c) $x(t)=-1.5+\displaystyle\sum_{k=1,\ k\ \text{odd}}^{\infty}\dfrac{1}{k\pi}\cos\left(k\ 2\pi\times10^3\ t +\pi\!/\!_2\right)$

9.9. Plot the magnitude and phase spectrum of the following DTFS representations of $x[n]$:

(a) $x[n]=0.6150\ \sin\left[\left(\dfrac{2\pi}{10}\right)n\right]-0.2749\ \sin\left[2\left(\dfrac{2\pi}{10}\right)n\right]$

$+0.1451\ \sin\left[3\left(\dfrac{2\pi}{10}\right)n\right]-0.0649\ \sin\left[4\left(\dfrac{2\pi}{10}\right)n\right]$

(b) $x[n]=1+\cos\left[\left(\dfrac{2\pi}{8}\right)n\right]+\sin\left[\left(\dfrac{2\pi}{8}\right)n\right]+\cos\left[2\left(\dfrac{2\pi}{8}\right)n\right]+\sin\left[2\left(\dfrac{2\pi}{8}\right)n\right]$

$-\cos\left[3\left(\dfrac{2\pi}{8}\right)n\right]-\sin\left[3\left(\dfrac{2\pi}{8}\right)n\right]$

9.10. A DTFS representation expressed in complex form is given by

$x[n]=1+0.2e^{j\left[\left(\frac{2\pi}{8}\right)n-\frac{\pi}{4}\right]}+0.3e^{j\left[3\left(\frac{2\pi}{8}\right)n+\frac{\pi}{3}\right]}+0.3e^{j\left[5\left(\frac{2\pi}{8}\right)n-\frac{\pi}{3}\right]}+0.2e^{j\left[7\left(\frac{2\pi}{8}\right)n+\frac{\pi}{4}\right]}$

Express $x[n]$ in trigonometric form. Plot the magnitude and phase spectra.

9.11. A DTFS representation expressed in complex form is given by

$x[n]=1+(1.8-j1.9)e^{j\left[2\left(\frac{2\pi}{8}\right)n\right]}+(0.75+j0.25)e^{j\left[3\left(\frac{2\pi}{8}\right)n\right]}$

$+(0.75-j0.25)e^{j\left[5\left(\frac{2\pi}{8}\right)n\right]}+(1.8+j1.9)e^{j\left[6\left(\frac{2\pi}{8}\right)n\right]}$

Express $x[n]$ in trigonometric form and plot the corresponding magnitude and phase spectra.

9.12. Derive the polar form of Eq. (9.43) from (9.42).

9.13. Sketch the periodic extension for the following sequence of points:

(a) [0, 0.7071, 1.0, 0.7071, 0.0, –0.7071, –1.0000, –0.7071]

(b) [0, 0.7071, 1.0, 0.7071, 0.0, –0.7071, –1.0, –0.7071, 0.0, 0.7071]

9.14. Using the FFT algorithm in MATLAB or some other software package, verify your answers to Problems 9.4, 9.5, and 9.6.

9.15. Given

$$x[n] = 0.25 + 0.5 \; \cos\left[\left(\frac{2\pi}{8}\right)n\right] + 0.1 \; \sin\left[\left(\frac{2\pi}{8}\right)n\right] + 2.1 \; \cos\left[2\left(\frac{2\pi}{8}\right)n\right]$$

$$- 0.9 \; \cos\left[3\left(\frac{2\pi}{8}\right)n\right] - 0.1 \; \sin\left[3\left(\frac{2\pi}{8}\right)n\right]$$

Express $x[n]$ in complex form.
Using MATLAB, write a script that samples $x[n]$ for $n = 0,1,\ldots,7$.
Compute the FFT of the samples found in part (b) and write the corresponding DTFS representation in complex form. How does it compare with $x[n]$ found in part (a)?

9.16. Given

$$x[n] = (0.2 - j0.4)e^{j\left[\left(\frac{2\pi}{10}\right)n\right]} + (0.25 + j0.25)e^{j\left[3\left(\frac{2\pi}{10}\right)n\right]}$$

$$+ (0.25 - j0.25)e^{j\left[7\left(\frac{2\pi}{10}\right)n\right]} + (0.2 + j0.4)e^{j\left[9\left(\frac{2\pi}{10}\right)n\right]}$$

(a) Express $x[n]$ in trigonometric form.
(b) Using MATLAB, write a script that samples $x[n]$ for $n = 0,1,\ldots, 9$.
(c) Compute the FFT of the samples found in part (b) and write the corresponding DTFS representation in trigonometric form. How does it compare with $x[n]$ found in part (a)?

9.17. Evaluate $x[n] = n^3 - 2n^2 - 2n + 1$ for $n = 0,1,\ldots,15$. Using MATLAB, compute the FFT of the 16 samples of $x[n]$ and determine the corresponding $\{c_k\}$ and $\{\phi_k\}$ spectral coefficients. Express the magnitude coefficients in RMS form.

9.18. Investigate the effects of increasing the observation window on the spectrum of a non-coherent sinusoidal signal. Consider generating a sinusoidal signal using parameters $A = 1$, $\phi = 0$, $M = 9.9$, and $N = 64$. Next, compare the magnitude spectrum of this signal when the following number of samples are collected: (a) 64 samples, (b) 512 samples, (c) 1024 samples, (d) 8192 samples. In all cases, estimate the amplitude of the sinusoidal signal.

9.19. An observation window consists of 64 points. Plot the frequency response of (a) rectangular window, (b) Hanning window, (c) Blackman window, and (d) Kasier ($\phi = 10$) window all on the same graph. *Hint*: Determine the frequency response of a window function using the Fourier transform definition of Eq. (9.47).

9.20. Compute the window shape factors for the rectangular, Hanning, Blackman, and Kasier ($\beta = 10$) windows using the window function coefficients.

9.21. Repeat Problem 9.18 but view the data first through a Blackman window. Estimate the amplitude of the sinusoidal signal.

9.22. Repeat Problem 9.18 but view the data first through a Kasier ($\beta = 10$) window. Estimate the amplitude of the sinusoidal signal.

9.23. A signal has a period of 128 μs and is sampled at a rate of 1 MHz. If 128 samples are collected, what is the frequency resolution of the resulting FFT? If the number of samples collected increases to 8192 samples, what is the frequency resolution of the FFT?

9.24. Repeat Example 9.9, but this time use a Blackman window. By what factor does the accuracy of the calculation improve over the rectangular window?

9.25. Repeat Example 9.10, but this time use a Blackman window.

9.26. Using the FFT and IFFT routines found in MATLAB, together with the samples described in Exercise 9.12, verify that $IFFT(FFT(x)) = x$.

9.27. Using the trigonometric identities described in Eq. (9.23), derive the trigonometric form of Parseval's theorem given in Eq. (9.76).

9.28. Using the trigonometric form of Parseval's theorem in Eq. (9.76) as a starting point, derive the corresponding complex form of the theorem given in Eq. (9.77).

9.29. The complex coefficients of a spectrum of a sampled signal are

$$\{X_k\} = \{0.5, 0.2 - j0.4, 0, 0.25 + j0.25, 0, 0.25 - j0.25, 0, 0.2 + j0.4\}$$

What is the RMS value of this signal?

9.30. The magnitude coefficients of a spectrum of a sampled signal are

$$\{c_k\} = \{0.1, 0.3, 0, 0.05, 0, 0.001\}$$

What is the RMS value of this signal?

9.31. Find the coherent sample set of $x[n]$ using MATLAB's IFFT routine assuming its spectrum is described by the following:

(a) $x[n] = 0.25 + 0.5 \ \cos\left[\left(\dfrac{2\pi}{8}\right)n\right] + 0.1 \ \sin\left[\left(\dfrac{2\pi}{8}\right)n\right] + 2.1 \ \cos\left[2\left(\dfrac{2\pi}{8}\right)n\right]$

$\qquad -0.9 \ \cos\left[3\left(\dfrac{2\pi}{8}\right)n\right] - 0.1 \ \sin\left[3\left(\dfrac{2\pi}{8}\right)n\right]$

(b) $x[n] = 1 + 0.2e^{j\left[\left(\frac{2\pi}{8}\right)n - \frac{\pi}{4}\right]} + 0.3e^{j\left[3\left(\frac{2\pi}{8}\right)n + \frac{\pi}{3}\right]} + 0.3e^{j\left[5\left(\frac{2\pi}{8}\right)n - \frac{\pi}{3}\right]} + 0.2e^{j\left[7\left(\frac{2\pi}{8}\right)n + \frac{\pi}{4}\right]}$

(c) $x[n] = (0.2 - j0.4)e^{j\left[\left(\frac{2\pi}{10}\right)n\right]} + (0.25 + j0.25)e^{j\left[3\left(\frac{2\pi}{10}\right)n\right]}$

$\qquad + (0.25 - j0.25)e^{j\left[7\left(\frac{2\pi}{10}\right)n\right]} + (0.2 + j0.4)e^{j\left[9\left(\frac{2\pi}{10}\right)n\right]}$

9.32. Verify the samples in Problem 9.31 by evaluating the function at each sampling instant.

9.33. A signal with a DTFS representation given by

$$x[n] = 0.1 + 2 \ \cos\left[\left(\frac{2\pi}{16}\right)n - \frac{\pi}{5}\right] + 0.5 \ \cos\left[5\left(\frac{2\pi}{16}\right)n + \frac{\pi}{5}\right]$$

passes through a system with the following transfer function

$$H(z) = \frac{0.0020 + 0.00402z^1 + 0.0020z^2}{1 - 1.5610z^1 + 0.64135z^2}$$

What is the DTFS representation of the output signal?

9.34. A signal with noise is described by the following DTFS representation:

$$x[n] = 0.25 + 2 \ \cos\left[\left(\frac{2\pi}{8}\right)n\right] + 10^{-5}\cos\left[2\left(\frac{2\pi}{8}\right)n\right]$$

$$- 10^{-7}\cos\left[3\left(\frac{2\pi}{8}\right)n\right] - 10^{-5}\sin\left[3\left(\frac{2\pi}{8}\right)n\right] + 10^{-5}\cos\left[4\left(\frac{2\pi}{8}\right)n\right]$$

What is the RMS value of the noise signal that appears between bins 2 and 4?

REFERENCES

1. A. V. Oppenheim and R. W. Schafer, *Discrete-Time Signal Processing*, Prentice Hall, Englewood Cliffs, NJ, March 1989, ISBN 013216292X.
2. A. V. Oppenheim et al., *Signals and Systems*, Prentice Hall, Englewood Cliffs, NJ, 1997, ISBN 0138147574.
3. W. McC. Siebert, *Circuits, Signals, and Systems*, The MIT Press, Cambridge, MA, 1985, ISBN 0262192292.
4. R. W. Ramirez, *The FFT Fundamentals and Concepts*, Prentice Hall, Englewood Cliffs, NJ, 1985, ISBN 0133143864.
5. J. G. Proakis and D. G. Manolakis, *Digital Signal Processing: Principles, Algorithms, and Applications,* 3rd Edition, Prentice Hall, Englewood Cliffs, NJ, 1996, ISBN 0133737624.
6. F. J. Harris., *On the use of windows for harmonic analysis with the discrete fourier transform,* *Proceedings of the IEEE,* **66** (1), pp. 51–83, January 1978.
7. M. Mahoney, *Tutorial DSP-Based Testing of Analog and Mixed-Signal Circuits,* The Computer Society of the IEEE, Washington, D.C., 1987, ISBN 0818607858.

CHAPTER 10

Analog Channel Testing

This chapter shows how basic AC channel tests can be performed economically using DSP-based testing. This chapter covers only nonsampled channels, consisting of combinations of op amps, analog filters, programmable gain amplifiers and other continuous-time circuits. Such AC channel tests include gain and level, frequency response related to magnitude, phase and group delay behavior, distortion, signal rejection, and noise.

10.1 OVERVIEW

10.1.1 Types of Analog Channels

Analog channels include any nonsampled circuit with analog inputs and analog outputs. Examples of analog channels include continuous-time filters, amplifiers, analog buffers, programmable gain amplifiers (PGAs), single-ended to differential converters, differential to single-ended converters, and cascaded combinations of these circuits. Channels including ADCs, DACs, switched capacitor filters, and other sampling circuits will be discussed in Chapter 11, "Sampled Channel Testing." However, we will see in Chapter 16, "Design for Test (DfT)," that a sampled channel may be broken into subsections using DfT test modes, as illustrated in Figure 10.1.

DfT allows the filter and PGA in Figure 10.1 to be isolated from the rest of the sampled channel for more thorough testing. Since DfT allows portions of sampled channels to be reduced to analog channel subsections, analog channel testing is extremely common in mixed-signal device testing.

It is critically important for the mixed-signal test engineer to gain a solid understanding of analog channel testing. The analog tests described in this chapter will represent at least half of many mixed-signal test programs. Although many analog channel tests can be measured without DSP-based techniques, this chapter will concentrate only on DSP-based methods of channel testing.

As previously explained in Chapter 9, DSP-based testing is the primary technique used in high-volume production testing of mixed-signal devices. A principal advantage of DSP-based

Figure 10.1. Analog bus DfT for a DAC mixed-signal channel.

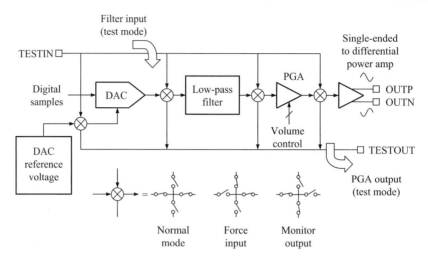

testing is that many of the parameters described in this chapter can be measured simultaneously. Simultaneous measurements save test time and thereby reduce production costs.

10.1.2 Types of AC Parametric Tests

Analog and sampled channels share many AC parametric test specifications. Most of these specifications fall into a few general categories, including gain, phase, distortion, signal rejection, and noise. Each of these categories will be discussed in the sections that follow. We will examine the definition of each type of test, common test conditions, the causes and effects of parametric failures, common test techniques, and common measurement units (volts, decibels, etc.) for each test. Finally, an example of each test will be presented to clarify test definitions and techniques.

The test definitions in this chapter are all based on the assumption that the signals to be sourced or measured are voltages. Some circuits operate with currents rather than voltages. ATE testers are typically unable to measure or source AC currents directly. Some form of voltage-to-current or current-to-voltage circuit will be needed in cases where the DUT produces AC current outputs or requires AC current inputs.

In this chapter, we will assume that the tester's digitizer and AWG are perfect, having no gain errors at any frequency. This is a naive assumption at best, but one that will suffice for now. To make accurate AC measurements with general-purpose digitizers and arbitrary waveform generators (AWGs), proper use of software calibration is often required. Focused calibrations can be used to compensate for the various measurement errors inherent in the general purpose AWGs, digitizers, and other instruments in an ATE tester.

10.2 GAIN AND LEVEL TESTS

10.2.1 Absolute Voltage Levels

Absolute voltage levels are perhaps the simplest AC parameters to understand, but they can be among the most difficult parameters to measure accurately using a general-purpose digitizer. Most electrical engineers are familiar with the use of bench equipment such as an oscilloscope or an AC voltmeter. The absolute voltage of a test tone is simply the RMS voltage of the signal under test, evaluated at the test tone's frequency. Energy at other frequencies is eliminated from the measurement. DSP-based measurement techniques allow noise, distortion, and other test tones to be easily eliminated from the RMS measurement. RMS voltmeters and oscilloscopes, by contrast, measure the total signal RMS, including noise, distortion, and other test tones. Spectrum analyzers offer a more frequency-selective voltage measurement capability, but they are not always as accurate as RMS voltmeters in measuring the absolute voltage level of a pure sinusoidal signal.

Absolute level specifications can be applied to any single-tone or multitone signal. The purpose of an absolute level test is to detect first order defects in a circuit, such as resistor or capacitor mismatch, DC reference voltage errors, and grotesque clipping or other distortion. For example, if the DC voltage reference for a DAC exhibits a 5% error, then the DAC's AC output amplitude will likely show a 5% absolute voltage level error as well. As a second example, consider a low-pass filter having a defective op amp (Figure 10.2). A clean sine wave input may become clipped very badly by the defective filter, resulting in an absolute voltage level at the filter output that is totally wrong. Obviously, it will also exhibit very high harmonic distortion as well. Absolute voltage level tests are a good way to find grossly defective circuits very quickly.

Loading conditions can be very important to many AC and DC parametric measurements, including absolute voltage level tests. If a buffer amplifier is designed to drive 32 Ω in parallel with 500 pF, then it makes no sense to test its absolute level in an unloaded condition. Generally, the test engineer must determine the worst-case loading conditions for a given output and test AC parameters using that loading condition. Device data sheets usually list a specific loading condition or a worst-case loading condition, which the output must drive during an absolute level test. The test engineer must design this load into the device interface board (DIB). In most cases, the load must be removable so that tests like continuity and leakage can also be performed on the DUT output. Electromechanical relays are often added to the DIB board to facilitate the removal of loads from DUT outputs.

Units of measure for absolute level tests vary somewhat, but they usually fall into a few common categories. Absolute levels may be specified in RMS volts, peak volts, peak-to-peak volts, dBV (decibels relative to 1.0 V RMS), and dBm (decibels relative to 1.0 mW at a specified load impedance). When dealing with differential inputs or outputs, each of these measurement units can be defined from either a single-ended or differential perspective (Figure 10.3). It is critical for a test engineer to be able to communicate these units of measure without ambiguity. It is aggravating to ask an engineer to measure the voltage at the output of a differential circuit, only to get the reply "1.2 V." Does the answer refer to a single-ended or differential measurement? Is it peak, peak-to-peak, or RMS?

When referring to a single-ended signal, it is acceptable to drop the single-ended/differential notation since there is no ambiguity. But when referring to a differential signal, it is imperative to specify the measurement type. Many times, correlation errors between the bench and ATE tester turn out to be simple misunderstandings about signal level definitions. Design errors can even be introduced if a customer specifies a differential signal level in the data sheet and the design engineer misinterprets it as a single-ended specification. Absolute voltage levels must be specified using a clear level definition, such as 1 V RMS, differential or 1 V peak, single-ended.

Figure 10.2. Absolute voltage level test detects gross circuit defects quickly.

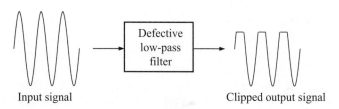

Input signal   Clipped output signal

Figure 10.3. Equivalent voltage measurements for single-ended and differential signals.

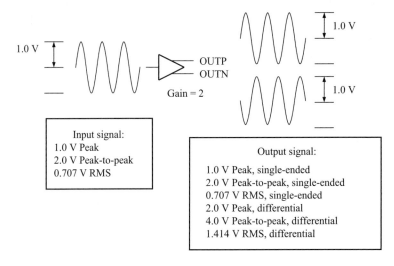

Decibel units can be abused as well. A signal level of +3.7 dB is meaningless, because it has no point of reference. What voltage level corresponds to 0 dB? The decibel unit represents a ratio of values, and as such it is inappropriate to refer to an absolute voltage level using decibels. Decibel units, when used to specify absolute levels, must include a definition of the 0-dB reference level. A common point of reference is 1 V RMS. When this definition of 0 dB is used, the measurement is expressed in dBV, or decibels relative to 1 V RMS. Differential reference levels are always used when measuring differential signals, and single-ended reference levels are always used when measuring single-ended signals. Therefore, decibel units do not require a single-ended/differential notation. Nevertheless, we shall specify the measurement type explicitly in this text to avoid confusion.

EXAMPLE 10.1

The positive side of a differential sine wave has a peak amplitude of 500 mV, as shown in Figure 10.4. Assuming the negative side is perfectly matched in amplitude and is 180 degrees out of phase, calculate the signal amplitude in dBV, differential.

Figure 10.4. Differential sine wave at 500 mV peak, single-ended.

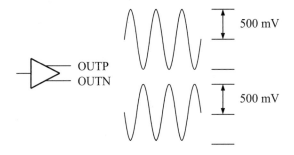

Solution:

Since we need to compare this signal to 1.0-V RMS, differential, we start by converting the single-ended peak signal into differential RMS units. We rely on the fact that a sine wave always has a peak-to-RMS ratio of 1.414 (square root of two)

$$500 \text{ mV peak, single-ended} = 1.0 \text{ V peak, differential}$$
$$= 0.707 \text{ V RMS, differential}$$

Next, we convert RMS volts to dBV using the equation

$$\text{signal level (dBV)} = 20 \ \log_{10} \left| \frac{\text{RMS signal level}}{1.0 \text{ V RMS}} \right|$$

resulting in

$$\text{signal level (dBV)} = 20 \ \log_{10} \left| \frac{0.707 \text{ V RMS}}{1.0 \text{ V RMS}} \right| = -3.01 \text{ dBV}$$

Another commonly used absolute unit is the dBm (decibels relative to 1.0 mW). It is a measurement of power P dissipated in some load impedance and is expressed as

$$\text{signal level (dBm)} = 10 \ \log_{10} \left(\frac{P}{1 \text{ mW}} \right) \qquad (10.1)$$

As ADCs are generally voltage-sensitive, the captured data will be is expressed in units of volts. With a known load impedance R, we can convert a voltage level to a power level through the general expression $P = V^2/R$, resulting in

$$\text{signal level (dBm)} = 10 \ \log_{10} \left(\frac{V^2/R}{1 \text{ mW}} \right) \qquad (10.2)$$

EXAMPLE 10.2

Convert a 250-mV single-ended RMS measurement into dBm units at 600 Ω.

Solution:

Using Eq. (10.2), we write

$$\text{signal level (dBm)} = 10 \ \log_{10} \left[\frac{(250 \text{ mV})/600}{1 \text{ mW}} \right] = -9.823 \text{ dBm}$$

Exercises

10.1.	Convert a 1-V peak, single-ended signal into dBV units.	ANS. −3.01 dBV.

10.2.	Convert a 1-V peak-to-peak, differential signal into differential-ended dBV units.	ANS. −9.03 dBV.

10.3.	Convert a 1-V RMS, differential signal into differential dBm units at 50 Ω.	ANS. +13.01 dBm

---

10.4.	A DTFS analysis reveals that a particular tone has a complex spectral coefficient of −0.2866 − j0.133643 V. What is the amplitude of this tone? What is its RMS value? Express this value in dBV units.	ANS. 0.3162 V; 0.2236 V; −13.01 dBV.

On a mixed-signal ATE tester, absolute voltage levels are usually measured using a general-purpose digitizer in conjunction with Fourier analysis (i.e., DSP-based testing). Digitizers are often capable of measuring either single-ended signals or differential signals. To measure differential signals, the digitizer uses an instrumentation amplifier at its front end. The instrumentation amplifier converts a differential input signal into a single-ended signal before it is measured. If a digitizer lacks the capability to measure differential signals, then the test engineer must capture each side of the signal separately, using either two digitizers or using the same digitizer twice. The two signals must then be combined mathematically by subtracting the negative signal from the positive signal.

10.2.2 Absolute Gain and Gain Error

In analog channels, absolute gain is simply a ratio of output AC signal level divided by input signal level at a specified frequency:

$$G = \frac{V_{out}}{V_{in}} \tag{10.3}$$

Absolute gain is frequently converted from V/V units to decibel units using the formula

$$G\,(\text{dB}) = 20\ \log_{10}\left|G\,(\text{V/V})\right| \tag{10.4}$$

Often a channel's gain is specified using a minimum and maximum absolute gain. Sometimes, though, a channel's gain is specified in terms of its error relative to the ideal absolute gain. This parameter is called *gain error*.

Gain error ΔG is defined as the actual (measured) gain of a channel, in volts/volt, divided by its ideal (expected) gain. When working with decibels, gain error is defined as the actual gain in dB minus the ideal gain G_{IDEAL} in dB (since subtracting logarithms is equivalent to dividing ratios).

$$\Delta G\,(\text{dB}) = G\,(\text{dB}) - G_{IDEAL}\,(\text{dB}) \tag{10.5}$$

For example, a channel may have an absolute gain of 12.35 dB (4.145 V/V), but its ideal gain should be 12.04 dB (4.0 V/V). Its gain error is therefore 12.35-dB − 12.04 dB = 0.31 dB (or equivalently, 4.145 V/V / 4.0 V/V = 1.036 V/V). While absolute gain and gain error can be specified using either V/V or decibels, gain error is usually specified in decibels.

Gain errors are frequently the result of component mismatch in the DUT. For example, mismatched resistors in an op amp gain circuit lead to gain errors. Excessive gain errors can lead to a number of system-level problems. In audio circuits, gain error can result in volumes that are too loud or too soft. Extreme gain errors can also lead to clipped (distorted) analog signals. In data transmission channels, the distortion caused by gain errors can lead to corrupted data bits. Absolute gain and gain error tests are well-suited to the detection of gross functionality errors like dead transistors or incomplete signal paths.

Gain tests are commonly performed at a signal level below the maximum allowed signal level in a channel. The reason a full-scale signal is not used is that it might cause clipping (harmonic distortion). Distortion can be introduced if the gain error or offset of the signal causes the circuit under test to clip either the top or bottom of the test signal. Distortion, in turn, causes an error in the absolute gain reading, since energy from the test signal is transferred into distortion components. Since we want to be able to distinguish between gain errors and distortion errors, gain measurements are often performed a few decibels below full scale (typically 1 to 3 decibels below full scale). A separate distortion test can be performed at full scale to determine the extent of clipping near the full-scale signal range.

Gain and gain error tests are often tested with a single test tone, rather than a multitone signal. In audio circuits, a 1-kHz test tone is very commonly specified in the data sheet, since the average human ear is maximally sensitive near 1 kHz.

EXAMPLE 10.3

A tester's AWG sources a sine wave to a low-pass filter with a single-ended input and differential output. A digitizer sampling at 8 kHz captures 256 samples of the sine wave at the input to the filter (Figure 10.5). The FFT of the captured waveform shows a signal amplitude of 1.25 V RMS at the 37th FFT spectral bin. The digitizer is then connected to the output of the filter using electromechanical relays (Figure 10.6). The digitizer captures 1024 samples of the output of the filter (differentially) using a 16-kHz sampling rate. The output FFT shows a signal amplitude of 1.025 V RMS in one of the spectral bins. What is the frequency of the test signal? Which spectral bin in the FFT of the output signal most likely showed the 1.025-V RMS signal level? Assuming the digitizer is perfectly accurate in both sampling configurations, what is the gain (in decibels) of the low-pass filter at this frequency? The ideal filter gain at this frequency is −1.50 dB. What is the gain error of the filter at this frequency? Is the filter output too high or too low?

Figure 10.5. Sampling system configuration during LPF input measurement.

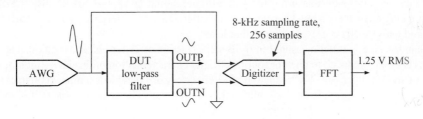

Figure 10.6. Sampling system configuration during LPF output measurement.

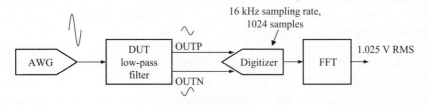

Solution:
First we have to understand our sampling systems. When sampling the input signal, the digitizer samples at 8 kHz and captures 256 samples. This results in a fundamental frequency of 8 kHz/256 = 31.25 Hz. The test frequency is therefore 37 × 31.25 Hz = 1156.25 Hz. Since the output of the filter must occur at the same frequency as the input, the output signal should also appear at 1156.25 Hz. When sampling the output signal, the digitizer samples at 16 kHz and captures 1024 samples. This sampling system results in a fundamental frequency of 16 kHz/1024 = 15.625 Hz. The FFT spectral bin containing this signal energy must therefore be located at 1156.25 Hz/15.625 Hz = spectral bin 74. Any signal energy falling in any other FFT spectral bin is therefore either noise or distortion.

The gain is calculated using a logarithmic calculation as follows

$$G(dB) = 20 \, \log_{10} \left| \frac{V_{out}}{V_{in}} \right| = 20 \, \log_{10} \left| \frac{1.025 \text{ V}}{1.25 \text{ V}} \right| = -1.724 \text{ dB}$$

The gain error is therefore equal to −1.724 dB−(−1.50 dB) = −0.224 dB. This means that the filter output is lower than it should be.

10.2.3 Gain Tracking Error

Gain tracking, or gain tracking error, is defined as the variation in the gain G (expressed in dB) of a channel with respect to a reference gain G_{REF} (also expressed in dB) as the signal level V_{in} changes.

$$\Delta G(dB) = G(dB) - G_{REF}(dB) \quad \text{versus} \quad V_{in} \tag{10.6}$$

Ideally, a channel should have a constant gain, regardless of the signal level (unless of course the signal is high enough to cause clipping). A perfectly linear analog channel has no gain tracking error. But small amounts of nonlinearity and other subtle circuit defects can lead to slight differences in gain at different signal levels. Gain tracking error is also introduced by the quantization errors in a DAC or ADC channel. As the signal level in a DAC or ADC quantized channel falls, the quantization errors become a larger percentage of the signal. Thus gain tracking errors in a quantized channel are most severe at low signal levels.

Gain tracking is calculated by measuring the gain at a reference level, usually the 0-dB level of the channel, and then measuring the gain at other signal levels (+3 dB, -6 dB, -12 dB, etc.). Gain tracking error at each level is calculated by subtracting the reference gain G_{REF} (dB) from the measured gain G (dB) corresponding to that level.

Gain tracking is often measured in 6-dB steps (factors of 2) for characterization. The number of steps is usually reduced to three or four levels after characterization of the device identifies which levels are most problematic. The reduction of levels saves considerable test time. Gain tracking error is almost always specified in decibels, although V/V would also be an acceptable unit of measure.

10.2.4 PGA Gain Tests

A programmable gain amplifier (PGA) can be set to multiple gain settings using a digital control signal. PGAs are commonly used as volume control circuits in cellular telephones, televisions, and so on. The absolute gain at each step in a PGA's gain curve is often less important than the difference in gain between adjacent steps. PGAs are often specified with an absolute gain at the first setting, a total gain difference between the highest and lowest setting, and the gain step size from each gain setting to the next.

EXAMPLE 10.4

The data sheet for a single-ended analog voltage follower defines a 0 dB reference level of 500 mV at the voltage follower input. The gain tracking specification for this device calls for a gain tracking error of ±0.25 dB at a +3.0-dB signal level and ±0.05-dB gain tracking error from 0.0-dB to −54.0-dB signal level. In this data sheet, gain tracking error is referenced to the gain at the 0-dB signal level. The gain error of the voltage follower is measured at each of the signal levels in Table 10.1, resulting in a series of absolute gain values. Calculate the gain tracking errors at each level and determine whether or not this device passes the gain tracking test. Calculate the input signal level at −54-dB.

Table 10.1. Absolute Gains for Gain Tracking Measurement

Signal Level	Absolute Gain	Signal Level	Absolute Gain
0 dB (500 mV)	0.02 dB (reference gain)		
+3 dB	−0.18 dB (slight clipping causes test tone attenuation)	−30 dB	−0.02 dB
−6 dB	0.02 dB	−36 dB	−0.03 dB
−12 dB	0.01 dB	−42 dB	−0.04 dB
−18 dB	0.00 dB	−48 dB	−0.05 dB
−24 dB	−0.01 dB	−54 dB	−0.07 dB

Solution:

We convert each absolute gain measurement into a gain tracking error measurement using the absolute gain at 0-dB as the reference level. To convert each absolute gain into gain tracking error, we subtract the reference gain (0.02 dB) from the absolute gain at each level. The gain tracking errors at each level are listed in Table 10.2.

Table 10.2. Gain Tracking Errors for Voltage Follower

Signal Level	Gain Error ΔG	Signal Level	Gain Error ΔG
0 dB (500 mV)	0.00 dB (by definition)		
+3 dB	−0.20 dB	−30 dB	−0.04 dB
−6 dB	0.00 dB	−36 dB	−0.05 dB
−12 dB	−0.01 dB	−42 dB	−0.06 dB
−18 dB	−0.02 dB	−48 dB	−0.07 dB
−24 dB	−0.03 dB	−54 dB	−0.09 dB

This voltage follower fails the specification limits of ±0.05-dB at all signal levels lower than −30 dB. The signal level at −54 dB is calculated using the formula:

$$-54 \text{ dB signal level} = 0 \text{ dB level} \times 10^{-54 \text{ dB} / 20 \text{ dB}} = 500 \text{ mV RMS} \times 10^{-54 \text{ dB} / 20}$$
$$= 0.998 \text{ mV RMS}$$

For example, consider a 32-step PGA that has an ideal step size of 1.5 dB from each gain step to the next (Figure 10.7). If its lowest gain setting is 0 dB, then it should ideally be programmable to any of 32 different gains: 0 dB, 1.5 dB, 3.0 dB, 4.5 dB, . . . , 46.5 dB. The gain range is defined as the highest gain (dB) minus the lowest gain (dB). Ideally, the gain range in this example should be 46.5 dB. PGA gain specifications, like other gain specifications, are usually tested at a particular frequency, such as 1 kHz. The absolute gain, gain step size, and gain range can all be measured by setting the PGA into each of its 32 settings and measuring the output voltage level divided by the input voltage level.

The absolute gain of each step can be measured by leaving the input signal unchanged and observing the change in output voltage. This eliminates the need for a focused calibration process, since the change in output level is equal to the gain step size regardless of any gain errors in the digitizer or AWG. Although it is possible to measure gain steps in this manner, it is probably best to adjust the input level at each step to produce a fairly constant output level at least 3 dB lower than the full-scale output level. This second technique avoids clipping while producing a strong

---

Exercises

10.5. The gain of an amplifier is measured with a 1-kHz tone test. An FFT analysis reveals that the input and output signals have spectral coefficients of $-0.2866 - j0.133643$ and $0.313150 - j0.044010$ V, respectively. What is the gain of this amplifier?

ANS. 1 V/V.

10.6. The gain of an analog channel is assumed to be described by the following equation

$$G = 0.9 + 0.1\,V_{in} - 0.01\,V_{in}^2$$

What is the gain error of the channel at an input level of 3.0-V RMS if the ideal gain is 1 V/V? What is the gain error when the input is increased to 5.0 V RMS?

ANS. 0.11 V/V; 0.15 V/V.

10.7. Using the gain expression in Exercise 10.6, determine the gain tracking errors at input levels of 3 dB, 0 dB, −3 dB, −6 dB, and −12 dB, when the 0-dB reference level corresponds to a 100-mV RMS input level?

ANS. 0.0383 dB, 0 dB, −0.0274 dB, −0.0470 dB, −0.0709 dB.

---

output signal level at all gain settings. Remember that strong signals are less susceptible to noise, yielding better repeatability in the measurement.

It is worth noting that the 32 gain measurements could potentially be reduced to only 6 measurements, assuming that the PGA is well-designed. If the PGA is designed with 32 individual resistors in the op amp gain circuit, then each step must be measured individually (since any one resistor may be defective). However, if the PGA's gain in dB is controlled by a sum of five binary-weighted resistors, each controlled by one of the PGA's control bits, then only the op amp and the five resistor paths need to be tested and all other levels can be deduced from this primitive set. We saw an example of this back in Chapter 6, "DAC Testing" where a major carrier model of the DAC behavior was derived in Section 6.3.6. This same approach can be extended to the PGA. In this case the PGA model would be one that relates the gain of the PGA expressed in decibels to the state of the D-bit control switches, For example,

$$\text{Gain}_{PGA} = b_0 W_0 + b_1 W_1 + \cdots + b_{D-1} W_{D-1} + \text{base value} \tag{10.7}$$

For the 5-bit example described here, the model would be

$$\text{Gain}_{PGA} = b_0 W_0 + b_1 W_1 + b_2 W_2 + b_3 W_3 + b_4 W_4 + \text{base value} \tag{10.8}$$

By cycling through five independent control words, say, 00000, 00001, 00010, 00100, 01000, and 10000, we can solve for the model parameters: $W_0, W_1, \ldots, W_4$ and the base value.
Figure 10.7. Programmable gain amplifier gain curve.

The complete gain curve can be calculated by adding the gains in a binary-weighted fashion. This partial testing approach requires that superposition be proven to be a valid assumption through characterization of the PGA. For example, assume that superposition is shown to be valid in a 3-bit PGA (8 gain stages) with 1.5-dB gain steps. Measuring the gain at each major transition yields the gain measurements in Table 10.3.

The remaining gains can be calculated using superposition of the measured gain steps according to the general expression of Eq. (10.7). The resulting calculations can be seen listed in Table 10.4.

The binary-weighted PGA is an example of a subtle form of design for test (DfT). By choosing a design architecture that is based on a weighted structure with excellent superposition characteristics, the IC design engineer can reduce the test requirements from eight measurements to four. The remaining four measurements can be calculated in a fraction of the time it would take to measure them explicitly. The test engineer should get involved early in the design process to suggest such architectural features, or at least make the test impact of such design choices known to the design engineers.

Figure 10.7. Programmable gain amplifier gain curve.

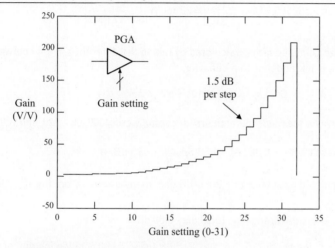

Table 10.3. Measured PGA Gains

Gain Setting	Model Parameter	Measured Gain
0-dB setting (000)	Base Value	0.01 dB
1.5-dB setting (001)	W_0	1.48 dB
3.0-dB setting (010)	W_1	3.25 dB
6.0-dB setting (100)	W_2	6.01 dB

Table 10.4. Calculated PGA Gains

Gain Setting	Calculated Gain $\text{Gain}_{PGA} = \text{Base Value} + b_0 W_0 + b_1 W_1 + b_2 W_2$
4.5-dB setting (011)	Gain = 0.01 dB + 1.48 dB + 3.25 dB = 4.74 dB
7.5-dB setting (101)	Gain = 0.01 dB + 1.48 dB + 6.01 dB = 7.5 dB
9.0-dB setting (110)	Gain = 0.01 dB + 3.25 dB + 6.01 dB = 9.27 dB
10.5-dB setting (111)	Gain = 0.01 dB + 1.48 dB + 3.25 dB + 6.01 dB = 10.75 dB

Table 10.5. PGA Output Voltages with 100-mV Input

Gain Setting	Output Voltage	Gain Setting	Output Voltage
0 dB	100.00 mV RMS	6.0 dB	200.45 mV RMS
1.5 dB	119.26 mV RMS	7.5 dB	239.06 mV RMS
3.0 dB	140.44 mV RMS	9.0 dB	280.87 mV RMS
4.5dB	167.50 mV RMS	10.5dB	334.96 mV RMS

EXAMPLE 10.5

A single-ended 3-bit PGA is stimulated with a constant 100-mV RMS sine wave input at 1 kHz. The output response at each of the eight gain settings is listed in Table 10-5. Calculate the absolute gain at each gain setting. Calculate the gain step size at each transition. Calculate the gain range. Is superposition a valid assumption for this PGA? Is it safe to modify the test program to measure only four gains and calculate the other four using a binary-weighted mathematical approach?

Solution:

Table 10-6 lists the absolute gain at each PGA step, calculated using the formula

$$G(dB) = 20 \ \log_{10} \left| \frac{V_{out}}{V_{in}} \right| = 20 \ \log_{10} \left| \frac{V_{out}}{100 \ mV \ RMS} \right|$$

Table 10-7 lists the gain step sizes, calculated by taking the difference in gain between adjacent steps. Assuming a 3-bit model of gain behavior,

$$Gain_{PGA} = \text{Base Value} + b_0 W_0 + b_1 W_1 + b_2 W_2$$

we can directly solve for the model parameters for control words, 000, 001, 010, and 100, as follows:

$$\text{Base Value} = 0 \ dB, \ W_0 = 1.53 \ dB, \ W_1 = 2.95 \ dB, \ W_2 = 6.04 \ dB$$

Gain range is calculated by subtracting the 0-dB gain measurement from the 10.5-dB gain measurement, that is,

$$\text{gain range} = 10.50 \ dB - 0.00 \ dB = 10.50 \ dB$$

If we calculate the gains at 4.5, 7.5, 9.0, and 10.5 dB using superposition instead of actual measurements, as we did in Table 10-4, we get the results in Table 10-8.

Table 10.6. PGA Measured Gains

Gain Setting	Measured Gain	Gain Setting	Measured Gain
0 dB (000)	0.0 dB	6.0 dB (100)	6.04 dB
1.5 dB (001)	1.53 dB	7.5 dB (101)	7.57 dB
3.0 dB (010)	2.95 dB	9.0 dB (110)	8.97 dB
4.5 dB (011)	4.48 dB	10.5 dB (111)	10.50 dB

Table 10.7. PGA Gain Step Sizes

PGA Transition	Gain Step Size
0 dB to 1.5 dB	1.53 dB − 0.00 dB = 1.53 dB
1.5 dB to 3.0 dB	2.95 dB − 1.53 dB = 1.42 dB
3.0 dB to 4.5 dB	4.48 dB − 2.95 dB = 1.53 dB
4.5 dB to 6.0 dB	6.04 dB − 4.48 dB = 1.56 dB
6.0 dB to 7.5 dB	7.57 dB − 6.04 dB = 1.53 dB
7.5 dB to 9.0 dB	8.97 dB − 7.57 dB = 1.44 dB
9.0 dB to 10.5 dB	10.50 dB − 8.97 dB = 1.53 dB

Table 10.8. Comparison of Measured Gains with Calculated Gains

		Calculated Gain using Superposition $\text{Gain}_{PGA} = \text{Base Value} + b_0 W_0 + b_1 W_1 + b_2 W_2$	
Gain Setting	Actual Gain		Error
4.5 dB	4.48 dB	0.00 dB + 1.53 dB + 2.95 dB = 4.48 dB	0.00 dB
7.5 dB	7.57 dB	0.00 dB + 1.53 dB + 6.04 dB = 7.57 dB	0.00 dB
9.0 dB	8.97 dB	0.00 dB + 2.95 dB + 6.04 dB = 8.99 dB	0.02 dB
10.5 dB	10.50 dB	0.00 dB + 1.53 dB + 2.95 dB + 6.04 dB = 10.52 dB	0.02 dB

The question of whether or not superposition holds and whether or not we can change to a superposition calculation instead of a full measurement process is a trick question for two reasons. First, we have not specified the test limits. Do we have test limits that are tight or loose relative to the behavior of the typical DUT? If the limits are loose, we can probably tolerate the 0.02-dB errors in the superposition calculations. We can account for the errors by tightening our guardbands (tightening the normal test limits by 0.02 dB, for example). If the average device performance is very close to the test limits, however, then we cannot tolerate as much error and the superposition technique may not be acceptable.

Another reason this is a trick question is that we are looking at the results from a *single* DUT. A good engineer should *never* draw broad conclusions about device characteristics from a single DUT or even a limited sample of DUTs. This is one of the most common mistakes a novice test engineer makes. We would have to see superposition hold over at least three production lots (thousands of devices) before we would have confidence that we could change the test program to the faster superposition methodology. Also, we would need to confirm with the design engineer that the PGA is designed in such a way that superposition should be a valid assumption. This gives us confidence that we are making a sensible decision about the expected characteristics of the DUT.

Exercises

10.8. An analog channel with a differential input and differential output has an ideal gain of 6.02 dB. What is the gain error of the channel if a 202.43-mV RMS signal appears at the output when a 100-mV RMS sine wave is applied to its input?   ANS. +0.1055 dB.

10.9. An analog channel is excited by a 100-mV single-ended sinusoidal signal. A digitizer captures 512 samples using a 16-kHz sampling rate. An FFT analysis reveals a single peak value of 0.043 V RMS in the 111th spectral bin. What is the frequency of the test signal? What is the absolute gain of the channel in V/V?   ANS. 3.46875 kHz, 0.43 V/V.

10.2.5 Frequency Response

Frequency response is similar to gain tracking in the sense that it is a measurement of gain under varying signal conditions, relative to a reference gain. Frequency response is most commonly used to measure the transfer function of a filter. Sometimes a frequency response test is used to measure the bandwidth of a circuit such as an op-amp gain circuit to verify that its gain/bandwidth product is acceptable.

If a filter's transfer function is not within specifications, the system-level consequences depend on the filter's purpose. For example, a low-pass antialias filter removes high-frequency components from a digital audio recorder's ADC input. If the antialias filter transfer characteristics are not correct, the result can be unpleasant alias tones in the audio signal once it is reconstructed with a digital audio playback DAC.

Unlike gain tracking tests, in which the signal amplitude is varied, a frequency response test measures the variation in the gain of a circuit as the signal frequency is varied. One signal frequency is chosen as the reference frequency, and the gain of the circuit at that frequency is the reference gain. All other gains are measured relative to the reference gain:

$$\Delta G\,(\text{dB}) = G\,(\text{dB}) - G_{REF}\,(\text{dB}) \quad \text{versus} \quad \text{frequency} \qquad \textbf{(10.9)}$$

For this reason, the gains computed according to Eq. (10.9) are called *relative gains*. Sometimes, the reference frequency is 0 Hz, meaning that all gains are measured relative to the circuit's DC gain. When DC is used as the reference frequency, it is often possible to measure the gain at a very low frequency (say, 100 Hz) rather than making a separate DC gain test as described in Chapter 3. This approach allows a single-pass DSP-based test, which saves test time. Sometimes the reference gain is defined as the midpoint between the highest and lowest gain in the frequency response curve. For example, if a filter's absolute gain varies from +0.25 dB to −0.31 dB across its in-band frequency range, then the reference gain G_{REF} is the average of these maximum and minimum gains

$$G_{REF} = \frac{(0.25 - 0.31)}{2} = -0.03 \text{ dB}$$

Frequency response is usually measured using a coherent multitone signal so that all signal frequencies can be measured simultaneously, saving test time. Sometimes the test must be broken into two parts, an in-band test and an out-of-band test. The reason it must sometimes be split is that the out-of-band components at the output of a filter may be extremely low in amplitude compared to the in-band components.

The out-of-band components must sometimes be amplified by the front-end circuitry of the ATE digitizer or by a local gain circuit on the DIB board before they can be measured accurately. This amplification technique is especially useful if the nonamplified out-of-band components would otherwise fall below the digitizer's quantization noise floor. Applying the same amplification to the in-band components might result in clipped signals. If so, the in-band and out-of-band components must be measured separately, at two different amplification settings. Ideally, though, the in-band and out-of-band components should be measured simultaneously to save test time.

Frequency response is usually measured with equal-level multitone signals in which the RMS amplitudes of the test tones are set equal to one another. To achieve a desired signal level, the amplitude of each tone is set to the desired total signal RMS amplitude divided by the square root of the number of test tones. For example, to produce a four-tone multitone signal with 100-mV RMS signal level, each tone must be set to $100 \text{ mV RMS}/\sqrt{4}$, or 50 mV RMS.

The frequencies of the tones should be chosen so that they do not produce harmonic or intermodulation distortion overlaps. The phase of each tone is randomly selected to produce a signal with an acceptable peak-to-RMS ratio. (For a review of these and other considerations in making a multitone measurement, refer to Chapter 8, "Sampling Theory" and Chapter 9, "DSP-Based Testing.")

As in all AC measurements, the DUT and tester must be allowed to settle to a stable state before valid data can be collected. Therefore, the tester's AWG must begin sending the input signal to the DUT for several milliseconds before the digitizer begins collecting samples. This precollection time is referred to as *settling time*. The settling time of an AC measurement is related to the signal frequency of interest and the filter characteristics of the DUT, AWG, and digitizer. In general, the lower the frequency being tested, the longer it takes to settle.

For example, a 10-Hz high-pass filter is difficult to test in production because it takes many tens or hundreds of milliseconds to settle to a steady state. Also, the higher the order of the filter, the longer it takes to settle. A band-pass filter with a quality factor (Q) of 10 takes much longer to settle than one with a Q of 1. For this reason, it is a good idea to get the DUT and AWG settling process started as early as possible in each test to reduce the test program's settling time overhead.

Once the filter has settled and its output has been digitized, frequency response is simple to calculate. Frequency response is calculated by first performing Fourier analysis (i.e., a DFT or FFT) on the waveforms collected at the DUT input and output. The gain at each frequency is then calculated by dividing the FFT's response to the DUT output signal by the DUT's input signal. The FFT allows a separate gain calculation at each frequency of interest. The reference gain (in dB) is then subtracted from each of the other gains to normalize them to the gain at the reference frequency. The absolute gain of the filter at the reference frequency is usually tested as a separate specification to guarantee that the filter's overall absolute gain is within specifications.

EXAMPLE 10.6

A bandpass filter is formed by cascading a 60-Hz high-pass filter with a 3.4-kHz low-pass filter. The filter's frequency response specification is shown in Table 10.9. Specifications at intermediate points are determined by linear interpolation between the specified points, on a log/log scale. These upper and lower gain limits form the filter's gain mask (Figure 10.8).

The data sheet defines 1 kHz as the reference frequency. The data sheet also specifies that the −3.01-dB gain points must occur between 170 and 190 Hz (high-pass cutoff) and between 3550 and 3650 Hz (low-pass cutoff). Measure the frequency response of the filter, using a multitone signal. Calculate the signal level of each input tone that will result in a combined signal level of 1.0 V RMS.

Table 10.9. Filter Gain Mask Inflection Points

Freq.	Lower Limit	Upper Limit	Freq.	Lower Limit	Upper Limit
<10 Hz	None	−30 dB	300 Hz	−0.5 dB	+0.5 dB
50 Hz	None	−25 dB	3000 Hz	−0.5 dB	+0.5 dB
60 Hz	None	−23 dB	3400 Hz	−1.35 dB	0.0 dB
200 Hz	−1.28 dB	0.0 dB	4000 Hz	None	−14.0 dB
			>4600 Hz	None	−32.0 dB

Solution:
This specification points out a number of problems with frequency response testing. First, the in-band response has to be guaranteed over a range of frequencies, yet the data sheet does not specify which subset of frequencies should be measured in production. Obviously, we could sweep the frequency from 0 to 8 kHz in 1-Hz steps, but that would be unacceptably time-consuming for production testing. A limited number of tones must be chosen. Second, the 3-dB cutoff frequencies are specified. We do not have time to search for the exact frequency that results in a relative gain of −3.01 dB; thus we have to find a compromise.

The solution to the first problem is to determine the frequencies that are most likely to cause the filter to fail the upper and lower frequency-response limits of the filter through a detailed Monte

Carlo or sensitivity analysis. This is usually a procedure that the design engineer has performed prior to releasing the design into production. We can see the potential problem areas by looking at a magnified view of the ideal frequency response plotted against the gain mask (Figure 10.9).

Figure 10.8. Ideal filter frequency response and gain mask.

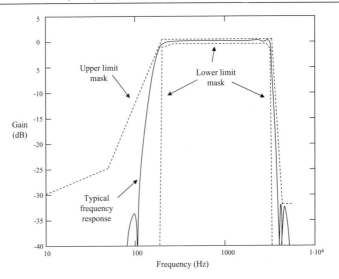

Figure 10.9. Problem areas in the filter frequency response.

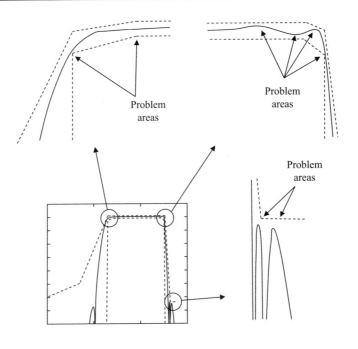

We begin by choosing the problematic frequencies in the passband, corresponding to the peaks and valleys of the ideal frequency-response curve. In this example, the peaks and valleys of the in-band ripple are expected at 850, 1600, 2310, 2860, and 3150 Hz. Next we select frequencies at the center of the out-of-band side lobes, located at 100, 4430, and 4860 Hz. The gains at these frequencies are the most likely to cross the upper gain mask.

One solution to the −3-dB frequency problem is to place two frequencies at the upper and lower limits of the −3-dB points (170, 190, 3460, and 3560 Hz). We can then interpolate between the gains at these frequencies (using a log/log interpolation) to find the approximate location of the −3-dB frequency. Log/Log interpolation is simply one where we fit two gain measurement values expressed in decibels at two different frequency locations to the equation $dB = m \times \log_{10} f + b$ and solve for model parameters m and b; subsequently, we obtain the interpolated gain value at the desired frequency value f, where f lies between the two known frequency values used previously. Alternatively, we can simply require that one gain must be greater than −3.01 dB and the other must be less than −3.01 dB. The interpolation technique is more accurate, but the pass/fail technique is a bit faster.

Next we choose frequencies at the inflection points of the upper and lower masks. These correspond to the frequencies in Table 10.9. Since a 10-Hz tone would require a minimum of 100 ms of data collection time to capture just one cycle, and since the ideal characteristics are so far from the specified mask, we will eliminate the 10- and 50-Hz tests from our frequency list. This is another example where a robust design allows us to avoid costly testing. Of course, we should be prepared to verify that the average device does have this characteristic, but we can perform that characterization in a separate test and later eliminate it from the production test list.

We now have the following list of frequencies corresponding to the in-band and out-of-band regions of the filter: 60, 100, 170, 190, 200, 300, 850, 1000 (reference frequency), 1600, 2310, 2860, 3000, 3150, 3400, 3460, 3560, 4000, 4430, 4600, and 4860 Hz. Clearly the choice of a limited set of frequencies is a complicated one. The test engineer needs to work with the design team to predict which tones are most likely to cause filter response failures. These "problem areas" are the logical choice for production testing.

Of course, we have to adjust the frequencies of the actual test tones slightly to accommodate coherent DSP-based testing, as explained in Chapters 8 and 9. When correlating the ATE tester results to measurements made with bench equipment, the test engineer must communicate the exact frequencies used on the tester. Otherwise, differences in test conditions may introduce correlation errors between the tester and bench.

We can either test the filter with this 20-tone multitone signal all at once or we can split it into two tests. The decision is based on the repeatability we can achieve with a single-pass test. For this example a single-pass test is probably acceptable, since the lowest out-of-band gain specification is −32 dB. If the lowest gain were instead −80 dB, then we might worry about the measurement of such a small signal in the presence of the much larger in-band signals.

To calculate the desired signal level of each tone, we divide the total signal RMS by the square root of the number of tones. Each tone should therefore be set to 1.0 V RMS/$\sqrt{20}$ or 223.6-mV RMS to achieve a total signal amplitude of 1.0 V RMS at the filter's input.

Figure 10.10 shows the 20-tone input signal, digitized at a sampling rate of 16 kHz and 2048 captured samples. Figure 10.11 shows the magnitude of the spectrum of the input signal. Figure 10.12 shows the magnitude of the spectrum of the digitized output. Both plots have been converted to a logarithmic scale (dBV) so that we can see the low-amplitude signal components (including noise) as well as the high ones.

The absolute gains are computed using the formula

$$G(\text{dB}) = 20 \, \log_{10} \left| \frac{V_{out}}{V_{in}} \right| \qquad (10.10)$$

Figure 10.10. Filter input voltage versus time.

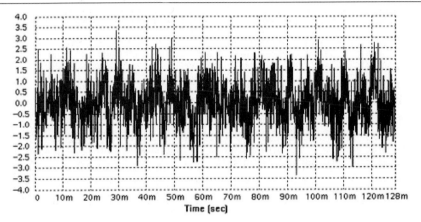

Figure 10.11. Filter input voltage (dBV) versus frequency.

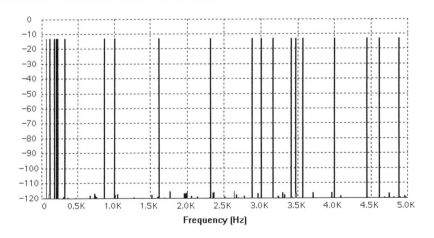

Figure 10.12. Filter output voltage (dBV) versus frequency.

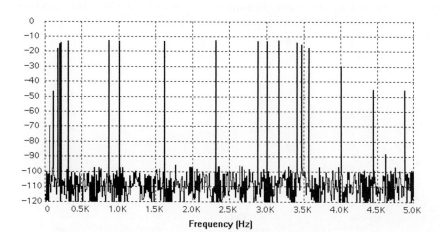

The computed absolute gains and the frequency response (gains relative to the gain at 1 kHz) are listed in Table 10.10. Note that this example filter failed in two places, corresponding to the lower mask problem areas in Figure 10.9. We can see from the enlarged view of the problem areas in the filter mask that the gains at these frequencies were going to be very close to failure.

In addition to the traditional multitone technique, applying a narrow pulse, or impulse, to the circuit under test and observing the filter's impulse response can also measure frequency response. The Fourier transform of the impulse response is the filter frequency response. The advantage of this approach is that it gives the full frequency response, at all frequencies in the FFT spectrum. The problem with the impulse response approach is that it is very difficult to measure the gain at any one frequency with any degree of accuracy, since the energy contained in a narrow impulse is very small. The small signal level makes the measurement very susceptible to noise. Also, an impulse contains energy at many frequencies, which interact with each other through distortion processes. Therefore, the gain at any one frequency may be corrupted by distortion components from other frequencies.

Another technique for frequency response testing is to assume a perfect filter with a known mathematical frequency response. If the mathematical frequency response has a limited number of independent variables that control its behavior, then the gain only needs to be measured at a few frequencies. Using N equations in N unknowns, the full filter transfer function can be estimated from a limited number of gain measurements. This technique has the advantage that only a few tones need to be measured to predict the complete transfer curve of the filter. This technique works well for simple filters with wide design margins, but may not work as well with more complex filters having very tight specifications. Although there are several different ways to approach frequency response testing, coherent DSP-based testing is still the most common methodology for measuring the frequency response characteristics of a filter or other circuit in high-volume production testing.

Table 10.10. Bandpass Filter Absolute and Relative Gain Measurements

Frequency	Absolute Gain	Relative Gain (Frequency Response)	Lower Limit (dB)	Upper Limit (dB)
1000	0.168	0.00	−0.50	+0.50
60	−57.44	−57.61	NA	−23
100	−33.62	−33.79	NA	−13.24 (interpolated)
170	−4.87	−5.04	NA	−3.01
190	−1.89	−2.06	−3.01	NA
200	−1.128	−1.296 **Fail**	−1.280	0.0
300	0.013	−0.155	−0.50	+0.50
850	0.181	0.013	−0.50	+0.50
1600	0.017	−0.151	−0.50	+0.50
2310	0.266	0.098	−0.50	+0.50
2860	0.018	−0.150	−0.50	+0.50
3000	0.071	−0.097	−0.50	+0.50
3150	0.170	0.002	−0.853	+0.292
3400	−1.264	−1.432 **Fail**	−1.35	0.0
3460	−2.29	−2.46	−3.01	NA
3560	−4.55	−4.72	NA	−3.01
4000	−16.82	−16.99	NA	−14.0
4430	−32.21	−32.38	NA	−32.0
4600	−75.23	−75.40	NA	−32.0
4860	−32.52	−32.69	NA	−32.0

Exercises

10.10. Using log/log interpolation, determine the −3-dB frequency of a filter corresponding to a measured gain of −1.56 dB at 3000 Hz and −4.32 dB at 3400 Hz.

ANS. 3202.4 Hz.

10.11. What is the total RMS value of a multitone signal consisting of 25 tones having an RMS value of 100 mV each?

ANS. 0.5 V RMS.

10.12. An eight-tone multitone signal having a combined RMS value of 250 mV is required to perform a frequency response test. What is the peak amplitude of each tone if they are all equal in magnitude?

ANS. 125 mV peak.

10.13. The frequency response behavior of an amplifier is measured with a multitone signal consisting of four tones. An FFT analysis of the input and output samples reveal the following complex spectral coefficients:

Tone	Input	Output
1.1	−0.8402 + j0.5424	−0.1102 + j0.6636
2.1	0.6286 + j0.7778	0.4181 − j0.1002
3.1	−0.9180 − j0.3966	−0.2024 + j0.2308
4.1	−0.0134 + j0.9999	0.2294 + j0.0592

ANS. 3.888 dB, 0 dB, −2.9252 dB, and −5.1747 dB.

What is the relative gain (dB) of the amplifier at each frequency if the reference gain is based on the gain at 2.1 kHz?

10.3 PHASE TESTS

10.3.1 Phase Response

The transfer function (frequency response) of a filter or other analog channel is defined not only by the gain variations over frequency (magnitude response) but also by the phase shift variations (phase response). In analog and mixed-signal testing, the frequency response test often measures only the magnitude response of a circuit. The phase information contained in the FFT results is frequently discarded because phase response is often an unspecified parameter. If phase response is specified in the data sheet, however, it can be calculated using the FFT results collected during the frequency response test.

Assuming that the tester's FFT routine returns the complex coefficients of the discrete-time Fourier series in the form of a vector X, then according to the development in Chapter 9, the amplitude of the kth tone is calculated according to

$$
\tilde{c}_k = \begin{cases} \sqrt{\operatorname{Re}\{X(k)\}^2 + \operatorname{Im}\{X(k)\}^2}, & k = 0, N/2 \\ 2\sqrt{\operatorname{Re}\{X(k)\}^2 + \operatorname{Im}\{X(k)\}^2}, & k = 1, \dots, N/2 - 1 \end{cases} \tag{10.11}
$$

where $\operatorname{Re}\{X(k)\}$ and $\operatorname{Im}\{X(k)\}$ denote the real and imaginary parts of the kth element of vector X. Correspondingly, the phase shift of the kth tone can be calculated using the phase formula:

$$
\tilde{\phi}_k = \begin{cases} -\tilde{\phi}_o & \text{if } \operatorname{Re}[X(k)] \geq 0, \operatorname{Im}[X(k)] \geq 0 \\ \tilde{\phi}_o & \text{if } \operatorname{Re}[X(k)] \geq 0, \operatorname{Im}[X(k)] < 0 \\ \tilde{\phi}_o - \pi & \text{if } \operatorname{Re}[X(k)] < 0, \operatorname{Im}[X(k)] \geq 0 \\ \pi - \tilde{\phi}_o & \text{if } \operatorname{Re}[X(k)] < 0, \operatorname{Im}[X(k)] < 0 \end{cases} \tag{10.12}
$$

where

$$
\tilde{\phi}_o = \tan^{-1}\left\{ \frac{\left|\operatorname{Im}[X(k)]\right|}{\left|\operatorname{Re}[X(k)]\right|} \right\}
$$

To aid the user, most testers include a DSP routine to convert the results of an FFT into polar notation (magnitude and phase). The polar conversion routines perform whatever corrections are necessary to compute a correct phase shift. Although the built-in polar conversion approach is easier than doing the conversion manually, it can be less efficient and a source of confusion. Why should we calculate the magnitude and phases of all 511 complex spectral coefficients in a 1024-point FFT if only 10 of the phases are of interest? Moreover, the phase angle of a bin containing noise has no relationship to the phase relationship of a signal; hence, one should not attempt to correlate these phase terms (or as we shall learn later, apply correction factors to them). The end result is that phase calculations should only be performed on bins containing signals of interest; *all others should be ignored.*

While the concept of the phase of a signal is relatively straightforward, the convention that one uses can be a source of confusion. Figure 10.13 depicts a diagram of two sine waves (denoted A and B) separated in phase by the angle ϕ. Here the phase angle is measured at the point at which each waveform crosses the time axis with a positive slope. There are two ways in which

Figure 10.13 Illustrating an interpretation of a phase angle between two sinusoidal signals. Signal A is said to lead signal B by the phase angle ϕ, or alternatively, signal B is said to lag signal A by the phase angle ϕ.

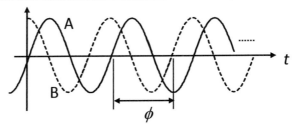

Figure 10.14. Analog delay line producing a -90 degree phase shift (phase lag).

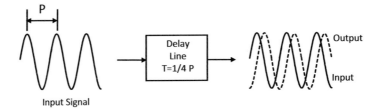

to describe the angle between these two signals. One is to refer to signal A with respect to signal B, and the other is to refer to signal B with respect to signal A. For the situation depicted in Figure 10.13, it is customary to say that signal A *leads* signal B by the phase angle ϕ, since the zero-crossing point of signal A occurs before that of signal B. However, it is equivalent to say that signal B *lags* signal A by the phase angle ϕ. We can also say that signal A lags signal B by the phase angle $-\phi$. As a matter of convention, all phase angles described in this textbook have been defined in terms of a phase lag.

Using multitone DSP-based testing, we can calculate the real and imaginary components of the input and output signal of a circuit under test. All phase shifts can thus be measured with a single test, using the same data collected while measuring the magnitude portion of the frequency response. The phase shift at each frequency in a multitone test signal is calculated by subtracting the input signal phase shift from the output signal phase shift and then correcting for any 360-degree wrap-around effects that result from the arctangent function. A positive phase shift indicates a positive time delay (i.e., the output lags the input), while a negative phase shift indicates a negative time delay (i.e., the output leads the input). Figure 10.14 shows an analog delay line with a time shift of $P/4$ producing a phase lag of 90 degrees at a frequency of $1/P$.

Polar notation is limited to a phase shift of at most ±180 degrees. For example, a phase shift of -190 degrees translates into a shift of $+170$ degrees. The test engineer has to account for this "wrapping" effect by adding or subtracting integer multiples of 360 degrees to the polar conversion results. The idea is to eliminate jump discontinuities in the phase behavior of the device because these rarely, if ever, occur in practice. This process can get confusing; thus we shall look at an example to illustrate the phase measurement process.

EXAMPLE 10.7

Measure the phase response of a 100-µs analog delay line in 1-kHz increments from 900 Hz to 9.9 kHz (10 tones). For simplicity, use a 1024-point sampling system with a 100-Hz fundamental frequency. Use odd harmonics so the time-domain signal will be symmetrical about the x axis (i.e., use spectral bins 9, 19, 29, 39, etc.). Use random phase shifts to produce a peak to RMS ratio of approximately 3.35:1.

Solution:

Figure 10.15 shows the digitized representation of the input signal to the delay line. The phases of the tones in this input signal were chosen to produce a peak to RMS ratio of 3.35:1. The real and imaginary spectral coefficients from an FFT analysis of the 10 digitized input tones are listed in Table 10.11, along with the phase calculated using Eq. (10.12). The digitized output signal is shown in Figure 10.16. The real and imaginary FFT results for the 10 digitized output tones are listed in Table 10.12, along with the calculated phases.

Figure 10.15. 100-µs delay line input voltage versus time.

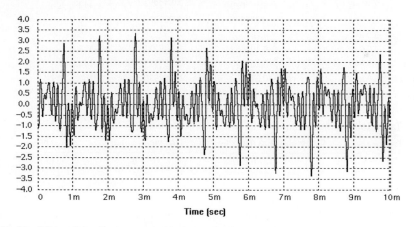

Figure 10.16. 100-µs delay line output voltage versus time.

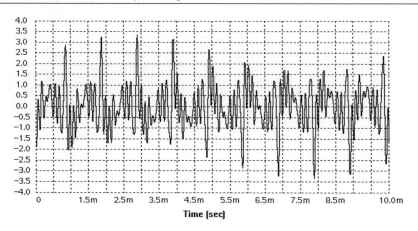

Table 10.11. Calculated Signal Phases for Digitized Input

Frequency	Real Part	Imaginary Part	Phase (deg)
900 Hz	−0.199009	−0.245755	129.0
1.9 kHz	−0.223607	−0.223608	135.0
2.9 kHz	0.238660	−0.207465	41.0
3.9 kHz	0.286600	0.133643	−25.0
4.9 kHz	−0.308123	0.071136	−167.0
5.9 kHz	−0.190311	−0.252550	127.0
6.9 kHz	0.313150	−0.044010	8.0
7.9 kHz	−0.215668	0.231275	−133.0
8.9 kHz	−0.071136	0.308123	−103.0
9.9 kHz	−0.308123	0.071136	−167.0

Table 10.12. Calculated Signal Phases for Digitized Output

Frequency	Real Part	Imaginary Part	Phase (deg)
900 Hz	−0.298347	−0.104827	160.6
1.9 kHz	−0.293618	0.117421	−158.2
2.9 kHz	−0.252395	−0.190517	143.0
3.9 kHz	−0.119020	−0.292974	112.1
4.9 kHz	0.314893	−0.033096	6.0
5.9 kHz	0.285237	0.133008	−25.0
6.9 kHz	−0.105152	0.298233	−109.4
7.9 kHz	−0.258195	−0.182579	144.7
8.9 kHz	−0.274162	0.157590	−150.1
9.9 kHz	−0.316175	0.005777	−179.0

Table 10.13. Phase Lag Calculations and Corrections

Frequency	Phase Shift (Out − In)	±360 Phase Adjustment	Actual Phase
900 Hz	160.6 − 129.0 = 31.6	0	31.6
1.9 kHz	−158.2 − 135.0 = −293.2	−360	66.8
2.9 kHz	143.0 − 41.0 = 102.0	0	102.0
3.9 kHz	112.1 − (−25.0) = 137.1	0	137.1
4.9 kHz	6.0 − (−167.0) = 173.0	0	173.0
5.9 kHz	−25.0 − 127.0= −152.0	−360	208.0
6.9 kHz	−109.4 − 8.0 = -117.4	−360	242.6
7.9 kHz	144.7 − (−133.0) = 277.7	0	277.7
8.9 kHz	−150.1 − (−103.0) = −47.1	−360	312.9
9.9 kHz	−179.0 − (−167.0) = −12.0	−360	348.0

Subtracting input phases from output phases should yield the phase lag at each frequency. Unfortunately, the polar conversion cannot distinguish between a shift of 360 degrees and a shift of 0 degrees. For this reason, we have to correct the output-input phase calculations in a two-step process. The raw output-minus-input phase shift calculations and the results of the correction steps are shown in Table 10.13.

In the first correction step (column three in Table 10.13), we have to account for the 360 degree ambiguity of the phase subtraction operation by first converting each phase shift to a value between −180 and +180 degrees. To do this, we either add 360 degrees or subtract 360 degrees from each phase number to limit our phase shift answers from −180 to +180 degrees. Then, we can see the discontinuity in the calculated phase shift between 4.9 and 5.9 kHz.

Observing the phase changes in the second column of Table 10.13, we see that there are several discontinuities in the calculated phase shift—in particular, between the frequencies of 1.9 and 2.9 kHz, 5.9 and 6.9 kHz, 6.9 and 7.9 kHz, and 8.9 and 9.9 kHz. These discontinuities are caused by a mathematical wrap-around effect, since the polar calculations cannot distinguish between a 360-degree shift and a 0-degree shift.

To compensate for this 360-degree discontinuity, we either add or subtract 360 degrees from the wrapped result as needed to make the phase shift measurements continue in the correct direction. The results of this correction process are shown in the last column of Table 10.13. A plot illustrating the wrapping effect on the system phase lag response and the phase after the correction factors are included is illustrated in Figure 10.17. As expected the system exhibits a straight-line phase lag response increasing with increasing frequency.

Sometimes phase calculations are specified not in degrees or radians, but in seconds or in fractions of a cycle. To convert phase shift in degrees or radians to phase shift in fractions of a cycle, we use the formula

$$\text{phase shift (fractions of cycle)} = \frac{\phi(\text{degrees})}{360 \text{ degrees}} \tag{10.13}$$

or equivalently

$$\text{phase shift (fractions of cycle)} = \frac{\phi(\text{radians})}{2\pi \text{ radians}} \tag{10.14}$$

To convert phase shift in degrees to phase shift in seconds, we use a similar calculation, taking the period $(1/f)$ into account

$$\text{phase shift (sec)} = \text{period (sec)} \times \frac{\phi(\text{degrees})}{360 \text{ degrees}} = \frac{1}{f\,(\text{Hz})} \times \frac{\phi(\text{degrees})}{360 \text{ degrees}} \tag{10.15}$$

or equivalently

$$\text{phase shift (sec)} = \text{period (sec)} \times \frac{\phi(\text{radians})}{2\pi \text{ radians}} = \frac{1}{f\,(\text{Hz})} \times \frac{\phi(\text{radians})}{2\pi \text{ radians}} \tag{10.16}$$

10.3.2 Group Delay and Group Delay Distortion

Group delay is a measurement of time shift versus frequency. It is commonly denoted as $\tau(f)$. Group delay is defined as the change in phase shift (fractions of a cycle) divided by the change in frequency (Hz), that is,

$$\tau(f) = \frac{\Delta\text{phase shift (fractions of cycle)}}{\Delta f} = \frac{1}{360} \frac{\Delta\phi(\text{deg})}{\Delta f} = \frac{1}{2\pi} \frac{\Delta\phi(\text{rad})}{\Delta f} \tag{10.17}$$

where $\Delta\phi$ is the change in phase shift expressed in either degrees or radians. Group delay is expressed in units of time. Strictly speaking, group delay is defined as the derivative of phase with respect to frequency, $\tau(f) = (1/360)(d\phi/df)$. In reality, it is extremely difficult to resolve tiny changes in phase called for by the derivative operation.

Instead, we have to measure the phase at two points that are sufficiently separated in frequency to allow an accurate measurement of phase change. Group delay is typically measured with tone pairs centered around each frequency of interest. Many tone pairs can be measured simultaneously using DSP-based testing.

In a simple delay line, the phase shift through a circuit is directly proportional to frequency. The group delay of a delay line is therefore equal to the time delay through the circuit. A constant group delay indicates a circuit that shifts each signal component by a constant amount of time. This leaves the relative time shifts of the various signal components unchanged and therefore results in a signal that is identical in shape but shifted in time (either delayed or advanced).

Figure 10.17. System phase lag behavior before and after phase correction factors included.

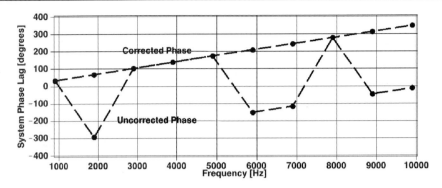

If, on the other hand, group delay varies over frequency, then the circuit will shift the various signal components relative to one another. This results in a change in shape as well as a shift in time. The variation in group delay over frequency is called *group delay distortion*. It is defined as the group delay minus the midpoint between the maximum and minimum group delay in the frequency range of interest, that is,

$$\tau_{distortion}\left(f\right)=\tau\left(f\right)-\frac{\tau_{max}+\tau_{min}}{2} \tag{10.18}$$

Group delay distortion may or may not be a problem in the system-level application. Signal clipping is one potential problem with a circuit that exhibits poor group delay characteristics. If different frequencies are shifted relative to each other, then the peak to RMS ratio may change enough to cause extreme peaks, which clip against the power supply rails of the analog channel.

Another problem that can arise from group delay distortion is incorrect data transmission in data communication channels, such as those used in modems, hard disk drive read/write channels, and cellular telephones. Since phase carries important information in many data communication protocols, group delay errors can lead to corrupted data.

EXAMPLE 10.8

Calculate the group delay and group delay distortion of the 100-µs delay line from the data gathered in the previous example.

Solution:
Group delay and group delay distortion are simple calculations once we have calculated the phase shift at each frequency. The group delay and group delay distortion for each change in frequency is listed in Table 10.14. Frequency change is a constant 1000 Hz.

Table 10.14. 100-µs Delay Line Group Delay and Group Delay Distortion Calculations

Test Tone Pairs	Phase Change (deg)	Group Delay	Group Delay Distortion
900 Hz and 1.9 kHz	66.8−(31.6) = 35.2	97.67 µs	−245 ns
1.9 kHz and 2.9 kHz	102.0−(66.8) = 35.2	97.78 µs	−135 ns
2.9 kHz and 3.9 kHz	137.1−(102.0) = 35.1	97.50 µs	−415 ns
3.9 kHz and 4.9 kHz	173.0−(137.1) = 35.9	99.72 µs	+1.8 µs
4.9 kHz and 5.9 kHz	208.0−(173.0) = 35.0	97.22 µs	−694 ns
5.9 kHz and 6.9 kHz	242.6−(208.0) = 34.6	96.11 µs	−1.8 µs
6.9 kHz and 7.9 kHz	277.7−(242.6) = 35.1	97.50 µs	−415 ns
7.9 kHz and 8.9 kHz	312.9−(277.7) = 35.2	97.78 µs	−135 ns
8.9 kHz and 9.9 kHz	348.0−(312.9) = 35.1	97.50 µs	−415 ns

Figure 10.18. Group delay versus frequency. The difference between any point on the group delay curve and the average group delay is group delay distortion.

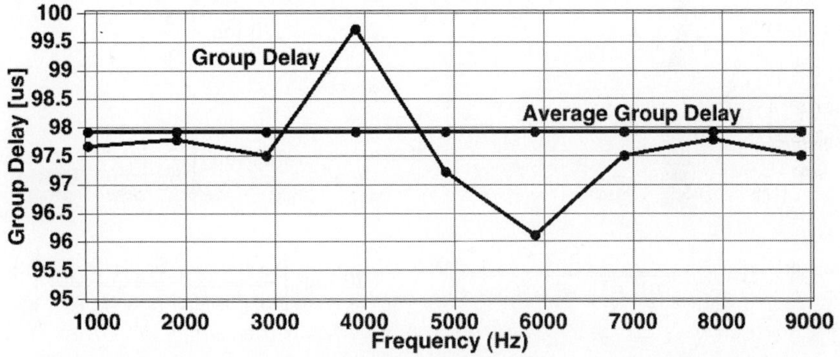

Group delay is equal to the change in phase (deg) divided by 360 × 1000 deg/s. Group delay distortion at each frequency is equal to the group delay minus the midpoint between the maximum and minimum group delay. The midpoint is calculated as the average between 99.72 and 96.11 µs, (99.72 + 96.11)/2 = 97.915 µs. A plot of group delay and its average is shown in Figure 10.18.

10.4 DISTORTION TESTS

10.4.1 Signal-to-Harmonic Distortion

Harmonic distortion arises when a signal passes through a nonlinear circuit. The spectrum of the output of a nonlinear circuit includes not only the frequency components that appeared at the input, but also integer multiples (harmonics) of the input frequency components. Harmonic distortion is often measured with a single-tone test signal, that is, a sine wave at a particular frequency (specified in the data sheet). To save test time, distortion can be measured in parallel with absolute gain using the FFT results from the gain test.

When passing a single test tone through the circuit under test, the harmonic distortion components appear at integer multiples of the test tone's frequency, F_t. F_t is often referred to as the *fundamental tone* (not to be confused with the fundamental frequency of the sampling system, F_f). Distortion that is symmetrical about the x axis gives rise to only odd harmonics ($3F_t$, $5F_t$, $7F_t$, etc.). Asymmetrical distortion, such as clipping on only the upper or lower portion of the waveform, gives rise to both odd harmonics and even harmonics ($2F_t$, $4F_t$, $6F_t$, etc.).

Signal to total harmonic distortion is defined as the ratio of the RMS signal level of the test tone divided by the total RMS of the odd and even harmonic distortion components. Signal to distortion is often expressed in decibel units, similar to gain. Since there are an infinite number of possible harmonics, the data sheet often calls out only a signal to second harmonic distortion and a signal to third harmonic distortion test. Also, the data sheet may call out a signal to total noise plus total harmonic distortion specification, which covers all harmonic distortion components and all noise components simultaneously (any spectral component that is not signal is either noise or distortion). To enable complete characterization, the test engineer will often write a test program that reports all these permutations of signal to noise and distortion measurements.

Table 10.15. Various Signal to Noise and Distortion formulae.[a]

Distortion Metric	Expression	
	V/V	**dB**
Signal to second harmonic distortion (S/2nd)	$\dfrac{S}{H_2}$	$20 \log_{10}\left(\dfrac{S}{H_2}\right)$
Signal to third harmonic distortion (S/3rd)	$\dfrac{S}{H_3}$	$20 \log_{10}\left(\dfrac{S}{H_3}\right)$
Signal to total harmonic distortion (S/THD)	$\dfrac{S}{\sqrt{H_2^2 + H_3^2 + H_4^2 + H_5^2 + \cdots}}$	$20 \log_{10}\left(\dfrac{S}{\sqrt{H_2^2 + H_3^2 + H_4^2 + H_5^2 + \cdots}}\right)$
Signal to noise (S/N)	$\dfrac{S}{N}$	$20 \log_{10}\left(\dfrac{S}{N}\right)$
Signal to total harmonic distortion plus noise (S/THD+N)	$\dfrac{S}{\sqrt{H_2^2 + H_3^2 + H_4^2 + H_5^2 + \cdots + N^2}}$ or $\dfrac{S}{\sqrt{(\text{total signal RMS})^2 - S^2}}$	$20 \log_{10}\left(\dfrac{S}{\sqrt{H_2^2 + H_3^2 + H_4^2 + H_5^2 + \cdots + N^2}}\right)$ or $20 \log_{10}\left(\dfrac{S}{\sqrt{(\text{total signal RMS})^2 - S^2}}\right)$

[a] Individual spectral components are expressed in terms of RMS value. Note that DC offset is not included in any of these formulae.

The definitions of the various signal to noise and distortion parameters are listed in Table 10.15. The symbol S denotes the fundamental signal component expressed in RMS volts, H_2 the RMS value of the second harmonic, H_3 the RMS value of the third harmonic, and so on, and N is the RMS value of all the nonharmonic bins combined (noise will be explained in a later section). Note that we add RMS signal levels using a square-root-of-sum-of-squares calculation. This is a direct result of Parseval's theorem described in Section 9.4.2. Also note that the numbers will typically be positive since the signal is always larger than the distortion (unless the DUT is completely defective).

Sometimes the data sheet will call out specifications in negative decibels, which simply means that the test engineer swaps the numerator and denominator in the log calculations or, equivalently, changes the sign of the dB number reported. In the final row of the table, the S/(THD+N) calculation is defined in two different ways. Both are equivalent, although, the second definition is sometimes faster to compute than pulling all the harmonics apart from one another only to recombine them later. Testers usually include a very efficient RMS routine that can calculate the total signal RMS quickly. It should also be noted that the abbreviation S/(THD+N) is often replaced by the equivalent expression, SINAD, which stands for signal to noise and distortion.

Table 10.16. Voltage Follower Output Levels at Fundamental and Second through Fifth Harmonics

FFT Spectral Bin	Frequency	RMS Voltage
51	1 kHz (fundamental tone)	1.025 V
102	2 kHz (second harmonic)	1.23 mV
153	3 kHz (third harmonic)	2.54 mV
204	4 kHz (fourth harmonic)	0.78 mV
255	5 kHz (fifth harmonic)	0.32 mV

Repeatable measurements of low-level distortion components can be extremely difficult and time-consuming to perform. For instance, if the specified distortion level is −85 dB, then the distortion component of interest may be very close to the noise floor of the DUT and/or ATE measurement hardware. This will lead to unrepeatable measurements of distortion that may or may not be tolerable.

The only way to improve the repeatability is to average or collect more samples with the ATE digitizer. The end result in either case is that data collection time (i.e., the number of samples collected) must quadruple to drop the nonrepeatability in half. The extra collection and DSP processing time obviously adds test time and drives up the cost of testing. Therefore, very low levels of distortion are inherently very costly to test, especially when they are close to failing test limits.

EXAMPLE 10.9

A 1-kHz sine wave passes through a voltage follower and a digitizer captures 512 samples of the output signal at a sampling rate of 10 kHz. The fundamental frequency is equal to the sampling rate divided by number of samples, or 10 kHz/512 = 19.531 Hz. To achieve coherent testing, the 1-kHz sine wave is actually generated by an AWG at 51 times the digitizer's fundamental frequency, or 996.094 Hz. An FFT of the output signal shows several distortion components, listed in Table 10.16. Calculate S/2nd, S/3rd, and S/THD.

Solution:
Notice that in a coherent DSP-based measurement system, the harmonic distortion components always fall into single FFT spectral bins. These can easily be calculated by multiplying the FFT bin of the fundamental tone, 51, by 2, 3, 4, 5, and so on. Therefore, we do not have to multiply 996.094 Hz times 2, 3, 4, and 5 and then try to figure out which spectral bins they will fall into. Working with spectral bins instead of frequencies is therefore easier in some cases than working with frequencies. The signal-to-distortion results are listed in Table 10.17.

Table 10.17. Signal-to-Distortion Results

S/2nd	S/3rd	S/THD
$20 \log_{10}\left(\dfrac{1.025 \text{ V}}{1.23 \text{ mV}}\right)$ $= 58.4 \text{ dB}$	$20 \log_{10}\left(\dfrac{1.025 \text{ V}}{2.54 \text{ mV}}\right)$ $= 52.1 \text{ dB}$	$20 \log_{10}\left(\dfrac{1.025 \text{ V}}{\sqrt{\left(1.23 \text{ mV}\right)^2 + \left(2.54 \text{ mV}\right)^2 + \left(0.78 \text{ mV}\right)^2 + \left(0.32 \text{ mV}\right)^2}}\right)$ $= 50.8 \text{ dB}$

Figure 10.19. Second and third intermodulation frequencies for two-tone signal at F_1 and F_2.

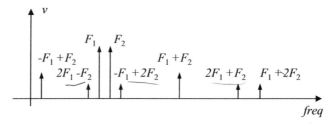

10.4.2 Intermodulation Distortion

Intermodulation distortion is very similar to harmonic distortion except that two or more tones are supplied to the DUT at once. The details of intermodulation tests vary widely from one type of device to another. Telecommunications and audio products usually specify a two- or four-tone test, while digital subscriber line (DSL) testing may require hundreds of test tones. Distortion components may appear at any sum or difference of any multiple of the test tones. Given any two test tones in a multitone signal, F_1 and F_2, there may be distortion components at any intermodulation frequency $F = |p \times F_1 \pm q \times F_2|$, where p and q may be any positive integers. Second-order intermodulation components are those for which $p + q = 2$. Third-order intermodulation components are those for which $p + q = 3$, and so on. Second- and third-order intermodulation components for two frequencies, F_1 and F_2, are shown in Figure 10.19.

Intermodulation distortion is expressed as a ratio of the signal RMS of any one test tone, denoted by, say, S_1, to the signal RMS of the intermodulation component(s):

$$20 \; \log_{10} \left(\frac{S_1}{\sqrt{IMD_1^2 + IMD_2^2 + IMD_3^2 + IMD_4^2 + \cdots}} \right) \tag{10.19}$$

where IMD_1, IMD_2, and so on., correspond to the appropriate intermodulation distortion components.

The signal to intermodulation distortion calculations are usually specified with a limited number of distortion combinations—for example, sum of all third-order intermodulation components. Alternatively, a signal to total noise plus distortion specification may be listed in the data sheet.

When calculating specific combinations, the test engineer adds or subtracts test tone FFT bin numbers and their multiples to determine which FFT bins are likely to contain intermodulation components.

EXAMPLE 10.10

A multitone test signal consists of a sum of two 1.0-V RMS sine waves, one at 1 kHz and the other at 1.1 kHz. Calculate the frequencies of the second-, third-, and fourth-order intermodulation components. The signal RMS at 100 Hz is 193 µV and the signal RMS at 2.1 kHz is 232 µV. Calculate the signal to second-order intermodulation distortion ratio, in dB.

Solution:
The second-order intermodulation components occur at

$$|1.0 \text{ kHz} \pm 1.1 \text{ kHz}| = 100 \text{ Hz and } 2.1 \text{ kHz}$$

The third-order intermodulation components occur at

$$|2 \times 1.0 \text{ kHz} \pm 1.1 \text{ kHz}| = 900 \text{ Hz and } 3.1 \text{ kHz}$$
$$|1.0 \text{ kHz} \pm 2 \times 1.1 \text{ kHz}| = 1.2 \text{ kHz and } 3.2 \text{ kHz}$$

The fourth-order intermodulation components occur at

$$|2 \times 1.0 \text{ kHz} \pm 2 \times 1.1 \text{ kHz}| = 200 \text{ Hz and } 4.2 \text{ kHz}$$
$$|3 \times 1.0 \text{ kHz} \pm 1.1 \text{ kHz}| = 1.9 \text{ kHz and } 4.1 \text{ kHz}$$
$$|1.0 \text{ kHz} \pm 3 \times 1.1 \text{ kHz}| = 2.3 \text{ kHz and } 4.3 \text{ kHz}$$

The signal to second-order intermodulation ratio is given by the equation

$$20 \log_{10}\left(\frac{1.0 \text{ V RMS}}{\sqrt{(193 \ \mu V)^2 + (232 \ \mu V)^2}} \right) = 70.4 \text{ dB}$$

10.4.3 Adjacent Channel and Noise Power Ratio Tests

A related concept to intermodulation distortion is the idea of adjacent channel and noise power ratio tests. These tests track the distortion behavior of a device over a continuous band of frequencies rather than at a set of discrete frequencies. Such tests provide a useful measure of nonlinear system operation as it tracks the mixing products associated with a complex input signal. Their use is growing in data communication systems and data converter applications.

Exercises

10.18. An FFT analysis of the output of an amplifier contains the following spectral amplitudes:

FFT Spectral Bin	RMS Voltage
31	0.9560 V
62 (2 × 31)	0.05 mV
93 (3 × 31)	1.64 mV
124 (4 × 31)	0.04 mV
155 (5 × 31)	1.04 mV

In addition, the total RMS value of the output signal is 0.95601 V. Calculate S/2nd, S/3rd, S/THD, and S/THD+N.

ANS. 85.63, 55.31, 53.84, 46.79 dB.

10.19. A multitone test signal consists of a sum of three 1.0-V RMS sine waves, one at 0.9 kHz, another at 2.1 kHz, and the third at 5.3 kHz. Determine the frequencies of all third-order intermodulation frequencies.

ANS. 0.3, 1.1, 2.3, 3.3, 3.5, 3.9, 4.1, 5.1, 6.5, 7.1, 8.3, 8.5, 9.5, 9.7, 11.5, 12.7 kHz.

Figure 10.20. Illustrating the input and output spectrum associated with: (a) an adjacent channel power ratio (ACPR) test; (b) noise power ratio (NPR) test.

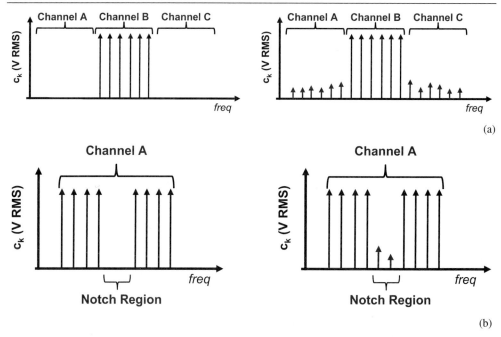

(a)

(b)

In the case of the adjacent channel power ratio (ACPR) test, the focus is on the distortion level created in a separate band of frequencies (i.e., such as an adjacent channel,) by a multitone input signal. Mathematically, ACPR is defined as

$$\text{ACPR} = \frac{\text{RMS level of input multitone signal}}{\text{RMS level in adjacent channel region}} \qquad (10.20)$$

The assumption here is that the adjacent channel occupies a different band of frequencies than the input multitone signal as depicted in Figure 10.20a for channels A, B, and C. In contrast, the noise power ratio (NPR) test considers the distortion that falls in some narrow bandwidth within the channel region of the device as illustrated in Figure 10.20b. In order to separate the distortion effect from the input signal, the input signal excludes any signals within this narrow frequency region. It is common practice to refer to this region as the notch region. Mathematically, NPR is defined as

$$\text{NPR} = \frac{\text{RMS level of input multitone signal}}{\text{RMS in notch region}} \qquad (10.21)$$

EXAMPLE 10.11

A multitone test signal is created using a 1024-point sample set with a sampling frequency of 16 MHz. One-hundred tones with an amplitude of 50 mV are used beginning with bin 101 and there-after except for bins 149, 150, and 151. A DUT was subject to this multitone signal and its output digitized at the same rate of 16 MHz. An FFT analysis reveals that bins 149, 150, and 151 had an RMS level of 1.1 mV, 1.5 mV, and 0.9 mV, respectively. What is the noise power ratio in decibels?

Solution:

As the input signal consists of 100 tones of 50-mV amplitude, the total RMS value of this input signal is

$$\text{RMS level of input} = \sqrt{100}\left(\frac{50\times10^{-3}}{\sqrt{2}}\right) = 0.3535 \text{ V}$$

The RMS level associated with the notch distortion is

$$\text{RMS in notch region} = \sqrt{\left(1.1\times10^{-3}\right)^2 + \left(1.5\times10^{-3}\right)^2 + \left(0.9\times10^{-3}\right)^2} = 2.0 \text{ mV}$$

Consequently, the NPR is

$$20\log_{10}\text{NPR} = 20\log_{10}\left(\frac{0.3535}{2.0\times10^{-3}}\right) = 44.6 \text{ dB}$$

10.5 SIGNAL REJECTION TESTS

10.5.1 Common-Mode Rejection Ratio

A number of signal rejection specifications are common to analog and sampled channel testing. Signal rejection tests are those that measure a channel's ability to prevent an undesired signal from propagating to the channel's output. The undesired signal may originate in the power supply, in another supposedly separate circuit, or in the channel itself.

One such signal rejection test, the common-mode rejection ratio (CMRR), is a measurement of how well a channel with a differential input can reject a common-mode signal. Ideally, a differential input circuit produces an output equal to GV_{diff}, where $V_{diff} = IN_P\_IN_N$ is the differential input voltage and G is the gain of the input circuit. Provided that IN_P and IN_N are exactly equal (i.e., if the input signal is purely common mode with no differential component), the circuit should produce zero output. However, due to mismatched components in the input circuit, a small amount of common-mode signal usually feeds through to the output.

In Chapter 3, we studied DC CMRR testing for differential circuits. AC CMRR for analog channels is defined in a similar manner to DC CMRR. AC CMRR is defined as the AC gain of the channel with a common-mode input divided by the gain of the channel with a normal, differential input

$$\text{CMRR}(f) \equiv \left|\frac{G_{cm}(f)}{G_{diff}(f)}\right| = \left|\frac{\left(\frac{V_{out}}{V_{cm}}\bigg|_{V_{diff}=0}\right)}{\left(\frac{V_{out}}{V_{diff}}\bigg|_{V_{cm}=\text{constant}}\right)}\right| \tag{10.22}$$

AC CMRR is a frequency-dependent parameter and is therefore denoted with a frequency argument.

The common-mode gain is measured by connecting the two inputs together and applying a single-ended signal to them. The common-mode gain is thus measured as if the channel had a single-ended input. Obviously, this gain should be very low, since IN_P and IN_N are equal and the output GV_{diff} should be zero. This very low gain is then divided by the differential AC gain of the channel to arrive at the CMRR value for the specified frequency. Often, the differential gain of the channel has already been measured at several frequencies during an absolute gain or frequency response test. In such a case, the differential gain results can be reused instead of repeating the same differential gain measurements again.

If high accuracy is not required, it is often acceptable to simply divide the measured common-mode gain by the ideal differential gain of the circuit, rather than measuring the differential gain at each frequency of interest. The measurement of CMRR is often performed at several frequencies, since a channel may have different characteristics at different frequencies. Multitone testing can be used to measure CMRR at several frequencies at once, saving test time.

Like many other AC parameters, CMRR is often expressed in decibel units rather than V/V. Since the gain of the circuit with a common-mode input is much smaller than the differential gain, CMRR should produce a negative decibel result. The calculation of CMRR for an analog channel is thus

$$\text{CMRR (dB)} \equiv 20 \ \log_{10}\left(\left|\frac{G_{cm}(f)}{G_{diff}(f)}\right|\right) \tag{10.23}$$

EXAMPLE 10.12

A differential gain circuit has a differential input, a differential output and an ideal gain of 6.02 dB (2.0 V/V). A differential multitone signal is applied to the input of the circuit at 300, 1020, and 3400 Hz. The level of each of the tones is 250 mV RMS, differential. The output of the circuit is digitized differentially, resulting in the following differential RMS output levels: 300 Hz: 510 mV RMS, 1020 Hz: 500 mV RMS, 3400 Hz: 480 mV RMS. Then the inputs are shorted together and a single-ended multitone is applied to the two inputs simultaneously.

The input signal level is 250 mV RMS, single-ended. The output of the channel is again digitized, resulting in the following differential RMS output levels: 300 Hz: 0.7 mV RMS, 1020 Hz: 0.8 mV RMS, 3400 Hz: 1.5 mV RMS. Calculate the CMRR at each frequency. If we observe the positive input during the differential gain measurement and then during the common-mode measurement, would the signal level change?

Solution:
The differential gain at each frequency is calculated as follows:

$$\textbf{300 Hz:} \quad \text{Gain} = \frac{510 \text{ mV}}{250 \text{ mV}} = 2.04 \text{ V/V}$$

$$\textbf{1020 Hz:} \quad \text{Gain} = \frac{500 \text{ mV}}{250 \text{ mV}} = 2.00 \text{ V/V}$$

$$\textbf{3400 Hz:} \quad \text{Gain} = \frac{480 \text{ mV}}{250 \text{ mV}} = 1.92 \text{ V/V}$$

The common-mode gain at each frequency is calculated as follows:

$$\textbf{300 Hz:} \quad \text{Gain} = \frac{0.7 \text{ mV}}{250 \text{ mV}} = 0.0028 \text{ V/V}$$

$$\textbf{1020 Hz:} \quad \text{Gain} = \frac{0.8 \text{ mV}}{250 \text{ mV}} = 0.0032 \text{ V/V}$$

$$\textbf{3400 Hz:} \quad \text{Gain} = \frac{1.5 \text{ mV}}{250 \text{ mV}} = 0.006 \text{ V/V}$$

The CMRR at each frequency is thus

$$\textbf{300 Hz:} \quad \text{CMRR} = 20 \log_{10}\left(\frac{0.0028 \text{ V/V}}{2.04 \text{ V/V}}\right) = -57.2 \text{ dB}$$

$$\textbf{1020 Hz:} \quad \text{CMRR} = 20 \log_{10}\left(\frac{0.0032 \text{ V/V}}{2.00 \text{ V/V}}\right) = -55.9 \text{ dB}$$

$$\textbf{3400 Hz:} \quad \text{CMRR} = 20 \log_{10}\left(\frac{0.006 \text{ V/V}}{1.92 \text{ V/V}}\right) = -50.1 \text{ dB}$$

Since the signal level during the differential test is 250 mV RMS differential (125 mV RMS, single-ended), we would see the signal level at the positive input increase by a factor of two during the common-mode test (250 mV RMS, single-ended).

10.5.2 Power Supply Rejection and Power Supply Rejection Ratio

Power supply rejection ratio is similar in nature to CMRR, except that the interference signal is applied through the power supply rather than through the normal inputs. In real-world applications, the power supply voltage is never perfect. It consists of a DC level plus AC variations caused by circuits pulling time-varying currents from the power supply circuits or from a battery.

PSRR is a measurement of a circuit's ability to reject a ripple signal added to its power supply voltage. PSRR is usually measured using a single tone or multitone signal, even though the device will seldom see a sinusoidal ripple on the power supply. PSRR is often specified for both the analog power supply and the digital power supply in mixed-signal devices. Usually, PSRR is much worse on the analog supply than on the digital supply, but not always.

Power supply rejection (PSR) is defined as the "gain" of a circuit with its input grounded or otherwise nonstimulated while an AC input signal is injected at a power supply pin. For instance, a test setup for a single-ended amplifier is shown in Figure 10.21. Here the PSR would be computed from the following formula:

$$PSR(f) \equiv \left| \frac{V_{out}}{V_{ac}} \bigg|_{V_{in} = V_{mid}} \right| \tag{10.24}$$

As with all AC test metrics, they are frequency-dependent and thus should be expressed as a function of frequency. Of course, the AC input signal at the power supply pin must include a DC offset that corresponds to the normal power supply voltage so that the circuit remains powered up.

Figure 10.21. PSR and PSRR test setup.

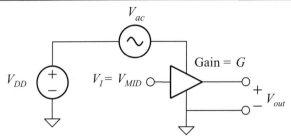

Power supply rejection ratio is defined as the PSR gain divided by the normal gain of the circuit according to

$$\text{PSRR}(f) \equiv \left| \frac{\text{PSR}(f)}{G(f)} \right| = \left| \frac{\left(\dfrac{V_{out}}{V_{ac}} \bigg|_{V_{in}=V_{mid}} \right)}{\left(\dfrac{V_{out}}{V_{in}} \bigg|_{V_{ac}=0} \right)} \right| \tag{10.25}$$

Both PSR and PSRR are usually specified in decibels, similar to CMRR. Like CMRR, the decibel results from a good DUT should be negative.

EXAMPLE 10.13

A differential multitone signal at 300, 1020, and 3400 Hz is applied to the input of the gain circuit in Example 10.12. Each of the tones is set to a level of 250 mV RMS, differential. The output of the circuit is digitized differentially, resulting in the following differential RMS output levels: 300 Hz: 510 mV RMS, 1020 Hz: 500 mV RMS, 3400 Hz: 480 mV RMS. Then the inputs are shorted to a DC midsupply voltage and a single-ended multitone is added to the power supply. The input signal level is 100-mV RMS, single ended. The output of the channel is again digitized, resulting in the following differential RMS output levels: 300 Hz: 0.12 mV RMS, 1020 Hz: 0.15 mV RMS, 3400 Hz: 0.20 mV RMS. Calculate the PSR and PSRR at each frequency.

Solution:
The differential gain at each frequency is the same as in the CMRR test

$$\textbf{300 Hz:} \quad \text{Gain} = \frac{510 \text{ mV}}{250 \text{ mV}} = 2.04 \text{ V/V}$$

$$\textbf{1020 Hz:} \quad \text{Gain} = \frac{500 \text{ mV}}{250 \text{ mV}} = 2.00 \text{ V/V}$$

$$\textbf{3400 Hz:} \quad \text{Gain} = \frac{480 \text{ mV}}{250 \text{ mV}} = 1.92 \text{ V/V}$$

The PSR at each frequ ency is calculated as follows

$$\textbf{300 Hz:} \quad \text{PSR} = 20\log_{10}\left(\frac{0.12 \text{ mV}}{100 \text{ mV}} \right) = -58.4 \text{ dB} \ \ (\text{or } 0.0012 \text{ V/V})$$

$$\textbf{1020 Hz:} \quad \text{PSR} = 20\log_{10}\left(\frac{0.15 \text{ mV}}{100 \text{ mV}} \right) = -56.4 \text{ dB} \ \ (\text{or } 0.0015 \text{ V/V})$$

$$\textbf{3400 Hz:} \quad \text{PSR} = 20\log_{10}\left(\frac{0.20\text{ mV}}{100\text{ mV}}\right) = -53.9\text{ dB } (\text{or } 0.0020\text{ V/V})$$

The PSRR at each frequency is thus

$$\textbf{300 Hz:} \quad \text{PSRR} = 20\log_{10}\left(\frac{0.0012\text{ V/V}}{2.04\text{ V/V}}\right) = -64.6\text{ dB}$$

$$\textbf{1020 Hz:} \quad \text{PSRR} = 20\log_{10}\left(\frac{0.0015\text{ V/V}}{2.00\text{ V/V}}\right) = -62.5\text{ dB}$$

$$\textbf{3400 Hz:} \quad \text{PSRR} = 20\log_{10}\left(\frac{0.0020\text{ V/V}}{1.92\text{ V/V}}\right) = -59.6\text{ dB}$$

10.5.3 Channel-to-Channel Crosstalk

Crosstalk is another common measurement in analog channels, though its exact definition can be very DUT-specific. Unlike CMRR or PSRR, crosstalk is a measurement with no exact definition. In general, crosstalk is the gain from one channel to a second supposedly independent channel. Ideally, of course, the channels should be perfectly isolated from one another so that there is no crosstalk.

In analog channels, crosstalk is often defined as the gain from one channel's input to another channel's output, divided by the gain of the second channel. Crosstalk is usually expressed in decibels, like CMRR and PSRR. Crosstalk may also be expressed as the gain from a channel's output to another channel's output divided by the second channel's gain. Other times, the crosstalk is not divided by the gain of the second channel at all. The test engineer has to clarify the definition of crosstalk in each case. For now, we will assume that crosstalk is defined as originally stated: the gain from one channel's input to another channel's output, divided by the gain of the second channel.

To aid the reader, a model of the channel-to-channel crosstalk is provided in Figure 10.22 where the crosstalk terms X_{rl} and X_{lr} are defined as

$$X_{rl}(f) \equiv \left|\frac{G_{rl}(f)}{G_r(f)}\right| = \frac{\left|\left(\dfrac{V_{out-r}}{V_l}\bigg|_{V_r=0}\right)\right|}{\left|\left(\dfrac{V_{out-r}}{V_r}\bigg|_{V_l=0}\right)\right|} \qquad X_{lr}(f) \equiv \left|\frac{G_{lr}(f)}{G_l(f)}\right| = \frac{\left|\left(\dfrac{V_{out-l}}{V_r}\bigg|_{V_l=0}\right)\right|}{\left|\left(\dfrac{V_{out-l}}{V_l}\bigg|_{V_r=0}\right)\right|} \qquad \textbf{(10.26)}$$

Figure 10.22. Modeling channel-to channel crosstalk.

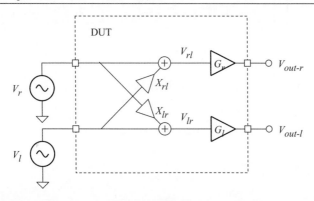

Crosstalk is often measured at several frequencies at once, using DSP-based multitone testing. The second channel's input is typically grounded (or connected to a midsupply voltage in single-supply circuits) during a crosstalk measurement. If crosstalk is specified from channel 1 to channel 2 and also from channel 2 to channel 1, however, then crosstalk can be measured from each channel to the other channel simultaneously to save test time.

To do this, the two channels have to be stimulated with slightly different frequencies so that each channel's crosstalk response can be isolated from its primary signal. The frequencies also have to be chosen so that the crosstalk components will not occur at the same frequencies as the harmonic and intermodulation distortion components of the primary signals. Otherwise, distortion will be misinterpreted as crosstalk.

EXAMPLE 10.14

A stereo audio channel consists of two identical analog signal paths (left and right). Each path is single ended (Figure 10.23). Using two digitizers and two AWGs, define a simultaneous sampling system that produces a multitone signal at approximately 300, 1020, and 3400 Hz on both the left and right channels. Use a 16-kHz sampling rate and 512 samples for the AWGs and digitizers. The actual test frequencies must be equal to the desired frequencies within a tolerance of plus or minus 10%.

Apply the three-tone equal-amplitude multitone with a total signal RMS of 500 mV. For simplicity, assume we have already measured a gain of exactly 6.02-dB through each channel. Using the following digitized signal levels, calculate crosstalk from L to R and from R to L. The frequencies below are approximate, since the exact frequencies should be slightly different for the left and right channels.

R Output:
          300 Hz: 0.07 mV RMS, 1020 Hz: 0.08 mV RMS, 3400 Hz: 0.22 mV RMS
L Output:
          300 Hz: 0.08 mV RMS, 1020 Hz: 0.09 mV RMS, 3400 Hz: 0.20 mV RMS

Figure 10.23. Simultaneous crosstalk measurements for stereo audio channels.

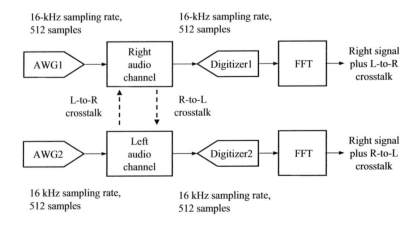

Solution:

First, we design the sampling system. We have two AWGs and two digitizers so we can perform two multitone crosstalk measurements simultaneously. The fundamental frequency for the 16-kHz sampling rate/512-sample system is 31.25 Hz. For the right channel, we can use spectral bins 11, 31, and 109, which gives frequencies of 343, 969, and 3406 Hz. For the left channel we can use spectral bins 9, 35, and 107, which gives frequencies of 281, 1103, and 3344 Hz. Note that we cannot use spectral bin 33 in the second multitone, even though it would give a frequency closer to 1020 Hz. If we tried to use bin 33, we would not be able to distinguish between third harmonic distortion from spectral bin 11 and crosstalk at spectral bin 33. These frequencies are within 10% of the desired frequencies. AWG #1 is set to produce the first three-tone multitone signal for the right channel, while AWG #2 is set to produce the second three-tone multitone for the left channel.

The two digitizers are set to capture the waveforms at the left and right channel outputs. Using the FFT from digitizer #1's waveform, we ignore the spectral components at spectral bins 11, 31, and 109 (the right channel's signal). Instead, we look for crosstalk components at 9, 33, and 107, which might result from the signal generated by AWG #2 (the left channel's signal). Similarly, we ignore spectral components at bins 9, 33, and 107 in the left channel's output, instead measuring the components at bins 11, 31, and 109, which might result from the right channel's signal.

Crosstalk is calculated as follows. The signal level of each input tone is equal to the total signal level divided by the square root of the number of tones (in this case, 3). Therefore, each tone's amplitude is 500 mV RMS/$\sqrt{3}$ or 288.68 mV RMS. Left-to-right crosstalk is defined as 20 $\log_{10}$(gain from L to R/gain in right channel) and right to left crosstalk is defined as 20 $\log_{10}$(gain from R to L/gain in left channel). The gain for both channels is exactly 6.02 dB, which is a gain of 2 V/V.

R-to-L crosstalk:

300 Hz: $\quad X_{rl} = 20 \log_{10}\left(\dfrac{0.08 \text{ mV}/288.68 \text{ mV}}{2}\right) = -77.2 \text{ dB}$

1020 Hz: $\quad X_{rl} = 20 \log_{10}\left(\dfrac{0.09 \text{ mV}/288.68 \text{ mV}}{2}\right) = -76.1 \text{ dB}$

3400 Hz: $\quad X_{rl} = 20 \log_{10}\left(\dfrac{0.20 \text{ mV}/288.68 \text{ mV}}{2}\right) = -69.2 \text{ dB}$

L-to-R crosstalk:

300 Hz: $\quad X_{lr} = 20 \log_{10}\left(\dfrac{0.07 \text{ mV}/288.68 \text{ mV}}{2}\right) = -78.3 \text{ dB}$

1020 Hz: $\quad X_{lr} = 20 \log_{10}\left(\dfrac{0.08 \text{ mV}/288.68 \text{ mV}}{2}\right) = -77.2 \text{ dB}$

3400 Hz: $\quad X_{lr} = 20 \log_{10}\left(\dfrac{0.22 \text{ mV}/288.68 \text{ mV}}{2}\right) = -68.4 \text{ dB}$

10.5.4 Clock and Data Feedthrough

The definition of clock and data feedthrough is even less standardized than crosstalk. Clock and data feedthrough is measured by digitizing the output of a channel and then applying one of several calculations to the resulting waveform. Digital feedthrough is usually very "spiky" in appearance. Clock feedthrough often has a signal bandwidth well into the megahertz range; thus a high-bandwidth digitizer is typically used to measure clock feedthrough.

If the clock and data feedthrough is coherent with the digitizer, then an FFT can be performed on the output signal. In this case, the feedthrough may be specified in terms of a maximum spurious tone, relative to the level of a carrier tone. When specified in this manner, the spurious tone is specified in dBc (dB relative to the carrier level). Spurious tones are often a major concern in communication devices, such as cellular telephones. Because the energy in a spurious tone is concentrated around a single frequency, it can cause electromagnetic compatibility problems. A cellular telephone that generates a spurious tone in its transmit channel might interfere with other cellular telephones operating at the same frequency as the spurious tone. Such a telephone would fail the Federal Communications Commission (FCC) compliance tests.

Clock and data feedthrough may instead be specified in terms of total RMS voltage, excluding DC offset. The removal of DC offset can be accomplished by applying a DC blocking capacitor or high-pass filter to the signal before measuring its RMS level. More often, the DC component is removed mathematically by subtracting the average of the digitized signal from each point in the captured waveform.

The test engineer will have to make sure the exact definition of clock feedthrough or glitch energy is unambiguously defined. Usually the systems engineers or the customers will be the only ones who can define their intentions. Unfortunately, specifications are often lifted from a competitor's data sheet, which does not clearly specify the test conditions or test definition (what data pattern is being sent to or from the DUT, etc.). Again, the systems engineers or customers will have to help clarify the test requirements.

Exercises

10.20. The gain of an instrumentation amplifier with differential inputs and outputs was measured at 1 kHz and 2 kHz in sequence with a 100-mV RMS input differential signal. The output at each frequency was found to be 1.1 V RMS and 0.97 V RMS, respectively. The differential inputs were then shorted together and the amplifier was again excited with a 100-mV RMS signal at 1 kHz and 2 kHz, The RMS level at the output at these two frequencies was 23 mV and 18 mV, respectively. Calculate the CMRR of this amplifier at each test frequency.

ANS. −33.59 dB, −34.6 dB.

10.21. A differential multitone signal at 400, 1160, and 3200 Hz is applied to the input of an instrumentation amplifier. Each of the tones is set to a level of 100 mV RMS, differential. The output of the circuit is measured and found to be 301 mV RMS, 521 mV RMS, and 423 mV RMS at 400, 1160, and 3200 Hz, respectively. The inputs are shorted to a DC midsupply voltage, and a single-ended multitone is added to the power supply. The input signal level is again 100 mV RMS, single-ended. The output of the channel is again digitized, resulting in the following differential RMS output levels: 400 Hz: 0.1 mV RMS, 1160 Hz: 0.15 mV RMS, 3200 Hz: 0.23 mV RMS. Calculate the PSRR at each frequency.

ANS. −69.5 dB, −70.8 dB, −65.3 dB.

Often, clock and data feedthrough is not specified as a separate parameter. It is simply considered part of the noise in a signal-to-noise test. Noise testing is another major category of analog and sampled channel testing. Noise tests can be among the most difficult, time-consuming measurements in a mixed-signal test program.

10.6 NOISE TESTS

10.6.1 Noise

Random noise is generated by every real-world circuit. It can be generated by thermal noise in the case of resistors, $1/f$ noise in the case of CMOS transistors, or quantization noise in the case of DACs and ADCs. Noise can also be injected into a circuit by external forces, such as light falling on a bare die or electromagnetic interference coupling into a circuit under test. Excessive noise can result in a hissing noise in audio circuits, corrupted data in a modem or cellular telephone, and many other system-level failure mechanisms. Noise is generally, but not always, an undesirable property of a circuit under test. Noise is one of the leading causes of long test time, since averaging or added measurements are needed to remove the nonrepeatability caused by random noise.

The spectral density of noise energy is often described using color analogies. White noise, like white light, contains energy that is evenly distributed across the frequency spectrum. White noise is noise whose RMS voltage is constant in any band of frequencies from F to $F + \Delta F$, regardless of the value of F. Pink noise, by contrast, is noise that is weighted more heavily at low frequencies.

Often the level of noise is assumed to exhibit a Gaussian (normal) statistical distribution. This is largely a result of the central limit theorem of large numbers. It is important to recognize, however, that the spectral properties of noise and its statistical distribution are separate concepts. They are combined for mathematical convenience.

Sometimes noise is defined as any signal component other than the primary test signal. (The 0 Hz, or DC component, is also excluded from the calculation of noise.) This definition of noise includes random noise as well as harmonic distortion, intermodulation distortion, clock feedthrough, sigma-delta converter self-tones, and so on. Since test engineering is frequently concerned with characterization and diagnosis of failure mechanisms, a good test program should isolate all the known failure mechanisms into separate measurements. It is therefore preferable to measure distortion components separately from clock feedthrough, separately from random noise, and so on. The signal to total noise, distortion, interference, and so on. can also be calculated separately for additional characterization information. Often a data sheet will call out such a signal to total noise plus distortion test as an overall measure of quality.

There are several different ways to measure noise. Idle channel noise is the RMS voltage variation with no input signal (grounded inputs or inputs connected to a DC midpoint voltage). Signal-to-noise and signal-to-noise plus distortion are other figures of merit. There are also other definitions of noise performance such as spurious free dynamic range. Each of these tests looks for a different noise-based weakness in the circuit under test.

10.6.2 Idle Channel Noise

Idle channel noise (ICN) is a measurement of noise generated by the circuit itself, plus noise injected from external circuits or signal sources through a variety of coupling mechanisms. During an idle channel noise test, the input to the circuit under test is either shorted to ground or otherwise disabled into a quiet state. Ideally, the output should also fall into a noise-free state, but of course all outputs exhibit some amount of noise.

Using DSP-based testing methodologies, this output noise is measured by digitizing the output of the circuit and performing a noise calculation on the captured samples. Idle channel noise can be expressed in many different units of measure. The most straightforward idle channel noise measurement is to simply calculate the RMS voltage level from the captured samples.

It is important to realize that the bandwidth of the digitizer during this measurement is extremely important. A digitizer with a wide bandwidth will see more RMS noise than a digitizer with a narrow bandwidth. Since white noise is spread evenly across the frequency spectrum, a wider bandwidth will allow more noise components to be added into the final calculation. For this reason, it is critical to express the noise in terms of RMS voltage over a specified bandwidth. For example, a data sheet may specify idle channel noise as <100 μV RMS from 100 Hz to 10 kHz.

The measurement of noise can be normalized by the bandwidth of the measurement using a unit of noise called the *root spectral density*, denoted S_n. This type of measurement is expressed in units of volts per root-Hz ($V/\sqrt{Hz}$) and it can be used to estimate RMS noise over other frequency bands. To convert a plain RMS voltage measurement into a $V/\sqrt{Hz}$ measurement, the RMS voltage is divided by the square root of the frequency span of the bandwidth B of the digitizer or RMS volt meter:

$$S_n\left(\frac{V}{\sqrt{Hz}}\right) = \frac{\text{noise RMS}}{\sqrt{B}} \qquad (10.27)$$

For example, if an RMS volt meter or digitizer allows only signal energy from 9 to 11 kHz to pass, and the RMS noise measurement is 100 μV RMS, then the noise can be expressed as $100\,\mu V/\sqrt{11\,kHz - 9\,kHz}$ or 2.236 $\mu V/\sqrt{Hz}$ from 9 to 11 kHz.

To estimate the RMS noise that would occur in the frequency span between 8 and 12 kHz, for example, one simply needs to multiply S_n by the square root of the frequency span (12 kHz–8 kHz = 4 kHz). Using the previous noise result, we would then estimate noise from 8 to 12 kHz to be $2.236\,\mu V/\sqrt{Hz} \times \sqrt{4\,kHz} = 141.41\,\mu V\,RMS$. However, since noise is sometimes unevenly distributed, it might or might not be appropriate to estimate the noise from 100 to 200 kHz using the root spectral density measured between 9 and 11 kHz. A separate measurement of root spectral density could be performed near 100—200 kHz for this range of frequencies.

Noise measurements can be converted to decibel units as follows:

$$\text{noise}\,(dB) = 20\,\log_{10}\left(\frac{\text{noise RMS}}{\text{reference RMS}}\right) \qquad (10.28)$$

Thus expressing noise in decibel units is fairly straightforward, except for the determination of the reference voltage. There are a variety of references that the test engineer will encounter. One reference is simply 1.0-V RMS. When using this reference the noise is expressed in dBV (decibels relative to 1 V RMS). Noise can also be specified in dBm, referencing it to a 1.0 mW signal level at a particular load impedance.

Idle channel noise may also be specified relative to a full-scale sine wave (full scale must be defined in the data sheet). For example, if a circuit generates 1-mV RMS idle channel noise and its full scale range is ±500 mV, then the noise is measured relative to a sine wave at 500-mV peak. This corresponds to $500\,mV/\sqrt{2} = 354\,mV\,RMS$. The idle channel noise would then be calculated as

$$\text{noise}\,(dB) = 20\,\log_{10}\left(\frac{1\text{-mV RMS}}{354\text{-mV RMS}}\right) = -51\,dB, \text{ relative to FS}$$

Idle channel noise can also be referenced to a 0-dB level, defined in the data sheet. For example, many central office telephone channels use a reference level that is about 3 dB lower than the full scale range of the channel. A sine wave at 0 dB produces a 0-dBm signal level at a specified point in the central office. Such a sine wave is said to be at a level of 0 dBm0. Idle channel noise measurements can also be referenced to this 0-dB level, resulting in the unit dBm0.

The dBm0 unit of measurement can be further refined by referencing the noise to a commonly accepted reference level of −90 dBm. This unit of measurement is called the dBrn (decibel referenced to noise). Referencing the measurement to the central office level of 0 dBm0 further gives us the dBrn0 unit. A dBrn0 measurement is therefore 90 dB higher than the equivalent dBm0 measurement. Finally, the noise can be weighted with a C-message filter before calculation of the RMS noise voltage, leading to the unit dBrnC0 (commonly pronounced duh-brink'-o). Weighting filters were discussed briefly in Chapter 9.

EXAMPLE 10.15

A CODEC (coder decoder) is a device that is used by the telephone company's central office to digitize and reconstruct analog voice signals during a telephone call. The digitized voice information is transmitted between two central offices as the two customers speak to one another. A CODEC DAC channel is sent a data sequence of all zeros (idle signal).

The DAC output is digitized with a bandwidth of 20 kHz. The resulting signal is filtered with a software C-message weighting filter that limits the bandwidth of the noise to 0 Hz to 3.4 kHz. After filtering, the noise level is calculated as 100 µV RMS. The 0-dBm0 reference level is 1.4 V RMS for the DAC channel. Calculate the idle channel noise, in dBrnC0 units.

Solution:
First we take the 100 µV RMS signal and calculate its level in dBmC0 units. This gives us

$$ICN(\text{dBmC0}) = 20 \ \log_{10}\left(\frac{100 \ \mu\text{V RMS}}{1.4 \ \text{V RMS}}\right) = -82.92 \ \text{dBmC0}$$

Conversion to dBrnC0 is accomplished by adding 90-dB to this result

$$ICN(dBrnC0) = -82.92 \ \text{dBmC0} + 90 \ \text{dB} = +7.08 \ \text{dBrnC0}$$

10.6.3 Signal to Noise, Signal to Noise and Distortion

Signal-to-noise ratio is another parameter that measures the noise performance of an analog or mixed-signal channel. It is different from idle channel noise in that it measures noise in the presence of a normal signal, usually a sine wave. When working with a purely analog channel, the presence of a signal should not change the amount of noise generated by the channel. However, in a DAC or ADC channel, quantization noise will not be present unless a signal is present. For this reason, it is necessary to measure not only idle channel noise, but also signal-to-noise ratio in mixed-signal channels. Signal-to-noise ratio is often measured using the same data collected during the gain and signal-to-distortion tests.

Signal-to-noise ratio can be defined as the ratio of the primary signal divided by all nonsignal AC components. However, this definition includes distortion components and other signal degradation factors that should be separated for characterization purposes. Therefore, signal-to-noise ratio is more commonly measured by excluding harmonic distortion components, or excluding selected harmonic distortion components. For example, it is common to measure signal to second distortion, signal to third distortion, and signal to total harmonic distortion plus noise (S/THD+N). Table 10.15 summarizes these typical noise and distortion measurements. Notice from this table that individual signal components are added using a square-root-of-sum-of-squares calculation. *Note: The DC offset component is always excluded from these calculations.*

To calculate the total noise in nonharmonic bins, the test engineer can set the spectral coefficients to zero for all signal and harmonic distortion bins that are to be excluded. Then the test engineer can either perform an inverse FFT and calculate RMS of the time-domain signal or, better yet, simply calculate the square root of the sum of squares of all the remaining FFT bins. These two approaches are mathematically equivalent. However, the avoidance of an inverse FFT can save quite a bit of test time.

Poor repeatability is one of the biggest problems with noise measurements. Random noise by its very nature is nonrepeatable. Also, any signal component that is near the noise level will be unrepeatable as well. For example, a distortion component at 100 µV RMS may result in a very unrepeatable measurement if the noise level is also 100 µV RMS. The only way to get a repeatable noise or low-level distortion measurement is to collect hundreds or thousands of individual samples for the FFT calculation. This can lead to extremely long test times, depending on the level of repeatability needed. This is one reason why devices that are designed very close to the specification limits are expensive to test. A design with 6 dB of margin between the typical device performance and the specified limit can allow a much less expensive test than a device with 1 dB of margin, since 6 dB of margin requires less accuracy and repeatability.

Exercises

10.22. Determine the root spectral density of a noise signal that has an RMS value of 250 µV over a frequency range of 1 to 4 kHz.

ANS. 4.564 $\mu V/\sqrt{Hz}$.

10.23. A digitizer sampling at 8 kHz captures 16 samples of a noise signal. An FFT analysis reveals the following spectral coefficients:

FFT Spectral Bin	RMS Voltage (mV)
0	0.0150 mV
1	0.2620 mV
2	0.4092 mV
3	0.5559 mV
4	0.1681 mV
5	0.7270 mV
6	0.4941 mV
7	0.2550 mV
8	0.2539 mV

Calculate the RMS level of the noise signal between 1 and 2.5 kHz. What is the corresponding root spectral density of the noise over this frequency interval? Repeat for the frequency range between 1 to 3 kHz.

ANS. 1 mV RMS, 26.24 $\mu V/\sqrt{Hz}$; 1.1 mV RMS, 25.27 $\mu V/\sqrt{Hz}$.

10.24. What is the idle channel noise level relative to a full-scale range of 2 V if the noise level is 1 mV RMS? What is the idle channel noise in dBm0 if the 0-dBm0 reference level is 500 mV RMS?

ANS. −57.0 dB, relative to FS; −54.0 dBm0.

EXAMPLE 10.16

256 samples of a 1-kHz sine wave are captured with a digitizer at a sampling rate of 16 kHz. An FFT analysis of the output reveals the magnitude of the spectrum shown in Figure 10.24. The magnitude of the spectrum shows the signal and five significant distortion components. The data sheet for this device defines noise as anything other than the fundamental signal, second-, and third harmonic distortion components.

The fundamental test tone, second-, and third-harmonic distortion components are set to zero, leaving the spectrum in Figure 10.25. Taking the square root of sum of squares of the modified spectrum gives an RMS noise level of 100 µV RMS. Calculate the signal-to-noise ratio for this signal.

Solution:

The fundamental tone was at 1 V RMS; so the signal-to-noise ratio is equal to

$$SNR(\text{dB}) = 20 \, \log_{10} \left(\frac{1 \, \text{V RMS}}{100 \, \mu\text{V RMS}} \right) = 80 \, \text{dB}$$

Figure 10.24. Magnitude spectrum (dBV) for 1-kHz test tone with noise and harmonic distortion.

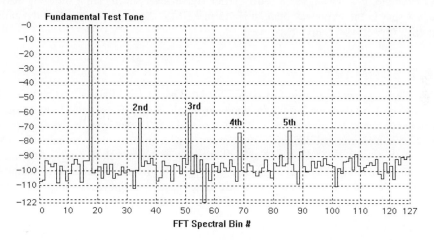

Figure 10.25. Magnitude spectrum with fundamental test tone, second and third distortion removed.

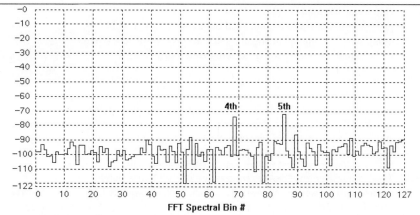

10.6.4 Spurious Free Dynamic Range

Spurious free dynamic range is a specification that is critical to audio circuits as well as telecommunication circuits that must pass FCC (or EC) certifications. A spur is defined as any nonsignal component that is confined to a single frequency. Spurs can be caused by harmonic and intermodulation distortion, clock feedthrough, sigma-delta converter self-tones, stray oscillations, or any of dozens of other undesirable processes.

Spurs are much more noticeable to the human ear than other types of noise. For this reason, sigma-delta converters sometimes include a randomizing circuit to inject random noise that spreads the inherent self-tones of the converter into many frequency bins. This degrades the signal-to-noise performance, but improves the spurious free dynamic range. The end result is less objectionable to the listener. In cellular telephone applications, spurs can be mixed into the transmitted signal, resulting in unwanted side lobes that might interfere with calls on other cellular telephones. For this reason, FCC compliance specifications limit unwanted spurs in transmitted signals.

A spur shows up in a magnitude FFT or on a spectrum analyzer display as a spike in the frequency domain. Spurious free dynamic range is often defined as the difference in decibels between the 0-dB signal level (the carrier level) and the maximum spur in the frequency domain. A spurious free dynamic range of 60 dBc would imply that no individual tone in the frequency domain is larger than 60 dB below the 0-dB carrier signal level as defined in the data sheet. For example, if a device has a 0-dB carrier signal level specification of 3.0 V RMS and the largest spur in the frequency domain is 1.5 mV RMS, then the spurious free dynamic range is $20 \log_{10}(3 \text{ V RMS}/1.5 \text{ mV RMS}) = 66$ dBc.

10.7 SUMMARY

This chapter provides a basic foundation for DSP-based mixed-signal testing, even though all the tests we have discussed so far are purely analog in nature. These same tests are performed on sampled channels, which may include DACs and ADCs. These channels require some slight differences in measurement definition. For example, we cannot calculate the gain of a DAC in volts per volt, because a DAC does not have a voltage input. Nevertheless, many of the measurement techniques are almost identical in nature to purely analog tests. In Chapter 11, "Sampled Channel Testing," we will continue developing DSP-based testing techniques as they apply to channels containing DACs, ADCs, switched capacitor filters, and other sampling circuits.

Exercises

10.25. A signal was digitized and analyzed for its spectral content. The signal was found to have a total RMS value of 0.867 V, a DC component of 0.500 V and a sinusoidal signal component of 1.00-V amplitude. The signal had insignificant levels of distortion. What is the signal-to-noise ratio associated with this signal? ANS. SNR = 24.7 dB.

10.26. A signal was digitized and analyzed for its spectral content. The signal was found to have a DC component of 0.500 V, a sinusoidal signal component of 1.00 V RMS, a distortion component of 0.01 V RMS and a noise component of 0.001 V RMS. What is the signal-to-distortion ratio associated with this signal? What is the signal-to-noise-plus-distortion ratio? ANS. SNR = 60 dB; SDR = 40 dB; SNDR = 39.96 dB.

10.27. What is the spurious free dynamic range of a system which has a maximum carrier signal level of ± 1 V and a signal spur at 1.01 kHz of 1.2 mV? ANS. SFDR = 55.4 dBc.

PROBLEMS

10.1. Perform the following signal conversions:

(a) 1.414-V peak, single-ended signal into dBV, differential.
(b) 0.5-V peak-to-peak, single-ended signal into dBm units at 600 Ω.
(c) 100-mV RMS, differential signal into dBV units.
(d) 700-mV RMS, differential signal into peak, single-ended signal.

10.2. An DTFS analysis reveals that a signal is present in the eleventh bin with a spectral coefficient described by $a_{11} = 0.707$ and $b_{11} = 0.100$ V. What is the amplitude of this signal? What is its phase? What is its RMS value? Express the signal amplitude in dBV units.

10.3. A DTFS analysis reveals that a particular tone has a complex spectral coefficient of $-0.5 - j0.5$ V. What is the amplitude of this tone? What is its phase? What is its RMS value? Express the signal amplitude in dBV units.

10.4. The small-signal gain of a channel is defined as the derivative of the output voltage with respect to the input voltage. Determine the gain of the following channels:

(a) $V_{out} = 0.1 + 0.99\ V_{in}$

(b) $V_{out} = 0.1 + 2V_{in} + 0.1\ V_{in}^2 + 0.01\ V_{in}^3$

(c) $V_{out} = a_0 + a_1 V_{in} + a_2 V_{in}^2 + a_3 V_{in}^3 + \cdots + a_N V_{in}^N$

(d) $V_{out} = 4\tan^{-1}\left(V_{in}\right)$

10.5. The gain of an analog channel as a function of the RMS signal level at its input is assumed to be described by the following equation,

$$G = 2.1 - 0.1\ V_{in-RMS} + 0.01\ V_{in-RMS}^2$$

(a) What is the gain error of the channel at an input level of 2.5 V RMS if the ideal gain is 2.0 V/V? What is the gain error when the input is increased to 4.0 V RMS?

(b) What are the gain tracking errors at input levels of 3, 0, –3, –6, and –12 dB, when the 0-dB reference level corresponds to a 3.0-V RMS input level? Plot the gain tracking error as a function of input level in decibels.

10.6. An analog channel is excited by a 1-V RMS single-ended sinusoidal signal from an arbitrary waveform generator. The output of the channel is digitized at a rate of 32 kHz and 1024 samples are collected and stored in memory. An FFT analysis reveals a signal amplitude of 2.1 V RMS in spectral bin 301. What is the frequency of the test signal? What is the absolute gain of the channel in V/V and in dB?

10.7. A digitizer captures 2048 samples of a signal with a sampling rate of 32 kHz. Signals are present in the following FFT bins: 31, 53, 54, 527, 544, 749, 1011. What are the frequencies of the corresponding signals?

10.8. What is the total RMS value of a multitone signal consisting of 16 tones having an RMS value of 25 mV each and 24 tones having an RMS value of 10 mV each?

10.9. A 15-tone multitone signal of 1 V RMS is required to perform a frequency response test. What is the amplitude of each tone if they are all equal in magnitude?

10.10. The frequency response behavior of an amplifier is measured with a 1-V RMS multitone signal consisting of eight equal-amplitude tones. An FFT analysis of the output samples reveal the following output signal amplitudes (RMS), in increasing frequency order: 353, 335, 314, 331, 349, 257, 158, 81 mV. What is the absolute gain of the channel at each frequency, in V/V? What is the relative gain (dB) of the amplifier at each frequency if the reference gain is based on the gain of the first tone?

10.11. A data sheet calls for the -1-dB gain point of a low-pass filter to occur between 19,500 and 20,000 Hz. A frequency response measurement was made and the gain at 19,486.3 Hz was found to be -0.01 dB and the gain at 20,001.97 Hz was -2.4 dB. Using log/log interpolation, determine the frequency at which the gain is -1 dB.

10.12. The frequency response behavior of a frequency selective analog channel is measured with a multitone signal consisting of eight tones. The data was collected at a 32 MHz sampling rate. An 64-point FFT analysis of the input and output samples reveal the following complex spectral coefficients (in V):

FFT Bin	Input	Output
0	$0.9609 - j0.2768$	$0.9609 - j0.2768$
1	$-0.0107 - j0.9999$	$-0.4896 - j0.8431$
2	$-0.9418 - j0.3363$	$-0.7132 + j0.6186$
3	$0.7735 - j0.6338$	$-0.8270 - j0.5488$
4	$-0.7078 - j0.7064$	$0.8208 - j0.4663$
5	$0.5466 - j0.8374$	$0.0004 + j0.0063$
6	$0.5371 + j0.8435$	$0.0309 - j0.0019$
7	$0.2319 + j0.9727$	$0.0228 - j0.00006$
8	$-0.7535 + j0.6575$	$0.0000 - j0.0000$

(a) What is the RMS value of the input and output signals as a function of frequency?
(b) What is the phase of the input and output signals as a function of frequency? Limit the range of the phase of each tone to ±180 degrees.
(c) What is absolute gain of this amplifier as a function of frequency?
(d) What is the relative gain of this channel as a function of frequency if the reference gain is based on the gain of the tone corresponding to the fourth FFT bin?
(e) What is the phase shift (in degrees) of the analog channel as a function of frequency? Provide an unwrapped description of the phase.
(f) What is the group delay of this channel? Also, determine the group delay distortion of this channel.

10.13. An FFT analysis of the output of an amplifier contains the following harmonically related spectral amplitudes:

FFT Spectral Bin	RMS Voltage
101	0.555 V
202	10 mV
303	1 mV
404	0.1 mV
505	0.01 mV

(a) What is the signal to second harmonic distortion ratio?
(b) What is the signal to third harmonic distortion ratio?
(c) What is the signal to total harmonic distortion ratio?
(d) If the RMS value of the noise component is 125 mV, calculate the signal to total harmonic distortion plus noise ratio.

10.14. An amplifier's input-output behavior can be described mathematically by the following third-order polynomial, $V_{out} = a_0 + a_1 V_{in} + a_2 V_{in}^2 + a_3 V_{in}^3$. What is the signal to third harmonic distortion ratio of this amplifier if the input sinusoidal signal is described by $V_{in}(t) = A\sin(2\pi ft)$.

10.15. A multitone test signal consists of a sum of four 1.0-V RMS sine waves, one at 1.3 kHz, another at 2.1 kHz, another at 3.2 kHz, and the fourth at 5.3 kHz. Determine the frequencies of second-order and third-order intermodulation frequencies.

10.16. An amplifier's input-output behavior can be described mathematically by the following third-order polynomial: $V_{out} = a_0 + a_1 V_{in} + a_2 V_{in}^2 + a_3 V_{in}^3$. What are the third-order intermodulation products (amplitude and frequency) produced by this amplifier if the input is described by $V_{in}(t) = A_1 \sin(2\pi f_1 t) + A_2 \sin(2\pi f_2 t)$.

10.17. A multitone test signal is created using a 512-point sample set with a sampling frequency of 32 MHz. One hundred tones with an amplitude of 75 mV are used beginning with bin 51 and thereafter except for bins 101, 102, 103, 104, 105, and 106. A DUT was subject to this multitone signal and its output digitized at the same rate of 32 MHz. An FFT analysis reveals that bins 101, 102, 103, 104, 105, and 106 had an RMS level of 1.1, 1.2, 0.81, 1.5, 1.1, and 0.9 mV, respectively. What is the noise power ratio in decibels?

10.18. The gain of an instrumentation ampifier with differential inputs and outputs was measured at 6 kHz and 8 kHz in sequence with a 250-mV RMS input differential signal. The output at each frequency was found to be 0.88 V RMS and 0.95 V RMS, respectively. The differential inputs were then shorted together and the amplifier was again excited with a 250-mV RMS signal at 6 kHz and 8 kHz. The RMS level at the output at these two frequencies was 12 mV and 11 mV, respectively. Calculate the CMRR of this amplifier at each test frequency.

10.19. A differential multitone signal at 400, 1160, and 3200 Hz is applied to the input of an instrumentation amplifier. Each of the tones is set to a level of 100 mV RMS, differential. The output of the circuit is measured and found to be 200 mV RMS, 400 mV RMS and 410 mV RMS at 400, 1160 and 3200 Hz, respectively. The input are shorted to a DC midsupply voltage and a single-ended multitone is added to the power supply. The input signal level is again 100 mV RMS, single ended. The output of the channel is again digitized, resulting in the following differential RMS output levels: 400 Hz: 0.01 mV RMS, 1160 Hz: 0.1 mV RMS, 3200 Hz: 0.13 mV RMS. Calculate the PSR and PSRR at each frequency.

10.20. An audio system consists of a two identical channels. The gain of each channel is measured using a three-tone equal-amplitude multitone signal with total RMS value of 250 mV. The gain of each channel was found to be exactly 1.0 V/V. Next, the crosstalk from channel 1 to channel 2 was measured using the previously used multitone signal and the following data measured at the output of channel 2:

100 Hz: 0.1 mV RMS, 920 Hz: 0.08 mV RMS, 3400 Hz: 0.12 mV RMS

The crosstalk from channel 2 to channel 1 was performed and the following data were measured at the output of channel 1:

100 Hz: 0.08 mV RMS, 920 Hz: 0.09 mV RMS, 3400 Hz: 0.16 mV RMS

Determine the crosstalk gain terms as a function of frequency.

10.21. A digitizer sampling at 4 kHz captures 16 samples of a 1-kHz sinusoidal signal corrupted by noise. An FFT analysis reveals the following spectral amplitudes:

FFT Spectral Bin	RMS Voltage
0	2300 μV
1	32 μV
2	14 μV
3	12 μV
4	0.707 V
5	31 μV
6	42 μV
7	11 μV
8	4 μV

(a) What is the total RMS level of the noise?

(b) Determine the signal-to-noise ratio of the channel.

(c) What is the root spectral density of the noise over the 2-kHz Nyquist interval.

10.22. Determine the root spectral density of a noise signal that has an RMS value of 543 µV evenly distributed over a frequency range of 19 to 23 kHz.

10.23. What is the RMS value of a noise signal measured over a 12-kHz bandwidth if its root spectral density is 10 $\mu V/\sqrt{Hz}$? Express this result in dBm units at 50 Ω.

10.24. What is the idle channel noise level in dBV if the noise level is equal to 2.5 mV RMS?

10.25. What is the idle channel noise level relative to a full-scale range of 0.5 V if the noise level is found to equal 2.5 mV RMS? What is the idle channel noise in dBm0 if the 0dBm0 reference level is 100 mV RMS?

10.26. What is the idle channel noise level relative to a full-scale range of 1.0 V if the noise level is found to equal 0.1 mV RMS? What is the idle channel noise in dBmC0 if the 0 dBm0 reference level is 500 mV RMS?

10.27. If a device has a 0-dB carrier signal level specification of 1.5 V RMS and the largest spur in the frequency domain is 0.95 mV RMS, what is the spurious free dynamic range asscoiated with this device?

CHAPTER 11

Sampled Channel Testing

This chapter explores many of the same AC tests as those performed in the previous chapter on Analog Channel Testing as they apply to sampled channels. Sample channels include DACs, ADCs, sample-and-hold circuits and amplifiers, and so on, whereby a clock controls the channel. The differences between analog and sampled channels will be highlighted below. The types of test that will be discussed include gain, frequency response, distortion, signal rejection and noise. Also included in this chapter is a discussion of undersampling and the modulo-time reshuffling algorithm as a means to extend the sampling capability of a digitizer to frequencies greater than the Nyquist frequency.

11.1 OVERVIEW

11.1.1 What Are Sampled Channels?

Sampled channels are similar to analog channels in many ways. Both channel types consist of a signal transmission path from one or more inputs to one or more outputs. Unlike analog channels, though, sampled channels operate on discrete waveforms rather than continuous ones. Discrete waveforms consist of a sequence of instantaneous voltage samples that are either represented as digital values or as sampled-and-held analog voltages.

Examples of sampled channels include digital-to-analog converters (DACs), analog-to-digital converters (ADCs), switched capacitor filters (SCFs), sample-and-hold (S/H) amplifiers, and cascaded combinations of these and other circuits. The test requirements for sampled channels are very similar to those described in Chapter 10, "Analog Channel Testing." However, the sampled nature of the signals transmitted by sampled channels forces some additional considerations in their test requirements. Because sampled circuits may be subject to quantization errors, aliased interference tones, and reconstruction image tones, they require additional test considerations that are not applicable to continuous analog channels.

Sampled channels are often tested for gain error, distortion, signal-to-noise ratio, PSRR, CMRR, and all the other tests common to analog channels. The similarities and differences between analog channel testing and sampled channel testing will be discussed later in this chapter. First, let us look at examples of sampled channels and how they are applied in system-level applications like cellular telephones and disk drive read channels.

11.1.2 Examples of Sampled Channels

Sampled channels form the interface between the physical world around us and the mathematical world of computers and digital signal processors. A digital cellular telephone (Figure 11.1) contains at least six sampled channels: three for the transmit channel and three for the receive channel. The transmit (XMIT) channel is the signal path that transmits the near-end speaker's voice, while the receive (RECV) channel is the signal path that receives the far-end speaker's voice. The transmit channel of a digital cellular telephone includes a number of ADCs, DACs, filters, and signal processing circuits that convert the speaker's voice into a modulated stream of digital data. The data stream is mixed with a high-frequency carrier signal so that it can be transmitted via radio waves to the cellular base station.

The first step in voice transmission is to digitize the speaker's voice using an ADC connected to the cellular telephone's microphone. The voice-band interface circuit is a sampled channel that contains a number of circuits that amplify, filter, and digitize the speaker's voice (Figure 11.2). The digitized voice signal is then routed to either a digital signal processor or a specialized modulator logic block, which converts the ones and zeros of the digitized voice samples into an amplitude/phase modulation protocol similar to that used in computer modems.

Unlike modems, which transmit data over telephone lines at audio frequencies, cellular telephones must transmit the data into an antenna using a radio frequency (RF) power amplifier. The RF transmission frequency is 900 MHz or higher, depending on the type of cellular telephone. Since it would be impossible to directly generate this 900-MHz modulated signal by feeding samples into a DAC (at a rate of 1.8 GHz or more), it is necessary to use an RF mixer circuit to upconvert an intermediate frequency (IF) signal to the RF frequency range. Since the IF signal is much lower in frequency than the RF signal, it can be generated using a DAC channel operating at a lower sampling frequency (Figure 11.3).

Figure 11.1. Digital cellular telephone block diagram.

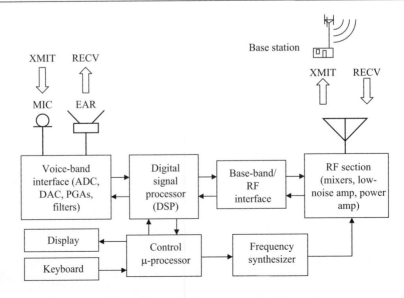

Figure 11.2. Voice-band XMIT (ADC) channel.

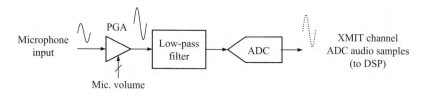

Figure 11.3. XMIT I-channel and Q-channel.

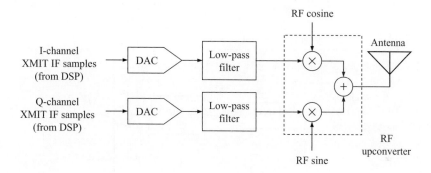

The digital signal processor converts each amplitude/phase pair of the modulation proto-col into a sine amplitude sample and a cosine amplitude sample. The sine and cosine amplitude samples are then converted into analog waveforms using two separate mixed-signal channels: the in-phase channel (I-channel) and the quadrature channel (Q-channel). The I-channel waveform is sent to an RF mixer circuit, controlling the amplitude of an RF cosine waveform. Similarly, the Q-channel waveform controls the amplitude of an RF sine waveform. By adding the RF cosine and sine waveforms together, the RF section of the cellular telephone reconstructs the ampli-tude/phase-modulated data waveform. This composite RF waveform is suitable for transmission through the cellular telephone's antenna.

The receive channel works in a very similar manner, except that the direction of the signal is reversed and the DACs are replaced by ADCs (Figures 11.4 and 11.5). The received data bits are downconverted to IF I and Q channel signals by the RF section, digitized by an IF ADC channel, demodulated into voice samples by the digital signal processor, and converted into audio voice waveforms by the voice-band interface circuit. Thus there are six separate sampled channels in this cellular telephone example: the voice-band interface transmit channel, the transmit I-channel, the transmit Q-channel, the receive I-channel, the receive Q-channel, and the voice-band interface receive channel.

Other examples of devices containing sampled channels are disk drive read channels, digital audio record and playback devices, and digital telephone answering devices (DTADs). A disk drive read channel is a sampled channel that is used to recover 1/0 data from the read coil of a disk drive read head. The read channel must digitize the high-frequency signal generated by the magnetic variations of the data stored on the spinning disk drive platter.

The read coil signal is typically noisy and distorted by the physical properties of the disk's magnetic storage media. The read channel must clean up and correct the signal using a variety

Figure 11.4. RECV I-channel and Q-channel.

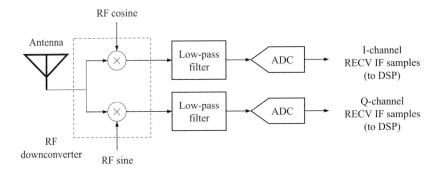

Figure 11.5. Voice-band RECV (DAC) channel.

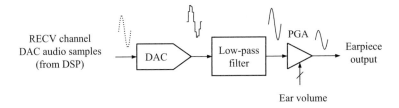

of analog and/or digital filters before data can be recovered. One of the major challenges in read channel testing is that the signals operate at extremely high frequencies. ATE testers are more adept at measuring and sourcing low-frequency signals with a high degree of accuracy. At higher frequencies, the signal source and measurement quality degrades, making the high-frequency read channel difficult to test.

Digital audio channels are similar to the voice-band interface transmit and receive channels of a cellular telephone. However, the sampled channels of a digital audio circuit must record and play back much higher-quality audio than that required for a telephone conversation. One of the major challenges in digital audio testing is that the signals are low in frequency but they have very tight signal-to-distortion and signal-to-noise specifications. Digital telephone answering devices, on the other hand, are more similar to telephones in that they do not need to record and play back especially high-quality audio signals.

Clearly, the uses for sampled channels are very diverse, and a wide variety of testing requirements are needed, depending on the end application of the channel. Despite the wide differences in functionality and quality requirements, most sampled channels have a series of common test specifications, including gain error, signal-to-noise, PSRR, and so on. Before we look at these common tests in detail, let us look at the different kinds of circuits that can be classified as sampled channels.

11.1.3 Types of Sampled Channels

Sampled channels fall into four basic categories: digital in/analog out, analog in/digital out, digital in/digital out, and analog in/analog out. For convenience, we will introduce the notation DIAO, AIDO, DIDO, and AIAO to refer to these four types of sampled channels. Let us look at some examples of each of the four categories of mixed-signal channels to see how they operate on continuous and sampled waveforms.

Figure 11.6. Digital input/digital output (DIDO) sampled channel (loopback test mode).

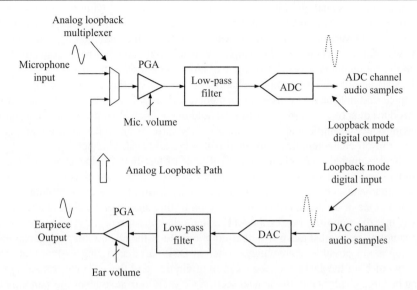

Falling into the digital in/analog out (DIAO) category are DACs and cascaded combinations of DACs and other circuits. DIAO channels are characterized by one or more digital signal inputs and one or more analog signal outputs. Note that a programmable gain amplifier does not fall into this category, even though it does have a digital input. A PGA's transmission channel consists of an analog input and an analog output. The digital control lines feeding a PGA are not generally used to transmit signal information; so we cannot really consider the PGA to have a digital *signal* input. Furthermore, the PGA does not sample or reconstruct its signals in any way; thus it is considered an analog channel rather than a sampled channel. A very high-speed digital line driver might also be considered a simple DIAO channel, as far as testing is concerned, since it converts a digital input into an output with analog characteristics (rise time, overshoot, undershoot, etc.).

ADCs and cascaded combinations of ADCs and other circuits fall into the second category, analog in/digital out (AIDO). AIDO circuits are characterized by one or more analog signal inputs and one or more digital signal outputs. Comparators and slicers fall into this category since they act as one-bit ADCs.

The third category is digital in/digital out (DIDO). While it may not be obvious that a DIDO circuit would have anything to do with a mixed-signal sampled channel, DIDO circuits are actually quite common in mixed-signal testing. Consider the analog loopback mode illustrated in Figure 11.6. In this test mode, a DIAO DAC channel's analog output is connected through a special test multiplexer to the analog input of an AIDO ADC channel. The cascaded circuit has a digital input and a digital output, yet it is a mixed-signal sampled channel. Another example of a DIDO sampled channel is a digital filter, which accepts digital samples at its input, filters the samples using a mathematical algorithm, and produces digital samples at its output. While a digital filter can be tested as a sampled channel by measuring its frequency response, it should be noted that it can be tested much more efficiently and thoroughly using traditional digital methodologies, since it does not contain any analog circuit elements.

An analog in/analog out (AIAO) sampled channel can be formed by placing a device in digital loopback mode, in which the digital output of an AIDO channel is looped back into a DIAO input. The resulting circuit has an analog input and output, but it may exhibit quantization errors, imaging, and aliasing, just like any other sampled circuit. Another example of AIAO sampled channels is the switched capacitor filter (SCF). Switched capacitor filters operate on sampled-and-held or continuous analog input signals, sampling them and applying a high-pass, low-pass, or other filter characteristic to the analog samples. The output of a switched capacitor filter is a sampled-and-held version of the filtered waveform. Since the switched capacitor output is a "stepped" waveform, it is considered a sampled channel, even though it never converts the analog signal into the digital domain. A third example of an AIAO sampled channel is the simple sample-and-hold (S/H) amplifier. This sampled channel converts a continuous analog input waveform into a sampled-and-held analog waveform. In theory, S/H amplifiers and switched capacitor filters do not introduce quantization errors, since they hold nonquantized voltages using capacitors. But since they introduce a sample-and-hold operation, they suffer from all the same imaging, aliasing, and $\sin(x)/x$ roll-off characteristics associated with ADC and DAC channels.

In general, sampled channels are more difficult to test than analog channels. Difficulties arises from a number of factors, such as quantization noise, image and alias tones, sampling rate synchronization headaches, and extra complexity in the form of coherent sampling loops in the digital pattern. One of the main differences between testing analog channels and testing sampled channels is that the sampling rate of the ATE tester's digital patterns, AWG sampling rates, and digitizer sampling rates must mesh with the sampling rate of the DUT. Otherwise, coherent DSP-based testing is not possible. Achieving a coherent DSP-based sampling system that meets the requirements of both the DUT's various sampling circuits and the ATE tester's various instruments can be quite challenging. In the next section we shall look at some of the sampling considerations we face as we design cost-effective tests for sampled channels. We will also look at the structure of the digital patterns that source and capture digital signal samples on the digital side of mixed-signal sampled channels.

11.2 SAMPLING CONSIDERATIONS

11.2.1 DUT Sampling Rate Constraints

When making a coherent DSP-based analog channel measurement, we only need to insure that the fundamental frequency of the AWG is related to the fundamental frequency of the digitizer by an integer ratio (usually a ratio of 1/1) and that the various Nyquist frequencies are above the maximum frequency of interest.[1] Other than these constraints, we are fairly free to choose whatever sampling frequencies we want. Once we begin testing sampled channels, however, we are often burdened with very specific sampling rate constraints placed upon us by the DUT specifications.

The data sheet for a mixed-signal DUT often requires a specific sampling frequency or list of sampling frequencies for each of the DUT's sampled channels. For example, the transmit (ADC) and receive (DAC) channels in a codec device may be specified at a sampling rate of exactly 8 kHz. The tester's sampling systems must mesh with this sampling rate so the waveforms generated and digitized by the DUT's sampled channels are coherent with the tester's AWGs, digitizers, and digital patterns. The most cost-effective testing for this type of device is simultaneous testing of both the transmit and receive channel at the same time. This type of test requires that all sampling rates must be coherent, including the transmit channel, receive channel, digital pattern frame syncs, digital source data rate, digital capture data rate, AWG, and digitizer.

In some cases, it may be acceptable to force-fit the sampling rate of a DUT into a sampling system that is more agreeable to the ATE tester's instruments. For example, we may find that a device whose master clock is specified at 38.88 MHz may actually fit the tester's constraints better if we shift it slightly to 38.879962 MHz. However, it is seldom acceptable to shift a clock by more than a fraction of a percent. If we shifted the 38.88 MHz clock to 40 MHz, for instance, we would run the risk of generating correlation errors between the tester and stand-alone bench equipment operating at 38.88 MHz. Correlation errors in mixed-signal tests are often the result of minor discrepancies in test conditions, such as sampling rates and output loading conditions. It is much safer to use the exact specified master clock and sampling rate combinations outlined in the device data sheet.

It would seem that a tester costing one million dollars or more would be able to produce any sampling rate the test engineer desires. Unfortunately, most testers have constraints that limit the clock frequencies that can be generated or utilized by each instrument. Even when a tester's clocks are highly programmable, certain frequencies may result in a better quality of test than others. For example, a digital pattern rate that is divided from the tester's master clock by a factor of 2^N may generate less sampling jitter than one generated using other divide ratios. The lower jitter may result in superior signal-to-noise ratio measurements. Of course, these constraints are highly tester dependent. A frequency that is not ideal on one type of tester may cause no problem on another. This is one of the reasons that it is so difficult to convert mixed-signal test programs from one tester platform to another.

Yet another headache in selecting sampling rates is the long settling time of low-jitter frequency generators and/or phase-locked loops (PLLs). These often take a long time (25–50 ms) to settle to a stable frequency after their frequency setting is changed. It is often better to select a single master clock frequency and let the tester's low-jitter master clock generator stabilize to this frequency. Ideally, all sampling rates in the test program can be derived from this single master clock frequency using digital clock divider circuits. Since digital divider circuits require no additional setting time, test time can be minimized using this approach. Unfortunately, this may or may not be possible. Sometimes it is necessary to switch the master clock back and forth between various frequencies, adding undesirable test time as the clock source stabilizes. Again, this constraint is highly tester-dependent and may not be an issue on some types of testers.

11.2.2 Digital Signal Source and Capture

When testing mixed-signal devices, the tester must apply digital signal samples to the DUT's inputs and collect digital signal samples from its outputs. The DUT usually requires these samples to be applied and captured at a particular sampling rate. A repeating digital pattern, called a *sampling frame*, is often required by the DUT to control the timing of the digital signal samples. Figure 11.7 shows an example of a digital pattern consisting of a repeating frame, digital signal inputs to a DAC channel, and digital signal outputs from an ADC channel. The pattern also includes a feature not found in purely digital patterns: source and capture shift and load commands. These pattern

Figure 11.7. Mixed-signal digital pattern example.

		pin RFID11 dig_src	pin RFID11 dig_cap	RFIA (H)	RFID (H)	R F I R D X S	R F I W R X S	P C M D C K S	
LABEL	COMMAND								COMMENT
	SET_LOOP 512	START	TRIG	.d0	.—	1	1	1	Prepare to collect 512 samples
ADC_DAC:				.—	.d0	1	1	0	Send Frame Sync (PCMDCK)
		SEND		.d8	.W	—	0	1	Source one DAC sample
				.—	.X	—	1	—	
			STORE	.d9	.—	0	—	—	Capture one ADC sample
	REPEAT 7			.—	.—	1	—	—	Finish the 12–cycle frame
	END_LOOP ADC_DAC			.—	.—	—	—	—	Loop back to ADC_DAC
	HALT			.—	.—	—	—	—	until 512 samples are collected

commands, called *microcode instructions*, determine when the data for source memory will be substituted in place of the normal pattern data. They also determine when the digital samples from the DUT will be stored into capture memory. If the DUT has a serial input or output, the micro-code SHIFT commands determine the time at which of each bit of the data word is shifted into or out of the DUT. Other microcode instructions initialize loop counters, determine loop endpoints, stop the pattern, and so on.

The waveforms stored in source memory are computed by the tester computer during an initialization run. Signals such as sine waves, multitones, and ramps can be stored into various locations in the source memory, ready to be inserted one sample at a time into the looping frame pattern. In Figure 11.7, a multibit parallel data word is written into the RFID pin group using a W character linked with the SEND command on the third vector. Likewise, the samples collected during the looping frame pattern are stored one sample at a time into capture memory using the STORE command on the fifth vector. The captured samples can then be moved into an array processor or tester computer for DSP analysis (DFTs, FFTs, average value, RMS value, etc). Naturally, the example pattern in Figure 11.7 is specific to a particular ATE tester. Different testers will use a variety of notations to specify the STORE and SEND operations, but all true mixed-signal testers provide a looped source and capture capability.

The sampling frame usually consists of one or more high-frequency clocks and one or more frame synchronization signals that determine the timing of the input and output sample data stream. For example, a digital audio device data sheet may specify that DAC channel samples are to be applied in parallel on DUT pins DAC7-DAC0 on the second rising edge of the master clock (MCLK) after the frame synchronization signal (FSYNC) goes low. Likewise, the data sheet may specify that the ADC channel samples will be available at ADC7-ADC0 on the third rising edge of the master clock after FSYNC goes low. Finally, the FSYNC frequency may be specified to run at a rate of MCLK divided by 16. Figure 11.8 shows these timing relationships.

In addition to the timing of the digital signal data inputs and outputs, the master clock and frame sync are usually required to operate at very specific frequencies. Like AWGs and digitizers, the digital source and capture instruments have their own fundamental frequencies, defined as the sampling rate divided by the number of samples sourced or captured. Coherent DSP-based testing requires that the fundamental frequency of the digital source, digital capture, AWG, and digitizer must all be equal (or at least related by an integer ratio).

Unlike digital circuits, mixed-signal channels are usually specified at a particular frequency, rather than a particular period. When making coherent DSP-based measurements, there is a huge difference between a 25-ns master clock period and a 38.88-MHz (25.72016460905 ... ns) period.

Figure 11.8. Sampling frame timing diagram.

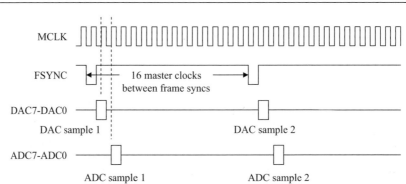

The frequency of these clocks (relative to the sampling frequencies of the tester's AWGs and digitizers) must be quite accurate to synchronize the fundamental frequencies of all the various sampling systems, including the DUT's sampling rates. A frequency accuracy of one part in a million or better is required to achieve acceptable signal-to-noise performance in coherent DSP-based tests. For this reason it is usually not acceptable to round off the clock periods to the nearest nanosecond or even the nearest picosecond. Digital circuits, by contrast, are often tested at frequencies higher or lower than the normal operating frequency. Their timing can often be rounded off to the nearest nanosecond to simplify the automated test pattern generation (ATPG) process.

Depending on the tester's architecture, it may or may not be possible to generate certain sampling rates. For example, a device with a master clock of 41.327 MHz may produce a fundamental frequency that cannot be easily matched by the tester's high-performance digitizer. It may be necessary to switch to a non-power-of-2 sample size or it may be necessary to shift the master clock slightly to allow a coherent measurement. When shifting the master clock to accommodate the tester's instruments, the test engineer should take care. The performance of the DUT may shift slightly as well. For example, the cutoff frequency of a switched capacitor filter changes with its master clock. Shifting the master clock by 1% will shift the 3-dB point of the filter by 1% as well. There are many more subtle problems that occur when the test conditions are shifted in this manner; thus it is best to test the DUT at exactly the specified frequencies to avoid correlation errors.

An interesting question arises when we try to test mixed-signal devices on digital testers: Do we really need source and capture memory to test mixed-signal devices? Let us look at the alternative. A mixed-signal DAC pattern could be written as a "flattened" series of samples with a repeating digital sample frame, as shown in Figure 11.9. Notice that the frame sync pattern is identical from one DUT sample to the next. If a frame consists of 1024 digital samples, each of which requires a 100 vector frame, then the pattern would require 102,400 vectors. Mixed-signal test programs may consist of dozens of these patterns. Testing mixed-signal devices using patterns such as the one shown Figure 11.9 would require many megabytes of vector memory, an obvious waste of digital pattern memory. Also, flattened patterns such as the one in Figure 11.9 pose

Figure 11.9. Source memory sample insertion.

LABEL	COMMAND	pin RFID11 dig_src	pin RFID11 dig_cap	RFIA (H)	RFID (H)	RFIRDRXX S	RFIWDRXX S	PCMDCK S	COMMENT
		START	TRIG	.d0	.−	−	−	−	Prepare to collect 512 samples
ADC_DAC				.−	.d0	1	1	0	
		SEND		.d8	.d94	1	0	1	Source 1st DAC sample
				.−	.X	1	1	1	
			STORE	.d9	.−	0	1	1	Capture 1st ADC sample
	REPEAT 8			.−	.−	1	1	1	Finish the 12−cycle frame
				.−	.d0	1	1	0	
		SEND		.d8	.dE89	1	0	1	Source 2nd one DAC sample
				.−	.X	1	1	1	
			STORE	.d9	.−	0	1	1	Capture 2nd ADC sample
	REPEAT 8			.−	.−	1	1	1	Finish the 12−cycle frame
				.−	.d0	1	1	0	
		SEND		.d8	.d299	1	0	1	Source 3rd DAC sample
				.−	.X	1	1	1	
			STORE	.d9	.−	0	1	1	Capture 3rd ADC sample
	REPEAT 8			.−	.−	1	1	1	Finish the 12−cycle frame

...Repeat this pattern for 512 samples

a debugging problem. What if we needed to quickly characterize a DAC's performance using a frequency of 2 kHz instead of 1 kHz? Likewise, what if we suddenly needed to try a frame of 102 vectors rather than a frame of 100 vectors? We would have to manually insert the new vectors or digital samples into the huge flattened vector set or use a cumbersome digital pattern compiler to recreate a totally new digital pattern. Fortunately, mixed-signal tester architectures provide source and capture memory to allow compact, easily modified digital patterns for mixed-signal tests.

A true mixed-signal tester uses source and capture memory to implement a type of vector compression that is ideally suited to mixed-signal sampling frames. Notice that Figure 11.9 consists of the same basic pattern of ones and zeros for each sample. Only the value of the DAC input data and ADC output data changes from one frame to the next. A mixed-signal tester allows digital samples to be inserted into or extracted from a repeating frame loop, as previously shown in Figure 11.7. The samples are specified with a digital signal sample notation, such as W or X instead of 1 or 0. The Ws and Xs are place holders for digital signal samples, which are either read from source memory or written to capture memory. Figure 11.10 shows how source memory and vector memory are combined to weave together the digital frame data with the digital samples. Similarly, Figure 11.11 shows how DUT output data is captured within a repeating capture frame.

Digital samples can be sourced to and captured from the DUT in either a parallel format or a serial format. In parallel format, the data for each sample are loaded into the device with a single clock cycle, as shown in the previous examples. Serial format, by contrast, involves a serial shifting operation that takes multiple data clock cycles. Versatile mixed-signal testers include built-in hardware features that ease the conversion from parallel to serial and serial to parallel data formats.

When supplying serial data to a DUT's digital signal input, data are loaded into source memory in a parallel format. The digital subsystem hardware performs a parallel to serial shift operation controlled by digital pattern microcode instructions such as SEND and SHIFT. Likewise, when capturing serial data from a DUT's digital signal output, data can be translated from serial format to parallel format before they are stored into capture memory. Using the hardware serial/parallel conversion features, the test engineer does not have to spend additional time writing software translation routines.

In addition to the parallel/serial data conversion, a mixed-signal tester should be capable of shifting the digital signal data into or out of the DUT with the most significant bit (MSB) first or the least significant bit (LSB) first. Mixed-signal testers accommodate the MSB-first/LSB-first translation using built-in digital logic in the digital subsystem.

Figure 11.10. Flattened mixed-signal DAC frames.

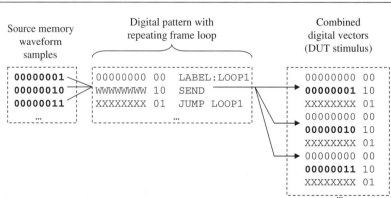

Figure 11.11. Capture memory sample extraction.

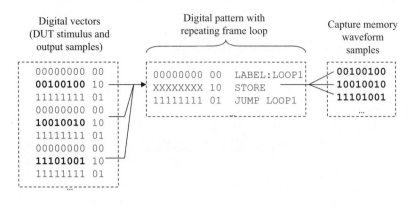

Figure 11.12. Scrambled bit order in a two-byte interface.

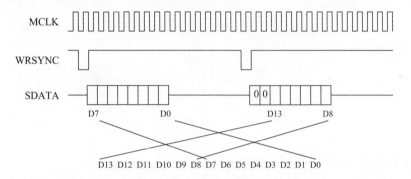

Unfortunately, mixed-signal ATE testers cannot predict every data format. Occasionally the test engineer will run across an "oddball" data format that requires an explicit software bit scrambling operation. For example, a device may have a 14-bit digital signal format that must be shifted into the device serially, with the least significant 8 bits shifted into the DUT MSB first, followed by the most significant 6 bits shifted into the DUT MSB first. Figure 11.12 shows why this is an odd format. Notice that the order of the bits is scrambled. Before we want to use the tester's parallel to serial shift hardware, we have to produce data samples that have the same scrambled bit order. This scrambling operation must be performed in software before the samples are loaded into the source memory. This is not a catastrophic problem; it just adds some overhead to the test development process and makes the code a little harder to follow.

11.2.3 Simultaneous DAC and ADC Channel Testing

When a DUT contains two or more channels that can be tested simultaneously, the test engineer will often test both channels at once to save test time. For example, a central office codec may have a transmit (ADC) channel and a receive (DAC) channel that both operate at a sampling rate of 8 kHz. The various parameters of these two channels can be tested in parallel, saving test time. For example, the absolute gain, distortion, and signal-to-noise of the DAC channel can be tested while the same tests are being performed on the ADC channel. The AC measurement system for simultaneous ADC and DAC testing is shown in Figure 11.13. In addition to the digital source and

Figure 11.13. Simultaneous DAC and ADC channel testing.

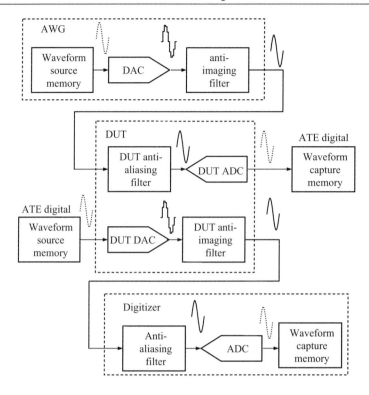

capture memories shown in Figure 11.13, the digital subsystem must also provide any necessary reset functions, initialization patterns, master clocks, frame syncs, and so on.

In Figure 11.13, we have shown four different sampling systems operating simultaneously. The AWG is one sampling system and the digitizer is another. The third sampling system is formed by the source memory and the DAC channel. The fourth sampling system consists of the ADC and the capture memory. In order to ensure that each sampling system forms a coherent DSP-based measurement, the sampling rate and the number of samples collected must be chosen in consideration with the other sampling systems.

Consider the situation involving the AWG and ADC, together with the capture memory. The AWG produces a test tone with frequency F_t that stimulates the ADC channel of the DUT. Assuming that the AWG is operating at a sampling rate of F_{s-AWG} and cycles through N_{AWG} samples, we can express the test tone frequency as

$$F_t = M_{AWG} \frac{F_{s-AWG}}{N_{AWG}} \tag{11.1}$$

where M_{AWG} is a spectral bin number for the AWG. We can also express Eq. (11.1) as

$$F_t = M_{AWG} F_{f-AWG} \tag{11.2}$$

where the fundamental frequency for the AWG is given by

$$F_{f-AWG} = \frac{F_{s-AWG}}{N_{AWG}}$$ (11.3)

As the ADC together with the capture memory forms another coherent sampling system with sampling rate F_{s-ADC} and N_{ADC} samples, we can also write the test tone frequency as

$$F_t = M_{ADC} \frac{F_{s-ADC}}{N_{ADC}}$$ (11.4)

or

$$F_t = M_{ADC} F_{f-ADC}$$ (11.5)

where M_{ADC} is a spectral bin number for the ADC and F_{f-ADC} is the fundamental frequency for the ADC given by

$$F_{f-ADC} = \frac{F_{s-ADC}}{N_{ADC}}$$ (11.6)

The number of samples collected in the capture memory N_{ADC} should be made a power of two in order for the sample set to be compatible with the FFT analysis that will follow.

Recognizing that Eqs. (11.2) and (11.5) are equal in a coherent sampling system allows us to write the following expression:

$$M_{AWG} F_{f-AWG} = M_{ADC} F_{f-ADC}$$ (11.7)

Any two sampling systems that satisfy this equation will be coherent with one another.

Following a similar development, we can state that the requirement for coherence between the DAC and digitizer is

$$M_{DAC} F_{f-DAC} = M_{DIGITIZER} F_{f-DIGITIZER}$$ (11.8)

Since the ADC and DAC samples are usually collected using a single digital pattern loop, their sampling rates and number of samples are not independent. The simplest sampling combination that meets all coherence requirements is one in which all four sampling systems have the same fundamental frequency. This requirement is met by any combination of sampling rates and number of samples in which

$$F_{f-AWG} = F_{f-ADC} = F_{f-DAC} = F_{f-DIGITIZER}$$ (11.9)

Of course, this also implies that the test frequency is equal in both the ADC and DAC channels as

$$M_{AWG} = M_{ADC} = M_{DAC} = M_{DIGITIZER}$$ (11.10)

In the 8-kHz codec example, the simplest sampling system is formed by choosing an 8-kHz sampling rate on all four systems. For example, if we want to collect 512 samples from the ADC channel while sending 512 samples to the DAC channel, we could use the following sampling system

$$F_{s-ADC} = F_{s-AWG} = 8 \text{ kHz}$$
$$N_{ADC} = N_{AWG} = 512$$
$$F_{s-DAC} = F_{s-DIGITIZER} = 8 \text{ kHz}$$
$$N_{DAC} = N_{DIGITIZER} = 512$$

In this case, each of the four sampling systems has a fundamental frequency of 8 kHz / 512 = 15.625 Hz. Often test engineers will double or quadruple the sampling rate of the digitizer and/or AWG to give a higher Nyquist frequency. A higher Nyquist frequency allows a digitizer to detect more frequency components in the output spectrum of a DAC. This is useful for testing the DAC's anti-imaging low-pass filter. Similarly, a higher Nyquist frequency allows an AWG to produce a cleaner test stimulus, free of images. To maintain fundamental frequency compatibility with these higher sampling rates, we have to double or quadruple the number of samples we collect with the digitizer or source with the AWG. For example, we can use the following sampling system to double the fundamental frequency of the tester's digitizer while quadrupling the fundamental frequency of the AWG:

ADC:

$$F_{s-ADC} = 8 \text{ kHz}$$
$$N_{ADC} = 512$$
$$F_{f-ADC} = 8 \text{ kHz} / 512 = 15.625 \text{ Hz}$$

AWG:

$$F_{s-AWG} = 8 \text{ kHz} \times 4 = 32 \text{ kHz}$$
$$N_{AWG} = 512 \times 4 = 2048$$
$$F_{f-AWG} = 32 \text{ kHz} / 2048 = 15.625 \text{ Hz (matches ADC)}$$

DAC:

$$F_{s-DAC} = 8 \text{ kHz}$$
$$N_{DAC} = 512$$
$$F_{f-DAC} = 8 \text{ kHz} / 512 = 15.625 \text{ Hz}$$

Digitizer:

$$F_{s-DIGITIZER} = 8 \text{ kHz} \times 2 = 16 \text{ kHz}$$
$$N_{DIGITIZER} = 512 \times 2 = 1024$$
$$F_{f-DIGITIZED} = 16 \text{ kHz} / 1024 = 15.625 \text{ Hz (matches DAC)}$$

11.2.4 Mismatched Fundamental Frequencies

The example of the previous subsection dealt with the situation where the fundamental frequencies were made equal. In this subsection, we shall investigate the fact that they need only be related by a ratio of two integers, as can be seen from Eq. (11.8) when rearranged.

$$\frac{F_{f-DIGITIZER}}{F_{f-DAC}} = \frac{M_{DAC}}{M_{DIGITIZER}} \tag{11.11}$$

For instance, in the previous example we could have used the following sampling system:

DAC:

$$F_{s-DAC} = 8 \text{ kHz}$$
$$N_{DAC} = 512$$
$$F_{f-DAC} = 8 \text{ kHz} / 512 = 15.625 \text{ Hz}$$

Digitizer:

$$F_{s-DIGITIZER} = 8 \text{ kHz} \times (3/2) = 12 \text{ kHz}$$
$$N_{DIGITIZER} = 512$$
$$F_{f-DIGITIZER} = 8 \text{ kHz} \times (3/2) / 512 = 15.625 \text{ Hz} \times (3/2) = 23.4375 \text{ Hz}$$

Clearly, the digitizer fundamental frequency is 3/2 times that of the fundamental frequency of the DAC. Therefore, the spectral bins in the DAC channel are also related to the spectral bins in the digitizer channel in much the same way, that is, $M_{DAC} = (3/2)M_{DIGITIZER}$. For example, if we use bin 9 in the DAC channel, it will produce a frequency of 9×15.625 Hz, or 140.625 Hz. Since the digitizer's fundamental frequency is $15.625 \text{ Hz} \times (3/2)$, this same frequency will appear at spectral bin 6 in the FFT of the digitizer's samples.

Note that this integer-ratio sampling approach will often force us into even-numbered spectral bins; so it is sometimes inferior to the more straightforward approach using equal fundamental frequencies. However, the test engineer will occasionally find that the use of an integer-ratio sampling approach is the best way to achieve a coherent sampling set given the constraints of the DUT and ATE tester.

Exercises

11.1. A codec consisting of a DAC and ADC is operating at a 11-MHz sampling rate. An AWG has a maximum operating frequency of 100 MHz and an allocated source memory capacity of 4096 samples. The digitizer has a maximum operating frequency of 300 MHz and an allocated captured memory capacity of 2048 samples. Select the appropriate test parameters so that the AWG-ADC and DAC-DIG sampling paths are coherent. What are the corresponding fundamental frequencies?

ANS. $F_{s,AWG} = F_{s,DIG} = 22$ MHz, $N_{ADC} = 2048$, $N_{DAC} = 1024$, $F_{f,AWG-ADC} = 5371.09$ Hz, $F_{f,DAC-DIG} = 10742.18$ Hz.

EXAMPLE 11.1

A DAC must be tested with a sampling rate of 5 MHz and a test tone frequency of approximately 5 kHz. The output of the DAC is sampled by a digitizer with a maximum sampling rate of 20 kHz. The ATE tester's source memory will be needed for other DAC tests; so we need to use the minimum number of DAC samples possible (no more than 1024 samples). Due to sampling constraints inherent to the tester, both the DAC sampling rate and the digitizer sampling rate must be integer multiples of 1 Hz. Find a coherent sampling system compatible with these constraints.

Solution:

First, let us first consider the DAC. The DAC operates at a sampling rate of approximately 5 MHz and is exercised by a tone at approximately 5 kHz. Hence we can write

$$\sim 5\ \text{kHz} = M_{DAC}\frac{\sim 5\ \text{MHz}}{N_{DAC}} \tag{11.12}$$

In order to determine values for the two unknowns (N_{DAC} and M_{DAC}), we shall consider that the test and sampling frequencies are exactly 5 kHz and 5 MHz, respectively, as we are attempting to approach these frequencies as closely as possible. In addition, we also know that N_{DAC} has to be less than 1024, due to source memory limitations. This can only be satisfied if $M_{DAC} = 1$, resulting in $N_{DAC} = 1000$. Thus, by storing 1000 samples of a single cycle of a sine wave in the source memory, we can produce a 5-kHz test tone to exercise the DAC.

Next, let us consider the digitizer. According to the information supplied, we can relate the sampling rate, number of samples, and test frequency according to

$$\sim 5\ \text{kHz} = M_{DIGITIZER}\frac{\sim 20\ \text{kHz}}{N_{DIGITIZER}} \tag{11.13}$$

Our first attempt at solving for the two unknowns ($N_{DIGITIZER}$ and $M_{DIGITIZER}$) is to consider setting the fundamental frequency of the digitizer equal to the fundamental frequency of the DAC, 5 MHz/1000 or 5 kHz. However, in doing so, we create a problem. A 20-kHz sampling rate would only allow us to collect only 4 samples with a fundamental frequency of 5 kHz. This is clearly not enough samples. One possible solution is to collect 256 samples at 20 kHz, and look for power in spectral bin 64 (64 × 20 kHz / 256 = 5 kHz). Unfortunately, this bin would result in the same samples over and over, because $N_{DIGITIZER}$ and $M_{DIGITIZER}$ are not mutually prime. So instead, we shall select $M_{DIGITIZER} = 67$, the nearest prime number, and $N_{DIGITIZER} = 256$. In turn, we must make an adjustment to the sampling rates F_{DAC} and $F_{DIGITIZER}$ in order for the DAC and digitizer to be coherent. This we obtain through the development in Section 11.2.3, where we found

$$M_{DAC}\frac{F_{s-DAC}}{N_{DAC}} = M_{DIGITIZER}\frac{F_{s-DIGITIZER}}{N_{DIGITIZER}} \tag{11.14}$$

Substituting known values, we can write

$$1 \times \frac{F_{s-DAC}}{1000} = 67 \times \frac{F_{s-DIGITIZER}}{256} \tag{11.15}$$

or

$$F_{s-DIGITIZER} = \frac{256}{67 \times 1000}F_{s-DAC} \tag{11.16}$$

With the DAC sampling rate set to 5 MHz, the digitizer's sampling rate would have to be 19,104.47 Hz to satisfy coherence.

Unfortunately, we are not done yet. The ATE tester imposes a further constraint whereby all sampling rates have to be integer multiples of 1 Hz. In other words, all sample rates must be integer numbers. In order to satisfy this constraint, the sample rates of the DAC and digitizer will have to be altered such that they are both integer numbers. To achieve this, we must first express the rational fraction in Eq. (11.16) as a product of prime numbers and eliminate any common terms. That is,

$$F_{s\text{-}DIGITIZER} = \frac{2^8}{67 \times 5^3 \times 2^3} F_{s\text{-}DAC} = \frac{2^5}{67 \times 5^3} F_{s\text{-}DAC} \tag{11.17}$$

Next, select $F_{s\text{-}DAC}$ as a multiple of 67×5^3 which is nearest the desired frequency of 5 MHz. That is,

$$F_{s\text{-}DAC} = 67 \times 5^3 \times n \tag{11.18}$$

where

$$n = \left\lceil \frac{5\text{ MHz}}{67 \times 5^3} \right\rceil = [597.015] = 597 \tag{11.19}$$

and [ ] indicates a rounding to the nearest integer operation. Therefore, $F_{s\text{-}DAC}$ = 4,999,875 Hz. Subsequently, from Eq. (11.17) we find $F_{s\text{-}DIGITIZER}$ = 19,104 Hz.

Summarizing, the final solution is

	DAC	Digitizer
Sampling rate	4,999,875 Hz	19,104 Hz
Number of samples	1000	256
Spectral bin	1	67
Fundamental frequency	4999.875 Hz	74.625 Hz
Test tone frequency	4999.875 Hz (approx. 5 kHz)	4999.875 Hz (matches DAC)

In conclusion, we used a different fundamental frequency for the DAC and digitizer, and we used a different Fourier spectral bin for each to achieve a coherent sampling system. Because there are numerous steps to follow to obtain coherence, test engineers will often devise software tools to aid in the selection of sampling rates, sample sizes, and so on, for a given set of DUT and tester constraints. This is particularly important when more than two sampling systems are required to be made coherent.

11.2.5 Reconstruction Effects in DACs, AWGs, and Other Sampled-Data Circuits

A common element of any sampled-data channel is the sample-and-hold operation associated with the ADC or DAC operation found along its signal path. While sampling and reconstruction was described in detail in Chapter 8, "Sampling Theory," this was done so with an ideal or impulse sampling process. In actual practice, such impulse-like signals are never realized; rather one is limited to signals that have a sampled-and-hold shape, such as that shown in Figure 11.14 and Figure 11.15 for the ADC and DAC processes. In the case of the ADC process shown in Figure 11.14, the sampling process is replaced by a sampled-and-hold (S/H) operation whereby the samples of the input signal are held constant over the duration of the sampling period T_s. During each period of the sampling clock, the ADC converts its input signal $v_{SH}(t)$ into a corresponding discrete-time signal $v[nT_s]$ synchronized with respect to the sampling clock frequency F_s. It is important to note here that the sampled-and-hold operation has no direct effect on the ADC output, because the ADC essentially re-samples the sampled-and-held signal with an impulse-train. Not shown, of course, is the effect of the quantization error introduced by the ADC. This will, in turn, lead to input–output signal differences.

Figure 11.14. Illustrating the effects of practical sampling (sampled-and-hold) process on the ADC operation. The output data sequence is unaffected by the sampled-and-hold operation (but essential).

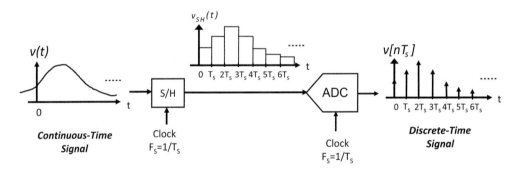

Figure 11.15. Illustrating the effects of practical sampling (sampled-and-hold) process on the DAC operation.

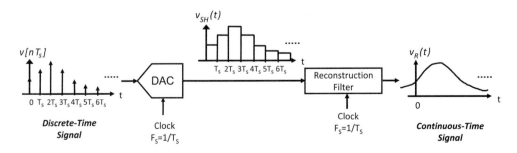

Exercises

11.2. Two sampling systems described in the following table are operating with a test tone of approximately 2000 Hz:

	System #1	System #2
Sampling rate	32,000 Hz	23,000 Hz
Number of samples	2000	1024
Spectral bin	125	89

Slightly adjust the sampling rate of each system so that the two sampling systems will become coherent. What are the sampling frequencies?

ANS. Coherence requires $F_{s1} = 1.390625 \times F_{s2}$. If $F_{s1} = 32$ kHz, then $F_{s2} = 23.011236$ kHz.

11.3. Repeat exercise 11.2, but this time limit the sampling frequencies to integer multiples of 1 Hz.

ANS. One possible solution: $F_{s1} = 31,951$ Hz and $F_{s2} = 22,976$ Hz.

In contrast, the DAC operation is affected by the sampled-and-hold operation introduced by the DAC itself. In Figure 11.15 a discrete-time signal is converted to a voltage level that is held constant over the duration of the sampling clock period. This signal is then applied to the reconstruction filter from which the continuous time signal $v_R(t)$ is obtained. As the sampled-and-held signal is applied directly to the reconstruction filter, the sampled-and-held signal will introduce artifacts into the output signal that is not associated with the discrete-time sample set. These artifacts, in turn, limit the quality of the signal produced by a DAC, or any other sampled-data circuit for that matter. In particular, we shall investigate the effects of $\sin(x)/x$ roll-off and images.

The impulse response behavior of a sample-and-hold operation can be described mathematically as the difference between two unit step functions normalized by the holding time $T_s = 1/F_s$ having the general form

$$g_{SH}(t) = \frac{u(t) - u(t - T_s)}{T_s} \tag{11.20}$$

Correspondingly, the transfer function $G_{SH}(s)$ of this operation can be obtain by taking the Laplace transform of (11.20), resulting in

$$G_{SH}(s) = \frac{1 - e^{-sT_s}}{sT_s} \tag{11.21}$$

Subsequently, by replacing s by $j2\pi f$, and rearranging, we obtain the frequency effect introduced by the sample-and-hold operation as

$$G_{SH}(f) = e^{-j\pi f/F_s} \frac{\sin\left(\pi \frac{f}{F_s}\right)}{\left(\pi \frac{f}{F_s}\right)} \tag{11.22}$$

A plot of the magnitude of $G_{SH}(f)$ is shown Figure 11.16. From this plot we see unity gain at DC and as the frequency increases toward F_s, the gain drops off to zero. Furthermore, above F_s, the magnitude response reduces asymptotically toward zero as the frequency tends toward infinity. With $x = \pi f / F_s$, the magnitude of Eq. (11.22) takes on the form $\sin(x)/x$, leading one to describe this filter behavior as a $\sin(x)/x$ roll-off.

The effects of the sample-and-hold operation can be corrected for by multiplying the spectrum of the discrete-time signal by the inverse of $\left| G_{SH}(f) \right|$. For example, the effects of the $\sin(x)/x$ roll-off on a single tone at frequency F_t can be corrected in software by boosting its amplitude by a correction factor given by

$$\text{correction factor} = \frac{1}{\left| G_{SH}(F_t) \right|} = \frac{\left(\pi \dfrac{F_t}{F_s} \right)}{\sin\left(\pi \dfrac{F_t}{F_s} \right)} \qquad (11.23)$$

Compensation can also be performed in hardware. In fact, some DAC channels include a $\sin(x)/x$ correction factor in the low-pass anti-imaging filter that automatically corrects for most of the roll-off. Therefore, one must know ahead of time whether $\sin(x)/x$ correction factors need to be included in the software description of the test samples.

Figure 11.16. Sample-and-hold gain versus frequency.

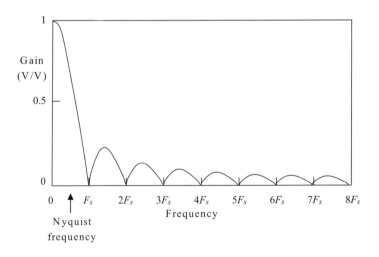

EXAMPLE 11.2

Calculate the sample set for a 3.0-V peak, 8-kHz sine wave that is to be reconstructed at 20 kHz using a conventional DAC. Boost the signal level to correct for sin(x)/x roll-off.

Solution:

First we calculate the correction factor:

$$\text{correction factor} = \left| \frac{\left(\pi \dfrac{8 \text{ kHz}}{20 \text{ kHz}} \right)}{\sin\left(\pi \dfrac{8 \text{ kHz}}{20 \text{ kHz}} \right)} \right| = 1.321$$

Next we multiply the peak value we want by the correction factor to get a sample set that will be attenuated to 3.0-V peak by $\sin(x)/x$ roll-off. An example MATLAB procedure is

```
% Correcting for DAC sin(x)/x effect
%
pi =3.14159265359;
N=2000; M=800; A=3.0; P=0.0;
x = pi*8e3/20e3;
correction_factor = x/sin(x);
for n=1:N,
    sinewave(n) = correction_factor * A * sin(2*pi*M/N*(n-1)+P);
end
```

These samples can be reconstructed using a sample-and-hold process (AWG or DAC) followed by a 10-kHz low-pass filter to remove the sampling images. The sample-and-hold process will attenuate the 8-kHz sine wave by 1.321, resulting in a 3.0-V peak waveform at 8 kHz.

As the signal spectrum of a digital signal is periodic in nature, the sampled-and-hold action of the DAC acts to attenuate the high-frequency signal content and removes the periodic nature of the signal spectrum. However, it does not remove all of the high-frequency content, and further filtering is required, i.e., purpose of the reconstruction filter. Regardless, no filter is perfect, and some energy associated with the images of the signal appears with the reconstructed signal. For a single tone with frequency F_t, the image tones, denoted by F_{image}, will appear in the reconstructed waveform at the following frequencies:

$$F_{image} = nF_s \pm F_t, \quad \text{where } n = 1, 2, 3, \ldots \tag{11.24}$$

Imaging is a one-to-many mapping process. For instance, a 5-kHz sine wave reconstructed at a 20-kHz sampling rate would produce images at 15, 25, 35, 45 kHz, ..., etc.

A common test performed on a DAC followed by a low-pass anti-image filter is to test the circuit with a test tone set very close to the Nyquist frequency of the channel. This is typically the worst-case condition, as the anti-imaging filter must pass the test tone at F_t undisturbed, but reject most of the image tone that appears very close to it at $F_s - F_t$. In other words, this single test acts to verify that the filter's pass-band and stop-band regions meet the maximum and minimum attenuation requirements, respectively.

Exercises

11.4. Samples from a 1-V RMS sine wave with a frequency of 1 kHz are reconstructed with a DAC at a frequency of 32 kHz. Sketch the RMS magnitude of the reconstructed waveform spectrum that includes the test tone plus the four lowest image tones.

ANS.

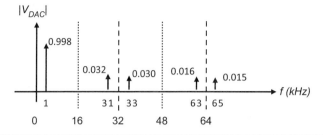

11.3 UNDERSAMPLING AND ALIASING

Undersampling is a technique that allows a digitizer or ADC to measure signals beyond the Nyquist frequency. A digitizer sampling at a frequency of F_s has a Nyquist frequency equal to $F_s/2$. Any input signal frequency, F_t, which is above the Nyquist frequency will appear as an alias component somewhere between 0 Hz and the Nyquist frequency. We normally try to filter the input signal so that it has no components above the Nyquist frequency. However, we may remove the filter if we want to allow our digitizer or DUT to collect samples from a signal that includes components above the Nyquist frequency. This technique is called *undersampling*. Undersampling can also be used to measure the out-of-band rejection of a low-pass anti-aliasing filter before an ADC as shown in Figure 11.17. Frequencies that extend beyond the Nyquist rate of the ADC may not be fully attenuated by the filter, and as a result they may appear as low-amplitude in-band alias tones. The amplitude of these alias components can be used to compute the filter cutoff performance at frequencies beyond the Nyquist frequency.

Undersampling is often used when we have a digitizer or a DUT ADC with a limited sampling rate, but we want to capture a signal with components beyond the Nyquist frequency. Provided that the bandwidth of the digitizer's front end is adequate, we can use undersampling to collect samples from the high frequency signal (Figure 11.18).

There are several things to consider when using undersampling. First, all the noise components from 0 Hz to the digitizer or ADC input bandwidth will be additively folded back into the range from 0 Hz to $F_s/2$. This means the signal-to-noise ratio of the aliased signal will probably be degraded compared to a fully sampled measurement.

Second, the digitizer's front end may be less linear or may have a gain error at the frequency of the test signal. These problems can usually be corrected using focused software calibration techniques. Finally, the aliased image of two or more tones in a multitone signal may fold back to the same frequency. Care must be taken when selecting frequencies so they each fall into a unique in-band FFT bin. To calculate the expected alias frequency of a test tone denoted F_{ta}, perform the following steps:

1. Repetitively subtract F_s from the test frequency F_t until the result is between 0 and F_s. Call this result, F'_{ta}. Mathematically, this is equivalent to

$$F'_{ta} = F_t - nF_s, \quad \text{where} \quad n = \left\lfloor \frac{F_t}{F_s} \right\rfloor \tag{11.25}$$

where [ ] indicates a rounding down to the nearest integer operation.

2. Next, check whether F'_n is above or below the Nyquist frequency $F_s/2$. If it is below the Nyquist frequency, then F'_{ta} is the frequency of the aliased tone. Otherwise, it is an image and the aliased tone will appear at frequency $F_s - F'_{ta}$. Mathematically, we can express these two conditions as

$$F_{ta} = \begin{cases} F'_{ta} & \text{if } 0 < F'_{ta} \leq F_s/2 \\ F_s - F'_{ta} & \text{if } F_s/2 < F'_{ta} \leq F_s \end{cases} \qquad (11.26)$$

Figure 11.17. Low-pass anti-aliasing filter and ADC.

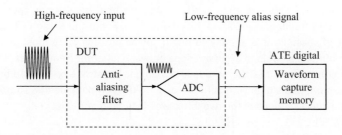

Figure 11.18. Undersampling a high-frequency sine wave.

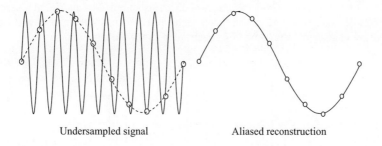

EXAMPLE 11.3

A sine wave with a frequency of 65 kHz is sampled by an ADC at a sampling rate F_s of 20 kHz. An FFT is performed on the samples collected by the ADC. At what frequency do we expect the 65-kHz tone to appear? Assuming N samples were collected, at what spectral bin will this frequency appear? Repeat the example with an input tone of 75 kHz.

Solution:
We subtract 20 kHz from 65 kHz until we get an answer less than 20 kHz

$$65 \text{ kHz} - 20 \text{ kHz} - 20 \text{ kHz} - 20 \text{ kHz} = 5 \text{ kHz}$$

This result is less than the Nyquist frequency of 10 kHz; so we would expect to see the aliased tone at 5 kHz. The spectral bin is calculated using the equation

$$5 \text{ kHz} = M \frac{20 \text{ kHz}}{N}$$

or rewriting

$$M = \frac{N \times 5 \text{ kHz}}{20 \text{ kHz}} = \frac{N}{4}$$

Repeating the example with a 75-kHz input tone, we keep subtracting 20 kHz until we get a number less than F_s: F'_{ta}=15 kHz. This frequency is above the Nyquist; thus we expect to see an aliased sine wave at 20 kHz – 15 kHz = 5 kHz. Again, this tone shows up in spectral bin $N/4$. Notice that both input frequencies (65 and 75 kHz) fold back into the same FFT spectral bin. This is the nature of aliasing—it is a many-to-one mapping process. Care must be taken when undersampling multitone signals to avoid overlaps between different frequencies. Each under-sampled tone must be selected so that it falls into a unique spectral bin. It would be a mistake to undersample a multitone signal with both 65- and 75-kHz components at a sampling rate of 20 kHz, since the aliased versions of the two tones would overlap at 5 kHz.

Exercises

11.5. A 1-V RMS sine wave with a frequency of 65 kHz is sampled by an ADC at a sampling rate of 32 kHz. Sketch an RMS magnitude spectrum that includes the four lowest frequencies that could alias into the same FFT spectral bin as the 65-kHz tone.

ANS.

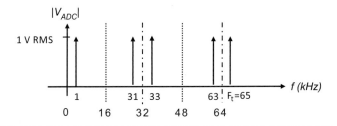

11.3.1 Reconstructing the High-Frequency Signal from the Aliased Sample Set

While undersampling is an effective method for capturing information about a high-frequency signal using a low-frequency sampling signal, it does so using samples from an aliased version of the high-frequency signal. Fortunately, these samples can be rearranged using a simple post-processing procedure to act as if the samples were taken directly from the original signal using a high-frequency sampling signal.

Consider a coherent test signal with frequency f_T is synthesized using the M and N principle according to

$$f_T = \frac{M}{N} F_s \qquad (11.27)$$

It is assumed that M and N are relatively prime. Let us further assume that $F_t \gg F_s$ so that the test signal is undersampled. Consequently, from Eq. (11.27), we see that $M \gg N$. Furthermore, let us assume that the test signal lies in an even image band of the frequency spectrum so that

$$M = M_{alias} + pN \qquad (11.28)$$

where p is some arbitrary positive value integer and $0 < M_{alias} < N/2$. Otherwise, the test signal lies in an odd image band where we would write

$$M = (N - M_{alias}) + pN \qquad (11.29)$$

Working with a test signal in the even band, we can substitute Eq. (11.28) into Eq. (11.27) and write

$$f_T = \left(\frac{M_{alias} + pN}{N} \right) F_s \qquad (11.30)$$

Multiplying the numerator and denominator by M_{alias}, we write

$$f_T = \frac{M_{alias}}{N} \left[\frac{(M_{alias} + pN)F_s}{M_{alias}} \right] \qquad (11.31)$$

Comparing Eq. (11.31) with (11.27), the latter term between the square brackets takes on the dimensions of a sampling frequency and is referred to as the effective sampling rate, that is,

$$F_{s,eff} \triangleq \left(\frac{M_{alias} + pN}{M_{alias}} \right) F_s \qquad (11.32)$$

Likewise, the term M_{alias} takes on a meaning of the number cycles that the sample set will complete over N sample points spaced $1/F_{s,eff}$ seconds apart. Of course this is equivalent to completing M_{alias} cycles of the original high-frequency signal. Correspondingly, we can express the test signal as

$$f_T = \frac{M}{N} F_s = \frac{M_{alias}}{N} F_{s,eff} \qquad (11.33)$$

Figure 11.19 illustrates the equivalence of the two sample sets, together with the appropriate timing relationships superimposed. This figure suggests the manner in which to reconstruct the original waveform from the aliased sample set. Simply retime the aliased data set using the sampling period $1/F_{s,eff}$.

Figure 11.19. Illustrating the sampling and timing relationships between the undersampled and reconstructed waveforms.

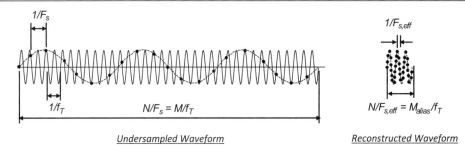

A second approach can also be envisioned from Eq. (11.30) above. Rather than normalize with respect to M_{alias}, we'll normalize with respect to unity and write

$$f_T = \frac{1}{N}\left[(M_{alias} + pN)F_s\right] \tag{11.34}$$

In this case, the data set is to complete 1 cycle of the original waveform over N points using an effective sampling rate of

$$F_{s,eff} \triangleq (M_{alias} + pN)F_s \tag{11.35}$$

Unlike previously, the data set is not in the correct order to display a single period of original signal. This is illustrated by the simple example shown in Figure 11.20. Here we see a three-cycle aliased waveform and the samples labeled consecutively. A waveform reconstructed over a single cycle of the original waveform requires a different order of samples as shown on the right of this figure. Fortunately, a simple algorithm can be devised to shuffle the data set so that they are in the right order. This algorithm is based on the modulo-time reshuffling algorithm[2] and acts to position the sample data with respect to the period of the test signal f_T normalized by effective sampling frequency $F_{s,eff}$, that is, N/M_{alias}.

The modulo-time reshuffling algorithm can be summaries as follows:

1. Map (n,v) pairs of signal samples to $(M_{alias} \times [n \bmod N/M_{alias}], v)$, where n is the sample index and v is the sample value.
2. Re-sort in ascending numerical order based on x-axis coordinate.

Here the $A \bmod B$ operation is equivalent to extracting the fractional part of the term A / B and multiplying the result by the factor B. Mathematically, we can write the mod operation as

$$A \bmod B = B \times f_r\left(\frac{A}{B}\right) \tag{11.36}$$

where $f_r(x)$ denotes the fractional part of x. This form of the mod operation eliminates any specific machine dependency especially when working with nonintegers. The following example will help to illustrate the use of this technique.

Figure 11.20. The order of the samples associated with the aliased waveform is different than those used to reconstruct a waveform over a single cycle of the original signal. This effect can be accounted for by using the modulo-time reshuffling algorithm

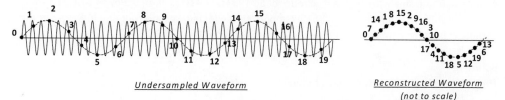

Undersampled Waveform　　　　　*Reconstructed Waveform*
(not to scale)

EXAMPLE 11.4

Reshuffle the following 8 points according to the modulo-time shuffling algorithm where $M_{alias} = 3$:

$$(0, 0), (1, 0.7071), (2, -1.0), (3, 0.7071), (4, 0), (5, -0.7071), (6, 1.0), (7, -0.7071)$$

Solution:

The x coordinate of each point is transformed according to $n \to N \times f_r\left(n \frac{M_{alias}}{N}\right)$, resulting in the following new data points:

$$(0, 0), \left(8 \cdot f_r\left(1 \times \frac{3}{8}\right), 0.7071\right), \left(8 \cdot f_r\left(2 \times \frac{3}{8}\right), -1.0\right), \left(8 \cdot f_r\left(3 \times \frac{3}{8}\right), 0.7071\right), \left(8 \cdot f_r\left(4 \times \frac{3}{8}\right), 0\right),$$

$$\left(8 \cdot f_r\left(5 \times \frac{3}{8}\right), -0.7071\right), \left(8 \cdot f_r\left(6 \times \frac{3}{8}\right), 1.0\right), \left(8 \cdot f_r\left(7 \times \frac{3}{8}\right), -0.7071\right)$$

As

$$f_r\left(1 \times \frac{3}{8}\right) = f_r(0.375) = 0.375, \ f_r\left(2 \times \frac{3}{8}\right) = f_r(0.75) = 0.75, \ f_r\left(3 \times \frac{3}{8}\right) = f_r(1.125) = 0.125,$$

$$f_r\left(4 \times \frac{3}{8}\right) = f_r(1.5) = 0.5, \ f_r\left(5 \times \frac{3}{8}\right) = f_r(1.875) = 0.875, \ f_r\left(6 \times \frac{3}{8}\right) = f_r(2.25) = 0.25 \text{ and}$$

$$f_r\left(7 \times \frac{3}{8}\right) = f_r(2.625) = 0.625$$

we obtain the new data set as

$$(0, 0), (3, 0.7071), (6, -1.0), (1, 0.7071), (4, 0), (7, -0.7071), (2, 1.0), (5, -0.7071)$$

Finally, we order each sample point in ascending x coordinate value and obtain the following shuffled data sequence:

$$(0, 0), (1, 0.7071), (2, 1.0), (3, 0.7071), (4, 0), (5, -0.7071), (6, -1.0), (7, -0.7071)$$

EXAMPLE 11.5

A unit amplitude sine wave with a frequency of 60.19531250 kHz is sampled by an ADC at a sampling rate of 20 kHz. Assuming that 512 samples are collected, reconstruct one full cycle of the original signal and compare with the sampled set captured by the ADC.

Solution:
The M value corresponding to this situation is found from

$$M = \frac{f_T}{F_s}N = \frac{60.19531250 \text{ kHz}}{20 \text{ kHz}} \times 512 = 1541$$

Next, we solve for M_{alias} by repeated subtract of N from M until the result is between 0 and 256:

$$M_{alias} = 1541 - 512 - 512 - 512 = 5$$

Subsequently, we compute the effective sample rate for this situation consisting of a single period of the original sampled signal:

$$F_{s,eff} = M \times F_s = 1541 \times 20 \times 10^3 = 30.82 \times 10^6 \text{ Hz}$$

The sample set captured by the ADC is shown plotted below corresponding to $N = 512$, $M = 1541$, $A = 1$ and $\Phi = 0$:

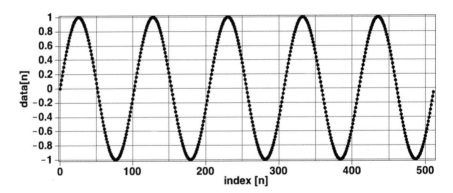

Reshuffling the data set using the modulo-time algorithm and spacing each data point $1/F_{s,eff} = 324.4646$ μs apart, we obtain the following plot of one full cycle of the original waveform referenced to absolute time:

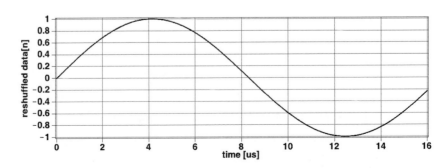

EXAMPLE 11.6

Undersampling applies to periodic signals of all types not just those involving sinusoids. The input to an amplifier was excited by a 1-V, 500-MHz square wave and the output was under-sampled using a digitizer operating at a sampling rate of 155.52855 MHz. 256 points were collected. Reconstruct the step response of the amplifier using the modulo-time shuffling algorithm and determine the maximum value of the overshoot and undershoot.

Solution:

The digitizer output was collected and plotted below:

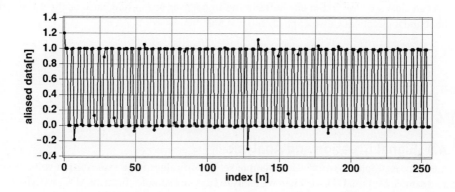

Clearly, not much can be seen from this plot. We will need to re-shuffle the data set and plot over one period of the input signal. According to the given information, f_T = 500 MHz, F_s = 155.52855 MHz, and N = 256. Hence, we compute the M value as

$$M = \frac{f_T}{F_s}N = \frac{500 \text{ MHz}}{155.52855 \text{ MHz}} \times 256 = 823$$

Subsequently, the aliased signal bin M_{alias} then becomes

$$M_{alias} = 823 - 256 - 256 - 256 = 55$$

The effective sampling frequency is then found to be

$$F_{s,eff} = M \times F_s = 823 \times 155.52855 \times 10^6 = 128 \text{ GHz}$$

Reshuffling the data set using the modulo-time algorithm and spacing each data point $1/F_{s,eff}$ = 7.8125 ps apart, we obtain the following plot of one full cycle of the original waveform referenced to absolute time:

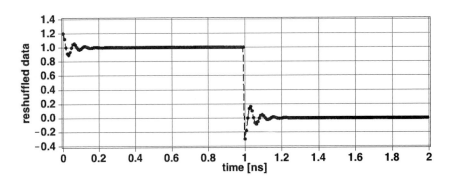

Here we can clearly establish the attributes of the signal over one period, such as the maximum overshoot and undershoot values. Specifically, we find

$$\text{max. overshoot} = 0.2 = 20\%$$
$$\text{max. undershoot} = 0.3 = 30\%$$

11.4 ENCODING AND DECODING

11.4.1 Signal Creation and Analysis

In a previous example, we produced a sample set whose values were expressed in volts. When we create a sample set for a DAC or when we analyze captured samples from an ADC, we have to work with LSB code steps (also called *quanta*). The term LSB is commonly used to refer to a single step in a DAC or ADC transfer curve, but this terminology can be a bit ambiguous. If we tell a design engineer that his ADC generates 3 LSBs of noise, do we mean 3 steps or do we mean that the 3 least significant bits of the ADC digital output are toggling randomly? If the three least significant bits (D2-D0) were toggling, that would correspond to 8 steps of noise. Unfortunately, the term "LSB" is much more common in data sheet specifications than the less ambiguous "quantum"; thus we shall use the term LSB when referring to a single division in a DAC or ADC transfer curve.

Exercises	
11.6. A test signal with frequency of 60.00292969 MHz is sampled by an ADC at a sampling rate of 1 MHz. Assuming that 1024 samples are collected, is the measurement coherent? If so, how many cycles of the original waveform is aliased and what is the effective sample period of this measurement setup?	ANS. Coherent, $M_{alias} = 3$, $T_{s,eff} = 16.27524698$ ps
11.7. Compute the following quantities: (a) 1013 mod 256, (b) 0 mod 10, (c) 25.1 mod 11.6, (d) 20 mod 10.	ANS. (a) 245, (b) 0, (c) 1.899, (d) 0.
11.8. A test signal is to be undersampled using a digitizer with a sampling frequency of 1 GHz. Assuming that the data sequence should complete 11 cycles in 1024 points, at which frequency should the input signal be set at? What is the corresponding effective sampling rate corresponding to the 11 cycles of the aliased waveform?	ANS. $F_t = 1.074218750 \times 10^7 + p10^9$, where p is any integer; for $p = 2$: $f_T = 2.010742188$ GHz, $F_{s,eff} = 187.1818096$ GHz.

To convert a series of desired voltages into a series of DAC codes, we have to know the DAC's ideal gain in bits per volt as well as its encoding format. This information is used to encode a series of floating point voltage samples into integer DAC samples. Similarly, before we can analyze the output of an ADC, we need to know its ideal gain and format. This information is used to convert the ADC output from integer values into floating point voltage samples. The encoding and decoding process depends on the encoding scheme of the DAC or ADC. In Chapter 6, "DAC Testing," we described several common digital encoding schemes, such as sign/magnitude, two's complement, and so on. If necessary, our readers can refer back to section 7.1.2 for further details.

11.4.2 Intrinsic (Quantization) Errors Associated with the DAC Operation

Whenever a sample set is encoded and then decoded (i.e., map from one digital number representation to another, and then back), quantization errors are added to the signal. In high-resolution encoding and decoding processes, these errors may be less than the errors generated by noise in the signal. But in low-resolution converters, or in signals that are very small relative to the full-scale range of the converter, the quantization errors can make a sine wave appear to be larger or smaller than it would otherwise be in a higher-resolution system.

These intrinsic errors can be compensated for in the final measured result by knowing ahead of time the gain error of a perfect ADC/DAC process as it encodes and decodes the signal under test. This is achieved by modeling the perfect DAC/ADC in software using, say, MATLAB. This process is made somewhat difficult, because these errors are dependent on the exact input signal characteristics, including signal level, frequency, offset, phase shift, and number of samples. All these parameters must be modeled correctly; otherwise the results will be incorrect. In the case of a DAC, the procedure is relatively straightforward. Unfortunately, the same cannot be said for an ADC.

Let us first consider the AWG encoding process of a single sinusoidal signal. One begins by writing a numerical routine that generates the samples using a floating-point number representation. Next, the samples are encoded into the corresponding format of the DAC found within the AWG. In the process of encoding the samples, they are normalized by the DAC gain due to the ideal LSB step change alone, that is, V_{LSB}, followed by a quantization or rounding operation, because only a finite number of bits can be used to represent each sample. Let us denote this particular gain term as $G_{DAC,IDEAL}$ to distinguish it from other DAC gains that we shall shortly describe. Subsequently, the samples are stored in the waveform source memory and passed to the DAC to be decoded (i.e., produce the output analog waveform). The DAC restores the samples to their original level, because the DAC has an ideal gain of $G_{DAC,IDEAL}$. We can model the encoding/decoding process with the block diagram shown in Figure 11.21a. Of particular interest is the quantization operation, denoted $Q(\cdot)$ in the figure. This block is the source of signal-dependent errors mentioned above. To quantify these errors, we first define the gain of the entire encoding process as the ratio of the output $V_{DAC,IDEAL}$ over the input V_{IN}. This gain is referred to as the intrinsic gain of the encoding process for a particular sample set and is defined as

$$G_{INTRINSIC} = \frac{V_{DAC,IDEAL}}{V_{IN}} \; \text{V/V} \tag{11.37}$$

Ideally, the entire encoding/decoding procedure should have a gain of unity (i.e. output equals input). Rather, the output contains intrinsic errors causing the output to differ from the input. Subsequently, the intrinsic gain error $\Delta G_{INTRINSIC}$ for a particular input becomes

$$\Delta G_{INTRINSIC} \equiv G_{INTRINSIC} - 1 = \frac{V_{DAC,IDEAL} - V_{IN}}{V_{IN}} \; \text{V/V} \tag{11.38}$$

Figure 11.21. Modeling the AWG waveform encoding/decoding process: (a) Nonlinear model; (b) Equivalent linear model for a particular sample set.

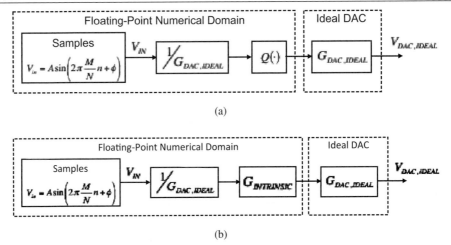

(a)

(b)

The fact that the output signal $V_{DAC,IDEAL}$ is related to the input V_{IN} by a gain constant $G_{INTRINSIC}$ suggests that we can model the nonlinear quantizer operation as a linear operation with gain $G_{INTRINSIC}$ as shown in Figure 11.21b. This representation is, of course, valid only for a particular sample set. If the sample set is changed, then a new $G_{INTRINSIC}$ gain constant must be derived.

For example, let us say that we want to generate a low level sine wave at the first Fourier spectral bin with no offset and 0 radians of phase shift using an 8-bit two's complement DAC. If we want to generate a very low level sine wave with an RMS amplitude of 8 LSBs, we could calculate the quantized/reconstructed sample set (normalized by $G_{DAC,IDEAL}$) shown in Figure 11.22 using the following MATLAB code:

```
% Coherent signal definition -x-
%
N=64; M=1; P=0;
A=sqrt(2)*8; % expressed in LSBs
for n=1:N,
    x(n) = A*sin(2*pi*M/N*(n-1) + P);
end
%
% Quantizer
%
GDAC=1; % gain expressed in volts per bits
for n=1:N,
    q(n) =GDAC *round(x(n)/GDAC);
end
%
```

If we perform an FFT on the output signal, we see that we get an RMS level of 7.861 LSBs, corresponding to an intrinsic error of (7.861–8.0) = –0.139 LSBs. Subsequently, the intrinsic gain and intrinsic gain error becomes 0.9826 and –0.0173 V/V, respectively. If we shift the phase of this sine wave by $\pi/3$ radians as shown in Figure 11.23, we get a different signal level of 8.026 LSBs, corresponding to an intrinsic error of +0.026 LSBs. The intrinsic gain and gain error will then be 1.00325 and 0.00325 V/V, respectively.

Figure 11.22. Quantized sine wave: intrinsic error = –0.139 LSBs.

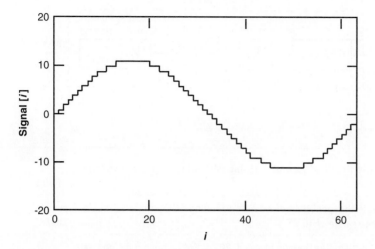

Figure 11.23. Quantized sine wave with phase shift: intrinsic error = +0.026 LSBs.

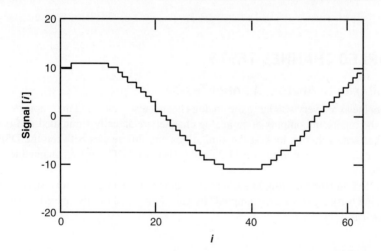

Attempting to apply this same approach to uncover the intrinsic errors associated with an ADC is much more difficult. A block diagram illustrating the decoding/encoding process for the digitizer is shown in Figure 11.24. Unlike the situation with the AWG, the input signal to the digitizer changes with each test and, hence, so does the quantization error. Thus, knowing the intrinsic error corresponding to a particular sample set provides no additional insight into the quantization errors that occur during a particular test.

Intrinsic error is the result of consistent quantization errors. In general, intrinsic error is less of a problem with higher-resolution converters and/or larger sample sizes. The intrinsic error of a DAC or ADC quickly approaches the noise floor of the measurement as the number of samples increases, as long as we use Fourier spectral bins that are mutually prime with respect to the sample size. Spectral bins that are not mutually prime will produce the same samples repeatedly, which in turn produces the same quantization errors over and over. This is another reason to use mutually prime spectral bins, since it tends to minimize intrinsic errors.

Figure 11.24. Modeling the digitizer/ADC waveform decoding/encoding process.

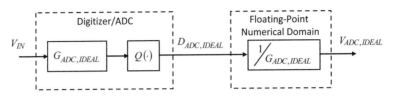

Exercises

11.9. What is the ideal gain of a 10-bit DAC with a full-scale voltage range of 1.5 V?

ANS. $G_{DAC,IDEAL} = 1.466$ mV/bit.

11.10. What is the ideal gain of a 12-bit rounding ADC with a full-scale voltage range of 10 V?

ANS. $G_{ADC,IDEAL} = 409.5$ bits/V.

11.11. A set of samples derived from a 0.707-V RMS sinusoidal signal is found to have an intrinsic gain error of 0.05 V/V. What is the RMS amplitude of the captured test signal?

ANS. 0.742 V.

11.5 SAMPLED CHANNEL TESTS

11.5.1 Similarity to Analog Channel Tests

Sampled channel tests are very similar to the analog channel tests described in Chapter 9. In DSP-based testing, the inputs and outputs of an analog channel are actually stimulated and measured using sampled channels. Let us look at the similarities and differences between the DSP-based analog channel gain test shown in Figure 11.25 and a DAC-to-ADC sampled channel test shown in Figure 11.26.

Figure 11.25 shows the full stimulus-to-analysis path for an analog channel test. A continuous mathematical sine wave is, in effect, "sampled" by calculating 512 evenly spaced values using a C code loop such as

```
int i;
float sample[512], pi=3.14159265359, A = 1.414 V;
for (i=0;i<512;i++)
  sample[i] = A * sin(2*pi*i/512);
  load_waveform_into_AWG(sample,512);
```

Whether the routine "load_waveform_into_AWG( )" is supplied by the ATE vendor as part of a library or whether the routine is written by the test engineer, it must perform a very important first step. Since most AWGs produce their waveforms by applying integer values to a DAC, the load_waveform_into_AWG( ) routine must first convert the continuous floating-point waveform, sample[ ], into a quantized integer waveform. It must also calculate the necessary AWG attenuation settings and mathematical scaling factors that make this waveform come out of the AWG at the proper voltage level. This scaled and quantized integer waveform is then compatible with the AWG's waveform memory.

Figure 11.25. DSP-based analog channel test.

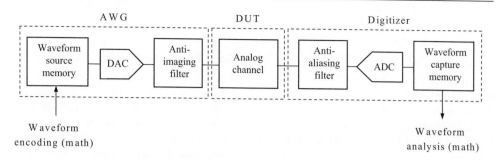

Figure 11.26. DAC-to-ADC sampled channel test.

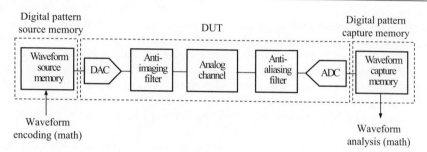

To create a continuous analog stimulus waveform, the quantized waveform is passed through the AWG's DAC, then through a reconstruction filter and various signal conditioning circuits (programmable attenuators, etc.). The resulting continuous waveform passes through the DUT's analog channel and into the input stage and low-pass anti-imaging filter of a digitizer. The digitizer's integer samples are stored into a bank of memory. These samples are then available for DSP operations, such as the FFT.

Now consider the sampled channel test in Figure 11.26. The same process occurs in this case, except the position of the components is shifted from the tester to the DUT. A continuous mathematical waveform is sampled by a mathematical process as before, but this time the integers are stored in the source memory of the digital pattern subsystem rather than being stored in the AWG waveform memory. The digital samples from the source memory are then shifted into the sampled channel's DAC input.

The DAC output is low-pass filtered, producing a continuous waveform. This waveform is then passed through an analog channel, such as a diagnostic loopback path inside the DUT. The continuous signal is then filtered and resampled by the DUT's ADC channel, producing digital output samples. These are captured and stored into the digital subsystem's capture memory, where they are available for DSP operations.

The sampled channel gain test is therefore almost identical to DSP-based analog channel testing. We could also show how DAC channels, ADC channels, switched capacitor filters, and any other sampled channel could be reduced to a similar measurement system. The only difference is that the location of DACs, ADCs, filters, and other signal conditioning circuits may move from the ATE tester to the DUT or vice versa. Unfortunately, this means that we have to apply more rigorous testing to sampled channels, since all the effects of sampling (aliasing, imaging, quantization errors, etc.) vary from one DUT to the next. These sampling effects are often a major failure mode for sampled channels. Let us look at each of the analog channel tests described in Chapter 10, and see how they must be modified as we apply them to sampled channels.

11.5.2 Absolute Level, Absolute Gain, Gain Error, and Gain Tracking

The process for measuring absolute level in DACs and other analog output sampled channels is identical to that for analog channels. The only difference is the possible compensation for intrinsic DAC errors. In this way, a measurement is made independent of the sample set used. Moreover, when compared to bench measurements made with noncoherent test equipment, better correlation is made possible. Otherwise, absolute voltage level measurements are performed the same way as any other AC output measurement. ADC absolute level is equally easy to measure. The difference is that we express the output measurement in terms of RMS LSBs (or RMS quanta, RMS bits, RMS codes, or whatever terminology is preferred) rather than RMS volts.

In some sampled channels, such as switched capacitor filters and sample-and-hold amplifiers, absolute gain is measured using the same voltage-in/voltage-out process as in analog channels. By contrast, measurement of absolute gain in mixed-signal channels is complicated by the fact that the input and output quantities are dissimilar. Gain in mixed-signal channels is defined not in volts per volt, but in bits per volt or volts per bit, where the term "bit" refers to the LSB step size. For example, if we have an 8-bit two's complement DAC with a full-scale range of -1 to $+1$ V, its ideal gain is 2.0/255 or 7.84 mV/bit. Notice that there are only 255 steps between codes in an 8-bit converter, not 256. Therefore one LSB is equal to 2.0 V/255 steps or 7.84 mV.

However, we sometimes see a data sheet that defines the upper voltage of a DAC as 1 LSB above the maximum valid DAC code. For example the data sheet in this 8-bit DAC example might define -1.0 V as code -128 and 1.0 V as code 128. Code 128 does not actually exist in an 8-bit two's complement DAC; thus it is an imaginary point on the DAC curve. In this case, we would calculate the ideal gain as 2.0/256 = 7.81 mV/bit. If the data sheet is not clear on this definition, the test engineer should request clarification. Similar issues exist on an ADC's gain specification. Data sheets do not always define the gain of a converter in bits per volt or volts per bit. Often they define the ideal LSB step size instead, which is equal to the gain in the case of a DAC or inverse of the gain in the case of an ADC.

Consider the DAC test setup shown in Figure 11.27a. The absolute gain of a DAC is expressed as the ratio of its actual output signal $V_{DAC,ACTUAL}$ divided by the input signal V'_{IN} (expressed in LSBs), that is,

$$G_{DAC,ACTUAL} = \frac{V_{DAC,ACTUAL}}{V'_{IN}} \ \text{V/bit} \tag{11.39}$$

As $V'_{IN} = \left(G_{INTRINSIC}/G_{DAC,IDEAL}\right)V_{IN}$, we can write the DAC's gain as

$$G_{DAC,ACTUAL} = \frac{G_{DAC,IDEAL}}{G_{INTRINSIC}} \frac{V_{DAC,ACTUAL}}{V_{IN}} \ \text{V/bit} \tag{11.40}$$

It is interesting to note that $G_{DAC,ACTUAL}$ also includes the $\sin(x)/x$ frequency roll-off of the sampled-and-hold action of the DAC. Thus $G_{DAC,ACTUAL}$ is a frequency-dependent parameter.

Figure 11.27. Modeling the test setup for anactual: (a) DAC and (b) ADC.

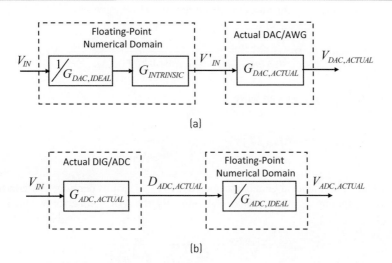

(a)

(b)

Similarly, as illustrated in Figure 11.27b, the absolute gain of an ADC without intrinsic error compensation is given by

$$G_{ADC,ACTUAL} = \frac{D_{ADC,ACTUAL}}{V_{IN}} \text{ bits/V} = G_{ADC,IDEAL} \times \frac{V_{ADC,ACTUAL}}{V_{IN}} \text{ bits/V} \qquad (11.41)$$

where $D_{ADC,ACTUAL}$ is the ADC digital output expressed in LSBs, $V_{ADC,ACTUAL}$ is the corresponding output signal level in volts, $G_{ADC,IDEAL}$ is the ideal ADC gain expressed in bits/V and V_{IN} is the ADC input voltage signal.

Converter gain cannot be specified in decibels, because it is a ratio of dissimilar quantities (e.g., volts per bit). Converter gain error, however, can be expressed in decibels. Gain error ΔG is equal to the actual gain G_{ACTUAL} divided by the ideal gain G_{IDEAL} (which includes any sampled-and-hold effects)

$$\Delta G = \frac{G_{ACTUAL}}{G_{IDEAL}} \text{ V/V} \qquad (11.42)$$

Either result can be converted into decibel units using the standard conversion expression

$$\Delta G = 20 \ \log_{10} \left(\frac{G_{ACTUAL}}{G_{IDEAL}} \right) \text{ dB} \qquad (11.43)$$

For example, the ideal gain of a DAC, which includes the sampled-and-hold effect, can be written as

$$G_{DAC,IDEAL,SH} \left(f \right) = G_{DAC,IDEAL} \times G_{SH} \left(f \right) \text{ V/bit} \qquad (11.44)$$

Substituting the above result into Eq. (11.42), where G_{ACTUAL} is replaced by $G_{DAC,ACTUAL}$ and G_{IDEAL} by $G_{DAC,IDEAL,SH}$, together with the measured result in Eq. (11.40), gives

$$\Delta G_{DAC}(f) = \frac{G_{DAC,ACTUAL}}{G_{DAC,IDEAL,SH}} = \frac{1}{G_{INTRINSIC}G_{SH}(f)} \frac{V_{DAC,ACTUAL}}{V_{IN}} \quad \text{V/V} \tag{11.45}$$

As $V_{DAC,IDEAL} = G_{INTRINSIC} V_{IN}$, we can write

$$\Delta G_{DAC}(f) = \frac{1}{G_{SH}(f)} \frac{V_{DAC,ACTUAL}}{V_{DAC,IDEAL}} \quad \text{V/V} \tag{11.46}$$

In the case of an ADC, the ideal gain is assumed to be $G_{ADC,IDEAL}$ bits per volt, as we have no accurate means of computing its intrinsic gain. Subsequently, the ADC gain error becomes

$$\Delta G_{ADC} = \frac{V_{ADC,ACTUAL}}{V_{IN}} = \frac{1}{G_{ADC,IDEAL}} \frac{D_{ADC,ACTUAL}}{V_{IN}} \quad \text{V/V} \tag{11.47}$$

EXAMPLE 11.7

An 8-bit two's complement DAC with a single-ended output has an ideal LSB size of 3.0 V/2^8 = 11.719 mV. The ideal output range is 1.0 to 4.0 V − 1 LSB, that is, the 4.0-V level corresponds to imaginary code +128. A sample set corresponding to a 1-kHz sine wave at 0.8 V RMS is desired. Assuming a perfect DAC, write a MATLAB routine that produces a 512-point sample set that will generate a 1-kHz sine wave at 800 mV RMS, at a DAC sampling rate of 16 kHz. If the DAC output is digitized and the actual RMS voltage is determined to be 780 mV RMS instead of 800 mV, what is the gain and gain error of the DAC at 1 kHz? Include sin(x)/x roll-off and intrinsic error in the gain and gain error calculations.

Solution:
First we need to calculate the sample set for the DAC. The peak amplitude of the tone is set at $\sqrt{2}$ × 0.8 V. We also have to compute the Fourier spectral bin for a 1-kHz tone with a 16-kHz sampling rate and 512 samples. The Fourier spectral bin is found by dividing 1 kHz by the fundamental frequency of the sample set: M = 1 kHz / (16 kHz / 512) = 32. Of course, 32 is a poor choice since it is not a mutually prime number and will generate excessive intrinsic error. We shift the spectral bin to 31 to achieve a prime bin. The resulting MATLAB code is therefore

```
% DAC Encoding/Decoding Procedure
%
D=8; % 8-bit DAC
VLSB=3.0/2^D; % least-significant bit using an imaginary bit at 4.0 V
%
% Coherent signal definition -x-
%
```

```
N=512; M=31; A=sqrt(2)*0.8; P=0;
for n=1:N,
x(n) = A*sin(2*pi*M/N*(n-1) + P);
end
%
% Quantize result and perform DAC operation
%
GDAC:=VLSB:
for n=1:N,
   q(n) = GDAC *round(x(n)/GDAC);
end
% end
```

We next need to calculate the absolute gain of a perfect DAC so we can compare our DAC's gain to the ideal gain. An FFT is performed on the scaled sample set. Using the FFT output, we calculate the voltage level at bin 31. It should be 800 mV RMS, but because of intrinsic quantization error, the sample set produces an RMS output of 800.127 mV.

Thus, according to Eq. (11.37), the sample set has an intrinsic gain of

$$G_{INTRINSIC} = 1.00016 \ \text{V/V}$$

Next, we need to consider the sampled-and-hold effect of the DAC on this sample set. As the frequency of the tone is F_t = 16 kHz / 512 × 31 or 968.75 Hz, and the sampling frequency is 16 kHz, we know from subsection 11.2.5 that the gain of the DAC at this frequency due to the sampled-and-hold operation is

$$G_{SH}\left(968.75\ \text{Hz}\right) = \left| \frac{\sin\left(\pi \dfrac{968.75\ \text{Hz}}{16\ \text{kHz}}\right)}{\left(\pi \dfrac{968.75\ \text{Hz}}{16\ \text{kHz}}\right)} \right| = 0.994 \ \text{V/V}$$

The expected RMS amplitude from the ideal DAC will then be 0.994 × 800.127 mV = 795.33 mV.

Finally, we load the sample set into digital source memory and start a digital pattern that sends the samples to the DAC at 16 kHz. A digitizer collects the output, and an FFT shows a signal level of 780-mV RMS. A perfect DAC would produce a sine wave at a signal level of 795.33 mV RMS. Since our DAC produced 780-mV RMS, the gain error according to Eq. (11.46) would be

$$\Delta G_{DAC} = \frac{780\ \text{mV}}{795.33\ \text{mV}} = 0.98072\,\text{V/V} = -0.1691\,\text{dB}$$

The absolute gain of a DAC is determined from Eq. (11.40) with intrinsic error compensation. Substituting the appropriate values, we obtain

$$G_{DAC,ACTUAL} \ \text{at 1 kHz} = \frac{3\ \text{V}}{256\ \text{bits}} \times \frac{780\ \text{mV}}{800\ \text{mV}} \times \frac{1}{1.00016\,\text{V/V}} = 11.42\ \text{mV/bit}$$

Fortunately, intrinsic error is usually very small; thus it is often ignored in gain and gain error calculations. However, intrinsic error may become much larger if a low-resolution converter is tested, if a low-level signal is to be tested, or if the number of samples is small. The test engineer should at least verify that intrinsic errors will be negligible before dropping intrinsic error correction from the gain calculations.

Intrinsic error does not apply to S/H amplifiers and switched capacitor filters, since these do not quantize the signal as they sample it. They are, however, subject to $\sin(x)/x$ roll-off since they produce a stepped version of the waveform samples. ADCs, on the other hand, quantize signals and are therefore subject to intrinsic error. However, ADCs collect instantaneous voltage values and therefore do not see $\sin(x)/x$ roll-off. Since many of the tests in this chapter are subject to intrinsic error and/or $\sin(x)/x$ effects, the test engineer should always keep these potential error sources in mind as the performance of sampled systems are measured. Often, the correlation errors between bench equipment and tester can be traced to a bench measurement that does not take such effects into account.

Exercises

11.12. An 8-bit unsigned binary formatted DAC has a full-scale range of 0.5 to 3.5 V. A sample set corresponding to 5-kHz sine wave at 0.75 V RMS is desired, assuming a DAC sampling rate of 32 kHz. An ideal analysis reveals an intrinsic gain error of –0.07 V/V. If the DAC output is digitized and the actual RMS output is found to be 0.81 V, what are the gain and gain error of the DAC at 5 kHz?   ANS. 14.2 mV/bit; 1.21 V/V (1.65 dB).

11.13. A 9-bit two's complement formatted ADC operating at an 8-kHz sampling rate has a full-scale range of 1 to 4 V. With a 1-V RMS sinusoidal signal at 3 kHz applied to its input, an analysis of the output codes indicates an RMS output value of 167.27 LSBs. Determine the gain and gain error of the ADC at 3 kHz.   ANS. 167.27 bits/V; 0.982 V/V (-0.16 dB).

Sampled channel gain tracking error is measured in a similar manner to analog channel gain tracking error. Intrinsic error is especially important in gain tracking measurements. It is important to realize that some of the variation in gain at different levels is caused by differences in quantization error. The intrinsic errors in each sample set should be extracted from the measurement of each level before calculating gain tracking error.

Intrinsic error and $\sin(x)/x$ corrections are usually not performed on any of the remaining tests in this chapter. These corrections are usually applied only to gain and absolute signal level measurements. In the remaining tests, $\sin(x)/x$ roll-off and intrinsic errors are usually considered part of the specified measurement. As usual, if there is any doubt about this issue, the test engineer should ask for clarification.

11.5.3 Frequency Response

Frequency response measurements of sampled channels differ from analog channel measurements mainly because of imaging and aliasing considerations. Since sampled channels often include an anti-imaging filter, the quality of this filter determines how much image energy is allowed to pass to the output of the channel. Frequency response tests in channels containing DACs, switched capacitor filters, and S/H amplifiers should be tested for out-of-band images that appear past the Nyquist frequency. In coherent DSP-based measurements, these images will appear at specific Fourier spectral bins, as explained in Section 11.2.5.

Notice that the digitizer used to measure these frequencies must sample at a high enough frequency to allow measurements past the Nyquist rate of the sampled channel. Also notice that each sampling process in a sampled channel has its own Nyquist frequency. An 8-kHz DAC followed by a 16-kHz switched capacitor filter has two Nyquist frequencies, one at 4 kHz and the other at 8 kHz. The images from the DAC must first be calculated. These images may themselves be imaged by the 16-kHz switched capacitor filter. Each of the primary test tones and the potential images should be measured.

The specification for a low-pass anti-imaging filter in a sampled channel may be stated in terms of the frequency response of the filter itself or it may be stated in terms of image attenuation of the total sampled channel. Since a sample-and-hold process introduces $\sin(x)/x$ roll-off, the images should appear even lower than the filter's gain curve would indicate. If image attenuation is specified rather than filter frequency response, then the test is a simple matter of comparing the amplitude of each in-band test tone with the amplitude of its image or images.

If the filter's frequency response is specified, then it can be measured in one of two ways. The best way is to provide design for test (DfT) access to the input and output of the filter and measure its frequency response using standard analog channel testing. The alternate test approach is to measure the attenuation of each test tone's first image compared to the ideal attenuation expected from $\sin(x)/x$ roll-off. The additional attenuation is due to the filter.

Unlike DAC channels, ADC channels do not suffer from $\sin(x)/x$ errors. Therefore, they can be measured without any additional compensation. However, ADC channels must be tested for

---

EXAMPLE 11.8

The 16-kHz DAC in the previous example is followed by a low-pass filter with a cutoff frequency of 8 kHz. The 1-kHz test tone is passed through the DAC and filter. Calculate the frequency of the first image of the 1-kHz test tone. A digitizer samples the filter output and sees a signal level of 780 mV at the 1-kHz frequency and a signal level of 5 mV at the out-of-band image frequency. Calculate the gain of the filter at the image frequency relative to the gain of the filter at 1 kHz.

Solution:

The image of the 1-kHz tone is located at $F_t = 16$ kHz − 968.75 Hz = 15031.25 Hz. Note that we will have to use a digitizer with a Nyquist frequency greater than 15031.25 Hz to see this frequency. If we used a digitizer with a 16-kHz sampling rate, then the digitizer would alias the 15031.25-Hz image back onto the test tone at 968.75 Hz, making the image measurement impossible. A logical choice for the digitizer sampling rate would be 32 or 64 kHz, since these are integer multiples of the DAC sampling rate.

Next, let us plot the spectral information that we are given in the problem, together with a block diagram that identifies each significant component of the AWG. This we do in Figure 11.28. The AWG/DAC combination is modeled with a frequency independent gain term $G'_{DAC,ACTUAL}$ and a sample-and-hold block, which accounts for the DAC's sampled-and-hold frequency dependency $G_{SH}(f)$. Of particular interest is the spectrum corresponding to the input to the DAC, denoted V'_{IN}. This particular spectrum is periodic; thus the test tone and all of its images will have equal amplitude, say, V_{RMS}. To keep the presentation relatively straightforward, we show only the test tone and its first image. Subsequently, the sample-and hold operation will modify the magnitude of the test tone and its image according to

$$V_{DAC,ACTUAL}\,(968.75\ \text{Hz}) = G_{SH}\,(968.75\ \text{Hz})G'_{DAC,ACTUAL}V_{RMS}$$

and

$$V_{DAC,ACTUAL}\left(15031.25\ \text{Hz}\right)=G_{SH}\left(15031.25\ \text{Hz}\right)G'_{DAC,ACTUAL}V_{RMS}$$

where the two gain factors are computed as follows

$$G_{SH}\left(968.75\ \text{Hz}\right)=\left|\frac{\sin\left(\pi\dfrac{968.75\ \text{Hz}}{16\ \text{kHz}}\right)}{\left(\pi\dfrac{968.75\ \text{Hz}}{16\ \text{kHz}}\right)}\right|=0.993981\ \text{V/V}=-0.052\ \text{dB}$$

and

$$G_{SH}\left(15031.25\ \text{Hz}\right)=\left|\frac{\sin\left(\pi\dfrac{15031.25\ \text{Hz}}{16\ \text{kHz}}\right)}{\left(\pi\dfrac{15031.25\ \text{Hz}}{16\ \text{kHz}}\right)}\right|=0.064061\ \text{V/V}=-23.868\ \text{dB}$$

$$V_{AWG}\left(15031.25\ \text{Hz}\right)=G_{filter}\left(15031.25\ \text{Hz}\right)G_{SH}\left(15031.25\ \text{Hz}\right)G'_{DAC,ACTUAL}V_{RMS}$$

Therefore, we can expect the relative gain of the first image amplitude to the test tone amplitude to be

$$\frac{V_{AWG}\left(15031.25\ \text{Hz}\right)}{V_{AWG}\left(968.75\ \text{Hz}\right)}=\frac{G_{filter}\left(15031.25\ \text{Hz}\right)G_{SH}\left(15031.25\ \text{Hz}\right)}{G_{filter}\left(968.75\ \text{Hz}\right)G_{SH}\left(968.75\ \text{Hz}\right)}$$

Subsequently, the gain of the filter at the image frequency relative to the gain of the filter at 1 kHz is derived from this equation to be

$$\frac{G_{filter}\left(15031.25\ \text{Hz}\right)}{G_{filter}\left(968.75\ \text{Hz}\right)}=\frac{V_{AWG}\left(15031.25\ \text{Hz}\right)G_{SH}\left(968.75\ \text{Hz}\right)}{V_{AWG}\left(968.75\ \text{Hz}\right)G_{SH}\left(15031.25\ \text{Hz}\right)}$$

Substituting known parameter values gives

$$\begin{aligned}\frac{G_{filter}\left(15031.25\ \text{Hz}\right)}{G_{filter}\left(968.75\ \text{Hz}\right)}&=\frac{V_{AWG}\left(15031.25\ \text{Hz}\right)G_{SH}\left(968.75\ \text{Hz}\right)}{V_{AWG}\left(968.75\ \text{Hz}\right)G_{SH}\left(15031.25\ \text{Hz}\right)}\\[4pt]&=\frac{5\ \text{mV}}{780\ \text{mV}}\ \frac{0.993981}{0.064041}\\[4pt]&=0.00994625\ \text{V/V}\\[4pt]&=-20.05\ \text{dB}\end{aligned}$$

Unfortunately, this answer may or may not correlate perfectly with a frequency response measurement of the continuous low-pass filter at these same frequencies. The potential error source is the shape of the DAC steps. If they are very sharp (i.e., very fast settling time), then the idealized sin(x)/x correction should be valid. However, if the steps of the DAC waveform are not sharp, there will be an additional low-pass filtering effect. The filter gain may not be accurately measurable using the sin(x)/x correction method in the previous example. This is the reason we usually prefer to measure both the absolute level of images relative to the reference tone, as well as measuring the true filter characteristics using a DfT test mode and analog channel test methodologies (see Chapter 16, "Design for Test"). This approach allows us to verify the filter characteristics separate from the DAC characteristics. The additional information gives us a more thorough characterization of the DAC/filter combination.

Figure 11.28. Modeling the frequency domain effects associated with the AWG.

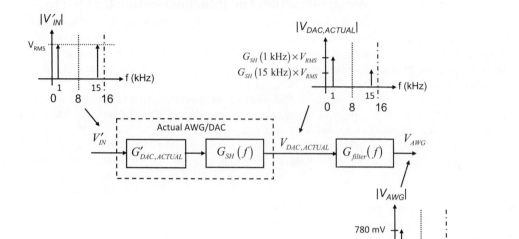

alias components rather than images. These alias components are likely to appear if the low-pass anitaliasing filter of the ADC channel is inadequate. Again, we prefer to measure both the filter in isolation (using a DfT test mode) and the alias components of the composite ADC/filter channel. The location of alias components in the ADC output spectrum is determined using the technique outlined in Section 11.3.

11.5.4 Phase Response (Absolute Phase Shift)

This is one of the more difficult parameters to measure in a mixed-signal channel (AIDO or DIAO). The problem with this measurement is that it is difficult to determine the exact phase relationship between analog signals and digital signals in most mixed-signal testers. The phase relationships are often not guaranteed to any acceptable level of accuracy. Also, the phase shifts through the analog reconstruction and anti-imaging filters of the AWGs and digitizers are not guaranteed by most ATE vendors. The solution to this problem is a complicated focused calibration process that is beyond the scope of this book. These problems are pointed out only as a warning to the new test engineer who might think that analog waveforms coming from an AWG or analog waveforms captured by a digitizer are exactly lined up with the digital samples coming out of a DUT or going into a DUT. Fortunately, phase response of mixed-signal channels is not a common specification.

Group delay and group delay distortion specifications are much more common; thus we will look more closely at these measurements.

11.5.5 Group Delay and Group Delay Distortion

These tests are much easier to measure than absolute phase shift, since they are based on a change-in-phase over change-in-frequency calculation. We can measure the phase shifts in a mixed-signal channel in the same way we measured them in the analog channel. The only difference between analog channel group delay measurements and mixed-signal channel measurements is a slight difference in the focused calibration process for this measurement.

11.5.6 Signal to Harmonic Distortion and Intermodulation Distortion

These tests are also nearly identical to the analog channel tests, except for the obvious requirement to work with digital waveforms rather than voltage waveforms. $\sin(x)/x$ attenuation is usually considered part of the measurement in distortion tests. In other words, if our third harmonic is down by an extra 2 dB because of $\sin(x)/x$ roll-off, then we consider the extra 2 dB to be part of the performance of the channel.

In ADC channels, we have to realize that some of the distortion components may fold back according to the rules of aliasing. We have to test these components just like any other distortion components. The following example shows how this is done.

When measuring DAC harmonic distortion frequencies, we do not have to worry about calculating alias frequencies as we did in the preceding ADC example. This is because the distortion

---

EXAMPLE 11.9

Five hundred twelve samples are collected from an ADC channel sampling at 8 kHz. A 3-kHz test tone (spectral bin 193) is applied to the input of the ADC. Calculate the frequency and spectral bin numbers of the second and third harmonics of this tone.

Solution:
The second and third harmonics would normally appear near 6 and 9 kHz. But since the Nyquist frequency of this channel is 4 kHz, we know these tones will fold back in band due to aliasing. Note that we may see distortion at these frequencies *even if the low-pass anti-aliasing filter is set to cut off everything above* 4 kHz. The reason for this is that the filter itself may be the source of the distortion, producing 6- and 9-kHz energy at its output which then gets aliased back into the 0 to 4 kHz band.

We could calculate alias frequencies using the traditional approach outlined in Section 11.3, but let us use a slightly different approach. Instead of converting all the tones from bin numbers into frequencies, let us work with bin numbers directly. We know our test tone is actually at a frequency slightly different from 3 kHz because we chose a prime bin number (bin 193). The Nyquist frequency exists at bin 256 (one half the sample size). The second and third harmonics would appear at bins $193 \times 2 = 386$ and $193 \times 3 = 579$, if these bins existed in a 512-point FFT (which they do not). We use a rule similar to the one in Section 11.3: Keep subtracting the number of samples from the bin number until the result is less than the number of samples. If the result is less than the Nyquist bin (number of samples divided by 2), the result is the alias bin number.

The second harmonic is at (nonexistent) bin 386, which is already less than 512 but greater than the Nyquist bin. Therefore, we subtract 386 from 512 to arrive at the alias bin of the second harmonic, bin 126. If there is any second harmonic distortion at the input to the ADC, or if the ADC itself generates second harmonic distortion, it will appear at bin 126, which corresponds to 1968.75 Hz. (If we were working with a multitone signal, we would need to make sure we did not have any other test tones at this frequency.)

We have to subtract 512 from the third harmonic once, to get an answer less than 512. The result is 579 – 512 = 67, which is less than the Nyquist bin; so we are done. The third harmonic should appear at bin 67. Bin 67 corresponds to 1046.875 Hz.

appears at the expected frequencies, rather than appearing at aliased frequencies. However, we do need to check to make sure we have no overlap between distortion components of one tone and images of other test tones caused by reconstruction.

11.5.7 Crosstalk

Crosstalk measurements in sampled systems are virtually identical to those in analog channels. The difference is that we have to worry about the exact definition of signal levels. If we have two identical DAC channels or two ADC channels, then we can say the crosstalk from one to the other is defined as the ratio of the output of the inactive channel divided by the output of the active channel. But what if the channels are dissimilar? If we have one DAC channel that has a differential output and it generates crosstalk into an ADC channel with a single-ended input, then what is the definition of crosstalk? Is it the single-ended level of the DAC divided by the ADC digital output, converted into equivalent input volts? Generally, crosstalk is defined by a ratio of voltages (converted to decibels). When working with digital samples, they are usually converted into volts using either the ideal gain of the converter or the actual gain of the converter. For example, the ratio of an ADC output, converted to RMS volts, relative to a DAC output in RMS volts, would be a good guess for the definition of DAC to ADC crosstalk. But this is not a solid rule to follow. The point is that the test engineer has to make sure that the data sheet clearly spells out the definition of crosstalk when dissimilar channels are involved.

One difference between analog channel crosstalk measurements and quantized channel crosstalk measurements relates to ADC quantization. A very low-level crosstalk signal may not be large enough by itself to toggle one LSB of a low-resolution ADC in a quiet state. For this reason, an ADC with an inactive DC input can mask low-level crosstalk. An AC dithering signal is sometimes applied to the ADC input rather than a DC signal. A low level sine wave added to the input to an ADC allows the very small crosstalk signal to appear as part of a multitone signal when it otherwise might be invisible by itself. However, a dithering signal is sometimes unnecessary. The noise in the DUT is often high enough to act as a dithering signal, overcoming the quantization masking effect. If the output of the ADC toggles by several LSBs because of noise, or if the crosstalk is well above 1 LSB in amplitude, then dithering is unnecessary. However, if the ADC output is only toggling between one or two LSBs, dithering may improve accuracy and repeatability of a crosstalk measurement.

DAC quantization, unlike ADC quantization, does not act as a hiding place for crosstalk signals. When measuring crosstalk into a DAC output, the DAC input is set to zero or midscale, and the crosstalk appears as a nonquantized continuous-time signal.

11.5.8 CMRR

DACs do not have differential inputs; thus there is no such thing as DAC CMRR. ADC channels with differential inputs, on the other hand, often have CMRR specifications. ADC CMRR is tested the same way as analog channel CMRR, except that the outputs are measured in RMS LSBs and gains are measured in bits per volt. Otherwise the calculations are identical. Like crosstalk, ADC CMRR tests may be affected by quantization masking effects. A dithering source can be used at the input to the ADC to uncover CMRR components below the 1 LSB level.

11.5.9 PSR and PSRR

Unlike analog channels, DAC and ADC channels do not have both PSR and PSRR specifications. A DAC has no analog input, and therefore no V/V gain. For this reason, it has PSR, but no PSRR. For similar reasons, ADCs have PSRR but no PSR. Unfortunately, data sheets usually list DAC channel PSRR, meaning PSR, but that is a minor semantic issue. Here is the definition of the ADC supply rejection

$$PSRR_{ADC} = 20 \ \log_{10}\left(\frac{D_{ADC}/V_{ripple}}{G_{ADC}}\right) dB \tag{11.48}$$

where

 V_{ripple} = RMS voltage added to power supply (typically around 50 to 100-mV RMS)

 D_{ADC} = ADC digital output expressed in RMS LSBs with V_{ripple} added to power supply

 ADC gain = gain of ADC from normal input to output, in bits per volt

The definition of DAC PSR is even simpler and is given by

$$PSR_{DAC} = 20 \ \log_{10}\left(\frac{V_{out}}{V_{ripple}}\right) dB \tag{11.49}$$

ADC PSRR is typically measured with the input grounded or otherwise set to a midscale DC level. However, like crosstalk, the ripple from a power supply may not be large enough to appear at the output of a low-resolution ADC with an inactive DC input. A dithering signal can be added to the input of the ADC to allow an accurate measurement of PSRR. DAC PSR is often measured with the DAC set to a static midscale value. However, it is important to realize that DACs may be more sensitive to supply ripple near one end of their scale, usually the most positive setting. PSR specs apply to worst-case conditions, which means that the DAC should be set to the DC level that produces the worst results. This level can be determined by characterization, combined with knowledge of the DAC architecture. ADCs may also suffer from worst results near one end of the scale or the other; thus they should be characterized as well.

11.5.10 Signal-to-Noise Ratio and ENOB

In sampled channels, signal-to-noise ratio (SNR) is again tested in a manner almost identical to that in analog channels. The output of the converter is captured using a digitizer or capture memory. The resulting waveform is analyzed using an FFT and the signal-to-noise ratio is calculated as in an analog channel. In this case, we do not care whether we are working with volts or LSBs, because SNR is a ratio of similar values.

 Excessive noise in an ADC or DAC can make it appear to have fewer bits of resolution than it actually has. For example, a 23-bit ADC that has only 98 dB of signal-to-noise ratio with a full-scale sine-wave input might as well have only 16 bits of resolution. This is because a perfect 16-bit converter has a SNR of about 98 dB. The apparent resolution of a converter based on its signal-to-noise ratio is specified by a calculation called the *equivalent number of bits,* or *effective number of bits* (ENOB). The ENOB is related to the SNR by the equation

$$ENOB = \frac{SNR\,(dB) - 1.761\,dB}{6.02\,dB} \qquad (11.50)$$

Exercises

11.14. An ADC channel is sampled at 16 kHz and 256 samples are collected. A 7.68-kHz test tone (spectral bin 123) is applied to the input of the ADC. Calculate the frequency and spectral bin numbers of the second and third harmonics of this tone.
ANS. 0.625 kHz, bin 10; 7.0625 kHz, bin 113.

11.15. A 2's complement formatted ADC has a nominal gain of 73.14 bits/V. With its input shorted to a DC midsupply voltage and a 100-mV RMS sinusoidal signal added to the power supply, the RMS digital output of the channel is found to be 1.15 LSBs. Determine the PSRR of the ADC.
ANS. −16.1 dB.

11.16. A sampled channel has 9.3 equivalent number of bits of resolution. What is the corresponding SNR of the channel?
ANS. 57.74 dB.

11.5.11 Idle Channel Noise

Idle channel noise (ICN) in DAC channels is measured the same way as in analog channels, except the DAC is set to midscale, positive full scale, or negative full scale, whichever produces the worst results. Usually there is not much difference in ICN results at different settings; so the DAC is simply set to midscale. Like analog channel ICN, DAC channel ICN is usually measured in RMS volts over a specified bandwidth. Often DAC ICN is specified with a specific weighting filter, as discussed in Chapter 9.

ICN testing of ADCs again involves quantization effects. Unfortunately, a dithering source in this case would introduce additional quantization noise, destroying the ICN measurement. If we instead apply a DC level, the ADC will produce different amounts of noise depending on the exact DC level and the ADC's own DC offset. If the ADC input is midway between two decision levels, we may get a fixed DC code out of the ADC and our ICN measurement will be zero LSBs RMS. If the input is equal to one of the ADC's decision levels, then we will get a random dithering between two levels, resulting in an unweighted ICN measurement of 1/2 LSBs RMS. So the exact DC offset will make the ICN measurement vary wildly. Despite this seeming flaw in test definition, this is how ADC ICN is measured.

Correlation can be a nightmare in ADC ICN tests. Extreme care must be taken to provide the exact DC input voltage specified in the data sheet during an ICN measurement. Otherwise the test results will be completely wrong. Because of the extreme sensitivity of an ADC ICN measurement to DC offset, some ADC channels include an auto-zero or squelch function to reduce the ICN of a DC input to zero regardless of the input offset. ICN in these devices is zero by design, as long as the auto-zero or squelch function is operational.

11.6 SUMMARY

DSP-based measurements of sampled channels are very similar to the equivalent tests in analog channels. The most striking differences relate to bit/volt gains and scaling factors, quantization effects, aliasing, and imaging. We also have to deal with a new set of sampling constraints, since the DUT must now be synchronized with the ATE tester's sampling system. Coherent testing requires that we interweave the DUT's various sampling rates with the sampling rates of the ATE tester instruments. Often this represents one of the biggest challenges in setting up an efficient mixed-signal test program.

Another difference between analog channel tests and sampled channel tests is in the focused calibration process, which we have only mentioned briefly. Focused calibrations provide the additional accuracy that the ATE vendor may not include with an off-the-shelf, general-purpose ATE tester. While focused calibrations are sometimes unnecessary in measurements requiring limited accuracy, a good command of focused calibration techniques is a must for the professional test engineer. Readers can find more on this subject in the first edition of this textbook. Space constraints forced its removal from this edition.

PROBLEMS

11.1. A codec consisting of a DAC and ADC is operating at a 32-kHz sampling rate. An AWG has a maximum operating frequency of 100 kHz and an allocated source memory capacity of 4096 samples. The digitizer has a maximum operating frequency of 200 kHz and an allocated captured memory capacity of 2048 samples. Select the appropriate test parameters so that the AWG and the ADC are coherent. Likewise, determine the test parameters of the DAC and the digitizer.

11.2. A DAC and digitizer are arranged according to the values listed in the following table to work with a test tone of approximately 6.15 kHz:

	DAC	Digitizer
Sampling rate	24,000 Hz	44,000 Hz
Number of samples	2000	1024
Spectral bin	513	143

Slightly adjust the sampling rate of each system so that the two sampling systems are coherent. Due to sampling constraints inherent to the tester, both the DAC and digitizer sampling rates must be integer multiples of 1 Hz. Determine the new sampling rates of the DAC and the digitizer.

11.3. An AWG and ADC are arranged according to the values listed in the following table to work with a test tone of approximately 15 kHz:

	AWG	ADC
Sampling rate	128,000 Hz	44,000 Hz
Number of samples	1500	1024
Spectral bin	175	349

Slightly adjust the sampling rate of each system so that the two sampling systems are coherent. Due to sampling constraints inherent to the tester, both the AWG and ADC sampling rates must be integer multiples of 3 Hz. Determine the new sampling rates of the AWG and the ADC.

11.4. For the codec test setup shown in Figure 11.13, the ADC must be tested with a sampling rate of 16 kHz and a sine wave of approximately 4.2 kHz. The AWG has a memory of only 1024 samples. What should the AWG sample rate be to establish coherence with the ADC if the sample rate must be a multiple of 1 Hz? How many samples should be collected by the waveform capture memory?

11.5. A 1-V RMS tone of 6.5 kHz is reconstructed with a DAC at a sampling rate of 16 kHz. Determine the RMS amplitude of the in-band test tone. What gain factor should be used to correct for the sampled-and-hold operation?

11.6. Samples from a 0.3-V RMS sine wave with a frequency of 5 kHz is reconstructed with a DAC operating a sampling rate of 12 kHz. What is the RMS amplitude of the in-band test tone and the RMS amplitude of the three lowest-frequency images?

11.7. Samples from a 2.5-V RMS sine wave with a frequency of 1 kHz are reconstructed with a DAC operating a sampling rate of 32 kHz, followed by a first-order low-pass filter having the following frequency response,

$$G(f) = \frac{1}{\sqrt{1 + (f/1000)^2}}$$

What is the RMS amplitude of the in-band test tone and the RMS amplitude of the lowest-frequency image?

11.8. A 1-V RMS sine wave with a frequency of 55 kHz is sampled by an ADC at a sampling rate of 24 kHz. Sketch the magnitude of the sampled spectrum that includes the six lowest frequencies related to this test tone.

11.9. A 1-V RMS sine wave with a frequency of 63 kHz is sampled by an ADC at a sampling rate of 24 kHz. Sketch the magnitude of the sampled spectrum that includes the six lowest frequencies related to this test tone.

11.10. A 1-V RMS two-tone multitone signal with frequencies of 55 and 63 kHz is sampled by an ADC at a sampling rate of 24 kHz. Sketch the RMS magnitude of the spectrum that includes the six lowest frequencies that could alias into the same spectral bin as the test tones. Do these two test frequencies overlap below the Nyquist frequency? Ignore any harmonic or intermodulation effects. Do the test tones overlap if the test tones are shifted to 55 and 65 kHz?

11.11. A test signal with frequency of 160.5625 MHz is sampled by an ADC at a sampling rate of 32 MHz. Assuming that 512 samples are collected, is the measurement coherent? If so, how many cycles of the original waveform is aliased and what is the effective sample period of this measurement setup?

11.12. A test signal with frequency of 91 MHz is sampled by an ADC at a sampling rate of 32 MHz. Assuming that 512 samples are collected, is the measurement coherent? If so, what how many cycles of the original waveform are aliased and what is the effective sample period of this measurement setup?

11.13. A test signal with frequency of 63.4 MHz is sampled by an ADC at a sampling rate of 32 MHz. Assuming that 512 samples are collected, is the measurement coherent? If so, what how many cycles of the original waveform is aliased and what is the effective sample period of this measurement setup?

11.14. A test signal is to be undersampled using a digitizer with a sampling frequency of 32 MHz. Assuming that the data sequence should complete 60 cycles in 512 points, at which frequency should the input signal be set? What is the corresponding effective sampling rate if the data set is used to reconstruct one cycle of the orginal high-frequency signal?

11.15. Compute the following quantities:

(a) 514 mod 256, (b) 103455 mod 2, (c) 33.2 mod 32, (d) 2033 mod 1024.

11.16. A sinusoidal test signal with frequency of 63.5 MHz is sampled by an ADC at a sampling rate of 32 MHz. Assuming 512 samples are collected, reconstruct the sampled signal over one period of the test signal.

11.17. A sinusoidal test signal with frequency of 61.5 MHz is sampled by an ADC at a sampling rate of 32 MHz. Assuming that 512 samples are collected, reconstruct the sampled signal over one period of the test signal.

11.18. A test signal consisting of 1-V amplitude tones, one at 61.5 MHz and another at 63.5 MHz is sampled by an ADC at a sampling rate of 32 MHz. Assuming that 512 samples are collected, reconstruct the sampled signal over the 1/61.5 μs period of the test signal.

11.19. Using MATLAB or some other software determine the intrinsic error of a quantized low-level sine wave with an RMS amplitude of 4 LSBs. Assume the sine wave is described by parameters $M = 1$, $N = 32$, $A = 4\sqrt{2}$, and $P = 0$. Repeat for $M = 8$.

11.20. A set of samples derived from a 0.6-V RMS sinusoidal signal is found to have an intrinsic gain error of -0.045 V/V. What is the actual amplitude of the test signal?

11.21. Assuming a perfect DAC, write a MATLAB routine, or some other equivalent software routine, that produces a 512-point sample set that will generate a 4-kHz sine wave at 500- mV RMS, at a DAC sampling rate of 16 kHz. What is the intrinsic gain error of this sampling set?

11.22. A sample set corresponding to an approximately 3-kHz sine wave at 0.5 V RMS is desired. Assuming a perfect DAC operating at a sampling rate of 16 kHz, write a MATLAB routine, or some other equivalent software routine, that produces a 1024-point sample set that will generate this desired signal. What is the intrinsic gain of this sampling set?

11.23. An 8-bit sign/magnitude formatted DAC has a full-scale range of 1.0 to 4.0 V. A sample set corresponding to a 12-kHz sine wave at 0.65 V RMS is desired, assuming a DAC sampling rate of 32 kHz. An ideal analysis reveals an intrinsic gain error of –0.045 V/V. If the DAC output is digitized and the actual RMS output is found to be 0.71 V, what are the gain and gain error (in dB) of the DAC at 12 kHz?

11.24. A 10-bit 2's-complement formatted ADC has a full-scale range of 0 to 5 V. A signal level of 1-V RMS is applied to the ADC input at a frequency of 4.2 kHz. The ADC output is measured and the actual RMS output is found to be 203.5 LSBs. What are the gain and gain error of the ADC at 4.2 kHz?

11.25. A 64-kHz DAC is followed by a low-pass filter with a cutoff frequency of 32 kHz. A 15-kHz test tone is passed through the DAC and filter. Calculate the frequency of the second image of the 15-kHz test tone. A digitizer samples the filter output and sees a signal level of 250 mV at the 15-kHz frequency and a signal level of 0.1 mV at the second image frequency. Calculate the gain of the filter at the second image frequency relative to the gain of the filter at 15 kHz.

11.26. A 12-kHz DAC is followed by a low-pass filter with a cutoff frequency of 6 kHz. A 15-kHz spurious signal appears at the input of the DAC. At what in-band frequency does the spurious tone appear?

11.27. An ADC channel is sampled at 16 kHz and 1024 samples are collected. A 2.5-kHz test tone (spectral bin 161) is applied to the input of the ADC. Calculate the frequency and spectral bin numbers of the fifteenth and twenty-third harmonics of this tone.

11.28. A two-tone multitone signal consisting of frequencies 3 kHz (spectral bin 191) and 3.2 kHz (spectral bin 205) is applied to the input of an ADC. The ADC channel is sampled at

8 kHz and 512 samples are collected. Calculate the frequency and spectral bin numbers of the second- and third-order intermodulation distortion components. Is their any frequency overlap between these distortion components and the images created by reconstruction?

11.29. A 1-V RMS sinusoidal signal at 1 kHz is applied to the input of an ADC. The output of the ADC is analyzed, resulting in an RMS digital output of 52.33 LSBs. Then the input is shorted to a DC midsupply voltage and a 100-mV RMS sinusoidal signal is added to the power supply. The output of the channel is again analyzed, resulting in an RMS digital output of 0.125 LSBs. Determine the gain of the ADC and its corresponding PSRR.

11.30. A digitizer sampling at 4 kHz captures 16 samples of a 0.5-kHz sinusoidal signal corrupted by noise from a DAC channel. An FFT analysis reveals the following spectral amplitudes in bins 0 to 8:

$$c_{RMS} = [2300, 32, 707000, 12, 15, 31, 12, 11, 4] \ \mu V$$

Calculate the total RMS level of the noise associated with this channel measurement. Determine the signal-to-noise ratio of the channel. Express this result in terms of its equivalent number of bits (ENOB).

REFERENCES

1. M. Mahoney, *DSP-Based Testing of Analog and Mixed-Signal Circuits*, The Computer Society of the IEEE, Washington, D.C., 1987, ISBN 0818607858, pp. 179–199.

2. F. H. Irons and D. M. Hummels, Modulo time plot—A useful data acquisition diagnostic tool, *IEEE Transactions on Instrumentation and Measurement*, **45**, pp. 734–738, June 1996.

CHAPTER 12

Fundamentals of RF Testing

In the first chapter of the RF portion of this textbook, we will review the physical basis of an RF signal so that we can better understand the RF test techniques introduced in the following chapter. We will start with the fundamentals of waves, which are the most efficient way of describing power transport within the RF frequency spectrum, followed by a vector description of an RF signal with scattering parameters.

12.1 INTRODUCTION TO RF TESTING

Radio transmission is over 100 years old, and the frequency bands for wireless transmission have been allocated and reallocated as technologies have progressed. In the beginning, there was only an AM modulated radio transmission, but in the last decades the technology has together with low cost manufacturing enabled a large variety of RF systems, like cellular telephones, pagers, satellite radio, or data services like WLAN, WiMAX, or UWB. Many of these applications had been previously served by wired application, but with the technology available, more and more of these services became wireless and commercially available in high volume. These and many other services use the electromagnetic spectrum depicted in Figure 12.1. A more detailed frequency allocation can be requested from national telecommunication organizations, like FCC in the US1 or ETSI2 in Europe, to name just a few.

With progressing technology, it was also commercially valid to build systems covering a wide power range for personal two-way radios. The low-power range had been accessible for a long time with radio receivers, while the high-power range had been used for transmitting radio stations. Modern personal two-way radio systems will continue to use the lower end of the power range, because the high end is limited due to (a) supply power restrictions and (b) as well as concerns over the health effects of high-power, high-frequency radio waves. The power range used for cellular phones or data services like WiMAX is in the range of a few watts.

There are multiple aspects the RF test engineer will have to consider when working on new devices. One is the extreme dynamic range covered with RF devices, ranging from noise measurement in the sub-attowatt range (10^{-18} W) up to a few watts when testing commercial power

Figure 12.1. Frequency spectrum of some commercial RF devices.

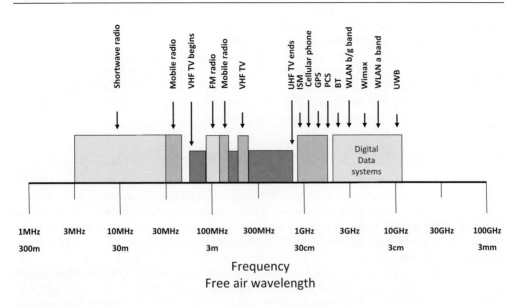

Frequency
Free air wavelength

Figure 12.2. Power spectrum of consumer and commercial RF devices.

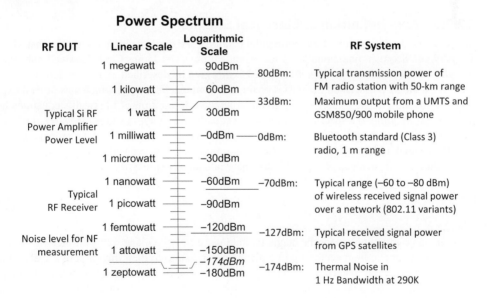

amplifiers, as depicted by the illustration in Figure 12.2. Commercial ATE systems typically cover a dynamic range as high as 170 dB without any external amplifier or attenuator.

A second RF differentiator to digital and mixed-signal testing is the huge bandwidth the test system needs to cover. Existing commercial ATE already cover the 12-GHz range for testing UWB devices. Recently, RF devices in the >50-GHz range have become commercially available for radar detection in automobiles. The authors assume that these frequency ranges will be around in the foreseeable time frame for data transmitting systems and other commercial products. In this chapter, we will limit our focus to test techniques in the sub-12-GHz range, however, the physics are still applicable to the high-gigahertz frequency range.

A third aspect that we will discuss in great depth is the physics of high frequency signals and its corresponding power. At a lower frequency, the power associated with an electrical signal can be expressed in terms of electron flow and electron position, while it will be more convenient at high frequency to describe the physical phenomena as waves. Later in this chapter, the scattering parameter will be explained based on the concept of waves.

12.2 SCALAR VERSUS VECTOR MEASURES

In RF test technology, we typically have two kinds of values: scalar and vector. Scalar values have by definition no direction and can be expressed by a single point in an n-dimensional space, while a vector has a magnitude and direction. Interesting enough, the magnitude of a vector can be described by a scalar value. In many cases in RF test technology, the magnitude of the wave is sufficient to describe the test metric, however, in some cases it will be necessary to take vector components into account.

Since time is a scalar (a nondirected value), we can assume that time-dependent performances, like magnitude over time as we have seen previously in Chapter 9, is a scalar measurement. However, it is best to describe multiple phenomena with waves (like the reflection we know from the propagation of light) in RF. These waves are not only time dependent, but also have a direction. Therefore we should treat electrical waves and their phenomena as vector values.

12.2.1 Wave Definition of Electrical Signals

Based on the definitions and the experimental results of the English chemist and physicist Michael Faraday, the Scottish mathematician and theoretical physicist James Clerk Maxwell published *A Treatise on Electricity and Magnetism* in 1873, in which he described all known dependencies of electricity. He also described the basics of the electromagnetic theory, which was proven experimentally in subsequent years. We will avoid a detailed description of Maxwell's equations here, and instead we jump to a specific result of these equations as it applies to plane waves (i.e., waves that have the same magnitude at any point in a plane perpendicular to the wave propagation).

Based on the Maxwell's equation, the equation for a plane wave with electric field E traveling in one direction in space in a homogeneous material can be expressed as

$$\frac{\partial^2 E}{\partial x^2} = \frac{1}{c^2} \frac{\partial^2 E}{\partial t^2} \tag{12.1}$$

The same form applies for the magnetic field wave B. For a plane wave, the electric field vector and the magnetic field vector are in a plane perpendicular to their direction of travel and they are also perpendicular to each other in this plane. In this discussion they are assumed to be traveling in the x direction. The symbol c represents the speed of the electromagnetic wave.

A solution to Eq. (12.1) for the electrical field E can easily be shown to be

$$E = E_m \sin\left(kx - \omega t\right) \tag{12.2}$$

and, similarly, for the magnetic field it can be shown as

$$B = B_m \sin\left(kx - \omega t\right) \tag{12.3}$$

Figure 12.3 is an illustration of a transverse electrical and magnetic wave (TEM) traveling in the *x*-direction as described by Eqs. (12.2) and (12.3). It is interesting to note that any linear combination of any number of sinusoids will also satisfy Eq. (12.1), hence, we can imagine through the concept of a Fourier series that plane waves can take on arbitrary shape.

12.2.1 Measures of Electrical Waves

Accepting the premise thaat energy is transported in RF systems in the form of propagating waves as described by Maxwell equation; we will now consider several attributes and metrics associated with RF signals.

A sinusoidal wave at some arbitrary point in space can be expressed in the time domain as

$$v_n(t) = A_n \sin(\omega_n t + \Theta_n) \tag{12.4}$$

or with complex notational form as

$$v_n(t) = \Re e \left\{ A_n \, e^{j(\omega_n t + \Theta_n)} \right\} \tag{12.5}$$

where A_n is the amplitude, $\omega_n = 2\pi f_n$ is the frequency and Θ_n is the phase lead relative to a defined reference. The wavelength λ of a propagating wave is defined as the distance in space over which the wave shape repeats as depicted in Fig. 12.4.

For the case of a sinusoidal wave of frequency f propagating in free space, the wavelength λ_0 is defined as

$$\lambda_0 = \frac{c_0}{f} = \frac{2.998 \times 10^8 \, [\text{m/s}]}{f \, [\text{Hz}]} \approx \frac{30 \, [\text{cm/s}]}{f \, [\text{GHz}]} \tag{12.6}$$

Figure 12.3. RF energy transport through space by perpendicular electrical and magnetic fields.

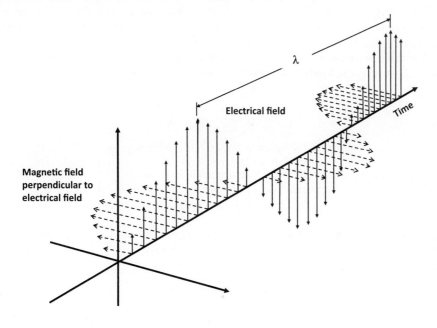

Figure 12.4. Wavelength λ is the distance in space between the start and the end of a complete sinusoidal wave.

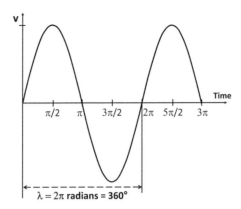

$\lambda = 2\pi$ **radians = 360°**

when c_0 is the phase velocity of an electromagnetic wave in vacuum. Using Eq. (12.6), we see that the wavelength of a 6-GHz WLAN signal in vacuum is about 5 cm.

If the same electromagnetic wave is propagating in an arbitrary nonmagnetic, homogeneous lossless dielectric medium, the propagation speed, denoted by v_p, will decrease with increasing relative dielectric constant ε_r value, as described by

$$v_p = \frac{c_0}{\sqrt{\varepsilon_r}} \tag{12.7}$$

The dielectric constant ε_r is the ratio of the electric field storage capability of the nonmagnetic, homogeneous and lossless material compared to that of free space, and it is a dimensionless number greater than 1. For optical waves, this dielectric constant ε_r is the square of the refractive index n. At this point we are using the dielectric constant ε_r as a constant; however, this parameter is highly dependent on frequency. It is for this reason that RF capacitor values need to be carefully chosen based on their operating frequency.

For nonhomogeneous media, like a microstrip or coplanar RF line, we will use the effective dielectric constant ε_{eff} instead of the dielectric constant ε_r. With this effective dielectric constant ε_{eff}, the effective wavelength λ_{eff} becomes

$$\lambda_{eff} = \frac{\lambda_0}{\sqrt{\varepsilon_{eff}}} \approx \frac{30[\text{cm/s}]}{f\,[\text{GHz}] \cdot \sqrt{\varepsilon_{eff}}} \tag{12.8}$$

An RF transmission line on FR4 board material might have an effective dielectric constant ε_{eff} of about 3.4. With this effective dielectric constant, the same WLAN 6-GHz signal wave, which had a wavelength of ~5 cm in free space, will have an effective wavelength of 2.7 cm (~1.06 in) on the board.

The fractional wavelength (FW) in percentage can be calculated using l, the physical wavelength of the line, and the effective wavelength λ_{eff} according to

$$FW = \frac{l}{\lambda_{eff}}100\% \quad [\%] \tag{12.9}$$

or it might be calculated in terms of phase Θ, expressed in degrees, as

$$\Theta = \frac{l}{\lambda_{eff}} 360° \quad [\text{degrees}] \tag{12.10}$$

A common rule of thumb is that a RF transmission line can be treated as a connection of two nodes with the same voltage at all points at any given time if the fractional wavelength is < 5%. For a transmission line on FR4 material with the $\varepsilon_{eff} = 3.4$ mentioned above, this would be given for frequencies up to 813 MHz for 10 mm (~0.39 in) connecting lines. For our 6-GHz WLAN signal the 5% rule would be equal to a line length of

$$l_{5\%} = FW \cdot \lambda_{eff} = FW \cdot \frac{30[\text{cm/s}]}{f[\text{GHz}] \cdot \sqrt{\varepsilon_{eff}}} = 0.05 \cdot \frac{30[\text{cm/s}]}{6[\text{GHz}] \cdot \sqrt{3.4}} = 1.35\,\text{mm} \tag{12.11}$$

For connecting lines longer than 1.35 mm (~0.053 in), the connecting line should be treated as a set of distributed elements rather than an lumped LCR equivalent.

EXAMPLE 12.1

What is the free space wavelength of an electromagnetic wave of a GPS signal with a frequency of 1575.42 MHz?

Solution:
Using Eq. (12.6),

$$\lambda_0 = \frac{c_0}{f} = \frac{2.998 \times 10^8 \,[\text{m/s}]}{1575.42 \times 10^6 \,[\text{Hz}]} = 190.3 \text{ mm}$$

The wavelength of a GPS signal in free space is 190.3 mm.

EXAMPLE 12.2

An RF transmission line on FR4 board material is used to delay a signal with a frequency of 1575.42 MHz by 90 degrees. The effective dielectric constant for this material is $\varepsilon_{eff} = 3.4$. What line length should be used?

Solution:
First, using Eq. (12.8), we calculate the wavelength of the signal as

$$\lambda_{eff} = \frac{1}{\sqrt{3.4}} \cdot \frac{2.998 \times 10^8 \,[\text{m/s}]}{1575.42 \times 10^6 \,[\text{Hz}]} = 103.2 \text{ mm}$$

For a 90-degree phase shift we require one-quarter of this wavelength, resulting in

$$l_{90\,deg} = \frac{103.2 \text{ mm}}{4} = 25.8 \text{ mm}$$

The board trace needs to be made 25.8 mm long.

Exercises

12.1. What is the frequency of a wave that has a wavelength of 10 cm in free space? What is the corresponding frequency if the wave is traveling in water with a relative dielectric constant of 80.1 at 20 degrees centigrade?

ANS. 2.998 GHz; 2.998 GHz (no change in frequency, because speed and wavelength change in equal proportions).

12.2. An RF transmission line on Roger's printed circuit board material is used to delay a signal with the frequency of 6 GHz by 45 degrees. The effective dielectric constant for this material is $\varepsilon_{eff} = 2.2$. What line length is required?

ANS. 4.2 mm.

12.3. An RF transmission line constructed on an FR4 printed circuit board is 10 mm in length. If the RF line is excited by a signal with frequency of 1 GHz do we treat this line as a distributed set of elements or as a lumped LRC equivalent assuming the 5% line length rule.

ANS. $l_{5\%} = 8.1$ mm; because the line length of 10 mm is greater than $l_{5\%}$, the line must be treated as a distributed line.

12.2.2 Power Definition

In the previous subsection, we described several attributes of a traveling plane wave: wavelength, frequency, and amplitude. This subsection will describe the concept of scalar power and voltage. The goal here is to gain an understanding how the shape of the waveform impacts a power measurement. It is important to understand expressions for average and peak power, as well as understand the role that they have when conducting a power measurement with an ATE.

In general, in a conservative field, voltage is defined as the line integral over an electrical field of an electromagnetic field. Sparing the reader the mathematical details, a voltage wave has a form of expression similar to that of Eq. (12.1); that being, a sinusoidal function of both time and space.

At low frequencies, the well-known Ohm's law relating a current signal to a voltage signal, that is, $v(t) = R \cdot i(t)$, provides a convenient shorthand solution to Maxwell's electromagnetic field equations. Maxwell's equation shows that the voltage and current are actually waves dependent on the frequency and the medium in which they are propagating through. At these low frequencies, the definition of power is the product of the instantaneous current and the instantaneous voltage or, together with Ohm's law, can be written in the following two ways:

$$p(t) = i(t) \cdot v(t) = \frac{v^2(t)}{R} \tag{12.12}$$

Often in RF circuit literature, the resistance of interest involved in the power transfer process is denoted with symbol Z_0. Since it is customary to use the capital letter Z for impedance, it is assumed that this impedance is real-valued. Using this notation, the power expression often appears as

$$p(t) = \frac{v^2(t)}{Z_0} \tag{12.13}$$

Figure 12.5 illustrates the instantaneous voltage and current waveforms corresponding to a voltage-source–resistive-load configuration. Here the voltage and current waveforms are in-phase. In contrast, Figure 12.6 illustrates the instantaneous voltage and current waveforms corresponding to a voltage-source–arbitrary-load configuration (i.e., one with resistive, capacitance and inductive components). Here we see the voltage and current waveforms are out-of-phase by some angle ϕ. For both cases, the instantaneous power $p(t)$ are seen superimposed on these two plots. In the case of the resistive load or in-phase situation, the instantaneous power is always positive in value, whereas, in the situation depicted in Figure 12.6 for the out-of-phase case, the instantaneous power goes negative at some points along the waveform. Positive instantaneous power refers to power being delivered by the source to the load, and negative instantaneous power refers to power being delivered to the source by the load. Since the load is passive, one would not expect a net flow of power to be delivered by the load to the source. A measure that describes this behavior is average power, P_{avg}.

Average power is defined as the average value of the instantaneous power waveform taken over some integration time, T_r. As the meaning of an average value depends on the integration time, it is important to standardize on this quantity. We encountered this same issue previously with DSP-based testing back in Chapter 9. There we spoke in terms of coherency; that is, input

Figure 12.5. Time dependent power as a function of in-phase voltage and current waves for a voltage source—resistive load configuration.

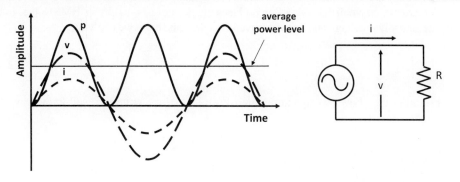

Figure 12.6. Time dependent power as a function of out of phase voltage and current waves for an arbitrary voltage source—arbitrary load configuration.

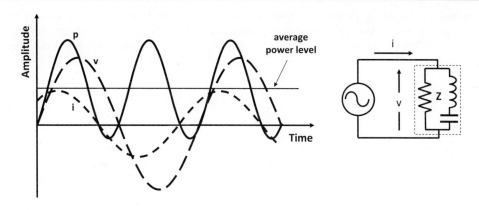

and output signals are coherent with respect to the unit test period (i.e., integration time). RF power measurements follow similar coherency constraints. For sinusoidal signals with period T_o, referred to as continuous wave (CW) sinusoidal signals in RF literature, as opposed to pulsed signals that also appear, we can state this same coherency condition by ensuring that the integration time is an integer multiple n of the period T_o. Assuming that the magnitude of the sinusoidal voltage wave is denoted by V_{peak} and magnitude of the sinusoidal current wave is described by I_{peak}, the average power is given by

$$P_{avg} = \frac{1}{nT_0} \int_0^{nT_o} V_{peak} \sin\left(\frac{2\pi}{T_0} t\right) \cdot I_{peak} \sin\left(\frac{2\pi}{T_0} t + \phi\right) dt \qquad (12.14)$$

where ϕ is the phase difference between the voltage and the current wave. Because n is an integer, the average power reduces to

$$P_{avg} = \frac{V_{peak} \cdot I_{peak}}{2} \cos(\phi) \qquad (12.15)$$

Here we see that P_{avg} is not a function of frequency and is dependent only on the phase difference ϕ, the magnitude of the voltage wave V_{peak}, and current wave I_{peak}. The average power for the in-phase and out-of-phase cases depicted in Figure 12.5 and Figure 12.6 is superimposed on each plot. As expected, the average power delivered to the load is positive in both cases.

As a matter of convenience, one often replaces the peak of a signal by its RMS value. In the case of the sinusoidal voltage and current waveforms describe here, we can write

$$V_{peak} = \sqrt{2}\, V_{RMS} \quad \text{and} \quad I_{peak} = \sqrt{2}\, I_{RMS} \qquad (12.16)$$

Substituting the above two quantities into Eq. (12.15), we write the well-known average power expression as

$$P_{avg} = V_{RMS} \cdot I_{RMS} \cos(\phi) \qquad (12.17)$$

In RF testing, signals can be more complex than a single-frequency sinusoidal wave, which requires a more general definition of the integration time. It is obvious that in the case of a superposition of multiple sinusoidal waves, the average for a power measurement needs to be taken over multiple periods of the lowest frequency involved. For example, in the case of the amplitude-modulated signal shown in Figure 12.7, the integration time is defined as an integer multiple of the period of the modulation signal T_m, not the carrier period T_c, as shown in Figure 12.7. We can then write the average power as

$$P_{avg} = \frac{1}{nT_m} \int_0^{nT_m} v(t) \cdot i(t) \, dt \qquad (12.18)$$

In modern digital RF transmission systems, the signals are often transmitted only in so called frames. In the time domain, the envelope of the signal power looks like a string of pulses as shown

Figure 12.7. Power of an amplitude-modulated signal.

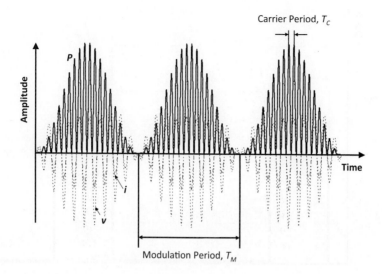

in Figure 12.8. Here the pulsed signal has a duration of τ seconds and a repetition period of T_r. For the system performance itself, only the power within the frame is of interest. One performance measure is the pulse power P_{Pulse}. It is defined as the average power in the pulse period τ according to

$$P_{Pulse} = \frac{1}{\tau} \int_0^{\tau} v(t) \cdot i(t) \, dt \tag{12.19}$$

In many cases, the pulse power can also be expressed as a function of the average power associated with the continuous wave and the duty cycle of the pulsing action; that is, consider rewriting Eq. (12.19) as

$$P_{Pulse} = \frac{T_r}{\tau} \frac{1}{T_r} \int_0^{\tau} v(t) \cdot i(t) \, dt = \frac{P_{avg}}{\tau/T_r} = \frac{P_{avg}}{DutyCycle} \tag{12.20}$$

As described above, it is important to consider the time dependence of the signal in order to be able to determine the correct power. It will often be beneficial to determine the signal in the time and frequency domains and then apply the correct power calculation or measurement method. For an appropriate ATE setting, it is important to understand the peak power P_{Peak} of the signal, even when only the average power is to be measured. This is especially important when there is a significant difference between the average and peak power, such as that which occurs with a pulse signal (see Figure 12.8). As we will discuss later in Chapter 13, it is critical to measure in the linear range of the ATE to guarantee that no stage of the measurement path enters the compression state at any time during the modulation cycles.

Figure 12.8. Power of a pulsed signal.

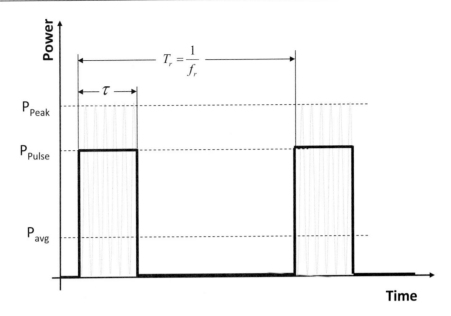

12.2.3 Crest Factor

In the previous section we have seen that the average power of a signal can have a complex relation to its peak power. For optimal RF system performance, like the test of an RF DUT with an ATE, it is essential to know not only the average power level but just as important the peak power used to source or measure the DUT, or even within the source and measurement path of the ATE. A description of the peak and average power used for complex signals is the signal dependent crest factor (CF). The CF is the ratio of the peak power P_{Peak} to the average power value P_{avg} of an instantaneous power waveform. CF can be written simply as

$$CF = \frac{P_{Peak}}{P_{avg}} \tag{12.21}$$

Exercises

12.4. The instantaneous voltage across a pair of terminals in a circuit can be described as $v(t) = V_{peak} \cos(2\pi f_0 \cdot t)$. Similarly the current in and out of each of these terminals can be described as $i(t) = I_{peak} \cos(2\pi f_0 \cdot t + \phi)$. Compute the instantaneous power associated flowing across these two terminals?

ANS. $p(t) = V_{peak} I_{peak} \cos(2\pi f_0 \cdot t + \phi)$
$\cos(2\pi f_0 \cdot t)$ W.

12.5. For the circuit situation described in Exercise 12.4 compute the average power.

ANS. $P_{avg} = f_0 \times \int_0^{1/f_0} p(t)dt$

$= \frac{1}{2} V_{peak} I_{peak} \cos(\phi)$ W.

12.6. Using the trigonometric identity, $\cos(\alpha)\cos(\beta) = \frac{1}{2}\cos(\alpha - \beta) + \cos(\alpha + \beta)$, write the instantaneous power expression found in Exercise 12.4 in terms of the sum and difference phase terms.

ANS.

$$p(t) = \frac{V_{peak} I_{peak}}{2}\cos(\phi)$$
$$+ \frac{V_{peak} I_{peak}}{2}\cos(4\pi f_o \cdot t + \phi)$$

12.7. Using the trigonometric identity, $\cos(\alpha + \beta) = \cos(\alpha)\cos(\beta) - \sin(\alpha)\sin(\alpha)$, rewrite the instantaneous power expression found in Exercise 12.6.

ANS.

$$p(t) = \frac{V_{peak} I_{peak}}{2}\cos(\phi) + \frac{V_{peak} I_{peak}}{2}$$
$$\cos(\phi)\cos(4\pi f_o \cdot t)$$
$$- \frac{V_{peak} I_{peak}}{2}\sin(\phi)\sin(4\pi f_o \cdot t)$$

12.8. Plot the total instantaneous total power in terms of the real and imaginary instantaneous power if $v(t) = 1.0\cos\left(2\pi \times 10^3 \cdot t\right)$ and $i(t) = 0.5\cos\left(2\pi \times 10^3 \cdot t + \pi/4\right)$.

Legend: solid = total power, dot = real power, dash = imaginary power.

ANS.

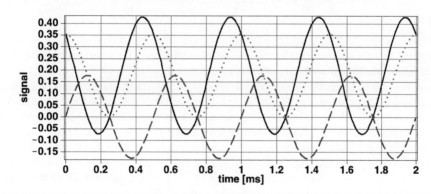

Often the crest factor is expressed in terms of decibels, where the conversion is done through the operation,

$$\left.\mathrm{CF}\right|_{dB} = 10\log_{10}\mathrm{CF} = 10\log_{10}\left(\frac{P_{Peak}}{P_{avg}}\right) \tag{12.22}$$

Back in Chapter 8, Section 8.3.3, we described the idea of a crest factor of a multitone voltage signal in terms of the peak-to-RMS ratio. This definition is often used in mixed-signal testing. However, for RF testing, the idea of a crest factor is related to a power waveform, and it should not be confused with parameters extracted from a voltage or current signal.

Referring to the power waveforms shown in Figure 12.5 we see that the peak value of this waveform is equal to twice the average power level, resulting in a crest factor of 2 or, equivalently, 3 dB.

For modulated RF signals, the crest factor is referred to as the peak value of the instantaneous power modulation envelope, instead of the peak value of the instantaneous power of the RF carrier signal. A frequency-modulated (FM) signal has a constant envelope and thus a crest factor, CF = 1 or 0 dB.

For signals with numerous uncorrelated sinusoidal voltages, we can bound the peak value of the instantaneous power waveform as the sum of individual voltage amplitudes, all squared, normalized by the appropriate impedance level, Z; that is,

$$P_{Peak} = \frac{\left(V_{Peak,1} + V_{Peak,2} + V_{Peak,3} + \cdots + V_{Peak,n}\right)^2}{Z} \tag{12.23}$$

Correspondingly, the average power of a multitone signal is simply the sum of the power of each sinusoidal term written as follows

$$P_{avg} = \frac{1}{2}\frac{V_{Peak,1}^2}{Z} + \frac{1}{2}\frac{V_{Peak,2}^2}{Z} + \frac{1}{2}\frac{V_{Peak,3}^2}{Z} + \cdots + \frac{1}{2}\frac{V_{Peak,n}^2}{Z} \tag{12.24}$$

Using Eqs. (12.23) and (12.24), the crest factor of multiple uncorrelated sinusoidal signals can be estimated to be

$$CF = \frac{P_{Peak}}{P_{avg}} \approx 2 \times \frac{\left(V_{Peak,1} + V_{Peak,2} + V_{Peak,3} + \cdots + V_{Peak,n}\right)^2}{V_{Peak,1}^2 + V_{Peak,2}^2 + V_{Peak,2}^2 + \cdots + V_{Peak,n}^2} \tag{12.25}$$

Figure 12.9 shows an example of the peak-to-average ratio for seven uncorrelated sinusoidal signals. Here we see the peak power exceeds the average signal power by a factor of about 11 times. If we assume the peak value each sine wave is 1 V, then according to Eq. (12.25) one would expect to see a CF of about $2 \times 7^2/7 = 14$. This is indeed what we see here.

The crest factor of other signal types are: noise has a crest factor of around 11 dB, an OFDM signal used in DAB, DVB-T and WLAN also have a crest factor of around 11 dB. The crest factor of CDMA2000 and UMTS mobile radio standards range up to 15 dB, but these could be reduced to 7–9 dB using special filtering techniques. Except during a burst, GSM signals have a constant envelope resulting from MSK modulation and thus have a crest factor of 0 dB. EDGE signals have a crest factor of 3.2 dB due to the filter function of the 8PSK modulation.

In the design of any RF system, the average power requirements are traded-off for lower peak power requirements. Crest factor is a parameter that relates these two quantities. Knowing any two parameters enables one to calculate the third. For instance, knowing the crest factor and the average power, the test engineer can calculate the expected peak power. This situation often arises in RF test when one is concerned about the maximum power a signal source can drive. The source will often go into compression at the peak power at any given time. In turn, source compression will set the maximum available power in a test system when using a specific modulation feature.

The complementary cumulative distribution function (CCDF) is the probability that a power is equal to or greater than a certain peak-to-average ratio as a function of such ratios. The higher the peak-to-average ratio, the lower the probability of reaching this point. The statistics of the signal determine the headroom required in the RF system. CCDF statistics are important in understanding digital modulated RF systems, because the statistics may vary. For instance, in CDMA systems, the statistics of the signal will vary with the number of codes that are used simultaneously.

Figure 12.9. Superposition of multiple sinusoidal waves resulting in high peak-power-to-average-power ratio defined as Crest Factor.

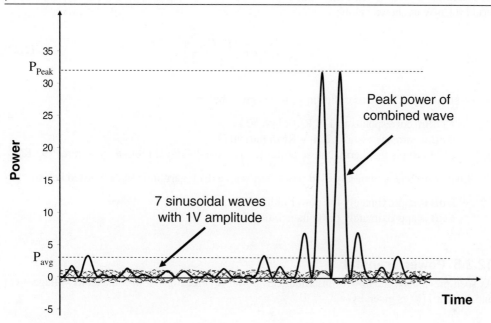

Since a couple of RF parameters like EVM and ACPR (to be introduced later in Chapter 13) are dependent on the peak-to-average ratio, it is important to understand their statistics to set up repeatable measurements.

Exercises

12.9. The crest factor of a continuous waveform is 7 dB. If the average power is 13 W, what is the peak power associated with this waveform?   ANS. 65.2 W.

12.10. A signal consisting of 4 uncorrelated sinusoids of 2 V. Estimate the crest factor associated with the composite waveform.   ANS. 9.03 dB.

12.2.4 Power in dBm

In actual RF systems, power is often defined for a (real) reference or characteristic impedance of Z_0. While any value can be selected for Z_0, most test equipment is built with a reference impedance Z_0 of 50 Ω. As such, a power measurement is equivalent to computing the square of the RMS value of the voltage and normalizing this value by 50 Ω, that is,

$$P_{avg} = \frac{V^2_{RMS}}{Z_0} = \frac{V^2_{RMS}}{50\,\Omega}\ \text{W}$$

(12.26)

It is a common practice in RF measurement and test to use the unit dBm instead of watt for a measurement of power. The dBm value is calculated by dividing the power transferred to a 50-Ω load by 1 mW as shown below:

$$P\big|_{dBm} = 10 \cdot \log\left(\frac{P_{avg}}{1\ \mathrm{mW}}\right) \tag{12.27}$$

A summary of some useful dBm values is given below:

13 dBm corresponds to 1-V RMS into 50 Ω
0 dBm corresponds to 0.224-V RMS into 50 Ω
−174 dBm is the thermal noise power in 1-Hz bandwidth at room temperature (293 K)

Likewise, some commonly used power ratios on a dB logarithmic scale is listed below:

3 dB is approximately the power ratio of 2
6 dB is approximately the voltage ratio of 2

12.2.5 Power Transfer

As seen previously, the flow of instantaneous power $p(t)$ for a given instantaneous voltage $v(t)$ and current $i(t)$ is given by

$$p(t) = v(t) \cdot i(t) \tag{12.28}$$

This instantaneous power represents the power that is dissipated, which is also known as real power, and the power that is bouncing back and forth between the reactive elements of the circuit. The latter power is known as the imaginary power. When the power is applied to an antenna, the real power is partially lost in dissipative elements of the antenna while the majority of it is transmitted (radiated) away from the antenna. The reactive power becomes the near-field reactive energy required to match boundary conditions of the antenna structure and, for the most part, is not radiated away from the antenna.

In the following, we will discuss the real power flow in an RF circuit. Here we will assume that the voltage and current signals are sinusoidal with some arbitrary phase angle difference Θ. If we assume that the voltage has the form $v(t) = V_{Peak}\sin(\omega t)$, then the current takes on the form $i(t) = I_{Peak}\sin(\omega t + \Theta)$. When working with RF circuits, it is preferable to make use of phasors, because it greatly simplifies their AC analysis. Consider representing the voltage and current signals using phasor notation. As mentioned previously, we write the voltage using complex notation as

$$v(t) = \Re\left\{V_{Peak}e^{j\omega t}\right\} \tag{12.29}$$

The leading term inside the brackets is the voltage phasor $\mathcal{V}$, which we shall denote using complex notation as

$$\mathcal{V} = V_{Peak}e^{j0} \tag{12.30}$$

Similarly, we can write the current signal using complex notation as

$$i(t) = \Re\left\{I_{Peak}e^{j(\omega t + \Theta)}\right\} = \Re\left\{\left(I_{Peak}e^{j\Theta}\right)e^{j\omega t}\right\} \qquad (12.31)$$

leading to the current phasor $\mathscr{I}$ as

$$V_L = I_{Peak}e^{j\Theta} \qquad (12.32)$$

The average power dissipated is then found from the expression

$$P_{avg} = \frac{1}{2}\cdot\Re\left(\mathscr{V}\cdot\mathscr{I}^*\right) = \frac{1}{2}\cdot V_{Peak}I_{Peak}\cos\left(\Theta\right) \qquad (12.33)$$

We can easily write the average power in terms of the RMS value of the voltage and current waveforms and get the familiar expression

$$P_{avg} = V_{RMS}I_{RMS}\cos\left(\Theta\right) \qquad (12.34)$$

The above expressions and their extension to the E and B fields are straightforward. Useful expressions for the power propagation in waveguides, transmission lines, and free space can be derived.

Maximum Power Transfer

When a voltage source is connected to a load, a flow of power from the source to the load results. As a matter of power transfer, the question related to the circuit conditions for maximum power transfer arises. Consider a voltage source with a real valued voltage level of $2V_A$ having a complex source impedance of $Z_S = R_S + jX_S$ connected to a load with impedance $Z_L = R_L + jX_L$. The corresponding circuit is shown in Figure 12.10. The voltage term V_A is called the *available voltage* from the source. This is not the maximum source voltage, as this would occur when the source is driven into an open circuit.

Through the applications of phasors, the average power delivered to the load is given by

$$P_{L,avg} = \frac{1}{2}\cdot\Re\left(V_L\cdot I_L^*\right) \qquad (12.35)$$

The load voltage and current phasors are found by straightforward circuit analysis resulting in

$$\mathscr{V}_L = \frac{Z_L}{Z_S + Z_L}2V_A = \frac{R_L + jX_L}{R_S + R_L + j\left(X_S + X_L\right)}2V_A$$

$$\mathscr{I}_L = \frac{2V_A}{Z_S + Z_L} = \frac{1}{R_S + R_L + j\left(X_S + X_L\right)}2V_A \qquad (12.36)$$

Figure 12.10. General network for power transfer calculations.

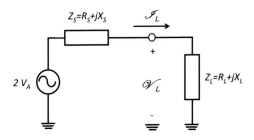

Substituting these two expressions back into Eq. (12.35), we obtain an expression for the average load power as

$$P_{L,avg} = 2|V_A|^2 \cdot \Re\left(\frac{R_L + jX_L}{R_S + R_L + j(X_S + X_L)} \cdot \frac{1}{R_S + R_L - j(X_S + X_L)}\right) \tag{12.37}$$

or we can write it more simply as

$$P_{L,avg} = 2|V_A|^2 \cdot \frac{R_L}{\left(R_S + R_L\right)^2 + \left(X_S + X_L\right)^2} \tag{12.38}$$

Assuming a fixed source impedance, together with some calculus (not shown), the maximum power transfer from source to load occurs when the following two conditions are met:

$$R_L = R_S$$
$$X_L = -X_S \tag{12.39}$$

Collectively, we can summarize the above two conditions and write $Z_L = Z_S^*$. This is known as the conjugate match condition. Under these matched conditions, the maximum power delivered to the load is referred to as the *maximum power available* from the source and is commonly denoted by the symbol P_A. Substituting the conditions listed in Eqs. (12.39) into (12.38), we find

$$P_A = \frac{1}{2}\frac{|V_A|^2}{R_S} \tag{12.40}$$

It is interesting to note that the source delivers twice this amount of power. However, the other half of this power is dissipated by the source resistance, R_S. Under general circuit conditions, the average power delivered by the source is given by

$$P_{S,avg} = 2|V_A|^2 \frac{\left(R_S + R_L\right)}{\left(R_S + R_L\right)^2 + \left(X_S + X_L\right)^2} \tag{12.41}$$

EXAMPLE 12.3

A voltage source with a voltage level described by phasor $100e^{j\pi/4}$ V and source impedance $Z_S = 54 + j6 \, \Omega$ drives load impedance of $Z_L = 25 + j25\Omega$. What is the average power delivered to the load and how does it compare to the maximum power available from the source? What is the average power delivered by the source?

Solution:
From Eq. (12.38), together with the relationships $2V_A = 100e^{j\pi/4}$, $R_S = \Re e(Z_S) = 54 \, \Omega$, $X_S = \text{Im}(Z_S) = 6 \, \Omega$, $R_L = \Re e(Z_L) = 25 \, \Omega$, and $X_L = \text{Im}(Z_L) = 25 \, \Omega$, we solve for the average load power as

$$P_{L,avg} = 2\left|\frac{100e^{j\pi/4}}{2}\right|^2 \cdot \frac{25}{(54+25)^2 + (6+25)^2} = 17.3 \text{ W}$$

The maximum available power can found from Eq. (12.40) as

$$P_A = \frac{1}{2}\left|\frac{100e^{j\pi/4}}{2}\right|^2 \frac{1}{54} = 23.1 \text{ W}$$

Using Eq. (12.41), we solve for the average source power as

$$P_{S,avg} = 2\left|\frac{100e^{j\pi/4}}{2}\right|^2 \cdot \frac{(54+25)}{(54+25)^2 + (6+25)^2} = 54.8 \text{ W}$$

EXAMPLE 12.4

Repeat Example 12.3 using a load impedance of $Z_L = 54 - j6 \, \Omega$ (conjugate matched to the source impedance).

Solution:
For the revised load, we write $R_L = \Re e(Z_L) = 54 \, \Omega$ and $X_L = \text{Im}(Z_L) = -6 \, \Omega$ and solve for the average load power as

$$P_{L,avg} = 2\left|\frac{100e^{j\pi/4}}{2}\right|^2 \cdot \frac{54}{(54+54)^2 + (6-6)^2} = 23.1 \text{ W}$$

The maximum available power can found from Eq. (12.40) as

$$P_A = \frac{1}{2}\left|\frac{100e^{j\pi/4}}{2}\right|^2 \frac{1}{54} = 23.1 \text{ W}$$

Using Eq. (12.41), we solve for the average source power as

$$P_{S,avg} = 2 \left| \frac{100e^{j\pi/4}}{2} \right|^2 \cdot \frac{(54+54)}{(54+54)^2 + (6+-6)^2} = 46.2 \text{ W}$$

As the source and load average power are equal, the circuit is operating under maximum power transfer conditions. Nothing can be done to have the load receive more power than ½ the source power.

12.2.6 Conjugate and Reflectionless Matching

Having just learned that the maximum available power at the load is half the source power, we must condition this statement with the fact that this is a best-case situation and only occurs when the load impedance is matched to the complex conjugate of the source impedance. In reality the source and the load are connected through a transmission line, whereby the line alters the load impedance seen by the source. If we denote the impedance seen by the source looking into the transmission line as Z_{TL}, one can show for a transmission line of length d having characteristic impedance Z_0 that

$$Z_{TL} = Z_0 \times \frac{Z_L + jZ_0 \tan(\beta \cdot d)}{Z_0 + jZ_L \tan(\beta \cdot d)} \tag{12.42}$$

Figure 12.11. (a) Complex conjugate match transmission line, (b) reflectionless matched transmission line.

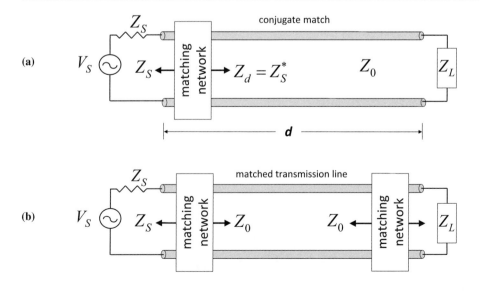

where $\beta = \dfrac{2\pi f}{v_p}$. As the impedance of Z_S and Z_L is generally targeted for 50 Ω, the maximum power transfer condition might not be met on account of the transmission line effect. To circumvent this situation, a matching network is inserted in the line at the source end as shown in Figure 12.11a. Here the matching network alters the impedance seen by the source and denoted as Z_d, such that the conjugate matched condition is met, that is, $Z_d = Z_S^*$. For a real-valued source impedance, that is, $Z_S = Z_S^*$, this condition ensures that no reflections occur at the source side of the transmission line. However, reflections will occur at the load side if $Z_L \neq Z_0^*$. In this case, an additional matching network is used to match the load to the transmission line as depicted in Figure 12.11b. Using two matching networks at each end is called reflectionless matching.

An interesting benefit arises with reflectionless matching. As the load and matching network combine to realize an impedance equal to the line impedance, Eq. (12.42) reveals that the impedance seen at the source end of the line is equal to Z_0 regardless of the line length or operating frequency f. This implies that the signaling will remain reflectionless with maximum power transfer. Moreover, this matched condition will be independent of line length and operating frequency. Section 5 of Chapter 15 will discuss in more detail the design of matching networks using Smith Charts.

12.2.7 Power Loss Metrics

Power transfer between source and load experiences various losses. In the following we shall describe two commonly used measures of power loss: insertion and transducer loss.

Insertion Loss

Insertion loss is the loss of signal power that result from the insertion of a device between the source and load connections as illustrated in Figure 12.12. Here a two-port network is shown inserted between the source and load. A two-network is a general term for an electrical network with two ports—for example an input and an output port. Two terminals constitute a port together with the condition that the current that enters one terminal of the port must equal the current leaving the other terminal. An n-port network is a network with n ports. A typical example of a two-port network is an amplifier. An example of a three-port is a mixer with an input RF port, an output IF port and a local oscillator port.

Figure 12.12. Two-step method to measure insertion and transducer loss: (a) power delivered to a load directly from source, (b) power deliver to load via a two-port network.

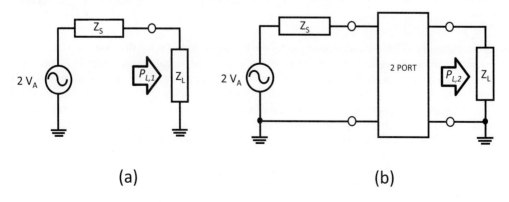

(a)   (b)

A two-step procedure to obtain the insertion loss of a two-port network is as follows:

1. A load is connected directly to the source and the average load power P_{L1} is measured as shown in Figure 12.12a. The load impedance can be, for example, the input impedance of a power meter.
2. A two-port is placed between the source and the load, as shown in Figure 12.12b, and again the average load power P_{L2} is measured.

The insertion loss IL is then calculated according to the formula

$$IL = \frac{P_{L1}}{P_{L2}} \tag{12.43}$$

Insertion loss IL can also be expressed in dB as

$$IL\big|_{dB} = 10\log\left(\frac{P_{L1}}{P_{L2}}\right) \tag{12.44}$$

As can be seen from Figure 12.12, an insertion loss measurement is dependent on the specific values of Z_L and Z_S. Without knowing these values exactly, the effect of inserting a two-port cannot be repeated at different times or locations. In the next chapter, we will discuss the insertion loss using S-parameters. This will provide more information about the test situation and allow for repeatable insertion loss measurements. The interested reader is recommended to study more details found in, for example, reference [5] or [6].

Transducer Loss

Transducer loss (TL) is the ratio of the maximum average power available from a source to the average power that the transducer delivers to a load, that is,

$$TL = \frac{P_A}{P_L} \tag{12.45}$$

or expressed in decibels as

$$TL\big|_{dB} = 10\log\left(\frac{P_A}{P_L}\right) \tag{12.46}$$

The maximum available source power can be derived from the test setup of Figure 12.12a with the load impedance conjugate matched to the source impedance. Because this power is equal to the maximum available source power, we can write $P_A = P_{L1,match}$, where $P_{L1,match}$ signifies the conjugate matched test condition. Note that the condition of conjugate matching makes this test setup different from that used in the insertion loss test. Next, the load power is measured with the transducer inserted in the circuit. This is identical to the test setup of Figure 12.12b where the load power is equal to P_{L2}. The transducer loss can then be expressed in terms of the two separate measurements as

$$TL = \frac{P_{L1,match}}{P_{L2}} \tag{12.47}$$

When the source and load impedances are complex conjugates of one other, insertion and transducer loss are equal. During a microwave test, the generator and load are commonly matched to a real impedance Z_0 known as the standard characteristic impedance (typical 50 Ω). By doing so, the conjugate match requirement will be satisfied.

When using this method to measure either insertion or transducer loss, it is important to remember that reflections should not compromise the test results. This can be achieved by adding attenuators to the source and load, which will minimize the impact of the reflected power. Another method would be with use of a network analyzer, which separates the incident and reflected wave with a directional coupler.

EXAMPLE 12.5

A 6-in. cable is used to connect a load to a source. A power measurement is made at the load and found to be equal to 1.2 W. It is assumed that this cable has negligible insertion loss. If a second cable is used instead with an insertion loss of 0.9 dB, what is the expected level of load power?

Solution:
Using Eq. (12.44), we write

$$IL\big|_{dB} = 10 \log\left(\frac{1.2 \text{ W}}{P_{L2}}\right) = 0.9 \text{ dB}$$

Solving for the unknown power level, PL2, we write

$$P_{L2} = \frac{1.2 \text{ W}}{10^{4.5/10}} = 0.97 \text{ W}$$

Therefore the expected level of load power is 0.97 W.

12.3 NOISE

The signal-to-noise ratio (SNR) at the output of RF systems is a very important quantity, if not the most important parameter. The SNR often sets the dynamic range limit of system operation. Listening or watching an analog audio RF system, like analog radio or television, in the presence of noise will have an adverse effect on the experience. Noise figure and sensitivity are other important system parameters, and are in most, if not all, cases the key figures of merit. For digital RF systems, the reliability of an RF system is often stated in terms of bit error rate (BER). The BER is related to the noise figure and the SNR in a nonlinear manner. For signals close to the noise floor (often referred to as low SNR signals), the noise level will make it impossible to decode such signals leading to transmission errors.

12.11. A 10-V source with source impedance $Z_S = 43 + j12\ \Omega$ drives a load impedance of $Z_L = 52 + j5\Omega$. What is the average power delivered to the load and how does it compare to the maximum power available from the source? What is the average power delivered by the source?

ANS. $P_L = 0.279\text{W}$, $P_A = 0.291$ W, $P_S = 0.51$ W.

12.12. A load is connected directly to a source where the load power was measured to be 2.4 W. The same load was then connected to the source through an RF connector and DIB trace. The load power was then measured to be 2.3 W. What is the insertion loss associated with the RF connector and DIB trace?

ANS. 0.185 dB.

12.13. The maximum available power from a source is 1 W. If a transducer with a 16-Ω load is connected to the source, what is the transducer loss if the load power is 0.9 W?

ANS. 0.46 dB.

A continuous sinusoidal wave (CW) can be described mathematically as

$$v(t) = A_0 \sin(\omega_0 t + \Theta_0)$$ (12.48)

where A_0 is the voltage amplitude of the sinusoidal wave, ω_0 is the angular frequency expressed in rad/s, and Θ_0 is the phase offset (also in radians). In practice, a CW signal will experience additive noise as it propagates through a channel. There are two types of noises that the wave will experience: amplitude and phase noise. We can model these two time-varying noise effects by including an additive amplitude noise term $a(t)$ and an additive phase noise term $\phi(t)$ in the expression above and write

$$v(t) = \left[A_0 + a(t)\right]\sin\left[\omega_0 t + \Theta_0 + \phi(t)\right]$$ (12.49)

Noise in RF test technology is a random variation of one or more characteristics of any entity such as voltage, current, phase, distribution, or spectral density. Noise effects are modeled in general as random processes. Typical measurements for RF noise effects are noise figure (NF) and phase noise ($\pounds(f)$). These measurements account for the random variations in magnitude and phase associated with the CW signal. Some system-like measurements such as EVM (error vector magnitude) and BER (bit error rate) also take into account noise effects while at the same time measuring system transmission capability.

In the following we will discuss the physics behind these amplitude and phase noise measurements.

12.3.1 Amplitude Noise

Generally, noise consists of spontaneous stochastically distributed fluctuations caused by ordinary phenomena in electronic circuits. These fluctuations can arise from different physical effects, but all lead to a random effect. The most important noise effects are due to:

- Thermal or Johnson–Nyquist noise
- Shot noise
- Flicker or $1/f$ noise

Thermal Noise

The thermal or Johnson–Nyquist noise arises from vibrations of the electrons and holes in conducting or semiconducting material due to their finite temperature. Some of the vibrations have a spectral content, but the magnitude of these variations for RF frequencies is nearly uniform. As such, we talk about the noise as being white, which is assumed to have a constant power as a function of the frequency. For a single resistor, the thermal noise power made available at its terminals is

$$P_N = kTB \qquad (12.50)$$

where k is the Boltzmann constant ($k = 1.38 \times 10^{-23} \, JK^{-1}$), T is the temperature in kelvin, and B is the bandwidth expressed in hertz. It is important to note that the thermal noise power available is independent of the system impedance. Using Eq. (12.50) we can calculate the noise power available at room temperature (say 20°C) for a 1 Hz bandwidth as follows

$$P_{N,1\mathrm{Hz}} = 1.38 \cdot 10^{-23} \, \mathrm{WsK}^{-1} \cdot (273 + 20)\mathrm{K} \cdot 1\mathrm{Hz} = 4.04 \times 10^{-21} \; \mathrm{W} \qquad (12.51)$$

Assuming this noise power is constant over a large range of frequencies, we can speak in terms of the power per 1-Hz bandwidth, or power spectral density (PSD), denoted by S_N, and write

$$S_N \triangleq \frac{P_{N,1\mathrm{Hz}}}{1\mathrm{Hz}} \left[\frac{\mathrm{W}}{\mathrm{Hz}} \right] \qquad (12.52)$$

In the case described here, the PSD would be $S_N = 4.04 \times 10^{-21}$ W/Hz . We can convert the PSD to a dBm/Hz scale by referencing to a 1-mW power level according to

$$S_N \big|_{\mathrm{dBm/Hz}} = 10 \log_{10} \left(\frac{S_N}{1\mathrm{mW}} \right) \qquad (12.53)$$

For the running example given here, the PSD expressed in dBm/Hz is found according to

$$S_N \big|_{\mathrm{dBm/Hz}} = 10 \log_{10} \left(\frac{4.04 \times 10^{-21}}{1\mathrm{mW}} \right) = -173.93 \frac{\mathrm{dBm}}{\mathrm{Hz}} \qquad (12.54)$$

If the noise power is measured with an instrument that has a resolution bandwidth other than 1 Hz, say instead equal to BW_{RES} in Hz, then the PSD will be increased by a factor equal to this BW_{RES} and falsely representing the PSD level. Therefore, to correct for this effect, we modify the PSD given in Eq. (12.52) according to

$$S_N \triangleq \frac{1}{BW_{RES}} \frac{P_{N,BW_{RES}}}{1\,\text{Hz}} \quad \left[\frac{W}{\text{Hz}} \right] \tag{12.55}$$

or in dBm/Hz as

$$S_N \big|_{\text{dBm/Hz}} = 10 \log_{10} \left(\frac{1}{BW_{RES}} \frac{P_{N,BW_{RES}}}{1\,\text{mW}} \right) \tag{12.56}$$

which can also be written as

$$S_N \big|_{\text{dBm/Hz}} = 10 \log_{10} \left(\frac{P_{N,BW_{RES}}}{1\,\text{mW}} \right) - 10 \log_{10} \left(BW_{RES} \right) \tag{12.57}$$

Here the left-hand side term represents the actual PSD of the signal, the first term on the right-hand side is the actual power level captured by the instrument, and the second term on the right is related to the instrument bandwidth. If the PSD of an input signal is -173.93 dBm/Hz at room temperature and an instrument with a 100-kHz resolution bandwidth measures it, then the capture power would display a power level equal to

$$10 \log_{10} \left(\frac{P_{N,BW_{RES}}}{1\,\text{mW}} \right) = -173.93 + 10 \log_{10} \left(100 \times 10^3 \right) = -173.93 + 50 \approx -124\,\text{dBm}$$

This example shows the impact of the bandwidth and the minimum detectable power. In this case, a noise power level of -124 dBm is displayed. If we want to measure lower amplitude signals, we need to reduce the measurement or receiver bandwidth.

Shot Noise

Shot or Schottky noise in electronic devices is caused by the random fluctuations of the number of charge carriers (electrons and holes) that cross a region of a conductor in a given amount of time. Shot noise is most pronounced in *pn*-junction-associated semiconductors. It is less of a concern in metal wires, as correlations between individual electrons remove these random fluctuations. The spectrum of shot noise is broadband and flat for RF frequencies, that is, white noise.

Shot noise is to be distinguished from thermal or Johnson–Nyquist noise. Thermal noise occurs without any applied voltage and without any current flow. This is not the case for shot noise. The noise power of shot noise can be described in most practical applications by

$$P_N = 2qIBR \tag{12.58}$$

where q is the elementary charge, B is the bandwidth in hertz over which the noise is measured, and I is the average current through the effective resistance R of the signal path.

Flicker Noise

Flicker noise is a type of electronic noise with a *1/f* frequency behavior or pink spectrum. It is therefore often referred to as *1/f* or pink noise, though these terms have wider definitions. It occurs in almost all electronic devices and results from a variety of effects, including (a) impurities in a conductive channel and (b) generation and recombination effects associated with a semiconductor material. In electronic devices, it is a low-frequency phenomenon, as the higher frequencies are overshadowed by white noise from other sources. In oscillators, however, the low-frequency noise is mixed up to frequencies close to the carrier that results in oscillator phase noise.

Active devices like MOSFETs, JFETs, and BJTs all suffer from flicker noise. Flicker noise is often characterized by the corner frequency f_c, the frequency at which the flicker noise meets the thermal noise level. MOSFETs have a higher than JFETs or bipolar transistors, which is usually below 2 kHz.

12.3.2 Noise Figure

The basic definition of noise factor (F) was introduced in the 1940s by Harold Friis as the ratio of SNR at a system input to the SNR at the system output.[3] It is written as

$$F = \frac{\text{SNR}_i}{\text{SNR}_o} = \frac{S_i/N_i}{S_o/N_o} \tag{12.59}$$

It is common to express the noise factor on a logarithmic scale in decibels with a new name, the noise figure (NF), according to

$$NF\big|_{\text{dB}} = 10\log_{10}\left(F\right) \tag{12.60}$$

A perfect system would add no noise to the signal, which would result in the output SNR being the same as that occurring at its input. In other words, the ideal system would have a noise factor or noise figure F of 1 or NF of 0 dB, respectively. Of course, real systems will add noise to the incoming signal (including the input noise) resulting in a reduction of the output SNR, in turn, leading to a noise factor F greater than one. Such a situation is illustrated in Figure 12.13 for an RF amplifier. Assuming that the amplifier corresponding to the data shown in Figure 12.13 is tested with a signal with a limited bandwidth, we can assume that we can read the signal power at the frequency of 2.45 GHz while the noise power can be read at, for example, 2.4 GHz. The error introduced is minimal since the noise power added to the signal at a frequency of 2.45 GHz is basically the same as the power around 2.4 GHz. Using Eqs. (12.59) and (12.60), we get

$$\text{NF} = \text{SNR}_i\big|_{\text{dB}} - \text{SNR}_o\big|_{\text{dB}} = \left[(-69)-(-100)\right]-\left[(-43)-(-64)\right]=10 \ \text{dB}$$

Assuming an RF system with a power gain $G = P_o/P_i$, where P_i and P_o is the input and output power, respectively, the noise factor F can be expressed in terms of the input noise level N_i as well as the additive amplifier noise N_a as follows:

$$F = \frac{S_i/N_i}{S_o/N_o}$$
$$= \frac{S_i/N_i}{GS_i/\left(N_a+GN_i\right)} \tag{12.61}$$
$$= \frac{N_a+GN_i}{GN_i}$$

Equation (12.61) shows the dependency of the noise at the input. Often this noise can be assumed to be thermal noise.

For a system with multiple gain stages, the system noise factor can be calculated by the Friis equation given by

$$F_{sys} = F_1 + \frac{F_2 - 1}{G_1} + \frac{F_3 - 1}{G_1 G_2} + \cdots + \frac{F_n - 1}{G_1 G_2 \ldots G_{n-1}}$$

(12.62)

where F_i and G_i are the noise factor and gain, respectively, of the ith amplifier stage expressed in linear magnitude form (rather than in dBm).

Figure 12.13. Typical signal and noise levels vs. frequency at an amplifier's input and at its output.

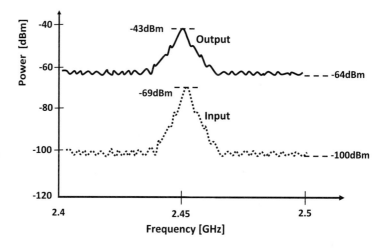

Figure 12.14. Illustrating the noise additions as a function of the number of amplifier stages.

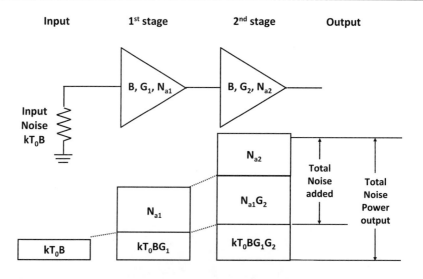

The basis of the Friis general expression of Eq. (12.62) can be better understood from the two-stage amplifier example shown at the top of Figure 12.14. Here the amplifier consists of two stages with gains G_1 and G_2. Each stage is assumed to have the same bandwidth B. Assume that the input noise to the amplifier is thermal noise arising from a resistor operating at a temperature of T_o with bandwidth B. The first stage will amplify this noise component by the gain of the first stage, G_1, as well contribute an additional noise component of N_{a1}, resulting in the output noise from the first stage as $kT_oBG_1 + N_{a1}$. This is depicted in the middle block in the lower figure of Figure 12.14. Subsequently, this noise component appears as the input to the second stage, whereby it is amplified with gain G_2 and combined with the noise of the second stage. The net result is the noise at the amplifier output becomes $kT_oBG_1G_2 + N_{a1}G_2 + N_{a2}$. As depicted by the lower sketch of Figure 12.14, the amplifier has contributed a noise component of $N_{a1}G_2 + N_{a2}$ that is not present at the input. As this additive noise component is the sum of two terms, where N_{a1} and N_{a2} are generally similar in magnitude, the first noise term $N_{a1}G_2$ dominates the sum. It also suggests that the noise of the first stage is the most important component affecting the amplifier's noise factor.

EXAMPLE 12.6

What is the combined noise figure of a two-stage amplifier if the first stage has a noise figure of 2 dB with a gain of 7 dB, and the second stage has a noise figure of 5 dB with a gain of 20 dB? What would be the overall noise figure if the amplifier cascade is rearranged so that the second stage becomes the first stage?

Solution:
Our first step is to convert all given terms into linear magnitude form, that is,

$$G_1 = 10^{G_1/10} = 10^{7/10} = 5.012, \quad G_2 = 10^{G_2/10} = 10^{20/10} = 100,$$
$$F_1 = 10^{NF_1/10} = 1.585, \qquad F_2 = 10^{NF_2/10} = 3.162$$

Next, using Friis general noise equation (i.e., Eq. (12.62)), we calculate

$$F_{sys} = F_1 + \frac{F_2 - 1}{G_1} = 1.585 + \frac{3.162 - 1}{5.012} = 2.016$$

Finally, the overall system noise figure in decibels is

$$NF = 10\log_{10}(2.016) = 3.046 \text{ dB}$$

Now, if the second stage is placed in front of the first stage, the noise factor becomes

$$F_{sys} = F_2 + \frac{F_1 - 1}{G_1} = 3.162 + \frac{1.585 - 1}{5.012} = 3.168$$

and is expressed in decibels as

$$NF = 10\log_{10}(3.168) = 5.008 \text{ dB}$$

Clearly, the arrangement of the first stage gain of 7 dB followed by the second stage gain of 20 dB results in the lower noise figure of 3.046 dB.

12.3.3 Phase Noise

One of the most significant parameters impacting the overall performance of an RF system is phase noise. Phase noise often becomes the limiting factor with respect to system sensitivity, maximum data rate, and bit error rate, to name just a few. This subsection will outline the fundamentals of phase noise, its sources, and how jitter in digital systems is related to phase noise such as that described in reference [10].

Exercises

12.14. A 1-kΩ resistor is connected across a 1-µF capacitor, what is the available noise power due to thermal noise at the terminals of this resistor at 300 K?   ANS. −151.8 dBm.

12.15. If a spectral analyzer reveals a power level of -150 dBm at 500 MHz using a resolution bandwidth of 50 kHz, what is the actual PSD appearing at the input to this instrument?   ANS. −147.0 dBm/Hz.

12.16. If a given amplifier has a 4 dB noise figure at 300 kelvin, a noise bandwidth of 500 kHz and an input resistance of 50 Ω, what is the output signal-to-noise ratio if the input signal RMS level is 1 mV.   ANS. 65.9 dB.

12.17. What is the combined noise figure of three-stage amplifier? The first stage has a noise figure of 2.1 dB and a gain of 8 dB, the second stage has a noise figure of 3 dB and a gain of 10 dB, and the third stage has a noise figure of 5 dB and a gain of 14 dB.   ANS. 2.59 dB.

Fundamentals

Phase noise, which falls into the wider category of frequency stability, is the degree to which an oscillating source produces the same single frequency value throughout a given time. This definition suggests that the frequency stability of an oscillator that produces a signal other than a perfect sinusoidal will be less than perfect.

A continuous sinusoidal wave subject to phase noise can be described as

$$v(t) = A_c \sin\left[\omega_c t + \Theta_0 + \phi(t)\right] \tag{12.63}$$

where A_c is the voltage amplitude of the sinusoidal wave, ω_c is the angular frequency expressed in rad/s, and Θ_0 is the phase offset (in radians) and $\phi(t)$ is a noise signal that modulates the phase of the carrier, also expressed in radians. Figure 12.15a illustrates the behavior of a sinusoidal signal whose phase is modulated by a noise signal in the time domain. One effect of the noise signal is to alter the zero-crossing values of the sinusoidal wave. Because noise affects the zero crossing values in a random manner, it is customary to model the attributes of this noise signal as a random process with specific types of statistics (e.g., Gaussian statistics). Alternatively, these phase fluctuations will also appear as changes in the period of the signal and thus show up as variations in the frequency of the signal. One can describe these frequency variations by a power spectral density (PSD) function as illustrated in Figure 12.15b.

Figure 12.15. (a) Impact of phase noise on sinusoidal signal in time domain. (b) PSD of sinusoidal signal in the frequency domain.

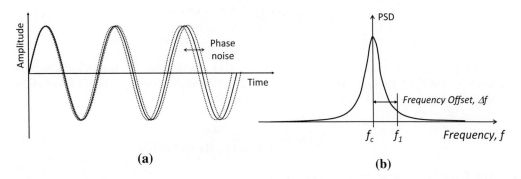

(a)   (b)

Spectral Density of Phase Fluctuations

As just mentioned, phase noise can be described in the time or frequency domain, depending on the application. For digital signals, a time domain description is used whereby a statistical model of the jitter introduced. For RF systems, a frequency domain description is more commonly used, because most measurement techniques are frequency-domain based.

The double-sided power spectral density (PSD) of a sinusoidal signal with phase noise can be described in the frequency domain along the positive frequency axis as having a frequency component concentrated at the carrier frequency f_c combined with a noise component that generally falls off on each side of the carrier at a rate of $\sim 1/(f - f_c)^2$ as illustrated in Figure 12.15b. Mathematically, one can describe the positive components of the frequency distribution in general terms as

$$S_v(f) = a \cdot \delta(f - f_c) + \frac{a \cdot b}{\pi^2 b^2 + (f - f_c)^2} \frac{V^2}{Hz} \tag{12.64}$$

where a and b are arbitrary positive constants. Defining $\Delta f = f - f_c$ as the frequency offset from the carrier, the general form of the PSD for $v(t)$ can be described as

$$S_v(f_c + \Delta f) = a \cdot \delta(\Delta f) + \frac{a \cdot b}{\pi^2 b^2 + \Delta f^2} \frac{V^2}{Hz} \tag{12.65}$$

The rightmost term in the above PSD is known as a Lorentzian distribution and often appears in oscillator studies.

While the above PSD was for the power associated with the signal $v(t)$, one can extract a single-sided PSD that reflects the RMS variations in the instantaneous phase difference φ between the carrier signal located at f_c and a signal occupying some bandwidth BW offset from the carrier by some frequency Δf. Assuming the RMS value of the phase signal extracted at each Δf is Φ_{RMS}, the PSD of the phase difference can be defined as

$$S_\phi(\Delta f) = \frac{\Phi_{RMS}^2}{BW} \frac{rad^2}{Hz} \tag{12.66}$$

Such a measurement would be carried out using a phase demodulator technique rather than a spectrum analyzer (see Chapter 13, Section 13.4).

Through some rigor, this phase PSD can be approximated in terms of the original waveform PSD according to

$$S_\phi(\Delta f) \approx 2 \frac{S_v(f_c + \Delta f)}{S_v(f_c)} \quad \frac{\text{rad}^2}{\text{Hz}} \qquad (12.67)$$

where $S_v(f_c + \Delta f)$ represents the PSD of the signal $v(t)$ at a distance Δf from the carrier frequency f_c and $S_v(f_c)$ represents the PSD at the carrier frequency (a power term). In terms of the PSD defined by Eq. (12.65), $S_\phi(\Delta f)$ would be written as

$$S_\phi(\Delta f) = \frac{2b}{\pi^2 b^2 + \Delta f^2} \frac{\text{rad}^2}{\text{Hz}} \qquad (12.68)$$

Often $S_\phi(\Delta f)$ is expressed in logarithmic terms as

$$S_\phi(\Delta f)\Big|_{\text{dBm/Hz}} = 10\log_{10}\left[S_\phi(\Delta f)\right] = 10\log_{10}(2b) - 10\log_{10}\left(\pi^2 b^2 + \Delta f^2\right) \frac{\text{dBm}}{\text{Hz}} \qquad (12.69)$$

or it can be rewritten in a piecewise linear fashion as

$$S_\phi(\Delta f)\Big|_{\text{dBm/Hz}} \approx \begin{cases} 10\log_{10}\left(2/\pi^2 b\right), & \Delta f < \pi b \\ 10\log_{10}(2b) - 20\log_{10}\Delta f, & \Delta f > \pi b \end{cases} \qquad (12.70)$$

As is evident from the above expression, for $\Delta f > \pi b$ the $S_\phi(\Delta f)$ will roll off at a rate of -20 dBm/Hz per decade or -6 dBm/Hz per octave.

It is important to note at this point in the discussion that not all oscillators behave according to a Lorentzian distribution. Often their behaviors can deviate quite significantly from this simple first-order model.

In the RF literature one often speaks about the quantity called *phase noise*. While multiple definitions of phase noise exist, here we make use of the definition provided in the 1139 Institute of Electrical and Electronic Engineers (IEEE) standard where the single sideband phase noise with symbol $\pounds(\Delta f)$ is defined as

$$\pounds(\Delta f) \overset{\Delta}{=} \frac{1}{2} S_\phi(\Delta f) \frac{\text{rad}^2}{\text{Hz}} \qquad (12.71)$$

Combining Eq. (12.67) with (12.71), we can write the phase noise as

$$\pounds(\Delta f) = \frac{S_v(f_c + \Delta f)}{S_v(f_c)} \frac{\text{rad}^2}{\text{Hz}} \qquad (12.72)$$

Taking the logarithm of each side of Eq. (12.72) and multiplying by 10, we can write the above phase noise expression as

$$10\log_{10}\left[\pounds(\Delta f)\right] = 10\log_{10}\left[S_v(f_c + \Delta f)\right] - 10\log_{10}\left[S_v(f_c)\right] \qquad (12.73)$$

While the units of the two right-hand terms are well-defined—that is, the term $10 \log_{10}[S_v(f_c + \Delta f)]$ is expressed in dBm/Hz and the term $10 \log_{10}[S_v(f_c)]$ is expressed in dBm—the left-hand side term is less clear. The units of the phase noise $10 \log_{10}[\pounds(\Delta f)]$ is simply $10 \log_{10}(\text{rad}^2/\text{Hz})$, which is conveniently written as dBc/Hz. The dBc is stated as "decibels with respect to carrier."

Often the phase noise is measured with an instrument that has a bandwidth other than 1 Hz, say, a spectrum analyzer with resolution bandwidth BW. As such, the PSD displayed must be normalized by the instrument bandwidth. The phase noise would then be computed according to

$$\pounds\big|_{\text{dBc/Hz}}(\Delta f) = P_{SSB}\big|_{\text{dBm}} - P_{carrier}\big|_{\text{dBm}} \tag{12.74}$$

where we define

$$
\begin{aligned}
\pounds\big|_{\text{dBc/Hz}}(\Delta f) &\triangleq 10 \log_{10}\big[\pounds(\Delta f)\big] \; \frac{\text{dBc}}{\text{Hz}} \\
P_{carrier}\big|_{\text{dBm}} &\triangleq 10 \log_{10}\big[S_v(f_c)\big] \;\; \text{dBm} \\
P_{SSB}\big|_{\text{dBm}} &\triangleq 10 \log_{10}\big[S_v(f_c + \Delta f)\big] - 10 \log_{10}(\text{BW}) \;\; \text{dBm}
\end{aligned}
\tag{12.75}
$$

Figure 12.16. Illustrating the definition of phase noise in terms of the carrier power and single sideband power at an offset of Δf.

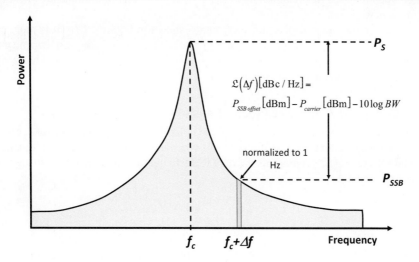

Figure 12.17. Single-sided phase noise $£(f)$ shown as function of frequency offset.

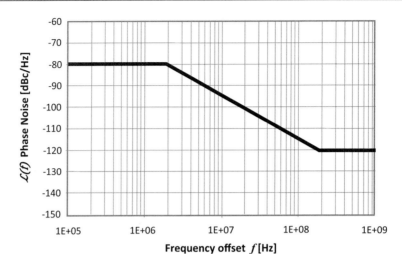

Figure 12.16 illustrates the relationship between the phase noise definition and power attributes of PSD of the original sinusoidal waveform defined in Eq. (12.63). If the PSD of a signal has been captured on a spectrum analyzer, then the phase noise is simply the difference in power of the carrier $P_{carrier}$ and the power at the some offset frequency Δf from the carrier frequency f_c normalized over 1 Hz bandwidth, denoted P_{SSB}.

While in this subsection we define the frequency offset with the explicit frequency term Δf, one commonly sees phase noise written without the delta symbol as $£(f)$. It should be clear to the reader when they see this notation that the frequency f refers to a frequency offset from the carrier frequency.

In practice, rather than display the spectrum as shown in Figure 12.16, it is more common in RF discussions to display the phase noise $£(f)$ as a single-sided spectrum dependent on the frequency offset from the carrier as shown in Figure 12.17. Both the vertical and horizontal axes are plotted on a log scale, that is, dBc/Hz versus Hz.

Spectral Density of Frequency Fluctuations

Another PSD that one may encounter with a phase noise description is the spectral density of frequency fluctuations, denoted $S_f(\Delta f)$, that results when measuring phase noise with a frequency discriminator. Here $S_f(\Delta f)$ represents the total RMS frequency variation Ψ_{RMS} in a given bandwidth, BW, defined as

$$S_f(\Delta f) \triangleq \frac{\Psi^2_{RMS}}{BW} \frac{Hz^2}{Hz} \tag{12.76}$$

Because instantaneous frequency is the derivative of the instantaneous phase of a signal, that is, $f = d\phi/dt$ the PSD of the frequency fluctuations is simply related to the PSD of the phase according to

$$S_f(\Delta f)^2 = (\Delta f)^2 S_\phi(\Delta f) \tag{12.77}$$

Integrated Frequency and Phase Noise

In modern communication systems with signals using multiple carriers, it is of interest to have a measure of short-term instabilities integrated over the channel bandwidth or the integrated noise power in the communication channel. For example, in an FM radio, the integrated frequency noise (measured in RMS-Hz) over the channel bandwidth is important. In digital communication systems, the integrated phase fluctuations in RMS-radians or RMS-degrees can be useful for analyzing the system performance.

Integrated noise over any bandwidth of interest is easily determined from using spectral density functions. Integrated frequency noise, commonly called residual FM, can be calculated by integrating the spectral density function of frequency fluctuations $S_f(\Delta f)$ over the frequency band $BW = f_2 - f_1$ to arrive at

$$\text{Residual FM} = \Psi_{RMS} = \sqrt{\int_{f_1}^{f_2} S_f(\Delta f) d\Delta f} \quad [\text{RMS-Hz}] \qquad (12.78)$$

The integrated phase noise (referred to here as the residual PM) can also be calculated by integrating the spectral density function of phase noise fluctuations $S_\phi(\Delta f)$ over a similar bandwidth according to

$$\text{Residual PM} = \Phi_{RMS} = \sqrt{\int_{f_1}^{f_2} S_\phi(\Delta f) d\Delta f} \quad [\text{RMS-radians}] \qquad (12.79)$$

or in rms-degrees as

$$\text{Residual PM} = \frac{180}{\pi} \times \sqrt{\int_{f_1}^{f_2} S_\phi(\Delta f) d\Delta f} \quad [\text{RMS-degrees}] \qquad (12.80)$$

Jitter and Integrated Phase Noise

Jitter is a common term in digital testing, as will be explained in greater detail in Chapter 13 as well as in Chapter 14. Jitter is the measure of the variation of the instantaneous frequency of an oscillator, which is not constant but varies slightly around an average value, thus creating an uncertainty in the frequency at any given point in time. This frequency change can be viewed as a change in the time of the waveform edge from the nominal frequency edge. We often speak about variation in the zero-crossing point as defined by the waveform edge.

Jitter can be measured in the time domain as a statistical variation in the timing of the edges. For instance, one statistical measure is to compute the RMS value of the signal period (denoted by T_n for the nth period) and expressed in radians according to

$$\sigma_\phi \big|_{\text{rad}} = \sqrt{\lim_{N \to \infty} \left(\frac{1}{N} \sum_{n=1}^{N} (T_n - \mu_T)^2 \right)} \qquad (12.81)$$

where μ_T is the average period given by

$$\mu_T = \lim_{N \to \infty} \left(\frac{1}{N} \sum_{n=1}^{N} T_n \right) \qquad (12.82)$$

We can also relate the integrated phase noise quantity to this RMS quantity (in radians) through the following expression, by integrating the area underneath the phase noise curve over some frequency range, f_1 and f_2, according to

$$\sigma_\phi\big|_{rad} = \sqrt{\text{Residual PM}} = \sqrt{\int_{f_1}^{f_2} S_\phi(\Delta f)d\Delta f} \qquad (12.83)$$

which is sometimes more conveniently expressed in terms of the phase noise as

$$\sigma_\phi\big|_{rad} = \sqrt{2\int_{f_1}^{f_2} \pounds(\Delta f)\,d\Delta f} \qquad (12.84)$$

The integrated phase noise can also be expressed in degrees according to

$$\sigma_\phi\big|_{deg} = \frac{180}{\pi}\sigma_\phi\big|_{rad} \qquad (12.85)$$

Likewise, we can express this RMS quantity in seconds by multiplying the average signal period, μ_T, normalized by 2π, according to

$$\sigma_\phi\big|_{sec} = \frac{\mu_T}{2\pi}\sigma_\phi\big|_{rad} \qquad (12.86)$$

The above four equations were derived assuming that there is no 1/*f* noise or burst noise present with the signal.

EXAMPLE 12.7

The phase noise of a VCO with an oscillation frequency of 1 MHz has the phase noise spectrum shown in Figure 12.18. What is the RMS value of the phase jitter over a frequency range of 100 kHz to 10 MHz? Express your answer in degrees, radians, and seconds.

Solution:
As is evident from the phase noise $\pounds(\Delta f)$ plot, the phase noise is equal to -80 dBc/Hz from 100 kHz to 10 MHz. This is equivalent to 10^{-8} rad²/Hz over the same frequency range. Using Eq. (12.84), the integrated phase noise is computed according to

$$\sigma_\phi\big|_{rad} = \sqrt{2\times\int_{10^5}^{10^7} 10^{-8}d\Delta f} = \sqrt{2\times(10^7 - 10^5)} = 0.447\,rad$$

We can convert this to degrees simply by multiplying $\sigma_\phi\big|_{rad}$ by 180/π and get

$$\sigma_\phi\big|_{deg} = \frac{180}{\pi}\times 0.447 = 25.5 \text{ degrees}$$

Similarly, we can express the RMS value in seconds through the application of Eq. (12.86) where the average period of the oscillator is

$$\mu_T = \frac{1}{f_{VCO}} = \frac{1}{10^6} = 10^{-6} \text{ s}$$

leading to

$$\sigma_\phi\big|_{sec} = \frac{10^{-6}}{2\pi} \times 0.447 = 71.1 \text{ ns}$$

Figure 12.18. Phase noise plot for Example 12.7. Area under curve from 100 kHz to 10 MHz represents integrated phase noise.

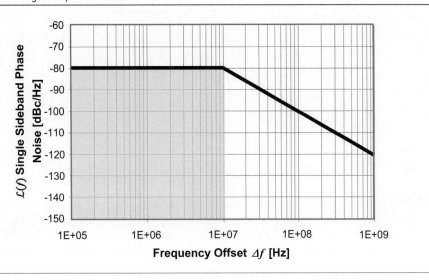

12.4 *S*-PARAMETERS

The performance and functionality of any RF circuit can be described by the signals going in and coming out of the circuit. In this case, the circuit can be viewed as an *n*-port. These *n*-ports are representatives for linear or nonlinear networks of the circuit. For signals sufficiently small to cause only a linear response, an *n*-port can be described via a set of parameters measured at the network ports. Once these parameters have been determined, the behavior in any external environment can be predicted without any detailed knowledge of the content of the *n*-port network. One set of parameters named *S*-parameters, also referred to as scattering parameters, has found widespread application for describing *n*-port networks. This was first published[4] by K. Kurokawa in 1965 and is used extensively in most RF work. In this section we will learn how *S*-parameters can help us better understand RF circuits.

12.4.1 Principles of *S*-Parameters of a Two-Port Network

S-parameters are commonly used to design and describe microwave circuits. They are easy to measure and are better suited to high-frequency applications unlike conventional two-port parameters

Figure 12.19. Two-port network with incident and reflected port waves: (a) waves associated with each port. (b) voltages and currents associated with each port.

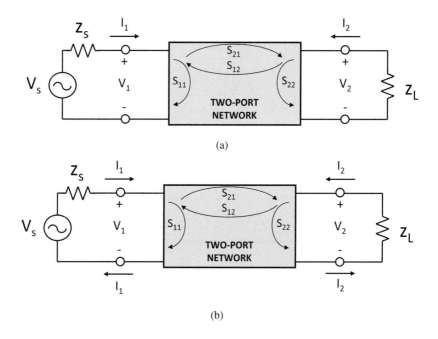

(a)

(b)

like Y-, T- or h-parameters, which one may have encountered in an introductory circuit analysis course. They are conceptually simple and give deep insight into RF circuit behavior and their properties. Here the fundamentals of S-parameters will be discussed.

Although an arbitrary network might have any number of ports, we will begin with a two-port network like one shown in Figure 12.19a. We begin by identifying the incident and reflected waves a_i and b_i at each port of the network. These waves are normalized by the square root of either the source or load impedance, depending on which branch the wave is associated with.

The general form of the incident voltage wave a_i can be expressed as

$$a_i = \frac{V_i - Z_i^* I_i}{2\sqrt{\left|\Re e\left\{Z_i\right\}\right|}}$$ (12.87)

and the reflected wave, b_i as

$$b_i = \frac{V_i - Z_i^* I_i}{2\sqrt{\left|\Re e\left\{Z_i\right\}\right|}}$$ (12.88)

In both cases, V_i is the terminal voltage at the ith port of the n-port network with terminal current I_i and complex impedance Z_i as highlighted in Figure 12.19b. The asterisk in Eq. (12.88) denotes the complex conjugate.

Assuming all impedances z_i are positive and real with value Z_0, Eqs. (12.87) and (12.88) can be expressed for a two-port network as

$$a_1 = \frac{V_1 + Z_0 I_1}{2\sqrt{Z_0}} = \frac{\text{voltage wave incident on port 1}}{\sqrt{Z_0}} \qquad (12.89)$$

$$a_2 = \frac{V_2 + Z_0 I_2}{2\sqrt{Z_0}} = \frac{\text{voltage wave incident on port 2}}{\sqrt{Z_0}} \qquad (12.90)$$

$$b_1 = \frac{V_1 - Z_0 I_1}{2\sqrt{Z_0}} = \frac{\text{voltage wave reflected from port 1}}{\sqrt{Z_0}} \qquad (12.91)$$

$$b_2 = \frac{V_2 - Z_0 I_2}{2\sqrt{Z_0}} = \frac{\text{voltage wave reflected from port 2}}{\sqrt{Z_0}} \qquad (12.92)$$

The following set of linear equations are used to describe the interrelationship of all the waves associated with the two-port network, in much the same way that nodal voltages and branch current interrelate in any electrical circuit:

$$b_1 = S_{11} a_1 + S_{12} a_2 \qquad (12.93)$$

$$b_2 = S_{21} a_1 + S_{22} a_2 \qquad (12.94)$$

The coefficients, S_{11}, S_{12}, S_{21}, and S_{22}, are referred to as the S-parameters of the two-port network. Sometimes, the above equations are lumped together into a matrix formulation as follows:

$$\begin{bmatrix} b_1 \\ b_2 \end{bmatrix} = \begin{bmatrix} S_{11} & S_{12} \\ S_{21} & S_{22} \end{bmatrix} \begin{bmatrix} a_1 \\ a_2 \end{bmatrix} \qquad (12.95)$$

Based of the above formulation, we can identify each S-parameter according to the following:

S_{11}: The input reflection coefficient with the output port terminated by a matched load ($Z_L = Z_0$) resulting in $a_2 = 0$;

$$S_{11} = \left. \frac{b_1}{a_1} \right|_{a_2 = 0} \qquad (12.96)$$

S_{22}: The output reflection coefficient with the input port terminated by a matched load ($Z_S = Z_0$ and $V_S = 0$) resulting in $a_1 = 0$;

$$S_{22} = \left. \frac{b_2}{a_2} \right|_{a_1 = 0} \qquad (12.97)$$

S_{21}: The forward transmission (insertion) gain with the output port terminated by a matched load ($Z_L = Z_0$) resulting in $a_2 = 0$;

$$S_{21} = \frac{b_2}{a_1}\bigg|_{a_2=0} \qquad\qquad (12.98)$$

S_{12}: The reverse transmission (isolation) gain with the input port terminated by a matched ($Z_S = Z_0$ and $V_S = 0$) resulting in $a_1 = 0$;

$$S_{12} = \frac{b_1}{a_2}\bigg|_{a_1=0} \qquad\qquad (12.99)$$

These S-parameters can be used to determine the percentage of the power going into a device and the amount reflected as a function of the impedance of the two-port network and the impedance of the termination. This termination might be the impedance of the source supplying an RF signal or the impedance of the test equipment measuring the RF power.

Equations (12.89)–(12.92) describe the incident voltage waves a_i and the reflected voltage waves b_i as effective (RMS) values and not peak values. Since the two-port network described with the S-parameter matrix,

$$S = \begin{bmatrix} S_{11} & S_{12} \\ S_{21} & S_{22} \end{bmatrix}_{Z_0} \qquad\qquad (12.100)$$

is embedded in a environment with a characteristic impedance of Z_0, these waves can be interpreted in terms of normalized voltage or current waves. This relation is explained below. In most test situations, a real characteristic impedance $Z_0 = (50 + j0)\Omega$ is selected. For such cases, the RF measurement equipment including all cables will be designed for 50 Ω. As such, the environment will be operating under reflectionless and conjugate matched conditions. Natural reflections within the measurement environment will not exist. Nonetheless, the test engineer must concern himself or herself with matching the DUT to the test environment. The general situation is depicted in Figure 12.20 where the middle block consisting of the S-matrix represents the DUT.

Let us look at one example involving S_{11}. From Eq. (12.96), we recognize that S_{11} relates the reflected wave to the incident wave at port 1. It is therefore referred to as the reflection coefficient for port 1. Of course S_{22} has the same meaning but applies to port 2. Let us consider some further

Figure 12.20. Two-port network embedded into test environment with characteristic impedance Z_0.

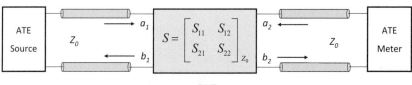

DUT

meaning for S_{11}. Assuming matched terminations, according to Eq. (12.96), together with Eqs. (12.89) and (12.91), we can express S_{11} as

$$S_{11} = \frac{b_1}{a_1} = \frac{\dfrac{V_1}{I_1} - Z_0}{\dfrac{V_1}{I_1} + Z_0} \qquad (12.101)$$

If we define the input impedance looking into port 1 as Z_1, we can write

$$Z_1 = \frac{V_1}{I_1} \qquad (12.102)$$

and then we can write the reflection coefficient as

$$S_{11} = \frac{Z_1 - Z_0}{Z_1 + Z_0} \qquad (12.103)$$

Clearly, if $Z_1 = Z_0$, then the reflection coefficient is zero and no reflected wave results. If $Z_1 = 0$ then the reflection coefficient is -1 and the reflected wave is equal but opposite to the incident wave.

Conversely, we can also rearrange this expression and write the input impedance of a two-port network as a function of the reflection coefficient S_{11} as

$$Z_1 = Z_0 \frac{1 + S_{11}}{1 - S_{11}} \qquad (12.104)$$

Of course, we could develop similar looking expressions for the reflection coefficient S_{22} at port 2 in terms of the impedance $Z_2 = V_2/I_2$ and find

$$S_{22} = \frac{Z_2 - Z_0}{Z_2 + Z_0} \qquad (12.105)$$

This relationship between the reflection coefficient and the port impedances is the basis of the Smith Chart transmission line technique. Consequently, the reflection coefficients S_{11} and S_{22} can be plotted on a Smith Chart (see Section 15.5). They can be converted directly into impedances; they can also be easily manipulated to determine a matching network to transform the port impedance so that the circuit operates more efficiently.

The remaining two S-parameters S_{12} and S_{21} are not reflection coefficients; rather they behave more as transmission coefficients from one port to the other. For instance, S_{21} can be shown to be equal to

$$S_{21} = \frac{I_2}{I_1} \left(\frac{Z_2 - Z_0}{Z_1 + Z_0} \right) \qquad (12.106)$$

and S_{12} as

$$S_{12} = \frac{I_1}{I_2}\left(\frac{Z_1 - Z_0}{Z_2 + Z_0}\right)$$

(12.107)

The magnitude of the S-parameters are expressed in one of two ways, linear magnitude or in decibels. To convert between the two, we simply use the following:

$$S_{11}\big|_{dB} = 20\log_{10}|S_{11}|, \quad S_{12}\big|_{dB} = 20\log_{10}|S_{12}|$$
$$S_{21}\big|_{dB} = 20\log_{10}|S_{21}|, \quad S_{22}\big|_{dB} = 20\log_{10}|S_{22}|$$

(12.108)

12.4.2 Scalar Representation of S-Parameters

It is a common practice to use scalar representatives for some of the S-parameters. The return loss, mismatch loss, VSWR, and mismatch uncertainty are scalar measures of different power parameters, which can be used to describe some RF circuit behavior, especially when formulating a test environment with a signal source, DUT, and measurement path. These scalar parameters are calculated using the magnitude of the complex value of the reflection coefficients.

Voltage Standing Wave Ratio

The concept of reflections can be best visualized by describing the reaction of an open-circuited lossless transmission line when a voltage pulse propagates along the transmission line and hits the open end. This is illustrated in Figure 12.21. The pulse will propagate as depicted in sketches 1 to 4, its voltage wave undiminished with the distance when the transmission line is assumed to be lossless (nondissipative and no radiative losses). When the pulse reaches the open end of the transmission line, all of the incident power must be reflected, since it can neither be radiated nor stored there. In practice some of the energy will radiate like an antenna. Since all the power needs to be reflected at the open end of the transmission line, the pulse will bounce back. For an open circuit, the current will need to be zero at the termination to satisfy this condition and the reflected current wave must be directed opposite to the incident current. This requires that the associated reflected voltage have the same polarity as the incident voltage, which results in a doubling of the voltage at the end of the line, as shown in Figure 12.21, sketches 5 to 7. The total voltage on the line is equal to the sum of the incident and reflected pulses. The current pulse at the end of the line is the sum of the incident and reflected pulse, resulting in a net zero. After the pulse is reflected at the end, it will propagate back in the direction of the source as shown in sketches 8 to 10, where the incident and reflected pulse are again separated in time. If the line impedance is different from the source impedance, the pulse will again reflect at the source end and this progress repeats itself until the energy of the pulse is all used up.

This pulse example is useful to demonstrate and visualize the effects of reflection. For one, it shows that the total voltage can vary on a transmission line due to reflections; on the other side it shows that a complex description is required to describe the effect of reflection. In the example above, we have seen that an open circuit, which requires having the condition of a zero current, causes the resulting pulse with twice the voltage amplitude to be reflected back. We can also consider the case of a short-circuited end as the other extreme. In this case, the voltage wave must be zero, and the current wave will be doubled, which will lead to an extinction of the voltage wave at the shorted end.

Just as pulses on a transmission line that is open-circuited produce a doubling of the incident voltage, so too do continuous sinusoidal waves (CW) propagating on a transmission line with

Figure 12.21. Pulse distribution on a loss-less transmission line and its reflections caused by the open circuit termination.

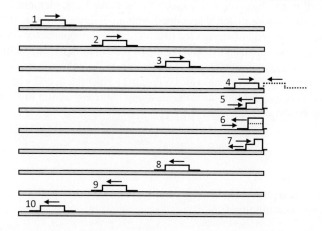

Figure 12.22. Sinusoidal CW incident, reflected and composite waves on a transmission line as a result of a mismatch in the line impedance at the far right hand side at distance 0.

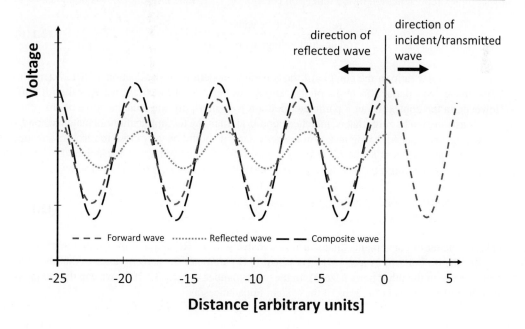

similar termination conditions. This situation is illustrated in Figure 12.22, where the impedance mismatch appears of the far right-hand side where the distance is 0. The reflected voltage wave will interfere constructively with the incident wave on the transmission line, generating voltage maxima and minima along the length of the transmission line. To physically separate these two waves, a device referred to as a directional coupler is used.

In the early years of network analyzers for microwave frequencies, high-performance directional couplers were unavailable; hence slotted transmission lines were used for return loss measurements. A slotted transmission line is a carefully fabricated line element with a longitudinal slot cut in its outer conducting material and fitted with an exterior carriage. Within the carriage, a small wire probe is mounted and connected to a rectifying semiconductor crystal. The crystal's rectified voltage output is connected to a voltage measurement unit, like an oscilloscope. With this test, the minimum and maximum voltage across the line could be measured. Based on this measurement, the voltage standing wave ratio (VSWR) can be calculated; and with the assumption of a loss-free line, the maxima and minima voltages along the transmission line due to constructive interference vary according to

$$\text{VSWR} = \frac{V_{max}}{V_{min}} = \frac{V_{incident} + V_{reflected}}{V_{incident} - V_{reflected}} \tag{12.109}$$

The VSWR is a scalar measure since V_{max} and V_{min} are themselves scalar quantities and VSWR is bounded between 1 and infinity.

Reflection Coefficient

For an expression of the fraction of the voltage wave reflected at the ith port of an n-port network, the complex representation of the reflection coefficient Γ_i can be introduced. It is defined as

$$\Gamma_i = \frac{\text{reflected voltage wave}}{\text{incident voltage wave}} = \frac{b_i}{a_i} = |\Gamma_i| e^{-j\Theta_i} \tag{12.110}$$

We saw a similar ratio in the context of the S-parameter definitions of Section 12.4.1, where the S_{ii} parameter was defined in terms of the ratio of the reflected-to-incident wave at the ith port. However, in the context of an S-parameter definition of an n-port, the incident wave on the other ports is assumed to be zero, that is, matched condition. In many test applications, a matched condition will be approached during test; if not exactly zero, it should be very close to zero. As a result, the reflection coefficient at the ith port can be used for S_{ii}. The reflection coefficient Γ_i can be expressed in polar coordinates with the phase angle Θ_i and the magnitude ρ_i as follows

$$\Gamma_i = \rho_i e^{-j\Theta_i} \tag{12.111}$$

This is a matter of convenience that is used extensively in the RF and microwave literature.

For the ith port, Γ_i is assumed to be equal to S-parameter S_{ii} (here we assume that the incident wave a_i on the other ports is zero). In the same manner as Eq. (12.103), we can then express the reflection coefficient in terms of the system impedances as

$$\Gamma_i = \frac{Z_i - Z_0}{Z_i + Z_0} \tag{12.112}$$

where Z_i is the impedance seen looking into the ith port and Z_0 is the impedance of the transmission line.

Often it is sufficient to represent an arbitrary reflection coefficient r with only its magnitude term ρ. The value of the reflection coefficient for a passive network is $0 \leq \rho \leq 1$. The value of $\rho = 1$ means that all power of the voltage wave at the port is reflected and the incident power is equal to the reflected power, while $\rho = 0$ means that no power is being reflected.

Figure 12.23. Definition of the refection coefficient in terms of the incident and reflective power.

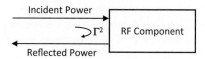

When working exclusively with power quantities, the magnitude of the reflection coefficient can be calculated as the ratio of the reflected to incident power as

$$\rho^2 = \frac{\text{reflected power}, P_r}{\text{incident power}, P_i} \tag{12.113}$$

Figure 12.23 depicts the relationship between the incident and reflect power at the interface of some arbitrary RF component. Here it is important to recognize that power (average power to be exact) is a scalar quantity and has no phase quantity associated with it. An alternative form for the magnitude of the reflection coefficient can be written in terms of the VSWR quantity as

$$\rho = \frac{\text{VSWR} - 1}{\text{VSWR} + 1} \tag{12.114}$$

On a lossless line, the respective magnitudes of the incident and reflected voltage waves do not change, thus the magnitude of the reflection coefficient $|\Gamma|$ does not change either, only the phase angle Θ changes as the two waves travel through each other in opposite directions. The phase angle Θ depends upon the relative phases of the incident and reflected voltage waves at the load, as well as the electrical distance from the load.

Return Loss

Return loss RL is defined as the ratio of the power incident to a discontinuity to that which is reflected. The return loss RL (often referred to as 'insertion return loss') for an RF port is equal to the reciprocal of the reflection coefficient according to

$$\text{RL} \triangleq \frac{\text{incident power}, P_i}{\text{reflected power}, P_r} = \frac{1}{\rho^2} \tag{12.115}$$

or expressed in decibels as

$$\text{RL}\big|_{dB} = 10\log_{10}\text{RL} = -20\log_{10}\rho \tag{12.116}$$

The return loss describes the difference between the incident wave and the reflected wave in decibels. The return loss itself is a scalar value and has no phase information. It is often used in production testing as a pass/fail criterion. A large RL value indicates little reflection; a small RL value means large levels of reflection. *Although RL is described as a loss quantity, its use is largely one to describe how well two devices match. The higher the value, the better the match.* Some literatures define return loss as the inverse of that which is given in Eq. (12.115). This typically results in a negative value for the RL in decibels. In this textbook, we make use of the positive value convention for return loss.

Passive circuits always have a return loss somewhere between 0 dB and infinity, that is, $0 \text{ dB} \leq RL \leq \infty$, while active circuits might also have a return gain, meaning that the reflected wave is greater than the incident wave, and as such, RL has a negative value.

Within the industry, VSWR and RL are used as alternative means for specifying match even though the slotted line is rarely used for measurements anymore. A perfect match of the load to the transmission line occurs when $Z_L = Z_0$. This condition results in all power being delivered to the load, and no reflected wave is produced. For this case, we get $\rho = 0$, VSWR = 1, and RL = ∞.

Mismatch Loss

Mismatch loss, ML, describes the amount of power loss at an impedance discontinuity due to reflections as shown in Figure 12.24. The path of the two-port is assumed lossless for the calculation of the mismatch loss. This condition can be expressed with S-parameters as $S_{21} = S_{12} = 1$. Physically, this describes a path without any dissipative loss or all power supplied into port one will exit the network at port two.

The mismatch loss ML is a computed parameter that describes the power attenuated (wasted) due to mismatch between two interconnects. It is a useful parameter that describes how much gain improvement could be achieved by an optimized matching.

For a two-port network, such as that shown in Figure 12.24 with a nonzero input reflection coefficient ($\Gamma \neq 0$), the normalized power delivered to the load as transmitted power is the difference between the available power from the source and the reflected power, that is, $1 - |\Gamma|^2$. The reciprocal of this amount is the fractional increase of power delivered to the load without reflection and is known as the mismatch loss, ML. It is defined in terms of the reflection coefficient Γ as follows

$$\text{ML} \triangleq \frac{\text{incident power}, P_i}{\text{transmitted power}, P_t} = \frac{1}{1 - \rho^2} \qquad (12.117)$$

Figure 12.24. Illustrating the power flow associated with mismatch loss.

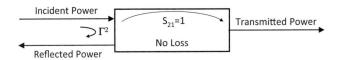

Figure 12.25. Illustrating the interfaces associated with power flow from source to load and their corresponding reflection coefficients Γ_1 and Γ_2.

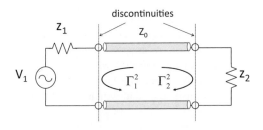

or expressed in decibels as

$$\text{ML}\big|_{\text{dB}} = 10\log_{10}\text{ML} = -10\log_{10}\left(1-\rho^2\right) \tag{12.118}$$

For passive circuits, such as the lossless network used here, ML expressed in decibels is a positive number (i.e., transmitted power is less than or equal to the incident power). If we extend the definition of ML to active circuits, then it is possible for ML expressed in decibels to be a negative quantity, that is, the transmit power is larger than the incident power. In such cases, one must be careful with the sign.

In the case of two discontinuities associated with signal transmission (such as the two that arise at the interface between each end of a transmission line with characteristic impedance Z_0 as depicted in Figure 12.25), the total mismatch loss can be shown to be described by

$$\text{ML} = \frac{\left|1-\Gamma_1\cdot\Gamma_2\right|^2}{\left(1-\left|\Gamma_1\right|^2\right)\left(1-\left|\Gamma_2\right|^2\right)} \tag{12.119}$$

or, when expressed in decibels, as

$$\text{ML}\big|_{\text{dB}} = 10\log_{10}\text{ML} \tag{12.120}$$

Here the reflection coefficients are described as

$$\Gamma_1 = \frac{Z_1-Z_0}{Z_1+Z_0} \text{ and } \Gamma_2 = \frac{Z_2-Z_0}{Z_2+Z_0} \tag{12.121}$$

The mismatch loss predicted by either Eq. (12.119) or Eq. (12.120) describes the power that is bouncing back and forth between the two unmatched ports of the network. As seen from the numerator of Eq. (12.119), both the magnitude and phase of each reflection coefficients Γ_1 and Γ_2 is required for this calculation.

Return loss and mismatch loss are frequently used to describe an RF circuit but often interpreted incorrectly. The return loss RL represents the ratio of the reflected power to incident power. This is sometimes stated as the difference in decibels between the reflected and the incident power. In general, it is desired to have a small positive return loss (large decibels value), indicating that the match at the interface is good. In contrast, mismatch loss represents the ratio of the power transmitted across an interface to the incident power, or the difference in decibels between the transmitted power and the power that is incident. This difference in decibels should be small. The following example will help to illustrate these differences.

Mismatch Uncertainty

In many cases, only the magnitude of the reflection coefficient is known, which may be due to the limitations of the measurement equipment used, like a power meter or a scalar network analyzer, or by simply calculating the reflection coefficient based on given VSWR values. As such, the precise mismatch loss cannot be calculated, however, upper and lower bounds can be established. The ratio of the upper to the lower bound of the mismatch loss is called mismatch uncertainty, *MU*.

EXAMPLE 12.8

Several power measurements were made on a two-port network. The incident power was found to be 1 W, the reflected power was 0.1 W and the power transmitted by the two-port was 0.9 W. Additional measurements indicate the two-port was lossless. The measurements are summarized in Figure 12.26. Compute the reflection coefficient, return loss, and mismatch loss associated with this two-port network.

Solution:
According to Eq. (12.113), we find the magnitude of the reflection coefficient as

$$\rho = \sqrt{\frac{P_r}{P_i}} = \sqrt{\frac{0.1\,\text{W}}{1.0\,\text{W}}} = \sqrt{0.1} = 0.316$$

Next, we find the return loss, RL, from Eqs. (12.115) and (12.116), as follows

$$RL\big|_{dB} = -20\log_{10}\rho = -20\log_{10}(0.316) = 10\ \text{dB}$$

Finally, we compute the mismatch loss ML for a single discontinuity using Eqs. (12.117) and (12.118), because the two-port network is assumed loss-less, resulting in

$$ML\big|_{dB} = -10\log_{10}\left(1-\rho^2\right) = 10\log_{10}\left(1-0.316^2\right) = 0.46\ \text{dB}$$

Figure 12.26. Power measurements associated with a loss-less two-port network.

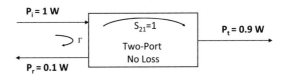

$P_i = 1\ \text{W}$

Γ

$S_{21} = 1$

Two-Port
No Loss

$P_t = 0.9\ \text{W}$

$P_r = 0.1\ \text{W}$

EXAMPLE 12.9

The reflection coefficients for a source and DUT were found to be 0.1778 and 0.3162, respectively. What is their corresponding VSWR?

Solution:
Rearranging Eq. (12.114), we can write

$$VSWR = \frac{1+\rho}{1-\rho}$$

leading to

$$\text{VSWR}_{SMA} = \frac{1+0.1778}{1-0.1778} = 1.432 \quad \text{and} \quad \text{VSWR}_{DUT} = \frac{1+0.3162}{1-0.3162} = 1.924$$

Exercises

12.18. The voltage wave incident at the interface of an RF component was measured with a network analyzer to be $0.9e^{j\pi/4}$ V, and the reflected wave was found to be $0.05e^{-j\pi/8}$ V. What is the reflection coefficient? What is the return and mismatch loss associated with this RF component?

ANS.
$\Gamma = 0.0213 - j0.0513 = 0.0556e^{-j1.1781}$,
RL = 25.1 dB, ML = 0.0134 dB.

12.19. An RF connector has a VSWR of 2.0, what is the reflection coefficient and return loss associated with this connector.

ANS. $\rho = 1/3$, RL = 9.54 dB.

The mismatch uncertainty will give the range of the power uncertainty at the input of the DUT. This power can be higher or lower than the source power. The mismatch uncertainty is in most cases the most significant contributor to the total error related to an RF measurement.

Consider the following situation where the magnitude of the reflection coefficients for two interfaces was found to be ρ_1 and ρ_2. Since the phase of each reflection coefficients is unknown, we can only assume that each reflection coefficient is either positive or negative in value, that is, $\Gamma_1 = \rho_1$ or $\Gamma_1 = -\rho_1$. Similarly for the other reflection coefficient, $\Gamma_2 = \rho_2$ or $\Gamma_2 = -\rho_2$. Under such uncertainty, we can bound the mismatch loss by substituting each of these reflection coefficients into Eq. (12.119) and identify the range of possible ML value as

$$\frac{(1-\rho_1 \cdot \rho_2)^2}{(1-\rho_1^2) \cdot (1-\rho_2^2)} \leq \text{ML} \leq \frac{(1+\rho_1 \cdot \rho_2)^2}{(1-\rho_1^2) \cdot (1-\rho_2^2)} \tag{12.122}$$

Expressing the mismatch loss in decibels, we can write Eq. (12.122) as

$$10\log_{10}\left[\frac{(1-\rho_1 \cdot \rho_2)^2}{(1-\rho_1^2) \cdot (1-\rho_2^2)}\right] \leq \text{ML}\big|_{dB} \leq 10\log_{10}\left[\frac{(1+\rho_1 \cdot \rho_2)^2}{(1-\rho_1^2) \cdot (1-\rho_2^2)}\right] \tag{12.123}$$

The difference between the limits of this bound is called mismatch uncertainty MU and is expressed mathematically as

$$\text{MU}\big|_{dB} \triangleq 10\log_{10}\left[\frac{(1+\rho_1 \cdot \rho_2)^2}{(1-\rho_1^2) \cdot (1-\rho_2^2)}\right] - 10\log_{10}\left[\frac{(1-\rho_1 \cdot \rho_2)^2}{(1-\rho_1^2) \cdot (1-\rho_2^2)}\right] \tag{12.124}$$

Since the denominator is the same for each term on the right hand side, the above expression reduces to

$$\left.MU\right|_{dB} = 20\log_{10}\left(1 + \rho_1 \cdot \rho_2\right) - 20\log_{10}\left(1 - \rho_1 \cdot \rho_2\right) \qquad (12.125)$$

If VSWR measurements are made instead of reflection coefficients measurements, Eq. (12.124) can be rewritten using the relationship between reflection coefficient and VSWR provided through Eq. (12.110). For instance, representing the VSWR of a RF source by $VSWR_{SRC}$ and that of the DUT by $VSWR_{DUT}$, the mismatch uncertainty can then be found from

$$\left.MU\right|_{dB} = 20\log\left(1 + \frac{VSWR_{SRC} - 1}{VSWR_{SRC} + 1} \cdot \frac{VSWR_{DUT} - 1}{VSWR_{DUT} + 1}\right) - 20\log\left(1 - \frac{VSWR_{SRC} - 1}{VSWR_{SRC} + 1} \cdot \frac{VSWR_{DUT} - 1}{VSWR_{DUT} + 1}\right)$$

$$(12.126)$$

The following example commonly found in RF testing applications involving an SMA connector, RF transmission line and a DUT will help to illustrate the application of the above theory.

EXAMPLE 12.10

A typical RF test setup involving a DIB is shown in Figure 12.27. Here a source is to be connected to a DUT through an SMA connector and a 50-Ω RF line fabricated on a PCB. If the SMA connector and DUT have return losses of 15 dB and 10 dB, respectively, what is the expected level of mismatch uncertainty associated with this test setup? Assume that the RF transmission line is perfect and has no loss.

Solution:
According to Eq. (12.116), we find the magnitude of the reflection coefficients in terms of the return loss as

$$\rho = |\Gamma| = 10^{-RL[dB]/20}$$

Correspondingly, the reflection coefficient for the SMA connector and DUT are found from above to be

$$\rho_{SMA} = 10^{-15/20} = 0.1778$$

and

$$\rho_{DUT} = 10^{-10/20} = 0.3162$$

According to Eq. (12.122), together with the knowledge of the magnitude of each reflection coefficient, we can bound the mismatch loss ML as

$$\frac{\left(1 - \rho_{SMA} \cdot \rho_{DUT}\right)^2}{\left(1 - \rho_{SMA}^2\right) \cdot \left(1 - \rho_{DUT}^2\right)} \leq ML \leq \frac{\left(1 + \rho_{SMA} \cdot \rho_{DUT}\right)^2}{\left(1 - \rho_{SMA}^2\right) \cdot \left(1 - \rho_{DUT}^2\right)}$$

leading to

$$1.022 \leq ML \leq 1.280$$

or in dB as

$$0.094 \text{ dB} \leq ML\big|_{dB} \leq 1.07 \text{ dB}$$

This suggest that the mismatch loss can range from 0.094 dB to 1.07 dB; a 0.978-dB variation. This, of course, is the mismatch uncertainty $MU\big|_{dB} = 0.978$ dB (see Eq. (12.124)).

Figure 12.27. RF DUT connected with a RF transmission line to SMA connector.

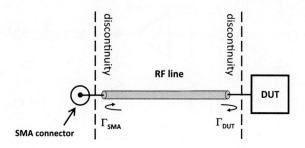

EXAMPLE 12.11

An RF source of an ATE system is coupled on a DIB to an amplifier (DUT) through an SMA connector with a return loss of 15 dB. The output of the amplifier is also coupled to a power meter in the ATE through another SMA connector with a return loss of $RL_M = 15$ dB. The DUT is known to have an input impedance of $Z_{IN} = 30$ Ω and an output return loss RL_{OUT} of 10 dB. The reverse isolation of the amplifier S_{12} is assumed to be high, so this DUT can be treated as a unilateral device. All RF lines are assumed to have a characteristic impedance of 50 Ω and are loss-free. An illustration of the physical configuration is shown in Figure 12.28a, and its corresponding electrical schematic is given in Figure 12.28b. Compute the return loss, mismatch loss, and mismatch uncertainty of the amplifier gain test.

Solution:
To start, we have to convert the input impedance using Eq. (12.101) into the reflection coefficient ρ. For a unilateral system, the reflection coefficient is equal to the input reflection coefficient S_{11}. Using Eq. (12.101), we get

Figure 12.28. DUT connected to the source and measurement path of an ATE. (a) Physical representation of the test setup. (b) Equivalent electrical wave representation.

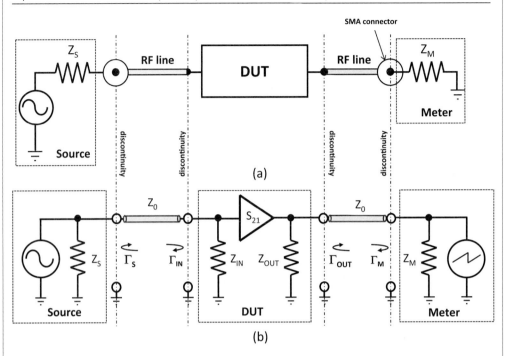

(a)

(b)

$$\rho_{IN} = |\Gamma_{IN}| = \left| \frac{Z_{IN} - Z_0}{Z_{IN} + Z_0} \right| = \left| \frac{30\,\Omega - 50\,\Omega}{30\,\Omega + 50\,\Omega} \right| = 0.250$$

and the return loss in decibels is

$$RL_{IN}\big|_{dB} = -20\log_{10}(\rho_{IN}) = -20\log_{10}(0.250) = 12.04\,dB$$

The source on the test board has a return loss RL_S of 15 dB. This is equal to a reflection coefficient Γ_S of

$$\rho_S = |\Gamma_S| = 10^{-RL_S/20} = 10^{-15/20} = 0.178$$

These two reflections on the single RF line leading into the DUT causes a mismatch loss ML that can be bounded between the upper and lower mismatch loss as

$$\frac{(1 - \rho_S \cdot \rho_{IN})^2}{(1 - \rho_S^2) \cdot (1 - \rho_{IN}^2)} \leq ML_{S-IN} \leq \frac{(1 + \rho_S \cdot \rho_{IN})^2}{(1 - \rho_S^2) \cdot (1 - \rho_{IN}^2)}$$

Substituting the appropriate values, we write

$$\frac{\left(1-0.250\cdot0.178\right)^2}{\left(1-0.250^2\right)\cdot\left(1-0.178^2\right)} \leq ML_{S-IN} \leq \frac{\left(1+0.250\cdot0.178\right)^2}{\left(1-0.250^2\right)\cdot\left(1-0.178^2\right)}$$

which after some algebra reduces to

$$1.005 \leq ML_{S-IN} \leq 1.202$$

We can express this mismatch loss in dB and write

$$0.025 \text{ dB} \leq ML_{S-IN}\big|_{dB} \leq 0.798 \text{ dB}$$

Here we see the mismatch loss varies between 0.025 dB to 0.798 dB. This is a 0.773 dB possible variation due to the standing wave on the front-end line on the DIB board. Hence, the mismatch uncertainty is

$$MU_{S-IN}\big|_{dB} = 0.773 \text{ dB}$$

For the output side of the DUT similar calculations can be performed. First, we compute the reflection coefficient related to the DUT output, i.e.,

$$\left|\Gamma_{OUT}\right| = \rho_{OUT} = 10^{-RL_{OUT}/20} = 10^{-10/20} = 0.316$$

and the one related to the power meter is found according to

$$\left|\Gamma_M\right| = \rho_M = 10^{-RL_M/20} = 10^{-15/20} = 0.178$$

Using the magnitude of these two reflection coefficients, we can bound the mismatch loss ML and write

$$\frac{\left(1-\rho_{OUT}\cdot\rho_M\right)^2}{\left(1-\rho_{OUT}^2\right)\cdot\left(1-\rho_M^2\right)} \leq ML_{OUT-M} \leq \frac{\left(1+\rho_{OUT}\cdot\rho_M\right)^2}{\left(1-\rho_{OUT}^2\right)\cdot\left(1-\rho_M^2\right)}$$

Substituting the appropriate values, wewrite

$$\frac{\left(1-0.316\cdot0.178\right)^2}{\left(1-0.316^2\right)\cdot\left(1-0.178^2\right)} \leq ML_{OUT-M} \leq \frac{\left(1+0.316\cdot0.178\right)^2}{\left(1-0.316^2\right)\cdot\left(1-0.178^2\right)}$$

leading to

$$1.022 \leq ML_{OUT-M} \leq 1.280$$

We can express this mismatch loss in decibels and write

$$0.094 \text{ dB} \leq ML_{OUT-M}|_{dB} \leq 1.072 \text{ dB}$$

The uncertainty of the mismatch loss is then

$$MU_{OUT-M}|_{dB} = 1.072 - 0.094 = 0.978 \text{ dB}$$

The output side of the ATE will be measured with an uncertainty of 0.978 dB. When combining the results from the input and output side of the device, that is,

$$MU_{ALL}|_{dB} = MU_{S-IN}|_{dB} + MU_{OUT-M}|_{dB}$$

in decibels, we get a mismatch uncertainty of ~1.75 dB. Again, this level of uncertainty is the result of the lack of phase information associated with each reflection coefficient. In the next sub-section, we shall consider a technique that can be used to reduce this uncertainty while working directly with scalar quantities for each reflection coefficient.

Exercises

12.20. The reflection coefficients between a source and DUT were found to be 0.21 and 0.25, respectively. What is the bound on the mismatch loss and the overall mismatch loss uncertainty?

ANS.
0.008 dB $\leq ML \leq$ 0.921 dB; MU = 0.913 dB.

12.21. The mismatch uncertainty associated with the input path to a source-DUT connection is 1 dB. The uncertainty associated with the output path is 0.5 dB. What is the total uncertainty associated with this measurement?

ANS. MU = 1.5 dB.

Reducing Measurement Uncertainty

To make reliable RF measurements, it is often necessary to reduce the measurement uncertainty caused by multiple discontinuities. This is often done in two steps. In the first step, the impedances of the source and the DUT are transformed with impedance transformers (also known as matching networks) to the line impedance. Techniques for this matching are provided in Section 15.5 of Chapter 15. In the second step, the impact of any residual mismatch can be further reduced

Figure 12.29. (a) Attenuators inserted at source and load side of test setup. (b) Resistive attenuator in T configuration. (c) Resistive attenuator in Π configuration.

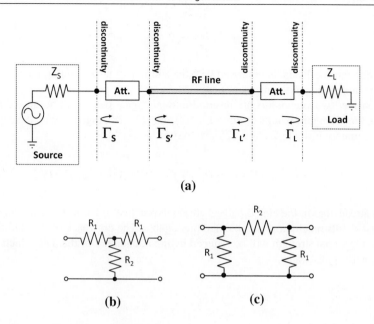

(a)

(b)   **(c)**

with the addition of an attenuator network or pad as depicted Figure 12.29a. In some cases the addition of this attenuator alone may be sufficient, especially when the impedance mismatch is not very large. Sometimes the attenuator network is combined with the matching network into a single network.

One simple realization of an attenuator is a T- or Π–shaped resistive network as shown in Figure 12.29b and 12.29c. Resistive attenuators typically have a wide bandwidth with a constant VSWR, as well as being easy to build using three resistors. The attenuator works on the principle of reducing the reflected wave back into the transmission line by twice its attenuation. To understand this, consider the following paragraph.

Let us assume that the reflection coefficient at an interface between the load and RF line is Γ_L. Recall that this reflection coefficients refers to the ratio of the reflected to incident power, that is, $P_L^- = \rho_L^2 P_L^+$. Now if an attenuator is inserted before the load interface, then the power incident at the interface is reduced, that is, $P_L^+ = A_{att}^2 P_{att}^+$, where P_{att}^+ is the power incident on the input side of the attenuator and A_{att} is the attenuation factor. The reflected wave at the load interface can then be described as $P_L^- = \rho_L^2 \cdot P_L^+ = \rho_L^2 \cdot A_{att}^2 \cdot P_{att}^+$. As this reflected wave passes back through the attenuator and gets attenuated by the same factor A_{att}, the reflected wave at the attenuator input is $P_{att}^- = A_{att}^4 \cdot \rho_L^2 \cdot P_{att}^+$. We can then define the reflection coefficient at the input of the attenuator as

$$\rho_{L'}^2 = \frac{P_{att}^-}{P_{att}^+} = A_{att}^4 \cdot \rho_L^2 \qquad (12.127)$$

Because A_{att} is less than one (i.e., gain loss), we see that the reflection coefficient at the input side of the attenuator has been reduced in magnitude by this factor A_{att}.

Following a similar argument for the attenuator placed right after the source interface, we can express the modified reflection coefficient at the source side as

$$\rho_{S'}^2 = \frac{P_{att}^-}{P_{att}^+} = A_{att}^4 \cdot \rho_S^2 \tag{12.128}$$

While the reflection coefficient is improved with the addition of an attenuator, it comes at a reduction in the dynamic range of the measurement.

If we consider two attenuators are inserted into the signal path, such because that shown in Figure 12.29, then using the results of Eqs. (12.127) and (12.128), together with the expression for mismatch uncertainty given in Eq. (12.125), we write

$$MU\big|_{dB} = 20\log_{10}\left(1 + A_{att}^4 \cdot \rho_S \cdot \rho_L\right) - 20\log_{10}\left(1 - A_{att}^4 \cdot \rho_S \cdot \rho_L\right) \tag{12.129}$$

Adding attenuators in the signal path is common practice in RF test engineering because it makes the test results much more reliable and predictable. The drawback of this additional attenuation is that the signal strength will be reduced by the attenuation of the pad, which can lead to a reduced signal-to-noise ratio.

EXAMPLE 12.12

The reflection coefficients for a source and DUT were found to be 0.1778 and 0.3162, respectively. If a 3-dB attenuator is placed at the source and DUT side in the signal path, what are the new reflection coefficients? What is the corresponding return loss with the attenuators inserted in the line and how much improvement was realized over the original arrangement? What impact does the attenuators have on the mismatch uncertainty of the measurement?

Solution:
A 3-dB attenuator has an linear magnitude gain given by

$$A_{att} = 10^{-3/20} = 0.707$$

Using Eqs. (12.127) and (12.128), we find the revised reflection coefficients as

$$\rho_{S'} = A_{att}^2 \rho_S = 0.707^2 \times 0.1778 = 0.0891$$
$$\rho_{DUT'} = A_{att}^2 \rho_{DUT} = 0.707^2 \times 0.3162 = 0.1585$$

Using Eq. (12.116), we find the return loss at the source side with and without the attenuators in the signal path as follows:

$$RL_S\big|_{dB} = -20\log_{10}\rho_S = -20\log_{10}(0.1778) = 15.0 \text{ dB}$$
$$RL_{S'}\big|_{dB} = -20\log_{10}\rho_{S'} = -20\log_{10}(0.0891) = 21.0 \text{ dB}$$

Similarly at the load side, we find

$$RL_{DUT}\big|_{dB} = -20\log_{10}\rho_{DUT} = -20\log_{10}(0.3162) = 10.0 \text{ dB}$$
$$RL_{DUT'}\big|_{dB} = -20\log_{10}\rho_{DUT'} = -20\log_{10}(0.1585) = 16.0 \text{ dB}$$

Here we see that the return loss has improved by twice the attenuator gain factor in decibels; that is, a 3-dB attenuator provides 6 dB improvement in return loss.

According to Eq. (12.125), we find the mismatch uncertainty without the attenuators present in the line as

$$
\begin{aligned}
MU\big|_{dB} &= 20\log_{10}(1+\rho_S\cdot\rho_{DUT}) - 20\log_{10}(1-\rho_S\cdot\rho_{DUT}) \\
&= 20\log_{10}(1+0.1778\cdot0.3162) - 20\log_{10}(1-0.1778\cdot0.3162) \\
&= 0.978 \text{ dB}
\end{aligned}
$$

Using the attenuators, we find the mismatch uncertainty as

$$
\begin{aligned}
MU\big|_{dB} &= 20\log_{10}(1+\rho_{S'}\cdot\rho_{DUT'}) - 20\log_{10}(1-\rho_{S'}\cdot\rho_{DUT'}) \\
&= 20\log_{10}(1+0.0891\times0.1585) - 20\log_{10}(1-0.0891\times0.1585) \\
&= 0.245 \text{ dB}
\end{aligned}
$$

The mismatch uncertainty has been reduced to 0.245 dB from 0.978 dB.

Exercises

12.22. The reflection coefficient for at a load is 0.3. How much attenuation is required to reduce the reflection coefficient to 0.1?

ANS. $A_{att} = 0.577, -4.8$ dB.

12.23. A 9-dB resistive pad is inserted in series with a source with a reflection coefficient of 0.4. What is the effective reflection coefficient of the source?

ANS. $\rho = 0.05$.

Insertion Loss

The reader first encountered insertion loss IL back in Section 12.2.7. There insertion loss was defined as the power loss associated with a two-port network inserted between the source and load. An alternative approach is one that involves the S-parameters of the two-port network. Assuming that the two-port network has been fully characterized, the insertion loss IL can be described as

$$IL\big|_{dB} = 10\log_{10}\left(\frac{|S_{21}|^2}{1-|S_{11}|^2}\right) \tag{12.130}$$

Figure 12.30. Highlighting the two-port transmission coefficient S_{21} for(a) insertion loss test setup, (b) mismatch loss test setup.

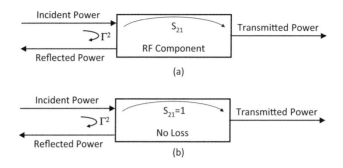

(a)

(b)

Here we adopt insertion loss as a loss instead of a gain (i.e., ratio of incident power to transmit power). We learned previously that $S_{11} = \Gamma$, so we can rewrite Eq. (12.130) as

$$IL\big|_{dB} = 10\log_{10}\left(\frac{|S_{21}|^2}{1-|\Gamma|^2}\right) \tag{12.131}$$

The reader may notice that the expression given here for insertion loss is very similar to that given in Eq. (12.117) for mismatch loss. The only difference is that insertion loss includes the term S_{21}. If one recalls the development of mismatch loss, the condition $S_{21} = 1$ was imposed on its definition.

While insertion loss and mismatch loss are very similar, they differ in terms of the transmission nature of the two-port network. To see this, compare the insertion loss expression given in Eq. (12.131) to the mismatch loss definition given in Eq. (12.117). On doing so, we see they are identical if and only if S_{21} is equal to unity. This leads us to conclude that insertion loss is a more general definition than mismatch loss; it is not restricted to the two-port condition of $S_{21} = 1$. To contrast these two loss definitions, Figure 12.30 summarizes their power flow and two-port conditions.

EXAMPLE 12.13

For a production test setup, an RF-DIB is connected to the ATE with an SMA RF cable. To characterize the losses, a network analyzer is first connected to the RF cable and the RF DIB as shown in Figure 12.31a. The RF cable is specified to have a forward transmission coefficient S_{21} equal to –0.3 dB. The DUT is replaced with a known 50-Ω load. The network analyzer measures S_{11} to be –15 dB. In production the network analyzer is replaced as shown in Figure 12.31b. What is the reflection coefficient Γ, VSWR, return loss, and the insertion loss of the RF cable-DIB combination? Also, what is the impedance seen by the network analyzer or ATE looking into the cable?

Figure 12.31. (a) A bench test setup for characterizing the interconnect associated with the RF line and DIB board. (b) Typical ATE production test setup for an RF DUT.

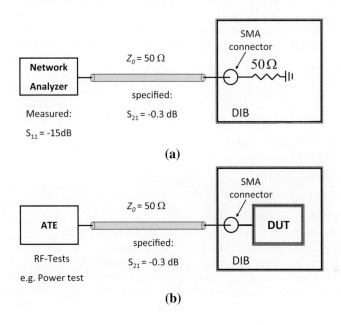

(a)

(b)

Solution:

The first step in solving this problem is to convert the two given S-parameters in decibels into voltage magnitude terms:

$$|S_{11}| = 10^{-15/20} = 0.1778 \text{ V}$$
$$|S_{21}| = 10^{-0.3/20} = 0.9661 \text{ V}$$

The magnitude of the reflection coefficient ρ associated with the RF line-DIB combination can be derived directly from the S_{11} measurement according to Eq. (12.96), leading to

$$\rho = |S_{11}| = 0.1778$$

Next, the VSWR is found from Eq. (12.113),

$$\text{VSWR} = \frac{1+0.1778}{1-0.1778} = 1.433$$

Similarly, the return loss is found from Eq. (12.116) to be

$$RL = -20\log_{10}(0.1778) = 15 \text{ dB}$$

The insertion loss of the DIB with the RF cable is then found from Eq. (12.131) with $\rho = |\Gamma|$ to be

$$IL = 10\log_{10}\left(\frac{|S_{21}|^2}{1-\rho^2}\right) = 10\log_{10}\left[\frac{0.9661^2}{1-0.1778^2}\right] = -0.161 \text{ dB}$$

Finally, the impedance seen by the network analyzer is found using Eq. (12.104) to be

$$Z_{IN} = Z_0 \frac{1+S_{11}}{1-S_{11}} = (50\ \Omega)\left(\frac{1+0.177e^{j0.125}}{1-0.177e^{j0.125}}\right) = 71.2 + j3.24\ \Omega$$

These numbers are showing a minimal impact on the test results while having a return loss of 15 dB. In Chapter 13 we will learn about a method in which to calibrate for this loss.

Exercises

12.24. An amplifier was characterized with the following S-parameters:

$S_{11} = -0.5638 - j0.2052,\quad S_{12} = 0.0433 + j0.0124$
$S_{21} = 2.165 + j1.250,\qquad S_{22} = -j0.5$

What is the insertion loss of this amplifier? What are the magnitudes of the reflection coefficients of the input and output ports of this amplifier? What is the input and output impedance of the amplifier, assuming that the *S*-parameters were obtained with ideal transmission lines with characteristic impedance of 50 Ω?

ANS.

IL = 9.9 dB, $\rho_{IN} = 0.6$, $\rho_{OUT} = 0.5$,

$Z_{IN} = 12.8 - j8.25\ \Omega$,

$Z_{OUT} = 30 - j40\ \Omega$

12.5 MODULATION

Modulation in mixed-signal or RF technology represents the process of varying a periodic waveform to transport a deterministic message. Typically, a sinusoidal waveform is used as a carrier signal for the modulation. The key parameters of a sinusoidal wave that can be modulated with a lower-frequency signal are the amplitude, the phase, and the frequency of the carrier signal. In the following sections, we will give a brief description of the fundamental modulations schemes.

12.5.1 Analog Modulation

Analog modulation is the transfer of an analog low-frequency signal—for example an audio signal for television over a high-frequency sinusoidal signal. In general, we can modulate the amplitude, frequency, and phase continuously, which results in amplitude modulation, frequency, modulation, and phase modulation, respectively. Below we will discuss the basics of these three analog modulation schemes.

Amplitude Modulation

Amplitude modulation (AM) had been the first technique used in telecommunication to transmit an audio message over a telephone line in the mid-1870s. Beginning with the Canadian inventor Reginald A. Fessenden in 1906, it was also the original method used for audio radio transmission and is still one of the most commonly used modulation schemes in modern telecommunication.

The general amplitude modulation formula is given by

$$s(t) = \left[1 + m \cdot s_m(t)\right] \cdot A_c \sin\left(2\pi f_c t + \phi_c\right)$$ (12.132)

where A_c, f_c, and ϕ_c are the parameters of the sinusoidal carrier wave, m is the modulation index, and $s_m(t)$ is the modulation signal. For a sinusoidal modulation signal,

$$s_m(t) = A_m \sin\left(2\pi f_m t + \phi_m\right)$$ (12.133)

with $\phi_c = \phi_m = 0$, Eq. (12.132) reduces to

$$
\begin{aligned}
s(t) &= \left[1 + m \cdot A_m \sin\left(2\pi f_m t\right)\right] \cdot A_c \sin\left(2\pi f_c t\right) \\
&= A_c \sin\left(2\pi f_c t\right) + m \cdot A_m \sin\left(2\pi f_m t\right) \cdot A_c \sin\left(2\pi f_c t\right)
\end{aligned}
$$ (12.134)

If we assign the leftmost term on the right-hand side as the carrier signal, i.e., $s_c(t) = A_c \sin\left(2\pi f_c t\right)$ and the rightmost term as the product of the carrier signal and the modulation signal, denoted by, $s_{mc}(t) = m \cdot A_m \sin\left(2\pi f_m t\right) \cdot A_c \sin\left(2\pi f_c t\right)$ then we can rewrite Eq. (12.134) as $s(t) = s_c(t) + s_{mc}(t)$. Using the trigonometric identity,

$$\sin\theta \sin\psi = \frac{1}{2}\cos\left(\theta - \psi\right) - \frac{1}{2}\cos\left(\theta + \psi\right)$$ (12.135)

the signal $S_{mc}(t)$ can be rewritten as

$$s_{mc}(t) = \frac{m \cdot A_m \cdot A_c}{2} \cos\left[2\pi\left(f_m - f_c\right)t\right] - \frac{m \cdot A_m \cdot A_c}{2} \cos\left[2\pi\left(f_m + f_c\right)t\right]$$ (12.136)

Equations (12.132) and (12.136) describe a general representation of an amplitude-modulated signal for the case in which the modulation signal is a sinusoidal signal. It consists of the term $s_c(t)$, which is independent of the modulated signal, and a signal on both sides of the carrier representing the modulated signal at the sum and difference frequencies. Figure 12.32 illustrates the time-domain representation of an amplitude modulated signal for two cases, m = 0.5 and m = 1.0. Here A_m is set equal to one. Also shown in this figure is the frequency domain representation of this signal consisting of three impulses in frequency; one at the carrier frequency f_c, and two sidebands at $f_c \pm f_m$. On account of the two sidebands, this type of modulation is called double-sideband amplitude modulation (DSB-AM). However, it should be noted that each sideband carries the same information; hence DSB-AM is inefficient in terms of power usage. At least two-thirds of the total RF power is used for the carrier and the second sideband, which carries no additional information beyond the fact that a signal is present. To optimize the power efficiency, the carrier will often be suppressed. This amplitude modulation is called a *double-sideband suppressed carrier* (DSBSC), but for demodulation, the carrier needs to be regenerated in the demodulator circuit for conventional demodulator techniques.

Figure 12.32. Time and frequency domains of an amplitude modulated sinusoidal wave.

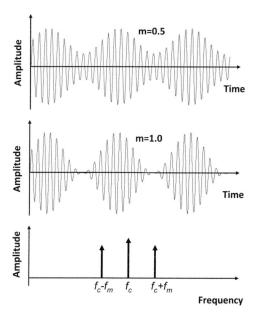

Frequency Modulation

Frequency modulation (FM) is well known as the broadcast signal format for FM radio. In contrast to AM, the modulation is achieved through the variance of the frequency of the carrier signal. This makes the modulation scheme more robust against amplitude noise.

Frequency modulation involves altering the frequency of the carrier according to some modulating signal. One might ask what is the frequency of a waveform. For a simple wave like a sine wave, the answer appears quite obvious, we can define the wave using an expression like

$$s(t) = A_c \cos\left[2\pi f_c t + \phi\right] \tag{12.137}$$

and identify the f_c as the wave's frequency. An alternative way is to represent such a wave as

$$s(t) = A_c \cos\left[\theta_i(t)\right] \tag{12.138}$$

where

$$\theta_i(t) = 2\pi f_c t + \phi \tag{12.139}$$

This term represents the instantaneous phase of the wave at time instant, t. For a simple sine wave, its frequency is constant and $\theta_i(t)$ increases steadily with time at the rate

$$\frac{d\theta_i(t)}{dt} = 2\pi f_c \tag{12.140}$$

We can define the FM wave, as one that results from the modulation of a carrier $s_c(t) = A_c \sin(2\pi f_c \cdot t)$ with signal $s_m(t)$ to be

$$s(t) = A_c \cos\left[2\pi f_i(t) \cdot t + \phi\right] \qquad (12.141)$$

where

$$f_i(t) = \frac{1}{2\pi}\frac{d\theta_i(t)}{dt} = f_c + k_f s_m(t) \qquad (12.142)$$

The term $f_i(t)$ represents the instantaneous frequency of the wave at the time instant, t. It is equal to the rate of change of the instantaneous phase $\theta_i(t)$ normalized by 2π. The term k_f is a constant whose value depends on the modulating system. It typically has units of hertz per volts.

The FM wave can now be used to convey information about the modulating signal in a manner very similar to that developed for AM modulation. Consider integrating Eq. (12.142) to get the instantaneous phase in terms of the modulating signal $S_m(t)$ as

$$\theta_i(t) = 2\pi f_c t + 2\pi k_f \int s_m(t)\, dt \qquad (12.143)$$

We can define the FM signal in terms of an arbitrary modulating signal by substituting Eq. (12.143) into Eq. (12.138) to obtain

$$s(t) = A_c \cos\left[2\pi f_c t + 2\pi k_f \int s_m(t)\, dt\right] \qquad (12.144)$$

To further our understanding of FM, consider the modulation signal as a simple sinusoidal signal as $s_m(t) = A_m \sin(2\pi f_m t)$. Substituting this signal into Eq. (12.142), we observe that the instantaneous FM signal frequency has the general form

$$f_i(t) f_c + k_f A_m \sin(2\pi f_m t) \qquad (12.145)$$

This expression indicates that the instantaneous frequency swings up and down around the carrier frequency f_c with amplitude $k_f A_m$. The range of this frequency change is commonly referred to as the peak frequency deviation value, denoted as

$$\Delta f = k_f A_m \qquad (12.146)$$

Combining the modulation signal $s_m(t) = A_m \sin(2\pi f_m t)$ together with Eq. (12.146) and Eq. (12.143), we can write the instantaneous phase as

$$\theta_i(t) = 2\pi f_c t + \frac{\Delta f}{f_m}\sin(2\pi f_m t) \qquad (12.147)$$

It is conventional to define a quantity called modulation index,

$$\beta = \frac{\Delta f}{f_m} \tag{12.148}$$

where we can write the FM wave in the form of

$$s(t) = A_c \cos\left[2\pi f_c t + \beta \sin\left(2\pi f_m t\right)\right] \tag{12.149}$$

This expression provides us with information about how the modulated signal varies with time. However, we often need to know the spectrum of the FM signal, as, for example, to determine the bandwidth of any filters, amplifiers, and so on, that the system requires.

Consider rewriting Eq. (12.149) using complex notation as follows

$$s(t) = A_c \Re e\left\{ e^{j\left[\omega_c t + \beta \sin(\omega_m t)\right]} \right\} \tag{12.150}$$

which, after several algebraic steps, leads to

$$s(t) = A_c \Re e\left\{ e^{j\left[w_c t + \beta \sin(\omega_m t)\right]} \right\} = A_c \Re e\left\{ e^{j\omega_c t} e^{j\beta \sin(\omega_m t)} \right\} \tag{12.151}$$

Using Bessel functional notation, we can write the above expression as

$$s(t) = A_c \Re e\left\{ e^{j\omega_c t} \sum_{k=-\infty}^{\infty} J_k(\beta) e^{j\beta k \omega_m t} \right\} = A_c \Re e\left\{ \sum_{k=-\infty}^{\infty} J_k(\beta) e^{j(\omega_c + k\omega_m)t} \right\} \tag{12.152}$$

where $J_k(\beta)$ is known as the Bessel function of the first kind of order k and argument β. The function $J_k(\beta)$ is a real number for real values of β (and any integer value for k). The reason for the introduction of the Bessel function representation is that Eq. (12.152) can be rewritten as a series of sinusoids (in much the same we do for a Fourier series) as follows

$$s(t) = A_c \sum_{k=-\infty}^{\infty} J_k(\beta) \cos\left[(\omega_c + k\omega_m)t\right] \tag{12.153}$$

Here we see an FM signal consists of an infinite number of sinusoidal components all positioned with respect to the carrier frequency ω_c provide $\beta \neq 0$. The side bands occur at multiples of the modulating frequency ω_m relative to the carrier frequency ω_c. As the series extends from minus infinity to plus infinity, FM signals have infinite bandwidth; albeit, the magnitude of these sidebands decrease with increasing index, k. This is a property of Bessel functions. For real systems, this requires limiting the bandwidth of the FM signal with a filter so that the signal stays within the channel bandwidth. An example of an FM signal is shown in Figure 12.33. The top plot illustrates the time domain FM signal for a sinusoidal modulating signal. The lower plot illustrates the multiple sidebands about the carrier signal.

Figure 12.33. Frequency modulated sinusoidal wave in the frequency and time domains.

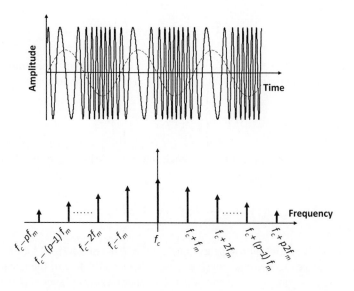

An FM signal created with a low modulation index (i.e., where $\Delta f \ll f_m$) is called a narrowband FM signal. Such an FM signal can be approximated as

$$s_{NB}(t) = A_c J_0(\beta)\cos(2\pi f_c t) + A_c J_1(\beta)\cos\left[2\pi(f_m - f_c)t\right] - A_c J_1(\beta)\cos\left[2\pi(f_m + f_c)t\right]$$

$$(12.154)$$

This narrowband FM signal is very similar to AM in that it has sideband components at $(f_c \pm f_m)$; hence narrowband FM signals only need a bandwidth of $2f_m$. It should be noted that the spectrum of a narrowband FM signal differs from AM in two ways. Firstly, the amplitude of an FM signal remains constant and secondly the phase of the sideband components are 180 degrees out of phase with respect to one another.

In the case of a high modulation index, an FM signal can be viewed as a sinusoid whose frequency is varied over a relatively large range. This implies that the bandwidth required for transmission is at least equal to the peak-to-peak frequency deviation or $2\Delta f$. A more refined analysis reveals the minimum bandwidth B required for FM signals is

$$B = 2(f_m + \Delta f)$$

$$(12.155)$$

This is known as Carson's rule for FM signals.

Phase Modulation

Phase modulation (PM) is very similar to frequency modulation, because any phase deviation can be interpreted as an instantaneous frequency variation. PM and FM are sometimes referred to as angular modulation. Consider a PM signal having the general form

$$s(t) = A_c \cos\left[\theta_i(t)\right]$$

$$(12.156)$$

where

$$\theta_i(t) = 2\pi f_c t + k_p s_m(t) \tag{12.157}$$

Here the term k_p is a constant whose value depends on the system. Combining Eqs. (12.156) and (12.157), we observe that the general form of a PM-modulated signal is

$$s(t) = A_c \cos\left[2\pi f_c t + k_p s_m(t)\right] \tag{12.158}$$

In the case of a sinusoidal modulating signal, that is, $s_m(t) = A_m \sin(2\pi f_m t)$, we can write the PM signal as

$$s(t) = A_c \cos\left[2\pi f_c t + k_p A_m \sin(2\pi f_m t)\right] \tag{12.159}$$

Although the above expression appears quite similar to the one derived for FM, it is important recognize the difference. To see this, consider that the time derivative of the instantaneous phase given in Eq. (12.157) is equal to the instantaneous angular frequency, such that we can write

$$
\begin{aligned}
f_i(t) &= \frac{1}{2\pi}\frac{d\theta_i(t)}{dt} \\
&= f_c + \frac{k_p A_m}{2\pi}\frac{d\sin(2\pi f_m t)}{dt} \\
&= f_c + f_m k_p A_m \cos(2\pi f_m t)
\end{aligned}
\tag{12.160}
$$

In comparison to the FM signal modulated by a similar signal, here we see the peak frequency deviation value is

$$\Delta f = f_m k_p A_m \tag{12.161}$$

which is dependent on both the amplitude and frequency of the modulating signal. This is quite different for the FM signal where we found $\Delta f = k_f A_m$ (see Eq. (12.146)). In terms of the modulation index, we find

$$\beta = \frac{\Delta f}{f_m} = k_p A_m \tag{12.162}$$

Substituting this result into Eq. (12.159), we write the PM signal subject to a sinusoidal modulating signal as

$$s(t) = A_c \cos\left[2\pi f_c t + \beta \sin(2\pi f_m t)\right] \tag{12.163}$$

We immediately recognize that this form of the PM signal is identical to that for the FM signal (i.e., see Eq. (12.149)). Hence, the frequency domain conclusions drawn previously for an FM signal applies equally to the PM signal (albeit under a sinusoidal modulation condition).

Even though PM is very similar to FM for a sinusoidal modulating signal, there is one important difference. In FM, the maximum frequency deviation Δf is proportional to the amplitude of the modulating signal $s_m(t)$ (see Eq. (12.146)). In PM, the maximum frequency deviation Δf depends on the amplitude A_m of the modulating signal $s_m(t)$ as well as its frequency f_m (Eq. (12.161)). Therefore, modulating signals with the same amplitude but different frequencies gives different values of maximum frequency deviation. The next example will illustrate this effect. In practice, PM is rarely used for analog modulation on account of this dependence.

EXAMPLE 12.14

A 100-mV, 1-kHz sinusoidal signal is used to modulate an FM and PM system with coefficients $k_f = 1$ kHz/V and $k_p = 1$ krad/V, respectively. The carrier frequency of both systems is 900 MHz. Compare peak-to-peak frequency deviation of these two systems. Repeat for a 100-mV, 10-kHz modulating signal.

Solution:
Using Eq. (12.146), at 1-kHz modulating frequency, we calculate the peak-to-peak-frequency deviation for the FM system as

$$\Delta f_{pp} = 2k_f A_m = 2 \times 1000 \, \frac{\text{Hz}}{\text{V}} \times 0.1 \, \text{V} = 200 \, \text{Hz}$$

Correspondingly, using Eq. (12.161), the peak-to-peak deviation of the PM system is

$$\Delta f_{pp} = 2f_m k_p A_m = 2 \times 1000 \, \text{Hz} \times 1000 \, \frac{\text{rad}}{\text{V}} \times 0.1 \, \text{V} = 200 \, \text{kHz} \cdot \text{rad} = 200 \, \text{kHz}$$

Repeating for a 10 kHz modulating signal, we find for the FM system remains the same as before, Δf_{pp}=200 Hz , but the PM system changes according to

$$\Delta f_{pp} = 2f_m k_p A_m = 2 \times 1000 \, \text{Hz} \times 1000 \, \frac{\text{rad}}{\text{V}} \times 0.1 \, \text{V} = 2 \, \text{MHz}$$

This example illustrates the subtle difference between FM and PM systems.

12.5.2 Digital Modulation

In contrast to the analog modulation, a digital signal with only a discrete number of valid states can be transmitted with a digital modulation scheme. Since the sinusoidal carrier signal is not modulated with an analog signal, but is shifted between discrete values, we refer to it as "shift keying." Like analog modulation, digital modulation schemes also vary in amplitude, phase, and frequency.

Figure 12.34. OOK (ASK) modulated signal.

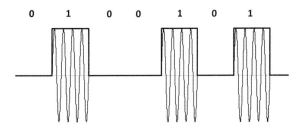

Figure 12.35. FSK modulation in the time and frequency domains.

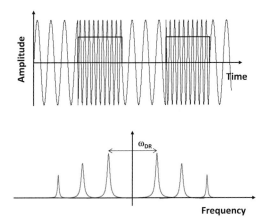

Amplitude Shift Keying

Amplitude shift keying (ASK) represents a simple form of digital modulation. The amplitude of the carrier $s_c(t) = A_c \cos(2\pi f_c t)$ varies in accordance with the bit stream, which keeps the frequency and phase constant. The level of the amplitude can be used to represent the binary values of 0s and 1s, by assigning different amplitude to these binary states. A special form of ASK is on–off keying (OOK) modulation where a zero amplitude sinewave is assigned to one of the binary states. In Figure 12.34 we illustrate OOK with logic 0 being assigned the zero amplitude sinewave.

Frequency Shift Keying

Frequency shift keying (FSK) is a modulation scheme similar to frequency modulation (FM). Instead of an analog modulating signal, a digital bit pattern is used to modulate a carrier wave. The simplest FSK is the binary FSK (often refer to as BFSK). BFSK transmits binary 0s and 1s information via the allocation of a single frequency per bit. This is captured in the time-frequency diagram shown in Figure 12.35. As the bits change, the frequency of the signal is shifts from one value to the opposite one.

The minimum frequency shift keying, or minimum shift keying (MSK), is a FSK modulation where the difference between the higher and the lower frequency is identical to half of the data (bit) rate. A variant of the MSK is the Gaussian MSK, or GMSK, which is used for cellular handsets using the global system for mobile communications (GSM) standard.

Figure 12.36. Phase shift modulated bit stream. The binary "1" signal has a phase offset of π radians with respect to the binary "0" logic value.

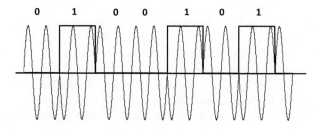

Figure 12.37. Constellation diagram for: (a) BPSK and (b) QPSK.

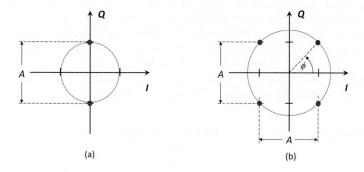

(a)   (b)

Phase Shift Keying

Phase shift keying (PSK) is a digital modulation scheme that conveys data by changing the phase of a carrier signal. The basic modulation scheme is the binary phase shift keying (BPSK), with exactly two defined phase states e.g.

$$s_0(t) = A\cos(2\pi f_c t) \text{ representing the binary "1"}$$

$$s_1(t) = A\cos(2\pi f_c t + \pi) \text{ representing the binary "0"}$$

A binary bit stream modulates the carrier waveform with constant amplitude, but alters the phase as a function of the binary signal to be transmitted, as seen at the beginning of every transmitted bit in Figure 12.36. A more general version of phase shift keying is the use of an array of $M = 2^n$ valid different phases, which represent multiple binary bits at one time.

Constellation Diagram

A convenient way of representing digital modulation schemes like PSK and ASK is on a constellation diagram. In a constellation diagram, the states of the modulation are shown in the in-phase-quadrature or IQ plane as shown in Figure 12.37 for two different signaling types: BPSK and QPSK. Here *I* represents the in-phase signal component and *Q* represents the quadrature or orthogonal signal component. In essence, the *IQ* plane is the familiar complex plane arising from complex number theory and is used to represent a transmitted signal using a complex number in rectangular form.

The constellation points in PSK are chosen such that the points are positioned with uniform angular spacing around a circle as shown in Figure 12.37. This gives maximum phase separation between adjacent points and thus gives the best immunity to a disturbance. As they are positioned around a circle, all points have the same amplitude, resulting in equal transmitted energy. For the BPSK signal shown in Figure 12.37a, two phases 180 degrees out of phase are used to represent the transmitted signal. In the case of a quadrature phase shift keying (QPSK) signal, each signal is placed 90 degrees out of phase of one another (i.e., occupies a quadrant of the complex plane).

12.5.3 Quadrature Amplitude Modulation

Quadrature amplitude modulation (QAM) is a modulation format that makes use of both PSK and ASK. The amplitude and phase of *M* sinusoidal waves are used to encode the in-coming digital data. The amplitude of each sine wave is uniformly distributed over some voltage range; also, the phase is evenly distributed over 180 degrees. For instance, for 2^M-QAM, where *M* is an integer, *M* sine waves having *M*/2 amplitudes and 180/*M* degrees out of phase with respect to one another are used to modulate *M*-bits of digital data. Commonly used signaling schemes are 16QAM and 64QAM. Figure 12.38 illustrates a 16QAM example. Here 4 bits of digital data are transmitted using two different amplitude sine waves with 90 degrees of phase separation.

12.5.4 Orthogonal Frequency Division Multiplexing

Orthogonal frequency division multiplexing (OFDM) is a frequently used modulation scheme for wideband digital communication. Wireless data communication systems like WLAN in the

Figure 12.38. 16QAM constellation diagram with 4 bit per symbol.

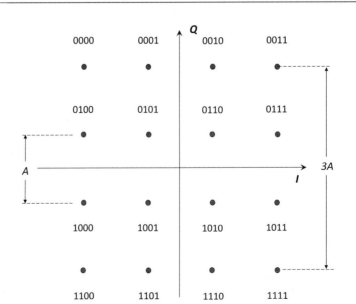

Figure 12.39. Orthogonal subcarrier frequency spectrum of an OFDM signal.

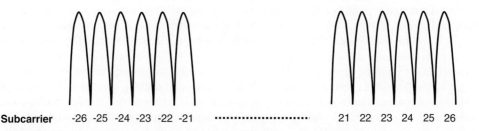

Subcarrier -26 -25 -24 -23 -22 -21 •••••••••••••••••• 21 22 23 24 25 26

Figure 12.40. Orthogonal signals for different modulation schemes within one time interval. The integral of the product of such signals over a time interval results in a value of 0.

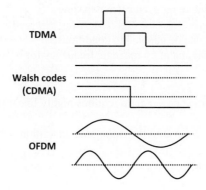

802.11a/b/g/n standard, or the WiMAX 802.16 standard are taking advantage on the high data rate combined with the robustness of data transmission. The high channel efficiency with OFDM modulated signals is utilized for digital radio systems like DAB, HD Radio or TV standards like DVB-T, DVB-H, as well as on wired standard for ADSL and VDSL broadband access and power line communication (PLC).

OFDM is a frequency division multiplexing (FDM) scheme, which utilizes a digital multi-carrier modulation method. A large number of closely spaced subcarriers are used to carry the data as shown in Figure 12.39. The data are divided into several parallel data streams, or channels, one for each subcarrier. Each subcarrier is modulated with a conventional modulation scheme, such as the QAM, and carries the data at a low symbol rate. The frequencies of these subcarriers are chosen so that the subcarriers are orthogonal to each other. Due to this orthogonally, the critical crosstalk between the subcarriers (subchannels) is eliminated, and inter-carrier guard bands are not required.

OFDM is addresses the problem with single carrier modulation (SCM) schemes. In a given environment, the symbol interval of a SCM becomes shorter than the delay spread associated with transmission. This effect becomes more acute with increased data rate. Multicarrier modulation (MCM) formats solve this problem by decreasing the symbol rate by increasing the number of carriers transmitted in parallel. The lower data rate within a single carrier of the MCM reduces the intersymbol interference (ISI) effect and reduces any multipath effects which might also occur. OFDM is one of the MCM formats and has a flexible modulation format, which can be scaled to meet the needs of multiple applications.

Another important consideration of OFDM is orthogonality (see Figure 12.40). Two signals are orthogonal in a given time interval if, when multiplied together and then integrated over that time interval, the result is zero. TDMA (time-domain multiple access) is not considered an orthogonal coding scheme. Over some time interval, the other signal is zero, so that the product of the two will be zero. Welch codes used in CDMA systems, like IS-95, are orthogonal and are probably the most common form of orthogonal signaling. OFDM uses Walsh codes instead of sinusoids. In a given period, the sinusoids will be orthogonally modulated over an integer number of cycles. Using multiple sinusoids instead of Walsh codes produces a spectrum, in which it is possible to assign a carrier frequency to a code channel. In other frequency division multiplexing systems (FDMA), the channel spacing is typically greater than the symbol rate to avoid an overlapping of the spectrum. In OFDM, the carriers are orthogonal and can overlap without interfering with one other.

Advantages of OFDM

- Increased efficiency due to reduced carrier spacing.
- Resistance to fading.
- Flexible data rate that can be adapted to conditions.
- Flexible frequency assignment up to single frequency for broadcast applications.

Disadvantages of OFDM

- The high peak-to-average power ratio requires highly linear transmit channels, especially a highly linear PA.
- The criticality of phase noise to the orthogonal condition require a tight phase noise and noise system.

12.6 SUMMARY

This chapter gave a brief introduction to the basics of RF theory as it applies to RF test. With the progression of semiconductor technology to higher bandwidths and higher operating frequencies to large volume applications, the need for the test engineer to expand his or her skill set with RF test has become a necessity.

This chapter began by introducing the reader to the concept of a propagating wave and the various means by which a wave is quantified (e.g., wavelength, velocity, etc.). Such metrics are important to the RF test engineer. For instance, understanding the wavelength of an RF signal is helpful when designing a device interface board and placing critical components on the test board. Knowledge of the wavelength will allow the engineer to find the best compromise between signal integrity, component size selection and placement, and position of the DUT.

Power measurements are the most common measurement in the RF test. Understanding its definition, as well as its related measures such as average power, peak power, power for framed signals, and so on, is critical to the power measurement itself. For instance, knowing the ratio of peak to average power commonly referred to as the crest factor enables the setting of the ATE to be optimized for maximum measurement accuracy.

A major portion of this chapter was dedicated to the theory of noise and phase noise. Like all electrical systems, noise is one of the limiting factors of system performance; the same holds true for RF systems. A brief description of various noise sources was introduced, as well as the concept of a system noise figure. Because most RF systems have a frequency translation device, such as a mixer and VCO, noise has a specific impact on their operation, which is quantified by a phase noise metric. The phase noise of the ATE reference source is often the limiting factor of any phase noise measurement, and it needs to be considered carefully in a production test.

A section of this chapter was dedicated to the concept of S-parameters because it applies to an n-port network, such as a two-port network. S-parameters are used to describe the small-signal performance of the network as seen from one port to another. S-parameters are useful descriptions of the how waves propagate and reflect at each interface of the network. Measures like reflection coefficient, mismatch loss, insertion and transducer loss, and various power gains can easily be defined in terms of these S-parameters. Most importantly, the idea of a mismatch uncertainty and its impact on a power measurement was introduced. Mismatch uncertainty is often the most significant contributor to measurement error. When estimating the quality of a test result, a careful discussion of the mismatch uncertainty will allow the engineer to estimate the tolerance in a measurement.

The final section of this chapter described several forms of modulation. We began with a set of traditional analog modulation schemes such as AM and FM used in early radio and television systems, followed with several digital modulation schemes used primarily for digital communication systems. This section concluded with a brief overview of an OFDM modulation as this scheme is used in many commercial RF communication systems.

Now that the reader has been introduced to the basics of RF theory they can move on to the next chapter where the specifics of various RF tests are described.

PROBLEMS

12.1. List four different RF applications and the corresponding frequency bands that they use.

12.2. List three major differences between mixed signal and RF test requirements.

12.3. Name the typical power level for noise, radio receive signals, data communication system transmit signals, and radio station transmit signals.

12.4. What is the difference between a scalar and a vector measure?

12.5. An RF transmission line on GETEK board material is used to delay a signal with a frequency of 2.4 GHz by 180 degrees. If the effective dielectric constant for this material is $\varepsilon_{\mathit{eff}} = 3.9$, what line length should be used?

12.6. An RF transmission line constructed on a Teflon printed circuit board is 10 mm in length. The effective dielectric constant for Teflon is $\varepsilon_{\mathit{eff}} = 2.2$. If the RF line is excited by a signal with frequency of 900 MHz do we treat this line as a distributed set of elements or as a lumped LRC equivalent assuming the 5% line length rule.

12.7. How is the wavelength changed when an electromagnetic wave is transferred from free space transmission to that guided by a microstripline?

12.8. Explain the superimposition principle.

12.9. Which circuits show power and voltage out of phase?

12.10. What is a universal definition of electrical power?

12.11. What is thereal and imaginary instantaneous power associated with a circuit if $v(t) = 0.25\cos\left(2\pi \times 10^4 \cdot t\right)$ and $i(t) = 0.25\cos\left(2\pi \times 10^4 \cdot t - \pi/5\right)$? *Hint:* See Exercise 12.7.

12.12. What is the average power associated with a circuit if $v(t) = 0.25\cos\left(2\pi \times 10^4 \cdot t\right)$ and $i(t) = 0.25\cos\left(2\pi \times 10^4 \cdot t - \pi/5\right)$?

12.13. What is the expected level of power associated with a pulsed signal if the continuous wave signal has an average power of 5 W and the pulsing action has a 20% duty cycle? Express your answer in terms of watts and in decibels.

12.14. Why is it important to take the crest factor of a signal into account when defining an RF system?

12.15. The crest factor of a continuous waveform is 12 dB. If the average power is 0.4 W, what is the peak power associated with this waveform?

12.16. A signal consisting of 16 uncorrelated sinusoids of 1.5 V. Estimate the crest factor associated with the composite waveform.

12.17. List the crest factors of three signals.

12.18. What is the formula to calculate power in dBm from the linear scale of Watt?

12.19. What is the ratio in decibels for doubling the power and doubling the voltage?

12.20. What load condition is required to transfer the maximum power to an electrical load?

12.21. What load condition is required to have no reflections?

12.22. A 1-V source with source impedance $Z_S = 100 + j21\ \Omega$ drives a load impedance of $Z_L = 50 + j10\Omega$. What is the average power delivered to the load and how does it compare to the maximum power available from the source? What is the average power delivered by the source?

12.23. A 1-V source with source impedance $Z_S = 50 + j10\Omega$ drives a load impedance of $Z_L = 100 + j21\ \Omega$. What is the average power delivered to the load and how does it compare to the maximum power available from the source? What is the average power delivered by the source?

12.24. Describe in your own words: (a) insertion loss, (b) transducer loss.

12.25. A source is terminated in a load where the load power was found to be 0.52 W. The same load was then connected to the end of a trace on a DIB board connected to the source through an SMA connector. The load power was then measured to be 0.42 dB. What is the insertion loss associated with the RF connector and DIB trace?

12.26. The maximum available power from a source is 5.0 W. If a transducer with a 50 Ω load is connected to the source, what is the transducer loss is the load power is 4.2 W?

12.27. What is the difference between a dissipative and reflection loss?

12.28. If a spectral analyzer reveals a power level of -92 dBm at 900 MHz using a resolution bandwidth of 80 kHz, what is the actual PSD appearing at the input to this instrument?

12.29. What method can be used to measure a signal below -174 dBm?

12.30. Why is it important to have a gain stage with a low noise figure as the first stage in an RF system?

12.31. If a given amplifier has a 3 dB noise figure at 300 K, a noise bandwidth of 10 MHz, and an input resistance of 50 Ω, what is the output signal-to-noise ratio if the input signal RMS level is 100 μV?

12.32. What is the combined noise figure of three-stage amplifier? The first stage has a noise figure of 5.2 dB and a gain of 10 dB, the second stage has a noise figure of 4 dB and a gain of 10 dB, and the third stage has a noise figure of 5 dB and a gain of 10 dB.

12.33. What is the principal definition of phase noise and how could it be measured?

12.34. What are the paramters S_{11}, S_{21}, S_{12}, and S_{22} describing in a two-port network?

12.35. For which case are the parameters S_{21} and S_{12} identical?

12.36. Describe in your own words VSWR, reflection coefficient, return loss, insertion loss, and mismatch loss.

12.37. The voltage wave incident at the interface of an RF component was measured with a network analyzer to be $3.5e^{-j\pi/16}$ V and the reflected wave was found to be $0.03e^{j\pi/4}$ V. What is the reflection coefficient? What is the return and mismatch loss associated with this RF component?

12.38. An RF connector has a VSWR of 1.45, what is the reflection coefficient and return loss associated with this connector.

12.39. What is the cause of mismatch uncertainty and what is the impact of mismatch uncertainty on a test result?

12.40. The reflection coefficients between a DUT and load were found to be 0.09 and 0.15, respectively. What is the bound on the mismatch loss and the overall mismatch loss uncertainty?

12.41. The mismatch uncertainty associated with the input path to a source-DUT connection is 0.43 dB. The uncertainty associated with the output path is 0.24 dB. What is the total uncertainty associated with this measurement?

12.42. A source is to be connected to a DUT through an SMA connector and a 50–Ω RF line fabricated on a PCB. If the SMA connector and DUT have return losses of 12 dB and 25 dB, respectively, what is the expected level of mismatch uncertainty associated with this test setup? Assume the RF transmission line is perfect and has no loss.

12.43. An RF source of an ATE system is coupled on a DIB to an amplifier (DUT) through an SMA connector with a return loss of 18 dB. The output of the amplifier is also coupled to a power meter in the ATE through another SMA connector with a return loss of $RL_M = 21 \text{ dB}$. The DUT is known to have an input impedance of $Z_{IN} = 42 \ \Omega$ and an output return loss RL_{OUT} of 18 dB. The reverse isolation of the amplifier S_{12} is assumed to be high, so this DUT can be treated as a unilateral device. All RF lines are assumed to have a characteristic impedance of 50 Ω and are loss free. Compute the return loss, mismatch loss, and mismatch uncertainty of the amplifier gain test.

12.44. The reflection coefficient for at a load is 0.2. How much attenuation is required to reduce the reflection coefficient to 0.1?

12.45. If a 6-dB attenuation pad is inserted in series with a source with a reflection coefficient of 0.5. What is the effective reflection coefficient of the source?

12.46. An amplifier was characterized with the following *S*-parameters:

$$S_{11} = 0.3e^{-j2.7925}, \quad S_{12} = 0.1e^{j0.2793}$$
$$S_{21} = 1.5e^{j0.5236}, \quad S_{22} = 0.9e^{-j1.5708}$$

What is the insertion loss of this amplifier? What are the magnitudes of the reflection coefficients of the input and output ports of this amplifier? What is the input and output impedance of the amplifier, assuming that the S-parameters were obtained with ideal transmission lines with characteristic impedance of 50 Ω?

12.47. Why is the frequency modulation more robust against noise compared to amplitude modulation?

12.48. How can the power efficiency improved for AM and FM signals?

12.49. What is the difference between GFSK and FSK and what is the advantage of GFSK?

REFERENCES

1. National Telecommunications and Information Administration, U.S. Department of Commerce.

2. European Telecommunications Standards Institute (ETSI).

3. T. H. Friis, Noise figures of radio receivers, Proceedings of the IRE, July 1944, pp. 419–422.

4. K. Kurokawa, Power waves and the scattering matrix, *IEEE Transactions Microwave Theory and Techniques*, Mar. 1965, pp. 194–202

5. L. Besser and R. Gilmore, *Practical RF Circuit Design for Modern Wireless Systems,* Vol. 1, *Passive Circuits an Systems*, Artech House, Boston, 2003.

6. J. F. White, *High Frequency Techniques—An Introduction to RF and Microwave Engineering*, Wiley Interscience, Hoboken, NJ, 2004.

7. K. Feher and Engineers of Hewlett Packard, *Telecommunications Measurements, Analysis, and Instrumentation*, Noble Publishing Corp., Raleigh, NC, 1997.

CHAPTER 13

RF Test Methods

This chapter will describe the principles of RF testing of electronic circuits using commercial automated test equipment (ATE). These principles are largely based on the physical concepts related to wave propagation outlined in Chapter 12. However, these techniques also make use of the sampling principles related to DSP-based testing developed in the earlier part of this textbook. In this chapter we shall focus on the most common types of RF tests for standard devices, like mixers, VCOs, and power amplifiers, to name just a few. For devices not covered in this work, the test principles provided should be easily extendable with only minor modifications to the device or test setup. Of course, the better one understands the fundamentals of RF test, the easier it will be to adjust a given test technique to a specific problem at hand.

In an RF test, many different test types are used to decide if a device is within specification and whether it should be shipped to the customer. Below is a list of some of these methods that are being considered for a production-oriented RF test by the research community:

- Structural test
- Functional test
- Observer (model)-based test, which includes testing against a regression
- Specification-based test
- Correlation-based test (alternate test)
- Defect-oriented test
- Build-in test (BIT) and Build-in self-test (BIST)
- Statistical-based test (e.g., BER, EVM)

While the research community continues to search for the most effective approach, this chapter is limited to a discussion related to functional or specification-based testing with some discussion about observer-based and statistical-based testing. In a later chapter, we will provide some discussion about BIST and BIT for RF devices, as well as correlation-based test. However, for the most part, these techniques are still early ideas and further work is required before they are declared ready for production test. Nonetheless, the test engineer needs to be exposed to early ideas in order to be better understand present-day test problems and possible directions from which solutions will appear.

One operation that is common to all product-oriented RF tests is the measurement of power and phase related to single or multiple sinusoidal waves. Generally, such measurements are made through the capture of the time-domain information using a digitizing operation, followed by an FFT to determine the frequency domain description of the signal. Our readers at this stage of the textbook should be able to fully understand this approach and the advantages it offers the test engineer. For those readers who skipped ahead to this chapter, they should consider reviewing Chapters 8–11 before reading this chapter.

This chapter will begin with a description of the general architecture of the RF section of the ATE. This will help provide a better understanding of how power measurements are made, as well as their limitations. Subsequently, we will describe various RF test metrics like power gain, intermodulation distortion, harmonics, noise figure, and phase noise. In this chapter we will also learn about some statistical test metrics that apply at the system level, like bit error rate (BER) and error vector magnitude (EVM) test measures.

When developing an understanding of RF test techniques, it is important to think of the RF portion of the ATE as an RF system unto itself. In this way, it is easier to encapsulate an understanding of the system operation based on maximum power limitations, linear power range, noise floor, dynamic range, gain, bandwidth and phase noise. These limitations are common to all RF systems and will continue to play a role in the RF test technique of subsequent chapters.

13.1 SCALAR MEASUREMENT METHODS

The scalar power measurement is the basis for a large number of RF test techniques used in production. In this section, we will discuss the principles and the limitations of a scalar power measurement made with an ATE, and we will compare them to the methods used in focus bench equipment found in characterization laboratories.

13.1.1 Principles of a Scalar Power Measurement

Let us look at the three common methods used to extract a power signal and, correspondingly, extract a power metric. This will include a discussion of the principles and limitations of a (a) calorimetric power sensor, (b) superheterodyne power sensor, and (c) zero-IF (intermediate frequency) with sampling power sensor.

Calorimetric Power Sensor

A power meter constructed with calorimetric and bolometric power sensors, also referred to as thermocouples, measure the true heating power of an RF signal. As shown in Figure 13.1, the RF power is transformed into an equivalent thermal power and detected with a temperature

Figure 13.1. Simple calorimetric set-up for accurate RF power measurement.

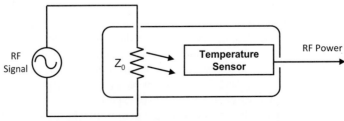

sensor by terminating the power source, here depicted as RF source, with a resistor with the real impedance Z_0. The signal from the temperature sensor is processed and displayed. These calorimetric sensors are very accurate and measure the RF power, including the noise floor, harmonics, spurious, and the signal itself in the measurement bandwidth of the power sensor. This technique is used in most RF power meters. These methods are very accurate with respect to the true RMS power of a signal, even for complex signal types. However, they have some serious limitations at low power levels as both the signal and the noise are combined into one power number. A related approach is based on computing the RMS value of a power signal electrically through an integration operation. Such instruments are limited in accuracy when measuring complex signals with a high crest factor or a complex duty cycle in contrast to calorimetric sensors.

Superheterodyne Power Sensor

Superheterodyne instruments measure power with a power sensor comparable to a spectrum analyzer. The principal architecture of these spectrum analyzers are based on the input stage of a traditional FM radio, which includes a mixer, an IFamplifier with an IFtracking filter, and a power detector as shown in Figure 13.2. While sweeping the center frequency of the IF-tracking filter with a sweeping generator over a predetermined range of frequencies, a spectrum analyzer is capable of measuring signal power as a function of the frequency (as opposed to power as a function of time). With the pre-selector filter together with the tracking filter, a spectrum analyzer is capable of limiting the bandwidth of the measured signal, which increases the measurement dynamic range significantly compared to a power meter. With this architecture, input bandwidths as high as 30 GHz can be measured, because only a small frequency range is measured at any one time. This has the added benefit that background noise levels appearing at the input will not compress or saturate the IF log amplifier and detector stages of the analyzer. The use of a logarithmic amplifier provides the instrument with the capability to separate signals that differ in signal strength by over 90 dB. On account of its versatility, the spectrum analyzer is standard equipment for bench characterization of various parameters requiring power measurements in the frequency domain.

Figure 13.2. Block diagram of a typical superheterodyne spectrum analyzer.

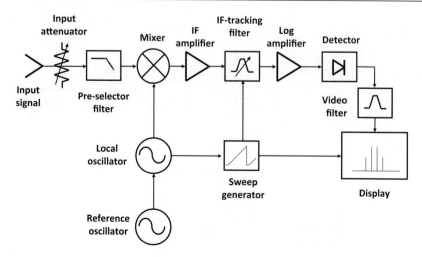

Zero-IF with Sampling Power Sensor

The architecture of a typical ATE power meter is much simpler than that seen for the superheterodyne spectrum analyzer. Most testers are built with what is referred to as a zero-IF (ZIF) architecture followed by an ADC to convert the IF signal into a digital signal as a function of time for further digital signal processing. A block diagram of the ZIF architecture can be seen in Figure 13.3. Some systems have a second IF stage to suppress spurious responses and allow for additional filtering. The mixer is a non-image reject mixer, one that generates both the in-band signal and its image. The overall bandwidth of the instrument is limited by the bandwidth of the front-end low-noise amplifier (LNA), as well by as the bandwidth of the IF stage and, possibly, the ADC. Typical input bandwidths found on commercial ATEs are about 6 GHz. It is important to note that the instrument described here based on the ZIF architecture performs measurements in the time domain on voltage samples presented to the ADC and converts this information into a power signal as a function of frequency using a FFT. This operation is clearly quite different from that which is performed by a spectrum analyzer.

Let us assume that an N-point data sequence $x[n]$ corresponds to the voltage signal appearing at the input to the ADC. Subsequently, an N-point FFT analysis is performed on this data set. As we learned in Chapter 10, the RMS value of the spectral coefficients c_k corresponding to this data set can be written in the usual way as

$$c_{k,RMS} = \begin{cases} |Y[k]|, & k = 0 \\ \sqrt{2}|Y[k]|, & k = 0,\ldots,N/2-1 \\ \dfrac{|Y[k]|}{\sqrt{2}}, & k = N/2 \end{cases} \tag{13.1}$$

where

$$Y = \frac{FFT\{x[n]\}}{N}$$

The power in each bin can then be found by simply squaring each $c_{k,RMS}$ term above and normalized by the appropriate real-valued input impedance level Z_0 associated with the ADC. Generally, Z_0 is

Figure 13.3. Block diagram of a typical ZIF ATE measurement path architecture.

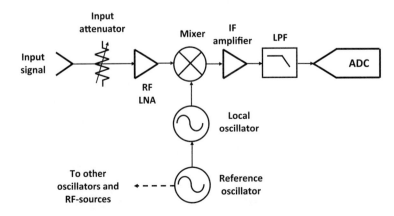

a standardized value of 50 Ω. For example, if the signal A is located in bin M_A and another signal B is located in bin M_B, then the power in dBm associated with each term is computed according to

$$P_{S_A}\Big|_{dBm} = 10 \log_{10} \left[\frac{C^2_{M_A, RMS}/Z_0}{1\,mW} \right] \tag{13.2}$$

and

$$P_{S_B}\Big|_{dBm} = 10 \log_{10} \left[\frac{C^2_{M_B, RMS}/Z_0}{1\,mW} \right] \tag{13.3}$$

An important limitation of the ZIF architecture is the requirement for a front-end LNA with large input signal bandwidth, a low-noise figure, a high compression point, and a low intermodulation distortion level (more on these two in moment). On account of the design difficulties of achieving all these specifications simultaneously, the LNA becomes the critical component of this instrument and often limits the smallest power level that can be measured. Another limitation of the ZIF architecture is the conversion range associated with the ADC. For small signals in the presence of a carrier, or some other large signal, the ADC must be capable of digitizing both the large and the small signals simultaneously. A similar argument can be made for measuring the phase noise associated with a signal; a large signal component carries a small signal in the form of a phase modulation. Of course, the ADC has a limited resolution, which ultimately limits the dynamic range of signals that it can convert. Another issue is, of course, the need for coherency between the measured IF signal and the sampling clock of the ADC. Generally, this is handled by setting the frequency of the local oscillator (LO) such that the measured signal frequency falls directly into a bin of the FFT. Since there is no settling time associated with the sampling process (as long as the signal is within the ADC bandwidth), the ZIF architecture leads to quicker test time than the spectrum analyzer architecture.

An important aspect of any power measurement regardless of the instrument's internal architecture is to establish the correct signal level at its input so that the system produces an output signal that is of the highest quality representation possible, i.e., one with highest accuracy. On the one hand, a large signal will maintain a large signal-to-noise ratio but will, simultaneously, result in a low signal-to-distortion ratio. The net result is poor measurement accuracy. On the other hand, a small signal will result in a low signal-to-noise ratio but a large signal-to-distortion ratio. Again, the net result is poor measurement accuracy. Somewhere in between a large signal input and a small signal input will result in the optimum input situation. It is the goal of the test engineer to establish these optimum signal conditions so each and every measurement is made with the right amount of accuracy. This is achieved with an attenuator at the front end of the instrument (see, for instance, Figure 13.3) whose gain is set by the test engineer according to the maximum expected level of power involved in the measurement.

In the context of an RF power measurement, we can see that these effects trade off in terms of the signal spectrum appearing at the output of the instrument. Let us assume that the input power signal consists of two independent tones. If the input peak power level is set too low through a large amount of attenuation, then the output of the instrument will consist of two very small-sized tones, positioned very close to the instrument noise floor, as illustrated in Figure 13.4a. If the attenuation is reduced, then higher level tones would be present at the instrument output, as seen in the spectrum of Figure 13.4b. At the optimum attenuation setting, some distortion caused by the instrument will just start to appear along side the two input tones produced by the DUT. If the attenuation is further reduced, then the two input tones will cause the instrument to compress, whereby the signal level of the two tones will decrease, the noise floor level will decrease and, in addition, higher distortion products will appear at the instrument output. This situation is depicted

Figure 13.4. Impact of expected power level setting on the signal, distortion and noise level: (a) expected power level set too high, (b) expected power level set correct, (c) expected power level set too low.

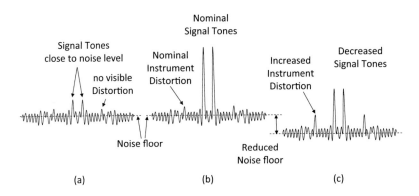

(a)   (b)   (c)

in Figure 13.4c. This problem is further complicated by the calibration procedure that follows the measurement. Since calibration factors are generated in terms of the attenuation level selected, if additional attenuation occurs through the instrument on account of compression, the calibration process will be in error, greatly reducing the accuracy of the measurement.

When testing the power of a known signal, it is important to know the full spectrum, as well as any time dependency, since out-of-band and unwanted signals can compress the measurement path in the same manner as the measurement signal. In the case of small signals, one can find that the wideband noise associated with the signal can cause the instrument to compress if the attenuator is not set properly. For large power signals, the crest factor of the signal, which includes multiple tones as well, must be taken into account. For example, in Section 13.2.4 we learned that the average or burst power in a given frame for a WLAN OFDM signal is 11 dB lower than its peak power. In this case, it is necessary to set the maximum expected power 11 dB higher than the burst power.

Commercial ATE RF power instruments are typically built according to the ZIF architecture shown in Figure 13.3. Like most ATE instruments, they are based on sampling and require coherent operation for best operation. In general, this might require synchronizing the DUT timing to the timing base of the ATE. Moreover, coherency requires that the incoming signal frequency to the ADC, denoted by f_{IF}, be related to the ADC sampling rate F_S according to

$$f_{IF} = \frac{M}{N} F_S \tag{13.4}$$

where N is the number of samples collected by the sampling process and M corresponds to the FFT bin in which the incoming signal will fall. Here we denote the ADC input frequency as that corresponding to the intermediate frequency associated with the mixer output, that is, f_{IF}. To have all signal power within one FFT bin, M must be an integer. When M is not an integer, some of the signal power will be distributed over multiple bins, causing frequency leakage and possible measurement error. This becomes even more problematic when testing for the phase noise of a signal, since the contribution of the phase noise of the signal and the power leakage cannot be separated. These two test situations are illustrated in the FFT plots shown in Figure 13.5. The top plot illustrates the noncoherent sampling situation and the bottom plot illustrates the desired coherent sampling situation.

Figure 13.5. A comparison of a non-coherent test situation (a) with a test condition that is coherent (b).

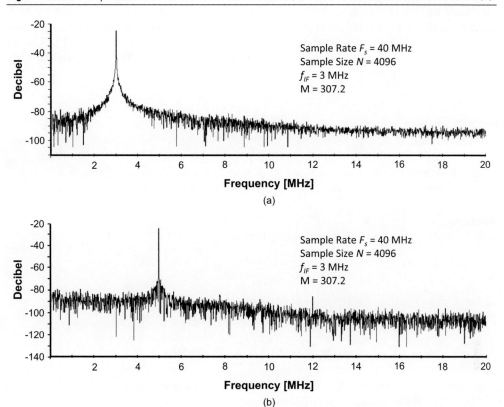

As the ADC input is derived from the output of the mixer, it is, in turn, related to the local oscillator frequency f_{LO}. The input frequencies to the mixer f_{RF} and f_{LO} are related to its output intermediate frequency f_{IF} according to

$$f_{IF} = f_{LO} - f_{RF} \tag{13.5}$$

or

$$f_{IF} = f_{RF} - f_{LO} \tag{13.6}$$

The case of Eq. (13.5) is low-side injection (the frequency of the RF signal is lower than the LO frequency), and for Eq. (13.6), it is high-side injection. Substituting Eq. (13.4) into Eq. (13.5) the LO frequency is

$$f_{LO} = \frac{M}{N} F_S + f_{RF} \tag{13.7}$$

or substituting Eq. (13.4) into Eq. (13.6), the LO frequency is

$$f_{LO} = \frac{M}{N} F_S - f_{RF} \tag{13.8}$$

Because the RF test signal frequency is, in general, arbitrary in value, we find from the above expression that the LO frequency will also have to be an arbitrary in value, even though the ADC input is constrained to a finite set of discrete frequencies. Fortunately, on ATEs, the LO is a continuous wave oscillator and is therefore not constrained to any finite set of frequencies. This, in turn, makes it a straightforward exercise to develop a frequency plan to have all single-frequency RF signals sampled coherently.

EXAMPLE 13.1

What IF frequency should be selected for a ZIF ATE architecture to be coherent with an ADC having a sampling rate of 20 MHz and 1024 samples are required for FFT processing.

Solution:
With $N = 1024$, the Fourier frequency will be given by

$$F_f = \frac{F_S}{N} = \frac{20\,\text{MHz}}{1024} = 19{,}531.250\,\text{Hz}$$

As the IF frequency must be a multiple of F_f, we select a value for M such that it shares no common factors with N; that is, let $M = 511$, then

$$f_{IF} = M \cdot \frac{F_S}{N} = 511 \cdot \frac{20\,\text{MHz}}{1024} = 9.9805\,\text{MHz}$$

Therefore an IF frequency of 9.9805 MHz is required for coherent sampling.

13.1.2 Gain Measurement

Device gain is one of the standard tests for RF devices. Figure 13.6 illustrates a typical ATE arrangement. Here, we assume that the more general term "gain" is used as a synonym for the available gain of the DUT, defined as follows:

$$G_A = \frac{\text{power available from output line, } P_{OUT}}{\text{power available from source, } P_A} \qquad (13.9)$$

where we replaced the word network by output line to be specific to the problem at hand. Here available power simply refers to the network situation where the power transfer is maximized, that is, source and load are conjugate matched. We learned in Section 12.4.2 that not all the power available from the source will make its way to the input of the DUT, because some of this power will be reflected and dissipated by the DUT interface board and the DUT itself. Also, the power leaving the DUT will experience similar losses as it makes it way to the power meter or load. It will be required to de-embed the DUT performance from the measured performance. Therefore, we modify the definition of available gain to account for this loss and write the desired device gain G_A' as

$$G_A' = \frac{\text{power available from DUT, } P_{OUT,DUT}}{\text{power available from input line, } P_{IN,DUT}} \qquad (13.10)$$

Exercises

13.1. What IF frequency should be selected for a ZIF ATE architecture to be coherent with an ADC having a sampling rate of 100 MHz, if 20ww48 samples are required for FFT processing and the output of the mixer ranges between 25 and 50 MHz?

ANS. For $M = 615$, $f_{IF} = 30.029$ MHz

13.2. A 1-GHz RF signal is to be down converted by a mixer and directed to a digitizer for digital conversion. If the digitizer is sampled at a clock rate of 100 MHz and 4096 samples are collected for digital signal processing using an FFT, what should be the local oscillator frequencyif the mixer has an output IF frequency range between 40 and 60 MHz?

ANS. $f_{LO} = 987.52$ MHz for $M = 511$.

By introducing the power gain of the interface board or, for that matter, any lossy RF line, we can define the ***input operating gain***, $G_{o,IN}$, of the source-DUT interconnect as the ratio of the power made available to the input of the DUT to the power available from the source, that is,

$$G_{o,IN} = \frac{\text{power available from input line, } P_{IN,DUT}}{\text{power available from source, } P_A} \tag{13.11}$$

In a similar manner, we can define the ***output operating gain***, $G_{o,IN}$, of the DUT-load interconnect as the ratio of the power made available at the load to the power made available by the DUT, that is,

$$G_{o,OUT} = \frac{\text{power available from output line, } P_{OUT}}{\text{power available from DUT, } P_{OUT,DUT}} \tag{13.12}$$

Collectively, these three power gains are combined according to

$$G_A = G_{o,IN} \cdot G'_A \cdot G_{o,OUT} \tag{13.13}$$

Expressing this relationship in terms of decibels, we write

$$G_A\big|_{dB} = G_{o,IN}\big|_{dB} + G'_A\big|_{dB} + G_{o,OUT}\big|_{dB} \tag{13.14}$$

Finally, our end goal is to determine G'_A; rearranging Eq. (13.14), we write

$$G_A\big|_{dB} = G_{o,IN}\big|_{dB} + G'_A\big|_{dB} + G_{o,OUT}\big|_{dB} \tag{13.15}$$

Because G'_A and $G_{o,OUT}$ are always less than one for passive lines, it is customary to speak in terms of a loss, so the two rightmost terms, including its sign, is encapsulated into a new term called the ***operating loss***, *OL*, expressed in decibels, as

$$G'_A\big|_{dB} = G_A\big|_{dB} + \text{OL}\big|_{dB} \tag{13.16}$$

Figure 13.6. ATE gain test setup with interface board.

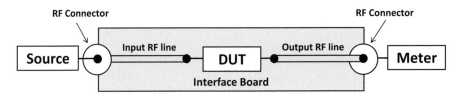

where

$$OL\big|_{dB} = -10 \cdot \log_{10}\left(G_{o,IN}\right) - 10 \cdot \log_{10}\left(G_{o,OUT}\right) \tag{13.17}$$

We can therefore conclude that to acquire the gain of a device with an ATE accurately, one must first account for the operating loss of the device interface board and any reflective effects associated with the DUT or meter.

One approach that is commonly used to estimate the operating losses is from the mismatch losses associated with the DUT. Using a calibrated network analyzer the input and output return loss of the DUT, together with the DIB, source, and meter, can be determined from which the mismatch loss associated with the input source-DUT signal path, denoted by ML_{S-DUT}, can be computed (see, for example, Eq. (12.117) in Chapter 12). Likewise, the mismatch loss associated with the DUT-meter signal path, denoted by ML_{DUT-M}, can be computed. The mismatch loss of the DUT, together with the any dissipative or other losses associated with the cables and traces on the input and output side of the DUT, denoted by $ML_{LINE-IN}$ and $ML_{LINE-OUT}$, respectively, can be approximately combined in a linear manner to provide an estimate of the operating losses of the test setup. Recalling that each mismatch loss is expressed in terms of a bound, we must write the operating loss in the following manner:

$$
\begin{aligned}
&\min\left\{ML_{S-DUT}\big|_{dB}\right\} + \min\left\{ML_{LINE-IN}\big|_{dB}\right\} + \\
&\min\left\{ML_{LINE-OUT}\big|_{dB}\right\} + \min\left\{ML_{DUT-M}\big|_{dB}\right\}
\end{aligned}
\le OL\big|_{dB} \le
\begin{aligned}
&\max\left\{ML_{S-DUT}\big|_{dB}\right\} + \max\left\{ML_{LINE-IN}\big|_{dB}\right\} + \\
&\max\left\{ML_{LINE-OUT}\big|_{dB}\right\} + \max\left\{ML_{DUT-M}\big|_{dB}\right\}
\end{aligned}
\tag{13.18}
$$

Any uncertainty associated with the mismatch loss parameter would translate to uncertainty in the operating loss parameters, which, in turn, leads to uncertainty in the gain term given by

$$\Delta G'_A\big|_{dB} = MU_{S-DUT}\big|_{dB} + MU_{DUT-M}\big|_{dB} + MU_{LINE-IN}\big|_{dB} + MU_{LINE-OUT}\big|_{dB} \tag{13.19}$$

where each uncertainty term is given by

$$
\begin{aligned}
MU_{S-DUT}\big|_{dB} &= \max\left\{ML_{S-DUT}\big|_{dB}\right\} - \min\left\{ML_{S-DUT}\big|_{dB}\right\} \\
MU_{DUT-M}\big|_{dB} &= \max\left\{ML_{DUT-M}\big|_{dB}\right\} - \min\left\{ML_{DUT-M}\big|_{dB}\right\} \\
MU_{LINE-IN}\big|_{dB} &= \max\left\{ML_{LINE-IN}\big|_{dB}\right\} - \min\left\{ML_{LINE-IN}\big|_{dB}\right\} \\
MU_{LINE-OUT}\big|_{dB} &= \max\left\{ML_{LINE-OUT}\big|_{dB}\right\} - \min\left\{ML_{LINE-OUT}\big|_{dB}\right\}
\end{aligned}
\tag{13.20}
$$

EXAMPLE 13.2

A gain test was performed on a DUT using an ATE with an interface board as shown in Figure 13.6. The source and power meter is located in the ATE. The return loss of the source and meter are known to be 15 and 21 dB, respectively. A network analyzer was used to measure the return loss of the input and output port of the DUT to be 10 and 12 dB, respectively. The PCB designer found through simulation that the traces and connectors introduce losses of about 0.5 dB equally on both sides of the DUT. If the ratio of the output available power to input available power was found to be 12.4 dB, what is the device gain when the board and reflective losses are taken into account?

Solution:
The reflection coefficient of the source, power meter, and DUT is found according to

$$\rho_S = 10^{-RL[dB]/20} = 10^{-15/20} = 0.1778$$
$$\rho_M = 10^{-RL[dB]/20} = 10^{-21/20} = 0.0892$$
$$\rho_{IN} = 10^{-RL[dB]/20} = 10^{-10/20} = 0.3162$$
$$\rho_{OUT} = 10^{-RL[dB]/20} = 10^{-12/20} = 0.2512$$

The expected level of mismatch loss on the DUT input side of the interconnect can be bounded using the theory of Chapter 12, Section 12.4.2, repeated here for convenience as

$$\frac{\left(1-\rho_S \cdot \rho_{IN}\right)^2}{\left(1-\rho_S^2\right)\cdot\left(1-\rho_{IN}^2\right)} \leq ML_{S-DUT} \leq \frac{\left(1+\rho_S \cdot \rho_{IN}\right)^2}{\left(1-\rho_S^2\right)\cdot\left(1-\rho_{IN}^2\right)}$$

Substituting the appropriate values, we find

$$0.0944 \text{ dB} \leq ML_{S-DUT}\big|_{dB} \leq 1.072 \text{ dB}$$

Repeating for the interconnect on the output side of the DUT, we obtain

$$\frac{\left(1-\rho_{OUT} \cdot \rho_M\right)^2}{\left(1-\rho_{OUT}^2\right)\cdot\left(1-\rho_M^2\right)} \leq ML_{DUT-M} \leq \frac{\left(1+\rho_{OUT} \cdot \rho_M\right)^2}{\left(1-\rho_{OUT}^2\right)\cdot\left(1-\rho_M^2\right)}$$

or

$$0.121 \text{ dB} \leq ML_{DUT-M}\big|_{dB} \leq 0.510 \text{ dB}$$

Subsequently, we compute the operating loss using Eq. (13.18) with the appropriate values substituted according to

$$0.0944 \text{ dB} + 0.5 \text{ dB} + 0.5 \text{ dB} + 0.121 \text{ dB} \leq OL\big|_{dB} \leq 1.072 \text{ dB} + 0.5 \text{ dB} + 0.5 \text{ dB} + 0.510 \text{ dB}$$

which reduces to

$$1.215 \text{ dB} \leq OL\big|_{dB} \leq 2.582 \text{ dB}$$

Finally, using Eq. (13.16), we can estimate the actual DUT gain to be

$$12.4 \text{ dB} + 1.215 \text{ dB} \leq G'_A\big|_{dB} \leq 12.4 \text{ dB} + 2.582 \text{ dB}$$

which further reduces to

$$13.61\,dB \le G'_A\big|_{dB} \le 14.98\,dB$$

The available gain of the device is therefore bounded between 13.61 dB and 14.98 dB when the interface board and DUT reflection losses are account for.

An alternative approach is to characterize the lossy line with a set of S-parameters, in much the same way that any two-port network would be characterized. Consider that a source is connected to a load through a lossy RF line. In addition, we shall assume that the lossy line is physically separated from the source and the load by two other unspecified transmission lines as shown in Figure 13.7a. We can then model this situation with the electrical wave network shown in Figure 13.7b. Here the lossy RF line is modeled with a set of S-parameters, denoted by $\{S_{11}, S_{12}, S_{21}, S_{22}\}$ and a set of reflection coefficients $\{\Gamma_S, \Gamma_{IN}, \Gamma_{OUT}, \Gamma_L\}$ established at the plane of discontinuities at the input and output ports of the lossy RF line, as indicated in Figure 13.7b. The specifics of the two other transmission lines are not important, because their effects are fully accounted for by the source and load reflection coefficients.

In order to measure the waves at both the input and output port of the lossy RF line, connectors must be inserted into the signal path at each end in order for the network analyzer to gain access to these waves. However, in many test situations this may not be possible or even desirable. Instead, a method that makes measurements at one end of the lossy line where a connector is generally available is preferred. This method makes use of several carefully calibrated standard loads that are inserted into a socket at the load end of the line hence giving rise to the name ***in-socket calibration method***. These standards include a short circuit, open circuit and a 50 Ω load. Since passive networks are generally reciprocal—that is, operate exactly the same way if the source and load is interchanged—the forward and reverse transmission coefficients of the S-parameter description are equal, that is, $S_{12} = S_{21}$. This implies that the lossy RF line model consists of only three unknown parameters. Through the application of three separate measurements (involving

Figure 13.7. (a) Illustrating the physical separation between a source, lossy RF line and a load. (b) A two-port equivalent representation of the lossy RF line with corresponding reflection coefficients shown.

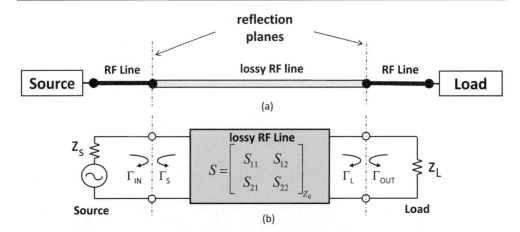

Figure 13.8. Making measurements at input port of lossy RF line using the in-socket calibration method: (a) The three terminating conditions for establishing the Sparameters of the lossy RF line. (b) Load with Z_{DUT}.

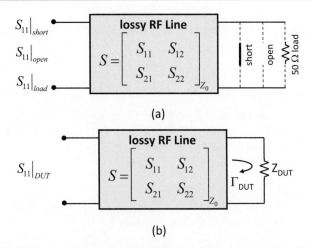

(a)

(b)

both magnitude and phase) at the input port of the lossy line corresponding to three separate termination conditions, as illustrated in Figure 13.8a, we obtain the following set of S-parameters, assuming the load is close to the matched condition (where the load reflection is small):

$$S_{11} = S_{11,load},$$

$$S_{12} = \sqrt{\left(\frac{\left(S_{11,load} - S_{11,short}\right)\left(2S_{11,open} - 2S_{11,load}\right)}{S_{11,open} - S_{11,short}}\right)}$$

$$S_{21} = \sqrt{\left(\frac{\left(S_{11,load} - S_{11,short}\right)\left(2S_{11,open} - 2S_{11,load}\right)}{S_{11,open} - S_{11,short}}\right)}, \quad S_{22} = \frac{S_{11,open} - 2S_{11,load} + S_{11,short}}{S_{11,open} - S_{11,short}}$$

$$(13.21)$$

A fourth measurement involving the DUT is performed in the exact same manner as that for the calibrated standard loads as shown in Figure 13.8b. This measurement provides information about the reflection loss as seen by the source, but it also provides information about what the lossy line sees at the load end of the line. Let us assume that the line is terminated with a device having input impedance Z_{DUT} and that the wave parameter made at the input of the line is $S_{11,DUT}$. In this case, the reflection loss seen by the source is simply

$$RL = -20 \cdot \log_{10} \left| S_{11,DUT} \right| \qquad (13.22)$$

and, equivalently, the mismatch loss ML can be computed from

$$ML = -10 \cdot \log_{10} \left(1 - \left| S_{11,DUT} \right|^2 \right) \qquad (13.23)$$

The load reflection coefficient Γ_{DUT} corresponding to the impedance seen looking into the load at the interface of the line and load can be computed according to

$$\Gamma_{DUT} = \frac{\left(S_{11,DUT} - S_{11,load}\right)}{S_{21} \cdot S_{12} + S_{22} \cdot \left(S_{11,DUT} - S_{11,load}\right)} \tag{13.24}$$

This, in turn, implies that the equivalent impedance seen by the lossy RF line toward the load side is

$$Z_{DUT} = Z_0 \left(\frac{1 + \Gamma_{DUT}}{1 - \Gamma_{DUT}}\right) \Omega \tag{13.25}$$

where Z_0 is generally assumed to be equal to 50 Ω.

Now that the lossy RF line has been fully characterized, we can quantify the input operating gain, $G_{o,IN}$, through a network analysis in terms of the S-parameters of the line, as well as the reflection coefficient at the DUT interface $G_{o,IN}$, to find

$$G_o = \frac{\left(1 - \left|\Gamma_{DUT}\right|^2\right)}{\left|1 - S_{22} \cdot \Gamma_{DUT}\right|^2} \left|S_{12} \cdot S_{21}\right| \tag{13.26}$$

The operating losses OL would then be computed from this operating gain according to Eq. (13.17).

Returning to the test setup shown in Figure 13.6, we can make use of the above theory and determine the operating loss of the input and output signal paths. In the case of the input path, we would insert the calibration standards into the DUT socket and perform the corresponding S_{11} measurements. Collectively, the S-parameters of the input signal path would then be determined through application of Eq. (13.21). Next, the S_{11} parameter corresponding to the line loaded with the input port of the DUT would be measured. Finally, we can put all the measurements together to arrive at the input path operating gain as

$$G_{o,IN} = \frac{\left(1 - \left|\Gamma_{IN,DUT}\right|^2\right)}{\left|1 - S_{22}^{IN} \cdot \Gamma_{IN,DUT}\right|^2} \left|S_{12}^{IN} \cdot S_{21}^{IN}\right| \tag{13.27}$$

where

$$\Gamma_{IN,DUT} = \frac{\left(S_{11,DUT}^{IN} - S_{11,load}^{IN}\right)}{S_{12}^{IN} \cdot S_{21}^{IN} + S_{22}^{IN} \cdot \left(S_{11,DUT}^{IN} - S_{11,load}^{IN}\right)} \tag{13.28}$$

Here we make use of the superscript *IN* as a reference to the specific S-parameters related to the input signal path. These superscripts are only used here to distinguish from the parameters related to the output signal path with the superscript OUT. Often, we will simply drop their use from our notation if their meaning is clear from the context (such as in Example 13.3).

In the case of the output signal path, the approach is very similar. The exception is that the test signal is driven into the output port of the load board where the meter would normally be connected. While the flow of power is opposite to that which the signal path normally sees, this prevents no difficulty to the characterization result, since the transmission line is reciprocal in nature. Also, we should note that output port of the DUT is used to load the line instead of the DUT input port. The DUT must be biased under normal operating conditions during this test. The output signal path gain would then be found from

$$G_{o,OUT} = \frac{\left(1 - \left|\Gamma_{OUT,DUT}\right|^2\right)}{\left|1 - S_{22}^{OUT} \cdot \Gamma_{OUT,DUT}\right|^2} \left|S_{12}^{OUT} \cdot S_{21}^{OUT}\right| \tag{13.29}$$

where

$$\Gamma_{OUT,DUT} = \frac{\left(S_{11,DUT}^{OUT} - S_{11,load}^{OUT}\right)}{S_{12}^{OUT} \cdot S_{21}^{OUT} + S_{22}^{OUT} \cdot \left(S_{11,DUT}^{OUT} - S_{11,load}^{OUT}\right)} \tag{13.30}$$

The in-socket calibration method of computing operating losses is more advantageous than the method based on estimated mismatch losses, because it is a true measurement of power loss. Moreover, this method includes any power loss due to mismatches in the signal path, any variations in the operating conditions of the device, and any dissipative loss associated with the board traces and matching components. Finally, the S-parameter-based measurement approach is not subject to the measurement uncertainty. This is a result of using both magnitude and phase information in all calculations.

EXAMPLE 13.3

What is the operating loss associated with the input side of the interface board shown in Figure 13.6, if the following measurements at the input port to the board were obtained using the in-socket calibration method, together with a DUT measurement:

$$S_{11,DUT} = -0.2 + j0.2, \quad S_{11,open} = -0.9 - j0.1,$$

$$S_{11,short} = 0.8 + j0.1, \quad S_{11,load} = -0.7 - j0.1$$

Solution:

The input side of the interface board shown in Figure 13.6 includes a lossy trace, an RF connector, and a socket for the DUT. Since the S_{11} port measurements include the effects of all three components in their result, the network diagram shown in Figure 13.7d correctly models this situation. We can then use the theory above to determine the operating loss of this interface board.

Using the above measured data, together with the S-parameter formulae seen listed in Eq. (13.21), we compute

$$S_{22} = \frac{S_{11,open} - 2S_{11,load} + S_{11,short}}{S_{11,open} - S_{11,short}}$$

$$= \frac{(-0.9 - j0.1) - 2(-0.7 - j0.1) + (0.8 + j0.1)}{(-0.9 - j0.1) - (0.8 + j0.1)}$$

$$= -0.7679 - j0.0273$$

and

$$S_{12} = S_{21} = \sqrt{\left(S_{11,load} - S_{11,short}\right) \cdot \left(\frac{2S_{11,open} - 2S_{11,load}}{S_{11,open} - S_{11,short}}\right)}$$

$$= \sqrt{\left[(-0.7 - j0.1) - (0.8 + j0.1)\right] \cdot \left[\frac{2(-0.9 - j0.1) - 2(-0.7 - j0.1)}{(-0.9 - 0.1) - (0.8 + j0.1)}\right]}$$

$$= 0.0046 - j0.5946$$

Next, using Eq. (13.27), we solve for the reflection coefficient at the input terminals of the DUT as

$$\Gamma_{IN,DUT} = \frac{\left(S_{11,DUT} - S_{11,load}\right)}{S_{21} \cdot S_{12} + S_{22} \cdot \left(S_{11,DUT} - S_{11,load}\right)}$$

$$= \frac{(-0.2 + j0.2) - (-0.7 - j0.1)}{(0.0046 - j0.5946)(0.0046 - j0.5946) + (-0.7679 - j0.0273)\left[(-0.2 + j0.2) - (-0.7 - j0.1)\right]}$$

$$= -0.740 - j0.158$$

Finally, we compute the operating loss using Eqs. (13.27) and (13.17) as follows:

$$G_{o,IN} = \frac{\left(1 - |\Gamma_{DUT}|^2\right)}{|1 - S_{22} \cdot \Gamma_{DUT}|^2} |S_{12} \cdot S_{21}|$$

$$= \frac{\left(1 - |-0.740 - j0.158|^2\right)| (0.0046 - j0.5946) \cdot (0.0046 - j0.5946)|}{|1 - (-0.7679 - j0.0273) \cdot (-0.740 - j0.158)|^2}$$

$$= 0.719$$

and

$$OL\big|_{dB} = -10 \cdot \log_{10}\left(G_{o,IN}\right) = -10 \cdot \log_{10}(0.719) = 1.434 \text{ dB}$$

The operating loss of the input RF line is then found to be 1.434 dB. It is interesting to note that the device is operating with an input impedance given by

$$Z_{DUT} = \left(\frac{1 + \Gamma_{DUT}}{1 - \Gamma_{DUT}}\right) \times 50 \text{ }\Omega$$

$$= \left[\frac{1 + (-0.740 - j0.158)}{1 - (-0.740 - j0.158)}\right] \times 50 \text{ }\Omega$$

$$= 7.01 - j5.187 \text{ }\Omega$$

Clearly the input impedance to the device is significantly different from the ideal situation of 50 Ω.

13.1.3 Scalar Power Measures Versus Time

A common measurement for dynamic systems is measure of how signal power varies over time. Typical examples are the power droop when turning on an amplifier or the settling time of a transceiver. Another common measurement is one that observes how the signal frequency changes over time, such as that related to the settling time of a VCO controlled by a PLL. Both measurements are executed in the same way by sampling of the RF power signal; the difference is in the analysis of the captured data.

To measure the time-dependent power such as that shown in Figure 13.9, it will be necessary to synchronize the digitizer with the device by utilizing the trigger bus of the test system and start digitizing the power at a known time point. This might be set within a digital pattern controlling the state of the device. The captured time signal can be processed to calculate the envelope using the ATE DSP or PC algorithm. The envelope is representative of the RF power over time, which can be extracted by calculating the peak power per period of the RF sinusoidal signal.

Exercises

13.3. A gain test was performed on a DUT and found to be 8.76 dB. The DUT is connected to an ATE through an interface board with an operating loss of 0.9 dB. What is the actual gain of the device?

ANS. 9.66 dB.

13.4. A gain test was performed on a DUT using an ATE with an interface board. The gain was found to be 8.76 dB. If the source, the meter, and the input and output ports of the DUT have return losses of 10, 12, 20, and 9 dB, respectively, and the traces introduce losses totaling 0.35 dB, what is actual gain of the DUT?

ANS. $9.39 \text{ dB} \leq G_A' \leq 11.49 \text{ dB}$.

13.5. What is the operating loss associated with the output side of an interface board shown in Figure 13.6, if the following measurements at the output port of the board were obtained using the in-socket calibration method, together with a single DUT measurement:

$S_{11,DUT} = 0.1 + j0.2$, $S_{11,open} = -0.8 - j0.2$, $S_{11,short} = 0.9 + j0.2$, and $S_{11,load} = -0.4 - j0.2$.

ANS. 1.67 dB.

13.6. For the conditions specified in Exercise 13.3, what is the output impedance of the DUT?

ANS. $9.0 - j7.8 \ \Omega$.

Figure 13.9. Time-dependent power with power mask.

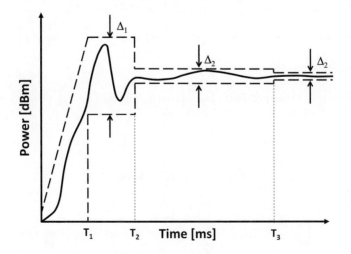

Similar to the measurement of the power over time, the measurement of the frequency components of a signal over time can also be captured. This is required when measuring the lock-in time of a VCO. Instead of calculating the envelope, the signal needs to be FM demodulated in the ATE DSP, where the demodulated signal will be the frequency envelope of the signal.

13.1.4 Intermodulation Measurement

The nth-order intercept point is a single figure of merit describing the nonlinear performance of an RF system or DUT. The intercept point is a purely mathematical concept and does not correspond to any practically occurring physical power level. In fact, in many cases, it exceeds the damage threshold of the device.

Intermodulation or intermodulation distortion (IMD) is the result of two or more signals of different frequencies being mixed together forming additional signals at frequencies that are not, in general, at harmonic frequencies of each other. Intermodulation is caused by the nonlinear behavior of the RF system.

The power output of a device or system can be expressed in terms of the input power level as a power series representation as follows:

$$P_{out} = a_0 + a_1 P_{in} + a_2 P_{in}^2 + a_3 P_{in}^3 + \cdots + a_n P_{in}^n \tag{13.31}$$

For a two-tone unit amplitude signal with frequencies ω_1 and ω_2, described as

$$P_{in} = \cos(\omega_1 t) + \cos(\omega_2 t) \tag{13.32}$$

the output signal of the device or system can be expressed as

$$\begin{aligned} P_{out} = a_0 &+ a_1 \left[\cos(\omega_1 t) + \cos(\omega_2 t)\right] + a_2 \left[\cos(\omega_1 t) + \cos(\omega_2 t)\right]^2 \\ &+ a_3 \left[\cos(\omega_1 t) + \cos(\omega_2 t)\right]^3 + \cdots + a_n \left[\cos(\omega_1 t) + \cos(\omega_2 t)\right]^n \end{aligned} \tag{13.33}$$

Equation (13.33) can be rewritten as

$$\begin{aligned} P_{out} = a_0 &+ a_1 \left\{\cos(\omega_1 t) + \cos(\omega_2 t)\right\} \\ &+ a_2 \left\{\cos^2(\omega_1 t) + \cos^2(\omega_2 t) + 2\cos(\omega_1 t) \cdot \cos(\omega_2 t)\right\} \\ &+ a_3 \left\{\cos^3(\omega_1 t) + \cos^3(\omega_2 t) + 3\cos^2(\omega_1 t) \cdot \cos(\omega_2 t) + 3\cos(\omega_1 t) \cdot \cos^2(\omega_2 t)\right\} + \cdots \end{aligned} \tag{13.34}$$

By using the trigonometric identities

$$\cos(x) \cdot \cos(y) = \frac{1}{2}\left\{\cos(x - y) + \cos(x + y)\right\}$$

$$\cos^2(x) = \frac{1}{2}\left\{1 + \cos(2x)\right\}$$

$$\cos^3(x) = \frac{1}{4}\left\{3\cos(x) + \cos(3x)\right\}$$

and

$$\cos(x) \cdot \cos^2(y) = \frac{1}{4}\left[2\cos(x) + \cos(2y - x) + \cos(x + 2y)\right]$$

Equation (13.34) can be further written in terms of cosine functions of first order. For instance, the third-order term of Eq. (13.34) can be written (ignoring the leading coefficient) as follows

$$P_3 \propto \cos\left(\omega_1 t\right) + \cos\left(\omega_2 t\right) + \cos\left(3\omega_1 t\right) + \cos\left(3\omega_2 t\right) + \cos\left(2\omega_1 t - \omega_2 t\right) + \cos\left(2\omega_1 t + \omega_2 t\right)$$ **(13.35)**
$$+ \cos\left(2\omega_2 t - \omega_1 t\right) + \cos\left(2\omega_2 t + \omega_1 t\right)$$

The above equation reveals some important system behavior. Here we see that the third-order term contains additional signals other than the inputs at frequencies at $3\omega_1$, $3\omega_2$, $2\omega_2 + \omega_1$, $2\omega_1 + \omega_2$, $2\omega_1 - \omega_2$, and $2\omega_2 - \omega_1$. A diagram depicting the third order distortion products, including its 2nd-order harmonics, is shown in Figure 13.10. The first four terms are relatively far away from the input frequencies (ω_1 and ω_2) and can easily be filtered out. However, the latter two terms, and, are very close to the two input frequencies. These two terms cannot be easily filtered out and are commonly referred to as the third-order intermodulation products. It is important to note that the intermodulation products will increase with increasing input power faster than the tones at the fundamental input frequencies. In dBm units, the third-order intermodulation signal will increase in power three times as fast as the power associated with the fundamental signals.

Figure 13.10. Frequency spectrum of a 2-tone signal driving a non-linear system.

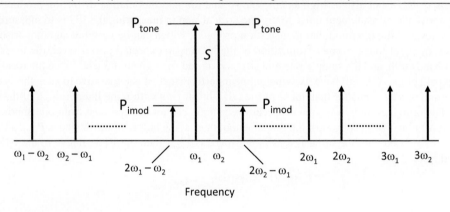

Figure 13.11. Illustration of the Third Order Intercept Point (IP3).

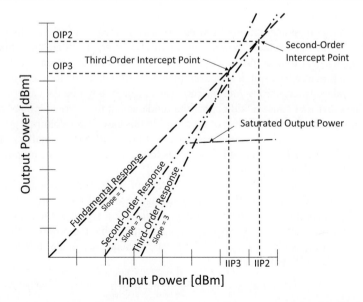

This is evident in the plot of a breakdown of the output power as a function of the input power shown in Figure 13.11. Here the powers of the fundamental, second-order and third-order intermodulation terms are plotted as a function of the input power level. We also noticed in this plot that the second-order and third-order response intersects with the fundamental response, at least the straight-line representation, at two separate points. In the case of the second-order response, the intersection point is called the second-order intercept (IP2) and, likewise, for the third-order response it is called the third-order intercept point (IP3). Since any point in a two-dimensional Cartesian plane is represented by two values, here the intercept points are presented by the input-referred and output referred value. In the case of the third-order intercept, we refer to the input-referred value as IIP3 and the output-referred value as OIP3. Intercept points are commonly found on the data sheet of an RF device.

To measure say the third-order intercept point, it is necessary to source a two-tone signal with equal tone powers to the RF system or DUT. The frequency spacing between each tone should be selected such that the third-order intermodulation distortion products are within the nominal bandwidth of the system or DUT and do not get filtered or attenuated other than that experienced by the input tones. The power level needs to be below the compression point of the system or DUT, but also as high as possible so that the intermodulation distortion products are well above the measurement noise floor. The critical step in measuring the IP3 is to measure the power level of the intermodulation distortion products, which can be very low in comparison to the strong input tones. Some optimization of the maximum expected power level, the measurement bandwidth, and the sample rate and size is required for a given IP3 test. Since the measurement path of the ATE will have its own nonlinear performance, it is imperative to keep the system distortion products smaller than those of the DUT by staying within the linear region of the system. Distortions from two separate sources, in general, cannot be separated from one another.

Assuming that the power of the fundamental tones at the output of the device is P_{OUT}, and the power associated with the in-band third-order tones is P_{imod}, then the IIP3 relative to the OIP3 is found from

$$\text{IIP3}\big|_{dBm} = \text{OIP3}\big|_{dBm} - \text{Gain}\big|_{dB} \qquad (13.36)$$

where OIP3 is given by

$$\text{OIP3}\big|_{dBm} = P_{tone}\big|_{dBm} + \frac{1}{2}\left(P_{tone}\big|_{dBm} - P_{imod}\big|_{dBm}\right) \qquad (13.37)$$

Theoretically, the two fundamental input power levels will be equal to one another, and the two intermodulation distortion products' power levels will also be equal. In general, this assumption is usually valid, so either tone could be taken to calculate the intercept point. However, in some cases, the test equipment will result in a measurement of different levels for these tones. For these cases, an investigation is required to find the root cause, which might be the calibration or a filter of the ATE. One other root cause could be that the fundamental tones are amplified differently in the DUT, causing an unequal power level of the tones. In this case, it is a common practice to measure all four tones and calculate the worst-case intercept point value (i.e., smallest IIP3 or OIP3). We should also mention that the gain of the device is found as the ratio of the output fundamental power to the input fundamental power.

In much the same manner, the second-order intercept point, IP2, is obtained by sourcing a two-tone signal to the RF system or DUT and measuring the second-order intermodulation distortion products. The second-order distortion products will have a signal at the following frequencies: DC, $2\omega_1$, $2\omega_2$, $\omega_1 + \omega_2$, $\omega_1 - \omega_2$, and $\omega_2 - \omega_1$. Assuming that the power of the fundamental tones at the output of the device is P_{OUT}, and the power associated with the in-band second-order tones is P_{imod}, then the OIP2 is found from

$$\text{OIP2}\big|_{dBm} = 2 \cdot P_{tone}\big|_{dBm} - P_{imod}\big|_{dBm} \qquad (13.38)$$

and the input referred second-order intercept point *IIP2* is found from

$$\left.IIP2\right|_{dBm} = \left.OIP2\right|_{dBm} - \left.Gain\right|_{dB}$$ (13.39)

The following three examples will help illustrate the application of these formulas to several different situations.

EXAMPLE 13.4

The intermodulation of an amplifier with a gain of 20 dB is measured. The intermodulation distortion measurements are made at a 15 dBm output-power level per fundamental tone. The intermodulation distortion products power level is at –55 dBm per tone. What is the third order intercept point?

Solution:
The third order intercept point can be calculated to

$$\left.OIP3\right|_{dBm} = 15\ dBm + \frac{15\ dBm - (-55\ dBm)}{2} = 50\ dBm$$

or

$$\left.IIP3\right|_{dBm} = \left.OIP3\right|_{dBm} - \left.Gain\right|_{dB} = 50\ dBm - 20\ dB = 30dBm$$

EXAMPLE 13.5

When testing the OIP3 of an amplifier with gain of $G = 30$ dB, we measure the spectrum shown below. What is the output OIP3 and input-referred IIP3?

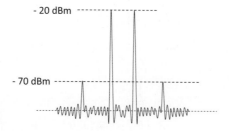

Solution:
Using Eq. (13.37), we write

$$\left.OIP3\right|_{dBm} = \left.P_{tone}\right|_{dBm} + \frac{1}{2}\left(\left.P_{tone}\right|_{dBm} - \left.P_{imod}\right|_{dBm}\right) = -20\ dBm + \frac{1}{2}\left[-20\ dBm - (-70\ dBm)\right] = 5\ dBm$$

and using Eq. (13.36), we compute

$$\left.IIP3\right|_{dBm} = \left.OIP3\right|_{dBm} - \left.Gain\right|_{dB} = 5\ dBm - 30\ dB = -25\ dBm$$

EXAMPLE 13.6

The spectrum below has been measured when supplying an amplifier with a −40 dBm per tone two-tone signal. What are the OIP3, IIP3, and the gain of this amplifier?

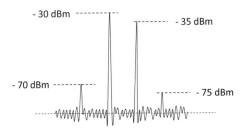

Solution:

Because the signal levels are all different, we must determine the worst-case IP3 point. As such, we compute all four possible OIP3 values using Eq. (13.37) as follows:

$$OIP3_1\big|_{dBm} = P_{tone1}\big|_{dBm} + \frac{P_{tone1}\big|_{dBm} - P_{imod1}\big|_{dBm}}{2} = -30 \text{ dBm} + \frac{-30 \text{ dBm} - (-70 \text{ dBm})}{2} = -10 \text{ dBm}$$

$$OIP3_2\big|_{dBm} = P_{tone2}\big|_{dBm} + \frac{P_{tone2}\big|_{dBm} - P_{imod2}\big|_{dBm}}{2} = -35 \text{ dBm} + \frac{-35 \text{ dBm} - (-75 \text{ dBm})}{2} = -15 \text{ dBm}$$

$$OIP3_3\big|_{dBm} = P_{tone2}\big|_{dBm} + \frac{P_{tone2}\big|_{dBm} - P_{imod1}\big|_{dBm}}{2} = -35 \text{ dBm} + \frac{-35 \text{ dBm} - (-70 \text{ dBm})}{2} = -17.5 \text{ dBm}$$

$$OIP3_4\big|_{dBm} = P_{tone1}\big|_{dBm} + \frac{P_{tone1}\big|_{dBm} - P_{imod2}\big|_{dBm}}{2} = -30 \text{ dBm} + \frac{-30 \text{ dBm} - (-75 \text{ dBm})}{2} = -7.5 \text{ dBm}$$

Because the smallest or lowest OIP3 value represents the worst-case value, we note

$$OIP3_{w/c}\big|_{dBm} = -17.5 \text{ dBm}$$

Next, the gain of the amplifier is found as the ratio of the output tone power to the input power level (or, equivalently, the difference in dBm). Since there are two different output levels corresponding to the fundamental frequencies, we compute the gain of each tone as

$$G_1\big|_{dB} = -30 \text{ dBm} - (-40 \text{ dBm}) = 10 \text{ dB}$$
$$G_2\big|_{dB} = -35 \text{ dBm} - (-40 \text{ dBm}) = 5 \text{ dB}$$

Finally, we compute the worst-case IIP3 using Eq. (13.36) as

$$IIP3_1\big|_{dBm} = OIP3_{w/c}\big|_{dBm} - G_1\big|_{dB} = -17.5 \text{ dBm} - 10 \text{ dB} = -27.5 \text{ dBm}$$
$$IIP3_2\big|_{dBm} = OIP3_{w/c}\big|_{dBm} - G_2\big|_{dB} = -17.5 \text{ dBm} - 5 \text{ dB} = -22.5 \text{ dBm}$$

The worst-case IIP3 (lowest value) is therefore

$$IIP3_{w/c}\big|_{dBm} = -27.5 \text{ dBm}$$

Exercises

13.7. The intermodulation of an amplifier with a gain of 15 dB is measured. The intermodulation distortion measurements are made at a –21-dBm output-power level per fundamental tone. The intermodulation distortion products power level is at –73 dBm per tone. What is the second-order intercept point?

ANS. IIP2 = 16 dBm, OIP2 = 31 dBm.

13.8. The spectrum below has been measured when supplying an amplifier with a –30 dBm per tone two-tone signal. What is the OIP3, the IIP3, and the gain of this amplifier?

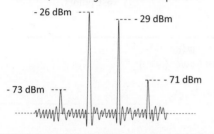

ANS. IIP3 = –12 dBm, OIP3 = –8 dBm, G = 4 dB (all worst-case).

13.1.5 Compression Point Measurement

The dynamic range of an amplifier is determined by its noise figure and 1-dB compression point. In multicarrier applications, the third-order intercept point also plays a role in its dynamic range. The compression point, denoted by the parameter P1dB, is the input power level at which the gain will deviate 1 dB from its small signal gain as illustrated in Figure 13.12. The value of 1 dB is commonly used but any other value could be specified, like, for instance, a 0.1-dB compression point. In this subsection, we will describe four different methods to test for the 1-dB compression point of a DUT. All four methods are based on the same approach: Sweep the input power level, calculate the gain of the device, and then decide where the 1-dB compression point is located.

Go–No-Go Method

The exact 1-dB compression point (P1dB) is not measured with the Go–No-Go method. This method is more useful for production tests, which have a pass–fail criterion. The gain of the device is tested well below the P1dB, and a second gain is tested at the power level specified for the device in the data sheet as the minimum P1dB. If the gain difference between the two measurements is less than 1 dB, the device will pass the test, because the actual P1dB point will occur at a higher input level than that specified in the data sheet. Since the measurement error for both gain tests can add up, it is necessary to minimize the measurement error by optimizing the ATE setting (e.g., establish the optimum power level), as well average the nominal small signal gain measurement over multiple runs to improve its measurement accuracy.

Binary Search Method

Another method for measuring the P1dB is to vary the input power and to apply a binary search routine (see further details in Chapter 3, Section 3.11) such as that illustrated in Figure 13.13 for an eight-step search. Since the gain is nonlinear around the P1dB, it will take a couple of iterations to be within a predefined band around the actual P1dB. Nonetheless, the search is quite fast when compared to a step search. A binary search will not work on multiple devices connected in parallel to a single source

Figure 13.12. Go–no-go method to test for P1dB specification.

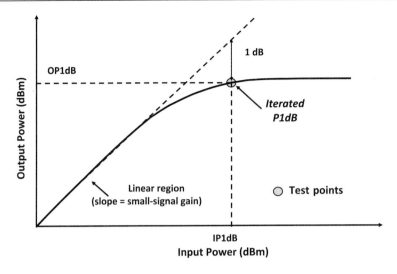

Figure 13.13. Binary search method to measure the P1dB.

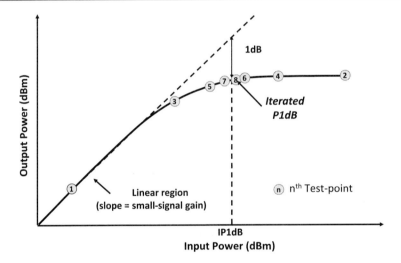

for the simple reason that it operates by adjusting the input power level such that a desired output is achieved. Since multiple devices have multiple outputs, the algorithm simply does not apply. Instead, one must resort to searching the P1dB one device at a time. In addition, because the search process is different for each device, the number of iterations required can also vary. This, in turn, can create complications with the data analysis tools associated with some testers. Also, one must be aware that the source and measurement error can accumulate and cause the search process to go longer than expected. Even more seriously is the potential for the search process to oscillate about a region that includes the P1dB. It is therefore important that the search algorithm take the instrument errors into account.

Interpolation Method

With the interpolation method, a predefined number of gain tests are taken over a range of input power levels, which includes the input power level corresponding to the expected 1-dB compression point. Such a situation is depicted in Figure 13.14. The power levels are usually selected with

Figure 13.14. Interpolation method applied to gain vs. input level to determine the P1dB.

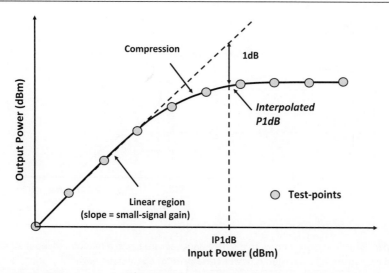

a constant power difference (e.g., 2 dB). These measurements can be analyzed in several ways to find the P1dB of the DUT. Two methods are described below:

1. The actual gain curve can be reconstructed with a high-order polynomial approximation such as using the Lagrange's interpolation method. For an Ψ-point data set, $[(P_{in,1}, G_1),\dots,(P_{in,k}, G_k),\dots,(P_{in,\Psi}, G_\Psi)]$, one can write the interpolation equation for the gain at any power level for an input power range defined between $P_{in,1}$ and $P_{in,\Psi}$ as

$$G(P_{in}) = \sum_{i=1}^{\Psi} L(i, P_{in}) G_i \qquad (13.40)$$

where

$$L(i, P_{in}) = \frac{(P_{in} - P_{in,1})\dots(P_{in} - P_{in,i-1})(P_{in} - P_{in,i+1})\dots(P_{in} - P_{in,\Psi})}{(P_{in,i} - P_{in,1})\dots(P_{in,i} - P_{in,i-1})(P_{in,i} - P_{in,i+1})\dots(P_{in,i} - P_{in,\Psi})}$$

Once the model of the power gain is formulated in terms of the input power level, the P1dB point can be determined through some mathematical root finding procedure such as those described earlier in Chapter 3, Section 3.11. Specifically, we solve for the P1dB from the expression

$$G(\text{P1dB}) = G_{nom} - 1 \qquad (13.41)$$

where G_{nom} is the nominal small-signal gain found under low power conditions. Since the search process is performed in software on a mathematical function rather than through a control loop involving the instruments interacting with the DUT, this approach can be a very time efficient. We leave the details of this procedure to Exercises 13.9 and 13.10.

2. A search routine can be implemented in the test program to find the closest two power levels that bound the 1-dB compression point. Let's say that this occurs between the $(k{-}1)$th and kth input power step. Using with the nominal small-signal gain G_{nom} found in a separate measurement, together with the gains corresponding to the $(k{-}1)$th and kth step, denoted G_{k-1} and G_k, together with the two corresponding input power levels, $P_{in,k-1}$ and $P_{in,k}$, we can estimate the 1-dB compression point from the following straight-line approximation:

$$\text{P1dB}_{est} = P_{in,k-1} + \left[\frac{G_{k-1} - G_{nom} + 1}{G_{k-1} - G_k}\right]\left[P_{in,k} - P_{in,k-1}\right] \qquad (13.42)$$

Figure 13.15. Interpolation between two points that encompass the 1-dB compression point.

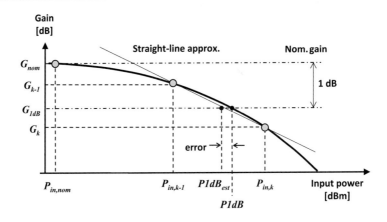

Figure 13.16. P1dB measurement with a single data capture using a modulated input signal: (a) a segmented waveform, (b) a saw-tooth power waveform for continuous power variation.

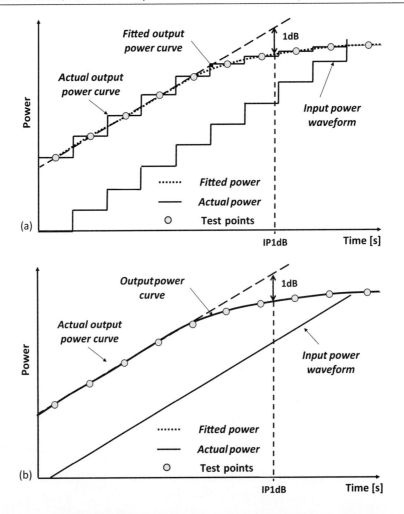

Figure 13.15 illustrates the idea behind this approximation. The inherent error of this method is small for reasonable input power step size.

The advantage of either of these two interpolation methods is that the individual gain measurements across a lot of devices can be compared, since all are taken under the same input conditions. This information can be quite useful during the test debug phase.

Modulation Method

The fourth method is similar to the interpolation method just described, except that the gain is not measured at individual input power levels using a continuous-wave stimulus, but rather with a modulated input signal containing the various power levels over different sections of the modulated waveform, such as that shown in Figure 13.16a. When driving the DUT with this waveform and capturing the output power over time, the individual power levels can be extracted and calculated from a single

EXAMPLE 13.7

The measurements below are taken on a PA to test the P1dB point. The small signal gain had been measured previously and found to be equal to 19.60 dB. Estimate the P1dB point with the given data (which contain a realistic repeatability and accuracy error).

Step	Input [dBm]	Gain [db]
1	0	19.566
2	1	19.405
3	2	19.237
4	3	19.027
5	4	18.653
6	5	18.352
7	6	17.875
8	7	17.423
9	8	16.831
10	9	16.313

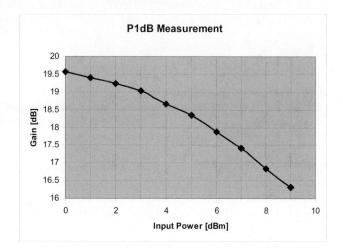

Solution:
With a nominal gain of 19.60 dB, the 1-dB compression point occurs at 18.60 dB. Therefore, we observe from the tabulated data that the 1-dB compression point occurs between an input power level of 4 and 5 dBm (step 5 and 6). Using Eq. (13.42), we get

$$P1dB_{est} = P_{in,5} + \left[\frac{G_5 - G_{nom} + 1}{G_5 - G_6} \right] \left[P_{in,6} - P_{in,5} \right]$$

$$= 4\,dBm + \left[\frac{18.653\,dB - 19.6\,dB + 1}{18.653\,dB - 18.352\,dB} \right] \left[5\,dBm - 4\,dBm \right]$$

$$= 4.176\,dBm$$

Therefore we estimate the 1-dB compression point at 4.176 dBm.

captured array. These individual gain values can be processed in the same way as was done with the interpolation method. Instead of setting up the waveform with separate power levels, the power level can be steadily increased in a sawtooth-like manner, such as that shown in Figure 13.16b. The reconstructed gain curve can then be used to compute the 1-dB compression point.

The advantage of the modulated method is that the 1-dB compression point can be accurately measured with a single data capture. The data can be analyzed in the ATE DSP, or as a background process while the ATE is executing another test. The disadvantage is that the requirement to create a complex input waveform and that the ATE needs to have a source with modulation capability. Another advantage of sourcing a modulated signal is the high relative power accuracy. The reconstruction of the output power curve also minimizes the measurement error by a weighted averaging of the measured power, which can be typically seen as a smooth regression curve.

13.2 *S*-PARAMETER MEASUREMENTS

A special application of a power measurement is one that involves capturing the *S*-parameters of a two-port network. With the addition of some hardware in the RF ATE measurement and source path, the magnitude and phase of the incident and reflected wave can be obtained, allowing the calculation of the corresponding *S*-parameters (see Section 12.4). To be able to measure the *S*-parameter, the ATE needs to be capable of separating the incident and the reflected wave of the signals at the source and measurement side of the DUT. This separation can be done with an RF coupler, which is built into most commercial RF ATE. In the following two subsections, we will discuss the principles of the directional coupler, their physical constraints, and the *S*-parameter measurement technique.

13.2.1 Principles of a Directional Coupler

A directional coupler is a passive four-port device, which separates the forward and reverse propagating wave. A symbolic representation of a directional coupler is shown in Figure 13.17. A part of the power transmitted from the *input* to the *transmitted* port will be coupled to the *coupled* port, while a part of the power supplied to the transmitted port will be coupled to the *isolated* port.

The directional coupler can be defined as a four-port device with a corresponding set of *S*-parameters (e.g, S_{11}, S_{12}, S_{13}, S_{14}, etc.). For a backward wave coupler (a forward wave coupler would have the coupled port 4 next to the transmitted port 2), port 4 is the coupled port, and port 3 is the isolated port as shown in Figure 13.17. Power supplied to port 1 will ideally appear at port 2, and a portion will be coupled to port 4, but no power will be transmitted to port 3. The insertion loss IL of an ideal coupler is the loss of the through line between *input* port 1 and *transmitted* port 2 and can be defined as the ratio of the corresponding powers, that is,

$$IL\big|_{dB} = -10\log_{10}\left(\frac{P_2}{P_1}\right) = 10\log_{10}\left(1 - \frac{P_4}{P_1}\right) = -20\log_{10}|s_{21}| \qquad \textbf{(13.43)}$$

Figure 13.17. Schematic symbol of a directional coupler as a four-port device.

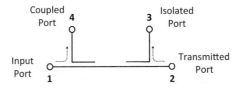

Exercises

13.9. The minimum P1dB is listed in a device data sheet to be 0 dBm. A gain test is performed for an input of –30 dBm and the gain is found to be 22.3 dB. A second test is performed at 0 dBm and the gain is found to be 21.87 dB. Does this device pass its 1-dB compression point test?

ANS. Yes, as the gain difference is less than 1 dB.

13.10. Using the capture data seen listed in Example 13.7, determine the corresponding Lagrange interpolation function. What is the value of this function at an input power level of 3.5 dBm?

$$G\left(P_{in}\big|_{dBm}\right)\bigg|_{dB} = 19.566 + 2.9160P_{in} - 7.9809P_{in}^2$$
$$+ 8.0681P_{in}^3 - 4.2441P_{in}^4$$
$$+ 1.2905P_{in}^5 - 0.2355P_{in}^6$$
$$+ 0.0254P_{in}^7 - 0.0015P_{in}^8$$
$$+ 3.71 \times 10^{-5} P_{in}^9$$

ANS. $G(3.5) = 18.82$ dBm.

13.11. Using the binary search method described in Chapter 3, determine the 1-dB compression point for the amplifier described in Exercise 13.9 if the nominal gain is 20.0 dB?

ANS. P1dB = 3.070 dBm.

The coupling factor C is the ratio in decibels of the power that appears at the *coupled* port 4 to that which is applied to the *input* port 1 when all other ports are terminated by an ideal reflection free termination, defined as follows:

$$C\big|_{dB} = -10\log_{10}\left(\frac{P_4}{P_1}\right) = -20\log_{10}|s_{41}| \tag{13.44}$$

Real couplers will have some power leaking through to port 3. A quantity used to characterize the power leakage is the isolation factor, I. It is defined as the ratio of the power that appears at the isolated port to the input power at port 1,

$$I\big|_{dB} = -10\log_{10}\left(\frac{P_3}{P_1}\right) = -10\log_{10}\left(\frac{P_4}{P_2}\right) = -20\log_{10}|s_{31}| = -20\log|s_{42}| \tag{13.45}$$

The final metric of interest is the directivity factor D, which is equal to the difference in decibels between the power that appears at the *coupled* port in the forward direction when the coupled port is terminated to the power that appears at the terminated *isolation* port when driven in the reverse direction (which becomes the coupled port in the reverse direction). It is defined as

$$D\big|_{dB} = 10\log_{10}\frac{P_4}{P_1} - 10\log_{10}\frac{P_3}{P_1} \tag{13.46}$$

which reduces to

$$D\big|_{dB} = -10\log_{10}\left(\frac{P_3}{P_4}\right) = -20\log_{10}|s_{34}| \tag{13.47}$$

Practically, the directivity factor should be very high, and this makes it very difficult to measure directly. Instead, using the coupling and isolation definitions of Eq. (13.44) and Eq. (13.45), the directivity factor seen listed in Eq. (13.46) can also be computed as

$$D\big|_{dB} = I\big|_{dB} - C\big|_{dB} \tag{13.48}$$

13.2.2 Directional Couplers on an ATE

With the definitions of S-parameters provided in Section 4 of Chapter 12, S-parameter-type measurements can be made with an ATE so long as it is equipped with directional couplers, sources and some digitizers. Figure 13.18 illustrates the general test set up for extracting the S-parameters of a DUT. Two ATE sources are required to drive each side of the DUT. The separation of the incident and reflected wave is done in the hardware with the directional coupler. Each digitizer is used to extract both the magnitude and phase of the voltage signal appearing at the appropriate port of the directional coupler. Even when the ATE has only one digitizer, the incident and reflected waves can still be extracted, albeit each wave a_1, a_2, b_1, and b_2, will be extracted sequentially. Of course, it is assumed here that the signals are stationary and their statistics do not change with time.

With the addition of each directional coupler, signal gain from the source to each interface port must be calibrated for its frequency dependencies. This gain dependency must be incorporated into a set of calibration factors, which will be used internally in the ATE to correct the measured magnitude and phase of the captured value. For the situation when the S-parameters of the DUT alone is desired, the impact of the load board and socket needs to be measured to calculate an error model. This error model will be used to de-embed the measured S-parameter of the load board, socket, and DUT to the DUT-specific S-parameter.

The understanding behind the theory behind the S-parameter measurement makes this test extremely powerful when describing the performance of an RF DUT.

Figure 13.18. Two-port S-parameter implementation on an ATE.

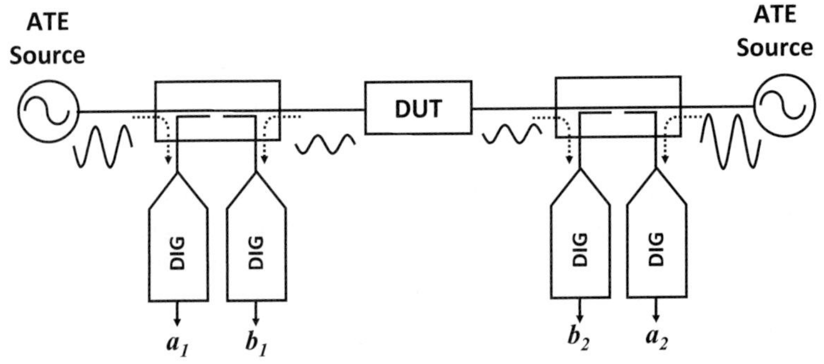

13.3 NOISE FIGURE AND NOISE FACTOR

In Section 12.3 we described (a) the theory behind noise and noise figure and (b) its importance as an RF system parameter. In this subsection, we will explain two different techniques for measuring noise figure-specifically, the Y-factor method and the cold noise method. While there are additional noise measurement techniques commonly found on a bench setup, such as the 3-dB power increase method, the majority of ATE systems make use of the Y-factor and cold-noise methods. We shall therefore limit our discussion here to these two methods.

13.3.1 Noise Figure and Noise Factor Definition

It is common to all noise figure tests that thermal noise is considered the dominant component. This simplifies the analysis significantly because thermal noise is also frequency-independent, as outlined in Section 12.3. Recall that the noise factor F of a system or device is defined as the ratio of the signal-to-noise ratio at the input to that found at its output, that is,

$$F = \frac{\text{SNR}_i}{\text{SNR}_o} \qquad (13.49)$$

whereby it is assumed that the system or device is operating at a reference temperature of T_0. Noise factor is also expressed in terms of decibels, but given the new name of noise figure NF defined as

$$\text{NF} = F\big|_{dB} = 10 \log_{10}(F) \qquad (13.50)$$

Often 290 K is taken as the reference temperature, because it is declared in one of the IEEE standards, however, the noise figure and noise factor can be defined for any temperature. Since noise is a function of temperature, that is,

$$N = kTB \qquad (13.51)$$

device temperature must be declared alongside a noise figure or noise factor in order to compare one device to another.

The noise of an RF system or device can consist of thermal noise or any other noise-generating process, as described in Section 12.3. The individual noise contributions can be added together and regarded as the total noise power. This combined noise power can then be described with a *noise temperature* T_N, defined as the temperature generating the equivalent thermal noise P_N in a defined noise bandwidth B_N given by

$$T_N = \frac{P_N}{k B_N} \qquad (13.52)$$

For systems or devices with the same bandwidth, B_N, noise temperature and noise power are proportional to one another, hence noise temperature can be combined in much the same way that noise power is combined-that is, as a linear sum. To see this more clearly, consider the RF system shown in Figure 13.19. Here the noise power associated with the DUT is modeled as a noise component $N_e = kBT_e$ and added to the incoming signal with noise component $N_s = kBT_s$. Moreover, the DUT is now assumed to be noiseless with noise figure $NF = 0$ dB. For a DUT with power gain G, the noise appearing at the output can be described as

$$N_o = G(N_S + N_e) \qquad (13.53)$$

Figure 13.19. Noise temperature to describe the total noise power.

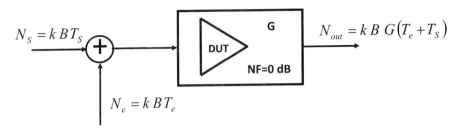

which can be rewritten as

$$N_o = G\left(kBT_S + kBT_e\right) = GkB\left(T_S + T_e\right) \tag{13.54}$$

As is evident from above, the noise at the DUT output is proportional to the linear sum of the noise temperature corresponding to its input and the noise temperature of the DUT itself.

It is worth mentioning that the noise we refer to here is the average noise power over a predefined bandwidth. Moreover, this noise metric does not include any signals like spurious or other unwanted signals like the harmonics associated with the power supply.

To measure the noise figure of a device, a noise source is often used. This source generates a noise power having specific statistical quantities as a function of a DC input voltage. Most of the commercially available noise sources have two defined states, one for a 0-V input and another typically found at a 28-V input. These two states give rise to two distinct noise power levels, referred to as the hot and cold noise levels (N_{hot} and N_{cold}) or as the on and off noise levels (N_{on} and N_{off}). Equivalently, and not surprising, the hot and cold noise levels can be described with two temperatures T_{hot} and T_{cold}. It is a common practice to quantify the function of the noise source with the parameter *excess noise ratio* or *ENR* value. The *ENR* value of a noise source describes the noise output of a source compared to the off-state of the noise source, given by

$$\text{ENR} = \frac{\text{noise power difference between hot and cold noise source}}{\text{noise power at } T_0} \tag{13.55}$$

or with the definition of the thermal noise and the noise temperature we write

$$\text{ENR} = \frac{k\left(T_{hot} - T_{cold}\right)B}{kT_0 B} = \frac{T_{hot} - T_{cold}}{T_0} \tag{13.56}$$

Here T_{hot} is the equivalent noise temperature of the noise source in the on (hot) state in kelvin, T_{cold} is the equivalent noise temperature for the off (cold) state, and T_0 is the reference temperature of 290 K. Most of the time, the reference temperature T_0 is the same as the cold noise temperature T_{cold}, which reduces Eq. (13.56) to

$$\text{ENR} = \frac{T_{hot}}{T_0} - 1 \tag{13.57}$$

or on a logarithmic scale in decibels as

$$\text{ENR}\big|_{dB} = 10\log\left(\frac{T_{hot}}{T_0} - 1\right) \tag{13.58}$$

The typical ENR value of noise sources varies between 5 dB and 20 dB. Sources with a high ENR value will generate a higher noise power output, as Eq. (13.58) shows, which makes the noise detection with the ATE easier for low gain devices. However, the disadvantage of high ENR noise sources is that the output impedance is not as stable, which will generate some mismatch error due to a variation in DUT loading. A low ENR source can be built by combining a high ENR source with an attenuator. Noise sources need calibration, like any other RF signal source, to guarantee NIST traceability of the test signals.

13.3.2 Noise Measurement Technique with the *Y*-Factor Method

The *Y*-factor method is a popular way of measuring noise figure. Most commercial ATE systems incorporate this approach for measuring noise figure. To use the *Y*-factor method, a noise source with a known *ENR* value is connected to the DUT as shown in Figure 13.20 and the output of the DUT is delivered to a power meter specifically tailored to low noise measurements, commonly referred to as a noise figure meter. Sometimes an RF signal is supplied to the internal mixer of the DUT to down-convert its output before reaching the power meter. The ATE or noise figure meter generates a 28-V DC pulse signal to drive the noise source, which, in turn, generates the noise signal used to drive the DUT. Since the ratio of the two input noise levels is known (i.e., noise source ENR), the noise figure of the DUT can be derived from the noise measurement made at the DUT output. To see this, consider that the noise factor of a DUT is expressed as the ratio of the input SNR to the output SNR according to

$$F = \frac{\text{SNR}_i}{\text{SNR}_0} \tag{13.59}$$

Assuming that the input signal to the DUT is S_i with noise component kBT_0 at reference temperature T_0 and the gain of the device is G with additive noise component kBT_e, then we can express the noise factor above as

$$F = \frac{S_i \big/ kT_0}{GS_i \big/ Gk(T_0 + T_e)} \tag{13.60}$$

which reduces further to the final result of

$$F = 1 + \frac{T_e}{T_0} \tag{13.61}$$

Figure 13.20. Testing the noise figure of a DUT with a noise source.

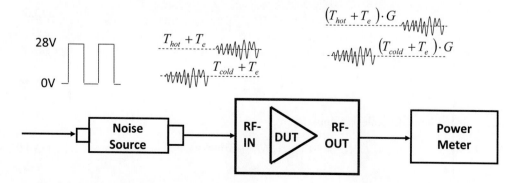

The goal of the Y-factor method is to determine the noise factor F, which, in turn, is related to the ratio of terms T_e to T_0. Since T_0 is related to the input source, it is a known quantity. However, T_e is device-dependent and is the missing quantity required to compute the noise factor of a DUT. To obtain this parameter, we make use of the relationship shown in Eq. (13.54), together with the hot and cold noise inputs, to arrive at two measurements of the output noise level as

$$N_{o,hot} = GkB\left(T_{hot} + T_e\right) \tag{13.62}$$

and

$$N_{o,cold} = GkB\left(T_{cold} + T_e\right) \tag{13.63}$$

Recognizing that we have two equations in two unknowns, specifically, GkB and T_e, we solve using straightforward algebra to arrive at

$$GkB = \frac{N_{o,hot} - N_{o,cold}}{T_{hot} - T_{cold}} \tag{13.64}$$

and

$$T_e = \frac{N_{o,cold}T_{hot} - N_{o,hot}T_{cold}}{N_{o,hot} - N_{o,cold}} \tag{13.65}$$

If we define the Y-factor as the ratio of the hot and cold noise power measurements, we write

$$Y = \frac{N_{o,hot}}{N_{o,cold}} \tag{13.66}$$

then Eq. (13.65) can be rewritten as

$$T_e = \frac{T_{hot} - Y \cdot T_{cold}}{Y - 1} \tag{13.67}$$

Replacing this expression back into Eq. (13.61), we write the noise factor as

$$F = 1 + \frac{T_{hot}/T_0 - Y \cdot T_{cold}/T_0}{Y - 1} \tag{13.68}$$

As before, let us select the cold temperature of the noise source as the reference temperature, that is, $T_0 = T_{cold}$; then the noise factor becomes

$$F = 1 + \frac{T_{hot}/T_0 - Y}{Y - 1} \tag{13.69}$$

With the definition of the ENR value given in Eq. (13.57), we can rewrite the noise factor as

$$F = \frac{ENR}{Y - 1} \tag{13.70}$$

or on a logarithmic dB scale as a noise figure,

$$NF = 10\log_{10}\left(ENR\right) - 10\log_{10}\left(Y - 1\right) \tag{13.71}$$

For $Y \gg 1$, which applies in most practical situations, Eq. (13.71) will simplify to

$$\text{NF} = \text{ENR}\big|_{dB} - N_{o,hot}\big|_{dBm} + N_{o,cold}\big|_{dBm} \tag{13.72}$$

Both Eqs. (13.71) and (13.72) are commonly used in extracting the noise figure from an ATE with a built-in noise source or a noise source placed on the load board. In some cases, the noise source is replaced with an AWG associated with the ATE, which drives a pseudorandom noise signal with two different power levels. This method has the advantage of having a variable ENR, which can be used to optimize individual test cases with respect of the input power level and dynamic range of the ATE.

13.3.3 Noise Measurement Technique with the Cold Noise Method

The cold noise method or gain method is another technique frequently used for measuring the noise figure of a device with a production ATE. This method relies on measuring the cold noise power of a DUT, while terminating the input with a 50-Ω load. This method requires a DUT with a high gain, which often applies to RF-to-base band devices with a high IF gain. Additionally, the method requires a two-step measurement process: First, the gain G of the device is measured, then, as the second step, the noise power $N_{o,cold}$ of the DUT is measured with the input terminated in 50 Ω. Collecting both pieces of information, the noise factor F can be calculated from

$$F = \frac{N_{o,cold}}{kTBG} \tag{13.73}$$

or, in decibels, as

$$\text{NF} = N_{o,cold}\big|_{dBm} - 10\log_{10}(kT) - 10\log_{10}(B) - G\big|_{dB} \tag{13.74}$$

where B is the bandwidth over which the noise power $N_{o,cold}$ is measured. To simplify its use, we recognize that the thermal noise spectral density kT at 290 K is approximately -174 dBm. Hence, we can rewrite Eq. (13.74) as

$$\text{NF} = N_{o,cold}\big|_{dBm} + 174 \ \text{dBm} - 10\log_{10}\left(\frac{T}{290 \ \text{K}}\right) - 10\log_{10}(B) - G\big|_{dB} \tag{13.75}$$

13.3.4 Comparison of the Noise Figure Test Methods

The Y-factor and the cold noise methods are commonly used to measure the noise figure of a device in production, however, both methods have their advantages and disadvantages.

The Y-factor method only measures the ratio of the noise power for two states. Since this is a relative measurement, the absolute accuracy of the test equipment is of less concern. The primary disadvantage of the Y-factor method is that the noise source will terminate the DUT with two slightly different input impedances, which can cause some test uncertainty. This disadvantage can be overcome with low ENR noise sources, but this is only applicable to devices with low gain and low noise figure.

The cold noise method has the advantage that it does not require any additional ATE hardware, because it uses no noise source other than a 50-Ω terminating resistor of the turned off RF source. However, this approach is limited to devices with a high gain or a high noise figure, or both. Table 13.1 summarizes the advantages and disadvantages of these two noise figure methods.

Independent of the test method, the noise power associated with a system or device is generally a low power level, which can easily be masked by measurement error. It is important to consider

Table 13.1. Comparison of Cold Noise to Y Factor Noise Figure Test Method

Method	Suitable Applications	Advantage	Disadvantage
Cold noise method	High gain or high NF	Accurate at measuring high NF, suitable for a wide frequency range. Easy to implement.	Error can be large when measuring DUT with low gain and low NF. Limited by ATE noise floor.
Y-factor Method	Wide range of NF	Suitable for DUT with a wide range of NF and frequency regardless of gain.	Error can be large when measuring DUT with high NF.

EXAMPLE 13.8

What is the maximum noise figure that can be tested with a noise source with an ENR = 10 dB?

Solution:
Using Eq. (13.72) and repeated below, we obtain

$$\text{NF} = \text{ENR}\big|_{dB} - N_{o,hot}\big|_{dBm} + N_{o,cold}\big|_{dBm}$$

As is evident from the right-hand side of this expression, the NF is maximized when the term on the right-hand side between the brackets is minimized. This occurs when $N_{o,hot} = N_{o,cold}$. As such, the noise figure is equal to ENR, the maximum theoretical limit. However, for a reliable measurement, $N_{o,hot}$ needs to be significantly greater than $N_{o,cold}$.

EXAMPLE 13.9

What is the maximum power level expected along the measurement path of an ATE when testing the noise figure of a device. The DUT has a NF of 4 dB, the noise source has an ENR = 28 dB and the effective bandwidth of the measurement path is 1 MHz. The cold noise is assumed to be equal to thermal noise at $T = 273$ K.

Solution:
Using Eqs. (13.72), together with the given information, we can write

$$\text{NF} = \text{ENR}\big|_{dB} - N_{o,hot}\big|_{dBm} + N_{o,cold}\big|_{dBm} \Rightarrow 4 = 28 - N_{o,hot}\big|_{dBm} + N_{o,cold}\big|_{dBm}$$

We are also told that the cold noise is equal to thermal noise at 273 K over a bandwidth of 1 MHz, hence we find that the cold noise power in dBm is equal to

$$N_{o,cold} = 10\log_{10}\left(\frac{kTB}{1\,\text{mW}}\right) = 10\log_{10}\left(\frac{1.38 \times 10^{-23} \times 273 \times 1 \times 10^{6}}{1 \times 10^{-3}}\right) = -114.2 \text{ dBm}$$

Substituting this cold noise power into the expression written above, we solve

$$N_{o,hot} = 28 - 4 - 114.2 = -90.2 \text{ dBm}$$

Accountingfor the random noise crest factor corresponding to 11 dB, we estimate the peak power to be about –90.2 dBm +11 dB or –79.2 dBm.

Exercises

13.12. What is the NF of a DUT if the noise source has an ENR of 32 dB, and the hot and cold noise power is –90.2 and –120.3 dBm, respectively?   ANS. 1.9 dB.

13.13. What is the noise factor *F* of a DUT if the noise source has an ENR of 32 dB and the *Y*-factor is 900?   ANS. 1.76.

13.14. What is the peak power level expected at the output of a DUT if it has an NF of 3 dB, the noise source has an ENR = 31 dB, and the effective bandwidth of the measurement path is 10 MHz. The cold noise is assumed to be equal to thermal noise at *T* = 225 K.   ANS. –66.1 dBm.

---

that the noise figure of the ATE will be tested with an ATE with its own noise floor, which needs to be calibrated out. In the cold noise method the device input is terminated and the thermal noise is measured. At the output, a power measurement is performed with a calibrated power meter. For *Y*-factor it's a relative measurement and the measurement path needs to be treated as second stage in the deFriis equation. The other caveat is that noise signals will have a high crest factor, which makes the measurement on top of the low level unrepeatable when selecting a short integration time. Averaging the noise power is good practice in order to have a stable reading with a good repeatability.

13.4 PHASE NOISE

Phase noise is a critical performance parameter of RF systems and devices. It is related to the power spectral density of noise around a carrier and is typically expressed in rad^2/Hz or on a logarithmic scale as dBc/Hz. Here dBc refers to the ratio of power of a signal or noise relative to the power in a carrier expressed in decibels. The theoretical background of phase noise was described previously in Section 12.3.2. Here we will discuss methods for measuring the phase noise of an RF device and discuss their limitations.

A frequency synthesizer constructed from a reference oscillator and PLL is part of all RF systems and is included in most RF devices. The reference oscillator is in most test applications bypassed with a reference signal from the ATE to synchronize the timing system of the DUT with the ATE. The PLL is used to translate a reference frequency to an RF frequency so that a specific signal-processing task, like mixing, can be performed. Understanding the functionality of the PLL is beneficial for understanding phase noise test techniques.

Figure 13.21 illustrates a basic block diagram of a frequency synthesizer. Here a reference signal is applied to the input of PLL whose output generates an RF signal. The PLL consists of a two-input phase/frequency detector (PFD), followed by a charge pump, loop filter, and voltage-controlled oscillator (VCO). The loop filter acts to smooth out any quick changes appearing in the phase of the input reference signal, thereby maintaining a steady control signal at the input of the VCO. The output of the VCO is then feedback via a divider circuit to the second input of the phase detector. Collectively, the components of the PLL form a negative feedback loop, whereby the phase and frequency of the divider output is adjusted such that the phase difference between the reference and the feedback signal is a constant. This results in the frequency of the feedback

Figure 13.21. Principle block diagram of a PLL with phases/frequency detector (PFD), charge pump, loop filter, VCO and frequency divider.

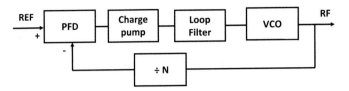

signal to be exactly equal to the reference frequency. As a result, the output of the VCO (denoted f_{RF}) will be N_{PLL} times the input reference frequency (denoted f_{Ref}), that is,

$$f_{RF} = N_{PLL} \times f_{REF} \qquad (13.76)$$

where N_{PLL} is the divider ratio of the divider block.

Not only will the reference frequency f_{REF} be transformed to the RF signal of the VCO, but so too will the phase noise of the reference signal. For the situation where the reference signal is the sole phase noise contributor, the phase noise of the RF signal can be assumed to be

$$\pounds_{RF}\left(f\right)\big|_{\text{dBc/Hz}} = \pounds_{REF}\left(f\right)\big|_{\text{dBc/Hz}} + 20\log_{10}\left(N_{PLL}\right) \qquad (13.77)$$

The dynamics of the PLL is a function of all elements of the PLL but is generally dominated by the dynamics of the loop filter. Since the loop filter is designed for a small bandwidth operation, the PLL has a relatively slow settling and lock time. The phase noise characteristics of the PLL are dependent on the characteristics of the loop filter and also on the noise characteristics associated with the reference input and the VCO. It should be noted that any frequency error associated with the reference signal translates directly to the PLL output, however, the PLL itself will not introduce any frequency errors because its input–output behavior is governed by Eq. (13.76).

For frequencies within the loop filter bandwidth, the phase PSD of the PLL is dominated by the noise of the reference signal. However, for frequencies outside the loop filter bandwidth, the PLL output PSD is dominated by the noise of the VCO. For this reason, the spurious free dynamic range is often part of the required set of tests performed on a synthesizer. Common sources for these spurs are the reference signal with its harmonics and the divider.

13.4.1 Measuring Phase Noise Using Spectral Analysis

One of the most common ways of measuring the phase noise of a device with an ATE is through an FFT analysis of a time-sampled signal. As such, the phase noise is defined in terms of the spectrum of a signal captured by a digitizer. Specifically, phase noise is defined as the ratio of the power in a 1-Hz bandwidth offset from the carrier frequency f_c by a frequency f denoted by P_{SSB} to the power in the tone at f_c, denoted by P_S, that is,

$$\pounds\left(f\right)\big|_{\text{dBc/Hz}} = 10\log_{10}\left(\frac{P_{SSB}}{P_S}\right) \qquad (13.78)$$

A diagram illustrating the components of this measurement is shown in Figure 13.22. Often, the offset power component P_{SSB} is measured with a bandwidth other than 1 Hz. This is particularly true when working with an FFT analysis where the resolution bandwidth of an N-point FFT is F_s/N. As such, Eq. (13.78) is modified to account for this bandwidth difference, resulting in the phase noise being described as

EXAMPLE 13.10

The phase noise of the output of a PLL at an offset of 100 kHz from the carrier is equal to –100 dBc/Hz. Write a mathematical description of the phase noise as a function of frequency, and plot the result from 1 to 200 kHz on a log-linear scale.

Solution:

Assuming that thermal noise is solely responsible for the phase noise from a PLL, a Lorentzian process can be used to describe the PSD. Using Eq. (13.72), we write

$$\mathcal{L}_{RF}(f) = \frac{1}{2} S_{\phi,RF}(f) = \frac{a_{PLL}}{\pi^2 a_{PLL}^2 + f^2} \ \text{rad}^2/\text{Hz}$$

Using the given information, $\mathcal{L}_{RF}(100 \text{ kHz})\big|_{\text{dBc/Hz}} = -100 \text{ dBc/Hz}$, we solve for the phase noise in rad^2/Hz and find

$$\mathcal{L}_{RF}(100 \text{ kHz}) = 10^{\left(\mathcal{L}_{PLL}|_{\text{dBc/Hz}}\right)/10} = 10^{-10} \ \text{rad}^2/\text{Hz}$$

Substituting this result into the expression listed above, with $f = 100 \times 10^3$ Hz, we write

$$10^{-10} = \frac{a_{PLL}}{\pi^2 a_{PLL}^2 + \left(100 \times 10^3\right)^2}$$

This leads to the coefficient $a_{PPL} = 1$ or 10^9, because either result will model the phase noise correctly. Selecting the former value, we find the following mathematical formula for the phase noise as

$$\mathcal{L}_{RF}(f) = \frac{1}{\pi^2 + f^2} \ \text{rad}^2/\text{Hz}$$

Expressing the above in dBc/Hz, we can write

$$\mathcal{L}_{RF}(f)\big|_{\text{dBc/Hz}} = 10\log_{10}\left(\mathcal{L}_{RF}(f)\right) = 10\log_{10}\left(\frac{1}{\pi^2 + f^2}\right)$$

Approximating the phase noise as two straight lines, we can write

$$\mathcal{L}_{RF}(f)\big|_{\text{dBc/Hz}} \approx \begin{cases} -20\log_{10}(\pi), & f < \pi \\ -20\log_{10}(f), & f > \pi \end{cases}$$

A plot of the phase noise in dBc/Hz over a 200 kHz frequency interval is shown below:

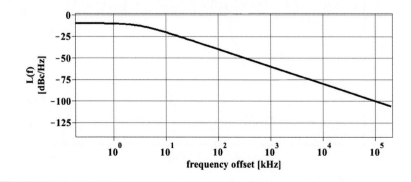

EXAMPLE 13.11

The phase noise of a VCO needs to be tested. A 20-MHz reference signal is sourced from an ATE to a PLL with a feedback divider ratio of 100. If the ATE reference has a phase noise of −120 dBc/Hz, what effect does this noise have on the PLL output?

Solution:

To measure the phase noise of the VCO, we need to make sure that the impact of the phase noise of the reference source is significantly lower than that of the VCO. One typically assumes that the source phase noise is about 10 dB less than the contribution from the VCO. Using Eq. (13.77), we compute the PLL RF output phase noise due to the reference as

$$\pounds_{RF}\left(f\right)\Big|_{dBc/Hz} = \pounds_{REF}\left(f\right)\Big|_{dBc/Hz} + 20\log_{10}\left(N_{PLL}\right) = -120 \text{ dBc/Hz} + 20\log_{10}\left(\frac{2 \text{ GHz}}{20 \text{ MHz}}\right) = -80 \text{ dBc/Hz}$$

Clearly, the effect of the reference source phase noise is increased by 40 dBc/Hz on account of the multiplicative frequency effect of the PLL.

$$\pounds\left(f\right)\Big|_{dBc/Hz} = 10\log_{10}\left(\frac{P_{SSB}}{P_S}\right) - 10\log_{10}\left(\frac{F_s}{N}\right) \tag{13.79}$$

or as

$$\pounds\left(f\right)\Big|_{dBc/Hz} = P_{SSB}\Big|_{dBm} - P_S\Big|_{dBm} - 10\log_{10}\left(\frac{F_s}{N}\right) \tag{13.80}$$

In terms of the FFT analysis, here $P_s\big|_{dBm}$ represents the magnitude of the power in dBm in the FFT bin corresponding to the carrier and $P_{ss}\big|_{dBm}$ represents the magnitude of the power in the FFT bin at some integer multiple of F_s/N offset from the carrier. In terms of the spectral coefficients

Figure 13.22. Illustrating the components of a phase noise metric from a spectrum analysis of a signal.

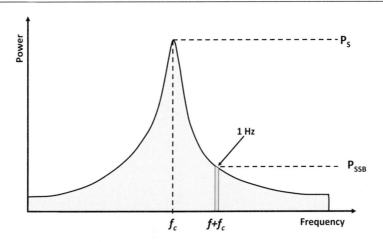

derived from an N-point FFT analysis of the captured data set [recall Eq. (13.1)], these two power terms would be computed according to

$$P_S\big|_{dBm} = 10\log_{10}\left(\frac{c_{M_S,RMS}^2/Z_0}{1\ mW}\right)$$

(13.81)

and

$$P_{SSB}\big|_{dBm} = 10\log_{10}\left(\frac{c_{M_{SSB},RMS}^2/Z_0}{1\ mW}\right)$$

(13.82)

where it is assumed that the carrier signal is located in bin M_S, the offset spectral component of interest is located in bin M_{SSB}, and the ADC is assumed to be terminated in real-valued impedance Z_0.

For this measurement it is critical to have the carrier signal coherent with the digitizer-sampling frequency F_s; otherwise, power will leak across different spectral bins. This will require synchronizing the DUT with the timing system of the ATE by applying a reference signal to the PLL. Since the phase noise of the reference signal can have an impact on the measured phase noise, it is necessary that the reference signal have a better phase noise than the PLL itself. This low phase noise can be achieved directly by using an ATE reference source with a very low phase noise component, or indirectly, by using a higher-frequency source whose output is divided down in frequency, which, in turn, reduces its effective phase noise.

For a reference signal with frequency f_{RF} and phase noise $\pounds(f_{Reference})$ applied to the input of a PLL with output frequency f_{RF}, the phase noise seen by the DUT can be described by

$$\pounds\left(f_{RF}\right) = \pounds\left(f_{reference}\right) + 20\log_{10}\left(\frac{f_{RF}}{f_{reference}}\right)$$

(13.83)

Here we see that the latter term can magnify the effect of the phase noise of the reference signal. As such, the phase noise of the reference source is a critical parameter of the ATE.

Another important consideration when testing the phase noise of a device is the measurement dynamic range of the ATE. ATE instruments must be capable of measuring both the power of the carrier P_S and the power at some offset from the carrier, P_{SSB}. Since these two powers may differ in magnitude by several orders of magnitude, a critical review on the ATE noise floor and measurement path compression point is necessary before undertaking this measurement.

Sometimes a performance metric related to the integration of the phase noise over the channel bandwidth is desired. This is called the integrated phase noise and is expressed in various formats. For a channel defined between frequencies f_l and f_u, we can define the integrated phase noise expressed in squared radians as

$$\Phi_{PN} = 2 \times \int_{f_l}^{f_u} \pounds\left(f\right) df$$

(13.84)

In terms of the FFT data, the integrated phase noise expressed in square radians would be computed according to

$$\Phi_{PN} = \frac{F_s}{N} \times \frac{\sum_{k=M_l}^{M_u} c_{k,RMS}^2}{c_{M_S,RMS}^2}$$

(13.85)

where M_l and M_u represents the lower and upper FFT bins corresponding to the frequency edges of the channel. Often we speak in terms of the RMS phase error. This is simply the square root of the integrated phase noise,

$$\sigma_\phi\big|_{\text{rad}} = \sqrt{\Phi_{PN}} \qquad (13.86)$$

To convert this quantity to degrees, we follow

$$\sigma_\phi\big|_{\text{degrees}} = \frac{180}{\pi}\sqrt{\Phi_{PN}} \qquad (13.87)$$

Sometimes one is interested in the RMS frequency error or residual FM. The residual FM is expressed in Hz and is defined as

$$\sigma_{FM}\big|_{\text{Hz}} = \sqrt{2 \times \int_{f_l}^{f_u} \pounds(f) \cdot f^2 \, df} \qquad (13.88)$$

The integrated phase noise test is easy to implement in production, however, it has some limitations. The first is inherent in the definition itself. The measured power is the combined power of phase noise and amplitude noise. For many cases, the phase noise contribution is dominant so that we can assume that the amplitude noise portion is negligible. The other limitation is a result of the performance limitations of the ATE. The inherent phase noise of the ATE LO synthesizer and reference source can limit the minimum measurable phase noise. Furthermore, this method requires a large measurement dynamic range, as the carrier is always present during this measurement. Few commercial ATE can block specific frequencies in the captured bandwidth. Hence, the carrier will establish the highest power level in the measurement path of the ATE.

EXAMPLE 13.12

A frequency synthesizer produces an output frequency at 1.20117 GHz. What is the phase noise of this synthesizer at a frequency offset of 4.8828 MHz if the output is digitized at a sampling rate of 5 GHz using an ADC with an input impedance of 50 Ω. Subsequently, a 2048-point FFT analysis reveals the following information:

$$\big|FFT\{x[n]\}\big| = \begin{cases} 1.11 \times 10^{-6} & \text{Bin} = 0 \\ \vdots & \\ 1.21 & \text{Bin} = 492 \\ \vdots & \\ 6.23 \times 10^{-3} & \text{Bin} = 494 \\ \vdots & \end{cases}$$

Solution:
According to the given information, the expect bin for the carrier is in FFT bin

$$M_c = \frac{f_c}{F_s}N = \frac{1.2012 \text{ GHz}}{5 \text{ GHz}} \times 2048 = 492$$

Likewise, the offset bin relative to the carrier is found from the following:

$$M_{c+\text{offset}} = \left(\frac{f_c + f_{\text{offset}}}{F_s}\right)N = \left(\frac{1.2012 \text{ GHz} + 4.9928 \text{ MHz}}{5 \text{ GHz}}\right) \times 2048 = 494$$

According to the given FFT data, we can compute the spectral coefficients using Eq. (13.1), to arrive at

$$c_{492,rms} = \frac{\sqrt{2}}{2048} \times 1.21 = 835.55 \times 10^{-6}$$

$$c_{494,rms} = \frac{\sqrt{2}}{2048} \times 1 \times 10^{-5} = 4.30 \times 10^{-6}$$

Finally, the power associated with each term is found, by application of Eqs. (13.81), and (13.82) to be

$$P_S\big|_{dBm} = 10\log_{10}\left(\frac{c_{492,RMS}^2 / 50\ \Omega}{1\ mW}\right) = -48.55\ dBm$$

$$P_{SSB}\big|_{dBm} = 10\log_{10}\left(\frac{c_{494,RMS}^2 / 50\ \Omega}{1\ mW}\right) = -94.31\ dBm$$

Finally, the phase noise at a frequency offset of 4.9928 MHz is found, using Eq. (13.80), to be

$$£(4.9928\ MHz)\big|_{dBc/Hz} = P_{SSB}\big|_{dBm} - P_S\big|_{dBm} - 10\log_{10}\left(\frac{F_s}{N}\right) = -94.31\ dBm - (-48.55\ dBm) - 10\log_{10}\left(\frac{5 \times 10^9}{2048}\right),$$

reducing to

$$£(4.9928\ MHz)\big|_{dBc/Hz} = -109.64\ dBc/Hz$$

13.4.2 PLL-Based Phase Noise Test Method

The spectral analysis approach to phase noise extraction is prone to AM modulation interference. Two techniques that avoid this problem are the PLL-based and delay line phase noise test methods. A drawback to these methods, though, is that they require load board circuits to complete the test setup.

The basic concept of the PLL-based phase noise test method is to mix the output of a DUT using a quadrature mixer acting as a phase discriminator as depicted in Figure 13.23. This requires a low phase noise reference source, which is phase locked to the DUT. The PLL will force the feedback signal to be at 90° out of phase with respect to the DUT signal. This biases the phase detector in its linear region at the most sensitive point. In this method, the short-term phase fluctuations of the DUT are translated directly to low-frequency voltage fluctuations, which can then be measured by a low-frequency analyzer. We assume that the signal from the DUT can be described as $v_{DUT}(t) = \sin[2\pi f_c t + \phi_{DUT}(t)]$, where $\phi_{DUT}(t)$ is the phase signal of interest. Also, due to the feedback action associated with the PLL, the signal being fed back to the other input of the phase detector is $v_{Ref}(t) = \sin = (2\pi f_c t - \pi/2) - \cos(2\pi f_c t)$. Then at the output of the phase detector we would see

$$v_{PD}(t) = -k_{PD}\sin\left[2\pi f_c t + \phi_{DUT}(t)\right]\cos\left(2\pi f_c t\right) \approx -\frac{k_{PD}}{2}\left[\sin\left[2 \cdot 2\pi f_c t + \phi_{DUT}(t)\right] + \sin\left[\phi_{DUT}(t)\right]\right]$$

$$\textbf{(13.89)}$$

Passing the mixer signal through a low-pass filter removes the double frequency term, resulting in the filter output signal as

$$v_{LPF}(t) = -\frac{k_{PD}}{2}\sin\left[\phi_{DUT}(t)\right] \qquad \textbf{(13.90)}$$

Figure 13.23. Phase detector (PD) method to measure phase noise.

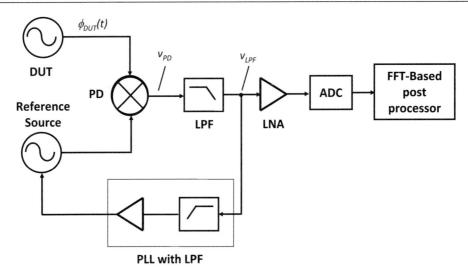

Assuming that the phase variation is small, the filtered output of the phase detector can be approximated by

$$v_{LPF}(t) \approx -\frac{k_{PD}}{2} \cdot \phi_{DUT}(t) \tag{13.91}$$

Taking the Laplace transform of the above expression leads to

$$V_{LPF}(s) \approx -\frac{k_{PD}}{2} \cdot \Phi_{DUT}(s) \tag{13.92}$$

where $\Phi_{DUT}(s)$ is the Laplace transform of $\phi_{DUT}(t)$ and $V_{LPF}(s)$ is the Laplace transform of $v_{LPF}(t)$. The magnitude of the transfer function from the output of the DUT to the output of the low-pass filter can then be written as

$$\left|H(jf)\right| = \frac{\left|V_{LPF}(jf)\right|}{\left|\Phi_{DUT}(jf)\right|} = \frac{k_{PD}}{2} \tag{13.93}$$

Consequently, the spectrum of the voltage signal that appears at the output of the LPF can be written in terms of the phase spectrum of the DUT as

$$S_{v,LPF}(f) = \left|H(jf)\right|^2 S_{\phi,DUT}(f) \tag{13.94}$$

which can just as easily be written in terms of the phase noise of the DUT as

$$\pounds_{DUT}(f) = \frac{4}{k_{PD}^2} \pounds_{v,LPF}(f) \tag{13.95}$$

Except for the scale factor pertaining to the phase detector, the measurement of the phase noise using the PLL-based approach is carried out in the exact same way using the FFT spectral analysis approach of the previous method. Since the phase noise is dependent on the phase detector gain

k_{PD}, this quantity must be accurately determined through a calibration procedure during the load board setup time.

For large phase noise signals, especially at close-in offset frequencies, the linear assumption pertaining to Eq. (13.91) will be wrong. Moreover, for offset frequencies within the loop bandwidth of the PLL, the PLL tracks the phase signal of the DUT and cancel it. This will limit this PLL-based method to frequencies greater than the loop bandwidth. The advantage of this method is that the carrier itself is canceled due to the quadrature condition of the mixer, which improves the dynamic range of the signal processing, including the LNA and ADC. The other advantage as noted before is that this method separates the AM component from the phase noise component, which allows testing of the true phase noise. The PLL-based phase noise test technique is by far the most accurate method, but also the most expensive in terms of equipment and implementation cost.

13.4.3 Delay-Line Phase Noise Test Method

The other approach that measures the true phase of the DUT is the delay-line phase noise test method.[6] Figure 13.24 shows the delay line/mixer implementation of a phase discriminator to measure the phase noise of a DUT. The DUT signal is split into two channels: One channel is applied directly via a phase shifter to the double balanced mixer, the other channel is delayed through a time delay element and then applied to the mixer. The tunable phase shifter provides a means to adjust the output signal for maximum sensitivity. In the time domain, the output of the phase detector/LFP can be shown[6] to be equal to

$$v_{LPF}(t) = k_{PD}\left[\phi_{DUT}(t) - \phi_{DUT}(t - \tau_d)\right] \qquad (13.96)$$

where $\phi_{DUT}(t)$ the instantaneous phase signal associated with the DUT and τ_d is the propagation delay of the delay line minus the delay introduced by the phase shifter. In essence, this circuit converts phase fluctuations into voltage fluctuations, thereby removing the carrier from the measurement path. The noise floor of this system is typically not critical since the carrier does not limit the dynamic range of this method and an LNA with a decent noise figure can significantly improve the measurement sensitivity. Furthermore, the output signal from the LNA is digitized and an FFT is used to extract the phase noise of the DUT. To quantify this, consider taking the Laplace transform of Eq. (13.96) according to

$$V_{LPF}(s) = k_{PD}\left(1 - e^{-s\tau_d}\right)\Phi_{DUT}(s) \qquad (13.97)$$

where $\Phi_{DUT}(s)$ is the Laplace transform of $\phi_{DUT}(t)$. The magnitude of the transfer function from the output of the DUT to the output of the low-pass filter can then be written as

$$\left|H(jf)\right| = \frac{\left|V_{LPF}(jf)\right|}{\left|\Phi_{DUT}(jf)\right|} = k_{PD}\left|1 - e^{-j2\pi f\tau_d}\right| = 2k_{PD}\sin(\pi f\tau_d) \qquad (13.98)$$

Consequently, the spectrum of the voltage signal that appears at the output of the LPF can be written in terms of the phase spectrum of the DUT as

$$S_{v,LPF}(f) = \left|H(jf)\right|^2 S_{\phi,DUT}(f) \qquad (13.99)$$

which can just as easily be written in terms of the phase noise of the DUT as

$$\pounds_{DUT}(f) = \frac{1}{4k_{PD}^2 \sin^2(\pi f\tau_d)}\pounds_{v,LPF}(f) \qquad (13.100)$$

Figure 13.24. Delay line method to measure phase noise.

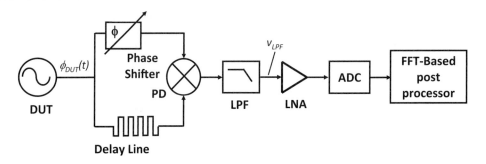

Table 13.2. Comparison of different methods for measuring phase noise.

Phase Noise Test Method	Advantage	Disadvantage
Spectrum analysis method	• Easy setup with RF ATE • No calibration required • No external circuits required on load board	• AM and phase noise can not be separated • No carrier suppression require high dynamic range • Measurement accuracy limited by ATE LO and reference source phase noise
Phase detector Method	• AM and phase noise separated • Carrier suppression allow high dynamic range	• Test method require external circuits on load board • calibration required • complicated setup with calibration
Delay line method	• Suitable for drifting oscillators • No reference oscillator required • AM suppression • Carrier suppression allow high dynamic range	• Complicated setup requiring external components on load board • Complicated calibration procedure • Restricted measurement range

Once again, we see that the phase noise of the DUT is simply obtained through an FFT analysis of the signal captured by the ADC with the appropriate calibration factors incorporated—for example, phase detector gain, delay, LNA and ADC measurement path gain, and so on.

Comparison of the Phase Noise Test Methods

Table 13.2 compares the three phase noise test methods described earlier. It is obvious that the spectrum analysis method is the easiest to set up with an ATE, because no external circuits are required. However, it has the disadvantage that the carrier is not suppressed, which limits the dynamic range of the measurement. Furthermore, this method does not separate AM signal effects from the phase noise. Some ATEs have a built-in phase noise module with a delay line, which significantly improves the measurement dynamic range. The carrier will be suppressed and the method produces a signal proportional to the phase noise without any AM effects. It also does not require an expensive low phase noise reference source.

13.5 VECTOR SIGNAL ANALYSIS

Vector signal analysis (VSA) is the measurement of complex dynamic signals, which might be time-varying or complex-modulated. Time-varying signals, like burst, gated, pulsed, or transient

Exercises

13.15. The phase noise of an oscillator at an offset of 500 kHz from the carrier is equal to −90 dBc/Hz. Assuming that the noise process is best described as Lorentzian, write a general description of the phase noise as a function of frequency.

ANS. $\mathcal{L}_{OSC}(f)\big|_{dBc/Hz} = 10\log_{10}\left(\dfrac{250}{250^2\pi^2 + f^2}\right)$

13.16. The phase noise of a VCO needs to be tested. A 320-MHz reference signal is sourced from an ATE to a PLL with a feedback divider ratio of 16. If the ATE reference has a phase noise of −105 dBc/Hz, what is the minimum level of VCO phase noise that can be measured with this setup?

ANS. −70.92 dBc/Hz; assuming 10 dB margin.

13.17. The phase noise of a VCO within a PLL with a feedback divider ratio of 32 is to be tested using a delay-line phase noise test setup. The VCO is to be tuned to a frequency within the ISM frequency band at 2.450 GHz, and the phase noise is to be measured at approximately 1 MHz offset. What frequency should be sourced to the PLL from the ATE, and what sampling frequency should the digitizer use to collect 4096 samples? What is the exact offset frequency from the VCO output?

ANS. $F_s = 5$ GHz; $F_{REF,ATE} = 76.5625$ MHz; $f_{offset} = 1.2207$ MHz.

13.18. The phase noise of an oscillator is to be tested using a PLL-based phase noise test setup. The oscillator is tuned to a frequency of 1.401367 GHz. What is the phase noise of this device at a frequency offset of 4.8828 MHz if the output is digitized at a sampling rate of 5 GHz using an ADC with an input impedance of 50 Ω. Subsequently, a 2048-point FFT analysis reveals the following information:

$$FFT\{x[n]\} = \begin{cases} 3.42 \times 10^{-6} & \text{Bin} = 0 \\ \vdots \\ 1.51 & \text{Bin} = 287 \\ 8.23 \times 10^{-3} & \text{Bin} = 288 \\ 6.23 \times 10^{-3} & \text{Bin} = 289 \\ \vdots \end{cases}$$

ANS. −112.16 dBc/Hz

signals, change their properties during the measurement sweep. Complex modulated signals cannot solely be described in terms of a simple analog modulation (see Chapter 12, Section 12.5), but instead require more elaborate descriptions such as quadrature amplitude modulation (QAM). When measuring the power with a power meter or spectrum analyzer, it is necessary that the power will be stable during the measurement time. A traditional spectrum analyzer sweeps a narrow-band filter across a frequency range, sequentially measuring one frequency at a time. This requires a repetitive or constant signal. Furthermore, this technique provides only a scalar measurement, such as the magnitude of a signal. In contrast to this, a vector analysis measures both the magnitude and phase of a signal as a function of time. By applying an FFT function, this result can be transformed into the frequency domain. Figure 13.25 shows the block diagram of a typical vector signal analyzer.[1,2]

When comparing the block diagram in Figure 13.25 with a typical ATE architecture (see Chapter 2), it appears that the hardware implementation of a vector signal analyzer is similar to the ATE architecture. The DSP function can be built as firmware in the operating system or as part of the test program. This makes an ATE a good solution for vector analysis of complex RF signals. The only critical task will be the implementation of the required analysis software. In the following subsection, we will discuss the most common techniques used for vector analysis of RF signals.

13.5.1 In-Phase and Quadrature Signal Representation

RF circuits process information that is carried by complex-valued signals. Since a complex signal at any instant of time has two independent signal components, we can refer to such signals as phasors or, equivalently, as two-dimensional vectors. To gain an understanding of RF signals, let us define the concept of an analytic signal.

If the negative frequency components of a signal are discarded, such as through the process of quadrature modulation, then the modulated signal will take on complex values as a function of time. This is in contrast to the signal prior to modulation where it varies as a function of time as real-valued quantities. It is customary to describe base-band signals with real-value functions of time [e.g., $A\sin(2\pi f \cdot t)$], and bandpass signals with complex or analytic signals [e.g., $A\sin(2\pi f \cdot t) + j A\cos(2\pi f \cdot t)$]. In general, for any signal $x(t)$ the corresponding analytic representation is given by $x(t) + j \cdot \hat{x}(t)$, where $\hat{x}(t)$ is known as the Hilbert transform of the real-valued signal $x(t)$. The Hilbert transform is simply an operation that phase shifts $x(t)$ by 90 degrees over all positive frequencies. It is important for the reader to recognize that the actual j operator is not transmitted with the signal. Rather, it is implicit in the modulation scheme whereby the signals are made orthogonal to one another. As such, the signals can easily be separated from one another at the receiver end.

From the theory of complex numbers, a complex quantity can be represented by a rectangular formulation, such as $x(t) + j \cdot \hat{x}(t)$ or, equivalently in magnitude and phase form, as $\sqrt{x^2(t) + \hat{x}^2(t)} \cdot e^{j\phi(t)}$. At any instant in time, the signal will be complex in value. This complex number can be plotted in the complex plane using rectangular coordinates as shown in Figure 13.26. Here the horizontal axis is denoted as I and is equal to the real part of the signal, $x(t)$. Similarly, the vertical axis is denoted as Q and is equal to the imaginary part of the signal $\hat{x}(t)$. It is tradition to refer to I and Q as the in-phase and quadrature signals, respectively. In this textbook, we shall refer to the complex or analytic signal as the IQ signal.

13.5.2 Test of Relative Phase

Figure 13.27 shows a typical RF-to-base-band transceiver device, which might also be part of an RF-to-digital system-on-chip (SOC). For more complex digital modulation schemes, the phase of the signal (albeit a modulated signal) is important for obtaining measures of system performance.

Figure 13.25. Block diagram of a typical vector signal analyzer.

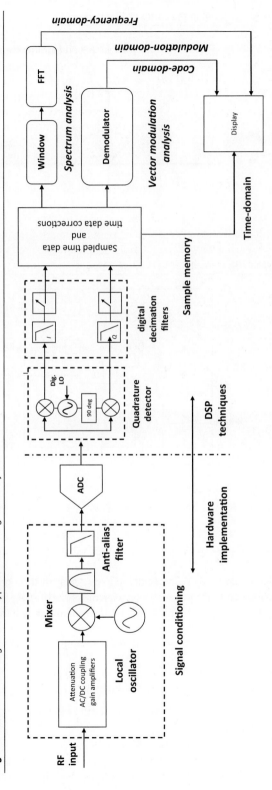

Figure 13.26. Arbitrary signal in the complex IQ-plane.

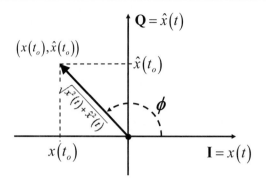

Figure 13.27. Typical RF transceiver DUT.

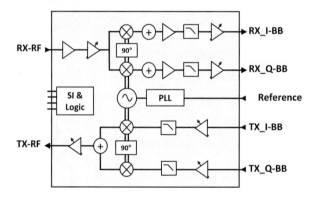

EXAMPLE 13.13

An ideal 1-GHz continuous wave sinusoidal signal is passed through a quadrature modulator, resulting in an IQ signal having the following form: $\sin(2\pi 10^9 \cdot t) + j \cdot \cos(2\pi 10^9 \cdot t)$. Plot the I and Q components of this complex signal. Also, plot the magnitude and instantaneous phase of this signal.

Solution:
The I and Q signals can be described as

$$I(t) = \sin\left(2\pi 10^9 \cdot t\right)$$
$$Q(t) = \cos\left(2\pi 10^9 \cdot t\right)$$

A plot of these two signals superimposed on the same time scale is shown below:

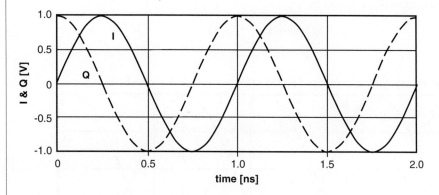

Expressing the complex signal in magnitude and phase form, we write

$$M(t) \cdot e^{j\phi(t)} = \sin\left(2\pi 10^9 \cdot t\right) + j \cdot \cos\left(2\pi 10^9 \cdot t\right)$$

where

$$M(t) = \sqrt{\sin^2\left(2\pi 10^9 \cdot t\right) + \cos^2\left(2\pi 10^9 \cdot t\right)} = 1$$

$$\phi(t) = \tan^{-1}\left[\frac{\cos\left(2\pi 10^9 \cdot t\right)}{\sin\left(2\pi 10^9 \cdot t\right)}\right] = 2\pi 10^9 \cdot t$$

A plot of these two components for the RF complex signal is shown below:

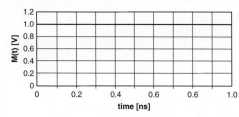
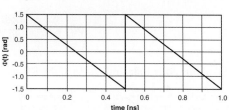

In terms of a vector in the complex IQ plane, we see from the results above that the vector associated with the ideal reference source has a constant magnitude of unity and rotates uniformly around the complex plane in a counterclockwise motion with frequency $2\pi 10^9$ radians per second.

EXAMPLE 13.14

A 1-GHz continuous wave sinusoidal signal is pass through a quadrature modulator resulting in an IQ signal having the following form: $\sin(2\pi 10^9 \cdot t - \pi/5) + j \cdot \cos(2\pi 10^9 \cdot t + \pi/8)$. Plot the I and Q components of this complex signal and the corresponding magnitude and phase of this signal. Also, what is the phase angle difference between the I and Q components of this signal?

Solution:

The I and Q signals can be described as

$$I(t) = \sin\left(2\pi 10^9 \cdot t - \pi/5\right)$$

$$Q(t) = \cos\left(2\pi 10^9 \cdot t + \pi/8\right)$$

A plot of these two signals superimposed on the same time scale is shown below:

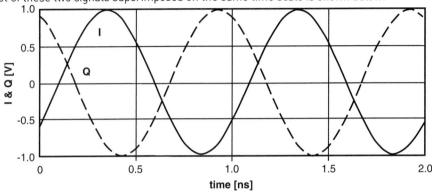

Expressing the complex signal in magnitude and phase form, we write

$$M(t) \cdot e^{j\phi(t)} = \sin\left(2\pi 10^9 \cdot t - \pi/5\right) + j \cdot \cos\left(2\pi 10^9 \cdot t + \pi/8\right)$$

where

$$M(t) = \sqrt{\sin^2\left(2\pi 10^9 \cdot t - \pi/5\right) + \cos^2\left(2\pi 10^9 \cdot t + \pi/8\right)}$$

$$\phi(t) = \tan^{-1}\left[\frac{\cos\left(2\pi 10^9 \cdot t + \pi/8\right)}{\sin\left(2\pi 10^9 \cdot t - \pi/5\right)}\right]$$

A plot of the magnitude and phase of the RF complex signal is shown below:

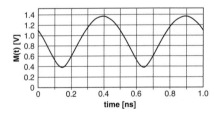 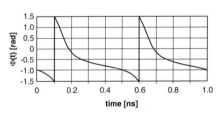

The phase difference between the I and Q components can be established by rewriting each component in terms of the same trigonometric function, that is, same phase reference point. Since $\cos(x) = \sin(x + \pi/2)$, we can write the I and Q signals as

$$I(t) = \sin\left(2\pi 10^9 \cdot t - \pi/5\right)$$

$$Q(t) = \cos\left(2\pi 10^9 \cdot t + \pi/8\right) = \sin\left(2\pi 10^9 \cdot t + 5\pi/8\right)$$

Hence, we conclude that the quadrature signal leads the in-phase signal by $33\pi/40$ radians rather than ideal amount of $\pi/2$.

Figure 13.28. IQ demodulation.

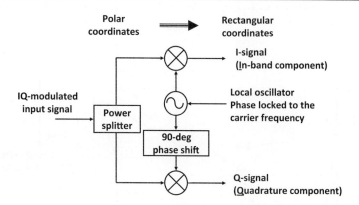

For the receive (RX) path, the phase between the RX_I-BB and RX_Q-BB can be measured. For the TX path, the DUT will be supplied with a perfect I and Q signal by the ATE, and the embedded IQ signal can be measured after demodulating the RF signal with the ATE.

To measure the phase difference of an RX I and Q signal, the DUT will be supplied with a CW sinusoidal signal to the RX input, and the output signal will be captured in the time domain. An FFT will calculate the power and the phase of the appropriate signals. The accuracy of the measured phase difference between the I and Q signal is dominated by the calibration accuracy of the ATE between both measurement channels.

To measure the IQ signals in TX mode, it will be necessary to supply the DUT with low frequency I and Q base-band signals and to analyze the RF signal by digital demodulation. Figure 13.28 shows the block diagram, which needs to be implemented into the test program, either as a firmware routine controlling a DSP engine or directly as a test routine in the program code. The RF signal needs to be captured in the time domain and multiplied with a sine signal with the carrier frequency ω_0 to generate the I signal, as well as with a cosine signal to generate the Q signal. An FFT can be applied to these signals to capture their corresponding magnitude and phase information.

13.5.3 Error Vector Magnitude Test Method

The most widely used figure of merit for modulation quality of digital communication systems is the *error vector magnitude* (EVM) metric. The EVM measurement is sensitive to any signal impairment that affects the magnitude and/or the phase trajectory of the signal. As the name already indicates, EVM is a vector analysis of the error at any given time between an ideal reference signal and a measured signal as shown in Figure 13.29. All the vectors are assumed to be rotating around the origin in a counterclockwise fashion at a rate set by the signal frequency. Here a snapshot of the measured and reference vectors are plotted at a particular point in time. Also shown is the error vector connecting the ends of the measured and reference vectors. It has a magnitude and a phase, Φ. It is the magnitude of this error vector that is related to the EVM metric mentioned above. Also indicated on the drawing is the magnitude and phase difference between the IQ measured vector and the IQ reference vector. Both the magnitude and angle are different from that which is associated with the error vector. It is important that one does not confuse the "magnitude of the error vector" with the "magnitude error" or the "phase of the error vector" with the phase error, because these two phrases sound so similar.[3,4,5]

Figure 13.29. Error vector magnitude (EVM) with reference and measured vector signal at particular point in time, as the vectors are constantly rotating around in a counterclockwise motion at a rate determined by the signal frequency.

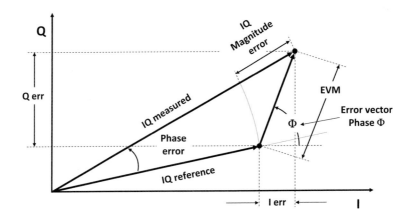

A complex-valued error vector for a particular constellation point can be defined in terms of the corresponding sampled sequence of I and Q components of the measured signal and reference signal as follows:

$$E[n] = \left[I_{measured}[n] - I_{REF}[n] \right] + j \left[Q_{measured}[n] - Q_{REF}[n] \right] \tag{13.101}$$

The EVM metric is defined as the root mean square of the error vector over N samples normalized by a reference power level (P_{REF}) as follows:

$$\text{EVM} = \frac{\sqrt{\dfrac{1}{N} \sum_{n=0}^{N-1} |E[n]|^2}}{\sqrt{P_{REF}}} \tag{13.102}$$

The reference power P_{REF} level is either the amplitude of the outermost symbol or the square root of the average symbol power. Substituting Eq. (13.101) into Eq. (13.102), we can write EVM as

$$\text{EVM} = \frac{\sqrt{\dfrac{1}{N} \sum_{n=0}^{N-1} \left\{ \left(I_{measured}[n] - I_{REF}[n] \right)^2 + \left(Q_{measured}[n] - Q_{REF}[n] \right)^2 \right\}}}{\sqrt{P_{REF}}} \tag{13.103}$$

There are two parts associated with an EVM measurement, the source, and the measurement path. This is depicted in the diagram shown in Figure 13.30. In the source path, the serial data are transformed into a parallel data stream for a multicarrier system. The data are mapped onto the appropriate IQ constellation points and then transformed into the frequency domain. This waveform can be used to modulate an RF source to drive the DUT. The time-domain capture of the DUT output signal is then transformed into a frequency domain signal from which the data are analyzed for its accuracy in regenerating the constellation set.

EXAMPLE 13.15

The following two *IQ* signals were described in Examples 13.13 and 13.14:

$$x_{measured}(t) = \sin(2\pi 10^9 \cdot t - \pi/5) + j \cdot \cos(2\pi 10^9 \cdot t + \pi/8)$$
$$x_{REF}(t) = \sin(2\pi 10^9 \cdot t) + j \cdot \cos(2\pi 10^9 \cdot t)$$

Plot the magnitude and phase of vector error sequence between the above two signals.

Solution:

As the *I* and *Q* signals of the measured and reference signal can be described as

$$I_{REF}(t) = \sin(2\pi 10^9 \cdot t), \quad I_{measured}(t) = \sin(2\pi 10^9 \cdot t - \pi/5)$$
$$Q_{REF}(t) = \cos(2\pi 10^9 \cdot t), \quad Q_{measured}(t) = \cos(2\pi 10^9 \cdot t + \pi/8)$$

We compute the vector error $E(t)$ as

$$E(t) = \left[I_{measured}(t) - I_{REF}(t) \right] + j \left[Q_{measured}(t) - Q_{REF}(t) \right]$$

which, on substituting the given information from above, we write

$$E(t) = \left[\sin(2\pi 10^9 \cdot t - \pi/5) - \sin(2\pi 10^9 \cdot t) \right] + j \left[\cos(2\pi 10^9 \cdot t + \pi/8) - \cos(2\pi 10^9 \cdot t) \right]$$

The magnitude and phase of this error signal is then calculated as

$$M(t) = \sqrt{ \left[\sin(2\pi 10^9 \cdot t - \pi/5) - \sin(2\pi 10^9 \cdot t) \right]^2 + \left[\cos(2\pi 10^9 \cdot t + \pi/8) - \cos(2\pi 10^9 \cdot t) \right]^2 }$$

$$\phi(t) = \tan^{-1} \left[\frac{ \cos(2\pi 10^9 \cdot t + \pi/8) - \cos(2\pi 10^9 \cdot t) }{ \sin(2\pi 10^9 \cdot t - \pi/5) - \sin(2\pi 10^9 \cdot t) } \right]$$

A plot of the magnitude and phase of the error vector is shown below:

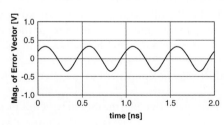
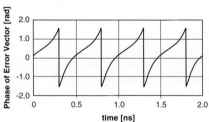

13.5.4 Adjacent Channel Power Tests

One other important measurement related to an RF DUT used in a digital communication system is the adjacent channel power (ACP) or the adjacent channel power ratio (ACPR). ACP and ACPR are measures of the leakage power in the adjacent channels normalized with respect to the power in the channel. While an EVM test describes the DUT performance in-band, ACPR is a test that specifies the out-of-band behavior of the DUT. Leakage power influences the system capacity

Figure 13.30. Source and measurement path to test EVM.

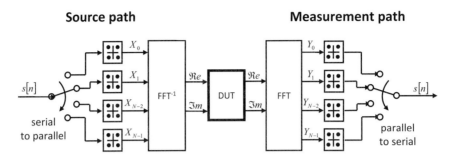

because it interferes with the transmission in the adjacent channels. Therefore, the adjacent channel power or the adjacent channel power ratio test is one of the most common test parameters for production test of an RF transmitter. The ACPR is a measure of the ratio of the wanted channel power to the power in the adjacent and alternate channel due to spectral re-growth. Even though the measurement path is a scalar power measurement, the ATE needs to source the DUT with an *IQ* signal.

Measuring the spectral re-growth of a wideband signal like a WLAN or CDMA signal is one of the most demanding measurements for an ATE. The spectral re-growth is the effect that generates signal power into the adjacent and alternate channel of a particular RF system with defined channels and channel bandwidth. In terms of signal statistics, having a signal leak into an adjacent channel is different than noise in this channel.

In Section 13.1.4, intermodulation distortion was introduced as a measure of the nonlinear performance of a DUT. This measure is still useful here, but the ACPR is a more valuable metric especially for RF systems using a digital modulation scheme. The third-order product of two sinusoidal signals could land in a neighboring channel and cause interference with another signal. Since the modulation becomes more complex, it becomes less obvious that the sinusoidal intermodulation analysis will be adequate for describing the system limitations for complex modulations. Measuring the ACPR with the nominal modulated signal will be the logical extension of the intermodulation distortion measurement. As described previously, IMD is defined as the ratio of the power in one of the third-order tones to that in one of the main tones. ACPR is defined as the ratio of the power in the bandwidth of the channel adjacent to the main signal to the power in the bandwidth within the main signal channel. Alternate channel power ratio (ACP) is also defined in a similar way. It refers to the ratio of the power in a bandwidth two channels away from the main signal to the power in the main channel with the same bandwidth as shown in Figure 13.31. In terms of IMD measurement, a higher-order product like the fifth-order product may be equivalent to the alternate channel power ratio.

While the ACPR measurement is simple to perform in principle, the difficulty of this measurement is due to the complexity of sourcing a defined RF signal into a DUT and measuring the output of the DUT with a high dynamic range without introducing new signal products by the ATE.

Modulated Sources

A key component of the ACPR measurement is the signal source, which needs to provide a standard conformance signal to the DUT. Most of the commercial ATE have a modulated RF source built in, and most of the ATE vendors provide a set of modulated signals for the most common wireless standards. Like IMD or any other distortion measurement, the level of distortion is a very strong

Figure 13.31. Adjacent and alternate channel power.

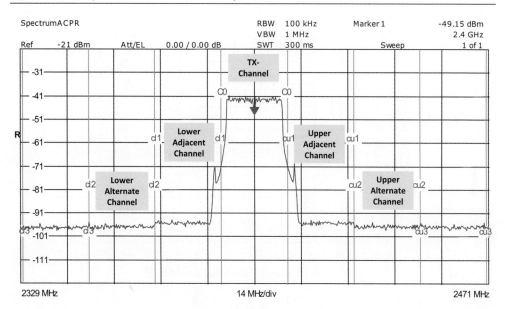

function of the source power level, which might already cause some distortion by the ATE internal source switch matrix and amplification stages. In addition to the power level, ACPR also varies strongly with the modulation format being employed due to the different crest factor associated with the different modulation schemes. It is good practice to verify the ACPR of the ATE sourced signals with bench equipment like a spectrum or signal analyzer with a known performance.

Modulation Measurements

The ATE measurement path architecture needs to be capable of measuring the integrated power in the transmit channel and the adjacent and alternate channels. This will not only require a receiver bandwidth which covers at least the channel bandwidth, but will also need to have a significant dynamic range, which covers the sum of the ACPR power, and crest factor of the modulated signal. In addition, it will be important that the tester's LO will not introduce phase-noise-related signal components.

The dynamic range of the ATE measurement path is at a premium, and the limits must be understood in order to select the optimum ATE settings. The primary constraints are (a) receiver nonlinearities at the high-power end and (b) the measurement path noise floor at the low power end. In addition, the LO phase noise provides a lower bound, but in many cases the measurement path noise floor itself is usually the lower limit. A comparison of the different components that contribute to a ACPR measurement is shown in Figure 13.32.

To measure ACPR, it will be important to follow the specified channel bandwidth and the frequency spacing between the main transmit channel and the adjacent or alternate channel and the filter. Since the ACPR value will vary with the power in the main transmit channel, the channel power will be given as a requirement in the DUT or test specification.

In conclusion, the ACPR measurement is one of the most challenging tests performed on an ATE. The source needs to be modulated with a waveform defined by the specific standard. The dynamic range of the source and measurement path of the ATE will require special attention when correlating the ATE result to bench measured results.

Figure 13.32. Parameter impacting the dynamic range of the measurement path of an ATE. The power level will vary for different ATE and ATE settings.

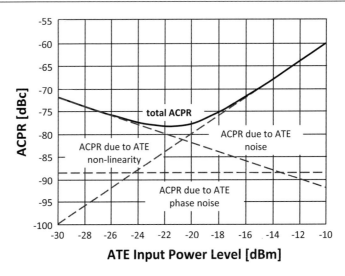

13.5.5 Transmit Mask

Spectral mask test is a good indicator of the deteriorating performance of wireless transmission systems for digital modulation schemes like WLAN or cellular phone systems. The DUT is sourced with a standard compliant digitally modulated signal, while the power at the transmit output is measured. The spectral mask test measures the distribution of the transmitted power spectral density similar to the ACPR test. This test makes sure that no unwanted spectral components will be transmitted, which can cause interference in other channels.

When transmitting a digitally modulated signal, all spectral content needs to be below a specified limit or mask. Figure 13.33 shows this mask for a WLAN transmitter.

The reference power level is taken as the peak power spectral density in the signal. The spectral mask applies to frequencies in the frequency band of the adjacent channels. For all offset frequencies, the power spectral density is calculated and compared relatively to the reference level. For all offset frequencies specified in the standard, the power spectral density is tested against a defined mask.

The challenge of this mask test is that the signal bandwidth is already wide for WLAN—for example close to 20 MHz, which is a typical receiver bandwidth of a commercial ATE system. To capture the full mask bandwidth, multiple captures need to be taken, which will increase the test time. The other challenge, which is similar to the ACPR test, is the required dynamic range, which requires a careful selection of the ATE power settings.

13.5.6 Bit Error Rate

Bit error rate (BER) is a fundamental system test for measuring a receiver performance, such as the sensitivity and selectivity. The BER is the percentage of erroneous bits received compared to the total number of bits sourced into the receiver during a defined measurement period. While the sensitivity for analog-modulated systems is often specified as the ratio of signal plus noise plus distortion (SINAD), for digital modulated systems and DUT, the sensitivity is specified as the maximum BER for a given sensitivity level. In addition to the power level of the modulated signal, the BER will also depend on in-band and out-of-band interferers and blockers. These interferers

Figure 13.33. Transmit spectrum mask for 802.11a.

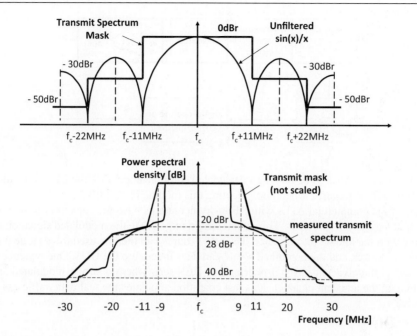

and blockers will simulate data traffic on an adjacent channel, or other strong signals out of band caused by other services. BER is also an efficient metric to specify the quality of the device for these critical parameters.

To perform a BER test, a standard compliant signal is supplied to the input of the receiver, which is modulated with a known pseudorandom bit sequence (PRBS). The analog or digital output will be captured with an analog digitizer, or digital capture using a digital pin.[6] The analog capture can be used to sample an analog as well as a digital bit stream. In the case of the capture of an analog signal, the signal will be digitally demodulated in the ATE DSP, or ATE computer. When capturing a digital bit stream with an analog digitizer, the bits need to be reconstructed and compared with the bit sequence used to modulate the input signal in order to calculate the BER. This method avoids synchronizing the device output with the input in terms of data rate, and thus delaying time caused by the DUT and ATE. Especially for devices without clock recovery circuitry, where the DUT can't trigger the ATE, this method might be the most robust one to implement. With a short known training sequence, the digitized bit stream can be synchronized to the known input signal using a correlation algorithm.

Another method used to measure BER is to capture the digital output of the DUT with the digital capture instrument of the ATE, as shown in Figure 13.34. This method has the advantage of being easily scalable for multisite parallel test setups. Another advantage is that the time delay does not need to be known, since the captured bit stream can be processed using the capture memory, which is synchronized with a matched loop.

The disadvantage of all methods is that the necessary number of bits will require a long bit sequence, which will cause significant test times, especially when data processing is required.

Other methods utilizing on-board test circuits can avoid some of the disadvantages. A built-in XOR circuit can detect failed bits on the flight while the bit sequence is measured by comparing with a known good bit sequence. Another on-board test method would be the implementation of

Figure 13.34. BER ATE test set-up.

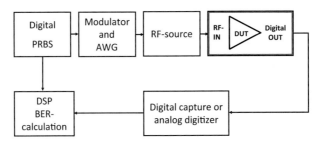

an FPGA. This FPGA could be built as a designated compute engine for the BER test. Additional debug effort will be required when implementing this module for the first time.

Most of the commercial ATEs with RF instrumentations have the capability to combine the signals of at least two RF sources. This enables them to generate the modulated signal on the first RF source with the known bit sequence. The other source can then be used to generate a defined interferer or blocker, and the combined signal can be sourced into the DUT. One typical example for this test is the adjacent channel interference BER test, where a strong modulated signal is transmitted on the adjacent channel. This test condition simulates the real data traffic in a digital RF system.

13.6 SUMMARY

In this chapter we reviewed three different types of RF scalar measurements: power, amplitude noise (thermal and $1/f$) and phase noise, as well as several vector-based tests. We began by describing a typical architecture of the RF portion of an ATE. This was followed by a discussion of the principles of a power measurement using an ATE. Issues relating to dynamic range, maximum power, noise floor, and phase noise were then introduced. A source of error in any RF system is the power loss due to impedance mismatches, transmission losses, and so on, associated with the DIB. To compensate for these errors, de-embedding techniques were introduced. This enables the DUT performance to be separated out from the operation of the DUT-DIB combination. The power measurement techniques can be applied to gain, P1dB, or intermodulation products, as well as the lock-in times of a VCO when its signal dependency with time is taken into account. Several power measurements made with the aid of directional couplers enables one to write a two-port S-parameter description of the DUT. The physics of a directional coupler was briefly described where it was shown how to measure the incident and reflected wave at the input and output port of a DUT.

Measurement using an ATE for both the noise associated with a power measurement as well as the phase noise associated with an oscillator output were described. The advantages of the Y-factor and cold noise method to measure the noise figure of a DUT were outlined. Phase noise measurements were shown to be made with methods like the PLL- and delay-line phase noise method instead of the commonly used bench-top approach using a spectrum analyzer.

In the last part of this chapter we studied the principles of a vector signal analysis. Since a standard RF ATE has all the building blocks of a vector signal analyzer, it was shown how an ATE could be used to measure more complex RF system parameters like EVM, ACPR, and BER. Common to all these tests is that the stimulus needs to be a modulated signal.

An important lesson derived from this chapter is that the DUT, the DIB, and the ATE must be treated as an RF system rather than a cascade of three independent electrical elements. This requires the test engineer to design the test system in much the same way that the system designer would design a new device or RF system. System parameters like dynamic range, maximum power, signal bandwidth, measurement bandwidth, noise figure, noise bandwidth, phase noise, compression point and gain distribution in the source and measurement path all need to be considered by the test engineer during the initial DIB design phase. It is the goal of the test engineer to develop an RF test program using an ATE such that the test results match the results of a dedicated RF measurement equipment, like a spectrum analyzer, RF synthesizer, vector signal analyzer, or network analyzer.

PROBLEMS

13.1. What are the major blocks of an spectrum analyzer?

13.2. What are the major blocks of an RF ATE measurement path with an zero IF architecture?

13.3. Why can the the dynamic range of a spectrum analyzer be large compared to a vector signal analyzer?

13.4. What can happen if the maximum expected power of an ATE measurement path is set too low?

13.5. What can happen if the maximum expected power of an ATE measurement path is set too high?

13.6. What condition is required for coherent sampling and what happens if this condition is not true for a phase noise measurement?

13.7. What IF frequency should be selected for a ZIF ATE architecture to be coherent with an ADC having a sampling rate of 50 MHz, if 1024 samples are required for FFT processing and the output of the mixer ranges between 15 and 20 MHz?

13.8. A 2.5-GHz RF signal is to be down converted by a mixer and directed to a digitizer for digital conversion. If the digitizer is sampled at a clock rate of 100 MHz and 2048 samples are collected for digital signal processing using an FFT, what should be the local oscillator frequency if the mixer has an output IF frequency range between 15 and 20 MHz?

13.9. A gain test was performed on a DUT and found to be 21.3 dB. The DUT is connected to an ATE through an interface board with an operating loss of 2.1 dB. What is the actual gain of the device?

13.1. A gain test was performed on a DUT using an ATE with an interface board. The gain was found to be 11.4 dB. If the source, the meter, and the input and output ports of the DUT have return losses of 10, 12, 13, and 12 dB, respectively, and the traces introduce losses totaling 0.45 dB, what is actual gain of the DUT?

13.2. What is the operating loss associated with the input-side of an interface board shown in Figure 13.6 if the following measurements at the input port of the board were obtain using the in-socket calibration method, together with a single DUT measurement:

$$S_{11,DUT} = 0.21 - j0.32, \quad S_{11,open} = -0.85 + j0.32$$
$$S_{11,short} = 0.85 - j0.12, \quad S_{11,load} = 0.53 - j0.31$$

13.3. For the measurements made in problem 13.11 above, what is the input impedance of the device under test?

13.4. What is the operating loss associated with the output side of an interface board shown in Figure 13.6 if the following measurements at the output port of the board were obtain using the in-socket calibration method, together with a single DUT measurement:

$$S_{11,DUT} = 0.21 + j0.22, \quad S_{11,open} = -0.89 + j0.13$$
$$S_{11,short} = 0.9 + j0.13, \quad S_{11,load} = 0.15 - j0.31$$

13.5. For the measurements made in Problem 13.13 above, what is the output impedance of the device under test?

13.6. A gain test was performed on a device under test and found to be 19.3 dB. The input port to the load board was measured using a vector network analyzer under four separate calibrated conditions, resulting in

$$S_{11,DUT} = 0.11 - j0.22, \quad S_{11,open} = 0.91 + j0.12$$
$$S_{11,short} = -0.54 + j0.21, \quad S_{11,load} = 0.35 - j0.13$$

What is the true gain of the device if the output signal path has zero losses? What is the input impedance of the DUT?

13.7. Why is the IP3 of an RF system so important?

13.8. What is the slope of the second- and third-order response in the input/output power diagram? What is the mathematical background for these slopes?

13.9. The intermodulation distortion measurements are made at a -10-dBm output-power level per fundamental tone. The intermodulation distortion products power level is at -73 dBm per tone, and the power level for the fundamental tones is -5 dBm. What is the third order intercept point?

13.10. The spectrum below has been measured when supplying an amplifier with a 20 dBm per tone two-tone signal. What is the OIP3, the IIP3, and the gain of this amplifier?

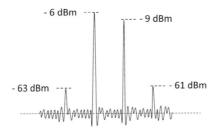

13.11. Name four methods used to measure the compression point of an RF system like an amplifier.

13.12. What advantages does the modulation method have when measuring the compression point?

13.13. The following measurments of amplifier gain expressed in decibels were found using 2 dBm increasing steps of input power beginning at -10 dBm: 11.92, 11.81, 11.35, 10.98, 10.13, 9.89, 9.35, 8.76. The small-signal gain was measured previously and found to be equal to 11.35 dB. What is the 1-dB compression point of this amplifier using interpolation with a a linear rule of gain behavior aaround the P1dB. How does this result compare with an interpolation method that uses a polynomial approximation of the data set?

13.14. The gain of an amplifier was found through interpolation to depend on its input power over a 0- to 10-dBm range according to

$$G = 0.1 + 10.2 \times P_{in} + 0.1 \times P_{in}^2$$

What is the 1-dB compression point associated with this amplifier if the nominal small-signal gain is 10 dB?

13.15. The minumum P1dB is listed in a device data sheet to be -10 dBm. A gain test is performed for an input of -30 dBm and the gain is found to be 15.3 dB. A second test is performed at –10 dBm and the gain is found to be 14.25 dB. Does this device pass its 1-dB compression point test?

13.16. Name the two most common methods used for measuring the noise figure with an ATE, and what are the advantages and disadvantage of each method. Why can the the dynamic range of a spectrum analyzer be large compared to a vector signal analyzer?

13.17. Why is it important to pay special attention to the load condition when using a noise source with a high ENR value?

13.18. What is the NF of a DUT if the noise source has an ENR of 21 dB and the hot and cold noise power is –101.2 dBm and –119.1 dBm, respectively?

13.19. What is the noise factor of a DUT if the noise source has an ENR of 34 dB and the Y-factor is 500?

13.20. What is the peak power level expected at the output of a DUT if it has an NF of 5 dB, the noise source has an ENR = 28 dB and the effective bandwidth of the measurement path is 0.75 MHz. The cold noise is assumed to be equal to thermal noise at T = 275 K.

13.21. The phase noise of an oscillator at an offset of 100 kHz from the carrier is equal to –96 dBc/Hz. Assuming that the noise process is best described as Lorentzian, write a general description of the phase noise as a function of frequency.

13.22. The phase noise of a VCO needs to be tested. A 320-MHz reference signal is sourced from an ATE to a PLL with a feedback divider ratio of 64. If the ATE reference has a phase noise of –125 dBc/Hz, what is the minimum level of VCO phase noise that can be measured with this setup? Assume that the VCO phase noise should be at least 10 dB greater than the reference noise contribution.

13.23. The phase noise of a VCO within a PLL with a feedback divider ratio of 64 is to be tested using a delay-line phase noise test setup. The VCO is to be tuned to a frequency within the ISM frequency band at 5.750 GHz, and the phase noise is to be measured at approximately 1-MHz offset. What frequency should be sourced to the PLL from the ATE, and what sampling frequency should the digitizer use to collect 4096 samples? What is the exact offset frequency from the VCO output?

13.24. The phase noise of an oscillator is to be tested using a PLL-based phase noise test setup. The oscillator is tuned to a frequency of 1.401367 GHz. What is the phase noise of this device at a frequency offset of 4.8828 MHz if the output is digitized at a sampling rate of 5 GHz using an ADC with an input impedance of 50 Ω. Subsequently, a 1024-point FFT analysis reveals the following information:

$$\text{FFT}\left\{x[n]\right\} = \begin{cases} 3.42 \times 10^{-4} & \text{Bin} = 0 \\ \vdots & \\ 2.32 & \text{Bin} = 4710 \\ 7.56 \times 10^{-3} & \text{Bin} = 4711 \\ 6.20 \times 10^{-3} & \text{Bin} = 4712 \\ \vdots & \end{cases}$$

13.25. What are the main blocks of a vector signal analyzer, and why is it similar to an RF ATE?

13.26. Why does the error vector magnitude (EVM) describe the transmit function of an RF system in a single figure of merit?

13.27. A 5-GHz continuous wave sinusoidal signal is passed through a quadrature modulator resulting in an IQ signal having the following form: $\sin(\pi 10^{10} \cdot t + \pi/12) + j \cdot \cos(\pi 10^{10} \cdot t - \pi/10)$.

What is the magnitude and phase of this signal. Also, what is the phase angle difference between the I and Q components of this signal?

13.28. The following IQ signal was generated by a quadrature modulation process $0.95\sin(2\pi f_o \cdot t - \pi/7) + j \cdot 1.1\cos(2\pi f_o \cdot t + \pi/8)$. If the reference is described by $\sin(2\pi f_o \cdot t) + j\cos(2\pi f_o \cdot t)$, what is the vector error associated with the signal? Plot its magnitude and phase as a function of time for f_o equal to 900 MHz.

REFERENCES

1. Agilent, *Vector Signal Analysis Basics*, Application Note 150-15, Agilent Technologies, Inc., 2004.
2. M. Hiebel, *Fundamentals of Vector Network Analysis*, Rohde & Schwarz, Beaverton, 2007.
3. G. Srinivasan, H-C. Chao, and F. Taenzler, *Octal-site EVM tests for WLAN transceivers on very low-cost ATE platforms*, in *Proceedings of the IEEE International Test Conference*, 2008.
4. 802.11a-1999. *Supplement to IEEE standard for information technology telecommunications and information exchange between systems-local and metropolitan area networks-specific requirements.*
5. Agilent Technologies Application Note 1380-1, *RF Testing of WLAN Products*.
6. K. B. Schaub and J. Kelly, *Production Testing of Rf and System-on-a-Chip Devices for Wireless Communications*, Artech House, Boston, 2004.

CHAPTER 14

Clock and Serial Data Communications Channel Measurements

M any of today's electronic devices make use of high-speed asynchronous serial links for data communications such as USB, Firewire, PCI-Express, XAUI, SONET, SAS, and so on. Such devices make use of a serializer-deserializer transmission scheme called SerDes. Older devices operated on a synchronous clocking system such as the recommended standard 232 serial bus, referred to as RS232, or the small computer system interface, known as SCSI. While such synchronous buses are being used less as a means to communicate between two separate devices, buses internal to most devices remain for the most part dependent on a synchronous clocking scheme. The goal of this chapter is describe the various data communication schemes used today and how such systems are characterized and tested in production.

This chapter will begin by describing the attributes of both synchronous clock signals and those signals transmitted asynchronous over a serial channel. In the case of synchronous clock signals, both time- and frequency-domain descriptions of performance are described. This includes various time-domain jitter metrics, like periodic jitter and cycle-to-cycle jitter, and frequency-domain metrics, like phase noise. Our readers encountered phase noise for RF systems back in Chapters 12 and 13. It is pretty much the same for clocks. For asynchronous systems, the ultimate measure of performance is bit error rate (BER). The student will learn how to calculate the necessary test time to assure a desired level of BER performance and learn about several techniques that are used in production to reduce this test time. The latter approaches are largely based on jitter decomposition methods, and this chapter will explore four common methods found in production test. Unique to this chapter is the extensive application of probability theory to quantify the use of these jitter decomposition methods. This chapter will conclude with a discussion of several DSP-based test techniques used to quantify jitter transmission from the system input to its output. This includes a discussion about jitter transfer function test and a jitter tolerance test.

14.1 SYNCHRONOUS AND ASYNCHRONOUS COMMUNICATIONS

When two devices exchange data, there is a flow of information between the two. Typically, this information is transmitted in the form of a set of binary bits referred to as a symbol. As this

Figure 14.1. Synchronous clock distribution between two D-type registers.

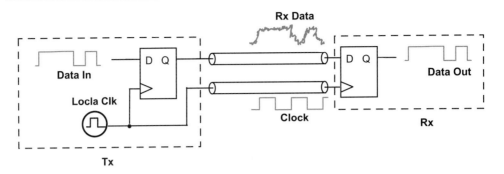

Figure 14.2. An asynchronous data communication link involving two D-type registers.

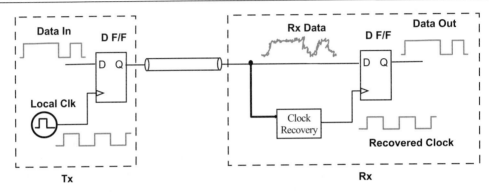

information is transmitted in sequence, the receiver must have the means to extract the individual symbols. As symbols arrive as a continuous stream of bits, one has to be able to separate one symbol from another. In asynchronous communications, each symbol is separated by the equivalent of a tag so that one knows exactly when the symbol arrives at its destination. In synchronous communications, both the sender and receiver are synchronized with a separate clock signal, thereby providing all necessary timing information at both ends.

Figure 14.1 illustrates a typical synchronous system involving two D-type registers physically separated from one another via a transmission line. One-bit data are exchanged between the two registers through the action of the falling or rising edge of the clock signal. The clock signal is used to specify when a specific bit is to be transmitted and received. Due to the physical distance between the sender and the receiver, and the fact that the clock signal travels at about one-half light speed on a PCB (equivalent to 3.3 ns/m or 2 ns/ft), the timing information is not the same at all locations. This results in timing differences between the clock signal at the transmit and receive ends, commonly referred to as clock skew. Ultimately, clock skew limits the maximum rate at which symbols can be exchanged between the transmitter and the receiver.

Conversely, an asynchronous system involving two D-type registers does not share a common clock signal (Figure 14.2). Rather, the timing information associated with the clock signal is embedded in the data stream and recovered by the receiver by a clock recovery (CR) circuit. While the system is more complicated than a synchronous system, no clock skew occurs. This enables faster data exchange rates. For this reason alone, asynchronous data communications is fast becoming the dominant method of exchanging information between electronic devices.

While asynchronous systems avoid the problem of clock skew, both synchronous and asynchronous systems suffer from the effects of circuit noise. While the source of noise is identical to that which we studied previously, here we are interested in the effect of noise on the timing information that is associated with a data communication link. Deviation from the reference signal is referred to as clock jitter.

14.2 TIME-DOMAIN ATTRIBUTES OF A CLOCK SIGNAL

A clock signal is used in both synchronous and asynchronous systems in one form or another. Deviations from its ideal behavior affect the overall performance of the data communication system.

Mathematically, we can define the transmit clock with frequency f_o as

$$c_{Tx}(t) = \text{sgn}\left[\cos(2\pi f_o t)\right] \tag{14.1}$$

where sgn( ) is the sign function and is defined as follows

$$\text{sgn}(x) = \begin{cases} 1, & x > 0 \\ 0, & x = 0 \\ -1, & x < 0 \end{cases} \tag{14.2}$$

The clock signal is assumed to be symmetrical with a zero DC offset and a 1-V signal amplitude. Of course, any offset and amplitude can be incorporated into the model of the clock by introducing the appropriate terms.

If the clock signal experiences a time delay as it travels from one point to another, say on a PCB, then we can model the received clock signal by incorporating a time delay T_D in Eq. (14.1) according to

$$c_{Rx}(t) = \text{sgn}\left\{\cos\left[2\pi f_o(t - T_D)\right]\right\} \tag{14.3}$$

From a system timing perspective, the time at which the rising edge of a clock signal crosses the logic level threshold, represented here by the 0-V level, can be deduced from Eq. (14.3) by equating it to 0 and solving for the zero crossing times. This is illustrated in Figure 14.3 for both the transmit and received clock signals. Following this logic, we can write

$$T_{ZC}(n) = \frac{2\pi n + 2\pi f_o T_D}{2\pi f_o} = \frac{n}{f_o} + T_D \tag{14.4}$$

where n represents the nth clock cycle associated with the transmit clock. In the manner written here, clock skew alters the zero crossing point by a constant amount T_D. However, in practice, T_D varies from system to system and from device to device, hence we must treat T_D as a random variable. Let us assume that T_D is a Gaussian random variable with mean value T_D and standard deviation σT_D; then we can write the mean value and standard deviation of the zero crossing point as

$$\mu_{ZC} = \mu_{T_D}$$
$$\sigma_{ZC} = \sigma_{T_D} \tag{14.5}$$

Figure 14.3. Zero crossing times associated with a clock signal.

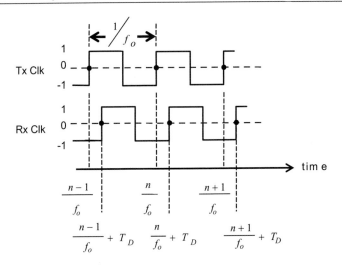

Figure 14.4. Illustrating the effects of noise on the received clock signal when compared against that which was transmitted.

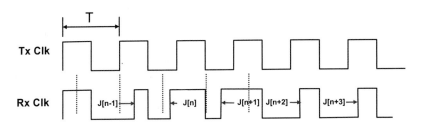

Since the standard deviation is nonzero, it indicates that the zero crossing times will vary and in some way act to reduce the amount of time available for the logic circuits to react to incoming signals, thereby increasing the probability of logic errors.

In addition to clock skew, noise is always present in any electronic circuit. While noise may arise from many different sources, the effect of noise on a synchronous or asynchronous system is to alter the zero crossing point. This is depicted in Figure 14.4, where a received signal is compared to a transmitted signal minus any physical delays for ease of comparison. Here we see that the zero crossing of the received clock varies with respect to that which was transmitted. If we denote the time difference between the transmit and received zero crossings as $J(n)$, then we can gather a set of data and perform some data analysis to better understand the underlying noise mechanism. For ease of discussion, we shall refer to $J(n)$ as the instantaneous time jitter associated with the received signal. Conversely, we can convert this jitter term into a phase-difference jitter term, that is,

$$\phi[n] = 2\pi f_o \times J[n] = 2\pi \times \frac{J[n]}{T} \tag{14.6}$$

In the study of jitter one finds several ways to interpret jitter behavior. The most common approach is to construct a histogram of the instantaneous jitter and observe the graphical behavior

Figure 14.5. Typical jitter histogram and its corresponding density function.

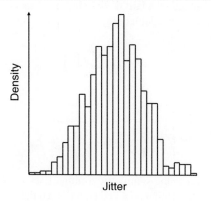

Figure 14.6. The histogram and PDF associated with a random noise and a sinusoidal signal.

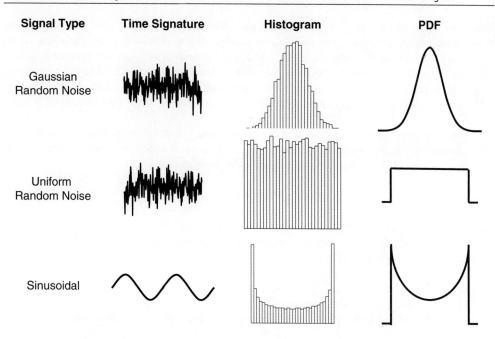

that results (see Figure 14.5). By comparing it to known shapes, one can deduce the nature of the underlying source of jitter, at least in a qualitative manner. For example, Figure 14.6 illustrates the histogram as well as its corresponding PDF for several known signal types such as a Gaussian random noise signal, a uniform random noise signal, and a sinusoidal signal. It is reasonable to assume that the distribution shown in Figure 14.5 was obtained from a noise source with a Gaussian distribution and not one from a signal involving a uniform random noise signal or a signal with a sinusoidal component.

To obtain a more quantitative measure of jitter, one can extract statistical measures[1] from the captured data such as the mean and standard deviation. For instance, if N samples of the instantaneous jitter is captured, then we can calculate the mean according to

$$\mu_J = \frac{1}{N} \sum_{n=1}^{N} J[n] \qquad (14.7)$$

and the standard deviation as

$$\sigma_J = \sqrt{\frac{1}{N} \sum_{n=1}^{N} \left\{ J[n] - \mu_J \right\}^2} \qquad (14.8)$$

Here the mean value would refer to the symmetric timing offset associated with the data set. If the data represent the jitter associated with a received signal, then the mean value represents the clock skew mentioned earlier. Often we are also interested in the peak-to-peak jitter, which is computed as follows:

$$pp_J = \max\left\{ J[1], J[2] \cdots J[N] \right\} - \min\left\{ J[1], J[2] \cdots J[N] \right\} \qquad (14.9)$$

Due to the statistical nature of a peak-to-peak estimator, it is always best to repeat the peak-to-peak measurement and average the set of values, rather than work with any one particular value. A peak-to-peak metric is a biased estimator and will increase in value as the number of points collected increases. This stems from the fact that a Gaussian random variable has a theoretically infinite peak-to-peak value, although obtaining such a value would require infinite samples. We often to refer to such random variables as being unbounded since they theoretically have no limit.

There are other types of analysis that can be performed directly on the time jitter sequence. For instance, one may be interested in how the period of the clock varies as a function of time. This removes any time offset or skew. Mathematically, this is defined as

$$J_{PER}[n] = J[n] - J[n-1] \qquad (14.10)$$

where we denote $J_{PER}[n]$ as the period jitter for the nth clock cycle. For a large enough sample set, we can calculate the mean and standard deviation of the period jitter. We can also define a cycle-to-cycle time jitter metric as follows:

$$J_{CC}[n] = J_{PER}[n] - J_{PER}[n-1] \qquad (14.11)$$

Substituting Eq.(14.10) into Eq.(14.11), we can write the cycle-to-cycle jitter in terms of the time jitter as

$$J_{CC}[n] = J[n] - 2J[n-1] + J[n-2] \qquad (14.12)$$

Mean and standard deviation can be computed when a significant number of samples of cycle-to-cycle jitter are captured.

Sometimes jitter metrics are used that go beyond comparing adjacent edges or periods and instead look at the difference between a multiple number of edges or periods that have passed.

Such metrics are referred to as N-period or N-cycle jitter. Let us assume that N is the number of periods of the clock signal that separate the edges or periods, then we can write the new metrics as

$$J_{N,PER}[n] = J[n+N-1] - J[n-1] \tag{14.13}$$

and

$$J_{N,CC}[n] = J_{PER}[n+N-1] - J_{PER}[n-1] \tag{14.14}$$

Phase, period, and cycle-to-cycle jitter provide information about clock behavior at a localized point in time. N-period or N-cycle jitter metrics track the effects of jitter that accumulate over time, at least, over the N period observation interval.

Time jitter, period jitter, and cycle-to-cycle jitter are related quantities. They are related through the backwards difference operator, a functionanalogous to differentiation for continuous functions. Assuming that time jitter has a uniform frequency distribution, period jitter will have a high-pass nature that increases at rate of 20 dB/decade across the frequency spectrum. Cycle-to-cycle jitter will also have a high-pass behavior but will increase at a much faster rate of 40 dB/dec. In both cases, low-frequency jitter components will be greatly attenuated and will not have much influence on the jitter metric. This may lead to incorrect decisions about jitter, so one must be careful in the use of these jitter metrics.

Metrics that provide insight into the low-frequency variation of jitter involve a summation or integration process as opposed to a differencing or differentiation operation. One such metric is known as accumulated jitter or long-term jitter. Accumulated jitter involves collecting the statistics of time jitteras a function of the number of cycles that has passed from the reference point. In order to identify the reference point, as well the time instant the time jitter is captured, we write $J[n,k]$ where n is the sampling instant and k is the number of clock delays that has passed from the reference point. Figure 14.7a illustrates the time jitter captured as a function of clock period delay, k. For each delay, we compute the statistics of the jitter, that is,

$$\mu_J[k] = \frac{1}{N}\sum_{n=0}^{N} J[n,k]$$

$$\sigma_J[k] = \sqrt{\frac{1}{N}\sum_{n=0}^{N}\left[J[n,k] - \mu_J(k)\right]^2} \qquad \text{for } k = 0,\ldots,\frac{N}{2} \tag{14.15}$$

Accumulated jitter refers to the behavior of the standard deviation $\sigma_j[k]$ as a function of the delay index, k, as illustrated in Figure 14.7b. This particular jitter accumulation plot would be typical of a PLL. It is useful for identifying low-frequency jitter trends associated with the clock signal. The concept of accumulated jitter is related to the autocorrelation function of random signals. As a word of caution, the length of each jitter sequence must be the same for each delay setting to ensure equal levels of uncertainty.Moreover, the length of each sequence must be no longer than the fastest time constant associated with the jitter sequence. For instance, if the jitter sequence has a noise bandwidth of 1 MHz, then the time over which a set of points are collected should be no longer than 1 μs.

Figure 14.7. Accumulated jitter: (a) illustrating the sampling instant as function of the number of clock period delays from the reference point. (b) Jitter standard deviation as a function of delay index, k.

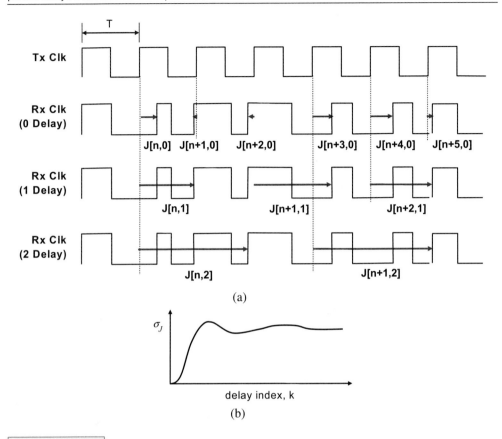

(a)

(b)

Exercises

14.1. A set of 8 samples was extracted from a received signal with respect to a reference clock having the following instantaneous time jitter values: [1 ps, 1 ps, -2 ps, 3 ps, 1 ps, 0 ps, -1 ps, 5 ps]. What is the mean, standard deviation and peak-to-peak values?

ANS. $\mu = 1 \times 10^{-12}$,
$\sigma = 2.06 \times 10^{-12}$,
$pp = 7 \times 10^{-12}$.

14.2. Compute the period jitter and cycle-to-cycle jitter from time jitter samples of exercise 15.1.

ANS.
$J_{PER} = [0\text{ ps}, -3.5\text{ ps}, -2\text{ ps}, -1\text{ ps}, -1.6\text{ ps}]$;
$J_{CC} = [-3.8\text{ ps}, -7\text{ ps}, 1\text{ ps}, 0\text{ ps}, 7\text{ ps}]$.

14.3 FREQUENCY-DOMAIN ATTRIBUTES OF A CLOCK SIGNAL

While time-domain measures of a clock signal are commonly used to describe its behavior, a power spectral density (PSD) description or phase noise $\mathcal{L}(\Delta f)$ is also used. The reader was introduced to this approach back in Sections 12.3.3 and 13.4 for RF circuits. The same approach extends to clock signals in almost the exact same way.[2,3]

Figure 14.8. (a) PSD of an ideal clock signal. (b) PSD of a jittered clock signal.

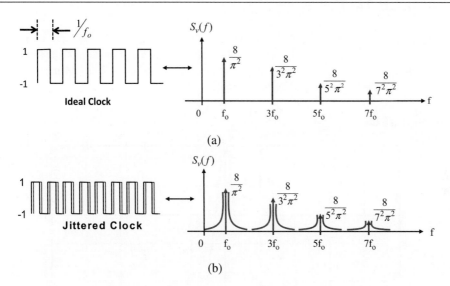

(a)

(b)

The PSD of an ideal clock signal $S_v(f)$ expressed in units of V^2 per Hz is illustrated in Figure 14.8a. Here the clock signal is decomposed into a set of monotonically decreasing sized impulses that are odd multiples of the clock frequency f_o. The impulses indicate that each harmonic contributes a constant level of power to the spectrum of the clock signal. When random jitter is present with the clock signal, it modifies the spectrum of the clock signal by adding sidebands about each harmonic as shown in Figure 14.8b. This spectrum is the result of noise in the clock circuitry phase modulating the output signal. We learned in Section 12.3.3 that the resulting PSD follows a Lorentzian distribution.

The instantaneous phase difference ϕ between the first harmonic of the clock signal located at f_o and a signal occupying a 1-Hz bandwidth offset from this harmonic by some frequency Δf has a PSD described by

$$S_\phi(\Delta f) = 2\frac{S_v(f_o + \Delta f)}{S_v(f_o)} \quad \frac{\text{rad}^2}{\text{Hz}} \tag{14.16}$$

As we learned in Section 12.3.3, $s_\phi(\Delta f)$ is related to the 1139IEEE standard definition for phase noise $\mathcal{L}(\Delta f)$ according to

$$\mathcal{L}(\Delta f)\big|_{\text{dBc/Hz}} \triangleq 10\log_{10}\left[\frac{S_\phi(\Delta f)}{2}\right] = 10\log_{10}\left[\frac{S_v(f_o + \Delta f)}{S_v(f_o)}\right] \tag{14.17}$$

The term dBc is a shorthand notation for "decibels with respect to carrier." In the context of clock signals, the carrier is the fundamental harmonic of the clock.

In many measurement situations involving a spectrum analyzer, the PSD displayed on the screen of the instrument is not in terms of V^2 per Hz but rather V^2 per resolution bandwidth in Hz. The resolution bandwidth (BW) represents the equivalent noise bandwidth of the front-end filter of the spectrum analyzer. To get the correct PSD level, one must scale the instrument PSD

EXAMPLE 14.1

Below is a PSD plot of the voltage level associated with a clock generator captured by a spectrum analyzer with a center frequency of 1.91 GHz and a resolution bandwidth of 200 kHz. Each frequency division represents 1-MHz span. What is the phase noise £(f) at a 1-MHz offset from the 1.91-GHz fundamental tone of this clock signal?

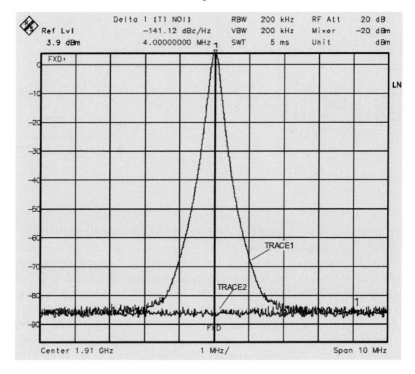

Solution:

As the carrier at 1.91 GHz has a signal level of 3.9 dBm and the PSD at 1 MHz offset from the carrier is –68 dBm, the phase noise metric [Eq. (14.19)] is computed as follows:

$$£ \text{ (1 MHz)} = -68 \text{ dBm} - 3.9 \text{ dBm} - 3.9 \text{ dBm} - 10\log_{10}(200 \times 10^3) = -124.9 \frac{dBc}{Hz}$$

Hence, the phase noise at a 1-MHz frequency offset is –124.9 dBc/Hz.

Exercises

14.3. The PSD of the voltage level of a 10-MHz clock signal is described by

$$S_v(f) = 10^{-8} / 10^4 + \left(f - 10^7\right)^2 + 0.5 \times \delta\left(f - 10^7\right)$$

in units of V2/Hz. What is the RMS noise level associated with this clock signal over a 1000-Hz bandwidth center around the first harmonic of the clock signal?

ANS.

$$V_{rms} = \sqrt{\int_{10MHz-500Hz}^{10MHz+500Hz} \frac{10^{-8}}{10^4 + \left(f - 10^7\right)^2} df} =$$

$$1.66 \times 10^{-5} \text{ V}$$

14.4. The PSD of the voltage level of a 10-MHz clock signal is described by $S_v(f) = 10^{-8}/10^4 + (f - 10^7)^2 + 0.5 \times \delta(f - 10^7)$ in units of V2/Hz. What is the phase noise of the clock at 1000-Hz offset in dBc/Hz?

ANS. $\mathcal{L}(1\,\mathrm{kHz}) = -137.0$ dBc/Hz.

by dividing by the factor BW. In terms of the phase noise metric, $\mathcal{L}(\Delta f)$, the expression would become

$$\mathcal{L}(\Delta f)\Big|_{\mathrm{dBc/Hz}} = 10\log_{10}\left[\frac{1}{BW}\frac{S_v(f_o + \Delta f)}{S_v(f_o)}\right] \tag{14.18}$$

More often than not, the data read off a spectrum analyzer is in terms of dBm, hence we convert Eq. (14.18) into the following equivalent form

$$\mathcal{L}(\Delta f)\Big|_{\mathrm{dBc/Hz}} = P_{SSB}\Big|_{\mathrm{dBm}} - P_{carrier}\Big|_{\mathrm{dBm}} \tag{14.19}$$

where we define $P_{carrier}\big|_{\mathrm{dBm}} = 10\log_{10}\left[S_v(f_o)\right]$ and $P_{SSB}\big|_{\mathrm{dBm}} = 10\log_{10}[S_v(f_o + \Delta f)] - 10\log_{10}(BW)$, both in units of dBm.

In the study of jitter, one often comes across different measures of the underlying noise mechanisms associated with a clock signal. In the previous section, measures such as a time-jitter sequence $J[n]$ or a period jitter sequence $J_{PER}[n]$ were introduced. In all cases, these sampled quantities are related to one another through a linear operation. For instance, from Eq. (14.6) the phase-difference sequence is related to the time-jitter sequence through the proportional constant $2\pi/T$, where T is the period of the clock signal. Correspondingly, the PSD of the time-jitter sequence can be expressed in terms of the PSD of the phase jitter according to

$$S_J(\Delta f)\Big|_{s^2/\mathrm{Hz}} = \left(\frac{T}{2\pi}\right)^2 S_\phi(\Delta f) \ \frac{s^2}{\mathrm{Hz}} \tag{14.20}$$

Normalizing by the period T, we can express the jitter PSD in terms of the unit interval (denoted as UI) and write Eq. (14.20) as

$$S_J(\Delta f)\Big|_{\mathrm{UI}^2/\mathrm{Hz}} = \left(\frac{1}{2\pi}\right)^2 S_\phi(\Delta f) \ \frac{\mathrm{UI}^2}{\mathrm{Hz}} \tag{14.21}$$

We can also relate the phase noise $\mathcal{L}(\Delta f)$ to the PSD of the jitter sequence expressed in s²/Hz or in UI²/Hz by substituting Eq. (14.20) or Eq.(14.21) into Eq.(14.16) above.

To obtain the PSD associated with the phase signal, we can perform an FFT analysis of the samples of the instantaneous phase jitter signal. Following the procedure outlined in Section 9.3.3, we first calculate the RMS value of the spectral coefficients of the DTFS representation of the phase error signal, that is,

$$c_{k,RMS} = \begin{cases} \left| \Phi\left[k\right] \right|, & k = 0 \\ \sqrt{2}\left| \Phi\left[k\right] \right|, & k = 1,\ldots,N/2-1 \\ \dfrac{\left| \Phi\left[k\right] \right|}{\sqrt{2}}, & k = N/2 \end{cases} \tag{14.22}$$

where the N-point variable Φ is obtain from the N-point FFT analysis from the phase-jitter sequence ϕ expressed in radians as follows:

$$\Phi = \frac{\mathrm{FFT}\left\{\phi\left[n\right]\right\}}{N} \tag{14.23}$$

As a phase signal may consist of both periodic signals (e.g., spurs) and noise, we must handle the two types of signals differently. Specifically, $S_\phi[k]$ of the random noise component expressed in rad²-per-Hz is given by

$$S_\phi\left[k\right] = c_{k,RMS}^2 \times \frac{N}{F_s} \; \frac{\mathrm{rad}^2}{\mathrm{Hz}} \tag{14.24}$$

whereas the power associated with any periodic component in the spectrum of the phase signal expressed in rad² is given by

$$S_\phi\left[k_{tone}\right] = c_{k,RMS}^2 \;\; \mathrm{rad}^2 \tag{14.25}$$

Here $k = k_{tone}$ represents the bin location of the tone. As these tonal components are not known *a priori*, the user must decide if a periodic component is present and its corresponding bin location. If the number of samples is increased, the PSD level of the random component will remain the same, however, a periodic component will increase in amplitude, thereby revealing its periodic nature. To summarize, the $S_\phi[k]$ of the phase signal can be described as

$$S_\phi\left[k\right] = \begin{cases} c_{k,RMS}^2 \times \dfrac{N}{F_s} \quad \dfrac{\mathrm{rad}^2}{\mathrm{Hz}}, & k = 0,\ldots,N/2, \;\; k \neq \text{tone BIN} \\ c_{k,RMS}^2 \quad\quad \mathrm{rad}^2, & k = \text{tone BIN} \end{cases} \tag{14.26}$$

Because the phase noise PSD sequence will change with different sample sets, it is customary to average K-sets of the PSD on a frequency-by-frequency basis and obtain an average PSD behavior defined as follows

$$\hat{S}_\phi\left[k\right] = \frac{1}{K}\sum_{i=1}^{K} S_{\phi,i}\left[k\right] \tag{14.27}$$

where the subscript i indicates the ith-PSD obtain from Eq. (14.26). These short-term PSDs are commonly referred to as periodograms.

EXAMPLE 14.2

The following plot of a 512-point phase-difference sequence was obtained from a heterodyning process for capturing the phase errors associated with a clock signal:

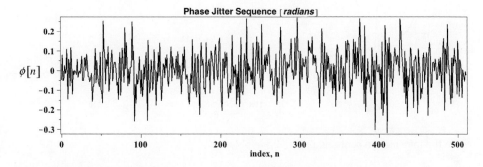

Phase Jitter Sequence [radians]

The samples were obtained with a digitizer operating at a 10 MHz sampling rate. The continuous-time phase signal passes through a 1-MHz low-pass filter prior to digitization. What is the phase noise $£(\Delta f)$ at a frequency offset of about 40 kHz for this particular sample set?

Solution:

Because the sampling rate is 10 MHz and 512 points have been collected, the frequency resolution of the PSD will be

$$\frac{F_s}{N} = \frac{10 \times 10^6}{512} = 19531.25 \text{ Hz}$$

Hence we can find the offset in the second bin of the PSD of phase error signal at about 39.0625 kHz. The PSD of the phase signal is found by sequencing through Eqs. (14.22) to (14.26), resulting in the following PSD plot of $S_\phi[k]$:

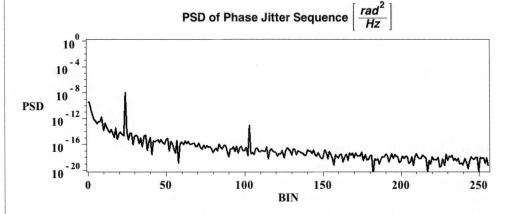

PSD of Phase Jitter Sequence $\left[\frac{rad^2}{Hz} \right]$

Here we see that spurious tones are present in bins 24 and 103 corresponding to frequencies of 468.75 kHz and 2.0117 MHz. In order to remove the statistical variation in the PSD, we repeated the sampling process an additional 10 times and then averaged the PSD. The resulting smoothed

PSD for the phase noise is as follows:

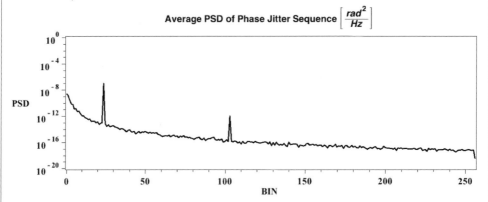

Average PSD of Phase Jitter Sequence $\left[\dfrac{rad^2}{Hz}\right]$

From the information contained in bin 2, we obtain

$$\hat{S}_\phi[2] = 2.66 \times 10^{-10} \quad \frac{rad^2}{Hz}$$

Subsequently, we compute the phase noise metric at a 39.0625 kHz offset from the carrier as follows:

$$\pounds(39.0625 \text{ kHz}) = 10\log_{10}\left[\frac{2.66 \times 10^{-10}}{2}\right] = -98.8 \quad \frac{dBc}{Hz}$$

Exercises

14.5. The PSD of the instantaneous phase of a 10 MHz clock signal is described by $S_\phi(f) = 10^{-8}/10^4 + f^2$ in units of rad²/Hz. What is the PSD of the corresponding jitter sequence in UI²/Hz?

ANS.

$$S_J(f) = \left(\frac{1}{2\pi}\right)^2 \frac{10^{-8}}{10^4 + f^2} \quad \frac{UI^2}{Hz}$$

14.6. The PSD of the instantaneous phase of a 10-MHz clock signal is described by $S_\phi(f) = 10^{-8}/10^4 + f^2$ in units of rad²/Hz. What is the phase noise of the clock at a 1000-Hz offset in dBc/Hz?

ANS. $\pounds = -143.1 \text{ dBc/Hz}.$

14.4 COMMUNICATING SERIALLY OVER A CHANNEL

Serial communication is the process of sending data one symbol at a time, sequentially, over a communication channel. In order to retrieve the symbols that were sent, some form of synchronization between the transmitter and the receiver is required. This is typically performed with the aid of a clock signal combined with the data. In this section, we shall create a model of the communication channel, provide measures of transmission quality, and describe various methods to test for system behavior. Many of the methods used to describe system behavior are extension of the methods introduced for clock signals.

14.4.1 Ideal Channel

A simple model of a serial communication system is shown in Figure 14.9 consisting of a transmitter, an ideal channel with T_D time delay, and a receiver. For sake of argument, we will assume that the symbol being transmitted is either a binary 0 or 1. With an ideal channel, the data that are transmitted is delivered to the receiver in the exact same form that it was transmitted. That is, if a binary 1 were sent at time instant TS_1, then at the receiver a binary 1 would arrive at time $TS_1 + T_D$. If another symbol were sent some time instant later, say TS_2, then this symbol would arrive at the receiver at time $TS_2 + T_D$. It is fair to say that at the receiver end, there is a 50% probability that the received symbol is a 1 if we assume that there is a 50% probability that a 1 was transmitted. Likewise, a binary 0 will arrive with identical probability since there are only two possible outcomes.

In practice, a signal cannot be sent in zero time, because this would require infinite bandwidth. A better model of system behavior is one where the transmitter sends a pulse of some finite width duration T and height 1 or 0, depending on the bit being sent, as depicted in Figure 14.10. At the receiver end, a process must be in place to decide whether a 0 or a 1 was transmitted at the appropriate time. To gain insight into this process, consider overlaying k durations of the received signal into one full bit period T, as illustrated in Figure 14.11a. It is customary to normalize the time axis by the bit duration and refer to the normalized bit period as the unit interval, UI. On doing so, the resulting diagram is called an *eye diagram* and is used extensively in data communication studies. We can just as easily express the horizontal axis of the eye diagram in terms of denormalized time as shown in Figure 14.11b. For our convenience throughout this chapter we shall work with the denormalized eye diagram.

An eye diagram is a two-dimensional diagram where the vertical axis represents the range of possible signals that can appear at the receiver end and the horizontal axis represents the range of times modulo the bit duration T that also appears at the receiver end. For any bit of information that is transmitted, two pieces of information must be transmitted: the bit level and the time at which bit is sent. At the receiver end, these two pieces of information must be extracted from the incoming signal. A decision-making process involving two decision thresholds performs this operation. One threshold V_{TH} is used to determine if the receive level is a logically high or low

Figure 14.9. Asynchronous communications through an ideal channel with ideal signaling.

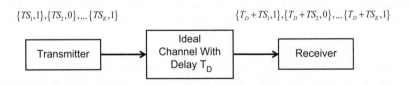

Figure 14.10. Asynchronous communications using finite width pulses.

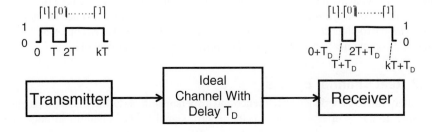

Figure 14.11. Illustrating the creation of an eye diagram from the received signal: (a) normalized eye diagram (b) denormalized eye diagram.

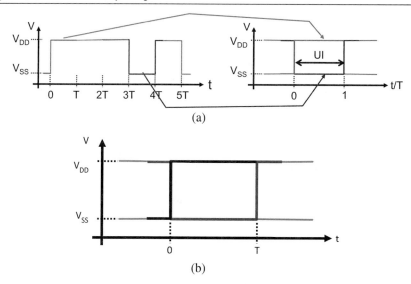

(a)

(b)

Figure 14.12. Illustrating the decision level thresholds associated with an eye diagram.

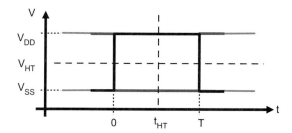

level and the other threshold t_{TH} is used to set the bit sample time (between 0 and T). To maximize the likelihood of making the correct decision, the decision point lies in the middle of the range of possible values as shown in the eye diagram of Figure 14.12.

At this point in our discussion it is interesting to make some general statement about the events that are taking place at the receiver. From the eye diagram we see that received signal crosses the logic threshold V_{TH} at two points: $t = 0$ and $t = T$. If we assume that a logical 0 or 1 is equally likely, then the probability that either a 0–1 transition or 1–0 transition occurs is also equally likely at a 50% probability level. Hence we can model the PDF of the time at which a signal crosses the logic threshold (also referred to as the zero-crossing) with two equal-sized delta functions at $t = 0$ and $t = T$, each having a magnitude of 50% as shown in Figure 14.13a. Likewise, we can model the PDF of the voltage level that appears at the receiver in a similar manner as shown in Figure 14.13b. These two plots correspond to the fact that perfect square waves appear at the receiver. In a real serial communication system, such perfection is a very long way off.

14.4.2 Real Channel Effects

Transmitted signals experience interference and dispersion effects as the signal propagates through the channel to the receiver. As a result, the receive signal contains additional noise not sent by

Figure 14.13. A PDF interpretation of the received signal: (a) as measured along the voltage decision axis defined by $V = V_{TH}$. (b) as measured along the time decision axis defined by $t = t_{TH}$.

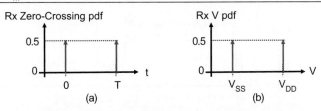

Figure 14.14. Real channel effects; (a) A pulse experiencing an additive noise effect. (b) A pulse experiencing a spreading effect due to the dispersion nature of the channel.

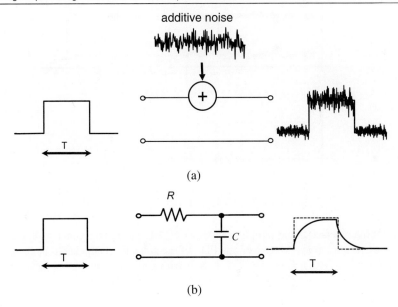

the transmitter, and the signal experiences a spreading effect due to the dispersion of the channel (e.g., high-frequency attenuation effects, as well as a delay dependent on frequency). We model the additive noise with a summing block placed in series with the channel, as illustrated in Figure 14.14a. A simple model that captures the basic idea of dispersion is that shown in Figure 14.14b. Here the channel is model as a simple RC circuit. A square pulse of duration T applied to one end of the RC line comes out of the other end as an exponential type pulse with a duration that extends beyond time T. It is this pulse spreading effect that can lead to a bit interfering with the next bit. When this occurs, it is referred to as *inter-symbol interference* (ISI).

Noisy Channel

Let us begin our investigation of real channels by modeling the effects of additive noise on the received signal. We shall assume that the noise is Gaussian distributed. A pulse train passing through a channel would experience additive noise and may appear as that shown in Figure 14.15a. The eye diagram corresponding to this pulse train would be that shown in Figure 14.15b. Also drawn alongside the eye diagram are the corresponding PDFs associated with each decision axis, that is, $V = V_{TH}$ and $t/T = UI_{TH}$. In contrast to the PDFs along the decision axis for the ideal channel shown in Figure 14.13, here we see that additive noise modified the PDF of the voltage

Figure 14.15. (a) Pulse train subject to additive channel noise. (b) Corresponding eye diagram and its PDFs along the decision-making axes ($V = V_{TH}$ and $t/T = t_{TH}$).

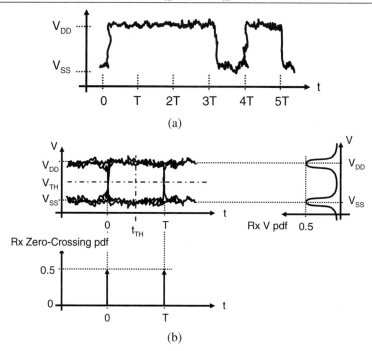

that appears at the received at the sampling instant UI_{TH} but has no effect on the PDF corresponding to the zero-crossing times. Moreover, the PDF of the noise level appears to be the convolution of the delta functions with a Gaussian distribution with zero mean and some arbitrary standard deviation, σ_N.

Dispersive Channel

In this subsection we shall consider the effects of a dispersive channel on the behavior of a transmitted square pulse train. The eye diagram corresponding to a pulse train subject to channel dispersive effects is illustrated in Figure 14.16. Here we see that the channel impairments alter the distributions of the received signal along both decision axes. We model these impairments as discrete lines in the PDF rather than as a continuous PDF because these effects are deterministic and dependent on the data pattern transmitted as well as the channel characteristics. This effect is commonly referred to as *data-dependent jitter* (DDJ). Due to the physical nature of a channel, the distribution along each decision axes is bounded by some peak-to-peak value. Unless the eye opening along the $t = UI_{TH}$ axis is closed, DDJ will have little effect on the bit error rate. However, when the effects of the channel noise is incorporated alongside the dispersive effects, we would find an increased bit error rate (over and above the bit error rate that would correspond to the additive noise effect alone). The reason for this stems from the way the different channel effects combine. From a probability point of view, assuming that the PDFs for the different channels impairments are independent, they would combine through a convolution operation as shown in Figure 14.18. The net effect is a wider PDF extending over a larger range on both decision axes, that is, closing the eye opening. Also, we note that the PDFs no longer follow a Gaussian distribution.

Figure 14.16. (a) Pulse train subject to channel dispersion or ISI. (b) Corresponding eye diagram and its PDFs along the decision-making axes ($V = V_{TH}$ and $t/T = t_{TH}$).

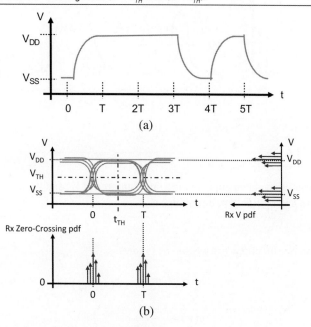

Figure 14.17. (a) Pulse train subject to channel dispersion and additive Gaussian noise. (b) Corresponding eye diagram and its PDFs along the decision-making axes ($V = V_{TH}$ and $t = t_{TH}$).

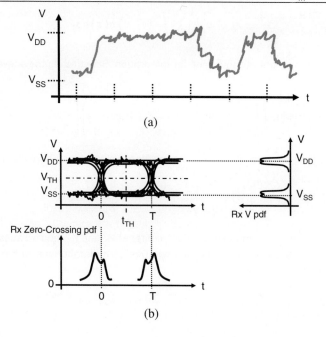

Figure 14.18. Independent PDFs are combined by a convolution operation.

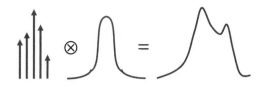

EXAMPLE 14.3

Determine the convolution of two impulse functions centered at V_{SS} and V_{DD} with a Gaussian function with zero mean and σ_N standard deviation.

Solution:
Mathematically, the two functions are written as follows:

$$f(v) = \frac{1}{2}\delta(v - V_{SS}) + \frac{1}{2}\delta(v - V_{DD})$$

$$g(v) = \frac{1}{\sigma_N\sqrt{2\pi}}e^{-v^2/2\sigma_N^2}$$

The convolution of two PDFs is defined by the following integral equation,

$$f(v) \otimes g(v) \triangleq \int_{-\infty}^{\infty} f(v - \tau)g(\tau)\,d\tau$$

where τ is just an immediate variable used for integration. Substituting the above two functions, we write

$$f(v) \otimes g(v) = \int_{-\infty}^{\infty} \left[\frac{1}{2}\delta(v - V_{SS} - \tau) + \frac{1}{2}\delta(v - V_{DD} - \tau)\right]\left[\frac{1}{\sigma_N\sqrt{2\pi}}e^{-\tau^2/2\sigma_N^2}\right]d\tau$$

Next, using the sifting property of the impulse function, we write the above integral as

$$f(v) \otimes g(v) = \frac{1}{\sigma_N 2\sqrt{2\pi}}e^{-(v-V_{SS})^2/2\sigma_N^2} + \frac{1}{\sigma_N 2\sqrt{2\pi}}e^{-(v-V_{DD})^2/2\sigma_N^2}$$

Hence, the convolution of two impulse functions center at V_{SS} and V_{DD}, and a Gaussian distribution with zero mean and standard deviation σ_N are two Guassian distributions with mean V_{SS} and V_{DD} having a standard deviation σ_N.

Figure 14.19. Illustrating the effects on the zero-crossing levels: (a) duty-cycle distortion and (b) periodic-induced jitter, without any noise present.

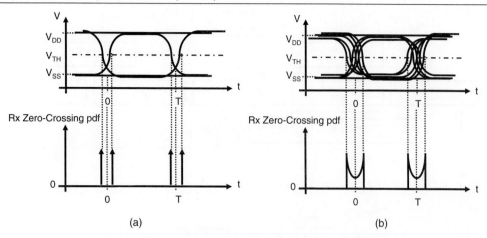

(a)   (b)

In many practical situations we encounter problems that involve the convolution of two Gaussian distributions. As such, one can easily show the convolution of two Gaussian distributions is another Gaussian distribution, whose mean value is the sum of individual mean values and whose variance is the sum of individual variances, that is,

$$\mu_T = \mu_1 + \mu_2$$
$$\sigma_T^2 = \sigma_1^2 + \sigma_2^2$$

(14.28)

where the two Gaussian distributions are described by parameters $N(\mu_1, \sigma^2_1)$ and $N(\mu_2, \sigma^2_2)$.

Transmitter Limitations

The transmitter also introduces signal impairments that show up at the receiver—specifically, duty-cycle distortion (DCD) and periodic induced jitter (PJ). Duty-cycle distortion is caused by the unsymmetrical rise and fall times of the driver located in the transmitter as represented by the eye diagram shown in Figure 14.19a. As can be seen from this figure, a logic 1 bit value has a shorter bit duration than a logic 0 bit value relative to the zero crossing point. In essence, DCD can be considered as a shift in the time of the rising and falling edge of the data bit. DCD is a deterministic effect because it depends only on the driver characteristic together with the logic pattern that is transmitted. Fortunately, its effect on the zero-crossing levels is small and bounded.

Sometimes one finds a periodic component of jitter showing up at the receiver end. This is typically caused by some periodic interference located at the transmitter or picked up via the channel. Such effects include crosstalk from adjacent power nets and noise from switching power supplies. Its effect on the zero-crossing levels is bounded. Figure 14.19b illustrates a single sine wave component and its effect on the zero-crossing PDF as shown in Figure 14.19b. Through the application of the Fourier series more complicated periodic components can be also be described.

Jitter Classifications

Figure 14.20 provides a quick summary of the breakdown of the various jitter components described in the previous subsection. At the top of the tree, we list the total jitter (TJ), which is

Figure 14.20. A summary of the different components of jitter.

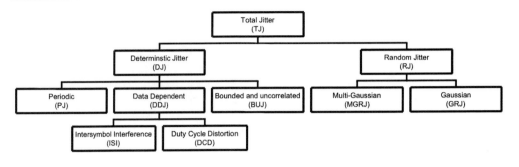

Figure 14.21. A simple receiver arrangement.

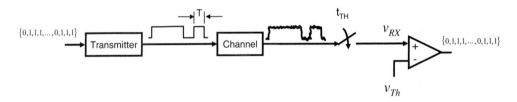

broken down into two parts: deterministic (DJ) and random (RJ) jitter. The deterministic jitter component is further broken down into period jitter PJ, data-dependent jitter DDJ, and a catch-all category called bounded and uncorrelated jitter denoted by BUJ. We further divide DDJ into intersymbol interference (ISI) and duty-cycle distortion (DCD) components. The BUJ component includes tones that are unrelated to the input data sequence, as well as uncorrelated noise-like signals. In essence, it is a catch-all quantity that accounts for unexplained effects. Random jitter is divided into single or multi-Gaussian distributions, denoted as MGRJ or GRJ, respectively.

The entire jitter classification is further divided between whether the jitter is bounded or unbounded—that is, varies with sample set. As a general rule, bounded parameters will be specified with a min–max value or a peak-to-peak value, whereas unbounded parameters have no limit so instead are specified with anRMS or standard deviation parameter.

14.4.3 Impact of Decision Levels on Receiver Performance

In this section we shall investigate the effects that jitter has on the performance of a simple receiver arrangement as a function of the decision levels t_{TH} and V_{TH}. Consider the simple receiver arrangement shown in Figure 14.21 consisting of a switch in series with the channel and a comparator with reference level V_{TH}. The series switch is used to sample the incoming signal at the appropriate time designated by t_{TH}. The resulting sampled level is then applied to a comparator where a decision of being logical 0 or 1 is made based on a comparison with the voltage threshold level V_{TH}. Let us consider the effect of each decision level—first separately, then combined.

Received Voltage Decision Level, VTH

Let us assume that the channel has corrupted the received signal v_{Rx} in such a way that a logical 0 level, denote V_{Logic0}, has a Gaussian distribution with mean value V_{Logic0} and standard deviation σ_N. Likewise, a logical 1 level, denoted V_{Logic1}, is also Gaussian distributed with mean value V_{Logic1} and

Figure 14.22. Modeling the PDF of the received voltage signal and identifying the regions of the PDF that contributes to the probability of error.

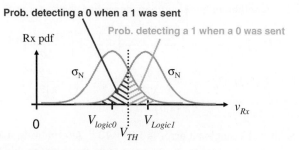

standard deviation σ_N. Superimposing these two distributions along the same axis we obtain the decision diagram shown in Figure 14.22. The dotted vertical line indicates the voltage threshold V_{TH}. The portion of the distribution centered around V_{Logic0} but above V_{TH} represents the probability that logic 1 is detected when logic 0 was sent. Mathematically, we write

$$P\left(V_{Rx} > V_{TH} \mid Tx = 0\right) = \int_{V_{TH}}^{\infty} pdf_{Rx,Logic0} dv_{Rx} = 1 - \Phi\left(\frac{V_{TH} - V_{Logic0}}{\sigma_N}\right) \tag{14.29}$$

Conversely, the portion of the distribution centered around V_{Logic1} but below V_{TH} represents the probability that logic 0 is detected when logic 1 was sent. This allows us to write

$$P\left(V_{Rx} < V_{TH} \mid Tx = 1\right) = \int_{-\infty}^{V_{TH}} pdf_{Rx,Logic1} dv_{Rx} = \Phi\left(\frac{V_{TH} - V_{Logic1}}{\sigma_N}\right) \tag{14.30}$$

Assuming the probability of transmitting a logic 0 and a logic 1 is 1/2, respectively, then we can write the total probability that a single bit is received in error as

$$P_e\left(V_{TH}\right) = \frac{1}{2} \times P\left(V_{Rx} > V_{TH} \mid Tx = 0\right) + \frac{1}{2} \times P\left(V_{Rx} < V_{TH} \mid Tx = 1\right) \tag{14.31}$$

Substituting Eqs. (14.29) and (14.30) into Eq.(14.31), we write

$$P_e\left(V_{TH}\right) = \frac{1}{2} - \frac{1}{2} \times \Phi\left(\frac{V_{TH} - V_{Logic0}}{\sigma_N}\right) + \frac{1}{2} \times \Phi\left(\frac{V_{TH} - V_{Logic1}}{\sigma_N}\right) \tag{14.32}$$

The following example will help illustrate the application of this formula.

EXAMPLE 14.4

A logic 0 is transmitted at a nominal level of 0 V and a logic 1 is transmitted at a nominal level of 2.0 V. Each logic level has equal probability of being transmitted. If a 150-mV RMS Gaussian noise signal is assumed to be present at the receiver, what is the probability of making an single trial bit error if the detector threshold is set at 1.0 V.

Solution:

According to Eq.(14.32), together with the numbers described above, we write

$$P_e\left(V_{TH}\right) = \frac{1}{2} - \frac{1}{2} \times \Phi\left(\frac{V_{TH} - V_{Logic0}}{\sigma_N}\right) + \frac{1}{2} \times \Phi\left(\frac{V_{TH} - V_{Logic1}}{\sigma_N}\right)$$

$$P_e\left(1.0\right) = \frac{1}{2} - \frac{1}{2} \times \Phi\left(\frac{1.0 - 0}{0.15}\right) + \frac{1}{2} \times \Phi\left(\frac{1.0 - 2.0}{0.15}\right)$$

$$\therefore P_e\left(1.0\right) = 1.31 \times 10^{-11}$$

Therefore the probabilitry of a single bit error is 1.31×10^{-11}. Alternatively, if 1.31×10^{11} bits are sent in one second, then one can expect about 1 error to be made during this transmisison.

EXAMPLE 14.5

For the system conditions described in Example 14.4, compute the probability of error as a function of the threshold voltage V_{TH} beginning at 0 V and extending to 2 V.

Solution:

Using Eq. (14.32), together with the data from Example 14.7, we can write the probability of error as a function of the threshold voltage V_{TH} as follows:

$$P_e\left(V_{TH}\right) = \frac{1}{2} - \frac{1}{2} \times \Phi\left(\frac{V_{TH} - 0}{0.15}\right) + \frac{1}{2} \times \Phi\left(\frac{V_{TH} - 2.0}{0.15}\right)$$

Numericially, we iterate V_{TH} through this formula beginning at 0 V and ending at 2 V with a 10 mV step, resulting in the P_e plot shown below:

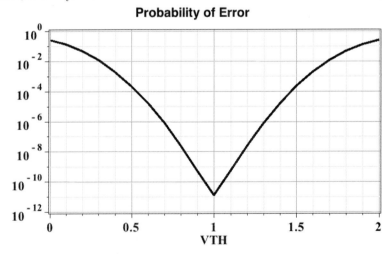

Here we the P_e plot is minumum at about the 1-V threshold level, a level midway between the logic 0 and logic 1 level. Also, we see the Pe is symmetrical about this same threhold level. The above plot is commonly referred to as a *bathtub plot* on account of its typical shape. (Albeit, this particular plot looks more like a valley associated with a mountain range than a bathtub. This is simply a function of the numerical values used for this example.)

Exercises

14.7. A logic 0 is transmitted at a nominal level of 0 V and a logic 1 is transmitted at a nominal level of 1.0 V. A logic 0 level is three times as likely to be transmitted as a logic 1 level. If a 100-mV RMS Gaussian noise signal is assumed to be present at the receiver, what is the probability of making an single trial bit error if the detector threshold is set at 0.6 V? What about if the detector threshold is set at 0.35 V?

ANS.

$P_e = 15.8 \times 10^{-6}$;

$P_e = 116.3 \times 10^{-6}$.

Up to this point in the discussion, we have assumed that the distribution of the received signal is modeled as a Gaussian random variable. We learned in the previous section that this is rarely the case on account of the channel dispersion effects as well as circuit asymmetries associated with the transmitter.From a mathematical perspective, this does not present any additional complication to quantifying the probability of error provided that the PDFs of the received signal is captured in some numerical or mathematical form described in the general way by the following expression:

$$P_e\left(V_{TH}\right) = \frac{1}{2} \times \int_{V_{TH}}^{\infty} pdf_{Rx,Logic0} dv_{Rx} + \frac{1}{2} \times \int_{-\infty}^{V_{TH}} pdf_{Rx,Logic1} dv_{Rx} \qquad \textbf{(14.33)}$$

Received Zero-Crossing Decision Level, t_{TH}

Let us now turn our attention to the performance of the receiver from the perspective of the normalized sampling instant represented by t_{TH}. Let us begin by assuming that the distribution of zero crossings around the $(n-1)$th-bit transition and the nth-bit transition are both Gaussian distributed with standard deviation σ_{ZC}. Such a situation is depicted in Figure 14.23. Once again, on account of channel effects (i.e., additive noise and dispersion), a zero-crossing edge intended for the $(n-1)$ th sampling instant may find itself occurring after the sampling instant defined by t_{TH}, resulting in a synchronization error that may lead to a bit error. Of course, the reverse is also possible whereby the intended zero-crossing corresponding to the nth-bit arrives before the sampling instant, resulting in a potential bit error. Following the same theoretical development pose earlier for the received voltage signal, we can state that the probability of error due to a sampling timing error is

$$P_e\left(t_{TH}\right) = P\left(\text{bit transition}\right) \times P\left(t > t_{TH}\middle|\text{bit}=n-1\right) + P\left(\text{bit transition}\right) \times P\left(t < t_{TH}\middle|\text{bit}=n\right) \textbf{(14.34)}$$

Figure 14.23. Modeling the PDF of the received zero-crossing time and identifying the regions of the PDF that contributes to the probability of error.

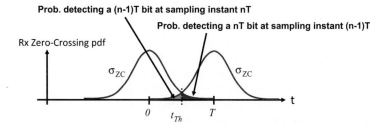

Since we assumed that a 0 and 1 are equally likely to occur, it is reasonable to assume that a bit transition is also equally likely, hence we write the probability of error as

$$P_e\left(t_{TH}\right) = \frac{1}{2} \times \int_{t_{TH}}^{\infty} pdf_{TJ,n-1} dt + \frac{1}{2} \times \int_{-\infty}^{t_{TH}} pdf_{TJ,n} dt \qquad (14.35)$$

where the subscript TJ, n signifies the PDF of the total jitter around the nth bit transition. Because the PDFs are Gaussian in nature, we can replace the integral expressionsby normalized Gaussian CDF functions and write

$$P_e\left(t_{TH}\right) = \frac{1}{2} - \frac{1}{2} \times \Phi\left(\frac{t_{TH} - 0}{\sigma_{ZC}}\right) + \frac{1}{2} \times \Phi\left(\frac{t_{TH} - T}{\sigma_{ZC}}\right) \qquad (14.36)$$

Interesting enough, the above formula has a very similar form as that for the received voltage signal probability of error shown in Eq. (14.32). Let us look at an example using this approach.

EXAMPLE 14.6

Data are transmitted to a receiver at a data rate of 1 Gbits/s through a channel that causes the zero crossings to vary according to a Gaussian distribution with zero mean and a 70-ps standard deviation. What is the probability of error of a single event if the sampling instant is set midway between bit transitions?

Solution:
As the data rate is 1 Gbits/s, the spacing between bit transistions is 1 ns. Hence the sampling instant will be set at 500 ps or 0.5 UI. From Eq. (14.36), the P_e is

$$P_e\left(t_{TH}\right) = \frac{1}{2} - \frac{1}{2} \times \Phi\left(\frac{t_{TH} - 0}{\sigma_{ZC}}\right) + \frac{1}{2} \times \Phi\left(\frac{t_{TH} - T}{\sigma_{ZC}}\right)$$

$$P_e\left(500 \text{ ps}\right) = \frac{1}{2} - \frac{1}{2} \times \Phi\left(\frac{500 \text{ ps} - 0}{70 \text{ ps}}\right) + \frac{1}{2} \times \Phi\left(\frac{500 \text{ ps} - 1000 \text{ ps}}{70 \text{ ps}}\right)$$

$$\therefore P_e\left(500 \text{ ps}\right) = 4.57 \times 10^{-13}$$

Therefore the probability of a single bit error is 4.57×10^{-13}.

Combined Effect of Varying Received Voltage Level Decision V_{TH} and Zero-Crossing Decision Level, t_{TH}

The previous two subsections look at the impact of the decision levels on the probability of error assuming Gaussian random variables. Let us combine these two effects assuming that the two sets of random variables are independent. In practice, one would expect that some correlation between the signal around the zero crossing and the voltage level appearing at the comparator input. However, away from the zero-crossing point, the signal saturates at one of its logic level; hence in this region one would expect little correlation between voltage level and the zero-crossing point.

14.8. Data are transmitted to a receiver at a data rate of 100 Mbits/s through a channel that causes the zero crossings to vary according to a Gaussian distribution with zero mean and a 100-ps standard deviation. What is the probability of error of a single event if the sampling instant is set at 0.3UI? Repeat for a threshold of 0.75UI?

ANS.
$P_e = 4.93 \times 10^{-10}$;
$P_e = 1.43 \times 10^{-7}$.

Regardless of whether or not we obtain complete independence, this analysis helps to illustrate the dependence of the decision levels on the eye opening. Consider the total probability of error as

$$P_e\left(t_{TH}, V_{TH}\right) = P_e\left(t_{TH}\right) + P_e\left(V_{TH}\right) \tag{14.37}$$

Substituting Eqs. (14.32) and (14.36), we write the above expression as

$$P_e\left(t_{TH}, V_{TH}\right) = 1 - \frac{1}{4} \times \Phi\left(\frac{V_{TH} - V_{Logic0}}{\sigma_N}\right) + \frac{1}{4} \times \Phi\left(\frac{V_{TH} - V_{Logic1}}{\sigma_N}\right)$$
$$- \frac{1}{4} \times \Phi\left(\frac{t_{TH} - 0}{\sigma_{ZC}}\right) + \frac{1}{4} \times \Phi\left(\frac{t_{TH} - T}{\sigma_{ZC}}\right) \tag{14.38}$$

14.5 BIT ERROR RATE MEASUREMENT

All communication systems experience some form of transmission error. We learned previously about the various ways in which errors (e.g., additive noise, dispersion, etc.) arise in a communication system. We modeled these effects as variations in the decision points using different probability density functions and used them to compute theprobability of a single error during transmission, denoted by P_e. In this section we shall consider a cumulative measure of transmission quality called the bit error ratio or which is sometimes referred to as the bit error rate. We shall use either term interchangeably and we will often use the acronym BER.

The bit error ratio (BER) is defined as the average number of transmission errors, denoted by μ_{N_e}, divided by the total number of bits sent in a specified time interval, denoted N_T, that is,

$$BER = \frac{\mu_{N_E}}{N_T} \tag{14.39}$$

According to our earlier development, if N_T bits were transmitted to a receiver, then according to our probability model, we would expect the average number of bit errors to given by

$$\mu_{N_E} = N_T P_e \tag{14.40}$$

which leads to the simple observation that BER is equivalent to P_e, that is,

$$BER = P_e \tag{14.41}$$

Figure 14.24. BER test setup.

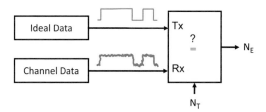

The test setup for measuring a BER is that shown in Figure 14.24. It consists of a block that compares the logic state of the data appearing at its two input port: Tx and Rx. One input contains the ideal data or the transmitted data, and the other input contains the received data. The compare block captures data over some specified time interval or over some total count N_T from which the number of transmission errors N_E is identified. While one may be tempted to compute the BER using the measured N_E, this would be incorrect as BER is defined in terms of the average N_E or as denoted in Eq. (14-39) by μ_{N_e}. Subsequently, a series of identical measurements must be made in order to extract the average value. However, we learned from Chapter 5, specifically Section 5.3, that extracting the average value from a finite-sized population of random variables will always have some level of uncertainty associated with it. If we were to assume that the error count is Gaussian distributed with parameters $N(\mu_{N_e}, \sigma_{N_e})$, then one could bound the variation in the BER value with a 99.7% confidence level as

$$\frac{\mu_{N_E} - 3\sigma_{N_E}}{N_T} \leq \text{BER} \leq \frac{\mu_{N_E} + 3\sigma_{N_E}}{N_T} \tag{14.42}$$

Clearly, the higher the number of bits transmitted (N_T), the smaller the expected range in possible measured BER values. While this conclusion is true in general, transmission bit errors do not obey Gaussian statistics, rather they follow more closely a binomial distribution. We saw this distribution back in Chapter 4, Section 4.3. There the binomial distribution is the discrete probability distribution of the number of successes in a sequence of N independent yes/no experiments, each of which yields a success with probability p. Following this train of thought, if N_T bits are transmitted, then the probability of having N_E errors in the received bit set assuming the errors are independent with probability of error P_e can be written as

$$P[X = N_E] = \begin{cases} \dfrac{N_T!}{N_E!(N_T - N_E)!} P_e^{N_E}(1 - P_e)^{N_T - N_E}, & N_E = 0, \ldots, N_T \\ 0, & \text{otherwise} \end{cases} \tag{14.43}$$

The binomial distribution has the following mean and standard deviation:

$$\mu = N_T P_e$$
$$\sigma = \sqrt{N_T P_e (1 - P_e)} \tag{14.44}$$

The cumulative distribution function CDF of the binomial distribution can be described by

$$P[X \leq N_E] = \sum_{k=0}^{N_E} P[X = k] = \sum_{k=0}^{N_E} \left(\frac{N_T!}{k!(N_T - k)!} \right) P_e^k (1 - P_e)^{N_T - k} \tag{14.45}$$

which can be further simplified when N_T BER < 10 using the Poisson approximation as

$$P[X \le N_E] \approx \sum_{k=0}^{N_E} \frac{1}{k!} (N_T \mathrm{BER})^k e^{-N_T \mathrm{BER}} \qquad (14.46)$$

The transmission test problem is one where we would like to verify that the system meets a certain BER level while at the same time we remain confident that the test results are repeatable to some statistical level of certainty.[4] Mathematically, we can state this as a conditional probability whereby we assume that the single bit error probability P_e is equal to the desired BER, and we further set the probability of N_E bit errors to a confidence parameter α as follows:

$$P[X \le N_E | P_e = \mathrm{BER}] = \alpha \qquad (14.47)$$

The meaning of what BER refers to should now be clear; every bit received has a probability of being in error equal to the BER. The parameter α represents the probability of receiving N_E errors or less. For a very large α, the probability of receiving N_E or fewer errors is very likely. However, a small α signifies the reverse, a very unlikely situation. Hence, focusing in on the unlikely situation provides us with a greater confidence that the assigned conditions will be met. It is customary to refer to $(1 - \alpha)$ as the confidence level (CL) expressed in percent, that is,

$$\mathrm{CL} = 1 - \alpha \qquad (14.48)$$

Combining Eq. (14.46) with Eq. (14.47), together with the appropriate parameter substitutions, we write

$$P[X \le N_E | P_e = \mathrm{BER}] = \sum_{k=0}^{N_E} \frac{1}{k!} (N_T \mathrm{BER})^k e^{-N_T \mathrm{BER}} = 1 - \mathrm{CL} \qquad (14.49)$$

To help the reader visualize the relationship provided by Eq.(14.49), Figure 14.25 provides a contour plot of the confidence levels corresponding to the probability of 2 errors or less as a function of BER and N_T. This plot provides important insight into the tradeoffs between CL, BER and bit length N_T. For example, if 10^{12} bits are transmitted and 2 or less data errors are received, then with 95% confidence level we can conclude that the BER $\le 10^{-11}$. Likewise, if after 10^{11} bits are received with 2 or less data errors, we can conclude with very little confidence (5%) that the BER $\le 10^{-11}$.

Exercises

14.9. Data are transmitted with a single-bit error probability of 10^{-12}. If 2×10^{12} bits are transmitted, what is the probability of no bit errors, one bit error, two bit errors, and five bit errors? What is the average number of errors expected?

ANS.

$P[N_E = 0] = 0.135$; $P[N_E = 1] = 0.271$; $P[N_E = 2] = 0.271$; $P[N_E = 5] = 0.036$; $\mu = 2$.

14.10. Data are transmitted at a rate of 10 Gbps over a channel with a single-bit error probability of 10^{-11}. If 10^{12} bits are transmitted, what is the probability of less than or equal to one bit error, less than or equal to two bit errors, and less than or equal to 10 bit errors.

ANS. $P[N_E \le 1] = 0.000499$; $P[N_E \le 2] = 0.00277$; $P[N_E \le 10] = 0.583$.

Figure 14.25. The confidence level for a probability of 2 errors or less as a function of BER and bit length, N_T.

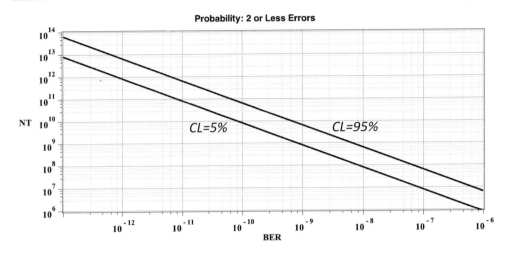

Equation (14.49) provides us with the opportunity to compute the number of bits N_T that need to be transmitted such that the desired level of BER is reached to a desired level of confidence CL. As N_T represents the minimum bit length that must pass with errors N_E or less, we shall designate this bit length as $N_{T,\min}$. Next, we rearrange Eq. (14.49) and solve for $N_{T,\min}$ as

$$N_{T,\min} = \frac{1}{\mathrm{BER}} \ln\left[\sum_{k=0}^{N_E} \frac{1}{k!}\left(N_T \mathrm{BER}\right)^k\right] - \frac{1}{\mathrm{BER}} \ln\left(1 - \mathrm{CL}\right) \qquad (14.50)$$

Using numerical methods, $N_{T,\min}$ can be solved for specific values of BER and CL. To achieve a high level of confidence, typically, one must collect at least 10 times the reciprocal of the desired BER. For example, if a BER of 10^{-12} is required, then one can expect that at least 10^{13} samples will be required.

It is interesting to note that Eq.(14.50) reveals a tradeoff between the confidence level of the test and the bit length $N_{T,\min}$, which in turn is related to test time, T_{test}. When Consider substituting $N_E = 0$ into Eq. (14.50), one finds

$$N_{T,\min} = -\frac{\ln\left(1 - \mathrm{CL}\right)}{\mathrm{BER}} \qquad (14.51)$$

Since the test time T_{test} is given by

$$T_{\mathrm{test}} = \frac{N_{T,\min}}{F_S} \qquad (14.52)$$

we find after substituting into Eq. (14.51),

$$T_{\mathrm{test}} = -\frac{\ln\left(1 - \mathrm{CL}\right)}{F_S \times \mathrm{BER}} \qquad (14.53)$$

The higher the confidence level CL, the longer the time required for completing the test.

EXAMPLE 14.7

A system transmission test is to be run whereby a BER < 10^{-10} is to be verified. How many samples should be collected such that the desired BER is met with a CL of 99% when no more than 2 errors are deemed acceptable. What is the total test time if the data rate is 2.5 Gbps?

Solution:
Using Eq. (14.50), we write

$$N_{T,min} = \frac{1}{10^{-10}} \ln\left[\sum_{k=0}^{2} \frac{1}{k!} \left(N_T 10^{-10}\right)^k \right] - \frac{1}{10^{-10}} \ln(1-0.99)$$

Here we have a transcendental expression in terms of $N_{T,min}$ only, as N_E was set to 2. Using a computer program, we solve for $N_{T,min}$ and get

$$N_{T,min} = \frac{1}{10^{-10}} \ln\left[1 + \frac{1}{1!}10^{-10}N_{T,min} + \frac{1}{2!}\left(10^{-10}N_{T,min}\right)^2 \right] - \frac{1}{10^{-10}}\ln(1-0.99) = 8.41 \times 10^{12} \text{ bits}$$

If after 8.41×10^{12} bits we have 2 bit errors or less, then we can conclude that BER < 10^{-10} with a confidence level of 99%. The time required to perform this test at a data rate of 2.5 Gpbs is

$$T_{test} = \frac{N_{T,min}}{F_S} = \frac{8.41 \times 10^{12} \text{ bits}}{2.5 \text{ Gbits/s}} = 3362.4 \text{ s}$$

Therefore, 3363.4 s is required to perfom this test. This is almost one hour of testing one part only!

BER tests are extremely time-consuming and very expensive to run in production. A second test limit can be derived that measures the confidence that the bit sequence will have more bit errors than the desired amount. Hence, once identified, the test can be terminated and declared a failure, thereby saving test time.

Consider the probability of achieving more bit errors than desired with confidence CL; that is, we write

$$P\left[X > N_E \middle| P_e = \text{BER} \right] = 1 - P\left[X \le N_E \middle| P_e = \text{BER} \right] = 1 - \text{CL} \qquad \textbf{(14.54)}$$

Substituting the appropriate CDF, we write

$$P\left[X > N_E \middle| P_e = \text{BER} \right] = 1 - \sum_{k=0}^{N_E} \frac{1}{k!}\left(N_T \text{BER}\right)^k e^{-N_T \text{BER}} = 1 - \text{CL} \qquad \textbf{(14.55)}$$

Following the same set of steps as before, we write a transcendental expression in terms of the bit length N_T. Because this bit length represents the maximum number of bits that can pass before exceeding a given number of errors, N_E, we designate this bit length as $N_{T,max}$ and write

$$N_{T,max} = \frac{1}{\text{BER}} \ln\left[\sum_{k=0}^{N_E} \frac{1}{k!}\left(N_T \text{BER}\right)^k \right] - \frac{1}{\text{BER}}\ln(\text{CL}) \qquad \textbf{(14.56)}$$

Figure 14.26. A flow diagram illustrating the test process for performing an efficient BER test.

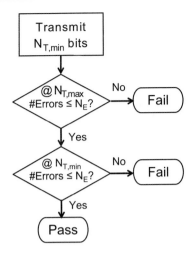

If the number of observed errors is greater than N_E after $N_{T,max}$ bits have been collected, the test can be declared a failure. Otherwise, we continue the test until all $N_{T,min}$ bits are transmitted. A flow diagram illustrating the test process is shown in Figure 14.26.

EXAMPLE 14.8

A system transmission test is to be run whereby a BER $< 10^{-10}$ is to be verified at a confidence level of 99% when 2 or less bit errors are deemed acccptable. What are the minumum and maximum bit lengths required for this test? Also what is the minimum and maximum time for this test if the data rate is 2.5 Gbps? Does testing for a fail part prior to the end of the test provide significant test time savings?

Solution:
From the previous example, we found the minumum bit length to be

$$N_{T,min} = \frac{1}{10^{-10}} \ln\left[1 + \frac{1}{1!}10^{-10}N_{T,min} + \frac{1}{2!}\left(10^{-10}N_{T,min}\right)^2 \right] - \frac{1}{10^{-10}}\ln(1-0.99) = 8.41 \times 10^{12} \text{ bits}$$

and the maximum bit length is found from

$$N_{T,max} = \frac{1}{10^{-10}} \ln\left[1 + \frac{1}{1!}10^{-10}N_{T,max} + \frac{1}{2!}\left(10^{-10}N_{T,max}\right)^2 \right] - \frac{1}{10^{-10}}\ln(0.99) = 4.36 \times 10^{11} \text{ bits}$$

The test time for any one test will be

$$T_{test,max} = \frac{N_{T,min}}{F_S} = \frac{8.41 \times 10^{12} \text{ bits}}{2.5 \text{ Gbits/s}} = 3362.4 \text{ s}$$

or

$$T_{test,min} = \frac{N_{T,max}}{F_S} = \frac{4.36 \times 10^{11} \text{ bits}}{2.5 \text{ Gbits/s}} = 1524.3 \text{ s}$$

Therefore we can save almost 50% of the test time on a bad part by introducing a second test.

14.11. A system transmission test is to be run whereby a BER < 10^{-11} is to be verified at a confidence level of 95% when 4 or less bit errors are deemed acceptable. What is the minimum bit length required for this test?

ANS. $N_{T,min} = 9.15 \times 10^{11}$.

14.12. A system transmission test is to be run whereby a BER < 10^{-11} is to be verified at a confidence level of 95% when 2 or less bit errors are deemed acceptable. What is the maximumbit length required for this test?

ANS. $N_{T,max} = 8.18 \times 10^{10}$.

14.5.1 PRBS Test Patterns

BER testing uses predetermined stress patterns comprising of a bit sequence of logical ones and zeros generated by a pseudorandom binary sequence (PRBS). The sequence is made up of 2^P bits that are generated from a core routine involving P-bits. In many PRBS sequences, the zero state is avoided, reducing the sequence to $2^P - 1$ unique patterns. There are many different pattern types, largely dependent on the system application and its data rates.[5,6] The bit sequence is pseudorandom because the pattern is not truly random, rather it is deterministic and repeats itself, unlike that which would be produced by a truly random source like white noise. The patterns must be long enough to simulate random data at the data rate being tested. If the pattern is too short, it will repeat rapidly and may confuse the clock recovery circuitry, which will prevent the receiver from synchronizing on the data.

Linear feedback shift registers (LFSRs) make extremely good pseudorandom pattern generators. An LFSR consists of a shift-register involving P flip-flops or stages, together with an exclusive-OR feedback function operating on the flip-flop outputs. Finite field mathematics in base 2 is used to derive PBRS patterns. Any LFSR can be represented as a polynomial in variable X referred to as the generator polynomial, $G(X)$, given by

$$G(X) = X^P + g_{P-1}X^{P-1} + g_{P-2}X^{P-2} + \cdots + g_2X^2 + g_1X^1 + 1 \tag{14.57}$$

The generalized Fibonacci implementation of an LFSR consists of a simple shift register in which a binary-weighted modulo-2 sum of the flip-flop outputs is fed back to the input as shown in Figure 14.27. (The modulo-2 sum of two 1-bit binary numbers yields 0 if the two numbers are identical, and 1 if they differ, e.g., $0 + 0 = 0$, $0 + 1 = 1$, $1 + 1 = 0$.) For any given flip-flop output except the first flip-flop, the weight g_i is either 0, meaning "no connection," or 1, meaning it is fed back. The 0th flip-flop weight is always 1 and the output of the XOR operation feeds into the D-input of the $(P - 1)$th stage. The most economical manner in which to realize an LFSR with a generator polynomial of Pth-order involves those Pth-order polynomials with minimum number of nonzero coefficients. In addition, the generator polynomial $G(X)$ must also be made primitive, that is, it cannot be factor into smaller terms. When this condition is met, the LFSR will produce a maximal length sequence, that is, a pattern with period equal to $2^P - 1$. Table 14.1 lists some primitive generator polynomials of degrees 2 to 33.

Figure 14.28 illustrates the implementation of a four-stage LFSR with generator polynomial,

$$G(X) = X^4 + X + 1 \tag{14.58}$$

Here the first and second flip-flop outputs are feedback to the shift register input through a single XOR operation. As each flip-flop delays their input by one clock cycle, we can write the following equations for each flip-flop output at time index n as

Figure 14.27. A generalized N-stage LFSR based on the Fibonacci implementation.

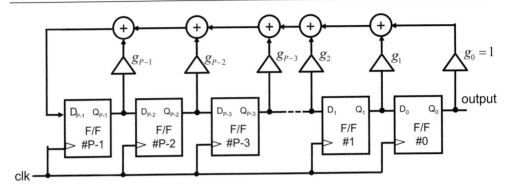

Figure 14.28. A 4-stage linear feedback shift register.

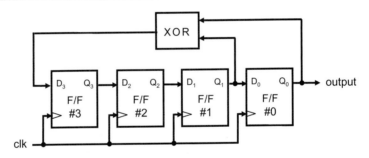

Table 14.1. Some Primitive LFSR Generator Polynomials of Degrees 2 to 33

Degree	Generator Polynomial	Degree	Generator Polynomial
2	X^2+X+1	18	$X^{18}+X^7+1$
3	X^3+X+1	19	$X^{19}+X^5+X^2+X+1$
4	X^4+X+1	20	$X^{20}+X^3+1$
5	X^5+X^2+1	21	$X^{21}+X^2+1$
6	X^6+X+1	22	$X^{22}+X+1$
7	X^7+X^3+1	23	$X^{23}+X^5+1$
8	$X^8+X^6+X^5+X^3+1$	24	$X^{24}+X^7+X^2+X+1$
9	X^9+X^4+1	25	$X^{25}+X^3+1$
10	$X^{10}+X^3+1$	26	$X^{26}+X^6+X^2+X+1$
11	$X^{11}+X^2+1$	27	$X^{27}+X^5+X^2+X+1$
12	$X^{12}+X^6+X^4+X+1$	28	$X^{28}+X^3+1$
13	$X^{13}+X^4+X^3+X+1$	29	$X^{29}+X^2+1$
14	$X^{14}+X^{10}+X^6+X+1$	30	$X^{30}+X^{23}+X^2+X+1$
15	$X^{15}+X+1$	31	$X^{31}+X^3+1$
16	$X^{16}+X^{12}+X^3+X+1$	32	$X^{32}+X^{22}+X^2+X+1$
17	$X^{17}+X^3+1$	33	$X^{33}+X^{13}+1$

$$F_3[n] = L_{XOR}[n-1]$$
$$F_2[n] = F_3[n-1]$$
$$F_1[n] = F_2[n-1] \tag{14.59}$$
$$F_0[n] = F_1[n-1]$$

where L_{XOR} is the output of the XOR gate. We can describe the XOR output as a summing operation over a finite field of mod 2 with inputs F_1 and F_0 as follows:

$$L_{XOR}[n] = F_1[n] + F_0[n] \mod 2 \tag{14.60}$$

Substituting the above equation into Eq. (14.59)and solving for the output $F_0[n]$, we write

$$F_0[n] = F_0[n-3] + F_0[n-4] \mod 2 \tag{14.61}$$

Including the initial register values, we further write

$$F_0[n] = \begin{cases} F_0[0], & n=0 \\ F_1[0], & n=1 \\ F_2[0], & n=2 \\ F_3[0], & n=3 \\ F_0[n-3] + F_0[n-4] \mod 2, & n=4,5,\cdots \end{cases} \tag{14.62}$$

The above time difference equation (mod 2) provides the complete pattern sequence from an initial state or seed. The following example will illustrate this procedure.

EXAMPLE 14.9

Derive the first 15 bits associated with the four-stage LFSR shown in Figure 14.28, assuming that the shift register is initialized with seed 1011 (from left to right).

Solution:
According to Eq. (14.62), we write

$$F_0[n] = \begin{cases} 1 & n=0 \\ 1 & n=1 \\ 0 & n=2 \\ 1 & n=3 \\ F_0[n-3] + F_0[n-4] \mod 2, & n=4,5,\ldots \end{cases}$$

Next, stepping through n from 4 to 19, we get the following 15 bits:

0, 1, 1, 1, 1, 0, 0, 0, 1, 0, 0, 1, 1, 0, 1

If the sequence is allowed to run longer, one would observe the following repeating pattern:

1, 1, 0, 1,0,   1, 1, 1, 1, 0, 0, 0, 1, 0, 0, 1, 1, 0, 1,   0, 1, 1, 1, 1, 0, 0, 0, 1, 0, 0, 1, 1, 0, 1,
0, 1, 1, 1, 1, 0, 0, 0, 1, 0, 0, 1, 1, 0, 1,   0, 1, 1, 1, 1, 0, 0, 0, 1, 0, 0, 1, 1, 0, 1, ...

In general, for any Pth-order generator polynomial described by Eq. (14.57) the output of the XOR gate can be written in terms of state of each flip-flop output at sampling instant n as

$$\text{XOR}\left[n\right] = g_{P-1}F_{P-1}\left[n\right] + g_{P-2}F_{P-2}\left[n\right] + \cdots + g_k F_k\left[n\right] + \cdots + g_1 F_1\left[n\right] + F_0\left[n\right] \qquad (14.63)$$

Correspondingly, we can write the state of the output of the 0th flip-flop in terms of its passed state as

$$F_0\left[n\right] = g_{P-1}F_0\left[n-1\right] + g_{P-2}F_0\left[n-2\right] + \cdots + g_k F_0\left[n-(P-k)\right] + \cdots + g_1 F_0\left[n-(P-1)\right] + F_0\left[n-P\right]$$

$$(14.64)$$

The next example will help to illustrate the use of these generalized equations, as well as describe how the initial seed is incorporated into the PRBS generation.

EXAMPLE 14.10

An LFSR of degree 8 is required. Write a short routine that generates the complete set of unique bits associated with this sequence. Initialize the LSFR using the seed 00000001, where the first bit corresponds to the state of the 7th flip-flop and the last bit is associated with the state of the 0th flip-flop.

Solution:
The generator polynomial for a PRBS of degree 8 (Table 14.1) is

$$G(X) = X^8 + X^6 + X^5 + X^3 + 1$$

Comparing this expression to the general form of the generating polynomial of degree $P = 8$, we determine

$$G(X) = X^8 + g_7 X^7 + g_6 X^6 + g_5 X^5 + g_4 X^4 + g_3 X^3 + g_2 X^2 + g_1 X^1 + 1$$

$$g_7 = 0, \ g_6 = 1, \ g_5 = 1, \ g_4 = 0, \ g_3 = 1, \ g_2 = 0, \ g_1 = 0$$

Subsequently, the recursive equation that governs the output behavior of the LSFR as desribed by Eq. (14.64) can be written as

$$F_0\left[n\right] = F_0\left[n-(8-6)\right] + F_0\left[n-(8-5)\right] + F_0\left[n-(8-3)\right] + F_0\left[n-8\right]$$

which reduces to

$$F_0\left[n\right] = F_0\left[n-2\right] + F_0\left[n-3\right] + F_0\left[n-5\right] + F_0\left[n-8\right]$$

Subsequently, we can write the following routine involving a programming for loop and the array varaible F as

```
# initialize the output using the given seed value
F[7] = 0 : F[6] = 0 : F[5] = 0 : F[4] = 0 : F[3] = 0 : F[2] = 0 : F[1] = 0 : F[0] = 1 :
# cycle through the leading flip-flop recursive equation
for n = 8 to 262,
    F[n] = F[n-2] + F[n-3] + F[n-5] + F[n-8]
end
```

Cycling through this routine will provide all 255 unique bits of this sequence corresponding to the initial conditions provided by the seed number.

Exercises

14.13. An LFSR has primitive generator polynomial $G(X) = X^{19} + X^5 + X^2 + X + 1$. How many bits are associated with this repeating pattern?

ANS. $2^{19} - 1$.

14.14. An LFSR has primitive generator polynomial $G(X) = X^{19} + X^5 + X^2 + X + 1$. Write the expression for the XOR operation assuming the output is taken from the 0th flip-flop.

ANS. XOR= $F_5 + F_2 + F_1 + F_0$; XOR is fed into the D-input of the #18 flip-flop.

14.15. Determine the first 15 unique bits associated with a four-stage LFSR with generator polynomial $G(X) = X^4 + X + 1$ with initial seed 0001.

ANS. 1, 0, 0, 1, 1, 0, 1, 0, 1, 1, 1, 1, 0, 0, 0.

14.6 METHODS TO SPEED UP BER TESTS IN PRODUCTION

A BER test generally takes a very long time to execute to any reasonable level of accuracy. There have been numerous attempts to reduce the amount of test time required to perform a BER test, such as the scan testapproach and the jitter decomposition technique. In all cases, these approaches reduce test time at the expense of accuracy. Their use in practice is therefore limited and application-dependent.

14.6.1 Amplitude-Based Scan Test

The amplitude-based scan test method attempts to estimate the BER of a system component in minutes rather than hours or days. It does so by measuring the amount of signal power andnoise power present in a system, assuming, the underlying statistics of the noise is Gaussian. In the statistical literature, this class of estimation is known as importance sampling. In the optical networking literature, the amplitude-based scan test method is referred to as the Q-factor test method.[11]

The amplitude-based scan test method further assumes that no periodic or pattern dependent jitter is present. Generally, a repeating 1010 pattern is used to excite the system rather than a PRBS pattern. This ensures that no ISI is present.

Recall from Section 14.4.3 that the probability of single bit error P_e as a function of the threshold voltage V_{TH} for a receiver is given by Eq. (14.32). Assuming that each bit has the same probability of error, we can set the BER as a function of V_{TH} and write

$$\text{BER}\left(V_{TH}\right) = \frac{1}{2} - \frac{1}{2} \times \Phi\left(\frac{V_{TH} - V_{Logic0}}{\sigma_N}\right) + \frac{1}{2} \times \Phi\left(\frac{V_{TH} - V_{Logic1}}{\sigma_N}\right) \qquad (14.65)$$

The minimum BER occurs when the threshold voltage V_{TH} is set equal to the average voltage representing the two logic levels, i.e.,

$$V^*_{TH} = \frac{V_{Logic0} + V_{Logic1}}{2} \qquad (14.66)$$

Hence the minimum BER is found by substituting Eq. (14.66) into (14.65), together with the mathematical identity,

$$\Phi\left(-x\right) = 1 - \Phi\left(x\right) \qquad (14.67)$$

and write Eq. (14.65) as

$$\text{BER}\left(V_{TH}^{*}\right) = 1 - \Phi\left(\frac{V_{Logic1} - V_{Logic0}}{\sigma_N}\right) \qquad (14.68)$$

The amplitude-based scan test strategy can now be identified. The goal is to identify the three unknown parameters of the BER expression found in Eq. (14.65) then using the ideal sampling point defined by Eq. (14.66), compute the corresponding BER. Moreover, as these three quantities are the same at any BER level, we can identify them at relatively large BER values (low performance) and save enormous test time in the process.

The procedure works by moving the receiver threshold level V_{TH} closer to the logic 1 voltage level such that the BER performanceis reduced, say, to some value between10^{-4} and 10^{-10} where the test time is short. Let us assume at this instant that the BER is set at some level near 10^{-9}. According to Eq. (14-65), with $V_{TH} = V_{TH,1}$, we can write

$$\text{BER}\left(V_{TH,1}\right) = \frac{1}{2} - \frac{1}{2} \times \Phi\left(\frac{V_{TH,1} - V_{Logic0}}{\sigma_N}\right) + \frac{1}{2} \times \Phi\left(\frac{V_{TH,1} - V_{Logic1}}{\sigma_N}\right) \qquad (14.69)$$

We further recognize that probability of error due to the noise centered on the logic 0 level is now insignificant at this BER level, hence we can write

$$\text{BER}\left(V_{TH,1}\right) \approx \frac{1}{2} \times \Phi\left(\frac{V_{TH,1} - V_{Logic1}}{\sigma_N}\right) \qquad (14.70)$$

Next, move the threshold level again such that the BER is reduced to say 10^{-6}. This means moving the threshold level even closer towards the logic 1 voltage level. We now have a second equation in terms of $V_{TH,2}$ and the revised BER level, i.e.,

$$\text{BER}\left(V_{TH,2}\right) \approx \frac{1}{2} \times \Phi\left(\frac{V_{TH,2} - V_{Logic1}}{\sigma_N}\right) \qquad (14.71)$$

We now have two linear equations in two unknowns, from which we can easilysolve $V_{Logic,1}$ and σ_N as follows:

$$V_{Logic1} = \frac{V_{TH,2} \times \Phi^{-1}\left[2\text{BER}\left(V_{TH,1}\right)\right] - V_{TH,1} \times \Phi^{-1}\left[2\text{BER}\left(V_{TH,2}\right)\right]}{\Phi^{-1}\left[2\text{BER}\left(V_{TH,2}\right)\right] - \Phi^{-1}\left[2\text{BER}\left(V_{TH,1}\right)\right]}$$

$$\sigma_N = \frac{V_{TH,2} - V_{TH,1}}{\Phi^{-1}\left[2\text{BER}\left(V_{TH,2}\right)\right] - \Phi^{-1}\left[2\text{BER}\left(V_{TH,1}\right)\right]} \qquad (14.72)$$

assuming

$$V_{TH,1} < V_{TH,2}$$

$$\text{BER}\left(V_{TH,1}\right) < \text{BER}\left(V_{TH,2}\right)$$

We repeat the above procedure, but this time we move closer to the other logic level using the same BER levels, and we obtain another set of equations assuming $V_{TH,4} < V_{TH,3}$ and BER $(V_{TH,3})$ < BER $(V_{TH,4})$, that is,

$$\text{BER}\left(V_{TH,3}\right) \approx \frac{1}{2} - \frac{1}{2} \times \Phi\left(\frac{V_{TH,3} - V_{Logic0}}{\sigma_N}\right)$$

$$\text{BER}\left(V_{TH,4}\right) \approx \frac{1}{2} - \frac{1}{2} \times \Phi\left(\frac{V_{TH,4} - V_{Logic0}}{\sigma_N}\right)$$

(14.73)

These two equations can then be solved for unknown parameters, $V_{Logic,0}$ and σ_N, as follows:

$$V_{Logic0} = \frac{V_{TH,3} \times \Phi^{-1}\left[1 - 2\text{BER}\left(V_{TH,4}\right)\right] - V_{TH,4} \times \Phi^{-1}\left[1 - 2\text{BER}\left(V_{TH,3}\right)\right]}{\Phi^{-1}\left[1 - 2\text{BER}\left(V_{TH,4}\right)\right] - \Phi^{-1}\left[1 - 2\text{BER}\left(V_{TH,3}\right)\right]}$$

$$\sigma_N = \frac{V_{TH,3} - V_{TH,4}}{\Phi^{-1}\left[1 - 2\text{BER}\left(V_{TH,4}\right)\right] - \Phi^{-1}\left[1 - 2\text{BER}\left(V_{TH,3}\right)\right]}$$

(14.74)

Here we have two expressions for the standard deviation of the noise. Based on our initial assumptions, these should both be equal. If they are not, we can simply average the two estimate of the standard deviation of the underlying noise process and use the average value in our estimates of this model parameter, that is,

$$\sigma_N = \frac{\sigma_{N,0} + \sigma_{N,1}}{2}$$

(14.75)

Conversely, we can create a new theory of BER performance based on two different Gaussian distributions with zero means and standard deviations of $\sigma_{N,0}$ and $\sigma_{N,1}$ and solve for the BER performance as

$$\text{BER}\left(V_{TH}\right) = \frac{1}{2} - \frac{1}{2} \times \Phi\left(\frac{V_{TH} - V_{Logic0}}{\sigma_{N,0}}\right) + \frac{1}{2} \times \Phi\left(\frac{V_{TH} - V_{Logic1}}{\sigma_{N,1}}\right)$$

(14.76)

The optimum threshold level V^*_{TH} tor minimum level of BER can then be found to occur at

$$V^*_{TH} = \frac{\sigma_{N,0}V_{Logic1} + \sigma_{N,1}V_{Logic0}}{\sigma_{N,0} + \sigma_{N,1}}$$

(14.77)

In order visualize the order of variables, Figure 14.29 illustrate the arrangement of threshold voltages with respect to the BER level. It is important that the reader maintains the correct order; otherwise the estimated BER will be incorrect.

To perform numerical calculations, we need to find the inverse value of the Gaussian CDF function for very small values. Appendix B, found at the back of the book, provides a short table of some of the important values of this inverse function. One can also make use of the approximation to the Gaussian CDF when x is large but negative:

$$\Phi\left(x\right) \approx \frac{1}{(-x)\sqrt{2\pi}} e^{-\frac{1}{2}x^2}, \quad x \ll 0$$

(14.78)

Figure 14.29. Highlighting the notation of the threshold voltages with respect to the BER level.

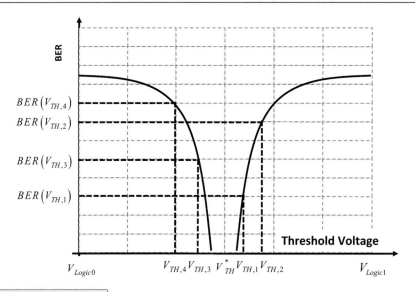

EXAMPLE 14.11

A system transmission test is to be run whereby a BER $< 10^{-12}$ is to be verified using the amplitude-based scan test method. The following BER measurements were made at the following four threshold levels:

Voltage Threshold Levels	BER Measurement
0.70	0.5 x 10^{-6}
0.90	0.5 x 10^{-9}
1.20	0.5 x 10^{-9}
1.35	0.5 x 10^{-6}

Does this system have the capability to meet the BER requirements and what is the optimum threshold level? Assume that the noise is asymmetrical.

Solution:
Following the reference system of equations we write

$$\text{BER}\,(V_{TH,1}) = 0.5 \times 10^{-9} \text{ @ } V_{TH,1} = 1.20, \qquad \text{BER}\,(V_{TH,2}) = 0.5 \times 10^{-6} \text{ @ } V_{TH,2} = 1.35,$$

$$\text{BER}\,(V_{TH,3}) = 0.5 \times 10^{-9} \text{ @ } V_{TH,3} = 0.90, \qquad \text{BER}\,(V_{TH,4}) = 0.5 \times 10^{-6} \text{ @ } V_{TH,4} = 0.70,$$

Substituting the above parameters into Eq. (14.72), we get

$$V_{Logic1} = \frac{1.35 \times \Phi^{-1}\left[10^{-9}\right] - 1.20 \times \Phi^{-1}\left[10^{-6}\right]}{\Phi^{-1}\left[10^{-6}\right] - \Phi^{-1}\left[10^{-9}\right]}$$

$$\sigma_{N,1} = \frac{1.35 - 1.20}{\Phi^{-1}\left[10^{-6}\right] - \Phi^{-1}\left[10^{-9}\right]}$$

Using the relationship for $\Phi^{-1}(x)$ from the table in Appendix B, specifically, $\Phi^{-1}(10^{-6}) = -4.7534$ and $\Phi^{-1}(10^{-9}) = -5.9978$, we find

$$V_{Logic1} = \frac{1.35 \times \Phi^{-1}\left[10^{-9}\right] - 1.20 \times \Phi^{-1}\left[10^{-6}\right]}{\Phi^{-1}\left[10^{-6}\right] - \Phi^{-1}\left[10^{-9}\right]} = \frac{1.35 \times (-5.9978) - 1.20 \times (-4.7534)}{(-4.7534) - (-5.9978)} = 1.923 \text{ V}$$

$$\sigma_{N,1} = \frac{1.35 - 1.20}{\Phi^{-1}\left[10^{-6}\right] - \Phi^{-1}\left[10^{-9}\right]} = \frac{1.35 - 1.20}{(-4.7534) - (-5.9978)} = 0.1205 \text{ V}$$

Likewise, for the other logic level parameters with $\Phi^{-1}(1 - 10^{-6}) = 4.7534$ and $\Phi^{-1}(1 - 10^{-9}) = 5.9978$, we write

$$V_{Logic0} = \frac{0.90 \times \Phi^{-1}\left[1 - 10^{-6}\right] - 0.70 \times \Phi^{-1}\left[1 - 10^{-9}\right]}{\Phi^{-1}\left[1 - 10^{-6}\right] - \Phi^{-1}\left[1 - 10^{-9}\right]} = \frac{0.90 \times 4.7534 - 0.70 \times 5.9978}{4.7534 - 5.9978} = -0.064 \text{ V}$$

$$\sigma_{N,0} = \frac{0.90 - 0.70}{\Phi^{-1}\left[1 - 10^{-6}\right] - \Phi^{-1}\left[1 - 10^{-9}\right]} = \frac{0.90 - 0.70}{4.7534 - 5.9978} = 0.1607 \text{ V}$$

The optimum threshold level is

$$V^*_{TH} = \frac{\sigma_{N,0}V_{Logic1} + \sigma_{N,1}V_{Logic0}}{\sigma_{N,0} + \sigma_{N,1}} = \frac{0.1607 \times 1.923 - 0.1205 \times (-0.064)}{0.1607 + 0.1205} = 1.071 \text{ V}$$

Finally, we compute the minimum BER from

$$BER\left(V^*_{TH}\right) = \frac{1}{2} - \frac{1}{2} \times \Phi\left(\frac{1.071 + 0.064}{0.1607}\right) + \frac{1}{2} \times \Phi\left(\frac{1.071 - 1.923}{0.1205}\right) = \frac{1}{2} - \frac{1}{2} \times \Phi\left(7.06\right) + \frac{1}{2} \times \Phi\left(-7.06\right)$$

Using the identity

$$\Phi\left(7.06\right) = 1 - \Phi\left(-7.06\right)$$

The above expression reduces to

$$BER\left(V^*_{TH}\right) = \Phi\left(-7.06\right)$$

Using the table in Appendix B, we note that is $\Phi(-7.06)$ is bounded by $\Phi(-7.06) = 10^{12}$; hence we estimate the minumum BER as

$$BER\left(V^*_{TH}\right) < 10^{-12}$$

Alternatively, we can make use of the approximation in Eq. (14.78) and estimate $\Phi(-7.06)$ as

$$\Phi\left(-7.06\right) \approx \frac{1}{7.06\sqrt{2\pi}} e^{-\frac{1}{2}(-7.06)^2} = 8.22 \times 10^{-13}$$

We can therefore conclude that the system meets the BER requirements (just barely, though).

A variant of the amplitude-based scan test technique is one that captures a histogram of the voltage level around each logic level at the center of the eye diagram. A digital sampler or digital sampling oscilloscope is often used to capture these histograms, from which its mean (logic voltage level) and standard deviation can be found. Subsequently, the optimum threshold and minimum BER can be found by combining Eq. (14.77) with Eq. (14.76). The following example will illustrate this approach.

EXAMPLE 14.12

A digital sampling oscilloscope obtained the following eye diagram while observing the characteristics of a 1-Gbps digital signal. Histograms were obtain using the built-in function of the scope at a sampling instant midway between transitions (i.e., at the maximum point of eye opening). Detailed analysis revealed that each histogram is Gaussian. One histogram has a mean value of 100 mV and a standard deviation of 50 mV. The other has a mean of 980 mV and a standard deviation of 75 mV. What is the theoretical minimum BER associated with this system?

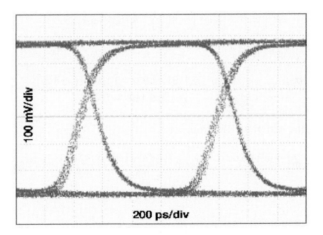

Solution:
Restating the given information in terms of the parameters used in the text above, we write

$$V_{Logic0} = 100 \text{ mV}, \qquad V_{Logic1} = 980 \text{ mV}$$
$$\sigma_{N,0} = 50 \text{ mV}, \qquad \sigma_{N,1} = 75 \text{ mV}$$

Substituting the above parameters into the expression for the optiumum thresold level given in Eq. (14.77), we write

$$V_{TH}^{*} = \frac{\sigma_{N,0}V_{Logic1} + \sigma_{N,1}V_{Logic0}}{\sigma_{N,0} + \sigma_{N,1}} = \frac{0.050 \times 0.980 + 0.075 \times 0.100}{0.050 + 0.075} = 0.452 \text{ V}$$

from which we write the minimum BER as

$$BER(V_{TH}^{*}) = \frac{1}{2} - \frac{1}{2} \times \Phi\left(\frac{0.452 - 0.100}{0.050}\right) + \frac{1}{2} \times \Phi\left(\frac{0.452 - 0.980}{0.075}\right)$$

resulting in

$$BER(V^*_{TH}) = \Phi(-7.04) = 9.61 \times 10^{-13}$$

Here we see that the theoretical minumum BER is slightly less than 10^{-12}.

Exercises

14.16. A system transmission test is to be run whereby a BER < 10^{-12} is to be verified using the amplitude-based scan test method. The threshold voltage of the receiver was adjusted to the following levels: 0.7, 1.0, 3.1 and 3.3 V. Also, the corresponding BER levels were measured: 5 x 10^{-6}, 5 x 10^{-9}, 5 x 10^{-9}, and 5 x 10^{-6}. What are the levels of the received logic values? Does the system meet the BER requirements if the receiver threshold is set at 2.0 V?

ANS. $V_{Logic,0} = -249.7$ mV;
$V_{Logic,1} = 3.933$ mV;
$BER(2.0\ V) = 1.3 \times 10^{-24}$.

14.17. A digital system is designed to operate at 25 Gbps and have a nominal logic 1 level of 0.9 V and a logic 0 at 0.1 V. If the threshold of the receiver is set at 0.6 V and the noise present at the receiver is 100 mV RMS, select the levels of the amplitude scan test method so that the system can be verified for a BER < 10^{-12} at a confidence level of 99.7% and that the total test time is less than 50 ms.

ANS.
$V_{TH,1} = 0.680$ V @ $BER_1 = 10^{-8}$,
$V_{TH,2} = 0.716$ V @ $BER_2 = 10^{-6}$,
Test Time = 46.9 ms.

14.18. A time jitter sequence is described by $s[n] = 1 \times \sin^2(14\pi/1024\ n)$ for $n = 0,1,..., 1023$. What is the mean, sigma and kurtosis of this data set? Is the data set Gaussian?

ANS.
$\mu = 0.5$, $\sigma = 0.354$, $k = 1.5$.
No, as $k \neq 3$.

The application of the amplitude-based scan test technique assumes that the histogram around each voltage level is Gaussian. One often needs to assure oneself that this assumption is realistic with the data set that are collected. Simply looking at the histogram plot to see if the distribution accurately follows a Gaussian is difficult because the subtleties in the tails are not easy to observe. A kurtosis metric or a normal probability plot can be used (see Section 4.2.4).

The validity of the amplitude-based scan test technique or Q-factor method has been called into question,[11] largely on account of its ability to accurately predict system behavior at very low BER from measurements made at much larger BER levels. While system impairments may be dominated by Gaussian noise behavior at large BER levels, one really doesn't know if such behavior continues at low BER levels unless one measures it. The converse is also true: When jitter impairments are present, the amplitude-based scan test technique predicts the minimum BER at a much higher level than a direct BER measurement would find—again, skewing the results. Even with these drawbacks, the amplitude-based scan test technique is a widely used method.

14.6.2 Time-Based Scan Test

The BER depends on the voltage noise present in the system as well as the noise associated with the sampling instant. We can repeat the BER analysis of the previous subsection, but this time let us write the BER in terms of the sampling instant assuming that the jitter distribution is Gaussian with zero mean and standard deviation σ_{RJ}. As previously stated for the amplitude-based scan test method, this test would be conducted with an altering 1010 pattern to ensure that no ISI is present.

In Section 14.4.3 we derived the BER in terms of the denormalized sampling instant as

$$\text{BER}\left(t_{TH}\right) = \frac{1}{2} - \frac{1}{2} \times \Phi\left(\frac{t_{TH} - 0}{\sigma_{RJ}}\right) + \frac{1}{2} \times \Phi\left(\frac{t_{TH} - T}{\sigma_{RJ}}\right) \tag{14.79}$$

Here we recognize that the BER expression contains one unknown σ_{RJ}; hence we can make a single measurement and solve for this unknown. Furthermore, we recognize that if we sample close to either edge of a bittransition, we can approximate Eq.(14.79) by the following:

$$\text{BER}\left(t_{TH}\right) \approx \begin{cases} \dfrac{1}{2} - \dfrac{1}{2} \times \Phi\left(\dfrac{t_{TH}}{\sigma_{RJ}}\right) & 0 \le t_{TH} \ll \dfrac{T}{2} \\[3mm] \dfrac{1}{2} \times \Phi\left(\dfrac{t_{TH} - T}{\sigma_{RJ}}\right) & \dfrac{T}{2} \ll t_{TH} \le T \end{cases} \tag{14.80}$$

Assuming that we set the sampling instant at $t_{TH,1}$ much less than $T/2$, then after making a single BER measurement we can write an expressionrelating this measurement as

$$\text{BER}\left(t_{TH,1}\right) = \frac{1}{2} - \frac{1}{2} \times \Phi\left(\frac{t_{TH,1}}{\sigma_{RJ}}\right) \tag{14.81}$$

We can then solve for σ_{RJ} and write

$$\sigma_{RJ} = \frac{t_{TH,1}}{\Phi^{-1}\left[1 - 2 \times \text{BER}\left(t_{TH,1}\right)\right]} \tag{14.82}$$

To estimate the location of the sampling instant, first assume a specific level of noise present and the level of BER that is expected, then rearrange Eq. (14.81) and write

$$t_{TH,1} = \sigma_{RJ} \times \Phi^{-1}\left[1 - 2 \times \text{BER}\left(t_{TH,1}\right)\right] \tag{14.83}$$

For instance, if we estimate that 10 ps of RJ is present on the digital signal and we want to make a BER measurement at a 10^{-6} level, we estimate the sampling time to be

$$t_{TH,1} = 10^{-11} \times \Phi^{-1}\left[1 - 2 \times 10^{-6}\right] \approx 46.1 \text{ ps}$$

Once σ_{RJ} is determined, we substitute this result back into Eq. (14.79) and solve for the system BER at the appropriate sampling instant, typically the middle of the data eye.

EXAMPLE 14.13

The BER performance of a digital reciever was measured to be 10^{-14} at a bit rate of 600 Mbps having a sampling instant at one-half the bit duration. Due to manufacturing errors, the sampling instant can vary by a $\pm20\%$ from its ideal position. Assuming an underlying noise component that is Gaussian in nature, what is the range of BER performance expected during production?

Solution:
As the ideal sampling instant is set at one-half the clock period, for a 600-MHz bit rate, this corresponds to a 0.833-ns sampling instant. Using Eq. (14.82), we can solve for the standard deviation of the underlying noise process as follows

$$BER\left(\frac{T}{2}\right) = 10^{-14} = \frac{1}{2} - \frac{1}{2} \times \Phi\left(\frac{T}{2\sigma_{RJ}}\right) + \frac{1}{2} \times \Phi\left(-\frac{T}{2\sigma_{RJ}}\right)$$

which reduces to the following when the identity is substituted:

$$BER\left(\frac{T}{2}\right) = 10^{-14} = \Phi\left(-\frac{T}{2\sigma_{RJ}}\right)$$

Rearranging, we write

$$\sigma_{RJ} = \frac{1/600 \times 10^6}{-2 \times \Phi^{-1}\left(10^{-14}\right)} = \frac{1/600 \times 10^6}{-2 \times -7.651} = 1.09 \times 10^{-10} \text{ s}$$

Next, given that the sampling instant can vary by $\pm20\%$ from its ideal position of 8.33×10^{-10} s, we write

$$\frac{1/600 \times 10^6}{2}(1 - 0.2) \le t_{TH} \le \frac{1/600 \times 10^6}{2}(1 + 0.2)$$

or when simplified as

$$6.66 \times 10^{-10} \le t_{TH} \le 1 \times 10^{-9}$$

As the BER performance varies in a symmetrical manner about the $T/2$ sampling instant, the BER performance level is the same at either extreme. Selecting $t_{TH} = 1 \times 10^{-9}$s, the BER performance we can expect is

$$BER\left(10^{-9}\right) = \frac{1}{2} - \frac{1}{2} \times \Phi\left(\frac{10^{-9} - 0}{1.09 \times 10^{-10}}\right) + \frac{1}{2} \times \Phi\left(\frac{10^{-9} - 1/600 \times 10^6}{1.09 \times 10^{-10}}\right) = \frac{1}{2} - \frac{1}{2} \times \Phi(9.18) - \Phi(-6.12) \approx \Phi(-6.12)$$

Through the application of a computer program (MATLAB, Excel, etc.) we find

$$BER\left(10^{-9}\right) = 2.33 \times 10^{-10}$$

Therefore during a production test, we can expect the BER performance to vary between 10^{-14} and 2.33×10^{-10}.

Exercises

14.19. A digital sampling oscilloscope was used to determine the zero crossings of an eye diagram. At 0 UI, the histogram is Gaussian distributed with zero mean and a standard deviation of 5 ps. At the unit interval, the histogram is again Gaussian distributed with zero mean value and a standard deviation of 8 ps. If the unit interval is equal to 100 ps, what is the expected BER associated with this system at a normalized sampling instant of 0.5 UI?

ANS. $BER = 1.03 \times 10^{-10}$

14.6.3 Dual-Dirac Jitter Decomposition Method

System transmission is often evaluated under various stressed conditions in order to better evaluate its behavior under typical operating conditions. System test specifications are often written to include BER tests for various types of PRBS input patterns. The amplitude or time-based scan test methods previously described are not suitable for this test, as the PRBS pattern introduces a data-dependent jitter component (as explained in Section 14.4.2). In other words, the underlying assumption that the noise associated with the received signal follows Gaussian statistics no longer holds. Instead, the method of jitter decomposition is used to evaluate the system transmission BER in production. The basic idea behind this method is to separate the random jitter (RJ) components from the deterministic jitter (DJ) components, write an expression for the BER in terms of these jitter terms, and then, in much the same way as the scan test approach, make measurements of BER at different sampling instants then solve for the unknown jitter terms.

Let us begin by assuming that the total jitter distribution is non-Gaussian and consisting of two parts: one random with PDF denoted by pdf_{RJ} and the other deterministic with PDF pdf_{DJ}. As we learned in Section 14.4.2, the PDF of the total jitter can be written as the convolution of the two separate parts, that is.,

$$pdf_{TJ} = pdf_{RJ} \otimes pdf_{DJ} \qquad \textbf{(14.84)}$$

We further learned that the DJ component could consist of many parts, including PJ, DDJ and DCD jitter components. One can extend the above theory for each of these separate noise components by expanding the convolution operation across all noise elements, for example,

$$pdf_{TJ} = pdf_{RJ} \otimes pdf_{PJ} \otimes pdf_{ISI} \otimes pdf_{DCD} \qquad \textbf{(14.85)}$$

Keeping the situation here more manageable, we shall consider the situation described by Eq. (14.84) and leave the more general case to some later examples.

We shall begin by describing the dual-Dirac jitter decomposition method[8] where the RJ noise component is assumed to follow Gaussian statistics with zero mean and standard deviation σ_{RJ} and the DJ component follows a distribution that has a dual-impulse or dual-Dirac behavior with parameter μ_{DJ}. Mathematically, we write these two PDFs as follows:

$$pdf_{RJ} = \frac{1}{\sigma_{RJ} 2\sqrt{2\pi}} e^{-t^2/2\sigma_{RJ}^2} \qquad \textbf{(14.86)}$$

$$pdf_{DJ} = \frac{1}{2}\delta\left(t - \frac{\mu_{DJ}}{2}\right) + \frac{1}{2}\delta\left(t + \frac{\mu_{DJ}}{2}\right) \qquad \textbf{(14.87)}$$

Figure 14.30. A multi-mode Gaussian representation of the total jitter distribution.

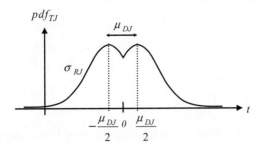

and denote the DJ and RJ measurement metrics as

$$DJ = \mu_{DJ}$$
$$RJ = \sigma_{RJ}$$

(14.88)

Through the convolution operation listed in Eq. (14.84), we can describe the total jitter PDF for any bit transition as

$$pdf_{TJ} = \frac{1}{2} \times \frac{1}{\sigma_{RJ}\sqrt{2\pi}} e^{-(t+\mu_{DJ}/2)^2 / 2\sigma_{RJ}^2} + \frac{1}{2} \times \frac{1}{\sigma_{RJ}\sqrt{2\pi}} e^{-(t-\mu_{DJ}/2)^2 / 2\sigma_{RJ}^2}$$

(14.89)

We illustrate this PDF in Figure 14.30, where it consists of two Gaussians distributions centered at $t = -\mu_D/2$ and $t = \mu_D/2$. It is important to recognize that the peaks associated with each Gaussian distribution are offset from one another by the parameter μ_{DJ}. Moreover, this offset is independent of the number of samples collected, as it is deterministic in nature. In practice, this fact is often used to identify the amount of DJ present with a jittery signal. In essence, the dual-Dirac modeling method is attempting to fit two Gaussian distributions with symmetrical mean values and equal standard deviation to the random portion of the digital signal.

Following the development of Section 14.4.3, the BER is computed as follows

$$BER\left(t_{TH}\right) = \frac{1}{2} \times \int_{(n-1)T+t_{TH}}^{\infty} pdf_{TJ}\big|_{t=(n-1)T} \, dt + \frac{1}{2} \times \int_{-\infty}^{nT-t_{TH}} pdf_{TJ}\big|_{t=nT} \, dt$$

(14.90)

where $pdf_{TJ}\big|_{t=(n-1)T}$ and $pdf_{TJ}\big|_{t=nT}$ represents the total jitter of $(n-1)$-th and nth consecutive bit transitions. t_{TH} is the sampling instance within the unit-interval bit period T. Substituting Eq. (14.89) into Eq. (14.90) and working through the integration, we find the BER can be written in terms of the Gaussian CDF and sampling instant t_{TH} as follows:

$$BER\left(t_{TH}\right) = \frac{1}{4} \times \left[2 - \Phi\left(\frac{t_{TH}+\mu_{DJ}/2}{\sigma_{RJ}}\right) - \Phi\left(\frac{t_{TH}-\mu_{DJ}/2}{\sigma_{RJ}}\right) \right]$$
$$+ \frac{1}{4} \times \left[\Phi\left(\frac{t_{TH}-T+\mu_{DJ}/2}{\sigma_{RJ}}\right) + \Phi\left(\frac{t_{TH}-T-\mu_{DJ}/2}{\sigma_{RJ}}\right) \right]$$

(14.91)

Because only two of the four terms contribute to the overall BER in a region, we can simplify the above equation and write

$$
\text{BER}\left(t_{TH}\right)=\begin{cases}\dfrac{1}{4}\times\left[2-\Phi\left(\dfrac{t_{TH}+\mu_{DJ}/2}{\sigma_{RJ}}\right)-\Phi\left(\dfrac{t_{TH}-\mu_{DJ}/2}{\sigma_{RJ}}\right)\right] & 0\le t_{TH}\ll T/2 \\[2em] \dfrac{1}{4}\times\left[\Phi\left(\dfrac{t_{TH}-T+\mu_{DJ}/2}{\sigma_{RJ}}\right)+\Phi\left(\dfrac{t_{TH}-T-\mu_{DJ}/2}{\sigma_{RJ}}\right)\right] & T/2\ll t_{TH}\le T\end{cases}
$$

(14.92)

The following example will be used to illustrate how the dual-Dirac jitter decomposition method is applied to a practical situation.

EXAMPLE 14.14

A digital system operates with a 1-GHz clock. How much RJ can be tolerated by the system if the BER is to be less than 10^{-10} and the DJ is no greater than 10 ps? Assume that the sampling instant is in the middle of the data eye having a unit time interval of 1 ns.

Solution:

Substituting the given information, that is, $T = 1000$ ps, $t_{TH} = 500$ ps, μ_{DJ} 10 ps, and BER (t_{TH}) 10^{-10} into Eq. (14.91), we write

$$
BER(t_{TH}) \le 10^{-10}
$$

or

$$
\frac{1}{4}\times\left[2-\Phi\left(\frac{500\text{ ps}+5\text{ ps}}{\sigma_{RJ}}\right)-\Phi\left(\frac{500\text{ ps}-5\text{ ps}}{\sigma_{RJ}}\right)\right]
$$
$$
+\frac{1}{4}\times\left[\Phi\left(\frac{500\text{ ps}-1000\text{ ps}+5\text{ ps}}{\sigma_{RJ}}\right)+\Phi\left(\frac{500\text{ ps}-1000\text{ ps}-5\text{ ps}}{\sigma_{RJ}}\right)\right]\le 10^{-10}
$$

which, when simplified, becomes

$$
\Phi\left(\frac{-495\text{ ps}}{\sigma_{RJ}}\right)+\Phi\left(\frac{-505\text{ ps}}{\sigma_{RJ}}\right)\le 2\times 10^{-10}
$$

Because this expression is transdental and nonlinear, we simply vary for σ_{RJ} until the left-hand side of the above expression is less than 10^{-10} and declare the largest value of σ_{RJ} as the largest RJ jitter that the system can tolerate. On doing this, we obtain the following listing of results:

σ_{RJ}(ps)	BER
60	4.93 x10^{-17}
70	5.19 x10^{-13}

σ_{RJ}(ps)	BER
78.5	1.00×10^{-10}
80	2.21×10^{-10}
90	1.45×10^{-8}
100	2.96×10^{-7}

As is evident from the table, the system can tolerate a random component of jitter of less having a standard deviaiton of less than 78.5 ps in order to ensure a BER of less than or equal to 10^{-10}.

Returning to Eq. (14.91), it is evident that the BER expression is a function of two unknown parameters, μ_{DJ} and σ_{RJ}. These two parameters can be identified by making two measurements of the BER under two separate sampling instants, say $t_{TH,1}$ and $t_{TH,2}$, both less than $T/2$, resulting in the following two equations:

$$\text{BER}\left(t_{TH,1}\right) = \frac{1}{4} \times \left[2 - \Phi\left(\frac{t_{TH,1} + \mu_{DJ}/2}{\sigma_{RJ}} \right) - \Phi\left(\frac{t_{TH,1} - \mu_{DJ}/2}{\sigma_{RJ}} \right) \right]$$

$$\text{BER}\left(t_{TH,2}\right) = \frac{1}{4} \times \left[2 - \Phi\left(\frac{t_{TH,2} + \mu_{DJ}/2}{\sigma_{RJ}} \right) - \Phi\left(\frac{t_{TH,2} - \mu_{DJ}/2}{\sigma_{RJ}} \right) \right]$$

(14.93)

Of course, we can also write two similar equations for the region $T/2 < tTH < T$. By selecting the sampling instants such that the expected BER level is not too small, a quick but accurate measurement of the BER can be made following the technique of Section 14.5. Once the two BER measurements are made, μ_{DJ} and σ_{RJ} can be solved using some form of numerical method.

Better yet, if the sampling instants are carefully selected so that we are on the tail of a single Gaussian distribution, as depicted in Figure 14.31 for the PDF of the zerocrossings, then only one of the two terms in each equation of Eq. (14-93) is significant, allowing us to write

$$\text{BER}\left(t_{TH,1}\right) \approx \frac{1}{4} \times \left[1 - \Phi\left(\frac{t_{TH,1} - \mu_{DJ}/2}{\sigma_{RJ}} \right) \right]$$

$$\text{BER}\left(t_{TH,2}\right) \approx \frac{1}{4} \times \left[1 - \Phi\left(\frac{t_{TH,2} - \mu_{DJ}/2}{\sigma_{RJ}} \right) \right]$$

(14.94)

The two unknown parameters can then be easily solved as follows:

$$\sigma_{RJ} = \frac{t_{TH,1} - t_{TH,2}}{\Phi^{-1}\left[1 - 4 \times \text{BER}\left(t_{TH,1}\right)\right] - \Phi^{-1}\left[1 - 4 \times \text{BER}\left(t_{TH,2}\right)\right]}$$

$$\mu_{DJ} = \frac{2 \times t_{TH,2} \times \Phi^{-1}\left[1 - 4 \times \text{BER}\left(t_{TH,1}\right)\right] - 2 \times t_{TH,1} \times \Phi^{-1}\left[1 - 4 \times \text{BER}\left(t_{TH,2}\right)\right]}{\Phi^{-1}\left[1 - 4 \times \text{BER}\left(t_{TH,1}\right)\right] - \Phi^{-1}\left[1 - 4 \times \text{BER}\left(t_{TH,2}\right)\right]}$$

(14.95)

The following example will help to illustrate the application of this theory.

Figure 14.31. (a) The zero-crossing distribution around two consecutive bit transitions; shaded area indicate bit error probability. (b) Individual Gaussian distribution around each bit transition. (c) BER as a function of sampling instant $(t = t_{TH})$.

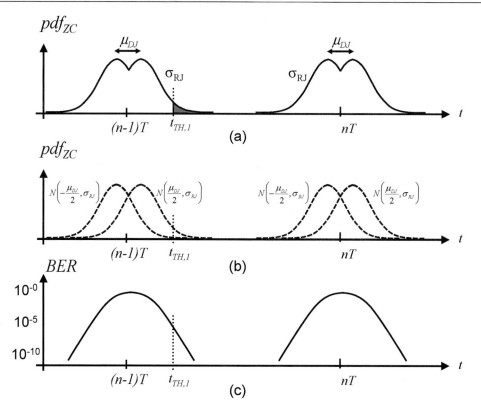

EXAMPLE 14.15

A system transmission test is to be run whereby a BER < 10^{-12} is to be verified using the jitter-decomposition test method. The system operates with a 1-GHz clock and the following BER measurements are at two different sampling instances: BER = 0.25 x 10^{-4} at 300 ps, and BER = 0.25 x 10^{-6} at 350 ps. Assume the digital system operates with a sampling instant in the middle of the data eye. Does the system meet spec?

Solution:
Following the reference system of equations, we write

$$BER(t_{TH,1}) = 0.25 \times 10^{-4} \text{ @ } t_{TH,1} = 300 \times 10^{-12}$$

$$BER(t_{TH,2}) = 0.25 \times 10^{-6} \text{ @ } t_{TH,2} = 350 \times 10^{-12}$$

Assuming the BERs were obtained on the tail of a single Gaussian, we substitute the above parameters into Eq.(14.95) and write

$$\sigma_{RJ} = \frac{300 \times 10^{-12} - 350 \times 10^{-12}}{\Phi^{-1}\left[1 - 10^{-4}\right] - \Phi^{-1}\left[1 - 10^{-6}\right]}$$

$$\mu_{DJ} = \frac{2 \times 350 \times 10^{-12} \times \Phi^{-1}\left[1 - 10^{-4}\right] - 2 \times 300 \times 10^{-12} \times \Phi^{-1}\left[1 - 10^{-6}\right]}{\Phi^{-1}\left[1 - 10^{-4}\right] - \Phi^{-1}\left[1 - 10^{-6}\right]}$$

Using the relationship for $\Phi^{-1}(1 - x)$ from the table in Appendix B, specifically, $\Phi^{-1}(1 - 10^{-4}) = 3.7190$, and $\Phi^{-1}(1 - 10^{-6}) = 4.7534$, we solve to obtain $\mu_{DJ} = 240.5$ ps and $\sigma_{RJ} = 48.3$. In order to ensure that our simiplified model of BER behavior is reasonable in the region where the samples of BER were obtained, we compare

$$BER(t_{TH}) = \frac{1}{4} \times \left[2 - \Phi\left(\frac{t_{TH} + 120.2 \text{ ps}}{48.3 \text{ ps}}\right) - \Phi\left(\frac{t_{TH} - 120.2 \text{ ps}}{48.3 \text{ ps}}\right)\right]$$

with

$$BER(t_{TH}) = \frac{1}{4} \times \left[1 - \Phi\left(\frac{t_{TH} - 120.2 \text{ ps}}{48.3 \text{ ps}}\right)\right]$$

as shown below:

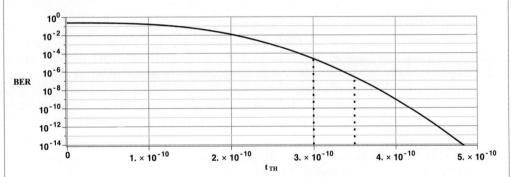

Clearly, we see that the two curves are identical in the region over the range of sampling instants chosen. Hence our initial assumption is correct and we are confident that our calculation of DJ and RJ are correct. Substituting these two parameters into Eq. (14.91) we obtain an expresson of the BER across the entire eye diagram as as a function of the sampling instant t_{TH} as

$$BER(t_{TH}) = \frac{1}{4} \times \left[1 - \Phi\left(\frac{t_{TH} + 240.5 \text{ ps}}{48.3 \text{ ps}}\right)\right] + \frac{1}{4} \times \left[1 - \Phi\left(\frac{t_{TH} - 240.5 \text{ ps}}{48.3 \text{ ps}}\right)\right]$$
$$+ \frac{1}{4} \times \Phi\left(\frac{t_{TH} - 1000 \text{ ps} + 240.5 \text{ ps}}{48.3 \text{ ps}}\right) + \frac{1}{4} \times \Phi\left(\frac{t_{TH} - 1000 \text{ ps} - 240.5 \text{ ps}}{48.3 \text{ ps}}\right)$$

Finally, we solve for the BER at a samping instant in the middle of the data eye, (i.e., t_{TH} = 0.5 ns), and obtain

$$BER (5 ns) = 9.8 \times 10^{-16}$$

Therefore we conclude the BER is less than 10^{-12} and the system meets spec. We should note that if the 1-Gaussian model of BER behavior differs from the 2-Gaussian behavior, then we must solve the RJ and DJ parameters directly from the 2-Gaussian model of BER behavior. We encourage our readers to attempt the exercise below.

Exercises

14.20. A system transmission test is to be run whereby a BER < 10^{-14} is to be verified using the dual-Dirac jitter-decomposition test method. The system operates with a 25-GHz clock and the following BER measurements are made at two different sampling instances: BER = 0.25 x 10^{-4} at 9 ps, and BER = 0.25 x 10^{-6} at 11 ps. Assume the digital system operates with a sampling instant in the middle of the data eye. Does the system meet spec?

ANS.

$BER(T/2) = 1.26 \times 10^{-21}$;
$RJ = 1.9$ ps; $DJ = 3.6$ ps.
Yes, the system meets spec.

Total Jitter Definition

Transmission specifications, such as fiber channel, Infini-band, and XAUI, often include a test metric called total jitter (denoted TJ), in addition to the test limits for DJ and RJ. While DJ and RJ are model parameters derive indirectly through the application of several BER measurements, TJ is derived directly from these two parameters.

To understand the definition of TJ, let us first consider the eye diagram shown in Figure 14.32a. Here we highlight the size of the eye opening along the time axis. Intuitively, the larger the

Figure 14.32. Illustrating the definition of total jitter (TJ) based on eye opening: (a) Eye diagram and (b) corresponding BER versus sampling instant $(t = t_{TH})$.

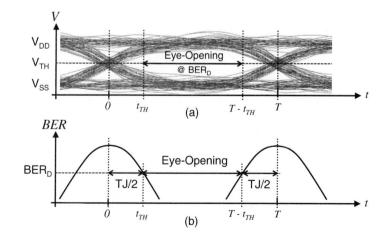

eye opening, the better the quality of the signal transmission, because there would be less chance of obtaining a bit error. Under ideal conditions, the maximum eye opening is equal to the bit duration, *T*. The difference between the ideal and actual eye opening is defined as the total jitter, TJ, that is,

$$\text{TJ} \triangleq \text{Ideal eye opening} - \text{Actual eye opening} \tag{14.96}$$

While we could define the eye opening based on the signal transitions that bound the open area encapsulatedby the data eye, we need to account for the random variations that will occur. Specifically, the edges of the eye opening along the time axis are defined based on an expected probability that a transition will go beyond a specific time instant, which we learned previously is equivalent to the BER. Figure 14.32(b) illustrates the BER as function of the sampling instant, and provides a manner in which to quantify TJ, that is,

$$\text{TJ} = 2 \times t_{TH} \tag{14.97}$$

At relative low BER levels, we can approximate the BER as

$$\text{BER}\left(t_{TH}\right) \approx \frac{1}{4} \times \left[1 - \Phi\left(\frac{t_{TH} - \mu_{DJ}/2}{\sigma_{RJ}}\right)\right] \tag{14.98}$$

Rearranging, we can write the sampling instant t_{TH} in terms of the desired BER, denoted by BER_D, as follows:

$$t_{TH} = \sigma_{RJ} \times \Phi^{-1}\left[1 - 4 \times \text{BER}_D\right] + \frac{\mu_{DJ}}{2} \tag{14.99}$$

Substituting the above expression into Eq. (14.97), we obtain the commonly used definition for TJ, that is,

$$\text{TJ} = 2 \times \sigma_{RJ} \times \Phi^{-1}\left[1 - 4 \times \text{BER}_D\right] + \mu_{DJ} \tag{14.100}$$

EXAMPLE 14.16

A system transmission test was run and the RJ and DJ components were found to be 13 ps and 64.6 ps, respectively. What is the TJ component at a BER level of 10^{-12}?

Solution:
Substituting RJ and DJ into Eq. (14.100), we write

$$\text{TJ} = 2 \times 13 \text{ ps} \times \Phi^{-1}\left[1 - 4 \times 10^{-12}\right] + 64.6 \text{ ps}$$

Using the data provided in the Table of Appendix B, we find $\Phi^{-1}[1 - 4 \times 10^{-12}]$, allowing us to write

$$\text{TJ} = 2 \times 13 \text{ ps} \times 6.839 + 64.6 \text{ ps} = 242.4 \text{ ps}$$

Exercises

14.21. A system transmission test was run where the RJ component was found to be 25 ps and the TJ component at a BER level of 10^{-14} was found to be 400 ps. What is DJ?   ANS. DJ = 26.48 ps.

14.22. A system transmission test was run where the DJ component was found to be 5 ps and the TJ component at a BER level of 10^{-12} was found to be 120 ps. What is RJ?   ANS. RJ = 8.4 ps.

14.6.4 Gaussian Mixture Jitter Decomposition Method

The dual-Dirac jitter composition method just described in the last subsection is a special case of fitting a set of independent Gaussian distributions to an arbitrary distribution, albeit without the physical understanding associated with the PDF convolution method. Assuming that the total jitter distribution consists of G independent Gaussian distributions with mean values of μ_i and standard deviation σ_i for $i = 1, ..., G$, we can write the PDF for the TJ as

$$pdf_{TJ} = \frac{\alpha_1}{\sigma_1\sqrt{2\pi}}e^{-(t-\mu_1)^2/2\sigma_1^2} + \frac{\alpha_2}{\sigma_2\sqrt{2\pi}}e^{-(t-\mu_2)^2/2\sigma_2^2} + \cdots + \frac{\alpha_{N_G}}{\sigma_G\sqrt{2\pi}}e^{-(t-\mu_G)^2/2\sigma_G^2} \tag{14.101}$$

where the terms α_i are weighting terms. Subsequently, we can substitute Eq.(14-101) into Eq. (14.90) and write the BER at sampling instant t_{TH} as

$$\begin{aligned}\text{BER}(t_{TH}) &= \frac{\alpha_1}{2}\times\left[1-\Phi\left(\frac{t_{TH}-\mu_1}{\sigma_1}\right)\right] + \frac{\alpha_2}{2}\times\left[1-\Phi\left(\frac{t_{TH}-\mu_2}{\sigma_2}\right)\right] + \cdots + \frac{\alpha_G}{2}\times\left[1-\Phi\left(\frac{t_{TH}-\mu_G}{\sigma_G}\right)\right]\\ &+ \frac{\alpha_1}{2}\times\Phi\left(\frac{t_{TH}-T-\mu_1}{\sigma_1}\right) + \frac{\alpha_2}{2}\times\Phi\left(\frac{t_{TH}-T-\mu_2}{\sigma_2}\right) + \cdots + \frac{\alpha_G}{2}\times\Phi\left(\frac{t_{TH}-T-\mu_G}{\sigma_G}\right)\end{aligned} \tag{14.102}$$

We depict a three-term Gaussian mixture in Figure 14.33. In part (a) we illustrate the total jitter distribution as a function of eye position, followed by the individual Gaussian mixture in part (b) and then the corresponding BER as a function of sampling instant in part (c). A specific sampling instance t_{TH} is identified in all three diagrams of Figure 14.33, highlighting a specific bit error probability and its impact on the BER. It is important to recognizefrom this figure that the distributions and BER function can be asymmetrical with respect to the ideal bit transition instant ($t = (n–1)T$ and $t = nT$).

For low BER levels, we can approximate Eq. (14.102) using two Gaussian distributions taken from the Gaussian mixture of Eq. (14.101) as follows:

$$\text{BER}(t_{TH}) \approx \begin{cases} \dfrac{\alpha_{tail+}}{2}\times\left[1-\Phi\left(\dfrac{t_{TH}-\mu_{tail+}}{\sigma_{tail+}}\right)\right], & 0 \ll t_{TH} \leq \dfrac{T}{2} \\[4mm] \dfrac{\alpha_{tail-}}{2}\times\Phi\left(\dfrac{t_{TH}-T-\mu_{tail-}}{\sigma_{tail-}}\right), & \dfrac{T}{2} \leq t_{TH} \ll T \end{cases} \tag{14.103}$$

where tail+ and tail– is a integer number from 1 to G, and tail+ can also be equal to tail–. Because the tails of a general distribution depends on both the mean values and the standard deviations of the Gaussian mixture, some effort must go into to identifying the dominant Gaussian function in each tail region of the total jitter distribution. A simple approach is to evaluate the BER contribution of each Gaussian term at some distance away from each edge transition of the eye diagram.

In terms of extracting the RJ and DJ metrics from the total jitter, we note that in the general case, DJ is defined as the distance between the two Gaussians that define the tail regions of the total distribution, that is,

$$\text{DJ} \triangleq \mu_{DJ} = \mu_{tail+} - \mu_{tail-} \tag{14.104}$$

Likewise, RJ is defined as the average value of the standard deviations associated with the tail distributions[9], that is,

$$\text{RJ} \triangleq \sigma_{RJ} = \frac{\sigma_{tail-} + \sigma_{tail+}}{2} \tag{14.105}$$

In addition to these two definitions, we can also write an equation for the TJ metric using the notation indicated in Figure 14.33c, specifically

$$\text{TJ} = T - \left(t_{TH,2} - t_{TH,1} \right) \tag{14.106}$$

where $t_{TH,1}$ and $t_{TH,2}$ correspond to the sampling instant at the desired BER level, BER_D, given by

$$t_{TH,1} = \sigma_{tail+} \times \Phi^{-1}\left[1 - \frac{2}{\alpha_{tail+}} \text{BER}_D \right] + \mu_{tail+}$$

$$t_{TH,2} = \sigma_{tail-} \times \Phi^{-1}\left[\frac{2}{\alpha_{tail-}} \text{BER}_D \right] + T + \mu_{tail-} \tag{14.107}$$

Substituting into Eq. (14.106) and simplifying, we obtain

$$\text{TJ} = \sigma_{tail+} \times \Phi^{-1}\left[1 - \frac{2}{\alpha_{tail+}} \text{BER}_D \right] + \sigma_{tail-} \times \Phi^{-1}\left[1 - \frac{2}{\alpha_{tail-}} \text{BER}_D \right] + \mu_{tail+} - \mu_{tail-} \tag{14.108}$$

Looking back at the dual-Dirac jitter decomposition methods of Section 14.6.3, these three definitions of DJ, RJ and TJ are consistent with the definitions given there when $\mu_{tail+} = -\mu_{tail-}$, $\sigma_{tail+} = \sigma_{tail-}$ and $\alpha_{tail+} = \alpha_{tail-} = 0.5$.

Finally, the random component of the jitter has a PDF that can be described by

$$pdf_{RJ} = \frac{1}{\sigma_{RJ}\sqrt{2\pi}} e^{-t^2/2\sigma_{RJ}^2} \tag{14.109}$$

The remaining task at hand is to determine the model parameters μ_{tail-}, μ_{tail+}, σ_{tail-}, σ_{tail+} and the weighting coefficients, α_{tail-} and α_{tail+}. While one may be tempted to go off and measure five different BER values at six different sampling times and attempt to solve for the six unknowns using some form of nonlinear optimization, much care is needed in selecting the sampling instants. It is imperative that samples are derived only from the outer tails of the total jitter distribution to avoid any contributions from the other Gaussians. This generally means sampling at very low BER levels, driving up test time. This approach is the basis of the Tail-Fit™ algorithm of Wavecrest™. At the heart of this method is the chi-square (χ^2) curve-fitting algorithm. Interested readers can refer to reference 14.

Exercises

14.23. The jitter distribution at the transmit side of a communication channel can be described by a Gaussian mixture model with parameters:

$$\mu_1 = -100 \text{ ps}, \quad \sigma_1 = 50 \text{ ps}, \quad \alpha_1 = 0.3$$
$$\mu_2 = -1 \text{ ps}, \quad \sigma_2 = 60 \text{ ps}, \quad \alpha_2 = 0.4$$
$$\mu_3 = 50 \text{ ps}, \quad \sigma_3 = 50 \text{ ps}, \quad \alpha_3 = 0.3$$

What is the RJ, DJ, and TJ at a BER of 10^{-14} assuming a bit rate of 100 Mbps? Which indiviudal distribution set the tail regions of the total jitter?

ANS.

RJ = 55 ps, DJ = 99 ps, TJ = 915.6 ps.

The positive tail is set by $N(\mu_2, \sigma_2)$ and the negative tail is set by $N(\mu_1, \sigma_1)$

Figure 14.33. (a) An asymmetrical zero-crossing distributions around two consecutive bit transitions; shaded area indicate bit error probability. (b) Individual Gaussian distribution around each bit transition. (c) BER as a function of sampling instant ($t = t_{TH}$).

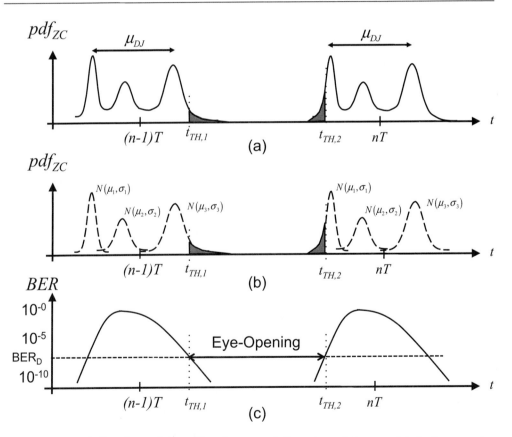

Another approach,[10] and one that tackles the general Gaussian mixture model, is the expectation maximum (EM) algorithm introduced back in Section 4.4.1. This algorithm assumes that G Gaussians are present and solves for the entire set of model parameters: $\mu_1, \mu_2, ..., \mu_G, \sigma_1, \sigma_2,..., \sigma_G, \alpha_1, \alpha_2, ..., \alpha_G$ using an iteration approach.[11,12] The next two examples will help to illustrate the application of the EM algorithm to the general jitter decomposition problem.

EXAMPLE 14.17

The jitter distribution at the receiver end of a communication channel is to be collected and analyzed for its DJ, RJ, and TJ corresponding to a BER level of 10^{-12}. In order to extract the model parameters, use the EM algorithm and fit three Gaussian functions to the jitter data set. In order to generate the data set, create a routine that synthesizes a Gaussian mixture consisting of three Gaussians with means –0.02 UI, 0 UI, and 0.04 UI and standard deviations of 0.022 UI, 0.005 UI, and 0.018 UI and weighting factors 0.4, 0.2, and 0.4, respectively. Provide a plot of the jitter histogram. Also, how does the extracted model parameters compare with the synthesized sample set model parameters?

Solution:

A short routine was written in MATLAB that synthesizes a Gaussian mixture with the above mentioned model parameters. The routine was run and 1000 samples were collected and organized into a histogram as shown below:

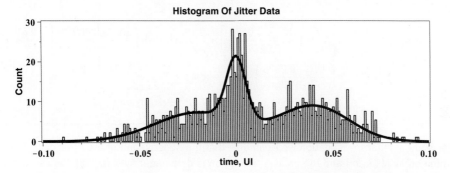

The PDF from which the samples were taken from is also superimposed on the above plot. We note that the general shape of this histogram reveals one Gaussian with a peak at about 0.04 UI, another with a peak at about 0, and the third may have a peak at about -0.03 UI. We estimate the standard deviation of the rightmost Gaussian at about 0.05 UI, the center one with about 0.01 UI, and the leftmost Gaussian with 0.05 UI. We will use these parameters as estimates for the Gaussian mixture curve fit algorithm. We will also assume that each Guassian component is equally important, and hence we assign a weight of 1/3 each. The better our estimate, the faster the algorithm will converge. A second MATLAB routinewas created based on the EM algorithm of Section 4.4.1. Using the above synthesized data, the Gaussian mixture curve-fitting algorithm was run. After 1000 iterations aiming for a root-mean-square error of less than 10^{-10}, the following model parameters were found:

$$\mu_1 = -0.018 \text{ UI}, \quad \sigma_1 = 0.0224 \text{ UI}, \quad \alpha_1 = 0.39$$
$$\mu_2 = 0.0007 \text{ UI}, \quad \sigma_2 = 0.0048 \text{ UI}, \quad \alpha_2 = 0.19$$
$$\mu_3 = 0.0398 \text{ UI}, \quad \sigma_3 = 0.0185 \text{ UI}, \quad \alpha_3 = 0.41$$

Substituting these parameters back into Eq. (14.101) and computing the value of this PDF over a range of –0.1 UI to 0.1 UI, we obtain the dashed line shown in the plot below. Also shown in the plot is a solid line that corresponds to the PDF from which the data set was derived. This is the same line as in the figure above superimposed over the histogram of the data set. It is quite evident from this plot that the estimated PDF corresponds quite closely to the orginal one. We also include in the plot below a dot-dashed line corresponding to the PDF of our initial guess. In this case, the initial and final PDFs are clearly quite different from one another; this highlights the effectiveness of the EM algorithm.

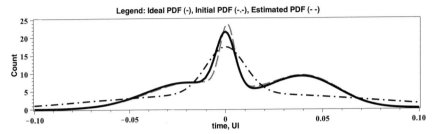

Legend: Ideal PDF (-), Initial PDF (-.-), Estimated PDF (- -)

We can write an expression for the BER as a function of the normalized sampling instant, UI_{TH} as follows

$$BER(\bar{t}_{TH}) = \frac{0.39}{2} \times \left[1 - \Phi\left(\frac{UI_{TH} + 0.018}{0.0224} \right) \right] + \frac{0.19}{2} \times \left[1 - \Phi\left(\frac{UI_{TH} - 0.0007}{0.0048} \right) \right] + \frac{0.41}{2}$$
$$\times \left[1 - \Phi\left(\frac{UI_{TH} - 0.0398}{0.0185} \right) \right] + \frac{0.39}{2} \times \Phi\left(\frac{UI_{TH} - 1 + 0.018}{0.0224} \right) + \frac{0.19}{2}$$
$$\times \Phi\left(\frac{UI_{TH} - 1 - 0.0007}{0.0048} \right) + \frac{0.41}{2} \times \Phi\left(\frac{UI_{TH} - 1 - 0.0398}{0.0185} \right)$$

After some further investigation we recongize that the negative tail of the jitter distrubution is dominated by $N(\mu_1, \sigma_1)$ and the postive tail region is dominated by $N(\mu_3, \sigma_3)$. We can therefore write the BER in the tail regions as

$$BER(\bar{t}_{TH}) \approx \begin{cases} \frac{0.41}{2} \times \left[1 - \Phi\left(\frac{UI_{TH} - 0.0398}{0.0185} \right) \right], & 0 \ll UI_{TH} < \frac{1}{2} \\ \frac{0.39}{2} \times \Phi\left(\frac{UI_{TH} - 1 + 0.018}{0.0224} \right), & \frac{1}{2} < UI_{TH} \ll 1 \end{cases}$$

Finally, we compute the test metrics DJ and RJ and TJ(10^{-12}) from Eqs. (14.104) and (14.105) as follows:

$$DJ = \mu_3 - \mu_1 = 0.0398 - (-0.018) = 0.058 \text{ UI}$$
$$RJ = \frac{\sigma_1 + \sigma_3}{2} = \frac{0.0224 + 0.0185}{2} = 0.020 \text{ UI}$$

and from Eq. (14.104), TJ at a desired BER of 10^{-12} can be written as

$$TJ = 0.0185 \times \Phi^{-1} \left[1 - \frac{2 \times 10^{-12}}{0.41} \right] + 0.0224 \times \Phi^{-1} \left[1 - \frac{2 \times 10^{-12}}{0.39} \right] + 0.0398 - (-0.018)$$

which is further reduced to

$$TJ = 0.0185 \times \Phi^{-1}[1 - 4.89 \times 10^{-12}] + 0.0224 \times \Phi^{-1}[1 - 5.05 \times 10^{-12}] + 0.0398 - (-0.018)$$

Substituting $\Phi^{-1}[1 - 4.89 \times 10^{-12}] = 6.8096$ and $\Phi^{-1}[1 - 5.05 \times 10^{-12}] = 6.8051$ as found by a built-in routine in MATLAB, we write TJ = 0.337 UI.

EXAMPLE 14.18

The jitter distribution at the receiver end of a communication channel consisting of 10,000 samples was modeled by a four-term Gaussian mixture using the EM algorithm, and the following model parameters were found:

$$\mu_1 = -0.03 \text{ UI}, \quad \sigma_1 = 0.02 \text{ UI}, \quad \alpha_1 = 0.4$$
$$\mu_2 = -0.02 \text{ UI}, \quad \sigma_2 = 0.05 \text{ UI}, \quad \alpha_2 = 0.2$$
$$\mu_3 = 0.03 \text{ UI}, \quad \sigma_3 = 0.01 \text{ UI}, \quad \alpha_3 = 0.1$$
$$\mu_4 = 0.04 \text{ UI}, \quad \sigma_4 = 0.04 \text{ UI}, \quad \alpha_4 = 0.3$$

Assuming that the unit interval is 1 ns and that the data eye sampling instant is half a UI, what is the BER associated with this system? Does the system behave with a BER less than 10^{-12}?

Solution:
Evaluating Eq. (14.102) a $UI_{TH} = 0.5$ UI and $T = 1$ UI, we write

$$BER(0.5UI) = \frac{0.04}{2} \times \left[1 - \Phi\left(\frac{0.5 - \mu_1}{\sigma_1}\right)\right] + \frac{0.2}{2} \times \left[1 - \Phi\left(\frac{0.5 - \mu_2}{\sigma_2}\right)\right] + \frac{0.1}{2} \times \left[1 - \Phi\left(\frac{0.5 - \mu_3}{\sigma_3}\right)\right] + \frac{0.3}{2} \times \left[1 - \Phi\left(\frac{0.5 - \mu_4}{\sigma_4}\right)\right]$$
$$+ \frac{0.04}{2} \times \Phi\left(\frac{0.5 - 1 - \mu_1}{\sigma_1}\right) + \frac{0.2}{2} \times \Phi\left(\frac{0.5 - 1 - \mu_2}{\sigma_2}\right) + \frac{0.1}{2} \times \Phi\left(\frac{0.5 - 1 - \mu_3}{\sigma_3}\right) + \frac{0.3}{2} \times \Phi\left(\frac{0.5 - 1 - \mu_4}{\sigma_4}\right)$$

Substituting the above Gaussian mixture parameters in normalized form, we find BER (0.5 UI) $\times 10^{-10}$. Therefore, we can expect that the system theoretically will not operate with a BER less than 10^{-12}.

Exercise

14.24. The jitter distribution at the receiver end of a communication channel consisting of 10,000 samples was modeled by a four-term Gaussian mixture using the EM algorithm and the following model parameters were found:

$$0.25 \times N(-10 \text{ ps}, 10 \text{ ps}) + 0.25 \times$$
$$N(-7 \text{ ps}, 7 \text{ ps}) + 0.25 \times N(4 \text{ ps}, 4 \text{ ps})$$
$$+ 0.25 \times N(11 \text{ ps}, 11 \text{ ps})$$

Assuming the unit interval is 0.2 ns and that the data eye sampling instant is half a UI, what is the BER associated with this system? Does the system behave with a BER less than 10^{-12}?

ANS. $BER(UI/2) = 3.7 \times 10^{-17}$.

14.7 DETERMINISTIC JITTER DECOMPOSITION

Using either the dual-Dirac or Gaussian mixture jitter decomposition method, we can determine the random and deterministic components of the jittery signal. In order to further decompose the deterministic jitter component into its constitutive parts, such as periodic or sinusoidal, data-dependent, and bounded uncorrelated jitter, additional data capturing and signal processing is required. More specifically, we analyze the jitter samples as a time series, attempting to find underlying structure in the data. As we have done throughout this text, we shall rely heavily on an FFT analysis to help us identify this structure. Once the DJ components are identified, the goal is to construct their corresponding PDF and use these PDFs for diagnostic purposes.

14.7.1 Period and Sinusoidal Jitter (PJ/SJ)

Period jitter (PJ) of a serial communication link is defined as the deviation in the period time of a clock-like output signal. Usually, a PJ test is conducted when the input to the channel is driven with a small-amplitude sinusoidal phase-modulated signal, but PJ also arises from internal sources unrelated to the input as well.

Let us assume that a sequence of N time difference values for the signal in question and a reference clock corresponding to the zero crossing times is captured and stored in vector $J[n]$ for $n = 1, \ldots, N$ (see Figure 14.4 for greater clarification). If we assume that the PJ components are dominant (i.e., PJ is greater than RJ), then the standard deviation σ_J of this data set represents the RMS value of the PJ components and can be computed as follows:

$$
\sigma_J = \sqrt{\frac{1}{N}\sum_{n=1}^{N}\left(J[n]-\frac{1}{N}\sum_{n=1}^{N}J[n]\right)^2}
\tag{14.110}
$$

Because DJ is expressed as a peak-to-peak quantity, PJ is described in the same peak-to-peak way by exploiting the relationship between the peak-to-peak value of a sine wave to its RMS value as follows

$$
PJ \triangleq 2\sqrt{2}\times\sigma_J
\tag{14.111}
$$

The user of Eq. (14.111) should be aware that this formula is only approximate, because it has no information about the underlying nature of the periodic jitter and how separate sinusoidal signals combine.

There is another definition of period jitter that one finds in the literature. Digital system designers often use this definition when describing jitter. We briefly spoke about it in Section 14.2. Specifically, period jitter is defined as the first-order time difference between adjacent bit transitions given by the equation

$$
J_{PER}[n]=J[n]-J[n-1]
\tag{14.112}
$$

The RMS value of this sequence (denoted by σ_{JPER}) would have a similar form to Eq. (14.110) above, except σ_{JPER} would replace σ_J. In the frequency domain, the period jitter sequence is related to the time jitter sequence by taking the z-transform of Eq. (14.112) and substituting for physical frequencies, that is, $z = e^{j2\pi fT}$, leading us to write

$$
J_{PER}\left(e^{j2\pi fT}\right)=\left(1-z^{-1}\right)J(z)\Big|_{z=e^{j2\pi fT}}=\left(1-e^{-j2\pi fT}\right)J\left(e^{j2\pi fT}\right)
\tag{14.113}
$$

Clearly, the frequency description of the period jitter sequence J_{PER} is proportional to the time jitter sequence J in the frequency domain. However, we note that the proportionality constant is also a function of the test frequency. This means that the PJ metric computed from the time jitter sequence will be different from that obtained using the period jitter sequence. This can be source of confusion for many. To avoid this conflict, for the remainder of this text, we shall rely exclusively on the time jitter sequence to establish the PJ quantity. Example 14.9 will help to demonstrate this effect.

A more refined approach for identifying the PJ component that makes no assumption of the relative strength of the PJ component relative to the RJ component is to perform an FFT analysis of the jitter time series J. If coherency is difficult to achieve, a Hann or Blackman window may be necessary to reduce frequency leakage effects. If we denote the period jitter frequency transform with vector X_J (using our usual notation) and identify the FFT bin containing the fundamental of the periodic jitter as bin M_{PJ} and the corresponding number of harmonics H_{PJ}, then we can determine the RMS value of the PJ component from the following expression

$$\sigma_{PJ} = \sqrt{\sum_{p=1}^{H_{PJ}} c_{p \times M_{PJ}, RMS}^2} \qquad (14.114)$$

where

$$c_{k,RMS} = \begin{cases} |X_J[k]|, & k = 0 \\ \sqrt{2}|X_J[k]|, & k = 1, \ldots, N/2 - 1 \\ \dfrac{|X_J[k]|}{\sqrt{2}}, & k = N/2 \end{cases}$$

If PJ consists of a single tone, then using Eq. (14.111) with σ_{PJ} replacing $\sigma_{J_{PER}}$, we can estimate the peak-to-peak value of the PJ component. If PJ consists of more than one tone, then we must account for the phase relationship of each tone in the calculation of its peak-to-peak value. One approach is to perform an inverse-FFT (IFFT) on only the PJ components (remove all others) and extract the peak-to-peak value directly from the time-domain signal. Mathematically, we write

$$PJ \cong \max\{\text{IFFT}(J_{PJ,only})\} - \min\{\text{IFFT}(J_{PJ,only})\} \qquad (14.115)$$

Correspondingly, we can also compute the samples of the time sequence $\text{IFFT}(J_{PJ,only})$ to construct a histogram of the PJ component. A mathematical formula can then be used to model the PDF for PJ. Because PJ is made up of harmonically related sinusoidal components, the PDF distribution will involve functions that appears in the general form as $1/\pi\sqrt{A^2 - t^2}$, where A is the amplitude of a single sine wave. In general, this is difficult to do and one general defers to thesingle tone case below.

As a special case of period jitter, if H_{PJ} is unity where only one tone is present, then PJ is sometimes referred to as sinusoidal jitter and is denoted with acronym SJ. Mathematically, we define SJ as

$$SJ \triangleq 2\sqrt{2} \times \sigma_{PJ} \qquad (14.116)$$

where the corresponding period jitter PDF would be written as

$$
pdf_{PJ} = \begin{cases} \dfrac{1}{\pi\sqrt{\left(\dfrac{SJ}{2}\right)^2 - t^2}}, & -\dfrac{SJ}{2} \le t \le \dfrac{SJ}{2} \\[4mm] 0, & \text{otherwise} \end{cases}
\tag{14.117}
$$

SJ plays an important role in jitter tolerance and transfer function tests of the next section.

EXAMPLE 14.19

A time jitter sequence $J[n]$ can be described by the following discrete-time equation:

$$
J[n] = 0.001 \sin\left(2\pi \frac{11}{1024} n\right)
$$

Here A_j is the amplitide of the phase modulated sequence and M_j is the number of cycles completed in N sample points. Compute the period jitter sequence and compare its amplitude to the time jitter sequence amplitude.

Solution:
The period jitter sequence is computed from time jitter sequence according to

$$
J_{PER}[n] = J[n] - J[n-1] = 0.001\sin\left[2\pi\frac{11}{1024}n\right] - 0.001\sin\left[2\pi\frac{11}{1024}(n-1)\right]
$$

Running through a short for loop, we obtain a plot of the period jitter together with the time jitter sequence below:

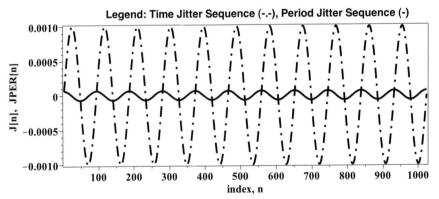

Legend: Time Jitter Sequence (-.-), Period Jitter Sequence (-)

The plot clearly reveals that the amplitudes of the two sequences are quite different. In fact, the analysis shows that the amplitude of the time jitter sequence is 10^{-3} and the amplitude of the period jitter sequence is 6.7×10^{-5}. In fact, the ratio of these two amplitudes is equal to

$$
\frac{\hat{J}_{PER}}{\hat{J}_p} = \left|1 - e^{-j2\pi^M/_N}\right| = 0.0067 \text{ with } M_j = 11 \text{ and } N = 1024.
$$

EXAMPLE 14.20

The SJ component of a jitter distribution was found to be 0.1 UI for a receiver operating at a bit rate of 1 GHz. Estimate the shape of the histogram that one would expect from this jitter component alone.

Solution:
Denormalizing the SJ component, we write its peak-to-peak value as $0.1/10^9$ or 10^{-10} s. Substituting this result back into Eq. (14.117), we write the PDF for this jitter component between –0.1 UI $< t < 0.1$ UI as

$$pdf_{SJ} = \frac{1}{\pi \sqrt{\left(\frac{10^{-10}}{2}\right)^2 - t^2}}$$

and everywhere else as 0. The following plot of this distrubtion would resemble the general shape of any particular histogram captured from this random distribution.

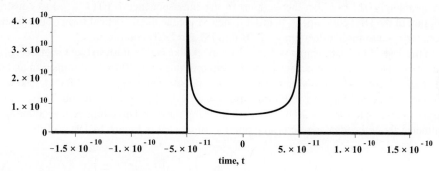

As is evident from the plot above, adual-Dirac like behavior results.

Exercises

14.25. A time jitter sequence can be described by the following discrete-time equation: $J[n] = 10 \sin (2\pi \, 13/1024 \, n)$. What is the probability density function for this sequence?

ANS. $f_J = \{1/\pi\sqrt{(10)^2 - J^2}$, $-10 \le J \le 10$ otherwise

14.7.2 Data-Dependent Jitter (DDJ)

When a serial communication link is driven by random data pattern, channel dispersion causes inter-symbol interference (ISI) to occur (see Figure 14.17). In addition, a transmitter may transmit a logic high value with different bit duration than a logic low value. Collectively, these two effects limitations give rise to data-dependent jitter (DDJ). Typically, a P-bit PRBS data pattern is used to excite a system by repeating this pattern throughout the duration of the test. Due to the repetition of this pattern, additional periodic jitter is produced that is harmonically related to this repletion frequency. For a P-bit PRBS pattern with bit duration T, the pattern repetition rate is $P \times T$ so that the fundamental frequency of the DDJ component will appear at $1/(P \times T)$ Hz. Assuming the harmonics of PJ and DDJ are different, then we can compute the RMS of the DDJ components using the exact same spectrum approach for PJ, except we drive the channel with a P-bit PRBS pattern. Assuming an N-point period jitter vector J, the fundamental component of DDJ will appear in bin $M_{DDJ} = N/P$, whence we write the RMS value of the DDJ component as

$$\sigma_{DDJ} = \sqrt{\sum_{p=1}^{H_{DDJ}} c^2_{p \times M_{DDJ}}} \tag{14.118}$$

where we assume that H_{DDJ} represent the number of harmonics associated with the DDJ jitter component. Once again, it is best to isolate the DDJ components so that an inverse FFT can be run and the peak-to-peak values extracted using the following expression:

$$DDJ \cong \max\{IFFT(J_{DDJ,only})\} - \min\{IFFT(J_{DDJ,only})\} \tag{14.119}$$

Correspondingly, we can also the samples of the time sequence $IFFT(J_{DDJ,only})$ to construct the histogram of the DDJ component. In order to determine the mathematical form of this jitter distribution, we need to further decompose DDJ into ISI and DCD components.

In the case of DCD, the input is driven by a clock-like data pattern so that the effects of ISI are eliminated. The time difference between the zero crossings for a 0-to-1 bit transition and a reference clock are collected from which the average time offset value $t_{DCD,01,av}$ is found. Likewise, this procedure is repeated for the 1–0 bit transitions and the $t_{DCD,10,av}$ value is found. These two results are then combined to form the PDF for the DCD using the dual-Dirac function, accounting for the asymmetry about the ideal zero crossing point as illustrated in Figure 14.19a, as follows

$$pdf_{DCD} = \frac{1}{2}\delta\left(t + \frac{t_{DCD,10,av}}{2}\right) + \frac{1}{2}\delta\left(t - \frac{t_{DCD,01,av}}{2}\right) \tag{14.120}$$

The DCD test metric expressed as a peak-to-peak quantity is defined as follows

$$DCD = t_{DCD,01,av} - t_{DCD,10,av} \tag{14.121}$$

When only the DCD metric is provided, we can approximate the PDF of the DCD as a dual-Dirac function as follows:

$$pdf_{DCD} \approx \frac{1}{2}\delta\left(t + \frac{DCD}{2}\right) + \frac{1}{2}\delta\left(t - \frac{DCD}{2}\right) \tag{14.122}$$

To identify the ISI distribution, a test very similar to DCD is performed but the input is driven with a specific data pattern other than a clock-like pattern. The zero crossings for the 0–1 bit transitions are collected and the extremes are identified, say $t_{ISI,01,min}$ and $t_{ISI,01,max}$. The process is repeated for the 1–0 zero crossings and, again, the extremes are identified as $t_{ISI,10,min}$ and $t_{ISI,10,max}$. The ISI test metric is then defined as the average of these two peak-to-peak values as follows

$$ISI \cong \frac{\left(t_{ISI,01,max} - t_{ISI,01,min}\right) + \left(t_{ISI,10,max} - t_{ISI,10,min}\right)}{2} \tag{14.123}$$

A first-order estimate of the PDF of the ISI can be written as another dual-Dirac function as follows:

$$pdf_{ISI} \approx \frac{1}{2}\delta\left(t + \frac{ISI}{2}\right) + \frac{1}{2}\delta\left(t - \frac{ISI}{2}\right)$$

(14.124)

As ISI and DCD convolve together to form the DDJ distribution, that is,

$$pdf_{DDJ} = pdf_{ISI} \otimes pdf_{DCD}$$

(14.125)

Substituting Eqs. (14.122) and (14.124) into Eq. (14.125), we write

$$pdf_{DDJ} = \left[\frac{1}{2}\delta\left(t + \frac{ISI}{2}\right) + \frac{1}{2}\delta\left(t - \frac{ISI}{2}\right)\right] \otimes \left[\frac{1}{2}\delta\left(t + \frac{DCD}{2}\right) + \frac{1}{2}\delta\left(t - \frac{DCD}{2}\right)\right]$$

(14.126)

Further expansion leads to the general form of the DDJ distribution as

$$pdf_{DDJ} = \frac{1}{4}\delta\left(t + \frac{DCD}{2} + \frac{ISI}{2}\right) + \frac{1}{4}\delta\left(t - \frac{DCD}{2} + \frac{ISI}{2}\right)$$
$$+ \frac{1}{4}\delta\left(t + \frac{DCD}{2} - \frac{ISI}{2}\right) + \frac{1}{4}\delta\left(t - \frac{DCD}{2} - \frac{ISI}{2}\right)$$

(14.127)

14.7.3 Bounded and Uncorrelated Jitter (BUJ)

During a data pattern test, additional tones unrelated to the test pattern can appear in the PSD of the jitter sequence; also the noise floor may rise. Firstly, for an alternating sequence of 1's and 0's, the PSD would contain the only the PJ and RJ components, as shown in Figure 14.34a. Here we illustrate the PJ component with a sine tone in bin M_{PJ}. When a PRBS pattern is applied to the channel input, theoretically, one would expect to continue to see the PJ component, as well as harmonics that are related to the repetition period of the PRBS pattern. In Figure 14.34b we depict this situation with three tones harmonically related to one another beginning with the fundamental in bin M_{DDJ}. In some situations, when testing with a PRBS sequence, additional tones may appear; also, the noise floor may increase. Because these signals have no theoretical relationship to the PRBS data pattern, these jitter components are classified as bounded and uncorrelated jitter (BUJ). It is essentially a catchall term that accounts for jitter that does not fit into PJ, DDJ, or RJ categories. DUJ generally arises from crosstalk and electromagnetic interference sources. It is important to note that BUJ jitter is bounded and its peak-to-peak value does not increase with more measurement samples.

Given measurements of the RMS value of PJ, DDJ, and RJ, together with the RMS value of the jitter sequence, denoted as σ_J, we can estimate the RMS value of the total BUJ component as

$$\sigma_{BUJ} = \sqrt{\sigma_J^2 - \sigma_{PJ}^2 - \sigma_{DDJ}^2 - \sigma_{RJ}^2}$$

(14.128)

where we assume that the powers of the individual components of BUJ add, that is, $(\sigma_{BUJ} = \sqrt{\sigma_{BUJ,1}^2 + \sigma_{BUJ,2}^2})$. We do not attempt to list the PDF for BUJ, because we need more information about the source of this jitter.

Figure 14.34. Power spectral density plot of a period jitter sequence with input: (a) alternating 1's and 0's pattern, (a) PRBS data sequence and (b) PRBS data sequence, but an additional tone and noise level is unaccounted for.

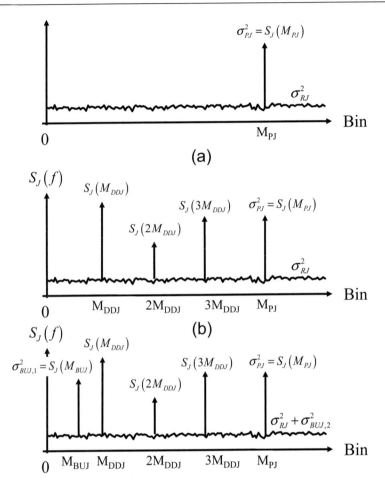

The following example will help to illustrate the deterministic jitter decomposition method described above.

EXAMPLE 14.21

A jitter sequence consisting of 1024 samples was captured from a serial I/O channel driven by two separate data patterns. In one case, an alternating sequence of 1's and 0's was used, in the other, a PRBS pattern was used. An FFT analysis performed on the two separate sets of jitter data, as well as with the PRBS sequence, resulted in the information below. Determine the test metrics: PJ, DDJ, BUJ.

Spectral Bin In Nyquist Interval	|XJ| Complex Spectral Coefficients (UI)		
	1010 clock-like Input	PRBS Input	
		Theory	Actual
0	0.001	0	6×10^{-4}
1, ..., 512 except	$\sim 1 \times 10^{-4}$	0	$\sim 2 \times 10^{-4}$
55	0	$0.021 - j\,0.028$	$0.02 - j\,0.03$
110	0	$0.002 + j\,0.003$	$0.001 + j\,0.002$
165	0	$0.006 - j\,0.007$	$0.005 - j\,0.006$
193	0	0	$0.01 - j\,0.02$
201	$0.02 + j\,0.05$	0	$0.018 + j\,0.053$

Solution:

Using the spectral data from the alternating 1010 data pattern, we find from the FFT data set above that a tone is present in bin 201 with an RMS value of

$$\sigma_{PJ} = \sqrt{2}\sqrt{0.02^2 + 0.05^2} = 0.076 \text{ UI}$$

For a single tone, the peak-to-peak value denoted by PJ or SJ is simply

$$PJ = SJ = 0.215 \text{ UI}$$

We also notice that from the spectral data for the clock-like input, the RJ noise component can be estimated to be

$$\sigma_{RJ} = \sqrt{\sum_{k=0}^{512} c_{k,RMS}^2 - c_{0,RMS}^2 - c_{201,RMS}^2} \approx \sqrt{512 \times \left(\sqrt{2} \times 10^{-4}\right)^2 - \sqrt{2}\left(0.02^2 + 0.05^2\right)} = 3.18 \times 10^{-3} \text{UI}$$

Next, as is evident from the middle column of data, through ideal simulation we expect to see harmonics related to the PRBS data pattern in bins 55, 110, and 165. Tones falling in any other bin is deemed unrelated to DDJ. Tonal components, other than those related to PJ, would be considered as being part of BUJ. Using the data from the rightmost column, we find that the RMS value of DDJ is

$$\sigma_{DDJ} = \sqrt{c_{55,RMS}^2 + c_{110,RMS}^2 + c_{165,RMS}^2} = \sqrt{2}\sqrt{\left(0.21^2 + 0.13^2\right) + \left(0.02^2 + 0.03^2\right) + \left(0.06^2 + 0.07^2\right)} = 50.6 \times 10^{-3}$$

To obtain the peak-to-peak value of DDJ, we notch out all bins other than those related to DDJ and perform an inverse FFT. This is achieved by setting bins 0–512 to zero except bins 55, 110, and 165 (as well as its first image). Next we scale these three bins by the factor N (=1024) and perform an inverse FFT, according to

$$d_{DDJ} = 1024 \times \text{IFFT}\left\{ \begin{bmatrix} 0,...,0,(0.02 - j0.03),0,...,0,(0.001 + j0.002),0,...,0,0,(0.005 - j0.006),0,...,0,0 \\ 0,...,0,(0.005 + j0.006),0,...,0,(0.001 - j0.002),0,...,0,0,(0.02 + j0.03),0,...,0 \end{bmatrix} \right\}$$

On doing this, the following DDJ jitter sequence results:

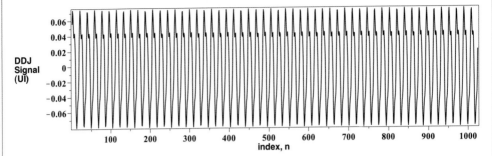

As is evident, the peak-to-peak value of DDJ is about

$$DDJ = 0.154 \text{ UI}$$

What remains now is the identification of the BUJ components. From the right-most column we recognize that a tone is present in bin 193 having an RMS value of

$$\sigma_{BUJ,1} = \sqrt{2}\sqrt{0.01^2 + 0.02^2} = 31.6 \times 10^{-3} \text{ UI}$$

We also recognize that the noise floor has increased when the data pattern changed. We attribute this increase to some bounded but uncorrelated jitter source. We quantify its value by taking the power difference between the two noise floors associated with the two different data patterns, that is,

$$\sigma_{BUJ,2} \approx \sqrt{\sum_{k=0}^{512} c_{k,RMS}^2 - c_{0,RMS}^2 - c_{55,RMS}^2 - c_{110,RMS}^2 - c_{165,RMS}^2 - c_{193,RMS}^2 - c_{201,RMS}^2 - \sigma_{RJ}^2} \equiv 5.4 \times 10^{-3} \text{UI}$$

The RMS value of all the BUJ jitter sources combined is then

$$\sigma_{BUJ} = \sqrt{\sigma_{BUJ,1}^2 + \sigma_{BUJ,2}^2} = \sqrt{\left(31.6 \times 10^{-3}\right)^2 + \left(5.4 \times 10^{-3}\right)^2} = 5.4 \times 10^{-3} \text{ UI}$$

To estimate the peak-to-peak value of the BUJ component, we note that the sinusoidal component is about four times larger in amplitude, so as a first estimate we obtain

$$BUJ \approx 2\sqrt{2}\sigma_{BUJ,1} = 2\sqrt{2} \times 31.6 \times 10^{-3} = 89.4 \times 10^{-3} \text{ UI}$$

Because this term does not include the correlated and bounded noise components evident in the noise floor, it underestimates the level of BUJ. A better estimate would be to repeat the test several times and average out the random noise leaving only the correlated signals.

EXAMPLE 14.22

A serial I/O channel was characterized for its RJ and DJ components resulting in the following table of measurements. Estimate the shape of the histogram of the total jitter distribution that would result? From this distribution, relate the peak-to-peak levels to the underlying test metrics found below. Also, plot the corresponding PDF of the zero-crossing distribution across two consectutive bit transitions assuming a bit duration of 100 ps.

	Jitter Metrics		
RJ (rms ps)	DJ		
	SJ (pp ps)	ISI (pp ps)	DCD (pp ps)
2	25	14	2

Solution:

Given the above metrics, we can represent each jitter distribution mathematically and picto-rially as follows:

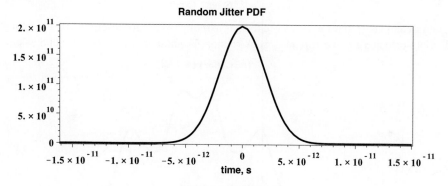

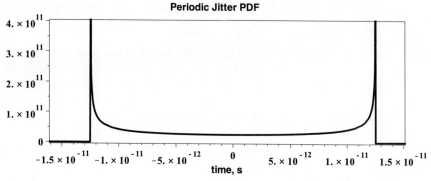

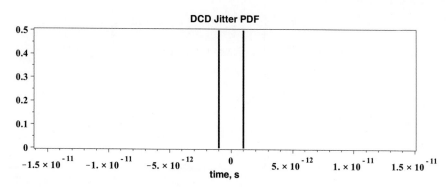

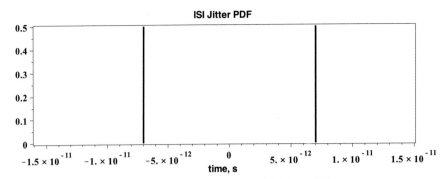

The total jitter distribution is a convolution of individual PDFs as follows:

$$pdf_{TJ} = pdf_{RJ} \otimes pdf_{PJ} \otimes pdf_{ISI} \otimes pdf_{DCD}$$

Using a computer program, we solved the above multiple kernel convolution for the total jitter distribution using a numerical integration routine. The numerical solution is found below:

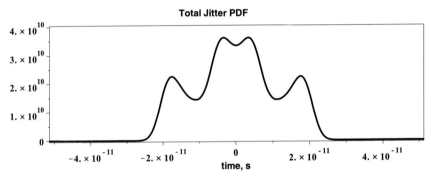

Here we see a multimodal distribution with four peaks, located at approximately -18 ps, -4 ps, 4 ps, and 18 ps. If we treat the sinusoidal distribution as a dual-Dirac density, then together with the other two Dirac distributions for ISI and DCD, we will end up with eight impulses located at ±12.5 ±7 ±1 ps or ±4.5, ±6.5, ±18.5 and ±20.5 ps. While these numbers are close to the peak locations, they do not coincide exactly, owing to the many interactions with the Gaussian distribution. Care must be exercised carefully when attempting to de-correlate jitter behavior from the total jitter distribution.

Finally, the PDF of the jitter distribution across two consecutive bit transitions separated by 100 ps is found by convolving the PDF of the total jitter with a dual-Dirac distribution with centers located at 0 and 100 ps, that is,

$$pdf_{ZC} = pdf_{TJ} \otimes \left[\frac{1}{2}\delta(t) + \frac{1}{2}\delta(t - 1 \times 10^{-10}) \right]$$

The resulting zero-crossing PDF is shown plotted below:

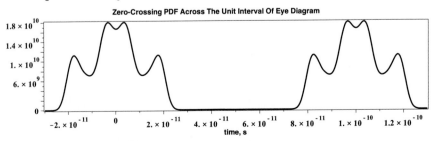

14.8 JITTER TRANSMISSION TESTS

Up to this point in this chapter we have been discussing the properties of jitter present at the transmitter or received output. In this section we shall address the sensitivity of a system to jitter imposed at its input and the effect that it has on its output. Two measures of jitter transmission are used: (1) jitter transfer and (2) jitter tolerance. We shall define these two terms and how to make their measurement below.

14.8.1 Jitter Transfer Test

Jitter transfer is a measurement of the amount of jitter present at the output of the digital channel relative to a particular amount of jitter applied to its input. The ratio is usually specified as a function of frequency, as the channel is generally frequency sensitive. Generally, a jitter transfer test uses a clock-like input data pattern to avoid ISI issues. In much the same way as we did for measuring the gain of an analog or sampled data channel, a sinusoidal phase modulated signal having specific amplitude and frequency is applied to the input of a device or system and the corresponding output sinusoidal phase modulated response is measured. Assuming the peak-to-peak value of the input phase modulated signal is SJ_{IN} and the peak-to-peak value of the output is SJ_{OUT} then we define the jitter transfer at frequency ω as

$$G_J(\omega) \triangleq \frac{SJ_{OUT}}{SJ_{IN}} \tag{14.129}$$

or, equivalently, in dB, as

$$G_{J,dB}(\omega) = 20 \log_{10} G_J(\omega) \tag{14.130}$$

Repeating this measurement for a range of input frequencies and comparing the jitter gain to a jitter transfer compliance mask, such as that shown in Figure 14.35, we would have a complete jitter transfer test. Any data point that falls inside of the hashed region would result in a failed test.

Figure 14.35. An example jitter transfer compliance mask with measured data superimposed. This particular data set passes the test as all the data points fall inside the acceptable region (as shown).

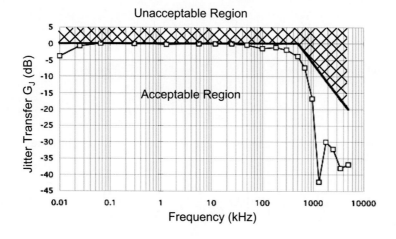

While we could measure both the input and output jitter level using the power spectral method described above for computing PJ, a more efficient method is to use coherent signal generation and sampling, in much the same way we did for the analog and sampled data channel measurements using the AWG and digitizer. Let us explain this more fully below.

Coherent Jitter Generation

As jitter transfer is based on injecting a known sinusoidal jitter signal at a particular frequency, we can use coherency to set up the test so that the spectrum of the samples of the output signal that is collected and processed with an N-point FFT falls in the appropriate bin. This avoids the need for any additional post-processing or windowing.

Mathematically, a clock-like digital signal can be described by a sinusoidal with unity amplitude and frequency f_c, followed by a squaring operation as follows

$$d(t) = \text{sgn} \left\{ \sin \left[2\pi f_c t \right] \right\} \tag{14.131}$$

where sgn{.} represents the signum function. Now, if we argument the phase of the sinusoidal function by adding a phase term J(t), to represent the jitter signal as a function of time, we would write the digital signal as follows

$$d(t) = \text{sgn} \left\{ \sin \left[2\pi f_c t + J(t) \right] \right\} \tag{14.132}$$

Furthermore, if we assume the phase function is sinusoidal, with amplitude A_J and frequency f_J, that is,

$$J(t) = A_J \sin \left(2\pi f_J t \right) \tag{14.133}$$

then Eq. (14.132) can be written as

$$d(t) = \text{sgn} \left\{ \sin \left[2\pi f_c t + A_J \sin \left(2\pi f_J t \right) \right] \right\} \tag{14.134}$$

The units of $J(t)$ are expressed in radians, but one often see this jitter signal expressed in terms of seconds,

$$J(t) \, [\text{seconds}] = \frac{1}{2\pi f_c} \times J(t) \, [\text{radians}] \tag{14.135}$$

or in terms of unit intervals (UI) as

$$J(t) \, [UI] = \frac{1}{2\pi} \times J(t) \, [\text{radians}] \tag{14.136}$$

If we assume that this signal is sampled at sampling rate F_s, then we can write the sampled form of Eq. (14.134) as

$$d[n] = \text{sgn} \left\{ \sin \left[2\pi \frac{f_c}{F_s} n + A_J \sin \left(2\pi \frac{f_J}{F_s} n \right) \right] \right\} \tag{14.137}$$

In order to maintain coherency, we impose the conditions

$$\frac{f_c}{F_S} = \frac{M_c}{N} \text{ and } \frac{f_J}{F_S} = \frac{M_J}{N}$$

(14.138)

where N, M_c, and M_f are integers. In the usually way, N represents the number of samples obtained from $d(t)$, and M_c and M_J represent the number of cycles the clock and jitter signal complete in N points, respectively. This allows us to rewrite Eq. (14.137) as

$$d[n] = \text{sgn}\left\{ \sin\left[2\pi \frac{M_c}{N} n + A_J \sin\left(2\pi \frac{M_J}{N} n \right) \right] \right\}$$

(14.139)

Deriving N points from Eq. (14.139) and storing them in the source memory of the arbitrary waveform generator (AWG) would provide a means to create a sinusoidal jitter signal with amplitude A_J riding on the clock-like data signal. A simple test setup illustrating this approach is shown in Figure 15.36a. For very-high-frequency data rates, an AWG would not be available to synthesis such waveforms directly. Instead, the digitally synthesized jitter signal would phase modulate a reference clock signal using some phase modulation technique such as the Armstrong phase modulator as shown in Figure 14.36b. A third technique is the method of data edge placement as shown in Figure 14.36c. This technique uses a digital-to-time converter (DTC) to place the data edges at specific points along the UI of the data eye according to an input digital code. The phase-lock loop removes high-frequency images, so it is essentially the reconstruction filter of this digital-to-time conversion process. A variant of this approach is one that makes use of a periodic sequence of bits together with a phase lock loop to synthesize a sinusoidal jitter signal.[14,15] The DTC is essentially realized with a set of digital codes.

Figure 14.36. Generating a phase modulated clock signal: (a) a direct digital synthesis method involving an AWG, (b) an indirect method using the Armstrong phase modulator technique and a low-frequency AWG, (c) edge-placement technique using a digital-to-time converter followed by a PLL.

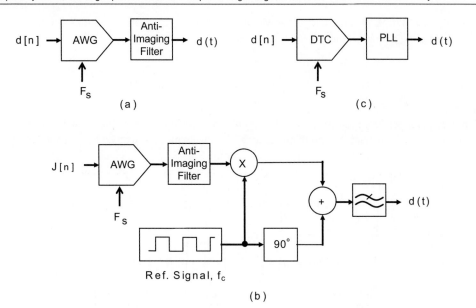

EXAMPLE 14.23

A digital receiver operates at a clock rate of 100 MHz. A jitter transfer test needs to be performed with a 10-kHz phase-modulated sinusoidal signal having an amplitude of 0.5 UI. Using an AWG with a sampling rate of 1 GHz, write a short routine that generates the AWG samples.

Solution:
Our first task is to detemine the number of samples and cycles used in the waveform sythesis. Substituting the given data in the coherency expressions, we write

$$\frac{100\,\text{MHz}}{1\,\text{GHz}} = \frac{M_c}{N} \quad \text{and} \quad \frac{10\,\text{kHz}}{1\,\text{GHz}} = \frac{M_J}{N}$$

Simplifying and isolating the two M terms, we write

$$M_c = \frac{1}{10} \times N \quad \text{and} \quad M_J = \frac{1}{10^5} \times N$$

As M_c, M_J and N are all integers, we immediately recongize that N must contain the factor 10^5. This makes it impossible for N to be expressed solely as a power of two. This leaves us with three options: (1) Alter the data rate f_c and/or the frequency of the modulating sinewave f_J such that N is a power of two, (2) work with a non-power-of-two vector length and use a window function together with the radix-2 FFT, or (3) work with a non-power-of-two vector length and assume that the jitter sequence consists of a single sine wave from which its RMS value is computed directly from the time sequence. We will select the latter two approaches and select $N = 3 \times 10^5$. Substituting N back into the above expressions, we obtain

$$M_c = 3 \times 10^4 \quad \text{and} \quad M_J = 3$$

As far as the jitter amplitude is concerned, we want a 0.5-UI amplitude but we need to convert between UI and radians. To do this, we use

$$A_J[\text{radians}] = 2\pi \times A_J[\text{UI}] = \pi$$

Using Eq. (14.137), M_J, M_c, and N, together with $A_J = \pi$ radians, the waveform synthesis routine can be written using a programming for loop as follows:

$$\text{for } n = 1 \text{ to } 3 \times 10^5$$
$$d[n] = \text{sign}\left\{ \sin\left[2\pi \frac{1}{10}(n-1) + 3.1415 \times \sin\left[2\pi \frac{1}{10^5}(n-1) \right] \right] \right\}$$
$$\text{end}$$

Running this procedure results in the following plot of a short portion of the sequence between $n = 15,000$ to $15,100$ where the modulating signal is near its peak, and again between $n = 50,000$ and $50,100$ where the modulating signal is near zero:

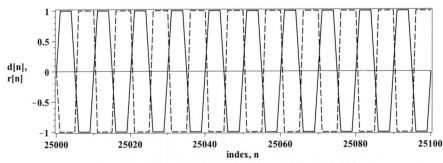

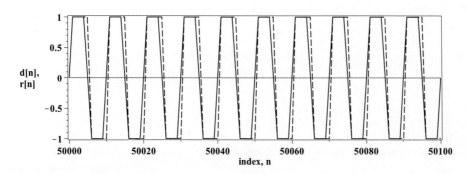

Clearly, the jitter sequence goes through a maximum phase shift of 3.14 radians or 0.5 UI as expected.

Exercises

14.26. A digital receiver operates at a clock rate of 600 MHz. A jitter transfer test needs to be performed with a 1-kHz phase-modulated sinusoidal signal having an amplitude of 10^{-8} s. Select the values of M_c, M_J and N such that the samples are coherent with a 4-GHz sampling rate.

ANS. $M_c = 1.8 \times 10^6$, $M_J = 3$, $N = 1.2 \times 10^7$.

Coherent Phase Extraction Using Analytic Signal Representation

In terms of extracting the sinusoidal phase-modulated response of the device or system, again, we can describe three approaches. The first method to be described is based on the analytic signal extraction method presented in references 16 and 17. This method uses a fast digitizer operating at a sampling rate of F_S, whereby the channel output signal is sampled and the data are collected and stored in vector d_{OUT}. In order to ensure that Shannon's sampling theorem is respected, the incoming digital signal must be band-limited so that only the fundamental components of the incoming signal are presented to the digitizer. Next, an analytic or complex signal representation of collected data d_{OUT} is created using the discrete Hilbert transform according to the following:

$$d_{OUT, analytic} = d_{OUT} + j\hat{d}_{OUT} \qquad (14.140)$$

where the discrete-time sequence $\hat{d}_{OUT}$ is related to the original time series through the discrete Hilbert transform[18] (DTH) as follows

$$\hat{d}_{OUT}[n] = \mathrm{DHT}\{d_{OUT}\} = \frac{2}{\pi} \begin{cases} \displaystyle\sum_{k=odd} \frac{d_{OUT}(k)}{n-k}, & n \text{ even} \\[2ex] \displaystyle\sum_{k=even} \frac{d_{OUT}(k)}{n-k}, & n \text{ odd} \end{cases} \qquad (14.141)$$

In general, the DHT defined above is computationally inefficient. Instead, one can work in the frequency domain using the spectral coefficients of d_{OUT}, that is,

$$D_{OUT}(k) = \mathrm{FFT}\{d_{OUT}\}, \quad k = 0,\ldots, N-1 \qquad (14.142)$$

and obtain the spectral coefficients of the analytic signal $D_{OUT,analytic}$ as

$$D_{OUT,analytic}(k) = \begin{cases} D_{OUT}(k), & k=0 \\ 2D_{OUT}(k), & k=1,2,\ldots,\dfrac{N}{2}-1 \\ 0, & k=\dfrac{N}{2},\ldots,N-1 \end{cases} \tag{14.143}$$

Through the application of the inverse FFT, we obtain the N-point Hilbert transform of the N-point sequence d_{OUT} as follows

$$\hat{d}_{OUT}[n] = \text{Im}\left\langle \text{IFFT}\left\{D_{OUT,analytic}\right\}\right\rangle, \quad n=0,\ldots,N-1 \tag{14.144}$$

Of course, the real part of $\left\langle \text{IFFT}\{D_{OUT,analytic}\}\right\rangle$ is the original data sequence d_{OUT}.

The discrete-time total instantaneous phase from the complex signal using the following recursive equation:

$$\theta_i[n] = \theta_i[n-1] + \tan^{-1}\left(\frac{\hat{d}_{OUT}[n]}{d_{OUT}[n]}\right) - \tan^{-1}\left(\frac{\hat{d}_{OUT}[n-1]}{d_{OUT}[n-1]}\right) \tag{14.145}$$

Because this phase term includes both the phase of the carrier and the jitter-injected signal, we need to determine the instantaneous phase of the carrier and subtract it from Eq. (14.145) to obtain the effect of the phase-modulated jitter signal alone. To begin, let us recognize that the sequence of phase samples corresponding to the reference signal varies proportionally with the sample time [argument term of Eq. (14.139) when A_j is zero], according to

$$\theta_{REF}[n] = 2\pi \frac{M_c}{N} n \tag{14.146}$$

or in recursive form as

$$\theta_{REF}[n] = \theta_{REF}[n-1] + 2\pi \frac{M_c}{N} \tag{14.147}$$

The phase-modulated jitter sequence would then be recovered by subtracting Eq. (14.146) from Eq. (14.145) to obtain

$$J[n] = \theta_i[n] - \theta_{REF}[n] \tag{14.148}$$

or simply as

$$J[n] = \theta_i[n] - 2\pi \frac{M_c}{N} n \tag{14.149}$$

Note that great care must be exercised when using the inverse-tangent operation associated with Eq. (14.145). Normally the built-in inverse-tangent operation of a tester or computer routine, such as MATLAB or Excel, limits its range to two quadrants of the complex plane, that is, $\pm\pi$.

Moreover, extension to four-quadrant operation does not solve the problem completely. We will address this issue shortly.

While the reference phase term is known at the AWG output, it may experience phase shift through the digital channel, so one should not rely on the generated signal properties alone. Instead, it is always best to extract the reference signal phase behavior from the measured data at the receiver end. There are at least two ways in which to do this. The first method is to fit a straight line through the instantaneous phase signal $\theta_i[n]$ using linear regression analysis (similar to the best-fit line method used in DAC/ADC testing), such as

$$\theta_{REF}[n] = \lambda n + \beta \tag{14.150}$$

Subsequently, the instantaneous excess phase sequence would then be found from Eq. (14.148).

A second approach is to work with the frequency information associated with the captured digital sequence d_{OUT}. Let us begin by taking the FFT of d_{OUT} and denote the result as D_{OUT}, that is,

$$D_{OUT} = FFT(d_{OUT}) \tag{14.151}$$

Next, set all bins to zero except bins M_c and $N\text{-}M_c$ and perform an inverse FFT according to

$$d_{REF} = IFFT\left\{[0,...,0,D_{OUT}(M_c),0,...,0,D_{OUT}(N-M_c),0,...,0]\right\} \tag{14.152}$$

Subsequently, perform a discrete Hilbert transform on d_{REF} and obtain the analytic signal representation of the reference signal as follows:

$$d_{REF,analytic} = d_{REF} + j\hat{d}_{REF} \tag{14.153}$$

The instantaneous phase behavior of the reference signal is then found using the following recursive equation:

$$\theta_{REF}[n] = \theta_{REF}[n-1] + \tan^{-1}\left(\frac{\hat{d}_{REF}[n]}{d_{REF}[n]}\right) - \tan^{-1}\left(\frac{\hat{d}_{REF}[n-1]}{d_{REF}[n-1]}\right) \tag{14.154}$$

The analytic signal representation for the reference signal in the time domain would then be described using the inverse FFT as follows:

$$d_{REF,analytic} = d_{REF} + j\hat{d}_{REF} = IFFT\left\{[0,...,0,2D_{OUT}(M_c),0,...,0]\right\} \tag{14.155}$$

Similarly, the analytic signal representation for the d_{OUT} sequence is found from

$$d_{OUT,analytic} = d_{OUT} + j\hat{d}_{OUT}$$
$$= IFFT\left\{[D_{OUT}(0),2D_{OUT}(1),...,2D_{OUT}(M_c),...,2D_{OUT}\left(\frac{N}{2}-1\right),0,0,...,0]\right\} \tag{14.156}$$

Before we move on to the calculation of jitter transfer gain, we need to close off the issue of unwrapping the phase function of the various complex signals. At the heart of this difficulty is the mathematical behavior of the inverse-tangent operation. While the trigonometric tangent maps one value to another, the inverse tangent operation maps one value to many values separated by multiples of $\pm 2\pi$. This creates discontinuities in the computed phase behavior of the complex signal and renders the analytic signal analysis method useless.

To get around the computational difficulties of unwrapping a phase function, we must keep track of the phase changes as a function of the sampling instant, in particular as we move from either the third or fourth quadrant of the complex place into the first or second quadrant, we must add 2π to the instantaneous phase function. Here we are assuming that a step change in phase is never greater than π. A condition that is met when the sampling process respects Shannon's sampling constraint (i.e., no less than 2 points per period on the highest frequency component). Following this process, assuming we have an analytic signal defined as $z[n] = x[n] + j\hat{x}[n]$, the instantaneous unwrapped phase function would be written as

$$
\theta_i[n] = \begin{cases} \theta_i[n-1] + 2\pi + \tan 4q^{-1}\left(\dfrac{\hat{x}[n]}{x[n]}\right) - \tan 4q^{-1}\left(\dfrac{\hat{x}[n-1]}{x[n-1]}\right) & \begin{array}{l} \hat{x}[n-1] < 0, x[n] > 0, \hat{x}[n] > 0 \\ \text{or} \\ x[n-1] > 0, \hat{x}[n-1] < 0, \hat{x}[n] > 0 \end{array} \\[4ex] \theta_i[n-1] + \tan 4q^{-1}\left(\dfrac{\hat{x}[n]}{x[n]}\right) - \tan 4q^{-1}\left(\dfrac{\hat{x}[n-1]}{x[n-1]}\right) & \text{otherwise} \end{cases}
$$

(14.157)

where the four-quadrant inverse-tangent function $\tan 4q^{-1}(.)$ is defined as follows

$$
\tan 4q^{-1}\left(\frac{y}{x}\right) = \begin{cases} \tan^{-1}\left(\dfrac{y}{x}\right), & x \geq 0, y \geq 0 \\[2ex] \pi - \tan^{-1}\left(\dfrac{y}{-x}\right), & x < 0, y \geq 0 \\[2ex] \pi + \tan^{-1}\left(\dfrac{-y}{-x}\right), & x < 0, y < 0 \\[2ex] 2\pi - \tan^{-1}\left(\dfrac{-y}{x}\right), & x \geq 0, y < 0 \end{cases}
$$

(14.158)

The following example will help to illustrate this jitter extraction technique on a multitone phase modulated signal.

EXAMPLE 14.24

A 4096-point data set is collected from the following two-tone unity-amplitude phase-modulated jitter sequence:

$$d_{OUT}[n] = \sin\left[2\pi\frac{1013}{4096}(n-1) + 1\times\sin\left[2\pi\frac{57}{4096}(n-1)\right] + 1\times\sin\left[2\pi\frac{23}{4096}(n-1)\right]\right]$$

Extract the jitter sequence using the analytic signal approach and compare this sequence to the ideal result.

Solution:
The first step to the analytic signal extraction method is to obtain the FFT of the sampled sequence. Let us denote the FFT of the data sequences as follows:

$$D_{OUT} = FFT(d_{OUT})$$

A plot of the magnitude response corresponding to the spectral coefficients of this data sequence is shown below over the Nyquist and its first image band:

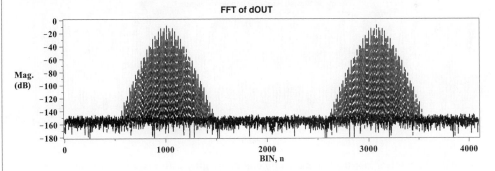

Here we see a very rich spectral plot with most of the power of the signal concentrated around the carrier located in bin 1013. We also notice that the magnitude response is symmetrical about the 2048 bin. Next, we perform the Hilbert transform of the data sequence in the frequency domain, resulting in the magnitude response shown below:

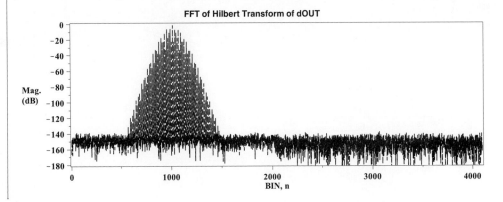

Here we see magnitude response in the Nyquist interval increase by 3 dB and the level of the first image reduced to the noise floor of the FFT. We can also extract the reference signal from the above spectral plot and obtain the spectral coefficients of the Hilbert transform of the reference signal as follows:

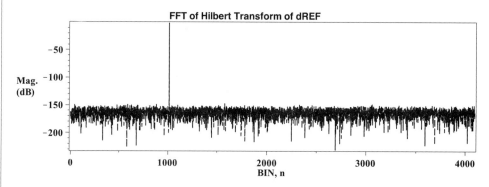

The inverse FFT of the Hilbert transform of D_{OUT} results in the complex sequence whose real and imaginary parts are plotted below for n between 0 and 150:

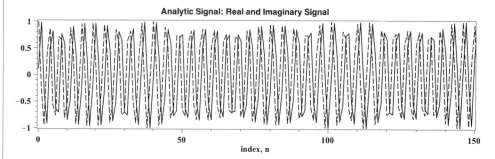

The reader can see two identical signals, one shifted slightly with respect to the other. In fact, the phase offset is exactly π/2, as defined by the Hilbert transform. Using the analytic signal representation for d_{OUT} and the reference, we extract the following two instantaneous phase sequences:

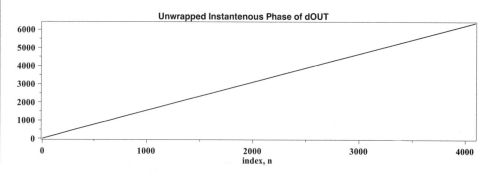

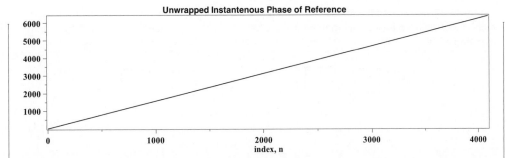

Unwrapped Instantenous Phase of Reference

Here both phase sequences increase monotonically with index, *n*. In fact, both plots look identical. However, taking their difference reveals the key information that we seek, that being the phase-modulated jitter sequence as shown below:

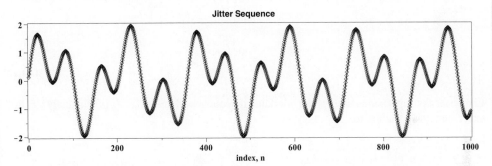

Jitter Sequence

Superimposed on this plot is the expected jitter sequence taken from the given phase-modulating data sequence (shown as circles).

We conclude that the analytic signal extraction method was effective in extracting the two-tone phase modulated signal. It is interesting to note from this example that if we take the FFT of the extracted jitter sequence we can obtain a spectral plot that separates the two tones in the usual DSP-based manner as follows:

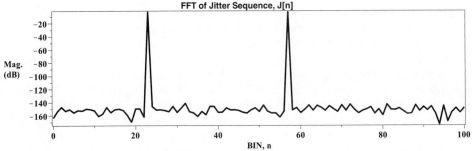

FFT of Jitter Sequence, J[n]

It should be clear from the above example that all DSP-based methods of previous chapters are directly applicable to a jitter sequence.

Final Comments

The objective of this section was to measure the sinusoidal phase-modulated response of a device or system so that the jitter transfer gain could be computed. We now are equipped with the DSP-based techniques in which to extract this information from a time sequence. If we define the jitter sequence at the output of the system as $J_{OUT}[n]$, then we can obtain the peak-to-peak value from the M_j bin using the spectral coefficient notation as follows:

Figure 14.37. Extracting the phase component of a digital signal: (a) indirect method involving a high-speed digitizer, (b) a direct method using a heterodyning technique, (c) another direct method involving a TDC.

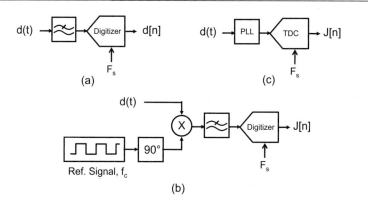

$$SJ_{OUT} = 2\sqrt{2} \times c_{M_J - RMS, J_{OUT}} \tag{14.159}$$

Likewise, the input sinusoidal phase-modulated sequence, denoted as $J_{IN}[n]$, is analyzed in the exact same way leading to

$$SJ_{IN} = 2\sqrt{2} \times c_{M_J - RMS, J_{IN}} \tag{14.160}$$

Finally, we combine Eqs. (14.159) and (14.160) and substitute into Eq. (14.129) to find the jitter transfer gain corresponding to Bin M_J as

$$G_J(M_J) = \frac{c_{M_J - RMS, J_{OUT}}}{c_{M_J - RMS, J_{IN}}} \tag{14.161}$$

The process can be repeated across a band of frequencies or one can create a multitone jitter signal and capture the entire jitter transfer over a band of frequencies.

To close this section, we point out through Figure 14.37 that there are several other ways in which to extract the jitter sequence of a phase-modulated signal. Figure 14.37a illustrates the high-speed sampling method just described. The setup in part Figure 14.37b illustrates a heterodyning approach whereby the system output is mixed with a reference signal and the low-frequency image is removed by filtering and then digitized. The test setup of Figure 14.37c makes use of a PLL and time-to-digital (TDC) converter arrangement to obtain the jitter sequence directly. A TDC essentially measures the time difference between the reference and the digital signal to produce $J[n]$. For the latter two test setups, regular FFT post-processing, together with Eq. (14.161), can be used to obtain the jitter transfer gain.

14.8.2 Jitter Tolerance Test

A jitter tolerance test measures the level of jitter that can be applied to the input to a device or system while maintaining a specific BER level at its output. This test generally requires a data source with controllable jitter, a means to verify the characteristics of the jittered input signal, and a means to detect whether the device or system is meeting its BER requirement. The general jitter

Figure 14.38. A general jitter tolerance test setup illustrating the different source conditions.

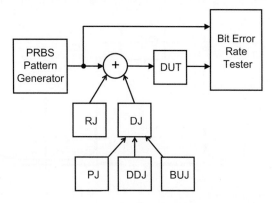

Figure 14.39. A jitter tolerance compliance mask for the SONET OC-12 specification.

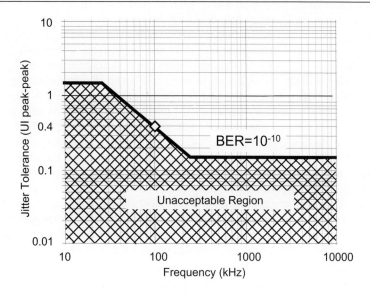

tolerance test setup is shown in Figure 14.38. Depending on the standard, the jitter on the input signal can take various forms. For example, SONET testing requires jitter in the form of a sinusoidal phase modulation, which is swept or stepped across a prescribed frequency band while the amplitude is adjusted so that a specific BER level is maintained. The input amplitude of the jitter is then compared against a frequency template, such as that shown in Figure 14.39 for the SONET OC-12 transmission specification. As an example, at an input sinusoidal frequency of 100 kHz, the device or system must be capable of handling an injected peak-to-peak sinusoidal jitter of no less than of 0.4 UI while maintaining a BER of 10^{-10} or smaller. It is important that the input signal have little RJ component, at least negligible when compared to the SJ component. In contrast to SONET testing, the SATA specification requires a test signal bearing a wide variety of both random and deterministic jitter down to a BER level of 10^{-12} or smaller. Great care must be exercised when selecting the jitter source, especially in light of the very high demands on its performance.

Figure 14.40. (a) Relating the input and output jitter behavior through a linear system function $G(j\omega)$. (b) Input-output PDFs for a sinusoidal input signal and its relationship to $G(j\omega)$.

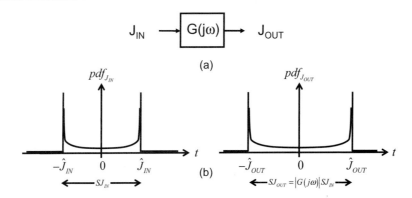

(a)

(b)

To gain a better understanding of a jitter tolerance test under sinusoidal input conditions, let us assume that the input and output phase or jitter behavior of a system is linearly related by some transfer function $G(j\omega)$ as depicted in Figure 14.40a. Such a transfer function arises naturally from the underlying mathematical description of a clock recovery circuit. If the input to the system is described by the equation $J_{IN} \sin(\hat{\omega} t)$, then the output can be described as $|G(j\omega)|\hat{J}_{IN} \sin(\omega t + \angle G(j\omega))$. Equally, we can relate the PDFs of the input and output distributions as shown in Figure 14.40b with peak-to-peak values of SJ_{IN} and $SJ_{OUT} = |G(j\omega)|SJ_{IN}$, respectively. Clearly, the larger the gain of the system at a particular frequency, the larger the output peak-to-peak jitter. Assuming that sinusoidal jitter is the only jitter present, then we can describe the PDF of the zero-crossings of two consecutive bits as shown in Figure 14.41a. Superimposed on this plot is the ideal sampling instant at $t_{SH} = T/2$. This threshold sits midway between the two ideal bit transitions. As is evident from this figure, the jitter is bounded and much less than the sampling threshold so that no transmission errors are expected. As the input peak-to-peak value increases, as shown in Figure 14.41b, the peak value of the output jitter distribution gets closer to the sampling instant, but again, no errors are expected. However, once the amplitude of the jitter exceeds the sampling threshold [i.e., $\hat{J}_{IN}|G(j\omega)| > T/2$], errors are expected as illustrated in Figure 14.41c. We can quantify the BER level by integrating the areas beyond the threshold on each side and write

$$BER = \frac{1}{2}P_e\bigg|_{t>t_{SH}} + \frac{1}{2}P_e\bigg|_{t<t_{SH}} = \frac{1}{2}\times\int_{T/2}^{\infty} pdf_{J_{OUT}}\bigg|_{t=0} dt + \frac{1}{2}\times\int_{-\infty}^{T/2} pdf_{J_{OUT}}\bigg|_{t=T} dt \qquad (14.162)$$

where

$$pdf_{J_{OUT}} = \begin{cases} \dfrac{1}{\pi\sqrt{\left(\dfrac{SJ_{OUT}}{2}\right)^2 - t^2}}, & -\dfrac{SJ_{OUT}}{2} \le t \le \dfrac{SJ_{OUT}}{2} \\ \\ 0 & \text{otherwise} \end{cases} \qquad (14.163)$$

Substituting and using the integral relationship $\int \dfrac{1}{\sqrt{a^2-x^2}}dx = -\cos^{-1}\left(\dfrac{x}{a}\right)$, $(a^2 > x^2)$, we write

$$BER = \int_{T/2}^{SJ_{OUT}/2} \dfrac{1}{\pi\sqrt{\left(\dfrac{SJ_{OUT}}{2}\right)^2 - t^2}} dt = \frac{1}{\pi}\cos^{-1}\left(\dfrac{T}{SJ_{OUT}}\right), \qquad SJ_{OUT} > T \qquad (14.164)$$

Figure 14.41. Illustrating the zero-crossing PDF for two consecutive bit transitions when the system is excited by a sinusoidal signal: (a) $\hat{J}_{OUT} < T/2$ no bit errors, (b) $\hat{J}_{OUT} < T/2$ but larger than situation in part (a) but again no bit errors, (c) $\hat{J}_{OUT} < T/2$ with bit errors.

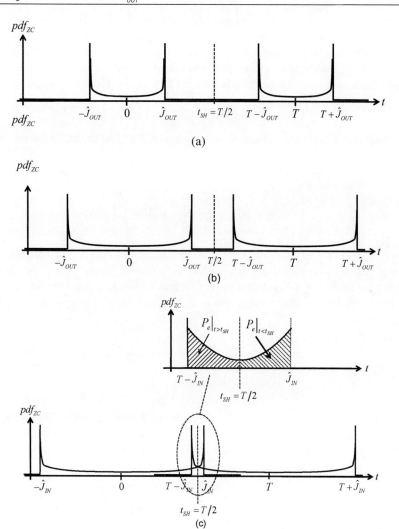

(a)

(b)

(c)

Rearranging and using the trigonometric power series expansion around $x = 0$ $\left(\text{i.e., } \cos x \approx 1 - \dfrac{x^2}{2} \right)$, we write the BER as

$$\text{BER} = \frac{2}{\pi} \sqrt{1 - \frac{T}{SJ_{OUT}}}, \qquad SJ_{OUT} > T \tag{14.165}$$

or expressing SJ_{OUT} in terms of UIs, we write

$$\text{BER} = \frac{2}{\pi} \sqrt{1 - \frac{1}{SJ_{OUT}}}, \qquad SJ_{OUT} > 1\ \text{UI} \tag{14.166}$$

Replacing the output SJ term by the input *SJ* term and system function $G(j\omega)$, we finally write

$$\text{BER} = \frac{2}{\pi}\sqrt{1 - \frac{1}{|G(j\omega)|SJ_{IN}}}, \quad |G(j\omega)|SJ_{IN} > 1 \text{ UI} \tag{14.167}$$

This equation provides the key insight into the jitter tolerance test. It shows how the peak-to-peak input jitter level, the system transfer function, and the BER performance are all interrelated. For instance, if $G(j\omega)$ has a low-pass nature whereby at DC, $G(j\omega) = 1$ and for $f > f_B$, $G(j\omega) = 0$, then the BER performance will have a value of $\frac{2}{\pi}\sqrt{1 - \frac{1}{SJ_{IN}}}$ for $f > f_B$ and a very large value of $2/\pi$ for $f > f_B$.

For a particular BER performance, we can rearrange Eq. (14.167) and write the input *SJ* as follows:

$$SJ_{IN} = \frac{1}{|G(j\omega)|\left[1 - \left(\frac{\pi}{2} \times \text{BER}\right)^2\right]} \tag{14.168}$$

The above expression reveals the nature of the mask that separates the acceptable region from the unacceptable region as illustrated in the jitter tolerance plot of Figure 14.39. The shape of this mask follows the inverse of the system transfer function $|G(j\omega)|^{-1}$. In order to clearly identify the points on this mask, let us denote these points as $SJ_{IN}|_{GOLDEN}$ and the corresponding system gain as $G_{GOLDEN}(j\omega)$, then we can write

$$SJ_{IN}|_{GOLDEN} = \frac{1}{|G_{GOLDEN}(j\omega)|\left[1 - \left(\frac{\pi}{2} \times \text{BER}\right)^2\right]} \tag{14.169}$$

Any measured SJ_{IN} that is above $SJ_{IN}|_{GOLDEN}$ will fall in the acceptable region. To understand why this is so, consider a situation where the input *SJ* at a particular frequency to a DUT is adjusted so that the BER at the output reaches the desired level. If $SJ_{IN}|_{DUT} > SJ_{IN}|_{GOLDEN}$, then we can write

$$\frac{1}{|G_{DUT}(j\omega)|\left[1 - \left(\frac{\pi}{2} \times \text{BER}\right)^2\right]} > \frac{1}{|G_{GOLDEN}(j\omega)|\left[1 - \left(\frac{\pi}{2} \times \text{BER}\right)^2\right]} \tag{14.170}$$

which leads to

$$|G_{GOLDEN}(j\omega)| > |G_{DUT}(j\omega)| \tag{14.171}$$

This expression suggests that the DUT gain is less than the golden device gain and hence can tolerate a greater level of jitter at its input before reaching the same output level or, equivalently, the same BER level. It should be noted here that the jitter tolerance test is much more than just a measurement of the gain of the system transfer function, because it includes nonlinear and noise effects not modeled by the above equations.

EXAMPLE 14.25

A jitter tolerance test was performed on a digital receiver that must conform to the SONET OC-12 specification shown in Figure 14.39. The following table lists the results of this test. Does the receiver meet the OC-12 specification? What is the gain of the measured device relative to the golden device gain at 300 kHz?

Input SJ (pp)	Input Frequency	Measured BER
2.2 UI	10 kHz	10^{-10}
2.3 UI	30 kHz	10^{-10}
1.1 UI	100 kHz	10^{-10}
0.11 UI	300 kHz	10^{-10}
0.21 UI	1,000 kHz	10^{-10}
0.32 UI	3,000 kHz	10^{-10}
0.32 UI	10,000 kHz	10^{-10}

Solution:
To see clearly the results of the test, we superimposed the data points from the table on a frequency plot with the OC-12 compliance mask superimposed. All points but one at 300 kHz appear above the BER = 10^{-10} line, hence this device does not pass the test.

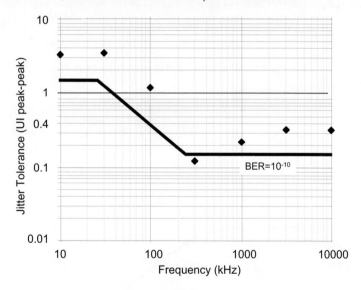

The gain of the DUT relative to the golden device gain is found from Eq. (14.167) as

$$\frac{2}{\pi}\sqrt{1-\frac{1}{\left|G_{GOLDEN}\left(j\omega\right)\right|SJ_{IN}\big|_{GOLDEN}}} = \frac{2}{\pi}\sqrt{1-\frac{1}{\left|G_{DUT}\left(j\omega\right)\right|SJ_{IN}\big|_{DUT}}}$$

which reduces to

$$\frac{\left|G_{DUT}\left(j\omega\right)\right|}{\left|G_{GOLDEN}\left(j\omega\right)\right|} = \frac{SJ_{IN}\big|_{GOLDEN}}{SJ_{IN}\big|_{DUT}}$$

Substituting the appropriate measured and golden device values leads to

$$\frac{\left|G_{DUT}\left(j2\pi \times 300 \text{ kHz}\right)\right|}{\left|G_{GOLDEN}\left(j2\pi \times 300 \text{ kHz}\right)\right|} = \frac{0.15 \text{ UI}}{0.11 \text{ UI}} = 1.36$$

Here we see the DUT has a gain that is 1.36 times larger than the golden device gain at 300 kHz.

EXAMPLE 14.26

A golden device operating at a bit rate of 100 MHz has an input–output phase transfer function described as follows:

$$G(s) = 50\frac{s+50}{s+5000}$$

Plot the expected jitter tolerance mask for a sinusoidal phase-modulated inputsignal with a BER of 10^{-10}.

Solution:
According to Eq. (14.169), the expected input SJ mask level can be written as

$$SJ_{IN}\big|_{GOLDEN}(f) = \frac{1}{\left|50\dfrac{j2\pi f+50}{j2\pi f+5000}\right|\left[1-\left(\dfrac{\pi}{2}\times 10^{-10}\right)^{2}\right]}$$

The corresponding jitter tolerance plot becomes

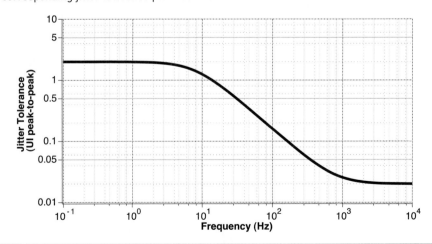

Exercises

14.27. Repeat the development of the BER expression shown in Eq. (14.162) for a sinusoidal jitter tolerance test but this time assume the PDF of output jitter is described by a uniform PDF from -SJ/2 to SJ/2.

ANS.
$$BER = \frac{1}{2}\left[1 - \frac{1}{|G(j\omega)|SJ_{IN}}\right],$$
$$|G(j\omega)|SJ_{IN} > 1\,UI$$

14.28. Repeat the development of the BER expression shown in Eq. (14.162) for a sinusoidal jitter tolerance test but this time assume the normalized sampling instant is an arbitray point between 0 and T.

ANS.
$$BER = \frac{1}{2}\left[1 + \frac{1}{\pi}\cos^{-1}\left(\frac{2UI_{TH}}{|G(j\omega)|SJ_{IN}}\right)\right.$$
$$\left. - \frac{1}{\pi}\cos^{-1}\left(\frac{2(UI_{TH} - 1)}{|G(j\omega)|SJ_{IN}}\right)\right],$$
$$|G(j\omega)|SJ_{IN} > 1\,UI$$

The above jitter tolerance test was developed from the perspective of a sinusoidal phase-modulated stimulus such as that used for SONET. One can just as easily develop the idea of a golden device behavior with respect to other types of input distributions. The reader can find several examples of this in the problem set at the end of this chapter.

During production a jitter tolerance test is conducted in a go/no-go fashion whereby the input DJ and RJ are first established by way of the transmission compliance mask, combined with the appropriate PRBS data pattern, and then applied to the device or system. The corresponding BER performance is then measured and compared to the expected level as specified in the communication standard (say BER_D). If the BER is less than or equal to BER_D over the entire frequency band, the device will pass the test. Owing to the number of BER measurements required during a jitter tolerance test, methods that speed up this test are often critical for a product's commercial success.

14.9 SUMMARY

In this chapter, we have reviewed the test principles related to clock and serial communication channel measurements. We began by studying the noise attributes of a clock signal in the time domain and then in the frequency domain. As with any noise process, statistical methods were used to quantify the level of noise present with the clock signal. This included some discussion on accumulated jitter, a concept related to the autocorrelation function of the jitter time sequence, and an aggregate power spectral density metric based on the average spectral behavior of the clock signal. Subsequently, the remainder of the chapter focused on measurement principles related to serial data communications. It began by introducing the concept of an eye diagram from which the idea of a zero crossing probability distribution was introduced. This was then expanded to include the effects of a real channel, such as noise and dispersion, leading to a classification of the various types of jitter one encounter in practice. Using a probabilistic argument, the probability of transmission error was introduced in terms of a probability density function for a jitter mechanism. The concept of bit error rate (BER) was introduced as the average probability of transmission error. Subsequently, the BER was modeled as a binomial distribution from which confidence intervals could be computed and related to test time. At this point in the chapter it became evident that a BER test requires excessive test time and the following section outlined different strategies to reduce the time of a BER test. All of these methods involved expressing the BER as a function of a model of the underlying jitter process and then solving for the parameters of this model at a low BER level where test time can be made short. A general method of modeling a Gaussian mixture was outlined and shown to be a powerful method of modeling various sources of jitter. Commonly used test metrics like random jitter (RJ), deterministic jitter (DJ), and total jitter (TJ) were defined. Methods to further decompose deterministic jitter were outlined. These methods were based on DSP techniques similar to the techniques used for analog and sampled-data channels. The chapter concluded with a short discussion on jitter transmission test whereby a jitter signal is injected into the channel and its effect on the output is quantified. This included a discussion on jitter transfer and jitter tolerance type tests.

PROBLEMS

14.1. A time jitter sequence is described by $J[n] = 0.001 \sin^2\left(14\pi/1024n\right) + 0.001$ for $n = 0,1,\ldots, 1023$. What is the mean, standard deviation, and peak-to-peak value of this time jitter sequence. Plot the histogram of this data sequence over a range of 0 to 0.002.

14.2. A time jitter sequence is described by $J[n] = 0.001 \sin\left(14\pi/1024n\right)$ for $n = 0,1,\ldots, 1023$. What is the period jitter and cycle-to-cycle jitter of this time jitter sequence.

14.3. A time jitter sequence is described by $J[n] = 0.001 \sin\left(14\pi/1024\,n\right) + 0.001 \times \left[1 - e^{-\left(\frac{3}{1024}n\right)}\right]$ for $n = 0,1,\ldots, 1023$. Plot and observe the behavior of the time, period, and cycle-to-cycle jitter. Comment on what you observe.

14.4. A time jitter sequence is described by $J[n] = 0.001 \sin\left(14\pi/64\,n\right) + 0.001 \times \left[1 - e^{-\left(\frac{3}{64}n\right)}\right]$ for all n. Plot and observe the behavior of the accumulated jitter sequence corresponding to this signal over a range of 64 delay instants using a sample set of 128.

14.5. A time jitter sequence is described by $J[n] = 0.001 \times \text{randn}() + 0.001 \times \left[1 - e^{-\left(\frac{3}{128}n\right)}\right]$ for all n. Here randn() represents a Gaussian random number with zero mean and unity standard deviation. Plot and observe the behavior of the accumulated jitter sequence for this signal over a range of 128 delay instants using a sample set of 256.

14.6. The PSD of the voltage level of a 100-MHz clock signal is described by $S_v(f) = 10^{-4}/10^4 + \left(f - 10^8\right) + 0.5 \times \delta(f - 10^{-8})$ in units of V²/Hz. What is the RMS noise level associated with this clock signal over a 1000-Hz bandwidth center around the first harmonic of the clock signal?

14.7. The PSD of the instantaneous phase of a 2.5-GHz clock signal is described by $S_\phi(f) = 4 \times 10^{-4}/10^4 + f^2$ in units of rad²/Hz. What is the RMS level associated with the phase of this clock signal over a 1000-Hz bandwidth?

14.8. The PSD of the voltage level of a 100-MHz clock signal is described by $S_v(f) = 10^{-4}/10^4 + (f - 10^8)^2 + 0.5 \times \delta(f - 10^8)$ in units of V²/Hz. What is the phase noise of the clock at a 10,000-Hz offset in dBc/Hz?

14.9. The PSD of the instantaneous phase of a 2.5-GHz clock signal is described by $S_\phi(f) = 4 \times 10^{-4}/10^4 + f^2$ in units of rad²/Hz. What is the phase noise of the clock at 10,000-Hz in dBc/Hz?

14.10. The PSD of the instantaneous phase of a 10-GHz clock signal is described by $S_\phi(f) = 10^{-5}/10^6 + f^2$ in units of rad²/Hz. What is the spectrum of a jitter sequence derived from this clock signal in s²/Hz? How about when expressed in UI²/Hz?

14.11. A phase jitter sequence expressed in radians can be described by the equation $\phi[n] = 0.1\sin(50\pi/1024n)$ for $n = 0,\ldots,1023$ was extracted from a 100-MHz clock signal using a 10-GHz sampling rate. Plot the phase spectrum in UI²/Hz.

14.12. A time jitter sequence expressed in nanoseconds can be described by the equation $J[n] = 0.1\sin(14\pi/1024n) + 0.1\sin 50\pi/1024n + 0.1\sin 124\pi/1024n$ for $n = 0,\ldots,1023$

was extracted from a 100-MHz clock signal using a 10-GHz sampling rate. Plot the phase spectrum in s²/Hz.

14.13. Determine the following list of convolutions:

(a) $\delta(t-a) \otimes \dfrac{1}{\sigma\sqrt{2\pi}} e^{-t^2/2\sigma^2}$,

(b) $\delta(t-a) \otimes \dfrac{1}{\sigma\sqrt{2\pi}} e^{-(t-\mu)^2/2\sigma^2}$,

(c) $\left[\dfrac{1}{2}\delta(t+a)+\dfrac{1}{2}\delta(t-a)\right] \otimes \left[\dfrac{1}{2}\delta(t+b)+\dfrac{1}{2}\delta(t-b)\right]$, and

(d) $\left[\dfrac{1}{4}\delta(f+a)+\dfrac{1}{4}\delta(f-a)+\dfrac{1}{4}\delta(f+2a)+\dfrac{1}{4}\delta(f-2a)\right] \otimes \left[\dfrac{1}{2}\delta(f+b)+\dfrac{1}{2}\delta(f-b)\right]$.

14.14. The PDF for the ideal zero crossings of an eye diagram can be described by $\dfrac{1}{2}\delta(t)+\dfrac{1}{2}\delta(t-10^{-8})$. If the PDF of the random jitter present at the receiver is $10^{-9}/\sqrt{2\pi}\, e^{-t^2/2\times10^{-18}}$, what is the PDF of the actual zero crossings?

14.15. The PDF for the zero crossings at a receiver due to inter-symbol interference can be described by $\dfrac{1}{4}\delta\left(t+10^{-9}\right)+\dfrac{1}{4}\delta\left(t-10^{-9}\right)+\dfrac{1}{4}\delta\left(t-10^{-8}+10^{-9}\right)+\dfrac{1}{4}\delta\left(t-10^{-8}-10^{-9}\right)$.
If the PDF of the random jitter present at the receiver is $10^{9}/\sqrt{2\pi}\, e^{-t^2/2\times10^{-18}}$, what is the PDF of the zero crossings?

14.16. In Chapter 5 we learned that the standard Gaussian CDF can be approximated by

$$\Phi(z) \approx \begin{cases} \left[1-\left(\dfrac{1}{(1-\alpha)z+\alpha\sqrt{z^2+\beta}}\right)\dfrac{1}{\sqrt{2\pi}}e^{\frac{-z^2}{2}}\right] & 0 \le z < \infty \\[4mm] \left[\left(\dfrac{1}{-(1-\alpha)z+\alpha\sqrt{z^2+\beta}}\right)\dfrac{1}{\sqrt{2\pi}}e^{\frac{-z^2}{2}}\right] & -\infty < z < 0 \end{cases}$$

where

$$\alpha = \dfrac{1}{\pi} \quad \text{and} \quad \beta = 2\pi$$

Compare the behavior of this function to the exact CDF function using a built-in routine found in MATLAB, Excel or elsewhere defined by

$$\Phi(z) = \dfrac{1}{\sqrt{2\pi}} \int_{-\infty}^{z} e^{\frac{-u^2}{2}}\, du$$

How well does these function compare for very small values of $\Phi(z)$? How about large values? Plot the difference between these two functions for -10 < z < 10. What is the maximum error in percent?

14.17. A logic 0 is transmitted at a nominal level of 0 V and a logic 1 is transmitted at a nominal level of 1.2 V. Each logic level has equal probability of being transmitted. If a 100-mV RMS Gaussian noise signal is assumed to be present at the receiver, what is the probability of making an single trial bit error if the detector threshold is set at 0.6 V. Repeat for a detector threshold of 0.75 V.

14.18. A logic 0 is transmitted at a nominal level of -1.0 V, and a logic 1 is transmitted at a nominal level of 1.0 V. If a 250-mV RMS Gaussian noise signal is assumed to be present at the receiver, what is the probability of making an single trial bit error if the detector threshold is set at 0.0 V. Assume that a logic one has twice the probability of being transmitted than a logic zero.

14.19. Data are transmitted to a receiver at a data rate of 2 Gbits/s through a channel that causes the zero crossings to vary according to a Gaussian distribution with zero mean and a 50-ps standard deviation. Assume that the 0 to 1 transitions are equally likely to occur as the 1 to 0 transitions. What is the probability of error of a single event if the sampling instant is set midway between bit transitions? What about if the sampling instant is set at 0.4 ns?

14.20. Describe the PDF of the zero crossings of an ideal receiver operating at a data rate of 1 Gbps when twice the number of logic 0's are transmitted than the logic 1's.

14.21. The set of zero crossings appearing on an oscilloscope corresponding to an eye diagram measurement is described by the following dual-Gaussian PDF:

$$pdf = \frac{1}{4}\frac{1}{\sigma\sqrt{2\pi}}e^{-t^2/2\sigma^2} + \frac{3}{4}\frac{1}{\sigma\sqrt{2\pi}}e^{1(t-1000\,ps)^2/2\sigma^2}$$

where $\sigma = 50$ ps. What is the probability of error of a single event if the normalized sampling instant is set at UI/4.

14.22. Data are transmitted to a receiver at a data rate of 2 Gbits/s through a channel that causes the zero crossings to vary according to a multi-Gaussian distribution described by

$$pdf = \frac{1}{4}\frac{1}{\sigma\sqrt{2\pi}}e^{-(t+5\,ps)^2/2\sigma^2} + \frac{1}{4}\frac{1}{\sigma\sqrt{2\pi}}e^{-(t-4\,ps)^2/2\sigma^2} + \frac{1}{4}\frac{1}{\sigma\sqrt{2\pi}}e^{-(t-500\,ps+5\,ps)^2/2\sigma^2}$$
$$+ \frac{1}{4}\frac{1}{\sigma\sqrt{2\pi}}e^{-(t-500\,ps-5\,ps)^2/2\sigma^2}$$

What is the probability of error of a single event if the sampling instant is set midway between bit transitions when $\sigma = 10$?

14.23. A series of 10 transmission measurements was conducted involving 10^{10} bits operating at a data rate of 600 Mbits/s resulting in the following list of transmisson errors: 5, 6, 4, 6, 3, 7, 3, 6, 1, 0. What is the 95% confidence interval for BER associated with this transmission system? How long in seconds did the test take to complete?

14.24. Data are transmitted at a rate of 5 Gbps over a channel with a single-bit error probability of 10^{-8}. If 10^9 bits are transmitted, what is the probability of no bit errors, one bit error, two bit errors, and ten bit errors. What is the average number of errors expected?

14.25. Data are transmitted at a rate of 10 Gbps over a channel with a single-bit error probability of 10^{-11}. If 10^{12} bits are transmitted, what is the probability of less than or equal to one bit error, less than or equal to two bit errors, and less than or equal to ten bit errors?

14.26. Demonstrate that the Possion approximation for the binomial distribution is valid over the range of N_T between 10^9 to 10^{14} for a BER of 10^{-12}:

$$\sum_{k=0}^{N_E}\left(\frac{N_T!}{k!(N_T-k)!}\right)BER^k(1-BER)^{N_T-k} \approx \sum_{k=0}^{N_E}\frac{1}{k!}(N_T BER)^k e^{-N_T BER}$$

14.27. A system transmission test is to be run whereby a BER $< 10^{-12}$ is to be verified. How many samples should be collected such that the desired BER is met with a CL of 97% when no more than 10 errors are deemed acceptable. What is the total test time if the data rate is 4 Gbps?

14.28. A system transmission test is to be run whereby a BER < 10^{-11} is to be verified at a confidence level of 99% when 3 or less bit errors are deemed acccptable. What are the minimum and maximum bit lengths required for this test? Also, what is the minimum and maximum time for this test if the data rate is 1 Gbps? Does testing for a fail part prior to the end of the test provide significant test time savings?

14.29. Draw the schematic diagram of a 5-stage LFSR that realizes a primitive generating polynomial.

14.30. Write a computer program to generate a maximum length sequence corresponding to a 5-stage LFSR. What is the expected length of the repeating pattern produced by this LFSR? Is the sequence periodic? How do you check?

14.31. Draw the schematic diagram of a 12-stage LFSR that realizes a primitive generating polynomial.

14.32. Write a computer program to generate a maximum length sequence corresponding to a 12-stage LFSR. What is the expect length of the repeating pattern produced by this LFSR? What is the value of the seed used to generate this sequence? Is the sequence periodic?

14.33. A PRBS sequence of degree 3 is required for a BER test. Determine the unique test pattern for a seed of 101. Compare this pattern to one generated with a seed of 001. Are there any common attributes?

14.34. A system transmission test is to be run whereby a BER < 10^{-12} is to be verified using the amplitude-based scan test method. The threshold voltage of the receiver was adjusted to the levels –200, –100, 100 and 200 mV, and the corresponding BER levels were measured: 5×10^{-6}, 5×10^{-9}, 5×10^{-9}, and 5×10^{-6}. What are the levels of the received logic values? Does the system meet the BER requirements if the reciever threshold is set at 10 mV?

14.35. A digital system is designed to operate at 2.5 Gbps and have a nominal logic 1 level of 3.3 V and a logic 0 at 0.1 V. If the threshold of the receiver is set at 1.65 V and the noise present at the receiver is 180 mV RMS, select the levels of the amplitude scan test method so that the system can be verified for a BER < 10^{-12} at a confidence level of 99.7% and that the total test time is less than 0.5 s.

14.36. A digital sampling oscilloscope obtained the following eye diagram while observing the characteristics of a 1-Gbps digital signal. Histograms were obtain using the built-in function of the scope at a sampling instant midway between transitions (i.e., at the maximum point of eye opening). Detailed analysis revealed that each histogram is Gaussian. One histogram has a mean value of 2.8 V and a standard deviation of 150 mV. The other has a mean of 120 mV and a standard deviation of 75 mV. Estimate the BER level when the threshold level is set at 2.0 V.

14.37. A digital sampling oscilloscope was used to determine the zero crossings of an eye diagram. At 0 UI, the histogram is Gaussian distributed with zero mean and a standard deviation of 80 ps. At unity UI, the histogram is again Gaussian distributed with zero mean value and a standard deviation of 45 ps. If the unit interval is equal to 1 ns, what is the expected BER associated with this system at a normalized sampling instant of 0.5 UI?

14.38. The BER performance of a digital receiver was measured to be 10^{-16} at a bit rate of 20 Gbps having a sampling instant at one-half the bit duration. If the sampling instant can vary by ±10% from its ideal position, what is the range of BER performance expected during production?

14.39. A digital system operates with a 20-GHz clock. How much RJ can be tolerated by the system if the BER is to be less than 10^{-14} and the DJ is no greater than 3 ps? Assume that the sampling instant is in the middle of the data eye and the total jitter is modeled as a dual-Dirac PDF.

14.40. A digital system operates with a 10-GHz clock. How much DJ can be tolerated by the system if the BER is to be less than 10^{-14} and the RJ is no greater than 5 ps? Assume that the sampling instant is in the middle of the data eye. Assume that the sampling instant is in the middle of the data eye and the total jitter is modeled as a dual-Dirac PDF.

14.41. A system transmission test is to be run whereby a BER $< 10^{-14}$ is to be verified using the dual-Dirac jitter-decomposition test method. The system operates with a 10-GHz clock, and the following BER measurements are at two different sampling instances: BER $= 0.25 \times 10^{-5}$ at 29 ps, and BER $= 0.25 \times 10^{-7}$ at 34 ps. Assume that the digital system operates with a sampling instant in the middle of the data eye. Does the system meet spec?

14.42. A system transmission test was run and the RJ and DJ components were found to be 100 ps and 75 ps, respectively. What is the TJ component at a BER level of 10^{-12}?

14.43. A system transmission test was run and the RJ and DJ components were found to be 100 ps and 75 ps, respectively. What is the TJ component at a BER level of 10^{-14}?

14.44. A system transmission test was run where the RJ component was found to be 10 ps and the TJ component at a BER level of 10^{-12} was found to be 150 ps. What is the DJ component?

14.45. The PDF for total jitter is described by the three-term Gaussian model as follows:

$$pdf = \frac{1}{4}\frac{1}{(2\,ps)\sqrt{2\pi}}e^{-(t+5\,ps)^2/2(2\,ps)^2} + \frac{1}{2}\frac{1}{(3\,ps)\sqrt{2\pi}}e^{-(t-1\,ps)^2/2(3\,ps)^2} + \frac{1}{4}\frac{1}{(4\,ps)\sqrt{2\pi}}e^{-(t-5\,ps)^2/2(4\,ps)^2}$$

For low BER levels, write an expression of the BER level as a function of the sampling instant. What is the BER at a sampling instant of 30 ps, assuming a 1-Gbps data rate?

14.46. The PDF for total jitter is described by the three-term Gaussian model as follows:

$$pdf = \frac{1}{4}\frac{1}{(2\,ps)\sqrt{2\pi}}e^{-(t+5\,ps)^2/2(2\,ps)^2} + \frac{1}{2}\frac{1}{(3\,ps)\sqrt{2\pi}}e^{-(t-1\,ps)^2/2(3\,ps)^2} + \frac{1}{4}\frac{1}{(4\,ps)\sqrt{2\pi}}e^{-(t-5\,ps)^2/2(4\,ps)^2}$$

What are the DJ, RJ, and TJ metrics at a desired BER of 10^{-12}?

14.47. The jitter distribution at the transmit side of a communication channel can be described by a four-term Gaussian mixture having the following parameters:

$$0.3 \times N\,(-100\ ps,\ 50\ ps) + 0.2 \times N\,(-10\ ps,\ 60\ ps) + 0.2 \times N\,(50\ ps,\ 60\ ps)$$
$$+ 0.3 \times N\,(120\ ps,\ 50\ ps)$$

What is the RJ, DJ, and TJ at a BER of 10^{-12} assuming a bit rate of 600 Mbps?

14.48. The jitter distribution at the receiver side of a communication channel can be described by a four-term Gaussian mixture having the following parameters:

$$0.2 \times N\,(-12\ ps,\ 5\ ps) + 0.3 \times N\,(-5\ ps,\ 6\ ps) + 0.3 \times N\,(4\ ps,\ 6\ ps) +$$
$$0.2 \times N\,(11\ ps,\ 7\ ps)$$

What is the RJ, DJ, and TJ at a BER of 10^{-12}, assuming a bit rate of 2 Gbps?

14.49. Extract a Gaussian mixture model using the EM algorithm from a data set synthesized from a distribution consisting of three Gaussians with means -0.04 UI, 0.01 UI, and 0.1 UI, standard deviations of 0.05 UI, 0.03 UI, and 0.04 UI and weighting factors 0.4, 0.2, and 0.4, respectively. How does extracted models compare with the original data set?

14.50. The jitter distribution at the receiver end of a communication channel consisting of 10,000 samples was modeled by a four-term Gaussian mixture using the EM algorithm and the following model parameters were found:

$$0.2 \times N\,\big(-12\ ps, 5\ ps\big) + 0.3 \times N\,\big(-5\ ps, 6\ ps\big) + 0.3 \times N\,\big(4\ ps, 6\ ps\big) + 0.2 \times N\,\big(11\ ps, 7\ ps\big)$$

Assuming that the unit interval is 0.12 ns and that the data eye sampling instant is half a UI, what is the BER associated with this system? Does the system behave with a BER less than 10^{-12}?

14.51. A time jitter sequence can be described by the following discrete-time equation: $J[n] = 2\sin(2\pi\, 51/1024n)$. Sketch the probability density function for this sequence.

14.52. A PRBS sequence of degree 5 is required to excite a digital communication channel. Using a sequence length of 1024 bits, together with a periodic PRBS sequence, what is the spectrum of this pattern? If any tonal behavior is present, what are the RMS values and in which bins of the FFT do these tones appear?

14.53. A jitter sequence consisting of 512 samples was captured from a serial I/O channel driven by two separate data patterns, in one case, an alternating sequence of 1's and 0's was used, and in the other a PRBS pattern was used. An FFT analysis performed on the two separate sets of jitter data, as well as with the PRBS sequence, resulted in the information below. Determine the test metrics: PJ, DDJ, BUJ.

Spectral Bin in Nyquist Interval	|XJ| Complex Spectral Coefficients (UI)		
	1010 Clock-like Input	PRBS Input	
		Theory	Actual
0	0.01	0	5×10^{-3}
1, ..., 256 except	$\sim 1 \times 10^{-5}$	0	$\sim 4 \times 10^{-4}$
33	0	$0.041 - j\,0.026$	$0.05 - j\,0.03$
91	0	$0.01 + j\,0.02$	$0.03 + j\,0.03$
153	0	$0.006 - j\,0.007$	$0.005 - j\,0.006$
211	0	0	$0.001 - j\,0.02$
251	$0.03 - j\,0.09$	0	$0.028 + j\,0.033$

14.54. The RJ, SJ, ISI, and DCD metrics are 5 ps rms, 20 ps pp, 6 ps pp, and 3 ps pp, respectively. Write an expression for the PDF for each jitter component. Provide a sketch of the total jitter distribution and the zero-crossing PDF assuming a bit duration of 200 ps.

14.55. The RJ, ISI, and DCD metrics are 5 ps rms, 6 ps pp, and 3 ps pp, respectively. Write an expression for the PDF of the total jitter distribution.

14.56. A digital receiver operates at a clock rate of 800 MHz. A jitter transfer test needs to be performed with a 3-kHz phase-modulated sinusoidal signal having an amplitude of 0.001 radians. Select the values of M_c, M_J, and N such that the samples are coherent with a 5-GHz sampling rate.

14.57. A digital receiver operates at a clock rate of 500 MHz. A jitter transfer test needs to be performed with a 5-kHz phase modulated sinusoidal signal having an amplitude of 1.0 UI. Using an AWG with a sampling rate of 10 GHz, write a short routine that generates the AWG samples.

14.58. Compute the discrete Hilbert transform of the sequence: $d[n] = 1 \times \sin(2\pi\, 11/1024)n$. Plot each sequence on the same time axis and compare.

14.59. A 2048-point data set is collected from the following two-tone unity-amplitude phase-modulated jitter sequence:

$$d_{OUT}[n] = \sin\left[2\pi\frac{601}{2048}(n-1) + 1 \times \sin\left[2\pi\frac{51}{2048}(n-1)\right] + 1 \times \sin\left[2\pi\frac{11}{2048}(n-1)\right]\right].$$

Extract the jitter sequence using the analytic signal approach and compare this sequence to the ideal result.

14.60. A golden device operating at a bit rate of 100 MHz has an input–output phase transfer function described as follows: $\Phi(s) = s(s + 18849.55)/s^2 + 18849.55s + 1.184352 \times 10^8$. Plot the expected jitter tolerance mask for a sinusoidal phase-modulated input signal with a BER of 10^{-12}.

REFERENCES

1. M. P. Li, *Jitter, Noise, and Signal Integrity at High-Speed*, Prentice Hall, Pearson Education, Boston, 2008.

2. Application Note, *Advanced Phase Noise and Transient Measurement Techniques*, Agilent Technologies, 2007, 5989-7273EN.

3. J. A. McNeil, *A simple method for relating time- and frequency-domain measures* of *oscillator performance*, in *2001 IEEE Southwest Symposium on Mixed Signal Design*, Austin, TX, February, pp. 7–12, 2001.

4. Maxim Application Note: Statistica; Confidence Levels for Estimating Error Probability, download, at http://pdfserv.maxim-ic.com/en/an/AN1095.pdf.

5. T. O. Dickson, E. Laskin, I. Khalid, R. Beerkens, J. Xie, B. Karajica, and S. P. Voinigescu, An 80-Gb/s 231-1 pseudo-random binary sequence generator in SiGe BiCMOS technology," *IEEE Journal of Solid-State Circuits*, **40**(12), pp. 2735–2745, December 2005.

6. M. Rowe, BER measurements reveal network health, *Test & Measurement World*, July 1, 2002.

7. K. Willox, Q factor: The wrong answer for service providers and equipment manufacturers, *IEEE Communications Magazine*, **41**(2), pp. S18–S21, February 2003.

8. Application Note, *Jitter Analysis: The Dual-Dirac Model, RJ/DJ, and Q-Scale*, Agilent Technologies, December 2004, http://cp.literature.agilent.com/litweb/pdf/5989-3206EN.pdf.

9. M. P. Li, J. Wilstru, R, Jessen, and D. Petrich, A new method for jitter decomposition through its distribution tail fitting, in *IEEE Proceedings, International Test Conference Proceedings*, pp. 788–794, October 1999.

10. A. P. Dempster, N. M. Laird, and D. B. Rubin, Maximum likelihood from incomplete data via the EM algorithm. *Journal of the Royal Statistical Society: Series B*, **39**(1), pp. 1–38, November 1977.

11. F. Nan, Y. Wang, F. Li, W. Yang, and X. Ma, Better method than tail-fitting algorithm for jitter separation based on gaussian mixture model, *Journal of Electronic Testing*, **25**(6), pp. 337–342, Dec. 2009.

12. N. Vlassis and A. Likas, A kurtosis-based dynamic approach to Gaussian mixture modeling, *IEEE Trans. on Systems, Man, and Cybernetics—Part A: Systems and Humans*, **29**(4), pp. 393–399, July 1999.

13. S. Aouini, *Gaussian Mixture Parameter Estimation Using the Expectation-Maximization Algorithm*, ECSE 509 Technical Report, McGill University, November 2004.

14. B. Veillette and G. W. Roberts, Reliable analog bandpass signal generation, in *proceedings of the IEEE International Symposium on Circuits and Systems*, Monterey, CA, June 1998.

15. S. Aouini, K. Chuai and G. W. Roberts, A low-cost ATE phase signal generation technique for test applications, in *Proceedings of the 2010 IEEE International Test Conference*, Austin, TX, November 2010.

16. C. M. Miller and D. J. McQuate, Jitter analysis of high-speed digital systems, *Hewlett-Packard Journal*, pp. 49–56, February 1995.

17. T. J. Yamaguchi, M. Soma, M. Ishida, T. Watanabe and T. Ohmi, Extraction of instantaneous and RMS sinusoidal jitter using an analytic signal method, *IEEE Transactions on Circuits and Systems II: Analog and Digital Signal Processing*, **50**(6), pp. 288–298, June 2003.

18. S. C. Kak, The discrete Hilbert transform, *Proceedings of the IEEE*, **58**(4), pp. 585–586, April 1970.

CHAPTER 15

Tester Interfacing—DIB Design

The device interface board (DIB) customizes the generic ATE to a specific device under test (DUT). With the wide variance of DUT, pure digital, pure analog, mixed signal up to high speed and RF devices, the DIB requirements lead to a high variance of design requirements. This chapter will give an overview on DIB design for a wide range of DUTs but discuss specific requirements for high-speed and high-frequency devices.

15.1 DIB BASICS

15.1.1 Purpose of a Device Interface Board

On any given day, a general-purpose ATE tester may be required to test a wide variety of device types. A mixed-signal tester may test video converters in the morning, modem chips in the afternoon, and standalone ADCs in the evening. Obviously, the electrical testing requirements of each type of device are unique to that device. Also, the mechanical requirements of each device are unique. The tester's various electrical resources must be connected to each of the DUT's pins, regardless of the mechanical configuration of the DUT package. For example, an 8-bit DAC might be available in several different packages such as the small outline IC (SOIC), quad flat pack (QFP), leadless chip carrier (LCC), and a chip scale package (CSP). Or this DAC can be part of a system on chip (SOC) in a ball grid array (BGA) of leadless quad flat pack (QFN). These package types are illustrated in Figure 15.1. Also, the tester needs to be connected to the bare die during wafer probing. Clearly, a general-purpose tester cannot be expected to provide all electrical resources and mechanical fixtures to test any arbitrary device type in any package.

The device interface board (DIB) provides a means of customizing the general-purpose tester to specific DUTs and families of DUTs. The DIB serves two main purposes. First, it gives the test engineer a place to mount DUT-specific circuitry that is not available in the ATE tester. This circuitry can be placed near the DUT to enhance electrical performance during critical tests. Second, the DIB provides a temporary electrical interface to each DUT during electrical performance testing. When testing packaged devices, the temporary connection is achieved using a hand-test

Figure 15.1. Common IC package types.

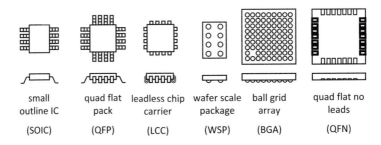

small outline IC	quad flat pack	leadless chip carrier	wafer scale package	ball grid array	quad flat no leads
(SOIC)	(QFP)	(LCC)	(WSP)	(BGA)	(QFN)

Figure 15.2. Electromechanical relays modify the DUT's electrical environment.

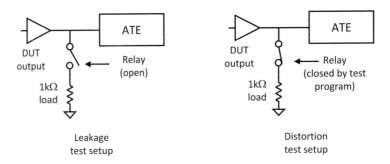

socket or a handler-specific mechanism called a *contactor assembly*. For this reason the DIB is called often also as HIB (handler interface board). Thus a DIB is often intended for use with only one type of DUT mounted in a particular mechanical package. When testing bare die on a wafer, the temporary DUT connection is made using the tiny probes of a probe card. A probe interface board (PIB) is usually required to interface the probe card to the tester's resources. Together, the PIB and probe card serve the same purpose as a DIB and contactor assembly. If the same device is offered in three different packages, then three different DIBs and a PIB may be required. Clearly, electromechanical hardware design represents a large portion of the test-engineering task.

DUTs that are purely digital in nature typically require a very simple DIB that simply provides point-to-point connectivity between the DUT pins and the tester's power supplies and digital pin card electronics. RF, analog and mixed-signal DUTs usually require much more elaborate DIBs. An RF DIB and PIB often require a matching circuit to adapt the tester impedance to the device impedance as well as decoupling circuits to provide a good device ground to accomplish the wave distribution of the stimulus and measure signals. A lower-frequency mixed-signal DIB often contains a variety of active and passive circuits that must be connected to or disconnected from various DUT pins as the test program progresses. For example, the harmonic distortion of an analog output may be specified with a 1-kΩ load connected between the analog output and ground. The same output may also have an off-state output leakage specification. The 1-kΩ load resistor must be disconnected during the leakage test to prevent current from leaking through the resistor to ground. Using electromechanical relays or solid state switches, the DIB can modify the DUTs electrical environment under test program control. The relays act as electrical switches that can be turned on and off by commands in the test program (Figure 15.2).

15.1.2 DIB Configurations

Thus far, we have talked about DIBs as if they are the same for each type of tester. In reality, the mechanical details of interface hardware vary widely from one tester type to another. Mechanical configurations may even vary within the same company, even when the various test development organizations all use the exact same tester. Figure 15.3 shows three possible interfacing schemes. The first is the simple DIB interfacing scheme. In this type of configuration, the DIB and contactor assembly form the entire interface between the tester and the DUT.

The second scheme shows a socket adapter/swap block stackup that is often used to test families of similar devices. In this configuration, the socket adapter (also called a *family board* or *mother board*) contains the support circuitry required to test a family of devices, such as video ADCs. The swap block (daughter board) provides the customization needed to test a particular video ADC in a particular type of package. For example, one swap block might be compatible with 64-pin LCCs while another is compatible with 64-pin QFPs. This scheme allows a relatively complex and expensive socket adapter to be reused for a family of similar DUTs. Of course, this scheme is not limited to only two layers of interfacing, but the attachment of a fully custom swap block to a semi-custom socket adapter is the most common multilayer configuration. The daughter board can be even mounted into a special machined mother board, so that the will become flat or as a stack-up as shown in Figure 15.3b.

A third possibility is the configuration board/DIB scheme. The configuration board is similar to the family board, except that it is located inside the test head. It is generally only intended to customize the tester for a particular organization's needs. Unlike the family board, which may

Figure 15.3. DIB interfacing schemes: (a) Simple DIB, (b) socket adapter/swap block, and (c) DIB and configuration board.

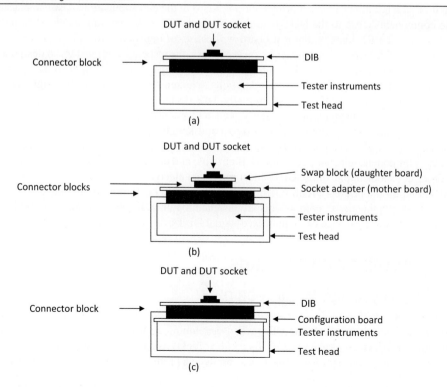

be changed daily, the configuration board is usually left in place for all device types. All DUT-specific circuitry is located on the DIB.

The terminology used to describe each layer of interface hardware varies widely from one ATE vendor to another. The terminology even varies from one vendor's tester to another. Throughout this book and for the remainder of this chapter, we will use the term DIB to refer to all the layers of custom and semi-custom interfacing hardware, including mother boards, daughter boards, swap blocks, and DIBs. In other words, we will treat the subject of DIB design as if all testers used the configuration illustrated in Figure 15.3.

15.1.3 Importance of Good DIB Design

One of the major causes of long test program development time is poor mixed-signal or RF DIB design and printed circuit board (PCB) layout. A DIB schematic shows only an idealized view of the DIB. Resistors are shown as ideal resistances, capacitors as ideal capacitances, and traces as perfect connections with no parasitic resistance, inductance, or capacitance. In reality, the exact mechanical layout of the components and traces on the DIB may make the difference between failing test results and passing results.

The performance of RF, analog, and mixed-signal devices is highly dependent on the quality of the surrounding circuit design. It is important to be able to distinguish between legitimate DUT failures and failures caused by poor design of the DIB. Consequently, a DIB should represent the best-case environment for the DUT, rather than a worst-case environment. With that said, the actual tests are more commonly designed to provide the worst-case conditions (e.g., supply voltage, input signal levels, input signal jitter, etc.) that the device is guaranteed to tolerate. Unfortunately, it is difficult to provide the DUT with a perfect environment using a general-purpose tester with bulky electromechanical interconnections. For example, the pins of the DUT socket will typically add more inductance and capacitance to the DUT's environment than the DUT will encounter when it is soldered directly onto a printed circuit board in the end application. Also the placement of the components close to the DUT is limited by the bulky socket causing a degradation of the performance. The RF lines on the test board will cause an impedance transformation when not matched to the DUT and ATE impedance. Nevertheless, the test engineer must try to design a DIB that does not present the DUT with unfair electrical handicaps.

There are so many performance considerations in mixed-signal and RF DIB design that many people consider it a mystical black art. Actually, DIB design is more of a "light gray" art, since many of the major considerations are fairly well understood. In this chapter, we will examine some of the main considerations in mixed-signal and RF DIB design such as transmission lines, matching circuits, power supply and grounding connections, shielding schemes, parasitic circuit elements, component selection, common DIB circuits, and common DIB mistakes.

First, let us look at one of the DIB's most important electrical components: the printed circuit board. Although the printed circuit board is often thought to be nothing more than a mechanical frame onto which the circuit components are fastened together, its physical construction is absolutely key to the performance of many mixed-signal DUTs.

15.2 PRINTED CIRCUIT BOARDS

15.2.1 Prototype DIBs Versus PCB DIBs

One of the common debates in test engineering is the choice between hand-wired prototype DIBs versus printed circuit board (PCB) DIBs when developing a DIB for lower-performing mixed-signal or digital devices. Hand-wired DIBs can be quickly constructed from prefabricated blank prototype boards. The alternate approach is to produce a production-worthy custom PCB version of the DIB without first building a hand-wired prototype. Each approach has advantages and disadvantages.

The hand-wired approach results in rapid turn-around at relatively low production cost. However, the resulting board is typically not very production-worthy, since the loose wires are easily broken. Also, hand-wired DIBs may not give the same high-quality electrical performance that can be achieved using PCB-based DIBs. When multiple DIBs are required or when the customer or the performance of the DUT requires, the PCB approach is usually the superior solution. PCB DIBs are easily manufactured in quantity, they are mechanically robust during debug and production, they provide superior electrical performance, and they provide good consistency (i.e., correlation) from one board to another. Correlation between hand-wired DIBs can be very problematic, since each is electrically unique depending on the exact length and physical layout of the wires on each board. At higher frequencies, hand-wired boards are often useless, since they can produce incorrect readings due to their inferior electrical characteristics. PCBs also have the advantage that commercial tools can be used to check their functionality with flying probes when being mass-produced.

The downside to PCB-based DIBs is primarily longer cycle time and higher initial cost. It may take several weeks to get a PCB DIB designed, laid out, and fabricated. Also, PCB DIBs are more expensive than hand-wired DIBs, at least in small quantities. However, assuming that the test engineer is skilled enough to produce a usable PCB DIB design on the first pass, the PCB DIB is actually a less expensive approach. After all, a PCB DIB will eventually be required for a robust production solution anyway.

Rapid turnaround is a problem that can be solved by good methodology. PCB-based DIBs can be designed, laid out, and fabricated in a matter of a week or two if the test engineer is skilled in the proper use of computer-aided design (CAD) tools. To achieve a rapid turnaround with minimal errors, a CAD-based design, layout, and fabrication approach must be established between the test organization and the PCB layout organization (which may either be an external vendor or an internal support group).

15.2.2 PCB CAD Tools

A streamlined PCB design and layout process requires the use of netlist-based CAD tools. A netlist is a database describing each interconnection in the circuit. For example, one line of a typical netlist file might tell the PCB layout tool that, for example, circuit node 55 interconnects resistor R1 pin 1, inductor L1 interconnects pin 2, and amplifier U37 interconnects pin 15. In addition to the point-to-point interconnection information, the netlist also includes such information as the footprint, or shape, of each component in the circuit as well as keeps out areas required for handler and prober docking interface. A footprint represents the mechanical specification of the component's package. Information such as (X,Y) pin locations, pad sizes, hole sizes, and package outline shapes to be printed on the finished PCB are included in the footprint description for each type of component.

Using a netlist-compatible schematic capture tool, the test engineer draws the circuit schematic on a computer workstation or PC. Then the schematic database (including the netlist) is transferred to the PCB designer for use in the DIB layout process. Once the netlist has been extracted from the database, the PCB designer begins laying out the DIB from a standard DIB template. The DIB template database represents a head start DIB design, which includes the shape of the DIB and its standard mechanical mounting holes as well as many preplaced standard components, such as tester connectors.

The netlist directs the PCB layout software to import all the required DIB components from a standard parts library. The PCB designer then places these components and connects them as shown in the schematic. The netlist prevents errors in point-to-point interconnections by refusing to let the layout designer place traces where they do not belong. The netlist also guarantees that none of the desired connections are mistakenly omitted. Once the DIB layout is completed, each layer of the design is plotted onto transparent film for use in PCB fabrication. These plots are commonly known as *Gerber* or *Gerber plots*, named after the company that pioneered some of the early plotting equipment (Gerber Scientific). Figure 15.4 illustrates the CAD-based DIB design, layout, and fabrication process.

15.2.3 Multilayer PCBs

Low-cost PCBs can be designed and fabricated using one or two layers of copper trace, as shown in Figure 15.5. Traces on opposite sides of a double-layer PCB can be connected using a copper plated through-hole called a *via*. Double-layer PCB fabrication starts with a blank PCB consisting of a sheet of insulator (e.g., fiberglass) plated with a thin layer of copper on both sides. The component lead holes and vias are drilled first. Then the holes are plated with copper to form the layer-to-layer interconnects. Finally, the traces are printed and etched using a photolithographic process similar to that used in IC fabrication. The via can be filled in a final step to have a solid surface and to avoid any problems when soldering components on the via pad.

Multilayer PCBs having four or more layers can be formed by stacking multiple two-layer boards together, as shown in Figure 15.6. The internal, or buried, layers are first printed and

Figure 15.4. CAD-based DIB design and fabrication process.

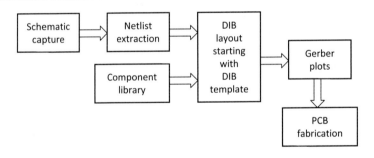

Figure 15.5. Single- and double-layer PCBs.

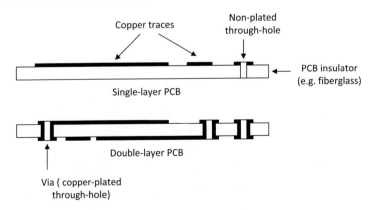

Figure 15.6. Multilayer PCBs.

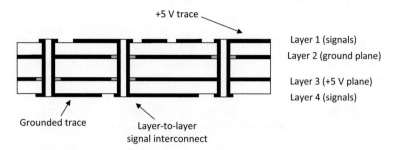

etched. The layers are all stacked and pressed together under heat to form a single board. Finally, the vias are drilled and plated and the outer layers are etched to form the finished PCB. In some instances, buried vias are added. These are vias that do not extend completely between the top and bottom layers, but rather go between sublayers of the PCB. These are often used to keep trace lengths short but add significant cost to the DIB.

Most mixed-signal DIBs are formed using 6- to 25-layer PCBs. The arrangement of layers in a PCB is known as the *stackup*. The stackup of a DIB may vary from one type of DUT to another, but some general guidelines are commonly followed. The internal layers are typically used for ground and power distribution, as well as for various noncritical signal traces. The outer layers are usually reserved for critical signals or those signal traces that might need to be modified after the DIB has been fabricated. External traces are also easier to access for observation during the test program debug process. If desired, test point vias can be added to a DIB to access buried signals for debugging purposes.

In addition to the trace layers and insulator layers in a PCB, the outer layers are usually coated with a material called a *solder mask*. This thin, nonconductive layer keeps solder from flowing all over the traces when the DIB components are soldered onto the PCB. The solder mask helps to prevent unwanted solder shorts between adjacent traces. Some test socket vendor require to have the solder mask excluded in the area the test socket is mounted onto the PCB.

A silkscreen pattern may also be printed on the outer layers of the PCB. The silkscreen patterns show the outline and reference numbers for all the DIB components, such as resistors, capacitors, relays, and connectors. The silkscreen patterns are quite useful during the DIB component assembly process, and they are equally useful during the test program debugging process.

15.2.4 PCB Materials

Printed circuit boards can be constructed using a variety of materials. The most common trace material is copper, due to its excellent electrical conductivity. The most common insulator material is FR4 (fire retardant, type 4) fiberglass. Fiberglass is an inexpensive material that exhibits good electrical properties up to several hundred megahertz. As frequencies approach 1 GHz, more exotic materials such as Teflon®* or cyanate ester may be the better choice.

Teflon® exhibits excellent RF characteristics, including low signal loss and an equally distributed dielectric constant. However, it suffers from poor mechanical stiffness. A DIB made exclusively of Teflon® insulator would be too weak to stand up to the force of DUT insertions by a handler. Cyanate ester is a material with reasonably good RF properties and yet it is stiff enough to withstand the mechanical stress of production testing. A hybrid stackup consisting of sandwiched

---

*Teflon® is a registered trademark of DuPont.

layers of Teflon® and cyanate ester provides a compromise between the good electrical properties of Teflon® and the good mechanical properties of cyanate ester.

15.3 DIB TRACES, SHIELDS, AND GUARDS

A trace on a PCB is the connection of two components. Other than in the schematics, however, this trace is not only a connection of two or more components, but rather needs to be treated as a component with parasitics itself, where the parasitics becomes more critical with higher frequencies. In the following section we discuss the basic parasitics like trace resistance, inductance, and capacitance; in a separate section we will discuss the trace as an RF component and its potential impact on the performance.

15.3.1 Trace Parasitics

One of the most important DIB components is the printed circuit board trace. It is easy to think that wires and traces are not components at all, but are instead represented by the connecting lines that appear in a schematic. However, PCB traces (and wires in general) are resistive, slightly inductive, and slightly capacitive in nature and have impedance, which can be calculated or simulated based on their geometry, proximity to another line, board thickness, and material constants. For RF boards these board traces become a component, which can even be used to match the DUT to the tester instruments.

The nonideal circuit characteristics are known as *parasitic elements*, though they are often simply referred to as *parasitics*. Often, trace parasitics can be ignored, especially when working with low frequencies and low to moderate current levels. Other times, the parasitics will have a significant effect on a circuit's behavior. The test engineer should always be aware of the potential problems that trace parasitics might pose.

Trace resistance on DIBs seldom exceeds a few ohms. Inductance can be anywhere from one or two nanohenries[*] (nH) to several microhenries (μH). Capacitance can range from one or two picofarads,[†] (pF) to tens of pF. Although these values are very approximate, they can be used as a thumbnail estimate to determine whether the parasitic elements might be large enough to affect the DUT's performance. To estimate trace parasitics with a little more accuracy, we need to review the equations for trace resistance, inductance, and capacitance.

15.3.2 Trace Resistance

The parasitic resistance of a PCB trace is directly proportional to the length of the trace, and inversely proportional to the height and width of the trace. The equation for resistance in a uniform conductive material with a rectangular cross section is

$$R = \frac{L_{Trace}}{\sigma W T} \tag{15.1}$$

where R is trace resistance, L_{TRACE} is trace length, W is trace width, T is trace thickness, and σ is the conductivity of the trace material. This resistance R is the DC resistance and not the same than the line impedance Z and is not considering the skin effect, which will have an effect at higher frequencies.

---

[*]Named in honor of the American scientist Joseph Henry (1797–1878), who discovered the electromagnetic phenomenon of self-inductance.
[†]Named in honor of the English chemist and physicist Michael Faraday (1791–1867), who contributed to the fields of electromagnetism and electrochemistry.

Most PCB traces are constructed using copper, which has a conductivity of about $5.7 \times 10^7 \, (\Omega \, \text{m})^{-1}$. The trace thickness is usually about 1 mil, although PCBs can be fabricated with a copper sheet thickness of 3 mils or more if desired. When working with equations such as Eq. (15.1), we will consistently convert all units of length to meters, since electrical units such as resistance, current, and voltage are metric units. Since the mil is an English unit (1 mil = 1/1000 in.), we will convert it to meters before using any of our electrical equations. The conversion factor is

$$1 \, \text{mil} = 25.4 \, \mu\text{m} \tag{15.2}$$

Often Eq. (15.2) is approximated by 1 mil = 1/39,000 meter.

EXAMPLE 15.1

Calculate the parasitic resistance of a PCB trace that is 15 in. long, 1 mil thick, and 20 mils wide.

Solution:
First we convert all units of length into meters

$$L_{TRACE} = 15 \, \text{in.} \times (1 \, \text{m} \, / \, 39 \, \text{in.}) = 0.385 \, \text{m}$$

$$T = 1 \, \text{mil} \times (1 \, \text{m} \, / \, 39{,}000 \, \text{mils}) = 2.56 \times 10^{-5} \, \text{m}$$

$$W = 20 \, \text{mil} \times (1 \, \text{m} \, / \, 39{,}000 \, \text{mils}) = 5.12 \times 10^{-4} \, \text{m}$$

Applying Eq. (15.1) to a copper trace with $\sigma = 5.7 \times 10^7 \, (\Omega \, \text{m})^{-1}$, we get a total parasitic trace resistance of

$$R = \frac{0.385}{5.7 \times 10^7 \times 5.12 \times 10^{-4} \times 2.56 \times 10^{-5}} = 515 \, \text{m}\Omega$$

15.3.3 Trace Inductance

The inductance of a DIB trace depends on the shape and size of the trace, as well as the geometry of the signal path through which the currents flow to and from the load impedance. Figure 15.7 shows a signal source feeding a load impedance through a pair of signal lines. In this example, the current is forced to return to the source through a dedicated current return line. The signal line and the current return line form a loop through which the load current flows. The larger the area of this loop, the higher the inductance of the signal path. This inductance can be modeled as a single inductor in series with the signal source, as shown in Figure 15.8.

A parasitic inductance such as the one in Figure 15.8 is generally an undesirable circuit component. We wish to minimize the effects of parasitic trace inductance on the DUT and DIB circuits. There are a number of ways to reduce this inductance. The first way is to minimize the area enclosed by the load current path. One easy way to do this is to lay a dedicated current return trace beside each signal trace. Of course, if we did this with every signal, we would have a very cluttered PCB layout. An easier way to obtain low inductance is to use one or more solid ground planes as current return paths.

By routing each signal trace over a solid ground plane, the load current can return underneath the trace along a path with very low cross-sectional area. The cross-sectional area can be minimized by placing the trace and ground plane very close together in the PCB layer stackup. Another way to reduce inductance is to make the trace as wide as is practical, since a wide trace over a

Figure 15.7. Signal source, load impedance, and current path.

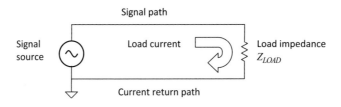

Figure 15.8. Schematic representation of signal path inductance.

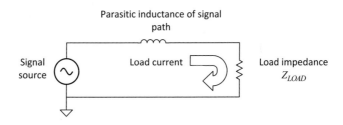

Figure 15.9. Cross section of a long trace over a ground plane (stripline).

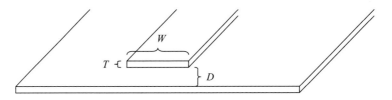

ground plane has minimal inductance. A thicker trace will also have somewhat less inductance, though PCBs are normally fabricated with trace thicknesses of 1 to 3 mils. We have less control over trace thickness than we have over trace width and layer spacing.

The inductance of a trace over a ground plane (examples are *microstrip* or *stripline* configurations) is dominated by the ratio of the trace-to-ground spacing, D, divided by the trace width, W (Figure 15.9). The parasitic inductance of a wide trace routed over a ground or power plane can be estimated using the equation

$$L_l = \mu_0 \mu_r \frac{D}{W} \tag{15.3}$$

where L_l is Inductance per unit length (H/m), μ_0 is magnetic permeability of free space (400π nH per meter), μ_r is magnetic permeability of the PCB material divided by μ_0, W is trace width, and D is separation between trace and ground plane.

Figure 15.10. Stripline trace inductance per meter versus D/W ratio ($\mu_r=1$).

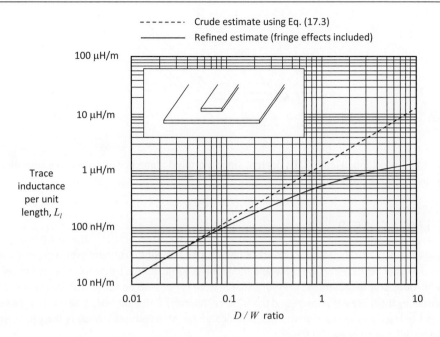

The value of μ_r is very nearly equal to 1.0 in all common PCB materials, so we can drop it from our calculations. The total inductance of the trace is directly proportional to the length of the trace

$$L = L_{TRACE} L_l \tag{15.4}$$

where L is total inductance and L_{TRACE} is trace length (meters).

Thus trace inductance increases as trace length increases and also increases as trace width decreases. Therefore, if we want to minimize parasitic inductance in PCB traces, we should make them as wide as possible, as short as possible, and as close to the ground or power plane as possible.

Unfortunately, Eq. (15.3) is only valid for traces in which $W \gg D$. In most PCB designs, the width of the trace is not much larger than the trace-to-ground spacing. In these cases, the magnetic fields between the trace and ground plane are not uniform, making Eq. (15.3) invalid. Figure 15.10 shows a more accurate relationship between the space-to-width ratio and the inductance per meter of a trace over a ground plane.[*] The dashed line represents the inductance per meter as estimated using Eq. (15.3). As we can see, the more accurate estimation converges with the estimations using Eq. (15.3) as the value of W becomes much larger than D.

It should be noted that the inductance of a stripline is approximately the same as the inductance of a trace over a second trace of equal size and shape. This is because most of the higher-frequency current returning through a stripline's ground plane returns directly underneath the stripline trace (i.e., the path of least inductance). However, the two-conductor configuration in

---

[*]The graph in Figure 15.10 was derived using a mathematical approximation, so its values should not be taken as absolutely accurate. However, the approximations are adequate for estimating the effects of parasitic elements on DIB circuits. The derivation of this graph and others in this chapter are based on electromagnetic field theory, a subject that is beyond the scope of this book.

Figure 15.11. Parallel traces on adjacent PCB layers.

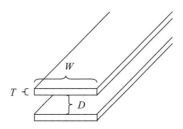

Figure 15.11 is seldom used in DIB design, since a ground plane permits a much easier means of achieving the same low inductance.

EXAMPLE 15.2

Calculate the parasitic inductance of a trace having a 16-mil width, running over a ground plane for 6 in. The spacing between the trace and the plane is 8 mils. If this trace is connected in series with a 50-Ω resistor to ground (as in Figure 15.12), what is the 3-dB bandwidth of the low-pass filter formed by the trace inductance and the 50-Ω resistance? Can a 50-MHz sine wave be passed through the trace to the resistor without significant loss of amplitude? What is the phase shift caused by the inductance at 50 MHz?

Figure 15.12. Low-pass filter formed by trace inductance and load resistance.

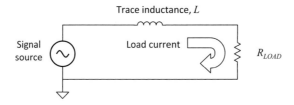

Solution:
Combining Eqs. (15.3) and (15.4), we write

$$L = 6 \text{ in.} \times \frac{1 \text{ m}}{39 \text{ in.}} \times \pi \times 400 \frac{\text{nH}}{\text{m}} \times 1.0 \times \frac{8 \text{ mils}}{16 \text{ mils}} = 97 \text{ nH}$$

However, we can see from the graph in Figure 15.11 that the actual inductance per meter at a D/W ratio of 0.5 is lower than that predicted from Eq. (15.3). If we use the more accurate value of inductance per meter from this graph (about 350 nH per meter), then we get a more accurate prediction of the trace inductance. The refined estimation of inductance is

$$L = 6 \text{ in.} \times \frac{1 \text{ m}}{39 \text{ in.}} \times 350 \frac{\text{nH}}{\text{m}} = 54 \text{ nH}$$

The trace inductance and load resistance form an *RL* low-pass filter, as shown in Figure 15.12. The 3-dB cutoff frequency, F_c, of this *RL* low-pass filter is given by

$$F_C = \frac{R_{LOAD}}{2\pi L} \qquad\qquad (15.5)$$

Thus the 3-dB bandwidth of the low-pass filter formed by the trace inductance and load resistor is equal to

$$F_C = \frac{50}{2\pi \cdot 54 \cdot 10^{-9}} = 147 \text{ MHz}$$

At a frequency *f*, the trace inductance and load resistance form a voltage divider having a transfer function equal to

$$H(f) = \frac{R_{LOAD}}{R_{LOAD} + Z_L(f)} \qquad\qquad (15.6)$$

where Z_L is the complex impedance of the trace inductance. Substituting $Z_L(f) = j2\pi f L$, calculating the magnitude of *H(f)*, and combining the result with Eq. (15.5) gives us the gain of the *RL* low-pass filter at any frequency *f*

$$\text{gain}(f) = |H(f)| = \left[\frac{1}{\sqrt{1 + \left(\dfrac{f}{F_C}\right)^2}}\right] \text{ V/V} \qquad\qquad (15.7)$$

At 50 MHz, the gain calculated using Eq. (15.7) is equal to = 0.947 V/V. Therefore, we get an attenuation due to the low-pass nature of the *RL* circuit that attenuates the 50-MHz sine wave by a factor of 0.947 V/V. This attenuation is probably unacceptable, unless we do not mind a 5% error in the amplitude of the signal at the load resistor.

The phase shift of the *RL* low-pass filter is given by

$$\Phi(f) = \angle H(f) = -\frac{180}{\partial}\tan^{-1}\left(\frac{f}{F_C}\right) \text{ degrees} \qquad\qquad (15.8)$$

Thus, at 50 MHz, the phase shift produced by the trace inductance and resistance is equal to −18.7 degrees. This is a fairly serious phase error. If we wanted to measure phase mismatch or group delay at a frequency near 50 MHz, the parasitic inductance of this example would be completely unacceptable.

If we want to achieve less attenuation and phase shift due to the trace inductance in Example 15.2, we must either shorten the trace, widen the trace, or reduce the spacing between the trace and ground. The spacing between the trace and ground in this example is 8 mils, which is about as thin as we can reliably fabricate a PCB. Rather than trying to fabricate a board with even thinner layer spacing, it is much easier to simply widen the trace.

As we will see in the following sections, widening the trace or reducing the trace-to-ground spacing has the unfortunate side effect of increasing the parasitic capacitance of the trace to ground. The extra capacitance may be just as undesirable as having too much inductance. Therefore, the

---

Exercises	

15.1. Calculate the parasitic resistance of a 5.7-in. PCB trace having a width of 2.5 mm and a thickness of 2 mils. If this trace feeds a 2.5-V DC signal to a 1-Ω load resistance, what will be the error of the voltage at the load as a percentage of the source voltage? (Assume a zero-resistance current return path.) How much power is dissipated by the trace?

ANS. $R = 18$ mΩ; $V_{ERR} = -44$ mV $= -1.77\%$; power $= 109$ mW.

15.2. Using Eq. (15.3), calculate the parasitic inductance of a 23 cm PCB trace having a width of 12 mils, and a spacing of 15 mils to the current-return ground plane. If this trace feeds a 125 kHz, 1.25-V RMS sinusoidal signal to a 10-Ω load resistance, what will be the error of the RMS voltage at the load as a percentage of the source voltage? (Assume zero trace resistance.) Compare your answers with those obtained using the refined inductance estimate of Figure 15.10.

ANS. Using Eq. (15.3), $L = 361.3$ nH and $V_{ERR} = -503$ μV $= -0.04\%$. Using Figure 15.10, $L = 115$ nH and $V_{ERR} = -50.9$ μV $= -0.004\%$. Eq. (15.3) yields significant error.

---

best solution is to keep traces as short as possible, since this reduces both the trace inductance and trace capacitance to ground.

15.3.4 Trace Capacitance

The capacitance between two parallel traces such as those in Figure 15.11 can be estimated using the standard parallel plate capacitance equation. The parasitic capacitance between two metal plates of area A is given by the equation

$$C = \varepsilon_r \varepsilon_0 \frac{A}{D} \tag{15.9}$$

where A is area of either plate ($L_{TRACE} \times W$ for rectangular traces), D is distance between the plates, ε_0 is electric constant* of free space (8.8542×10^{-12} F/m), and ε_r is relative dielectric constant† of the dielectric material between the plates.

The value of ε_r depends on the PCB insulator material. Air, for example, has a relative dielectric constant very near 1.0, while FR4 fiberglass has a relative dielectric constant of about $\varepsilon_r = 4.5$. Teflon®, by contrast, has a relative dielectric constant of about $\varepsilon_r = 2.7$. Therefore, Teflon® PCBs exhibit less capacitance per unit area than FR4 PCBs.

Equation (15.9) is only accurate for capacitor plates in which the length and width of the plates is much larger than the dielectric thickness, D. If W is about 10 times larger than D, then we can use Eq. (15.9) to estimate the capacitance per unit length of the trace

$$C_l = \frac{C}{L_{TRACE}} = \varepsilon_r \varepsilon_0 \frac{A/L_{TRACE}}{D} = \varepsilon_r \varepsilon_0 \frac{W}{D} \tag{15.10}$$

---

*The electric constant ε_0 is also known as vacuum permittivity or permittivity in free space.
†The (relative) dielectric constant ε_r is also known as relative permittivity.

Figure 15.13. Trace capacitance per meter versus D/W ratio ($\varepsilon_r = 4.5$).

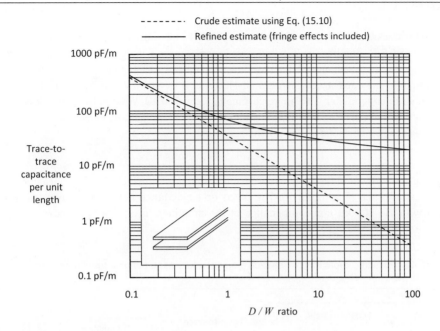

To calculate the total capacitance between two traces, we multiply the capacitance per unit length by the trace length. This is true for the configuration in Figure 15.11 as well as for any other configuration illustrated in this chapter.

$$C = L_{TRACE} \cdot C_l \tag{15.11}$$

When either the length or width is less than about 10 times the dielectric thickness, the so-called *fringe effects* in the electric field between the plates cause Eq. (15.10) to become inaccurate. Unfortunately, trace capacitance can seldom be accurately calculated using Eq. (15.10) since the width of the trace is often less than 10 times the trace to trace spacing. The graph in Figure 15.13 shows a more accurate estimation of the capacitance per meter between two parallel traces. The dashed line shows the capacitance per unit length as calculated by Eq. (15.10). Note that as the value of W becomes much larger than D, the refined estimation converges with the estimation from Eq. (15.10).

The chart in Figure 15.13 assumes a relative dielectric constant, ε_r, of $\varepsilon_r = 45$. If our PCB material has a different relative dielectric constant, ε', then we simply multiply the capacitance per unit length obtained from Figure 15.13 by the ratio of $\varepsilon'/4.5$ to calculate the correct capacitance per unit length. Equivalently, we can multiply the total capacitance (calculated using Eq. (15.11)) by $\varepsilon'/4.5$ to achieve the same result.

Example 15.3 shows how important it is to keep high-impedance nodes protected from potential sources of crosstalk. The best form of crosstalk prevention is to simply keep the sensitive trace as short as possible. Another method for reducing crosstalk is to place a ground plane underneath the critical signal traces, thus preventing layer-to-layer crosstalk such as that in Example 15.3. Each of the traces would then see a parasitic capacitance to ground, but the ground plane would block the trace-to-trace capacitance altogether. The effect of a ground plane on trace-to-trace capacitance is illustrated in Figure 15.15. The trace-to-trace capacitance is replaced by two parasitic capacitances to ground. This effectively shunts the offending source to ground so that it cannot inject its signal into the sensitive node.

EXAMPLE 15.3

An insulator thickness of 10 mils separates a pair of 12-mil-wide traces on adjacent layers in a multilayer FR4 DIB. One trace is 7 in. long, while the other is 5 in. long. The traces run directly over one another for a distance of 3 in., as shown in Figure 15.13, but do not cross each other at any other point. The upper trace carries a 10-MHz sine wave at 1.0 V RMS, while the lower trace is connected to an amplifier with an input impedance of 100 kΩ. What is the signal level of the crosstalk coupling from the 10-MHz signal into the input of the amplifier? Would a Teflon® PCB reducethe capacitance enough to give significantly better performance?

Solution:
First, we draw a model of the signal source and amplifier stage, including the parasitic trace-to-trace capacitance (Figure 15.14). The parasitic capacitance between the two traces will interact with the 100-kΩ input impedance of the amplifier to form a first-order high-pass filter.

Figure 15.14. Parasitic capacitance between a signal source and a high-impedance amplifier input.

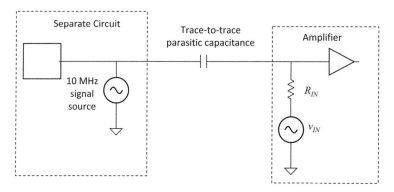

To estimate the value of the capacitance between the two traces, we first need to calculate the D/W ratio of this parasitic capacitor. The value of D is 10 mils, while the value of W is 12 mils. Thus the D/W ratio is equal to 0.833. From the graph in Figure 15.13 we can estimate a trace-to-trace capacitance of about 85 pF per meter. Applying Eq. (15.11), we calculate the total capacitance:

$$C = 3 \text{ in.} \cdot \frac{1 \text{m}}{39 \text{ in.}} \cdot 85 \text{pF/m} = 6.5 \text{ pF}$$

The 3-dB cutoff frequency of an RC high-pass filter is given by

$$F_c = \frac{1}{2\pi RC} \tag{15.12}$$

Using Eq. (15.12), we calculate the cutoff frequency

$$F_c = \frac{1}{2\pi \cdot 100 \times 10^3 \cdot 6.5 \times 10^{-12}} = 245 \text{ kHz}$$

Because the 10-MHz signal is well beyond the cutoff frequency, this would be a very bad DIB design. Using a Teflon® PCB rather than an FR4 fiberglass PCB, we would multiply the 6.5 pF capacitance by a factor of 2.7/4.5 (2.7 = relative permittivity of Teflon®, 4.5 = relative permittivity of FR4 fiberglass). This would result in a capacitance of 3.9 pF. The value of F_c would change to 408 kHz, which would not significantly reduce the crosstalk. To solve the crosstalk problem in this example, the traces must be moved farther away from one another. Also, the sensitive 100-kΩ line should be shortened to a fraction of an inch to minimize capacitive coupling from the 10-MHz signal source as well as any other potential sources of crosstalk.

The amount of capacitance between a trace and a ground plane (see Figure 15.9) is tricky to calculate. If the length and width of the trace are much larger than the trace-to-ground separation, then we can simply use Eq. (15.10) to calculate the capacitance per unit length. For most practical situations, though, the width of the trace is not much larger than the trace-to-ground separation. We again have to resort to a more accurate estimation, as shown in Figure 15.16. The dotted line shows the capacitance per unit length as calculated using Eq. (15.10). The solid line represents a more accurate calculation that takes the fringing effects of the electric fields into account. As expected, the two lines converge as W becomes much larger than D.

Next we consider the capacitance between two parallel traces on the same PCB layer (Figure 15.17). This configuration occurs very frequently in PCB designs, since many traces run parallel to each other for several inches on a typical DIB.

If the trace-to-trace spacing, S, is equal to or larger than the trace width, W, we can approximate this configuration as two circular wires having the same cross-sectional area as the traces and

Figure 15.15. Ground planes prevent layer-to-layer crosstalk.

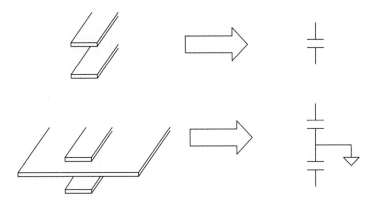

Figure 15.16. Trace capacitance per meter versus D/W ratio ($\varepsilon_r = 4.5$).

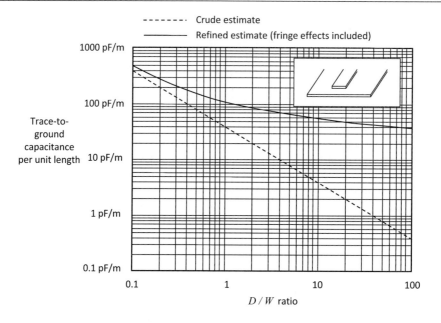

Figure 15.17. Coplanar PCB traces.

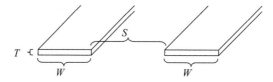

having a center-to-center spacing of $S+W$. The equation for the capacitance per unit length of two circular conductors having this geometry is given by[1]

$$C_l = \left[\frac{12.1 \times 10^{-12} \cdot \varepsilon_r}{\log_{10}\left[\frac{\left(\frac{S+W}{2}\right)}{\sqrt{\frac{T \cdot W}{\pi}}} + \sqrt{\left(\frac{\frac{S+W}{2}}{\frac{T \cdot W}{\pi}}\right)^2 - 1} \right]} \right] \tag{15.13}$$

where C_l is capacitance per unit length (F/m), ε_0 is electrical constant, ε_r is relative dielectric constant of the PCB material, W is width of the rectangular trace, and T is thickness of the rectangular trace.

Comparing this crude circular conductor approximation to a more accurate estimation based on flattened rectangular traces, we can see how closely the two approximations agree with one

Figure 15.18. Capacitance between two coplanar traces of equal width ($\varepsilon_r = 4.5$).

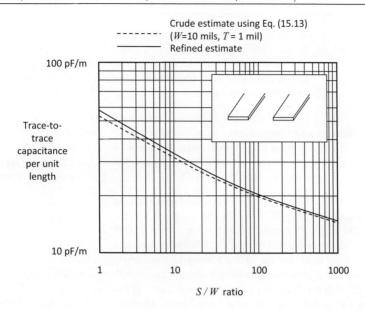

Figure 15.19. Coplanar traces over a ground plane.

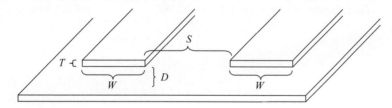

another (Figure 15.18). The dashed line shows the trace-to-trace capacitance per unit length using Eq. (15.13) while the solid line shows a more accurate calculation based on a flat trace geometry. These estimates are close enough to each other that Eq. (15.13) can probably be used in many cases as a reasonably good approximation.

We can reduce the effects of trace-to-trace crosstalk between coplanar traces using a ground plane. Figure 15.19 shows a pair of coplanar traces with a width of W separated from one another by a distance S and spaced a distance D over a ground plane.

The ground plane forms two parasitic capacitances to ground that serve to shunt the interference signal to ground (Figure 15.20). While this may not eliminate the crosstalk, it reduces it by a significant amount. From Figure 15.16 and Figure 15.18, we can see that the trace-to-ground capacitance will be several times larger than the trace-to-trace capacitance for values of $S > D$. Thus, if we lay out our traces so that the trace-to-trace spacing, S, is larger than our trace-to-ground spacing, D, we can make the shunt capacitance larger than the trace-to-trace coupling capacitor. This forms a capacitive voltage divider with good interference rejection.

When working with RF frequencies, a more prudent approach to quantifying the parasitics present with each interconnection is to simulate the structure using a CAD tool. This will not only

Exercises

15.3. Using Eq. (15.10), calculate the parasitic capacitance of a 14-cm-long, 25-mil-wide stripline trace with a spacing of 8 mils to the ground plane, fabricated on an FR4 PCB. Compare your answer with that obtained using the refined capacitance estimate of Figure 15.16.

ANS. Using Eq. (15.10), $C = 17.4$ pF; using Figure 15.16, $C = 28$ pF; Eq. (15.10) yields significant error.

15.4. Using Eq. (15.13), calculate the parasitic capacitance between two 4-cm-long, 16-mil-wide coplanar traces separated by a spacing of 30 mils, fabricated on an FR4 PCB. Compare your answer with that obtained using the refined capacitance estimate of Figure 15.18.

ANS. Using Eq. (15.13), $C = 1.61$ pF; using Figure 15.18, $C = 1.6$ pF; Eq. (15.13) agrees very well.

15.5. Using Eq. (15.10), calculate the parasitic capacitance of a 7.3-inch-long, 25-mil-wide stripline trace with a spacing of 10 mils to the ground plane, fabricated on a Teflon® PCB. This trace feeds a 50-kHz, 1.25-V RMS sinusoidal signal from a DUT output having a 100-kΩ output resistance to a buffer amplifier having an input capacitance of 2 pF. How much will the combined capacitance of the trace and buffer amplifier input capacitance attenuate the DUT signal? (Express your answer as a voltage gain in decibels.) Would this be an acceptable attenuation if the signal were the output of a gain test having ±0.8-dB limits? Compare your answer with that obtained using the refined capacitance estimate of Figure 15.16.

ANS. Using Eq. (15.10), $C = 11$ pF + 2 pF, G (50 kHz) = -0.69 dB, acceptable; using Figure 15.16, $C = 20$ pF + 2 pF, G(50 kHz) = −1.72 dB, not acceptable.

Figure 15.20. Equivalent circuit for coplanar traces over a ground plane.

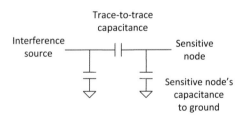

give better accuracy, but will also include effects caused by, for example, adjacent lines causing mutual inductance. We will have more to say about these CAD tools in a moment.

15.3.5 Shielding

Electrostatic shields can also be used to reduce coplanar trace-to-trace crosstalk. A shield is any conductor that shunts electric fields to ground (or a similar low-impedance node) so that the fields do not couple into a sensitive trace, causing crosstalk.[4] The electric fields can originate from external noise sources such as radio waves or 60-Hz power line radiation, or they can originate

Figure 15.21. Electrostatic shielding reduces trace-to-trace crosstalk.

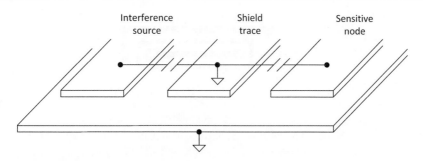

Figure 15.22. Shield trace routed around a sensitive node.

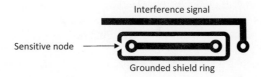

from other signals on the DIB. The ground plane in Figure 15.15 is one type of electrostatic shield. Ideally, a shield should completely enclose the sensitive node. A coaxial cable is one example of a fully shielded signal path. In most cases, it is impractical to completely shield every signal on a DIB using coaxial cables. However, we can achieve a close approximation of a fully shielded signal path by placing shield traces around sensitive signal traces. This configuration is called *coplanar shielding*.

Figure 15.21 illustrates how coplanar shielding can reduce crosstalk between a interference source and a sensitive DIB signal. The shield trace is connected to the ground plane to provide an extra level of protection for the sensitive node. Sometimes, a shield trace is routed all the way around a sensitive node, as illustrated in Figure 15.22. This type of shielding helps to reduce the coupling of electromagnetic interference from all directions.

15.3.6 Driven Guards

Electrostatic shields suffer from one small drawback. The shield forms a parasitic load capacitance between the sensitive signal and ground. The parasitic capacitance is both a blessing and a curse. It is a blessing because it shunts interference signals to ground, but a curse because it loads the sensitive node with undesirable capacitance. The capacitive loading problem can be largely eliminated using a driven guard instead of a shield. A driven guard is a shield that is driven to the same voltage as the sensitive signal. The guard is driven by a voltage follower connected to the sensitive node (Figure 15.23). The interference signal is shunted to the low-impedance output of the voltage follower, reducing its ability to couple into the sensitive signal node. A common mistake made by novice test engineers is to connect the tester's driven guards to analog ground. As seen in Figure 15.23, this is obviously a mistake.

The voltage follower drives the guard side of the parasitic load capacitance to the same voltage as the sensitive signal line. Since the parasitic load capacitance always sees a potential difference of 0 V, it never charges or discharges. Thus, the loading effects of the parasitic capacitance on the signal trace are eliminated by the voltage follower.

Figure 15.23. Driven guard—PCB layout and equivalent circuit.

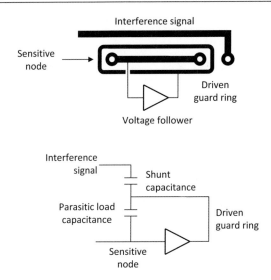

Of course, all voltage followers exhibit a finite bandwidth. Therefore, the parasitic capacitance can only be eliminated at frequencies within the voltage follower's bandwidth. For this reason, driven guards are typically used on relatively low frequency applications that cannot tolerate any crosstalk (e.g., high-performance audio circuits).

15.4 TRANSMISSION LINES

The previous section outlined the critical parameters of transmission lines for analog and mixed signal applications. This section will discuss the applications of transmission lines for RF circuits and device interface boards.

15.4.1 Various TEM Transmission Line Configurations

Here we will provide a brief summary of the most commonly used transmission line types for an RF DIB. This includes a coaxial line, a stripline, a microstrip line, a coplanar waveguide, and a coupled transmission line. The geometry of each line will offer its own advantages and disadvantages. A detail theoretical description is given in reference 5.

Coaxial Line

The coaxial line shown in Figure 15.24 provides perfect shielding, and its propagating energy does not radiate. This type of line is most commonly used to interface the ATE test head to the DIB. It generally consists of an SMA or SMP connector-type flexible cable assembly, which allows for flexible routing and easy access for debugging of RF circuits. The impedance of a coaxial line can be approximated by

$$Z = \frac{1}{2\pi}\sqrt{\frac{\mu}{\varepsilon}}\ln\left(\frac{D}{d}\right) \approx \frac{138\ \Omega}{\sqrt{\varepsilon_r}}\log\left(\frac{D}{d}\right) \tag{15.14}$$

Figure 15.24. Coaxial RF transmission line.

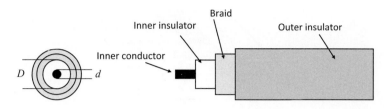

Figure 15.25. RF stripline with line width *W*, board thickness *D* and board dielectric constant ε_r.

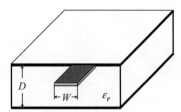

where *D* is the diameter of the braided conductor and *d* is the diameter of the inner conductor as shown in Figure 15.24.

Stripline

The stripline shown in Figure 15.25 is realized in printed circuit boards by placing the RF line symmetrically in between two parallel ground planes. These ground planes shield the RF lines when they are both connected through several vias. The stripline allows for low-cost integrated passive load board elements, like directional couplers and filters. The stripline configuration is most commonly used when high signal integrity is required and a well controlled via process is available to route the RF signal between the different board layers. A minor disadvantage is that striplines are not accessible for debugging and matching, and any access to this line will be made with a via, which is a discontinuity in itself.

Microstrip Line

Another variation of a stripline is as an open center conductor trace on the top layer of the PCB just above a ground plane. This is called a microstrip transmission line, and it is the most commonly used transmission line geometry for RF circuits (Figure 15.26). This configuration allows easy access for debugging purposes, and it allows for the addition of components to tune the RF performance when necessary if a GND plane is placed just below it.

One significant difference between the microstrip line and the stripline, and coaxial line configuration is that the dielectric medium through which the electromagnetic wave is traveling is not homogeneous, since parts of the EM wave will be in the PCB board with dielectric constant ε_r and the other part will be in the free air space above the line, which causes different phase velocities and wavelengths (see also Section 12.2.1). This effect is called dispersion. Since the EM wave propagates in two different dielectric media, it is necessary to calculate the effective dielectric constant $\varepsilon_{r,\,eff}$, which lies somewhere between the values of the relative dielectric constant of the PC board, and substrate and air. The effective dielectric constant is also dependent on the line

Figure 15.26. RF microstrip line with line width W, board thickness D, and board dielectric constant ε_r.

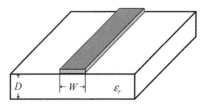

Figure 15.27. Coplanar waveguide with adjacent GND planes.

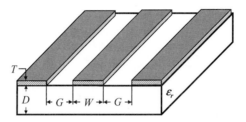

width W and the thickness of the dielectric material D. For higher frequencies, even the line thickness and surface roughness has an impact on the propagating parameter. There are multiple programs on the market for calculating or analyzing line parameters. Some of the most commonly used ones are LINECALC of the Agilent EESOF package, AWR's TXLINE, or Ansoft's TRL.

Coplanar Waveguide

Coplanar waveguide (CPW) was invented in 1969 by Cheng P. Wen, who worked for RCA Laboratories. CPW is formed from a conductor separated from a pair of adjacent ground planes as shown in Figure 15.27. For the ideal case, the thickness of the dielectric D is infinite, while in practice it needs to be thick enough so that the entire EM field will fit within the substrate. A variance of the CPW is a grounded coplanar waveguide (GCPW), where a ground plane is positioned on the opposite site of the substrate. The main advantage of CPW is that it provides an extremely high frequency response, because it uses no vias in the ground path. This is quite unlike that which occurs with striplines or microstrip lines where the presence of vias introduce parasitic discontinuities. A disadvantage of the CPW approach is that it requires substrate space, as the ground plane is placed on the component side of the board. The line impedance depends on the ratio of the strip width W to the ground plane spacing G, which allows for a given board thickness to control a wide variance of line impedances. Another advantage of CPW is that it offers a good transition to the coaxial board connector.

Coupled Transmission Lines

Coupled transmission lines are built by placing two uniform transmission lines close enough to each other so that their EM fields interact as shown in Figure 15.28. The interaction between each EM field is usually controlled by the separation distance S between each line. Typical applications for coupled transmission lines are directional couplers, or the lines connecting balanced

Figure 15.28. Coupled microstrip lines.

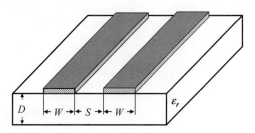

Figure 15.29. Coupled transmission line with (a) even mode configuration and (b) odd mode.

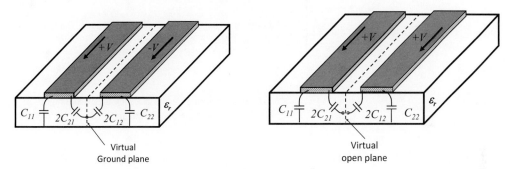

(differential) device ports. The line impedance is a function of the line width W and the board thickness D, as well as the separation distance between the two lines S.

The EM field of a coupled transmission line is somewhat more complex than the EM field of a single line, because the combined fields will have two different configurations. When the two lines are driven with the same in-phase signal as shown in Figure 15.29a, we refer to it as the even mode response. When both signals are out of phase as shown in Figure 15.29b, we refer to it as the odd mode response. Therefore, a coupled line has two sets of parameters, even and odd mode impedances, even and odd mode losses, and so forth. Only the electrical length will be the same, since it does not depend on the phase of the wave.

15.4.2 Transmission Line Discontinuities

Most RF DIBs will not use transmission line discontinuities as part of its design—for example, for a filter. It is important to understand the fundamental impact of line elements. Every time an EM wave hits a discontinuity in a transmission line, the RF current will crowd, which will behave like an inductor, or the electrical field will fringe, which makes it have a capacitance effect.

The most common discontinuities in transmission lines are shown below in Table 15.1. For the best and simplest design of an RF DIB, it is recommended to avoid any line discontinuities. As shown in Table 15.1, it should be obvious that RF lines cannot be connected in a tee or cross configuration without creating a new RF element. When designing the landing pattern for an RF-socket, it is recommended to avoid any open line stubs by following the recommendation provided by the specific socket vendor. Steps can be avoided by tapering the end, thereby making the transition of the line width smooth.

Table 15.1. Typical Discontinuities in Transmission Lines and Their Impact to the Wave Propagation

Open line stub		An open line stub or unterminated transmission line will cause a stray capacitance at the open end, as long the electrical length is less than 90°.
Shorted line stub		A shorted line stub behaves as an inductance, as long the electrical length is less than 90°.
Notch		A notch is built by a cut into the transmission line and can be represented by a series inductor.
Step		Two transmission lines with different line widths are connected to build a step. The current crowding takes place on both sides of the junction, and causes a stray capacitance on the wider line open edge.
Tee		Two lines are connected at a junction with a tee. The tee causes an excess capacitance to ground.
Cross		Two lines are crossing results in a complex discontinuity, which needs to be simulated with an EM simulation tool.
Bend		A line turning with a high angle leads to a significant discontinuity. The bend can be formed in a radius of at least three times the line width or by a mitered bend to avoid this.
Mitered bend		A mitered bend reduces the area of metallization of a bend, which leads to a reduction of excess capacitance. The mitered bend consumes less area of the DIB with similar RF performance, compared to a curved path.

15.4.3 Lumped- and Distributed-Element Models

In Example 15.2, we treated the PCB trace feeding the 50-Ω load resistor as if it had no capacitance to ground. In reality, the *RL* low-pass filter formed by the trace inductance and load resistance also includes a parasitic capacitance to ground, as shown in Figure 15.30. This simplistic model of a transmission line is known as a *lumped-element* model. The series inductance per unit length of the 6-in. trace in Example 15.2 was found to be 350 nH per meter. From the graph in Figure 15.30, we can determine that the capacitance per unit length of this trace is about 160 pF per meter, assuming FR4 PCB material. Therefore, the capacitance in Figure 15.30 is

Figure 15.30. Parasitic trace inductance and capacitance (lumped-element model).

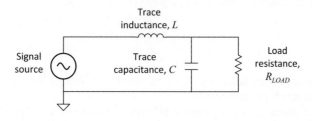

Figure 15.31. Distributed-element model of a transmission line.

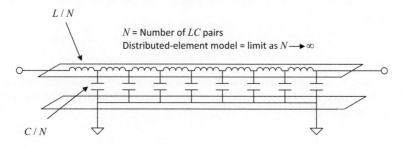

about 160 pF/m × [6 in. × (1 m/39 in.)] = 24.6 pF. At 50 MHz, this capacitance has an impedance of 130 Ω, which is not insignificant when placed in parallel with the 50-Ω load resistance. Therefore`, we should consider the trace inductance as well as the capacitance when evaluating the effects of trace parasitics on circuit performance.

Unfortunately, even the refined lumped-element model of Figure 15.30 becomes deficient at higher frequencies. In reality, the parasitic trace inductance and capacitance can only be modeled as a lumped inductance and capacitance at relatively low frequencies. At higher frequencies, we have to realize that the inductance and capacitance are distributed along the length of the trace. The effect of this distributed inductance and capacitance causes the true model of the trace to look more like an infinite series of infinitesimally small inductors and capacitors, as shown in Figure 15.31. This model is known as a *distributed-element* model. If we let the number of inductors and capacitors approach infinity as their values approach zero, the PCB trace becomes a circuit element known as a *transmission line*. The transmission line exhibits unique electrical properties, which the test engineer needs to understand.

As the voltage at the input to a transmission line changes, it forces current through the first inductor into the first capacitor. In turn, the rising voltage on the first capacitor forces current through the second inductor, into the second capacitor, and so on. The signal thus propagates from one *LC* pair to the next as a continuous flow of inductive currents and capacitive voltages. Notice that the transmission line is symmetrical in nature, meaning that signals can propagate in either direction through this same inductive/capacitive process.

Parallel trace pairs can form a transmission line as shown in Figure 15.31. These two traces can represent a stripline, a microstrip line, or a coaxial line. For the most part, these transmission lines are governed by the exact same equations. For example, one of the key parameters of a transmission line is its *characteristic impedance*, defined as

$$Z_0 = \sqrt{\frac{L_l}{C_l}} \qquad (15.15)$$

where Z_0 is characteristic impedance of the transmission line, L_l is trace inductance per unit length, and C_l is trace capacitance per unit length.

Notice that the characteristic impedance of a trace or cable is not dependent on its length. It is dependent only on the inductance per unit length and on the capacitance per unit length. Therefore, a 6-in. trace of a particular width and spacing to ground has the same characteristic impedance as one that is 10 ft long.

Signals injected into a transmission line travel down the line at a speed determined by the inductance and capacitance per unit length. The equation for the signal velocity v_{signal} is

$$v_{signal} = \sqrt{\frac{1}{L_l \cdot C_l}} \quad \text{m/s} \qquad (15.16)$$

The total time it takes a signal to travel down a transmission line is therefore equal to the length of the line divided by the signal velocity. This time is commonly called the transmission line's *propagation delay* T_d,

$$T_d = \frac{l_{line}}{v_{signal}} = l_{line}\sqrt{L_l \cdot C_l} \quad \text{s} \qquad (15.17)$$

Combining Eqs. (15.11), (15.15), and (15.17), we can find the total distributed capacitance of a transmission line as a function of its propagation delay and characteristic impedance:

$$C = \frac{T_d}{Z_0} \quad \text{F} \qquad (15.18)$$

The wavelength of a sine wave travelling along a transmission line is given by

$$\lambda_{signal} = \frac{v_{signal}}{f_{signal}} \quad \text{m/cycle} \qquad (15.19)$$

If the wavelength of a signal's *highest-frequency component of interest* is significantly larger than the length of the transmission line, we can use a lumped-element model, such as the one in Figure 15.30. Note that the highest-frequency component of a digital signal such as a square wave is determined by its rise and fall time, not by its period. A common rule of thumb for a digital signal is that the wavelength of the highest-frequency component must be 10 to 20 times the transmission line length before we can treat the line as a lumped-element model. Otherwise, we must treat the signal path as a transmission line with distributed rather than lumped parasitic elements. Another way to state this is that the period of the signal should be at least 10 times larger than the transmission line's propagation delay before we can treat the parasitic elements as lumped, rather than distributed. Another practical rule of thumb is that the highest-frequency component of interest in a digital signal is roughly equal to 1/3 the inverse of its rise or fall time.

In the case of RF devices transmitting analog information, most of the time we can assume sinusoidal signals with a single given frequency f_{signal}. For this case the wavelength can be calculated using Eq. (15.19). For RF systems and devices with multiple RF signals, the wavelength of the highest frequency needs to be considered when designing the PCB board as mentioned in Chapter 12, Section 2.1.

EXAMPLE 15.4

Determine the characteristic impedance of the PCB trace in Example 15.2 made from FR4 material. What is the velocity of a signal traveling along this transmission line? At what fraction of the speed of light does it travel? What is the propagation delay of this line? If we wish to transmit a 50-MHz sine wave along this trace, should we treat the parasitic capacitance and inductance as lumped elements, or should we treat the trace as a transmission line?

Solution:
Because the D/W ratio of this trace is 0.5, using Figure 15.10 we find that the inductance per unit length is 350 nH per meter. Similarly, using Figure 15.16, we find the capacitance per unit length to be 160 pF per meter. Using Eq. (15.15), the characteristic impedance of the trace is then found to be

$$Z_0 = \sqrt{\frac{350\frac{nH}{m}}{160\frac{pF}{m}}} = 46.77 \ \Omega$$

The velocity of a signal traveling along this line is given by Eq. (15.16) as

$$v_{signal} = \sqrt{\frac{1}{350\frac{nH}{m} \times 160\frac{pF}{m}}} = 133.63 \times 10^6 \ m/s$$

The speed of light, c_0, is 300×10^6 m/s. Therefore, signals travel down this transmission line at a speed of 133.63/300 times the speed of light, or 0.445c.

The propagation delay can be calculated using either of two methods. First we can divide the length of the transmission line by the signal velocity:

$$T_d = \frac{l_{line}}{v_{signal}} = \frac{6\,in. \times \frac{1\,m}{39\,in.}}{133.63 \times 10^6\,m/s} = 1.15\,ns$$

An alternative calculation uses Eq. (15.18) to find

$$T_d = C \cdot Z_0 = 24.6\,pF \times 46.77\,\Omega = 1.15\,ns$$

The wavelength of a 50-MHz signal, as calculated using Eq. (15.19), is

$$\lambda_{signal} = \frac{133.63 \times 10^6 \ m/s}{50\,MHz} = 2.67\,m$$

Since the length of the trace is only 6 in. and the wavelength of the 50-MHz signal is much larger (2.67 m × 39 in./m = 104 in.), we can safely treat this line as a lumped-element model.

15.4.4 Transmission Line Termination

Transmission lines can behave in a fairly complicated manner. We have discussed some of this behavior in Chapter 12 when defining the reflection coefficient for RF lines. Although their behavior is well-defined, a full study of transmission line theory is beyond the scope of this book, however, the theory required to design a good DIB will be explained in this subsection.

Fortunately, we can easily predict the basic behavior of a transmission line as long as we provide proper termination at one or both of its ends, as we have already seen in Chapter 12. Here we will apply a simple model sufficient for all mixed signal DIB designs.

To understand the purpose of transmission line termination, let us first examine the behavior of an unterminated line. An unterminated transmission line behaves as a sort of electronic echo chamber. If we transmit a stepped voltage down an unterminated transmission line, it will bounce back and forth between the ends of the line until its energy is dissipated and the echoes die out. The energy can be dissipated as electromagnetic radiation, as well as heat in the source resistance and parasitic resistance in the line. The resulting reflections appear as undesirable ringing on the stepped signal. Properly chosen termination resistors placed at either the source side or the load side of a transmission line cause it to behave in a much simpler manner than it would behave without termination. The purpose of termination resistors is to dissipate the energy in the transmitted signal so that reflections do not occur.

The simplest termination scheme to understand is the far-end termination scheme shown in Figure 15.32. As shown in this diagram, transmission lines are commonly drawn in circuit schematics as if they were coaxial cables, even if they are constructed using a PCB trace. This is because the basic behavior of a transmission line is dependent only on its characteristic impedance and propagation delay, rather than its physical construction.

If the termination resistor R_T is equal to the characteristic impedance of the transmission line, the transmitted signal will not reflect at all. The energy associated with the currents and voltages propagating along the transmission line is completely dissipated by the termination resistor. As far as the signal source is concerned, *a terminated transmission line looks just like a resistor whose value is equal to Z_0.* The distributed inductance and capacitance of the transmission line *completely disappear as far as the source is concerned.* The only difference between a purely resistive load and a terminated transmission line is that the signal reaching the termination resistor is

Figure 15.32. Terminated transmission line and resistive equivalent.

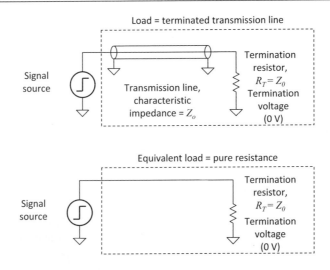

delayed by the propagation delay of the transmission line. It is also important to note that while the termination resistor is usually connected to ground (0 V), it can be set to any DC voltage, and the transmission line will still be properly terminated. The terminated line then appears to the source as a pure resistance connected to the DC termination voltage.

The ability to treat a terminated transmission line as a purely resistive element is very useful. Many tester instruments are connected to the DUT through a 50-Ω transmission line, which is terminated with a 50-Ω resistor at the instrument's input (see Figure 15.33). As far as the DUT is concerned, this instrument appears as a 50-Ω resistor attached between its output and ground. If the DUT output is unable to drive such a low impedance, we can add a resistor, R_s, between the DUT output and the terminated transmission line. The DUT output then sees a purely resistive load equal to $R_s + Z_0$. Although the signal amplitude is reduced by a factor of $Z_0/(Z_0 + R_s)$, we can compensate for this gain error using focused calibration (see Chapter 5).

If we observe the signals at the DUT output, the input to the transmission line, and the input to the tester instrument, we can see the effects of the resistive divider and the propagation delay of the transmission line as depicted in Figure 15.34. The signal is attenuated by the series resistor and termination resistor and is also delayed by a time equal to T_d.

Figure 15.33. Tester instrument with terminated transmission line.

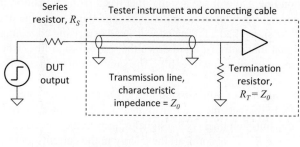

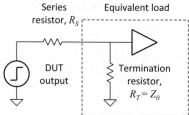

Figure 15.34. DUT output, transmission line input, and instrument input.

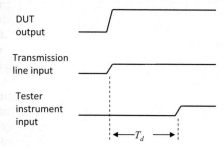

Figure 15.35. Transmission line with source termination.

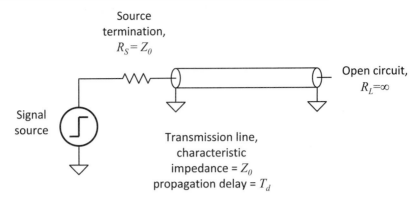

If we observe the voltage at a particular point along the transmission line, we will see a delayed version of the signal at the DUT output. For example, if we look at a point halfway down the transmission line, we will see a signal that has a rising transition that occurs halfway between the rising edge at the transmission line input and the rising edge of the transition at its output. Of course, if the signal had a falling edge instead of a rising edge, this same delay would occur. In fact *any* signal that is transmitted down the transmission line, whether it is a voltage step, sine wave, or complex signal, will exhibit the same attenuation and time delay described previously.

The next common method of transmission line termination is the source termination scheme shown in Figure 15.35. In this scheme, the transmitted signal is allowed to reflect off the unterminated far end of the transmission line. As the signal returns to the source end of the transmission line, the source resistance $R_S = Z_0$ absorbs all the energy in the currents and voltages of the transmitted signal. No further reflections occur. Source termination is used in the digital pin card driver electronics of most ATE testers to prevent ringing in the high-speed digital signals generated by the tester's digital subsystem.

While the signal propagates down the transmission line and back, the source cannot tell whether the far end is terminated or unterminated. *For a short period of time, the transmission line appears to the source as if it were a pure resistance of Z_0.* Therefore, during the period of time that the signal travels down the transmission line and back, the voltage at the transmission line input will be ½ that at the source output (since $R_S = Z_0$). Once the reflected signal returns to the source, the source resistor absorbs all the reflected energy and the voltage at the transmission line input becomes equal to the voltage at the output of the source. Therefore, the source only sees a load of $2 \times Z_0$ for a period of $2 \times T_d$. Afterwards, the source sees an open circuit.

At the far end of the transmission line, the voltage remains at the termination voltage until the incident signal arrives. The incident signal arrives at ½ the amplitude of the source signal, but it immediately adds to the reflected signal whose amplitude is also ½ the amplitude of the source signal. Therefore, the far end sees the unattenuated source signal with a delay of T_d. Figure 15.36 shows a stepped voltage as it appears at the source output, transmission line input, and transmission line output.

If we observe the voltage at a particular point along the source terminated transmission line, we will see a signal that is similar in shape to the signal at the transmission line input. However, the spacing between the first edge and the second edge will be closer together. At a point halfway down the transmission line, for instance, the edges will be spaced by a time equal to T_d, as illustrated in Figure 15.36. As shown, the edges observed at an intermediate point on the transmission line are always centered around the time the incident signal reflects off the unterminated transmission line output. Transmission lines can consist of multiple controlled-impedance segments, each having the same characteristic impedance. For example, the digital channel drivers from a tester are routed to the DIB through a series of cascaded coaxial cables and controlled impedance PCB traces (Figure 15.37). To create a cascaded controlled impedance transmission line, the DIB's traces must also exhibit the same characteristic impedance as the tester's transmission lines. If any of the impedances of the transmission line segments are not matched, then the point where they connect will generate signal reflections. Therefore, we have to make sure we lay out our DIB

Figure 15.36. Source output, transmission line input, and transmission line output.

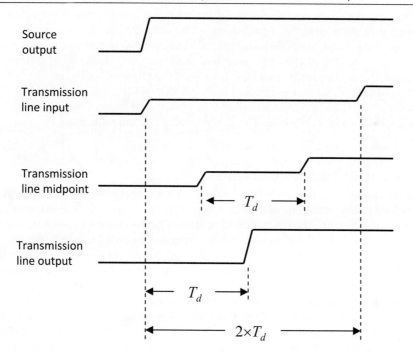

Figure 15.37. Cascaded transmission lines.

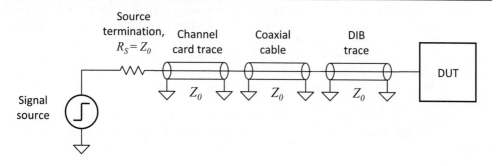

with a characteristic impedance equal to the tester instrument's characteristic impedance to avoid unwanted signal reflections.

One of the common mistakes made by novice test engineers is to observe the output of a digital channel at the point where the DIB connects to the test head. Such an observation point represents an intermediate point along the cascaded transmission line. As a result, a rising edge will appear as a pair of transitions such as those shown in Figure 15.36 rather than a single transition. The novice test engineer often thinks the tester driver is defective, when in fact it is working perfectly well in accordance with the laws of physics. The only way to see the expected DUT signal is to observe it at the DUT's input.

Notice that we can measure the propagation delay of a transmission line by measuring the time between the first and second step transitions at the source end of a source-terminated transmission line. This time is equal to $2 \times T_d$. We can divide the measured time by two to calculate the transmission line's propagation delay. This is how modern testers measure the propagation delays from the digital channel card drivers to the DUT's digital inputs. The tester can automatically compensate for the electrical delay in each transmission line, thereby removing timing skew from the digital signals. This measurement process is known as *time-domain reflectometry*, or TDR. On older testers, TDR deskew calibrations were not used. The timing edges of the digital channels were deskewed at a point inside the test head, and any delays caused by the various transmission lines were not taken into account. The test engineer had to lay out DIB traces that were equal in length to avoid channel-to-channel skew. This is the reason that DIBs are often round rather than square. The round board allows a radial layout, like spokes on a bicycle wheel. The spokes can be laid out with equal lengths, leading to matched delay times.

15.4.5 Parasitic Lumped Elements

So far we have only considered transmission lines that have been terminated with a pure resistance. Unfortunately, there are usually a few picofarads of parasitic capacitance, or a few nanohenries of parasitic inductance located at various points along a transmission line. These parasitic reactances arise from interconnections from one cascaded transmission line segment to another and from electromechanical relays located inside the tester instrument. There are also small amounts of inductance and capacitance at the ends of the transmission line. These are caused by the DUT's input or output parasitics, the DUT socket parasitics, or the pin card input and output parasitics. These parasitic elements are known as *lumped elements*, to distinguish them from the distributed inductance and capacitance of the transmission line itself. Lumped reactances can cause undesirable ringing, overshoot, and/or undershoot in our signals. Unfortunately, often there is not much we can do about the problem, other than to be aware of its existence.

Tester instruments are often specified using both lumped and distributed transmission line parameters. For example, a digital channel card is typically specified with a distributed capacitance and a lumped capacitance. The lumped capacitance is usually only a few picofarads, while the distributed capacitance (as defined by Eq. (15.18)) may be 30 pF or more. To accurately model the tester's load on a DUT output, we should use a model that includes both the transmission line and the lumped elements. We can use this type of model for simulations using software tools such as SPICE. Figure 15.38 shows a typical model for a tester's digital pin card electronics. The distributed capacitance listed in a tester's specifications may or may not include that associated with the DIB trace; thus the test engineer should read the channel card specifications carefully. A number of good books have been written on the subject of transmission lines and high-speed digital design. The reader is encouraged to refer to Johnson and Martin[2] and Wadell.[3]

Figure 15.38. Typical digital channel card model.

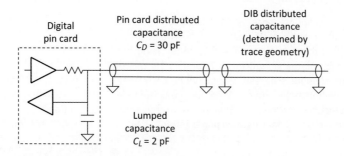

15.5 IMPEDANCE MATCHING TECHNIQUES FOR RF DIB

In many cases, the input and output impedance of RF DUT differsfrom that associated with the characteristic impedance of the ATE measurement path (normally 50 Ω). In these cases, the device will need to be matched by an impedance transformation in order to be able to measure the specified performance. In Chapter 12, we discussed the impact of mismatch resulting in a reflection coefficient other than zero, which also causes a mismatch loss and uncertainty. To be able to measure the true device performance accurately, it will be necessary to find the required matching circuit and implement it on the DIB.

The term *matching* implies performing an impedance transformation on some termination impedance. In the case of RF test, it usually refers to the transformation of the input impedance of a DUT to the port impedance of the ATE, which, in most cases, is 50 Ω. In principle there are three different kinds of matching:

1. Conjugate matches transform the load impedance Z_L to the conjugate match of the source impedance Z_S with $Z_S = Z_L^*$. This type of matching will maximize the power delivery to the load, but will not minimize the reflections, unless Z_S is real.
2. Load matching matches the load impedance Z_L to the line impedance Z_0. This matching minimizes the reflections, but does not maximize the delivered power, unless Z_0 is real. Load matching is best used at the ends of a transmission line.
3. Functional matching which will load the device with an impedance level that optimizesits system performance. This match is done, for example, when optimizing the noise figure of an LNA, since the best noise performance does not necessarily require the same load impedance as the best gain or power performance.

In the following, we will introduce the Smith Chart, an easy-to-use graphical tool to calculate matching components and topologies, which is well established and available as a paper chart, or as software tool from a number of vendors.

15.5.1 Introduction to the Smith Chart

In the previous section, we have seen that not only discrete RF components, like inductors and capacitors, need to be treated as RF circuits, but also RF lines connecting two points on a circuit board. In Table 15.1 we have seen that line elements, like stub lines, can even be used as matching components. Here we will describe the basics of a tool called the Smith Chart, which is used extensively by RF engineers. This tool is named after its inventor, Philip H. Smith (1905–1987), an engineer at the time with AT&T.

Exercises

15.6. A 12-in. stripline trace is fabricated on an FR4 PCB with a width of 15 mils. It is separated from its ground plane by a spacing of 12 mils. Using the refined estimates of Figures 15.10 and 15.16, calculate the stripline's parasitic capacitance per meter and inductance per meter. What is the stripline's characteristic impedance? Repeat the exercise for the same stripline fabricated using a Teflon® PCB.

ANS. FR4: $C_l = 110$ pF/m, $L_l = 500$ nH /m, $Z_0 = 67.4\ \Omega$; Teflon®; $C_l = 66$ pF/m, $L_l = 500$ nH /m, $Z_0 = 87.0\ \Omega$.

15.7. What is the velocity of a signal propagating along the stripline of Exercise 15.6? Express your answer in m/s and in a percentage of the speed of light. What is the stripline's propagation delay? Is the delay for the Teflon® PCB longer or shorter than the FR4 PCB?

ANS. FR4: $v_{signal} = 1.35 \times 10^8$ m/s, $0.45\ c$, $T_d = 2.28$ ns; Teflon®; $v_{signal} = 1.74 \times 10^8$ m/s, $0.58\ c$, $T_d = 1.77$ ns; Teflon® PCB propagation delay is shorter (signal velocity is higher) than FR4.

15.8. What is the wavelength, in inches, of a 500-MHz sine wave traveling along the FR4 stripline in Exercise 15.6? Can we treat the parasitic reactance of the stripline as lumped elements or do we have to treat the stripline as a transmission line at this frequency? Could we approximate the stripline using a lumped-element model if we used Teflon® instead?

ANS. FR4: $\lambda = 10.5$ in.—transmission line. Teflon®: $\lambda = 13.6$ in.—transmission line.

15.9. A 900-MHz sinusoidal signal is transmitted from a DUT output to tester digitizer along a 1.5-foot terminated 50-Ω coaxial cable having a signal velocity of $0.65 \cdot c_0$. What load resistance is presented to the DUT during the time the signal propagates? What resistance is presented to the DUT after the signal has settled? What is the distributed capacitance of this coaxial cable? What is the phase shift, in degrees, between the DUT output and the digitizer input? (The phase shift is equivalent to the propagation delay expressed in degrees, that is, $\phi = 360 \times f \times T_d$).

ANS. $R_{LOAD} = 50\ \Omega$, again $R_{LOAD} = 50\ \Omega$, $C = 47.3$ pF, phase shift = 766.9 degrees.

The Smith Chart is based on the principle of conformal transformation (mapping) associated with complex number theory. This theory says that any complex function, say the complex function $Z = x + jy$ in the x,y-plane, can be mapped into another function in another plane, say the u–v-plane, such that its angles are preserved. For example, the rectangle shown in the x–y plane on the left-hand side of Figure 15.39, when transformed to the u–v-plane using the appropriate conformal map, creates another rectangle rotated by about 45 degrees and doubled in size. The angles made between the various contours remain orthogonal to one another. This kind of mathematical transformation is used to map the complex right half of the impedance plane $z = r + jx$ into the complex plane of the reflection coefficient $\Gamma = \Gamma_r + \Gamma_i$ to generate what is referred to as the Smith Chart. This transformation is illustrated in Figure 15.40, where the entire right half pane of the impedance plane z is mapped inside the unit circle of the Γ-plane.

Figure 15.39. Mathematical transformation.

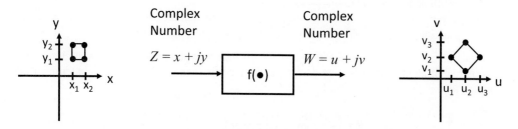

Figure 15.40. Bilinear conformal transformation of the right half of the complex impedance plane into the interior unit circle of the complex reflection coefficient Γ-plane.

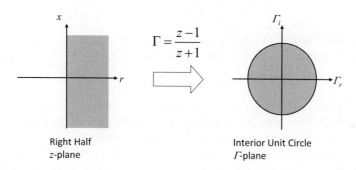

$$\Gamma = \frac{z-1}{z+1}$$

Right Half
z-plane

Interior Unit Circle
Γ-plane

15.5.2 Impedance Smith Chart

The Smith Chart is a polar plot of the reflection coefficient overlaid with an impedance or admittance grid. The rectangular coordinates of the impedance can be mapped with a mathematical bilinear conformal complex transformation given by

$$\Gamma = \frac{Z-Z_0}{Z+Z_0} = \frac{\dfrac{Z}{Z_0} - \dfrac{Z_0}{Z_0}}{\dfrac{Z}{Z_0} + \dfrac{Z_0}{Z_0}} = \frac{z-1}{z+1} \tag{15.20}$$

where $z \triangleq Z/Z_0$ is referred to as the normalized complex impedance. This impedance is typically expressed in terms of a real resistance component, r, and an imaginary reactance x whereby $z = r + jx$. Mathematically, this transformation expresses the following:

1. Every single point in the z-plane is mapped to a single point in the Γ-plane.
2. Circles in the z-plane map into circles in the Γ-plane, and vice versa. Radius and center point will be transformed. Straight lines can be seen as circles with an infinite radius. In the Γ-plane, constant-resistance circles (constant resistance with varying amount of reactance) have their center along the horizontal axis and are collinear, but not concentric. All circles have one common point at $z = \infty$ as shown in Figure 15.41.
3. The entire real positive half-plane is mapped inside of the circle of the Γ-plane. Since for all passive circuits Re{Z} ≥ 0, we can use the Smith Chart for all passive circuits without any limitations. For most active circuits, Re{Z} ≥ 0 is also valid, which underlines the power und usefulness of this tool.

4. This mapping is analytic. Therefore continuous contours in the z-plane are mapped into continuous contours in the Γ-plane.

5. The mapping is conformal. Contours in the z-plane that are orthogonal to each other will map into contours in the Γ-plane that are likewise orthogonal to each other.

The magnitude of the complex reflection coefficient seen at any point along a lossless transmission line of length l at a distance d from the load termination can be described in terms of the complex load reflection coefficient Γ_L as

$$\Gamma(d) = \frac{\text{reflected wave @ } d}{\text{incident wave @ } d} = \Gamma_L e^{-j4\pi d/\lambda_{\text{eff}}} \tag{15.21}$$

where λ_{eff} is the effective wavelength of the signal. Substituting $\Theta = 2\pi l/\lambda_{\text{eff}}$, the so-called electrical length parameter expressed in radians, we write

$$\Gamma(d) = \Gamma_L e^{-j2\Theta d/l} \tag{15.22}$$

It interesting to note that for an arbitrary load reflection coefficient described as $\Gamma_L = \rho e^{j\phi}$, the reflection coefficient at any point on the line can be expressed as

$$\Gamma(d) = \rho e^{j(\phi - 2\Theta d/l)} \tag{15.23}$$

As is clearly evident from Eq. (15.23) the magnitude of the reflection coefficient is constant along the transmission line, regardless where we are along the line, and equal to the magnitude of the load reflection coefficient. Therefore, on the complex Γ-plane of the Smith Chart, reflection coefficients are represented by constant circles (Figure 15.42).

For the full length of the line (i.e., $d = l$), the phase angle of the reflection coefficient seen at the input side of the line is rotated clockwise from ϕ by twice the electrical length of the line, 2Θ, that is,

$$\Gamma_{IN} = \Gamma_L e^{-j2\Theta} = \rho_L \angle (\phi - 2\Theta) \tag{15.24}$$

Figure 15.41. The Impedance Smith Chart as a bilinear transformation of the infinite right (positive) half of the z-plane to the inside of the circular Γ-plane. (a) Constant resistance vertical lines are mapped to constant resistance circles. (b) Constant reactance horizontal lines are mapped to constant reactance arcs.

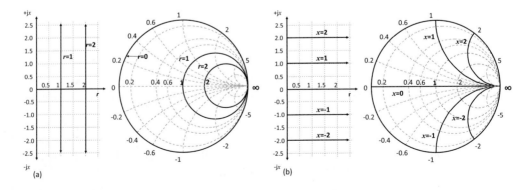

Figure 15.42. Smith Chart with constant reflection coefficient Γ circles. The shown reflection coefficient is Γ = 0.5∠50°. The phase θ can be read on the circular θ-scale, the magnitude of the reflection coefficient Γ on the Γ-scale below the Smith Chart.

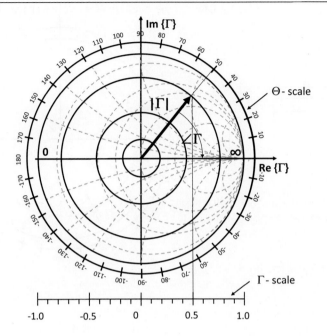

Because the phase angle is derived from the Smith Chart in degrees, the actual length of the line l is computed according to

$$l = \frac{\Theta|_{\text{deg}}}{360} \times \lambda_{\text{eff}} \qquad (15.25)$$

where Θ is expressed in degrees.

With the reflection coefficient known at any point on the lossless line, the value of the complex impedance seen looking into the line toward the load can be determined directly from the Smith Chart. Consider that the input impedance Z_{IN} of a lossless line with characteristic impedance Z_{TL} connecting a complex load impedance Z_L (Figure 15.43) is given by

$$Z_{IN} = Z_{TL}\left[\frac{Z_L + jZ_{TL}\tan\Theta}{Z_{TL} + jZ_L\tan\Theta}\right] \qquad (15.26)$$

Equation (15.26) defines the input impedance Z_{IN} as a function of the transmission line parameters including their electrical length Θ. This equation can be simplified by setting the source impedance equal to Z_{TL} and normalizing the circuit with respect to this quantity, thus obtaining $z_L = Z_L/Z_{TL}$ and $z_{TL} = Z_{TL}/Z_{TL} = 1$. If the transmission line is lossless, the magnitude of the amount of reflected power is constant throughout the entire length of the transmission line and only the phase of the reflected wave will change. The normalized form of Eq. (15.26) can then be written as

$$z_{IN} = \frac{z_L + j\tan\Theta}{1 + jz_L\tan\Theta} \qquad (15.27)$$

Figure 15.43. The Γ-plane representation of the load impedance Z_{TL} connected with a transmission line with the impedance Z_{TL} and the electrical length Θ.

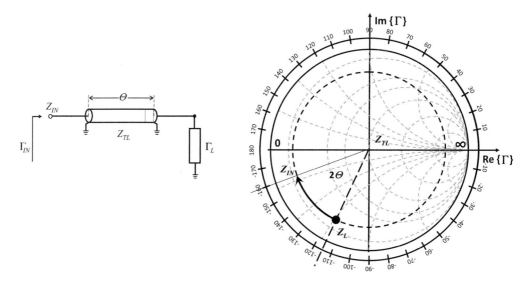

In a later subsection we will explain how this impedance transformation with transmission lines can be used to match a load to a given source impedance.

15.5.3 Admittance Smith Chart

When working with circuits that are parallel, it is often easier to use admittance instead of impedance. As such, we can rewrite the conformal transformation first seen in Eq. (15.20) from an admittance perspective as follows:

$$\Gamma = \frac{Z - Z_0}{Z + Z_0} = \frac{\dfrac{1}{Y} - \dfrac{1}{Y_0}}{\dfrac{1}{Y} + \dfrac{1}{Y_0}} = \frac{Y_0 - Y}{Y_0 + Y} = \frac{\dfrac{Y_0}{Y_0} - \dfrac{Y}{Y_0}}{\dfrac{Y_0}{Y_0} + \dfrac{Y}{Y_0}} = \frac{1 - y}{1 + y} \tag{15.28}$$

Here we make use of the normalized admittance, $y \triangleq Y/Y_0$. This admittance is normally expressed as $y = g + jb$, where the real part of y is the conductance g and the imaginary part is the susceptance, b. The denormalized admittance $Y = yY_0 = G + jB$ is measured in siemens.

Comparing Eq. (15.28) with Eq. (15.20), we see that that the general form differs by a negative sign. This implies that the admittance Smith Chart is rotated by 180 degrees with respect to its impedance form. In rectangular admittance coordinates, the vertical lines represent constant conductance, while the constant conductance in the Γ-plane is represented by collinear circles as illustrated in Figure 15.44a. Constant susceptance becomes constant semicircles or arcs as shown in Figure 15.44b. The open circuit point ($g = 0$) is as shown on the right side of the admittance chart, and the short circuit ($g = \infty$) is represented on the left side of the horizontal axis.

Figure 15.44. Admittance Smith Chart as bilinear transformation of the infinite right (positive) half y-plane to the inside of the circular Γ-plane: (a) Constant conductance vertical lines are mapped to constant conductance circles, and (b) constant susceptance horizontal lines are mapped to constant susceptance arcs.

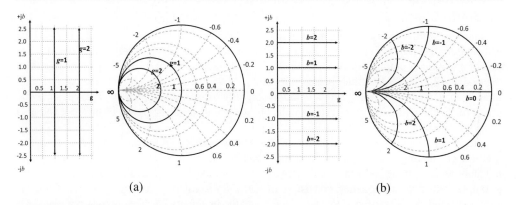

(a)   (b)

Figure 15.45. Immitance Smith Chart. The impedance grid is shown in black, and the admittance grid is shown in gray. Commercially available charts use two colors for easier usage.

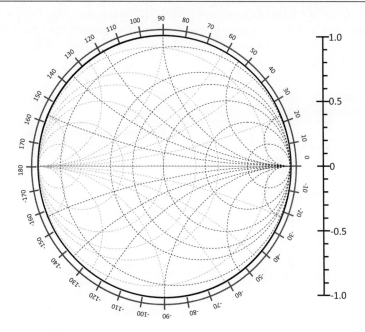

15.5.4 Immitance Smith Chart

To have a tool for developing a complex impedance transformation with parallel and serial components, both the impedance and the admittance Smith Charts are overlaid in one so-called immitance Smith Chart. The impedance and admittance parameters can be displayed simultaneously as shown in Figure 15.45. The immitance Smith Chart is typically colored, with the admittance part in another color. Here we use a black and gray scale to distinguish between impedance and admittance. In the next subsection, we will discuss how to use the Smith Chart for impedance transformations that match any impedance to a specified value.

15.5.5 Impedance Transformation with Discrete Components on Smith Chart

A common problem in the development of a test board, like a DIB, is to "match" the DUT to the test environment. The RF engineer understands the term "match" to be the impedance transformation of the input or output impedance of the DUT to the impedance of the ATE, or components on the test board, like a SAW filter. In this subsection we will demonstrate how to match a specific input or load condition using a circuit combination of two discrete components. We will show the use of the Smith Chart in a practical example, but we will also see the power of this tool and also some its limitations.

To aid the test engineer, Table 15.2 provides a brief overview of the most important formulas used for calculating reactance, susceptance, inductance and capacitance. These formulas come in handy when solving for a matching network. As a reminder, in general, impedance $Z(j\omega)$ is a complex function of frequency, consisting of real and imaginary parts. The real part of impedance is defined as the resistance $R = \mathrm{Re}\{Z(j\omega)\}$ and the imaginary part is known as the reactance $X = \mathrm{Im}\{Z(j\omega)\}$. Conversely, we can also speak in terms of admittance, defined as the inverse of the impedance, that is, $Y(j\omega) = \dfrac{1}{Z(j\omega)}$. Because admittance is also a complex function of frequency, the real part is referred to as the conductance, $G = \mathrm{Re}\{Y(j\omega)\}$, and the imaginary part is called the susceptance, $B = \mathrm{Im}\{Y(j\omega)\}$. Uppercase symbols represent denormalized values, and lowercase symbols represent normalized values, that is, $x = X/Z_0$ and $b = B/Y_0$, depending on the nature of the parameter.

Table 15.2. Summary of Basic Formulas to Calculate Lumped Components Reactance, Susceptance, Inductance and Capacitance

Denormalized Form	Normalized Form Using $Z_0 = 50\ \Omega$ and $Y_0 = 0.02$ S
Ideal Lumped Component Reactance and Susceptance	
$X_L = 6.283 \cdot f[\mathrm{GHz}] \cdot L[\mathrm{nH}]$	$x_L = 0.1257 \cdot f[\mathrm{GHz}] \cdot L[\mathrm{nH}]$
$B_L = -\dfrac{0.159}{f[\mathrm{GHz}] \cdot L[\mathrm{nH}]}$	$b_L = -\dfrac{7.96}{f[\mathrm{GHz}] \cdot L[\mathrm{nH}]}$
$X_C = -\dfrac{159}{f[\mathrm{GHz}] \cdot C[\mathrm{pF}]}$	$x_C = -\dfrac{3.183}{f[\mathrm{GHz}] \cdot C[\mathrm{pF}]}$
$B_C = 0.006283 \cdot f[\mathrm{GHz}] \cdot C[\mathrm{pF}]$	$b_C = 0.314 \cdot f[\mathrm{GHz}] \cdot C[\mathrm{pF}]$
Ideal Lumped Inductance and Capacitance in nH and pF	
$L[\mathrm{nH}] = \dfrac{0.159 X_L}{f[\mathrm{GHz}]} = \dfrac{0.159}{f[\mathrm{GHz}] \cdot (-B_L)}$	$L[\mathrm{nH}] = \dfrac{7.96 \cdot x_L}{f[\mathrm{GHz}]} = \dfrac{7.96}{f[\mathrm{GHz}] \cdot (-b_L)}$
$C[\mathrm{pF}] = \dfrac{159}{f[\mathrm{GHz}] \cdot (-X_C)} = \dfrac{159 B_C}{f[\mathrm{GHz}]}$	$C[\mathrm{pF}] = \dfrac{3.183}{f[\mathrm{GHz}] \cdot (-x_C)} = \dfrac{3.183\, b_C}{f[\mathrm{GHz}]}$

Figure 15.46. Adding a parallel (shunt) capacitor with admittance jb_2 to a one-port circuit with input admittance y_1.

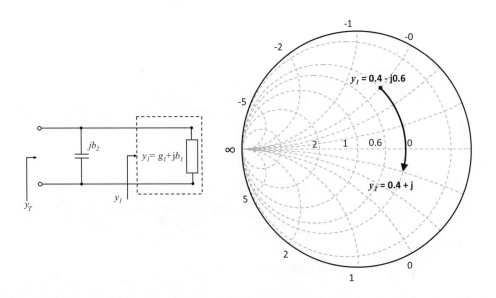

Parallel additions of components are most conveniently handled with the admittance Smith Chart, whereas series addition of components is most conveniently and efficiently handled using the impedance Smith Chart. Consider the parallel combination of an arbitrary admittance $y_1 = g_1 + jb_1$ and a lossless capacitor with admittance $y_2 = jb_2$ as illustrated in Figure 15.46. Beginning with a point in the admittance Smith Chart corresponding to y_1, the addition of the parallel capacitor shifts the admittance level along a clockwise path of constant conductance such that the total admittance point in the Smith Chart is equal to

$$y_T = y_1 + jb_2 = (g_1 + jb_1) + jb_2 = g_1 + j(b_1 + b_2) \qquad (15.29)$$

The specific value of parallel capacitance C_p placed in parallel to the admittance y_1 that gives rise to the susceptance of b_2 is simply found by rearranging the expression listed in Table 15.2 for a normalized impedance level of 50 Ω, that is,

$$C_p[\text{pF}] = \frac{b_2}{0.314 f[\text{GHz}]} \qquad (15.30)$$

Next, consider the series combination of arbitrary impedance $z_1 = r_1 + jx_1$ and a loss-less capacitor with impedance $z_2 = -jx_2$ as illustrated in Figure 15.47. Beginning with a point in the impedance Smith Chart corresponding to z_1, the addition of the series capacitor shifts the imped-ance level in a counterclockwise direction (more negative reactance) along a path of constant resistance such that the total impedance point in the Smith Chart is equal to

$$z_T = z_1 + z_2 = z_1 - jx_2 = r_1 + j(x_1 - x_2) \qquad (15.31)$$

Figure 15.47. Adding a series capacitor with impedance $-jx_2$ to a one-port circuit with input impedance z_1.

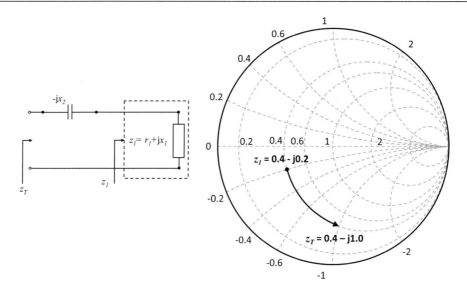

The series capacitance C_s can be calculated (as listed in Table 15.2) with negative reactance $-x_2$ at a normalized impedance level of 50 Ω corresponding to the following

$$C_s[\text{pF}] = \frac{3.183}{f[\text{GHz}] \cdot x_2} \qquad (15.32)$$

This same approach can easily be extended to the parallel or series addition of an inductor or resistor. The details are left for the reader to verify.

EXAMPLE 15.5

Assume that a capacitor with the susceptibility b_2 is connected in parallel to an admittance of $y_1 = 0.4 - j0.6$. The admittance of this parallel circuit can be read from the Smith Chart to be $y_T = 0.4 - j0.2$. What is the value of the capacitance placed in parallel with this admittance if the frequency is 2.0 GHz and the reference impedance level is 50 Ω?

Solution:
From Eq. (15.29), we write

$$y_T = y_1 - jb_2$$

allowing us to solve for b_2 according to

$$jb_2 = y_T - y_1 = (0.4 + j0.2) - (0.4 - j0.6) = j0.8$$

resulting in

$$b_2 = 0.8 \text{ units}$$

The actual capacitance that gives rise to this (normalized) susceptance is derived from Eq. (15.30) to be

$$C_p = \frac{b_2}{0.314 \cdot f} = \frac{0.8}{0.314 \cdot 2} = 1.274 \text{ pF}$$

The admittance for the one-port with and without the parallel capacitance included is highlighted in the Smith Chart shown on the right-hand side of Figure 15.46. Here it is shown how the initial one-port admittance of $y_1 = 0.4 - j0.6$ is transformed to a new admittance level of $y_T = 0.4 - j0.2$ along a path of constant conduction through the addition of a parallel capacitor with a normalized susceptance of 0.8 units.

15.5.6 Impedance Matching with a Series and Shunt Component Using the Immitance Smith Chart

In the previous subsection, we have seen how the Smith Chart was used to transform the impedance or admittance of a one-port circuit through the addition of a single passive element (i.e., resistor, capacitor, or inductor). In this subsection, we shall extend this approach to multiple additional passive elements so that a greater range of impedance or admittance transformation can be achieved. Throughout this subsection we will assume that the signal wavelength is long compared to the physical dimension of the components and RF lines. This enables one to neglect the propagation behavior of the connecting lines.

Since most matching networks require series and shunt elements, we shall make use of the immitance Smith Chart, because it combines both the impedance and admittance grids on the same graph. As before, all inductors and capacitors will be treated as reactance or susceptance and will thus follow the constant resistance or conductance circles of the Smith Chart. For serial components, we will use the impedance portion of the Immitance Smith Chart, as well as the admittance grid for parallel or shunt components. The goal will be to transform the impedance to be reflection-free, or moving the input impedance into the center point at $z = 1$, which is equivalent to a denormalized impedance level of $Z = 50 + j0\ \Omega$.

In order to illustrate the matching technique using the Smith Chart, let us assume that a DUT has an output impedance of, say $Z_{DUT} = 20 + j100\ \Omega$ at an operating frequency of 2 GHz. It is fair to assume that the test equipment has been designed with a nominal input impedance of $Z = 50\ \Omega$. It is our objective here to construct a matching network that connects between the DUT output and the test equipment input, such that the test equipment sees a matched condition, that is, an impedance level equal to 50 Ω at 2-GHz operation. This point is located at the center of the Smith Chart at $z_T = 1 + j0\ \Omega$.

We begin by first normalizing the DUT output impedance by 50 Ω, resulting in an initial normalized impedance level of $z_{DUT} = \dfrac{20 + j100}{50} = 0.4 + j2$ units. This point is then marked on the impedance Smith Chart, along with the desired impedance level, $z_T = 1 + j0$. Because z_{DUT} has a positive imaginary part (positive susceptance) corresponding to some inductive behavior, we will consider adding a series capacitor because it has a negative susceptance and will cause the impedance level to move toward zero. We also recognize that a series capacitor will cause the impedance level to move along the circle of constant resistance in a counterclockwise manner having a normalized level of 0.4 units. Recognizing that the final impedance point of $1 + j0$ lies

Figure 15.48. Immittance Smith Chart as tool to calculate the matching circuit for an impedance transformation involving multiple passive components.

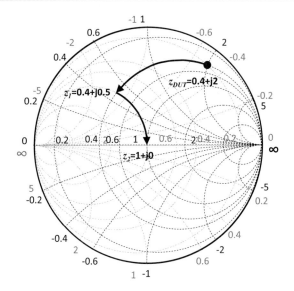

on a circle of constant admittance on the admittance Smith Chart of 0.5 units, we shall add a series capacitor C_s having an impedance of $-j1.5$ units, or equivalently, a capacitance value of

$$C_s = \frac{3.183}{f\,[\mathrm{GHz}]\cdot x_{C_s}} = \frac{3.183}{2\times 1.5} = 1.061\ \mathrm{pF}$$

The DUT impedance seen after the series capacitor is simply the sum of the initial DUT impedance and the impedance of the series capacitor, that is,

$$z_{T1} = z_{DUT} + z_{C_s} = 0.4 + j2 - j1.5 = 0.4 + j0.5\ \mathrm{units}$$

It is at this point that we convert the present impedance z_{T1} to an admittance, y_{T1}, so that we can make use of the admittance Smith Chart (gray scale in Figure 15.48), that is,

$$y_{T1} = \frac{1}{z_{T1}} = \frac{1}{0.4 + j0.5} = 0.976 - j1.22\ \mathrm{units}$$

Here we see that the admittance seen looking into the DUT-series-capacitor combination consists of a conductance very close to unity (error is due to limited accuracy of the Smith Chart) and a susceptance of -1.22 units (recall that a negative susceptance is caused by a shunting inductor). By shunting a capacitance with a susceptance of 1.22 units across the present network terminals, the imaginary component of the admittance can be eliminated, resulting in the desired admittance of

$$y_T = y_{T1} - y_{C_p} = (0.976 - j1.22) + j1.22 = 0.976 \cong 1\ \mathrm{units}$$

The value of the parallel capacitor C_p at 2 GHz having an admittance of $j1.22$ is then found from Eq. (15.30) to be

$$C_p = \frac{1.22}{0.314 \times 2} = 1.94\ \mathrm{pF}$$

Exercises

15.10. A DUT with a normalized output impedance $z_{DUT} = 0.4 + j2\ \Omega$ is to be matched with an LC network to 50 Ω at 2 GHz, where the matching network needs to be design in such a way that the device output will not be shorted to GND (inductors typically have a very low DC resistance).

ANS.

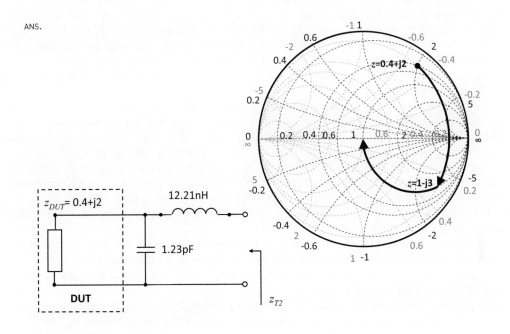

Figure 15.49. Matching network transforming the DUT impedance z_{DUT} to the test equipment impedance level.

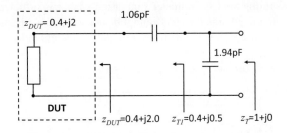

The final matching circuit is shown in Figure 15.49, together with the impedance levels at each node in the circuit.

15.6 GROUNDING AND POWER DISTRIBUTION

15.6.1 Grounding

The term "grounding" refers to the electrical interconnection and physical layout of the various ground nodes in an electronic system such as an ATE tester and DIB. In a circuit schematic, grounds are treated as perfect zero volt reference points exhibiting zero impedance. In a real system, there can only be one point that is defined as true ground. All other ground nodes are connected to true ground through resistive and inductive traces, wires, or ground planes. These parasitic resistors and inductors often play a significant role in the performance of the DUT and the ATE tester instruments.

Grounding is one of the more difficult DIB design subjects to master. Proper grounding is essential to a DUT's electrical performance. Poor grounding can lead to resistive voltage drops, inductive transients, capacitive crosstalk, interference, and a host of other electrical ailments that result in poor measurement accuracy, or noisy DUT performance. Grounding is not a subject taught in most college curricula. Instead, it is often learned through experience (i.e., bad experience!). Fortunately, a few books have been written on the subject of grounding and other related topics. Henry Ott's book,[4] *Noise Reduction Techniques in Electronic Systems*, provides an excellent introduction to the subjects of grounding, shielding, electromagnetic interference (EMI), and electromagnetic compatibility (EMC).

One way to achieve proper grounding is to pay close attention to the flow of currents through the traces, wires, and planes in the DIB and tester. The first thing we have to consider is DC measurement errors caused by resistive drops in ground connections. Figure 15.50 shows a simple test setup including a DC current source, a DC voltmeter, and a DUT (a simple load resistor in this case). We wish to measure the value of the resistor by forcing a current, I_{TEST}, and measuring the voltage drop across the resistor, V_{TEST}. The resistor's value is calculated by dividing V_{TEST} by I_{TEST}.

As we have seen in previous sections, the ground path from the DUT to the DC source will not be a 0-Ω path. If the ground path is several feet long, it may exhibit several ohms of series resistance, R_G. Therefore, the parasitic model illustrated in Figure 15.51 should be used to predict the measurement results obtained using the grounding scheme of Figure 15.50. Since the DC meter in this example is connected to the tester's internal ground, it measures the voltage across the series combination of the DUT resistance R_L and the ground interconnection resistance, R_G. Therefore, the measured resistance is equal to $R_L + R_G$, resulting in an error of several ohms. Notice that the value of R_S is unimportant, since the current source will force I_{TEST} through the DUT load resistance regardless of the value of R_S. Also, notice that the value of R_M is unimportant, since it carries little or no current (assuming the DC voltmeter's input impedance is sufficiently high).

Obviously, accurate mixed-signal testers cannot be constructed using such a simple grounding scheme. Instead, they use a signal, which we will call *device ground sense*, or DGS, to carry

Figure 15.50. Resistor measurement test setup: idealized model.

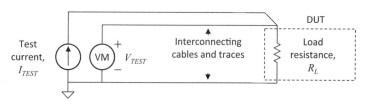

Figure 15.51. Resistor measurement test setup: parasitic model.

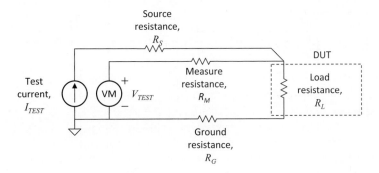

Figure 15.52. Resistor measurement test setup using device ground sense (DGS) line.

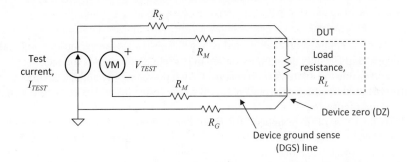

the DUT's 0-V reference back to each tester instrument. Since the DGS signal is carried on a network of zero-current wires, the series resistances of these wires do not result in voltage measurement errors. Figure 15.52 shows a resistor measurement test setup using the DGS reference signal. Note that any number of tester instruments can use DGS as a zero-volt reference, provided that they do not pull current through the DGS line. Consequently, each tester instrument typically contains a high input impedance voltage follower to buffer the voltage on DGS.

The DGS line is often routed all the way to a point near the DUT, which serves as the true 0-V reference point of the entire test system. This single point is known by several names, including *star ground* and *device zero*. We will use the term "device zero," or DZ, to refer to this point in the circuit. All measurement instruments should be referenced to the DIB's DZ voltage.

15.6.2 Power Distribution

As shown in Figure 15.53, the power supplies and sources in a mixed-signal tester are typically connected using a four-wire Kelvin connection (see Section 2.2.2). Each Kelvin connection includes a separate wire for high-force (HF), low-force or ground current return (LF), high-sense (HS), and low-sense (LS). The low-sense line is equivalent to DGS. Therefore all supplies may use a single DGS line as their low-sense reference, resulting in only three DIB connections per source.

All currents from a power supply must return through its low-force line, in accordance with Kirchhoff's current law. Therefore, we can control the path of DUT power supply currents by forcing them through separate DIB traces back to the low-force line of the supply. Separate return

Figure 15.53. Power and ground distribution in a mixed-signal ATE tester.

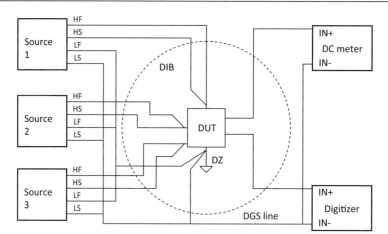

of each supply's current prevents unexpected voltage drops across the various current paths in the ground network. Some testers do not provide separate ground return lines for each supply. In these testers, it is impossible to isolate the currents from each power supply and DC source. Instead, the return currents are lumped together and returned to the tester through a common current return path, as shown in Figure 15.53.

The PCB traces in a DZ grounded DIB should be laid out as shown in Figure 15.54a. If the DGS line or the force and sense lines of a power supply are not connected properly, the full effectiveness of the zero-volt sense lines will not be realized.

In some situations, the DZ is placed on chip by routing ground force and sense traces through separate I/O pins of the IC to a common pad on the die. Power pins can also be connected to force and sense lines connected on chip in a similar fashion. In addition to improving the accuracy of DZ, this technique will cause a part to fail if either of the two wire bonds are left disconnected during assembly. The cost of joining power or ground Kelvin connections on the chip is extra package pins. In some cases this is not feasible, in others it is well worth the cost.

15.6.3 Power and Ground Planes

At this point in the chapter, the reader should be catching on that ground planes are a good idea for many reasons. The ground plane provides a low inductance connection between all the grounds on a DIB. Similarly, power supplies can be routed using power planes to reduce the series inductance between all power supply nodes. Power and ground planes can be divided into sections, forming what are known as *split planes*. Each section of a split plane can carry a different signal, such as +12 V, +5 V, and –5 V. This provides the electrical superiority of copper planes without requiring a separate PCB layer for each supply. Typically, power is applied through split planes while grounds are connected to solid (nonsplit) planes, but even the ground planes can be split if desired.

There are usually at least two separate ground planes in a mixed-signal DIB. One plane forms the ground for the transmission line traces carrying digital signals. This plane is subject to rapidly changing current flows from the digital signals, and therefore exhibits fairly large voltage spikes caused by the interactions of the currents with its own inductance. This ground plane is often called DGND (digital ground) in the DIB schematics. The second plane, AGND (analog ground), is for use by analog circuits. Ideally, this plane should carry only low-frequency, low-current signals that will not give rise to voltage spikes.

Figure 15.54. Physical layout of Kelvin connections and DZ grounding system: (a) Proper connections. (b) Improper connections.

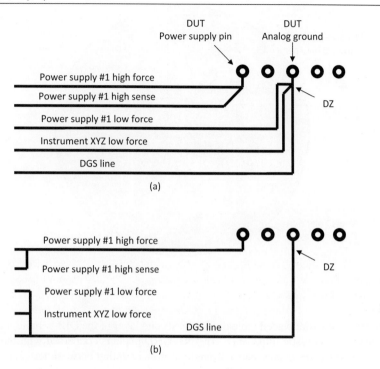

Figure 15.55. Three-plane grounding scheme.

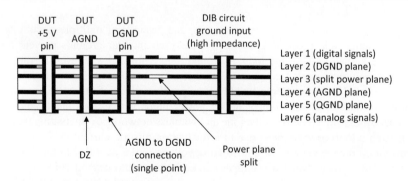

A third plane is occasionally used as a DIB-wide zero volt reference. This "quiet ground" plane (QGND) can be used by any analog circuits on the DIB that need a low noise ground reference. The QGND plane must be connected in such a way that it does not carry any currents exceeding a few milliamps. To guarantee this, the QGND plane should be tied only to the DZ node at a single point and to relatively high-impedance DUT pins and DIB circuit nodes. Often, the analog ground plane and the quiet ground plane are combined into a single plane, resulting in a DIB with only two ground planes (analog and digital). Figure 15.55 shows a cross section of a DIB with a three-plane grounding scheme.

Figure 15.56. DIB with RF-GND plane with trenches to improve site-to-site isolation.

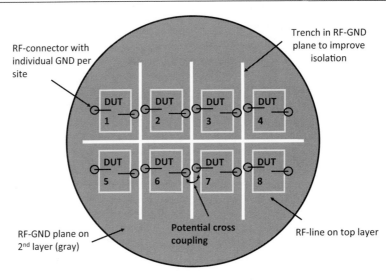

RF ground planes are placed typically on the second layer of the DIB to build microstrip lines with a defined characteristic impedance. This RF-GND plane is connected with the RF launcher and the DUT GND as solid as possible. To reduce cross coupling between test sites in a multisite test configuration, trenches in the GND plane can be build in between the test sites as shown in Figure 15.56.

15.6.4 Ground Loops

A star-grounding scheme is formed by connecting the grounds of multiple circuits to a single ground point, rather than connecting them in a daisy chain. Star grounds prevent a common grounding error known as a *ground loop*. A ground loop is formed whenever the metallic traces and wires in a ground network are connected so that a loop is formed (Figure 15.57). In a world without magnetic fields, ground loops would not be problematic. Unfortunately, fluctuating magnetic fields surround us in the form of 60-Hz power fields and various electromagnetic radio signals, such as those produced by cell phone and wireless networks. A fluctuating magnetic field passing through a loop of wire gives rise to a fluctuating electric current in the wire. The fluctuating current, in turn, gives rise to a fluctuating voltage in the wire due to the wire's parasitic resistance and inductance. Thus AC voltages can be induced into ground wires if we carelessly connect them in a loop. Ground loops are most commonly formed when we connect instruments such as oscilloscopes and spectrum analyzers to our DIB. The tester housing and its electrical ground must be connected to earth ground for safety reasons to prevent electrical shock. Likewise, an oscilloscope's housing and electronics must be connected to earth ground. When we connect a bench instrument's ground clip to the DIB ground, we form a large ground loop, as shown in Figure 15.57. Ground loops can also be formed when we attach test equipment to other external equipment such as handlers and probers.

Any fluctuating magnetic fields that intersect the area of the ground loop cause currents to flow around the loop. The parasitic resistance of the wires and cables in the loop cause voltage drops, which in turn cause unwanted noise in our grounding system, corrupting whatever

Figure 15.57. Bench instrumentation ground loop.

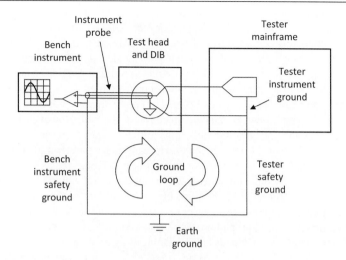

measurement we are trying to make. A ground loop often causes the tester's signals to appear terribly corrupted with 60-Hz power hum and other noise components. The test engineer has to realize that these signals are not present in the tester itself. They disappear as soon as we disconnect the bench instrument. Unfortunately, the instrumentation ground loop problem cannot be easily solved without violating company safety rules (e.g., by disconnecting the third prong of the oscilloscope's power cord from earth ground using a "cheater plug"). Ground loops caused by handlers and probers can sometimes be resolved by breaking unnecessary ground connections between the handler or prober and the test head.

15.7 DIB COMPONENTS

15.7.1 DUT Sockets and Contactor Assemblies

The DUT pins and the circuit traces on a DIB must be connected temporarily during test program execution. A hand-test socket or a handler contactor assembly makes the temporary connection. There are many different socket and contactor schemes, so we will not try to discuss them in any detail. The most important thing to note is that the metallic contacts of the socket or contactor assembly represent an extra resistance, inductance, and capacitance to ground that will not exist when the DUT is soldered directly to the PCB in the customer's system-level application. Sometimes the parasitic elements are unimportant to a device's operation, but other times, particularly at high frequencies, they can be extremely critical. When designing a DIB for an RF DUT it is important to compare the parasitics of the test socket especially the pin and GND inductance and compare with the device requirements. Often a good compromise needs to be found between the electrical, mechanical robustness and thermal performance. Often socket vendors provide the expected 1-dB or 3-dB bandwidth performance of their socket. This number is a first indicator of its high-frequency operation taken within a 50-Ω environment. It is a good practice to compare the parasitics and align them with the DUT requirement. Occasionally, the test engineer will find that socket or contactor pins are the cause of correlation errors between measurements made on the customer's application and measurements made on the tester.

15.7.2 Contact Pads, Pogo Pins, and Socket Pins

Contact pads are metal pads formed on the outer trace layers of a DIB PCB. They appear on DIB schematics as circles, black dots, or connector bars. These pads allow a relatively reliable, nonabrasive connection between one layer of interface hardware and the next. Two common uses for connector pads are pogo pin connections and DUT socket pin or contactor pin connections (Figure 15.58). A pogo pin is a spring-loaded gold-plated rod that provides a connection between two connector pads. Pogo pins may have blunted ends, pointed ends, or crown-shaped ends, depending on the connection requirements. Pointed or crown-shaped ends tend to dig into the pad surface, providing a reliable, low-resistance connection to the pad. However, the digging action may eventually destroy the pad. Blunted pogo pins are less abrasive to the pads, but are slightly less reliable since they do not dig into the pad surface. Contact pads can also be used to form a connection to DUT socket pins, such as illustrated in Figure 15.58.

Contact pads are usually plated with gold to prevent corrosion and to lower the contact resistance between the pad and the gold-plated pogo pin or socket pin. Without the gold plating, corrosion would lead to higher contact resistance, or even worse, it might result in a complete open circuit even when the pogo pin and pad are in physical contact. Since gold and copper are both soft metals, gold-on-copper contact pads can be damaged by repeated connect/disconnect cycles. To increase their hardness, the copper pads are often coated with a nickel alloy before the gold plating is applied.

A connection formed with a contact pad can be treated as an ideal zero-resistance connection in most cases. The connection adds a small amount of resistance, usually on the order of a few tens of milliohms. However, the resistance may change by 50% or more as the pad is connected and reconnected to the pogo pin or socket pin.

Pogo pins and socket pins also add inductance to the signal path. The inductance may be as little as a quarter of a nanohenry or as large as a few tens of nanohenries. Socket pins and pogo pins may also introduce pin-to-pin or pin-to-ground parasitic capacitance on the order of a few picofarads. In general, long thin pins add more inductance than short fat ones. One way to reduce the effects of pogo or socket pin inductance is to return all high-frequency currents through an adjacent pin. This minimizes the area of the loop through which the current must flow, reducing the inductance of the loop. Unfortunately, the shape and size of the current loops are often determined by the pinout of the device; thus the test engineer can do little to lower the inductance caused by socket pins. Of course, the best way to minimize parasitic socket pin inductance and

Figure 15.58. Contact pads, pogo pins, and DUT socket pins.

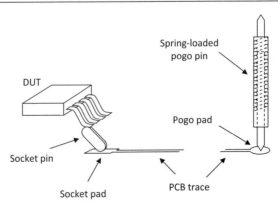

capacitance is to choose a socket with very short pins. Pogo pins can be also build with a defined impedance mainly 50 Ω, which allows us to minimize the reflection losses for devices with an equal input impedance.

In addition to the electrical performance, often the thermal performance is of interest. Most devices for the automotive applications will need to be tested at a specified elevated temperature, or power devices need to be cooled down to a specified ambient temperature. This will require a defined airflow during test as well as a thermal conductance bridging the thermal path from the device to the DIB. Socket vendors developed different approaches to this requirement: Some are building into their test socket a high number of pogo pins embedded in a solid socket housing, while others optimize the thermal performance with a solid metal (copper) insert bridging the thermal flow from the device to the test board.

Table 15.3 lists the specification of a RF socket and the impact that each parameter has on its operation. The parasitics of the test socket is a major concern during the selection of the socket, and its impact on the DUT performance needs to be considered carefully.

Table 15.3. Specification of a High-Performance RF Socket and Selection Criteria

Electrical Specification		Typical Value[a]	Measure or Indicator for
Electrical length (compressed height)		1.10 mm	potential if socket can be treated as lumped element and potential rotation of the reflection coefficient when not matched to Z_{IN}
Inductance	Self	0.23 nH	the impact to the device performance
	Mutual	0.14 nH	the efficiency to reduce inductance by placing multiple contact pins in parallel
Capacitance	Ground	0.16 pF	the impact to the device performance
	Mutual	0.05 pF	the efficiency to reduce inductance by placing multiple contact pins in parallel
S21 Insertion loss/bandwidth (ground–signal–ground)		−1 dB @ >40 GHz	the maximum frequency the socket can be used
S11 Return loss/bandwidth (ground–signal–ground)		−20 dB @ 14.5 GHz	potential reflections causing mismatch uncertainty
S41 Crosstalk/bandwidth (ground–signal–signal–ground)		−20 dB @ 32 GHz	cross coupling between two pins
Average contact dc resistance		50 mΩ	IR drop
Current carrying capability		3 A	current limit
Current leakage		<1 pA @ 10 v	leakage current measurement capability
Nearest decoupling area		1.25 mm	together with electrical length set the RF performance impact of the test socket

Mechanical Specifications	Typical Value[a]	Measure or Indicator for
Contact Compliance	0.175–0.200 mm	together with contact compliance a measure for acceptable mechanical tolerances
Contact wipe on pad	0.12 mm	cleaning effect
Contact force (per contact)	60 g	important by high pin count DUT
Contact tip coplanarity	0.05 mm	together with contact compliance a measure for allowable mechanical tolerances
Environmental	−40°C to 155°C	DUT test temperature range
Contact material	BeCuNiAu	together with roughness the capability to test specific lead materials

[a]Johnstech International, Product specification of ROL100 high-performance test socket.

15.7.3 Electromechanical Relays

One of the more common DIB components used in mixed-signal testing is the electromechanical relay. The relay is an electromagnetically controlled mechanical switch. Relays allow the DIB circuits to be appropriately reconfigured for each measurement in the test program. As one of the few moving parts on a DIB, relays represent a potential reliability problem in production. Very high-reliability relays must be chosen so that the DIB can operate through hundreds of millions of open/close cycles without failure.

The metal contacts in a relay are pulled open or closed using an electromagnetic field generated by a DC current passing through a coil of wire (Figure 15.59). The current is switched on and off under test program control as the test code is executed. As mentioned in Chapter 2, "Tester Hardware," flyback diodes are sometimes added to the DIB in parallel with each relay coil to prevent the coil's inductive kickback voltage from damaging the current-driving electronics located inside the tester.

In conventional relays such as the one in Figure 15.59, the moving armature is called the *wiper*. Since it pivots on its pole, it may eventually wear out and get stuck. A more reliable relay is the reed relay, which uses two springy metal reeds that become magnetized by the coil's electromagnetic field. They are attracted to each other by the induced magnetism. Since the reeds do not swing on any pivot, there are no parts to wear out other than the point of contact between the two reeds. Reed relays are often used on DIBs because of their superior reliability.

Occasionally, a resistive buildup can occur between the contacts of a relay, causing an open circuit. Relay damage can also result from poor DIB design. Care should be taken to avoid passing currents through a relay that exceed its rated current specifications. Damage can also be caused to the wiper and contacts by abrupt changes in the current passing through the contacts as they open and close. The high $\partial i/\partial t$ current changes can induce large inductive voltage spikes, leading to a spark that welds the contacts together. Care should be taken to avoid discharging capacitors directly through relays without a series resistance to limit the discharge current. Otherwise, the sudden surge in current from the capacitor may weld the relay contacts together.

Relays, like manually activated switches, are available in a variety of configurations. The most common versions are single-pole/single-throw (SPST), single-pole/double-throw (SPDT), double-pole/single-throw (DPST), and double-pole/double-throw (DPDT). The schematic representation of each of these configurations is shown in Figure 15.60.

The parasitic behavior of relays is fairly complicated. They may exhibit a number of possible nonideal characteristics, including series resistance through the wiper and posts (R_{WP}), series inductance through the wiper and contacts (L_W), capacitive coupling between the contacts and

Figure 15.59. Electromechanical relay (single-pole, double-throw).

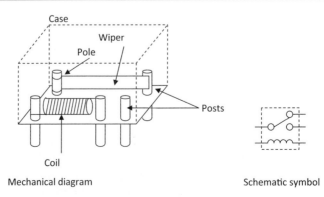

Mechanical diagram   Schematic symbol

Figure 15.60. Electromechanical relay schematic representations.

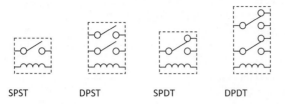

SPST   DPST   SPDT   DPDT

Figure 15.61. Parasitic model of SPDT relay.

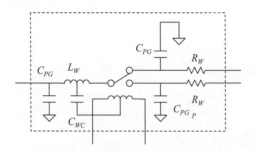

ground (C_{PG}), capacitive coupling between the wiper and the coil (C_{WC}), capacitive coupling from contact to contact, and mutual inductance between the coil and the wiper. A simplified model of a SPDT relay is shown in Figure 15.61. Series resistance is typically only a few hundred microhm, although the exact value changes from one closure to the next. Series inductance is often fairly high and may exceed 10 mH. Capacitance values are usually around 1 to 5 pF.

Some relays contain a built-in electrostatic shield that can be grounded to prevent capacitive coupling from external signals to the wiper. The shield helps to prevent electromagnetic interference from corrupting the signal passing through the relay. Similarly, controlled-impedance relays contain a shield conductor that can be grounded to form a continuation of a transmission line. This allows high-frequency signals to propagate smoothly through the relay without generating signal reflections. (Cascaded transmission lines were discussed in Section 15.4.4).

It is good design practice to connect relays so that they are in the most commonly desired position when they are not activated (i.e., when there is no current passing through the coil). For example, if a 1-kΩ resistor is to be connected from a DUT pin to ground during only one test, then it makes sense to connect the relay in the normally open configuration as shown in Figure 15.62a. Configuring the relay in this manner, the resistor is only connected when the test code sets the relay driver into the non-default state. This is a minor point, but it tends to save debug time since the test engineer does not have to explicitly add or remove the resistor during each of the remaining tests. Conversely, if the resistor is desired in all but one test, the relay should be connected in the normally closed configuration Figure 15.62b.

Good alternative solutions for mechanical relays are solid-state FET switches, or for very-high-frequency applications, MEMS switches. Solid-state FET switches are LED-driven MOS FET drivers that have an on-resistance as low as 1 Ω and an output capacitance of 1 pF or less. Since the solid-state FET switch has no mechanical parts, it can be constructed extremely small and can achieve hundreds of millions switching cycles. The principle structure is shown in Figure 15.63.

Figure 15.62. Relay default configurations: (a) Normally open. (b) Normally closed.

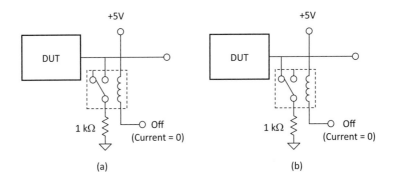

(a)   (b)

Figure 15.63. Solid-state FET switch.

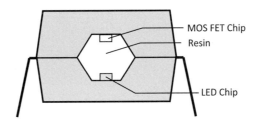

MEMS (micro-electro-mechanical systems) switches are typically used in RF DIB designs to utilize the low insertion loss over a wide frequency range. MEMS switches are built typically using a wafer process. Two electrodes within a DC electrostatic field actuate a thin membrane acting as the switch wiper. The insertion loss can be as low as 1dB or less, while the frequency characteristic have been reported as high as 40 GHz. A disadvantage of most MEMS switches is that they required high actuation voltage; also, they require a DC path through the switching electrodes.

15.7.4 Socket Pins

Since relays and active circuits such as op amps are subject to electrical or mechanical failures, they must be replaced from time to time. Although op amps and relays can be soldered directly onto the DIB PCB, replacement is far easier to perform if the relays are mounted in socket pins (Figure 15.64). Socket pins should ideally be used for any component having more than two or three leads.

Surface-mounted components with more than two pins are more difficult to unsolder without damaging the board. Unfortunately, the reduced pin inductance of surface-mount components is sometimes required for very-high-frequency testing. Also, the extra capacitance and inductance of socket pins may make a socket connection inferior at high frequencies. In such cases, surface-mounted relays, op amps, and other active devices may be the only viable alternative, even though they make the DIB more difficult to repair. Surface-mounted relays and active components should be used when needed for electrical performance.

Figure 15.64. Socket pins allow easy repair and maintenance of DIBs.

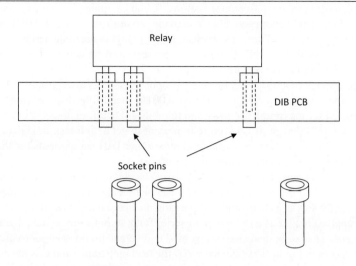

Figure 15.65. Common resistor packages.

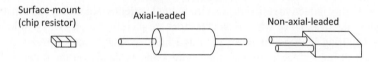

15.7.5 Resistors

Resistors are available in a variety of package types, including surface-mount, axial-leaded, and non-axial-leaded varieties (Figure 15.65). They can be constructed using a wide variety of resistive materials, most commonly carbon or metal (e.g., aluminum). Resistors can be constructed as either a solid core of resistive material, a coil of resistive wire, or a thin film of resistive carbon or metal. Carbon film and metal film resistors are constructed by depositing a thin film of the resistive material onto an insulator such as ceramic. The thin film gives a higher resistance than a solid core of the same material.

The choice of material and package type determines the power dissipation capabilities of the resistor, its accuracy, its stability over time and temperature, and its cost. Power dissipation is basically a function of the resistor's size. Larger resistors can typically dissipate more heat than small resistors. The test engineer should determine the maximum power that must be dissipated by each resistor on a DIB using the equation

$$\text{power} = \frac{V^2}{R} \text{ watt} \tag{15.33}$$

where V is the maximum RMS voltage dropped across the resistor at any point in the test program. The RMS voltage should include both DC and AC components.

The test engineer should also consider the accuracy requirements of each resistor. In general, metal film and wire-wound resistors are more accurate than carbon resistors. For tolerances of 5%

to 20%, solid carbon or carbon film resistors are typically acceptable. Tolerances of 1% generally require metal film resistors. Tolerances below 1% can be attained using a trimmed metal film resistor. Some component vendors are able to provide resistor tolerances of 0.01% or better using a trimmed metal film process. These resistors are not only highly accurate, but are also very stable over time and temperature. Extremely high accuracy can also be achieved using a wire-wound resistor constructed from a coil of wire.

In general, the cost of a resistor is inversely proportional to its accuracy. A 20% carbon resistor is far less expensive than a custom-trimmed 0.001% metal film resistor. However, DIBs are usually constructed for maximum accuracy and reliability. Therefore, the cost of components is a secondary issue in DIB design. It is far more important to get a high-test yield than to save a few dollars on DIBs. For this reason, resistors (and most other DIB components for that matter) are chosen based on their performance rather than their cost.

Performance is not only determined by a resistor's accuracy and power-handling capabilities, but is also determined by how closely the resistor can be modeled as a pure resistance. A resistor's material, shape, and size affect its nonideal performance characteristics. Resistors can be modeled to a first approximation as a resistance in series with an inductance and a parallel capacitor as shown in Figure 15.66a. In many cases a parallel equivalent circuit is easier to use; this parallel RLC model is shown in Figure 15.66b, where L is the lead inductance and C is the combination of parasitic capacitance which depends strongly on the resistor structure and build. At low frequencies, the inductance and capacitance may not be important, and the resistor can be modeled as a pure resistance. At higher frequencies, the series inductance and parallel capacitance may become a significant reactive element that affects circuit performance.

The most highly inductive resistors are wire-wound varieties, since they are constructed from a long, highly inductive coil of wire. These resistors are therefore used mainly in low-frequency applications requiring high-power dissipation and high accuracy. Leaded resistors suffer from a small amount of inductance caused by their wire leads, typically on the order of 1 to 10 nH. However, this inductance is often quite tolerable up to several tens of megahertz.

At higher frequencies, the smaller surface-mount packages typically provide the least series inductance. This is partly due to their lack of wire leads and partly due to the fact that they can be soldered directly onto the DIB without the use of a through-hole or via. Through-holes and vias add a small amount of series inductance as well as a few picofarads of parasitic capacitance to ground. Beyond the range of a few megahertz, the series inductance of vias, PCB traces, IC bond wires, and socket pins may become more of an issue than the inductance of the resistor itself.

For very-high frequency applications, resistors are normally selected based on maintaining their ratio of impedance-magnitude-to-resistance ($|Z|/R$) over a range of frequencies to be less than 1.2. A typical plot of this metric is shown in Figure 15.67 for several commercially available 50-Ω termination resistors. Another critical parameter of a resistor when used in high-frequency

Figure 15.66. Circuit model of a practical resistor: (a) Parallel-series RLC combination. (b) Equivalent parallel RLC circuit model representation.

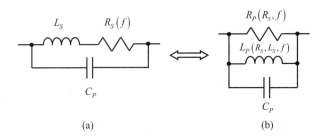

(a)   (b)

Figure 15.67. Typical |Z|/R-ratio for 50-Ω resistors.

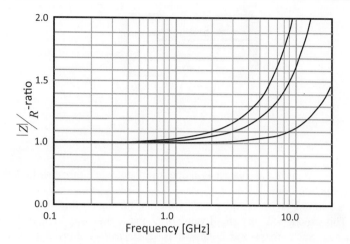

applications is its resonance frequency. Near the resonance frequency, minor changes in working frequency will cause major changes in impedance. We'll have more to say about this effect in a moment.

15.7.6 Capacitors

Like resistors, capacitors are also available in axial-leaded, non-axial-leaded, and surface-mount varieties. They can be constructed using a simple configuration of two parallel plates (called *electrodes*) separated by a nonconductive dielectric material, a sandwich of plates, or a pair of rolled foil plates. Capacitor performance is largely dependent on the shape and size of the capacitor as well as the type of dielectric material.

Small values of capacitance can be achieved using the simple two-plate arrangement in which capacitance is given by Eq. (15.9). The value A is the area of one of the plates, D is the distance separating the plates, and $\varepsilon = \varepsilon_r \, \varepsilon_0$ is the electrical permittivity of the dielectric material separating the plates. Small-value two-plate capacitors are often separated by a thin dielectric film of ceramic, mica, NPO, or even air. These capacitors are limited by physical constraints to a few hundred picofarads.

A larger value of capacitance requires either a larger area A, a smaller distance D, or a larger permittivity ε. Larger area can be achieved by stacking multiple layers of dielectric materials between parallel plates as illustrated in Figure 15.68. In effect, this forms a group of parallel capacitors in a small space, leading to a relatively large capacitor value in a compact package. High-value surface-mount capacitors up to 1 μF or more can be fabricated using this type of stacked configuration.

Various types of foil capacitors can be constructed using a pair of long metal foil strips separated by a dielectric film such as Mylar, polystyrene, or polypropylene. The long foil strips provide a large area A, but the strips and dielectric film must be rolled up and encapsulated in a tubular package to reduce the physical size of the capacitor.

To achieve dramatically larger values of capacitance, the dielectric film can be replaced by a permeable material such as paper soaked in an electrolytic liquid such as ammonia. The foil/liquid roll is sealed in an airtight package to prevent the liquid from evaporating. Then a small current is passed from one foil plate to the other through the electrolytic liquid. This process deposits an

Figure 15.68. Cross section of a multilayer chip capacitor.

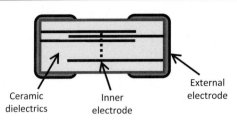

Ceramic dielectrics · Inner electrode · External electrode

extremely thin layer of insulating material on one foil plate, while the other plate remains in contact with the conducting liquid.

In effect, the liquid becomes one plate, while the extremely thin insulating deposit forms the dielectric. This produces a very small value of D and allows very high values of capacitance in a small package. Such a capacitor is known as an *electrolytic capacitor*. Aluminum electrolytic and tantalum electrolytic capacitors are constructed using foils of aluminum and tantalum, respectively.

Unfortunately, the deposition of the insulting layer in electrolytic capacitors is a reversible process. DC current passing through the capacitor in the wrong direction can lead to destruction of the insulating dielectric film, and the two plates of the capacitor can come into direct electrical contact through the electrolytic liquid. The large short circuit that results can lead to a rather destructive explosion. Therefore, electrolytic capacitors are marked with a polarity marker, typically a plus or minus sign or a black band representing a minus sign.

Electrolytic capacitors must never be reverse biased, although two of them can sometimes be connected in a back-to-back series connection to form a nonpolarized electrolytic capacitor. In this configuration, DC current is always blocked in one direction by the insulating film of one of the capacitors, regardless of the polarity of the applied voltage. Thus, neither capacitor's insulating layer can be damaged by a reversed DC current.

Ideally, a capacitor should lose no energy; however, an actual capacitor dissipates energy due to internal losses. These losses can be attributed to the series resistance of the internal and external electrodes and current leakage through the dielectric material. A model of a discrete capacitor is shown in Figure 15.69a, where R_e represents the resistance of the external electrode, R_d represents the effective energy loss associated with the polarization of the dielectric material, and R_{LEAK} represents the leakage resistance of dielectric material. R_{LEAK} is usually quite large, exceeding tens of mega-ohms in many cases. The highest leakage occurs in electrolytic capacitors, while the lowest leakage typically occurs in polystyrene or polypropylene capacitors. Smaller values of capacitance generally exhibit lower leakage currents than large ones simply because they generally have less plate and dielectric area. Also included in this model is the inductance L_s associated with magnetic flux self-coupling in the signal path through the capacitor. Because the makeup of each capacitor type is different, it is customary to approximate the high-frequency behavior of a capacitor with the series equivalent model shown in Figure 15.69b. This model includes an equivalent series resistance ESR, an equivalent series inductor ESL, and the nominal effective capacitance, C_{eff}. It is these three components that are specified in any technical literature made available from the component supplier.

The series inductance of a capacitor can vary widely from a fraction of one nanohenry to hundreds of nanohenries, depending on the shape and size of the capacitor. Typical values are given in Tables 15.4 and 15.5 for a select group of commercially available capacitors. In general, electrolytic capacitors exhibit very high series inductance, while surface-mounted chip capacitors

Figure 15.69. Parasitic model of a practical capacitor: (a) Individual parasitic components for each element of the capacitor. (b) Equivalent RLC series model.

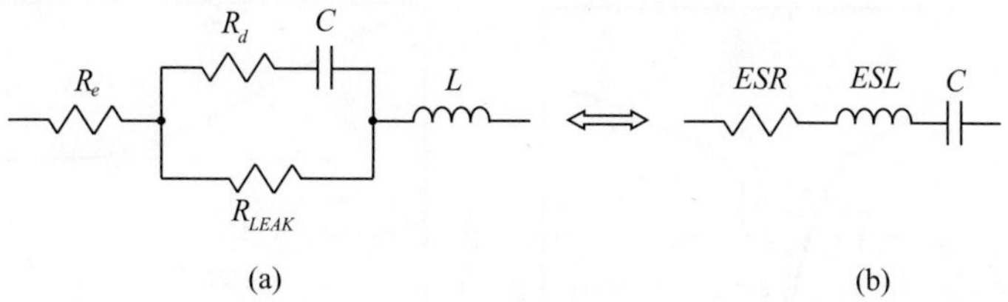

(a)   (b)

Table 15.4. Typical ESL Values for Common SMD Capacitor Sizes

Package	Size [mm]	ESL [pH]
0201	0.6×0.3	400
0402	1×0.5	550
0603	1.6×0.8	700
0805	2.0×1.25	800
1206	3.2×1.6	1250
0612	1.6×3.2	60

Table 15.5. Typical ESL Values for Leaded and Electrolytic Capacitors

Capacitor Type	Value	ESL (pH)
Leaded disc ceramic	0.01 mF	3000
Leaded monolithic ceramic	0.01 mF	1600
Leaded monolithic ceramic	0.1 µF	1900
SMD aluminum electrolytic	47 µF	6800
SMD tantalum electrolytic	47 µF	3400

exhibit very low series inductance. A typical self-resonance frequency of a leaded capacitor is in the range of 20 MHz, while for an SMD 1206 type capacitor it can be as low as 50 MHz. The exact value will be given in the data sheet of the individual capacitor series.

One important measure of a capacitor is the *quality factor Q*, where Q is defined as the ratio of the expected reactance of the capacitor (i.e., $|X_C| = 1/2\pi f C_{eff}$) to the equivalent series resistance ESR, given by

$$Q = \frac{|X_C|}{\text{ESR}} \tag{15.34}$$

An ideal capacitor has an ESR of 0, which leads to an infinite quality factor. Capacitors with very good Q factors are required in very-high-frequency applications such as microwave and RF systems. Surface-mounted NPO chip capacitors exhibit very low dielectric losses and series resistance, making them ideal for high-frequency applications. Also, surface-mounted ceramic capacitors are fairly well suited to high-frequency applications.

Figure15.70. Insertion loss frequency characteristic of an actual capacitor affected by (a) ESL, (b) ESR, and (c) ESL and ESR.

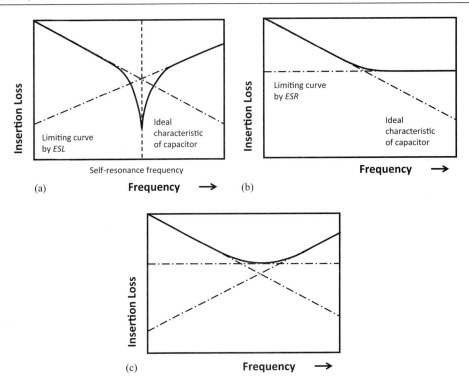

(a)

(b)

(c)

Another important metric of a capacitor is the *self-resonance frequency* SRF expressed in Hertz. It is defined in terms of the reactive components of the capacitor as follows:

$$SRF = \frac{1}{2\pi \sqrt{ESL \cdot C_{eff}}} \qquad (15.35)$$

The smaller the values of ESL, the larger SRF becomes; therefore, a small capacitor can be used to provide capacitive impedance for a higher frequency range. Figure 15.70 illustrates the impact of various parasitic components of the capacitor on its insertion loss (or equally impedance) as a function of frequency. Figure 15.70a illustrates the effect that ESL has on the insertion loss of an ideal capacitor. Instead of a decreasing value with frequency for all frequencies, at frequencies around the SRF, the insertion loss is dominated by the behavior of the parasitic inductor (ESL). In Figure 15.70b, we see the effect of the parasitic resistor (ESR) on the insertion loss. While effect of the ESR on insertion loss is the same across all frequencies, its effect on the capacitor is to limit its insertion loss to some minimum value. With both effects combined, we see from Figure 15.70c that the capacitor behaves as a capacitor only for frequencies below the SRF. By selecting the appropriate type of capacitor, this effect can be adjusted so that the required capacitance is realized over the appropriate frequencies. Table 15.6 lists the typical ESR values for a variety of capacitor types.

Table 15.6. Common Capacitor Parameter and Tradeoffs

Type	Picture	Capacitance Range	ESR	Leakage	Voltage Rating	Temperature Range	General Notes
Ceramic		pF to μF	Low	Medium	High	-55°C to +125°C	Multipurpose, inexpensive
MICA		pF to nF	Low 0.01Ω to 0.1Ω	Low	High	-55°C to +125°C	For RF filters, expensive, very stable
Plastic Film		few μF	Medium	Medium	High	varies	For low frequency, inexpensive
Tantalum		μF	High 0.5Ω to 5.0Ω	Low	Lowest	-55°C to +125°C	Expensive, nonlinear (limited for audio)
OSCON		μF	Low 0.01Ω to 0.5Ω	Low	Low	-55°C to +105°C	Best quality, highest price
Aluminum electrolytic		High μF	High 0.05Ω to 2.0Ω	Medium	Low	varies	For low to med frequencies, inexpensive, hold charge for long time, not for production test

15.7.7 Inductors and Ferrite Beads

Inductors are usually built using a length of wire, often looped into a coil to increase the magnetic coupling and, in turn, its inductance. Larger values of inductance can be achieved by wrapping the wire coil around a magnetic core, such as iron or a ceramic material with magnetic properties. Inductors are available in both surface-mount packages and leaded packages. However, the surface-mounted packages are typically limited to smaller values of inductance.

Inductors are occasionally required as part of a DUT circuit such as a voltage doubler. They may also be used as part of a passive load circuit that must be connected to a DUT output to simulate a speaker coil or similar system-level component. Inductors can also be used in conjunction with capacitors to simulate long transmission lines by building an *LC* network like the one in Figure 15.31. For RF DIB applications, inductors are frequently used as part of a matching circuit or as a choke to bias a DUT via an RF line. In matching circuits, the value of the inductor is of primary interest, however, for choke circuits, the self-resonant frequency is the key parameter of interest.

Figure 15.71. Parasitic model of a practical inductor: (a) Individual parasitic components for each element of the inductor. (b) Equivalent parallel RLC model.

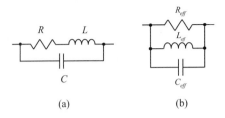

(a)   (b)

Figure 15.72. Typical inductance and impedance for a 100-nH wire wound inductor.

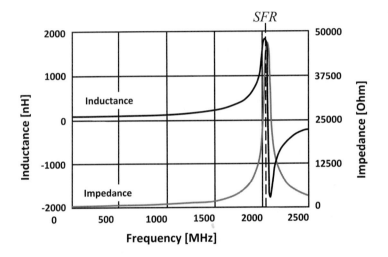

A practical inductor can be modeled by the combinational RLC circuit shown in Figure 15.71a. Here the inductor consists of an ideal inductor of value L, two parasitic elements, a series resistor R, and a shunting capacitor C. Because the specific circuit model varies with the device type, it is customary to model the inductor as an equivalent parallel RLC circuit shown in Figure 15.71b. In general, the series resistance R_{eff} of an inductor is quite low and often neglected. At low frequencies, the capacitor C_{eff} acts as an open circuit and the terminal behavior of the inductor is dominated by the behavior of inductor L_{eff}, that is, impedance increases linearly with frequency. At very high frequencies, the impedance of C_{eff} drops to very low levels, thereby shorting out the influence of the inductor. It is fair to say at very high frequencies the real inductor behaves like a capacitor. Somewhere in between these two extremes, both elements are contributing to the terminal behavior. An important frequency point within this range is the self-resonance frequency SRF expressed in Hertz, described by

$$SRF = \frac{1}{2\pi\sqrt{L_{eff} \cdot C_{eff}}} \tag{15.36}$$

A plot of the terminal impedance as a function of frequency is shown in Figure 15.72. Also superimposed on this plot is the effective inductance between the device terminals. As is evident from this plot, the impedance peaks around the SRF, as well, the effective inductance changes quickly from a positive to negative quantity (signifying capacitive behavior).

For impedance matching applications it is important to have constant inductance versus. frequency near the operating frequency. This requires selecting an inductor with an SRF well above the designated frequency of operation. A rule of thumb is to select an inductor with an SRF that is a decade higher than the operating frequency. Typically, the SRF is higher for lower inductance values. Also, tight inductance tolerance is desired. Wire-wound inductors typically achieve tighter tolerances than multilayer or thick film-type inductors.

If the inductor is to be used as a single-element high-frequency choke to supply a DC level via a RF terminal to the DUT, the inductor selection will be based on the highest reject frequency required. At the self-resonant frequency of an inductor, the series impedance is at its maximum. So, for a simple RF choke, the inductor selection will be based on finding an inductor whose SRF is very near the frequency where the choking is required. For a wider frequency range, a choke network with multiple inductors in series can be designed.

A high quality factor Q results in a narrow bandwidth, which is important if the inductor is to be used as part of a LC tank circuit of a VCO or a narrow band filter application. High Q also leads to a low insertion loss minimizing power consumption. The Q factor of an inductor is defined as

$$Q = \frac{\Im m\{Z\}}{\Re e\{Z\}} \tag{15.37}$$

where Z represents the impedance seen between the two terminals of the device. Typically, wire wound inductors have a much higher Q value than a multilayer inductor of the same size and value.

Other important parameters associated with an inductor are its current rating, DC resistance and temperature rating. While a small size is typically desired to keep the board space small and the operating frequency high, the laws of physics limit how small an inductor can be made for a given application.

Applications that require large current levels requires larger wire or more strands of the same wire size to keep losses and temperature rise to a minimum. Larger wires reduce the DC resistance and increase the Q factor, but will also increase the size and, in turn, reduce the self-resonance frequency. The current ratings for wire-wound inductors are typically better than the current rating for multilayer inductors of the same inductance value and size. The current capacity and lower DC resistance can be achieved by using a ferrite core inductor with a lower turns count. Ferrite, however, may introduce other limitations such as larger variations in the inductance value with temperature, looser tolerances, and a lower Q factor.

Inductor are often constructed with a magnetic core to increase its inductance. Depending on the magnetic core material used, the inductor may also exhibit lossy behavior at high frequencies, resulting in an apparent increase in the series resistance of the inductor. In fact, a class of inductors called *ferrite beads* is intentionally designed with very lossy core materials to achieve a component with near-zero resistance at low frequencies and higher resistance at higher frequencies. A typical ferrite bead impedance chart is shown in Figure 15.73. As shown, both the inductance $Z_L(f)$ and the series resistance $R(f)$ of the ferrite bead are functions of frequency.

Ferrite beads are useful for blocking high-frequency interference signals. They can reduce AC crosstalk from one circuit to another while allowing DC current to flow freely. For example, a ferrite bead can be placed in series with a power supply to prevent supply current spikes drawn by one circuit block from disturbing another circuit block, as shown in Figure 15.74. Both the inductance and resistance of the bead are near zero at DC; thus power supply current can flow freely

Figure 15.73. Typical ferrite bead impedance chart.

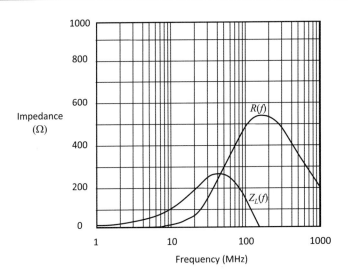

Figure 15.74. Ferrite beads in power supply connections.

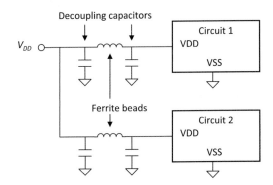

through the ferrite beads. Note that the schematic symbol for a ferrite bead is the same as that for an inductor. This is because a ferrite bead is an inductor with a lossy core.

15.7.8 Transformers and Power Splitters

Transformers are sometimes used on DIBs to translate a high-frequency single-ended signal into a differential signal or vice versa. Unfortunately, the transformer's frequency response is highly dependent on the output impedance of the transmitting circuit as well as on the input impedance of the receiving circuit. Consequently, the frequency response of the transformer is difficult to accurately calibrate, since it may change from one DUT to the next.

15.11. What is the self-resonance frequency of a 1-nF capacitor with a package size of 0402? Use the ESL data of Table 15.4.

ANS. 214.6 MHz.

15.12. A Vcc line of a 5.5-GHz circuit needs to be decoupled with a capacitor. A 0201-sized capacitor can be used. What maximum capacitor value can be selected, when the SRF should be 30% above the blocked frequency?

ANS. C = 1.239 pF.

15.13. What is the Q value of a 1-nF capacitor of the size 0402 if the ESR is per data sheet 0.1 Ω at 1 MHz?

ANS. 1592.

Power splitters are typically used in RF and microwave systems, but occasionally find use on mixed-signal DIBs. Their useful operation can be limited to a small range of frequencies, whereas resistive power splitters are generally able to handle a fairly wide range of frequencies.

Active circuits such as instrumentation amplifiers and other op-amp circuits are often superior to transformer circuits, since they can be accurately calibrated. Also, they present a consistent, high-impedance load to the DUT. Transformers and power splitters, by contrast, present a low-impedance inductive or resistive load to the DUT. Since many DUTs cannot drive low impedances, transformers and power splitters are often unusable. However, because they can pass frequencies well above those passed by active op-amp circuits, they are sometimes the only viable choice. Also, when attached to differential DUT inputs, transformers allow the DUT to set its own common-mode voltage at the differential input. By contrast, the active differential outputs of an op-amp single-ended to a differential converter forces the common-mode input voltage to a predetermined voltage, which may or may not be acceptable for a particular DUT input.

15.8 COMMON DIB CIRCUITS

15.8.1 Analog Buffers (Voltage Followers)

Sometimes a device output is incapable of driving the parasitic capacitance presented by the traces, cables, and relays leading to a tester instrument. An analog voltage follower with higher capacitive drive capability can be used to buffer the output signal before it is passed to the tester. The primary concerns with analog buffers are offset, signal bandwidth, and added noise from the amplifier. Generally, higher-bandwidth amplifiers generate more noise while low-noise amplifiers have a limited bandwidth. Offset and gain errors can be removed through a focused calibration process (see Chapter 5). However, noise generated from the buffer may or may not be removable through a calibration process. Figure 15.75 shows a simple op-amp buffer circuit including a relay for calibration and functional checking of the buffer.

Sometimes, an oscillating DUT amplifier can be stabilized using a small series resistor between the amplifier output and the tester. The resistor restores phase margin to the amplifier, eliminating the need for a buffer amplifier. This saves considerable complexity on the DIB.

15.8.2 Instrumentation Amplifiers

Another type of commonly used op-amp circuit is the differential to single-ended converter, also known as the *instrumentation amplifier*. Figure 15.76 shows an instrumentation amplifier

Figure 15.75. Buffer amplifier with calibration relay.

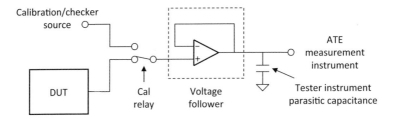

Figure 15.76. Op-amp differential to single-ended converter with calibration relays.

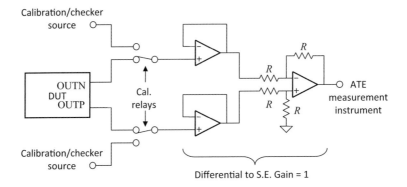

constructed using three op amps and four matched resistors. Today, such an arrangement can be found in a single IC with excellent performance specifications The two voltage followers at its input give the instrumentation amplifier a very high input impedance. Without these voltage followers, the resistors surrounding the third amplifier would present a load impedance of $2R$ to one of the DUT outputs and R to the other output. Of course, if the DUT outputs can drive these resistors without a problem, then the voltage followers are unnecessary. Like the previous analog buffer circuit, this circuit needs calibration relays to allow focused calibrations and checkers using a differential calibration/checker source.

The reason this type of circuit is sometimes needed on a DIB is that certain tester instruments are not capable of receiving a differential signal. In such instances, the differential DUT output must be converted to a single-ended signal before the tester can measure it.

A related problem arises with differential inputs that must be driven from an instrument having only a single-ended output. The simple single-ended to differential converter in Figure 15.77 uses an inverter to generate an inverted image of a single-ended input. The differential signal is centered around a common mode voltage, V_{CM}, which is set to the desired voltage level by a tester instrument or by a DUT V_{MID} output. Notice that this single-ended to differential converter has a gain of 2 V/V. Therefore, the signal level of the single-ended source must be attenuated to one-half of the desired differential signal level.

Since the inverter in Figure 15.77 may introduce a significant phase shift at higher frequencies, this circuit is not ideal for high-frequency differential circuits. A somewhat better circuit can

Figure 15.77. Single-ended to differential converter.

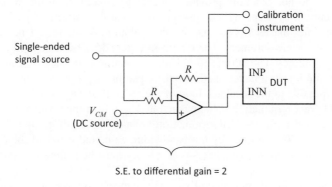

Figure 15.78. V_{MID} reference adder.

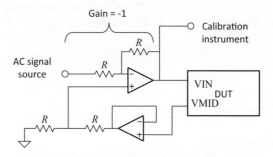

be built using a voltage follower in series with the noninverted signal to balance out the phase shift somewhat. Since the amplifiers are not in the same configuration, even the enhanced circuit produces some phase mismatch at very high frequencies. Fortunately, the circuit in Figure 15.77 is good enough for many DUTs having differential inputs.

15.8.3 V_{MID} Reference Adder

Sometimes, a DUT produces a V_{MID} voltage to which all input signals must be referenced. This type of input is commonly used in microphone inputs. Since microphones are basically differential signal generators, any noise or ripple present on the V_{MID} output gets canceled by the differential input circuits of the DUT. Therefore, an input signal generated by the tester has to fluctuate with any noise and ripple on the V_{MID} signal to simulate the differential nature of a microphone. A V_{MID} adder can be built using the simple op-amp circuit in Figure 15.78. The V_{MID} signal is added to the input signal from the tester, simulating the differential nature of a microphone. Since many DUTs are designed with a very weak V_{MID} output driver, a voltage follower is used to buffer the V_{MID} output before it is passed to the op-amp adder. Obviously, if the DUT's V_{MID} driver is strong enough to drive a load of $2R$, then this extra buffer is unnecessary.

The limitations of this circuit are its bandwidth and the small amount of noise generated by the op amps. The gain and offset of this circuit must be calibrated for maximum accuracy. A transformer might be used instead of the active op-amp circuit to work around noise and bandwidth limitations, but the transformer's frequency response would have to be carefully calibrated.

15.8.4 Current-to-Voltage and Voltage-to-Current Conversions

Tester instruments do not generally provide a means to directly measure AC currents. We can measure an AC current by dropping it across a resistor, but the parasitic capacitance of the tester instruments sometimes makes this an unacceptable solution. A low-impedance voltage output is a preferable signal for measurement. The circuit in Figure 15.79 can be used to convert current outputs to voltage outputs. The current-to-voltage translation is defined by Ohm's law ($V=I\,R$), although the op amp injects a factor of -1 into the equation ($V_{out}/I_{in}=-R$). The DUT sees a virtual ground at its output due to the feedback loop of the op amp. If the DUT is designed to drive its current into a different termination voltage (such as V_{MID}), then the noninverting input of the op amp can be connected to the desired termination voltage instead of ground. The simple I-to-V amplifier in Figure 15.79 is limited by the bandwidth, gain, and offset of the op amp. Its exact offset and voltage-over-current "gain" versus frequency must be calibrated using a focused calibration process.

Tester instruments also do not generally provide a means to force AC currents into the DUT. A transconductance amplifier circuit (Figure 15.80) can be used to make this conversion. In this circuit, the instrumentation amplifier senses the voltage drop across the source resistor, R_S, and feeds that voltage back to the inverting input of an op amp. The op amp adjusts its output voltage until the current forced across R_S generates a voltage drop equal to V_{in}. Of course, this transconductance amp must be calibrated at all frequencies of interest if maximum accuracy is to be achieved.

Figure 15.79. Current to voltage converter.

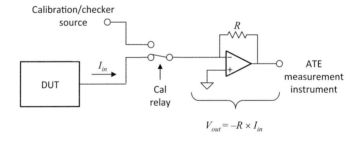

$$V_{out} = -R \times I_{in}$$

Figure 15.80. Voltage-to-current converter (transconductance amplifier).

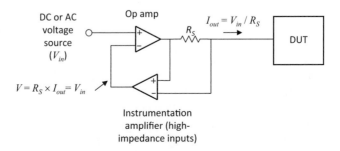

15.8.5 Power Supply Ripple Circuits

For some reason, mixed-signal testers have never included easily programmed ripple sources compatible with PSRR tests. PSRR tests require that we add a sinusoidal or multitone signal to one or more of the DUT's power supply voltages. This is not as easy as it might seem, since the output of power supplies include large bypass capacitors specifically designed to dampen ripple on the supply voltage. If the desired ripple cannot be provided by the tester itself, the test engineer can utilize any of a number of DIB circuits.

The simplest ripple injection approach takes advantage of the programmable DUT power supply's Kelvin sense line to force it to ripple its output. This ripple scheme is illustrated in Figure 15.81. The ripple source (a sine wave generator or arbitrary waveform generator) applies an AC signal to the sense line of the Kelvin connection through a resistor.

The power supply is forced to adjust its output to maintain a fixed voltage at its sense line. The power supply thus behaves as an op amp, with the sense line acting as a high-impedance inverting input and the programmed voltage level acting as a DC source at the op amp's noninverting input. The addition of the two resistors in the sense path forms an inverting gain stage, as shown in Figure 15.82. Once the Kelvin-connected DUT source is redrawn in this manner, it is easy to see how an AC signal can be injected into the power supply's output voltage. The output signal is given by

$$v_S = -\frac{R_2}{R_1}(v_R - V_P) \tag{15.38}$$

where v_S is the DUT power supply voltage (including its normal DC value and the injected AC ripple), v_R is the signal from the ripple source, and V_P is the programmed DC voltage level set by the test program.

Figure 15.81. Kelvin sense ripple injection circuit.

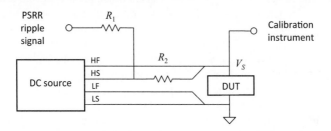

Figure 15.82. Equivalent op-amp inverting gain stage.

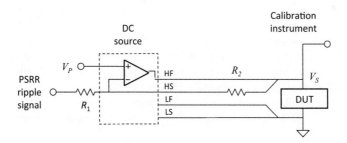

Figure 15.83. Equivalent ripple circuit with DC blocking capacitor.

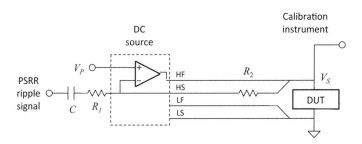

The values of R_2 and R_1 are chosen to give an attenuation, rather than a gain. This allows very small ripple voltages to be applied to the power supply using a fairly large ripple source amplitude. (Remember from Chapter 5 that large tester signal amplitudes are desirable because they are less susceptible to noise.) Values of $R_1 = 10$ kΩ and $R_2 = 1$ kΩ are commonly used, giving the ripple circuit a gain of 1/10.

If the circuit in Figure 15.82 is used, the ripple signal must include a DC offset equal to V_P. Otherwise, the ripple signal will introduce a DC offset into the power supply voltage. To use a ripple signal with no DC offset, a DC blocking capacitor can be added in series with the ripple source. Of course, this turns the ripple circuit into a first order high pass filter, as shown in Figure 15.83.

The cutoff frequency F_c of this filter is given by

$$F_C = \frac{1}{2\pi R_1 C} \quad \text{Hz} \tag{15.39}$$

Note that the circuit in Figure 15.83 includes a calibration path. This connection is absolutely necessary, since the frequency response of the Kelvin ripple circuit is not known. Each frequency in the injected signal must be calibrated during a focused calibration process to achieve acceptable accuracy in the injected power supply ripple signal. If the DUT needs a large DIB decoupling capacitor on the rippled power, the capacitor must be removed from the circuit temporarily (with a relay) to prevent it from damping the ripple signal. The Kelvin ripple circuit has one major drawback. It is impossible to ripple most power supplies at a frequency higher than a few kilohertz. At higher frequencies, a different approach must be taken.

One possibility is to use a DIB ripple buffer, which can provide the DC plus AC signals needed to drive the DUT during the supply ripple tests. This circuit is simply an op-amp adder circuit with a high-current buffer amplifier connected to its output (Figure 15.84). Again, a calibration path is added to improve the circuit's accuracy. This supply ripple circuit can be inserted into the power supply line using a SPDT relay, as shown.

Note that the large decoupling capacitor is automatically removed when the relay is thrown to the ripple circuit output. The large decoupling capacitor, typically a 10-µF electrolytic variety, provides relatively low-frequency currents to the DUT. Its own series inductance is many times that of the relay; thus we can safely connect and disconnect it using the relay. A smaller decoupling capacitor is often needed to provide higher-frequency currents to the DUT. It must be located very close to the DUT to minimize series inductance between the capacitor and the DUT power pin. Since the ripple circuit relay would provide too much series inductance, the small capacitor must be located next to the DUT. Therefore, the small high-frequency capacitor cannot be removed by the relay during the power supply ripple tests.

Figure 15.84. Buffer amp ripple injection circuit.

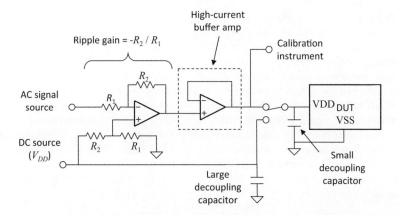

15.9 COMMON DIB MISTAKES

15.9.1 Poor Power Supply and Ground Layout

One of the most common sources of noise injection in mixed-signal DIBs is poor power and ground layout. The best way to avoid problems with power distribution and grounding is to use as many planes as needed, without regard to DIB cost. Although each layer in a multilayer DIB adds fabrication cost, the expense is fairly negligible compared to the production yield loss due to poor DIB performance. Therefore, a good DIB might include one or two layers dedicated to digital transmission line ground and noisy current returns, one layer dedicated to analog current returns, one layer dedicated to low-current analog ground (quiet ground serving the purpose of a zero-volt reference layer), and at least one layer dedicated to split power planes. Thus a 10-layer DIB may contain five or six layers dedicated to power and ground distribution.

15.9.2 Crosstalk

Another common problem on mixed-signal DIBs is crosstalk, especially between digital and analog signal lines. The digital-to-analog crosstalk problem can be dramatically reduced by placing analog signals on a separate PCB layer from digital signals, with an analog ground plane between the two signal layers. Analog-to-analog crosstalk can be minimized by simply realizing which signals are most susceptible (i.e., which signals have the highest impedance) and preventing high-frequency, high-amplitude signals from passing nearby. Also, the sensitive high-impedance nodes should be as short as possible to avoid crosstalk and coupling of external noise sources such as radio waves.

One of the most common sensitive nodes on a mixed-signal DUT is its current reference input. This input is typically tied to V_{DD} or ground through a very high-impedance bias resistor. The node between the bias resistor and the DUT is an extremely sensitive one. Noise injected into this node will translate directly into noise throughout the DUT. Therefore, current bias nodes should always be kept extremely short, preferably surrounded by a shield ring.

Another type of sensitive node is the DUT reference voltage. Reference voltages are typically driven by a low-impedance tester source and are therefore less susceptible to crosstalk than high-impedance bias nodes. However, any noise injected into the reference voltage will translate

---

Exercises

15.14 When placed in a particular test mode, a DUT routes a weak internal circuit node to an analog test pin for measurement. The signal at the test node contains frequencies from DC to 100 kHz. Its output impedance is 10 kΩ (purely resistive). We wish to measure the signal using a digitizer having a distributed cable capacitance of 350 pF plus a lumped input capacitance of 50 pF. Can this signal node be measured directly, or will a buffer amplifier be needed on the DIB?

ANS. At relatively low frequencies, the DUT output impedance forms a low-pass filter with the digitizer input capacitance. The 3-dB cutoff frequency of the filter is 3.98 kHz, preventing accurate measurement at 100 kHz. A local buffer amplifier, such as the one in Figure 15.75, will be needed on the DIB.

15.15 The Kelvin PSRR ripple circuit illustrated in is constructed using the values $R_2 = 1$ kΩ and $R_1 = 10$ kΩ. The power supply is programmed to 3.3 V DC. Describe a signal, $v(t)$, at the input to R_1 that would produce a DUT power supply ripple of 75 mV RMS at 1 kHz. (Assume no errors due to circuit bandwidth, component mismatch, etc.)

ANS. $v(t) = 3.3\text{ V} - \sqrt{2} \times 75\text{ mV} \times \dfrac{10\text{ k}\Omega}{1\text{ k}\Omega}\sin(2\pi f \cdot t)$

$= 3.3\text{ V} - 1.06\text{ V} \times \sin(2\pi \times 1000t)$

15.16 Repeat exercise 15.16 using the PSRR buffer circuit illustrated in Figure 15.84. (Assume V_{DD} is set to 3.3 V DC.)

ANS. $v(t) = -1.06\text{ V} \times \sin(2\pi \times 1000t)$

---

directly into noise in the DUT circuits. Therefore, reference voltage nodes should also be laid out as if they were extremely vulnerable to crosstalk.

15.9.3 Transmission Line Discontinuities

Small discontinuities in transmission lines can lead to glitches on the rising and falling edges of very fast digital signals. Such glitches can sometimes lead to timing errors or double-clocked logic in the DUT. The discontinuities are caused by lumped capacitance or inductance at transition points along the transmission line. For example, a lumped capacitance and/or inductance exists whenever a digital signal trace is routed between layers through a via or other through hole. It is best to avoid routing digital signals from one layer to another, unless absolutely necessary.

15.9.4 Resistive Drops in Circuit Traces

As we saw in Example 15.1, even relatively short traces may have a series resistance of several hundred millohms. If we try to force current through such a trace, we will get a voltage drop due to the parasitic resistance of the trace (this effect is also referred as IR drop). Sometimes these voltage drops are unimportant, but other times they can lead to errors nearly as large as the parameter we are trying to measure. The test engineer should always consider the effects of series resistance on each trace on the DIB. If the series resistance is serious enough to cause a problem, the trace can either be made wider, or a sensing circuit such as a Kelvin connection can be used to compensate for the resistive drops in the PCB trace.

15.9.5 Tester Instrument Parasitics

The various cables and wires that connect a DUT to a tester's instruments can present a significant capacitive load to the DUT's pins. Often, the loading is high enough to cause gain errors, phase shifts, or even DUT circuit oscillations. It is very important for a test engineer to ask the design engineer responsible for each DUT circuit what its capacitive drive capabilities will be.

If the output impedance of the DUT is incompatible with the tester's load capacitance, then a voltage follower will probably be needed on the DIB buffer the DUT's output. Even if the DUT is designed with a low output impedance, the load capacitance of some tester instruments may cause it to break into oscillations. The test engineer and design engineer should determine whether the unbuffered tester inputs might cause any DUT oscillations. If so, analog buffers must be added to the DIB design.

15.9.6 Oscillations in Active Circuits

Operational amplifiers used in buffer amplifiers or other DIB circuits may break into oscillations if they are not laid out properly. For example, the inverting and noninverting inputs to an op amp are extremely sensitive to parasitic capacitance. If these PCB traces are laid out so that they are more than a few tenths of an inch long, the amplifier will often break into oscillations. This problem is commonly seen in the nulling amplifier circuits described in Chapter 3, "DC and Parametric Measurements."

Another source of oscillations is poor power supply and decoupling capacitor layout. If the decoupling capacitors attached to an amplifier's power supplies are not positioned very close to the amplifier's power pins and ground plane, then the amplifier may break into oscillations. The oscillations are due to the extra parasitic inductance of the connecting PCB traces. This is especially true of high-bandwidth operational amplifiers. Decoupling capacitors in general should always be placed very close to their supply pin.

Oscillating amplifiers pose a particularly tricky problem. When measuring a DC offset using an oscillating buffer amplifier, the tester's DC voltmeter ignores the oscillation. Instead, it measures the average voltage level at the oscillating amplifier's output, which may or may not have any relation to the DUT signal to be buffered. Thus significant DC errors can be introduced by the oscillating amplifier.

15.9.7 Poor DIB Component Placement and PCB Layout

If the test engineer gets nothing else out of this chapter, at least one fact should come across loud and clear. The physical layout of the DIB is extremely critical to mixed-signal DUT performance. We have seen many cases throughout this chapter in which a short PCB trace is the ideal interconnection. Short traces have less parasitic resistance, inductance, and capacitance than long traces. The best way to achieve short PCB traces is to arrange the DIB components in a way that allows short traces, especially in critical nodes.

Component placement, power and ground schemes, trace layout, and other physical decisions must be made with knowledge of the DUT and DIB circuits and their required performance. This fact makes it very difficult for automatic routing software to lay out mixed-signal DIBs. In fact, it is very difficult to get a good DIB layout from a manual process, unless the test engineer sits with the PCB designer as the critical components and traces are placed on the DIB. This fact escapes many novice test engineers, who literally throw the DIB schematic into the PCB designer's lap and walk away, assuming all will turn out well.

15.10 SUMMARY

A good DIB is one of the most critical elements in a successful mixed-signal test solution. Without good DIB performance, the DUT may be unable to meet its specifications, regardless of the quality of the test code. Many things lead to good mixed-signal DIB design, including proper component selection and placement, proper power and ground layout, proper PCB stackup, and proper attention to parasitic components related to PCB traces and DIB components.

A good schematic design is essential as well as a good DIB layout. If the test engineer forgets to provide for an important connection between the tester and the DUT, then all the clever software routines in the world will not make the DIB usable. Also, if the test engineer does not provide for all the necessary hardware hooks to calibrate the DIB circuits, then it will be equally useless because it will not provide the necessary accuracy.

Often, the only way to produce a good DIB schematic is to have a complete test plan and data sheet to begin with. If the design specifications, DUT pinout, and test list are constantly changing, then it will be impossible to design a good DIB. Also, if the test engineer and design engineers do not work together closely, the DIB and DUT will often be incompatible with one another. It is critical for the test engineer to review his test plan and DIB design with the design engineers before the DIB is laid out and fabricated. Otherwise, the DIB may turn out to be an expensive but useless piece of test hardware, proving once again that concurrent engineering is critical to mixed-signal product development.

PROBLEMS

15.1. Calculate the parasitic resistance of a 13-in. PCB trace having a width of 20 mils and a thickness of 1 mil. A pair of these traces are used as the high-force and low-force lines of a Kelvin-connected voltage regulator located on the DIB. The regulator feeds a 3.3-V DC signal to a 5-Ω load resistance. How much current will flow through the four Kelvin lines? What will be the differential voltage between the high-force and low-force output of the voltage regulator, measured at the regulator side of the PCB traces?

15.2. Using Eq. (15.3), calculate the parasitic inductance per unit length and total inductance of a 3-in. stripline trace having a width of 50 mils and a spacing of 8 mils to the current-return ground plane. If this trace feeds a 5-MHz, 625-mV RMS sinusoidal signal to a 100-Ω load resistance, what will be the error of the RMS voltage at the load as a percentage of the source voltage? (Assume zero trace resistance and zero signal source impedance.) Compare your answers with those obtained using the refined inductance estimate of Figure 15.10.

15.3. Using Eq. (15.10), calculate the parasitic capacitance per unit length and total capacitance of a 3-cm-long, 35-mil-wide stripline trace with a spacing of 10 mils to the ground plane, fabricated on a Teflon® PCB. Compare your answer with that obtained using the refined capacitance estimate of Figure 15.16.

15.4. The stripline trace in Problem 15.3 is connected to the 50-kΩ source impedance of a DUT output. What are the gain and phase shift of the DUT output signal as a function of frequency, compared to its unloaded output (i.e., if it were not connected to the trace)? (Use the estimate of capacitance from Figure 15.16.)

15.5. Using Eq. (15.13), calculate the parasitic capacitance per unit length and total capacitance between two 8-in.-long, 10-mil-wide coplanar traces separated by a spacing of 20 mils with a thickness of 1 mil, fabricated on an FR4 PCB. Compare your answer with that obtained using the refined capacitance estimate of Figure 15.18.

15.6. Using Eq. (15.10), calculate the parasitic capacitance of an 11.5-in.-long, 15-mil-wide stripline trace with a spacing of 12 mils to the ground plane, fabricated on an FR4 PCB.

This trace feeds a 300-kHz, 1-V RMS sinusoidal signal from a DUT output having a 75-kΩ output resistance to an unterminated coaxial cable (tester instrument input) having a distributed capacitance of 35 pF. How much will the distributed capacitance of the trace and coaxial cable capacitance attenuate the DUT signal? (Express your answer as a voltage gain in decibels.) Would a DIB buffer amplifier be required for this output? Compare your answer with that obtained using the refined capacitance estimate of Figure 15.16.

15.7. An 8-in. stripline trace is fabricated on an FR4 PCB with a width of 24 mils. It is separated from its ground plane by a spacing of 16 mils. Using the refined estimates of Figure 15.10 and Figure 15.16, calculate the stripline's parasitic capacitance per meter and inductance per meter. What is the stripline's characteristic impedance? If we want to lower the characteristic impedance, would we make the layer spacing from trace to ground larger or smaller? If we were constrained to a layer spacing of 16 mils, would we make the trace wider or smaller? If we increased the length of the trace to 16 in., what would happen to the characteristic impedance?

15.8. What is the velocity of a signal propagating along the stipline of Problem 15.7? Express your answer in meters per second and in a percentage of the speed of light. What is the stipline's propagation delay? What is its distributed capacitance?

15.9. What is the wavelength, in meters, of a 20-MHz sine wave traveling along the FR4 stripline in Problem 15.7? Can we treat the parasitic reactances of the stripline as lumped elements, or do we have to treat the stripline as a transmission line at this frequency?

15.10. A 1.200-GHz sinusoidal signal is transmitted from a DUT output to tester digitizer along a 20-cm 50-Ω terminated coaxial cable having a signal velocity of $0.8c$. What is the wavelength of the transmitted signal as a percentage of the cable length? What is the distributed capacitance of this coaxial cable? What is the phase shift, in degrees, between the DUT output and the digitizer input? (Your answer may exceed 360 degrees.)

15.11. A DUT output signal under test contains frequencies from DC to 44 kHz. The DUT output impedance is guaranteed to fall between 50 and 100 Ω (purely resistive). We wish to measure the signal using a digitizer having a distributed cable capacitance of 120 pF plus a lumped input capacitance of 5 pF. Can this signal node be measured directly, or will a buffer amplifier be needed on the DIB?

15.12. A DUT with a normalized output impedance $z_{DUT} = 0.2 + j3.2$ Ω is to be matched with an LC network to 50 Ω at 1 GHz, where the matching network needs to be design in such a way that the device output will not be shorted to GND.

15.13. A DUT with a normalized output impedance $z_{DUT} = 0.7 + j3.2$ Ω is to be matched with an series-C shunt-C network to 50 Ω at 3 GHz. What are the values of these two capacitors?

15.14. Calculate the length and position for the open stub line matching of 50Ω with a load of $45 + j75$. What is the denomralized input admittance of the open-circuit stub line?

15.15. Calculate the length and position for the short-circuit stub line matching of 50Ω with a load of $65 + j150$ Ω. What is the normalized input admittance of the short-circuit stub line?

15.16. What is the self-resonance frequency of a 10-nF capacitor with a package size of 0603. Use the ESL data of Table 15.4.

15.17. What is the Q value of a 10-nF capacitor of the size 0603 if the ESR is per data sheet 0.15 Ω at 1 MHz?

15.18. The in-phase and quadrature outputs (IOUT and QOUT) of a cellular telephone DUT are produced by two supposedly identical amplifier circuits. According the the data sheet, the resistive output impedance of each amplifier circuit is specified at 75 Ω (min) to 125 Ω (max). We wish to measure the phase mismatch between these two outputs when a 70-kHz sine wave is driven from each output. Assuming a perfectly matched capacitive load of 300 pF from each of two digitizers, what is the worst-case phase mismatch caused by the tester loading? How could we eliminate the tester-induced phase error problem?

15.19. The Kelvin PSRR ripple circuit illustrated in Figure 15.81 is constructed using the values $R_2 = 1$ kΩ and $R_1 = 4.7$ kΩ. The power supply is programmed to 5.0 V DC. Describe a signal, $v(t)$, at the input to R_1 that would produce a DUT power supply ripple of 100 mV peak-to-peak at 2.4 kHz. (Assume no errors due to circuit bandwidth, component mismatch, etc.)

15.20. Repeat Problem 15.13 using the PSRR buffer circuit illustrated in Figure 15.84. (Assume V_{DD} is set to 5.0 V DC.)

15.21. Explain the dissipasion of a RF line and which lines are showing no dissipation.

REFERENCES

1. J. D. Kraus and K. R. Carver, *Electromagnetics*, 2nd edition, McGraw-Hill, New York, 1973, ISBN 0070353964, p. 85.

2. H. W. Johnson and G. Martin, *High Speed Digital Design: A Handbook of Black Magic*, Prentice Hall, Englewood Cliffs, NJ, April 1993, ISBN 0133957241.

3. B. Wadell, *Transmission Line Modelling Handbook (Artech House Microwave Library)*, Artech House, Boston, June 1991, ISBN 0890064369.

4. H. W. Ott, *Noise Reduction Techniques in Electronic Systems*, 2nd edition, John Wiley & Sons, New York, March 1998, ISBN 0471850683.

5. Y. Konishi, *Microwave Electronic Circuit Technology*, Marcel Dekker, New York, 1998.

CHAPTER 16

Design for Test (DfT)

Design for test (DfT, also DFT*) is a major topic of interest in the automated testing field. Any design methodology or circuit that results in a more easily or thoroughly testable product can be categorized as DfT. DfT, when properly implemented, can offer lower production costs and higher product quality by achieving a higher test coverage. Extensive literature exists on the subject of DfT,[1-3] though much of it pertains to purely digital circuits. Since so much digital DfT literature is already available elsewhere, this chapter will concentrate mostly on analog and mixed-signal DfT, and will give some outlook on RF DfT. Nevertheless, some of the more common digital concepts will be reviewed to give the reader a basic overview of digital DfT.

16.1 OVERVIEW

16.1.1 What Is DfT?

There are many types of DfT. Some DfT approaches are highly structured, using industry-defined standards. Other approaches are totally ad hoc, invented by the design engineer or test engineer to solve a specific test problem on a particular device or category of devices. Some DfT concepts are based on built-in circuits that allow easier or more complete testing. Other methodologies, such as increasing design margin to reduce test cost, are equally cost effective but may not be recognized as DfT concepts. In the end, the choice of DfT approach depends very much on (a) the specifics of the device under test (DUT) and (b) the demands placed on it by its system-level application.

In the past, design engineers were sometimes reluctant to add testability features to a device, since DfT added design cycle time, die area, and/or power consumption. Their reluctance was often reinforced by managers who judged design engineering performance based mainly on these

---

*This text discusses both design for test and the discrete Fourier transform. For clarity, the lowercase notation (DfT) is used throughout this text when referring to design for test while the uppercase notation (DFT) is used when referring to the discrete Fourier transform.

criteria rather than the overall cost effectiveness, marketability, and quality of the finished product. Fortunately, the attitude has changed in recent years from reluctance to enthusiasm as design engineers, managers, and customers have embraced the competitive advantages of DfT. Now DfT is seen as a major technological differentiator that can reduce production costs, enhance quality control, and even provide customers with value-added testability features for use in their system-level products.

Although we have only devoted a single chapter to the topic of DfT, this subject is of extreme importance to the semiconductor industry. Our light treatment of the topic is due to the broad, introductory nature of this text and to the fact that this book is targeted toward test engineers rather than design engineers. While test engineers are not typically expected to implement the DfT concepts suggested in this chapter, they are expected to participate in the DfT planning phase of new product development.

Test engineers also need to understand the types of DfT they may encounter when developing a test program for a new device. Our intention is to provide the test engineering professional a cursory introduction to digital and mixed-signal DfT rather than providing a design engineer with detailed knowledge to implement digital and mixed-signal DfT circuits.

16.1.2 Built-In Self-Test

Built-in self-test (BIST) circuits allow the DUT to evaluate its own quality without elaborate automated test equipment (ATE) support. Although BIST and DfT are often treated as if they were separate concepts, BIST is actually a type of DfT. Digital BIST circuits usually return a simple pass/fail bit or a multibit "signature" that allows the ATE tester to evaluate the quality of the device with a very simple (i.e., low cost) test. A BIST circuit may require little more than a power supply and a master clock from the tester. Since the DUT tests itself using BIST, a much less expensive ATE tester can be used. The limited tester resources required by BIST and the ability to perform parallel testing of multiple circuits on the DUT are key advantages of BIST-based testing methodologies.

Unfortunately, analog BIST technology has lagged behind digital BIST because of difficulties in guaranteeing the accuracy of signals generated and measured on-chip. Many books and technical papers have been written on digital BIST techniques,[4-6] but fewer have been written about analog and mixed-signal BIST.[7-11]

16.1.3 Differences Between Digital DfT and Analog DfT

DfT for purely digital designs has been extensively utilized for many years. Using a variety of software tools and industry standards, a digital design engineer can follow a well-defined path toward a testable design. Software tools can automatically insert the necessary DfT circuits into a digital design. The same tools can automatically generate the digital patterns to test the design.

For the most part, digital DfT test techniques are based on verifying the structural behavior of a circuit as opposed to its functional behavior (i.e., verify that all the transistors are present and can be switched between two logic states versus the device conforming to a set of system-level specifications listed in the device data sheet). While a structural-based test greatly simplifies the execution of the test, it is limited to identifying gross or spot defects in the IC as opposed to subtle changes in component behavior and its impact on the functional operation of the device.

Mixed-signal DfT is much less standardized because the testing requirements and failure mechanisms for the analog circuits in a mixed-signal device are often not particularly well understood or well defined. Digital circuits, for example, can be separated into subcircuits using a divide-and-conquer approach. The subcircuits are fairly independent from one another except for race conditions and other timing problems. To a large extent, these timing problems can be avoided using additional automated software tools. As a result, we can test the subsections of a digital circuit to guarantee the operation of the whole.

Mixed-signal circuits can also be subdivided, but the quality of the whole is seldom guaranteed by the quality of the parts. Analog circuits are frequently prone to obscure crosstalk problems, as well as other subtle interactions between circuit blocks. The divide-and-conquer approach is necessary for characterization and diagnosis of mixed-signal devices, but it may not be sufficient to guarantee the system-level specifications.

16.1.4 Why Should We Use DfT?

DfT circuits and methodologies offer a tremendous advantage in the marketplace. An IC designed without attention to testability may work perfectly well in the customer's application, but a competitor's IC may win in the marketplace, because of superior DfT features. The advantages of DfT include lower testing costs, higher product and process quality, ease in design diagnostics and characterization, ease in test program development, and enhanced diagnostic capabilities in the customer's system-level application. Let us look at some examples of each of these advantages.

16.2 ADVANTAGES OF DFT

16.2.1 Lower Cost of Test

Perhaps the most visible advantage of DfT is that it can lead to lower testing costs.[12] Consider the power-down logic block in Figure 16.1. This logic block controls the power-down status of all the circuit blocks in the device under test (DUT). The digital logic block accepts five digital inputs from a variety of external device pins and internal control register bits. It uses combinational logic to map the 32 possible input states to one of three valid power modes (normal mode, power-down mode, and standby mode). Each of the three power-down modes is expected to produce a unique combination of supply currents, I_{DDA} and I_{DDD}. The three power modes are determined by two power mode control lines, PWRMODE0 and PWRMODE1. The test engineer is required to verify the truth table of the power-down logic block, as well as to measure the I_{DDA} and I_{DDD} power supply currents in all three power modes.

From the design engineer's perspective, this circuit may appear to be well designed. The truth table for mapping the five input signals into three power-down states works perfectly with a minimum number of gates. The analog circuits power down as expected in each of the three modes. What could be wrong with this design?

Figure 16.1. Power-down logic block without DfT.

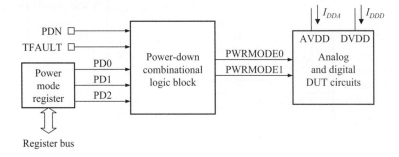

The problem is that the two outputs of the power-down logic block can only be observed by measuring the power supply current drawn by the DUT. Because the DUT requires power supply decoupling capacitors, the transition from one power state to another may require settling time as the decoupling capacitors charge and discharge. This is especially true when switching to the ultra-low-current power-down mode, since current discharges very slowly from the decoupling capacitors in this mode. In addition to the decoupling capacitors and other DIB circuits, the DUT and ATE meter may also require settling time. Due to the various settling time requirements, each measurement of I_{DDA} and I_{DDD} might take 5 ms to complete. The total test time for this logic block could therefore take as much as 320 ms (32 input state combinations times 2 supply currents times 5 ms). While 320 ms may sound like a reasonable test time, it is an eternity in production testing. Obviously, this is a ridiculous way to test a simple digital logic block. The problem is that the output bits of the power-down logic block can only be observed by making time-consuming analog measurements of I_{DDA} and I_{DDD}.

One possible solution to this problem is a very simple DfT circuit allowing the ATE tester to directly observe the power control bits of the logic block (Figure 16.2). This DfT circuit is implemented as a simple two-bit readback function using two unused bits in the existing power mode register (or two bits in a dedicated test register if there are no unused power mode register bits). Alternatively, the bits could be read back through a scan chain DfT structure, which we will discuss later. Using the readback DfT approach, all gates in the combinational logic block can be verified in less than a millisecond using a very fast digital pattern. After the digital logic has been verified, each of the three power supply current combinations only needs to be measured once. This DfT-based approach leads to a total test time of about 30 ms (3 input states times 2 supply current measurements times 5 ms) as opposed to the 320 ms of test time required for the non-DfT version of this design. The lower test cost resulting from the lower test time easily justifies the few logic gates required to add this type of readback DfT capability.

Another way in which DfT can reduce testing costs is by reducing the requirements of the ATE tester. A test that requires a 200 MHz digital pattern will cost more than a test operating at 50 MHz, because an ATE tester capable of running at high frequencies is generally more expensive than one that is only capable of lower frequencies. If the IC design engineer can find a way to test high-frequency signals using low-frequency stimulus and measurement hardware, the test cost savings can be substantial. The same comment applies to digital channel count. A device that requires a 64-channel tester will be much less expensive to test than a device that requires 512 or more channels. If the design engineer can find ways to reduce digital channel count (using multiplexed I/O pins, for example), then the test cost can be reduced significantly.

Figure 16.2. Power-down logic block with readback DfT.

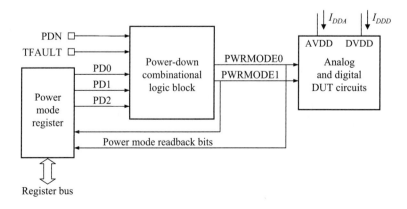

16.2.2 Increased Fault Coverage and Improved Process Control

Economic considerations are only one of the advantages of DfT. Another advantage is increased fault coverage. Fault coverage is defined as the percentage of possible failure modes that can be detected by a given test or series of tests. Therefore, increased fault coverage reduces the probability that defective devices will be shipped to the customer. While the economic advantage of lower test time is fairly easy to calculate, the economic advantage of happy customers is much harder to quantify financially. Many forms of DfT are designed to allow increased test coverage, with the understanding that the smiling face of a satisfied customer is well worth slightly higher test time and silicon cost. Devices going into the automotive market are expected to have 0 DPPM (defects parts per million). A typical automotive customer is expecting greater 98% percent test coverage for all digital and 100% test coverage for all analog and mixed signal circuits. This drives a special attention to the DfT and BIST techniques implemented into these devices.

I_{DDQ} testing is one such DfT methodology allowing detection of non-catastrophic defects in digital logic. An I_{DDQ} test configures all the gates in a CMOS device into a static digital state and then measures the tiny current leaking from power to ground. Excessive I_{DDQ} current indicates one or more resistive defects between power and ground that may or may not be detectable as a catastrophic failure in the operation of the DUT. When the I_{DDQ} tests suddenly begin rejecting many dies, it often indicates that the wafer fabrication process has gone awry for some reason, producing resistive shorts between circuit nodes. I_{DDQ} testing therefore allows the semiconductor wafer fab to monitor its process to detect and correct problems quickly.

16.2.3 Diagnostics and Characterization

When a design is first released to production, the new product undergoes a characterization process to determine whether or not it meets the customer's requirements. The first-pass design often has problems that must be corrected before a final version of the design can be released to production. To produce a production-worthy design in a short timeframe, we must be able to characterize the IC's performance and diagnose internal circuit problems very quickly. Lack of proper DfT observability circuits can make the diagnostic process extremely difficult or even impossible.

As an example of the diagnostic capabilities of DfT, consider a mixed-signal ADC channel (Figure 16.3). It may include several components, including an input amplifier, a programmable gain amplifier (PGA), a low-pass anti-aliasing filter, and the ADC itself. If a significant percentage of production devices fail the ADC channel's signal-to-distortion test, then the design needs to be corrected. Without DfT, it can be difficult to determine which of the four circuit blocks is introducing the distortion.

DfT can provide the necessary diagnostic capabilities to resolve problems like this. By providing test access points to each input and output node in the signal path, the test engineer can test each section of the circuit independently (Figure 16.4). The CMOS switch matrix at each circuit node is capable of three modes. First, it can pass the signal from one block to the next for normal

Figure 16.3. ADC channel without DfT.

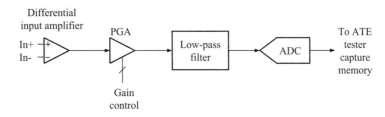

Figure 16.4. ADC channel with analog test access bus.

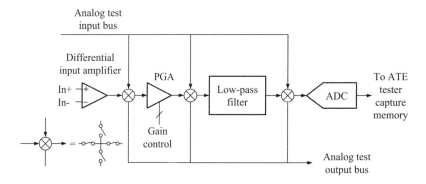

operation. Second, it can disconnect the output from one block and connect the input of the following block to the analog test input bus. The tester can then inject a test signal into the input of the circuit under test. Finally, the switch matrix allows observation of the output of the circuit under test through the analog test output bus.

The defective circuit can be isolated quickly by injecting signals into each block and observing the block's output. Once the defective circuit has been isolated, it can be redesigned to correct the problem. Observability and controllability test modes such as this are a major cornerstone of mixed-signal DfT. Similar observability/controllability circuits can be employed in digital circuits to accelerate the diagnosis of circuit problems. The CMOS switches are replaced by digital multiplexer circuits or scan cells to achieve the same results.

16.2.4 System-Level Diagnostics

A final advantage of DfT is that it often allows the customer to incorporate DfT into the end application (cellular phone, graphics card, etc.) more easily. Examples of IC-level DfT that are geared toward end application DfT include the IEEE 1149.1 and IEEE 1149.4 boundary scan standards. These standards, which allow chip-to-chip and circuit-to-circuit testing, will be reviewed in more detail later in this chapter. Customers may also request custom DfT test modes to allow easier integration of board-level DfT features.

16.3 DIGITAL SCAN

16.3.1 Scan Basics

Scan circuits allow a digital block to be isolated from surrounding circuits for the purpose of testing. Scan circuits facilitate a divide-and-conquer approach to testing that is exceptionally well suited to digital circuits. The scan circuits allow the normal inputs of the subcircuit under test to be replaced by tester-injected digital vectors. The tester injects the vectors into the subcircuit under test using a series of flip-flops called a *scan chain*.

A scan chain acts as both a serial-in, parallel-out (SIPO) shift register and a parallel-in, serial-out (PISO) shift register. The SIPO register allows a parallel stimulus vector to be shifted into the DUT through a dedicated serial input pin. A series of parallel stimulus vectors can be used to exercise the internal DUT circuits, or to verify interconnections from one IC to another on a finished printed circuit board. Scan chains would be somewhat useless if they did not allow observation of a circuit's response to the stimulus vectors; thus the PISO read-back capability is provided for capturing the output response from the circuit under test.

There are several different types of scan, including boundary scan (IEEE Std. 1149.1), full scan, and partial scan. Boundary scan is primarily directed toward the board-level chip-to-chip interconnection testing problem, although it is also extensible for testing internal circuits as well as in IEEE 1500. While 1149.1 boundary scan can be used to test internal circuits, full scan and partial scan methodologies are more commonly used for this purpose. These simpler forms of scan are somewhat more efficient in their use of extra circuitry than the more elaborate boundary scan architecture. Let us look at each of these methodologies very briefly to understand the advantages of scan over less-structured digital testing approaches.

16.3.2 IEEE Std. 1149.1 Standard Test Access Port and Boundary Scan

The IEEE Std. 1149.1 test access port and boundary scan standard[13] was developed by a consortium of industry participants from Europe and North America. The consortium, known as the Joint Test Action Group (JTAG, pronounced jay-tag), developed the standard to allow many different IC vendors to design chips compatible with a consistent board-level testing architecture. As a result, IEEE Std. 1149.1 is often referred to as JTAG 1149.1 boundary scan. JTAG 1149.1-compliant devices allow a system-level developer to test chip-to-chip interconnects on a finished printed circuit board. In addition, the system developer can reuse the production test vectors for each of the individual JTAG 1149.1 ICs to perform system-level diagnostics. In an 1149.1-based system, the test structures from multiple ICs can be tied together in a daisy chain configuration so that the entire system can be accessed through a single JTAG 1149.1 interface port. A simple illustration involving two ICs configured according to the JTAG 1149.1 protocol is shown in Figure 16.5.

A typical scan interface requires only four or five signals, which are connected to the ATE tester through dedicated test pins. In the JTAG 1149.1 standard, the serial scan interface makes use of a *test access port* (TAP) consisting of a four or five pin interface and a block of control logic called the TAP *controller*. The JTAG 1149.1 scan architecture uses four signals: TCK (test clock), TMS (test mode select), TDI (test data in), and TDO (test data out). The TAP controller controls the clocks, capture, shift/load and update signals, and mode control signals of the scan cell in Figure 16.7 using a state machine. The test-mode control signal, TMS, controls the operation of the state machine as it is clocked by the TCK signal. The state machine generates the appropriate capture, shift/load and update signals as needed. The state machine allows a sophisticated level of operation while using a minimum number of test-specific device pins. Additionally, the state machine can be used to initiate more advanced operations, such as initiating a BIST operation.

To facilitate an understanding of the boundary scan principle, consider the generic model of a synchronous sequential circuit consisting of combinational logic and a shift register (made from all flip-flops) in the feedback path as shown in Figure 16.6a. PI denotes the primary inputs to the digital system and PO denotes the primary outputs. In test mode, the shift register is divided into two equal-length registers, one called the scan chain register and the other the test stimulus register. At the start of the test, the scan chain register is initialized by shifting in a data vector through the TDI port of the test interface as depicted in Figure 16.6b. The number of test clock cycles required will be dependent on the length of the overall scan chain. Next, the contents of the scan chain register are shifted into the test stimulus register (Figure 16.6c). The contents of this vector will act as the pseudo primary inputs to the combinational logic during the next test cycle. Subsequently, the logic response vector (pseudo primary outputs) is captured in the scan chain register as shown in Figure 16.6d. The test cycle is repeated (Figure 16.6b–d), where the response vector is moved out of the scan chain while the next test vector is being loaded. Also, shifting the scan chain test vector into the test stimulus register can occur in parallel with capturing the response vector (i.e., overlap the operations depicted in Figure 16.6c and 16.6d).

Figure 16.5. Basic boundary scan architecture allows chip-to-chip interconnect testing through a single board-level interface.

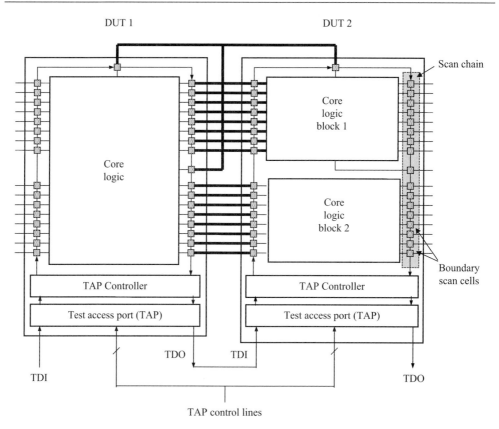

The logic element that enables the digital system to switch between normal and test operation is the *boundary scan cell* (BSC). Figure 16.7 shows a scan cell from the JTAG 1149.1 boundary scan standard. The Mode signal is used to select normal mode or test mode. In normal mode, the data D are passed directly to Q, bypassing the scan cell altogether. In test mode, Q gets its data from the test stimulus register. A load signal updates all the scan cell outputs in a scan chain simultaneously after all the serial bits of the test vector have been shifted into position. Similarly, a response vector from the circuit under test to the first pattern can be captured into the scan chain register and then shifted out while the second pattern is captured by scan register. While the next stimulus vector is shifted into the scan chain, the response vector is simultaneously shifted out to a dedicated serial output pin. It is possible for each shift and load cycle to apply one parallel stimulus vector to the circuit under test and then read back the previous parallel response vector. Chip-to-chip interconnects (highlighted in Figure 16.5) can be verified by applying vectors from one IC's boundary scan circuits and reading back the response from another IC's scan circuits.

The advantage of boundary scan chains is that a multibit test stimulus vector can be applied to any number of DUT circuit inputs using only a few device pins. Using JTAG 1149.1, the interface can be made consistent from one device type to another, allowing system-level test and diagnostics. Standardization allows a much more automated test generation process.

Figure 16.6. Scan-path operation: (a) Generic model of a synchronous sequential logic.
(b) Initializing scan chain register. (c) Shifting scan chain register contents into test stimulus register.
(d) Applying test vector to combinational logic and update response vector in scan chain register.

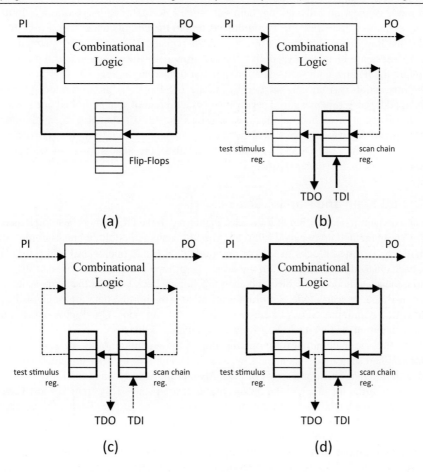

Figure 16.7. IEEE Std. 1149.1 boundary scan cell.

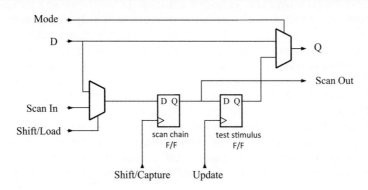

The state machine starts in a test-reset mode that disables all 1149.1 scan circuits, allowing normal operation of the device. The state of the TMS pin is used to direct the state machine through its various operational states. The first of these is the run-test idle state, which enables the test mode of the scan cells, taking them out of their normal (i.e., bypass) mode. In this state, the scan cells apply the parallel test stimulus vectors to the circuits under test. From the run-test idle state, the flow diagram continues to either a data register (DR) path or the instruction register (IR) path. These are identical except that the DR flow path shifts data into and out of the scan paths, while the IR flow path shifts data into and out of instruction register.

The instruction register can be used to initiate BIST operations, to implement proprietary scan chain addressing schemes, or to perform any of a myriad of standard or ad hoc testing operations. The inclusion of an instruction register allows the 1149.1 standard to enable much more powerful operations than a simpler data-only scan architecture could provide. The IR capability is a major forte of the 1149.1 standard, which should allow the standard to grow with advances in BIST and DfT for many years to come.

16.3.3 Full Scan and Partial Scan

As previously mentioned, testing of logic blocks internal to the DUT can be accomplished using a non-JTAG scan methodology called full scan. The full-scan methodology breaks complex circuits into small, easily tested blocks of simple combinational (i.e., nonclocked) logic. In a full-scan design, each clocked circuit element in the design (flip-flop or latch) serves a dual role. In normal operation, the scan mode is disabled and each flip-flop or latch behaves the same as its nonscannable counterpart. In scan mode, a multiplexer replaces the normal data input, D, of the clocked element with a scan input, SD. A buffered version of the flip-flop's Q output is passed out of the cell as a scan output, SQ. Figure 16.8a shows a scannable D flip-flop.

Connecting the scan output from one flip-flop to the scan input of the next, a scan chain can be formed. When scan mode is enabled, we can shift data into the first flip flop in a scan chain through a device pin. Shifting a series of data bits into the scan chains (one bit per clock pulse), we can preset all the flip-flops in the design into a desired state. This applies a test vector to all the combinational logic attached to the flip-flop outputs. The response from the combinational logic can then be latched into the flip-flops using a single clock cycle in normal mode. Then the captured response can be shifted out to a scan output pin using the scan chain.

Using the flip-flops of the scan chain, we can apply a series of parallel test vectors to all the combinational logic blocks in a design and read the response from the blocks using a very simple serial interface. Figure 16.8b shows a trivial example of a 4-bit scannable state machine. Using the flip-flops in scan mode, we can test the combinational logic without having to cycle the state machine through all its possible states.

Fortunately, we do not have to write scan vectors manually, because test vectors for simple combinational logic blocks can be easily calculated for worst-case process conditions. We can automatically generate high fault-coverage test vectors for a fully scannable device using an automated software tool. It is also worth noting that a scannable design can be generated automatically using a design methodology called *design synthesis*. In design synthesis, a high-level hardware description language such as VHDL (VLSI Hardware Description Language) can be used to describe the desired circuit functionality. Software can then synthesize the netlist for a scannable version of the circuit, and other software can generate test vectors to guarantee the circuit's quality.

Since a scannable flip-flop or latch requires more circuit area than a nonscannable equivalent, full scan adds about 10–15% to the area of a typical digital circuit. To reduce this area overhead, a modified version of full scan is sometimes used. The modified methodology, called *almost full scan*, prunes some of the scan circuits out of a full scan design. The result is a more area-efficient

Figure 16.8. Scan circuits: (a) Scannable D flip-flop. (b) Scannable 4-bit state machine.

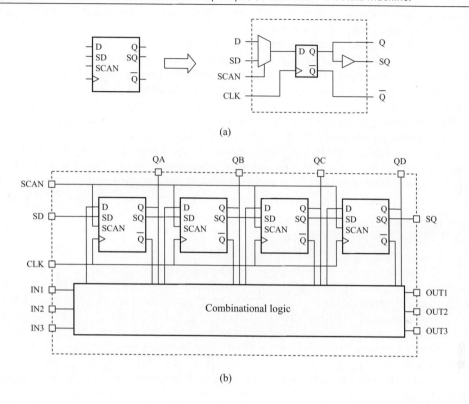

(a)

(b)

scannable design that can still be tested with fault coverage equivalent to the full-scan version of the circuit.

Another scan methodology, called *partial scan*, is similar to full scan, except that the scannable cells are added to a nonscan design until a desired level of fault coverage is attained. Using a partial scan methodology, we start with no scan capability and work our way toward a full-scan design, instead of starting with a full-scan design and working our way toward a design with no scan. As a result, the tools and methodologies for full scan and partial scan are different from one another.

A full treatment of scan circuits and scan methodologies are beyond the scope of this textbook. Although our coverage of scan has been very light, the subject is of extreme importance to design and test engineers. Fortunately, many books and papers have been written on the subject of scan. Rather than duplicating this information here, the reader is referred to the existing literature[1,13,14] for more in-depth information.

16.4 DIGITAL BIST

16.4.1 Pseudorandom BILBO Circuits

BILBO stands for built-in logic block observation. BILBO circuits are a form of BIST consisting of three parts: a pseudorandom data generator, a signature analyzer such as a cyclic redundancy checker (CRC), and a controller to synchronize the generator and analyzer. The pseudorandom data generator produces digital stimulus to be applied to the circuit under test, and the signature analyzer performs one of several mathematical operations (such as a check sum or CRC) to verify that the digital logic produced the correct sequence of outputs.

An example pseudorandom data generator is shown in Figure 16.9. This circuit is also known as a *linear feedback shift register* (LFSR). The pseudorandom generator circuit (Figure 16.9) sequences through all values from 1 to 511 in a pseudorandom sequence before repeating. (If initialized to all zeros, it will hang up in the all-zero state; thus it must be initialized to a nonzero value.)

The pseudorandom values can be passed through a digital circuit under test (Figure 16.10). The output of the digital circuit can be verified using a CRC circuit or other signature analyzer. A very simple example would be a simple checksum circuit, consisting of an 8-bit adder with no carry. During a BILBO BIST operation, all digital blocks are preset to a known state. A BILBO controller circuit then starts clocking pseudorandom data patterns through the circuit under test

Figure 16.9. LFSR pseudorandom number generator.

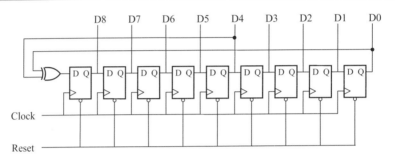

Figure 16.10. BIST testing of a circuit under test using BILBO.

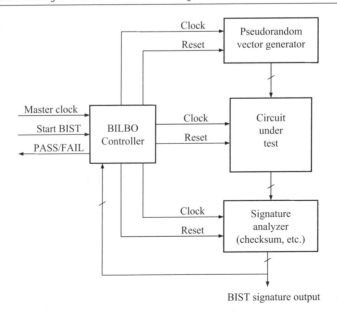

and into the signature analyzer. After a fixed number of clock cycles, the BILBO controller stops the process and the output of the signature analyzer is compared against its expected value. The comparison can be performed by the ATE tester or by the DUT itself. A defective circuit under test is highly likely to produce an incorrect signature.

One advantage of BILBO test circuits is that many of them can operate in parallel, saving test time. Another important advantage is that the circuit under test can be tested at its full digital clock rate without passing high-speed digital signals from the DUT into the ATE tester. This can allow very high-speed digital circuits to be tested on a slower (i.e., less expensive) tester. The tester only needs to supply a high-frequency clock for the BILBO circuit to operate at full speed.

16.4.2 Memory BIST

Memory BIST circuits are similar in nature to pseudorandom BILBO circuits, except that the data patterns are not generated by a pseudorandom algorithm. Instead, the bits of the memory are loaded with specific patterns, such as checkerboard and inverse checkerboard patterns (Figure 16.11), walking ones and zeros (Figure 16.12), and other standard patterns. The patterns in Figure 16.11 and Figure 16.12 represent the bits written into an 8x8 RAM array and then verified by a readback operation. Memory testers are designed to generate memory test patterns algorithmically (on-the-fly) rather than storing the repetitive test patterns in deep vector memory in the tester.

Rather than using an automated tester, BIST can be used to test memory circuits. On-chip BIST circuits can generate memory test patterns with a minimum number of gates. Alternatively, the patterns can be generated by a general-purpose microcontroller if one happens to be included as part of the DUT. Again, the advantages of memory BIST include the ability to test circuits at

Figure 16.11. Checkerboard and inverse checkerboard RAM test patterns.

```
10101010   11001100   11110000       01010101   00110011   00001111
01010101   11001100   11110000       10101010   00110011   00001111
10101010   00110011   11110000       01010101   11001100   00001111
01010101   00110011   11110000       10101010   11001100   00001111
10101010   11001100   00001111       01010101   00110011   11110000
01010101   11001100   00001111       10101010   00110011   11110000
10101010   00110011   00001111       01010101   11001100   11110000
01010101   00110011   00001111       10101010   11001100   11110000
```

(a) Checkerboard patterns   (b) Inverse checkerboard patterns

Figure 16.12. Walking ones and zeros RAM test patterns.

```
00000001   00000010   00000100       11111110   11111101   11111011
00000010   00000100   00001000       11111101   11111011   11110111
00000100   00001000   00010000       11111011   11110111   11101111
00001000   00010000   00100000       11110111   11101111   11011111
00010000   00100000   01000000       11101111   11011111   10111111
00100000   01000000   10000000       11011111   10111111   01111111
01000000   10000000   00000001       10111111   01111111   11111110
10000000   00000001   00000010       01111111   11111110   11111101
```

(a) Walking ones   (b) Walking zeros

full speed with minimal tester support and the ability to test the memory in parallel with other circuits to save test time. Memory testing is a topic unto itself, and will not be covered in detail in this book. Several good books and technical papers have been written on the subject of memory testing and memory BIST.[15–17]

16.4.3 Microcode BIST

If a DUT includes a microprocessor or microcontroller, it can be programmed to test itself using test-specific microcode instructions. For example, the microprocessor's arithmetic logic unit (ALU) can be verified by performing a series of mathematical operations, such as additions, subtractions, bitwise ANDs, bitwise XORs, and so on. The result of each operation is compared by the microprocessor against expected values. Microcode-based testing can also be performed on RAM blocks, I/O ports, and mixed-signal blocks such as ADCs and DACs.

The BIST instructions can be either hard coded into the microprocessor's ROM section, or it can be downloaded by the tester into the microprocessor's program RAM. The advantage of RAM-based BIST is that the BIST instructions do not occupy valuable program ROM space. However, RAM-based microcode BIST requires a longer test time, since the BIST instructions must be downloaded into the program RAM before the BIST testing can be performed. As always, tradeoffs between test time and silicon area must be considered.

16.5 DIGITAL DFT FOR MIXED-SIGNAL CIRCUITS

16.5.1 Partitioning

Highly complex digital circuits benefit from a structured testing approach, preferably using automated software tools to generate DfT structures and test vectors. However, simpler digital circuits such as those in many low-complexity mixed signal devices can be tested quite well without structured scan DfT techniques, saving the overhead of the structured approaches. While structured approaches with automated tools have become the rule rather than the exception as mixed-signal devices have become more complex, it is worth reviewing some of the common ad hoc DfT techniques that have been used on simpler circuits. It is interesting to note that the various structured approaches are based on many of the same concepts as the ad hoc methods that we will review in this section. For example, most structured testing approaches are based on the concept of circuit partitioning to reduce test time and increase circuit observability.

Digital circuits are particularly well suited to a divide-and-conquer approach to testing. In general it is possible to partition a complex digital circuit into pieces and test the pieces separately to guarantee the functionality of the whole. This approach gives us several advantages. First, the test time for exercising a complex circuit can be reduced significantly for certain types of circuits. Second, a complex circuit with many feedback paths may be difficult to force into each of the necessary logic states to guarantee good fault coverage. Partitioning allows the feedback paths of such a circuit to be broken, resulting in many simple circuits rather than one complex one. Finally, partitioning allows automated test generation software to produce test patterns for the simpler circuits that result from a divide-and-conquer approach. Clearly, full scan is the ultimate form of circuit partitioning, since it breaks a digital DUT into many simple subcircuits for quick and thorough testing.

As a very simple example of the test time reduction advantage of partitioning, consider the long divider chain illustrated in Figure 16.13. This circuit divides a 16.777 MHz input clock by 2^{24} to produce a 1-Hz clock. Testing this circuit in a straightforward manner requires that the divider chain step through each of its states, which would take 1 s of test time. Obviously, this is unacceptable. The simplest solution to this problem is to break the divider into three divide-by-256 sections

Figure 16.13. Divider chain without DfT.

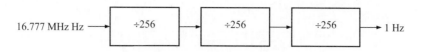

Figure 16.14. Partitioning of a digital divider chain for testability.

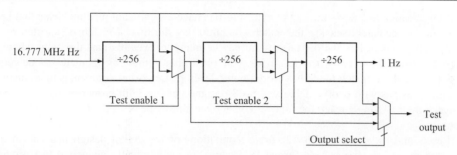

using a test mode (Figure 16.14). Each divider can be tested separately, guaranteeing the operation of the whole. The three separate tests can be executed in a fraction of the test time required to test the divider chain as a whole.

The multiplexers in Figure 16.14 can be controlled using bits in a test register. During normal mode the two multiplexers at the divider stage inputs are configured to pass the output from the previous stage into the following divider stage input. When configured into the test mode, the two input multiplexers replace the previous stage's output with the 16.777-MHz clock to speed up the countdown process for the following stage. Any of the three divider stage outputs can be selected for observation using the third multiplexer, which is controlled by another test-mode setting. The inputs and outputs of test multiplexers such as the ones in Figure 16.14 can be controlled and observed using register-based test modes and readback bits or they can be controlled using a multiplexed digital stimulus/observation bus connected to external device pins. Direct test access of internal circuit nodes through external pins is commonly referred to as parallel module testing, or PMT.

16.5.2 Digital Resets and Presets

One of the simplest but most critical DfT requirements in a digital design is the ability to reset or preset all register and flip-flop circuits into a known state before application of test vectors. In theory, this is usually unnecessary, since most testers can clock the device until it reaches the desired state. In practice, though, resets and presets allow faster testing and much easier test program development.

Consider the 4-bit ripple counter circuit in Figure 16.15. If the counter has no preset capability, it may come up in any of 16 states when power is applied to the DUT. Most testers have the ability to observe the four output bits of the counter and clock it until a desired output state is reached. The capability to search for a particular data pattern from the DUT is called *match mode*. Using match mode, a tester applies clocks to the counter until it sees a desired state, such as *0000*, at the DUT output. The match mode search process takes extra test time, which is not significant for the four-bit counter example. But if the circuit is more complex, it may take a very long time for the tester to find the desired state.

Figure 16.15. Four-bit counter without reset/preset.

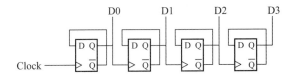

Complicating matters is the fact that most testers have a significant pipeline delay that keeps them from immediately sensing the match condition. By the time a match is detected by the tester's pattern controller, the digital pattern may be many test vectors past the point where the match occurred. The test engineer has to keep the pipelining in mind as the match mode code is developed. In short, match mode is a workaround developed by tester companies to compensate for digital designs with poor testability. A well-designed circuit with proper resets has no need for match mode. Using a reset, the four-bit counter can simply be reset to *0000* before testing its 16 states.

Resets and presets also allow accurate simulations of the digital design that can then be converted directly to the tester's format to produce an automatically generated test program. Simulation software seldom intentionally introduces random states at power up; so a simulation may appear to produce a predictable output pattern from the DUT even without resets. Since the actual nonresettable DUT does not behave predictably on power up, the simulations will not necessarily match the real DUT. Digital patterns that are automatically generated for such a design will be useless, since they are based on a simulation that starts in a nonrandom state.

16.5.3 Device-Driven Timing

Device-driven timing is another problematic issue for test engineers. In theory, it seems that a tester should be able to synchronize its digital pattern to a clock source or data strobe from the DUT. In practice, it is impractical to build a general-purpose tester with enough local circuitry to immediately respond to DUT outputs in real time. The pin card electronics in a typical ATE tester are located several inches if not several feet from the DUT because of mechanical constraints. The ATE tester's digital pattern generator and formatting circuits may be even farther away from the DUT. These are often placed inside the tester's mainframe cabinet, several feet away from the pin card drivers and comparators.

The delays caused by these paths are compounded by the pipelined architecture in the tester's high-speed digital pattern generator. It is common to see a pipeline depth of 60 or more digital pattern vectors between driven data from the tester and received data from the DUT. The tester's software compensates for the pipeline delay between the driven data in a pattern and the expect (compare) data in a pattern so that the test engineer normally does not need to worry about the pipeline delay. Match mode is one of the few instances where pipeline delay is not compensated by the tester software. Figure 16.16 shows a series of vectors with both driven and expected data. In reality, the drive data leave the pattern memory many cycles before the expect pass/fail result from the same vector arrives back at the pattern generator. The tester software takes this pipeline into account for all operations except match mode. This explains why the match mode has trouble immediately responding to a pattern from the DUT. By the time the tester senses the match, the pattern generator may have sent out dozens of additional cycles of drive data.

Because pipeline delay prevents the tester from immediately responding to device outputs, it is impossible for the tester to wait until it sees a sync pulse (data-ready signal) from the DUT to start clocking data into its capture memory. By the time the pattern generator sees the sync pulse, it is too late to begin generating the necessary signals to shift the data out of the DUT. If timing is

Figure 16.16. Pipeline delay in digital vectors.

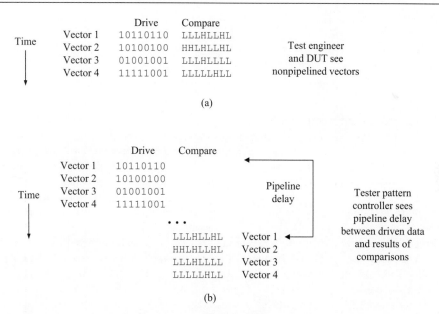

(a)

(b)

not critical, it may be possible to use match mode to learn where the next sync pulse is going to occur and then start clocking data out at that time whether a sync pulse occurs or not.

A subtler problem occurs when the device produces a sync pulse or clock whose timing relative to input or output data is very critical. In this case, it is very difficult to apply the input data or capture the output data at just the right time. The tester would have to learn exactly where the sync pulse or clock edge occurs relative to the digital pattern's bit rate, then adjust its own timing vernier circuits on the fly to find the exact position of the valid data. Testers are not able to do this easily or cost effectively.

It is much easier to tell the tester to accept data 2 ns after a rising edge of one of its own signals than to tell the tester to accept data 2 ns after the DUT suddenly decides to toggle a data-ready signal. Whenever possible, the DUT should be designed to provide data or accept data at a time specified by a tester signal. In cases where the DUT must define timing to meet system-level requirements, a separate test mode can be added that switches the DUT into a slave mode rather than a master mode; so that the ATE tester can define when events should occur.

As an example of DfT to allow ATE-driven timing, consider an on-chip clock generator that normally generates the master clock for a device. Oscillators, PLLs, and other master clock generating circuits should always include a bypass mode as shown in Figure 16.17 for testing purposes. Sometimes the same effect can be achieved by simply driving one side of the crystal with a digital clock. This is only acceptable if the DUT pin is truly a digital input, allowing the master clock to be stopped or single stepped as needed. AC-coupled input clocks are unacceptable, since they do not allow the DUT master clock to be reliably halted for static testing of I_{DDQ}, low-speed digital patterns, and so on.

The bypass mode allows the tester to control the timing of all digital events in the DUT rather than trying to let the DUT drive all timing. Allowing the tester to drive master clock timing is absolutely crucial in DSP-based mixed-signal testing, since the tester must have complete control of all sampling frequencies in the DUT to achieve coherent sampling (see Chapter 8, "Sampling Theory").

Figure 16.17. Oscillator bypass DfT.

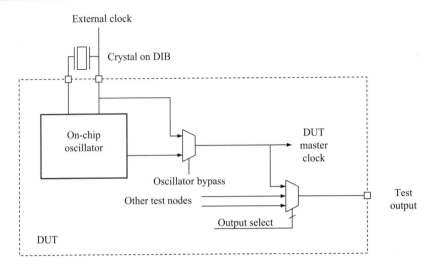

In addition to the bypass mode, the output of the on-chip clock generation circuitry must be observable, either directly or indirectly, say, through an on-chip clock divider circuit, to guarantee its functionality in the normal operational mode. Observability can be achieved either by using another test access point, as shown in Figure 16.17, or by observing a digital output whose operation depends on proper operation of the on-chip clock generator.

16.5.4 Lengthy Preamble

Another problem with complex digital circuits such as plug-and-play multimedia devices is that they cannot be enabled without a lengthy digital setup procedure, called a *preamble*. In a poorly designed device, the preamble must be executed every time a new measurement is performed. Since hundreds of AC channel tests may be performed on a stereo audio IC, the preamble must be executed hundreds of times, leading to needless test overhead. Whenever possible, a test mode should be provided to put the device directly into a test-ready state.

16.6 MIXED-SIGNAL BOUNDARY SCAN AND BIST

16.6.1 Mixed-Signal Boundary Scan (IEEE Std. 1149.4)

The IEEE Std. 1149.4 mixed-signal boundary scan standard[14] was developed by many companies and academic institutions around the world. IEEE 1149.4 is built upon the 1149.1 digital boundary scan standard. As an analog complement to the 1149.1 boundary scan for digital circuits, the 1149.4 standard allows chip-to-chip interconnect testing of analog signals. Optionally, it allows testing of internal circuit nodes. The 1149.4 standard provides a consistent interface for analog and mixed-signal tests for those signals that can tolerate the loading, series resistance, and cross-talk issues inherent in the physical implementation of the standard in the target IC process (e.g., CMOS).

The 1149.4 mixed-signal boundary scan standard is compliant with the 1149.1 digital TAP and boundary scan architecture. The major difference between 1149.4 and 1149.1 is that the

Figure 16.18. IEEE Std. 1149.4 analog boundary module. (Reproduced with permission from IEEE.)

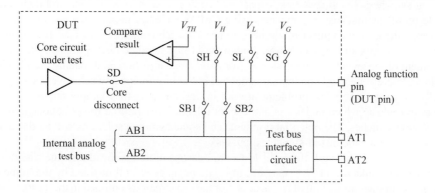

1149.4 standard includes some new test pins and analog switches for exercising nondigital circuits. Figure 16.18 shows the analog boundary module (ABM) for the 1149.4 standard. The ABM provides standardized access to analog input and output signals at the external device pins.

The 1149.4 standard allows a simple chip-to-chip interconnect verification scheme similar to that used in traditional digital boundary scan. A pair of switches at each analog input and output pin of the IC allows the pin's normal (analog) signal to be replaced by digital signal levels, V_H and V_L. V_H and V_L would typically be connected to V_{DD} and digital ground. In effect, the analog input or output becomes a simple digital driver. The interconnect between ICs can be tested by forcing either V_H or V_L from the pin and then checking the status of a receiver at the other end of the interconnection. The receiver, also part of the 1149.4 standard is an analog comparator tied to an 1149.1 digital boundary scan cell. It compares the incoming voltage against a threshold voltage, V_{TH}. In addition to the two logic level connections, the analog pin can also be connected to a quality ground, V_G (typically analog ground).

Although the 1149.4 standard is primarily targeted for chip-to-chip interconnect testing, it does include optional extensions for internal analog signal testing. For this purpose, the 1149.4 standard uses a pair of analog test buses, similar in nature to the analog test input and output buses in Figure 16.4. The analog switches are controlled by shifting control bits into the 1149.1 TAP, allowing a standardized method of setting up analog stimulus and measurement interconnects. The analog buses can be used for a variety of purposes, including internal testing as well as external (chip-to-chip) interconnect testing. As an example of external testing, engineers at Hewlett Packard and Ford Motor Company developed a method to use this structure to verify the interconnects between ICs and networks of passive components such as resistors, capacitors, and inductors.[18] This method forces DC or AC current through one test bus while measuring the voltage response through the other bus.

It should be noted that the switches defined by the 1149.4 standard do not necessarily need to be physical switches. For example, if the output of a particular circuit can be set to a high-impedance state, then it does not need to be disconnected using a switch. Similarly, if a circuit's output can be set to force a high level and a low level under 1149.1 digital control, then separate V_{DD} and ground switches are not needed. The switches defined by the standard are therefore behavioral in nature, rather than physical requirements. The advantage of eliminating switches when possible is twofold. First, the series impedance and/or capacitive loading of a CMOS transmission gate or other switching structure is not introduced into the signal path. Second, the silicon required to implement the 1149.4 standard can be minimized if the number of switches can be minimized.

The 1149.4 standard cannot be employed blindly to test internal signals without consideration of the effects of the standard on the analog circuits to be measured. Actually, it is not the standard that is the problem; it is the practical implementation of the standard using CMOS or other types of analog switches. Signal crosstalk, capacitive loading, and increased noise and distortion are possible problems that may occur when using CMOS switches in sensitive analog circuits. In some cases, the design engineer might need to use T-switch configurations (see 16.7.3) to minimize signal crosstalk and injected noise, though the 1149.4 standard does not specify the physical embodiment of the switches. The issues of crosstalk, noise injection, and loading are identical to those in the more general ad hoc mixed-signal test bus configurations, which have been used successfully for many years. The problems are not insurmountable; they simply require the design engineer to evaluate which nodes can and cannot tolerate the potential imperfections introduced by the analog switches. The potential problems and some common solutions will be discussed in a later section on ad hoc mixed-signal test busses.

Like the 1149.1 standard, the 1149.4 standard carries more overhead than the traditional ad hoc methods. But like the 1149.1 standard, the extra baggage is well justified by the tremendous enhancement in standardization of test access. For the same reasons outlined in the 1149.1 section, the overhead will eventually be much less of a problem as processing geometries shrink.

16.6.2 Analog and Mixed-Signal BIST

The IC industry has recently begun to apply BIST concepts to traditional specification-oriented parameters on mixed-signal circuits. Some of the more promising concepts have been presented at the International Test Conference in recent years, although many of these are designed for very focused test applications. The mixed-signal BIST designer faces some challenging problems that are not faced by digital BIST designers. Let us look at some of the more common challenges encountered when implementing mixed-signal BIST.

First, the more obvious implementations of analog BIST sometimes lack robust traceability to central standards such as those maintained by the NIST (the National Institute of Standards and Technology). It is usually easy to let the device wiggle its analog signals to see if it is basically functional or whether it is completely defective. Unfortunately, many parameters are very close to the specification limits, even on a good device. The use of uncalibrated on-chip analog stimulus and measurement circuits throws doubt into the accuracy of measurements, since there is a question about the quality of the signals generated and measured on a given DUT. Thus the analog BIST designer must define a calibration strategy for the analog circuits of the analog BIST structure.

For example, let us say we try to use an on-chip ADC to test the amplitude of a sine wave generated by a DAC (Figure 16.19). How do we know that the ADC gain on a given DUT is not

Figure 16.19. ADC/DAC loopback BIST.

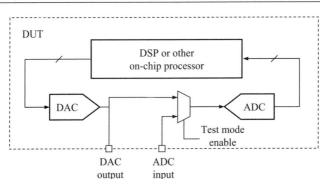

in error by –0.5 dB, canceling out a +0.5-dB amplitude error in its DAC? One possible solution is to provide a calibration signal of a known amplitude to the ADC and let the DUT calibrate itself. Certain parameters such as distortion do not tend to cancel, but are instead additive. These can be tested fairly effectively using ADC/DAC BIST without extra calibration.

Another issue with the DAC-and-ADC-based BIST is that the on-chip instrumentation is often inferior to the types of programmable equipment available on ATE equipment. These digitizers have programmable anti-aliasing filters and other features that allow a much more thorough evaluation of AC signals than can be achieved by an on-chip ADC. In the previous example, how do we know that the ADC output does not contain aliased signal components from the DAC's images? For that matter, how can we measure DAC images if the ADC samples at the same rate as the DAC? The Nyquist criterion becomes a problem.

Finally, the circuit overhead to implement a complete suite of production analog tests using BIST is often overwhelming, unless most of the circuits are already present in the design. In the previous ADC/DAC example, some kind of processor would be necessary to provide sine wave samples to the DAC and collect samples from the ADC. A useful BIST operation would then require the processor to perform an FFT on the results, evaluating *signal-to-noise* ratio, fundamental amplitude, distortion components, and so on. To truly perform on-chip BIST in this manner requires a fairly powerful processor such as a digital signal processor (DSP). If no processor exists on-chip, there is no straightforward way to let the device test itself for these types of parameters.

Despite this rather gloomy analysis of this particular BIST structure, there is a bright side to analog BIST. We have to keep in mind that one goal of BIST is to allow a customer to perform field diagnostics in the end equipment. For field testing, verification of basic functionality is perfectly adequate in many cases, giving analog BIST a very powerful advantage in system-level testing.

Another promising solution to the challenges faced in analog BIST is the use of defect-oriented testing (DOT). The problems outlined in the example are based on the assumption that we wish to perform traditional specification-oriented testing (SPOT). In other words, the problems are based on the fact that we are trying to measure system-level parameters such as gain, distortion, and noise that are very close to specification limits. However, if design margins and processing controls are maintained so that we do not need to measure these parameters with such extreme accuracy, then we can begin to take a defect-oriented approach to mixed signal testing. Inductive fault modeling is one example of defect-oriented testing that has been researched heavily in the past.[19–22]

In defect-oriented testing, we try to detect the cause of the failing parameter rather than the symptom. For example, we might measure the variation in a resistor value that ultimately results in frequency-response errors rather than trying to measure frequency response. Resistance is far easier to measure than frequency response; thus a much simpler BIST circuit might be used to detect this type of fault. Of course, this is a very simplistic example, but it serves to illustrate the thought process behind DOT. Detailed coverage of DOT is beyond the scope of this introductory text.

16.7 AD HOC MIXED-SIGNAL DFT

16.7.1 Common Concepts

Besides the IEEE 1149.4 analog boundary scan standard, there are few standardized approaches to mixed-signal DfT. Most of the more useful mixed-signal DfT concepts have been developed in an ad hoc manner, as needed for a particular application. These are very specific to the exact type of circuit under test. Some of the more common concepts are presented in the sections that follow.

16.7.2 Accessibility of Analog Signals

Accessibility of critical analog signals is one of the most important mixed-signal DfT concepts. DC voltage references, bias current generators, and other critical analog circuits should be accessible to the ATE tester, both for signal measurement and for insertion of signals from the tester. Let us look at a fictional cellular phone voice-band interface device to see how lack of analog test capability can hinder device debug and characterization. Remember that debug and characterization of the DUT is one of the more time-consuming tasks that frequently delays the release of a new product to market.

Figure 16.20 shows a portion of a cellular telephone voice-band interface device that converts received digital voice samples into an audio signal for the telephone's earpiece. This section of the device consists of a 16-bit DAC, an anti-imaging filter, a programmable gain amplifier (volume control), a power amplifier to drive the speaker, and a DC reference that sets the full-scale range of the DAC. The specifications for this particular device call for a gain error of 0 dB plus or minus 0.05 dB at 1 kHz, with a load of 32 Ω at the power amplifier output. A channel with 0-dB gain error is defined in the data sheet as one that produces 1 V RMS at the amplifier output while the DAC input is supplied with a digitized sine wave that is 3 dB below full scale. In addition to the gain error specification, the channel must have a harmonic distortion level at least 85 dB below the 1-kHz test tone.

The test engineer measures this device and discovers that the signal-to-distortion ratio is 75 dB, which fails by 10 dB. Also, the gain is –0.5 dB, failing by –0.45 dB. After the design engineers verify this result using bench equipment, they try to figure out which block is causing the distortion and gain error. Since there is no way to observe any voltages other than the power amplifier output, they can either (a) try to find the cause of the problem by running more elaborate simulations or (b) try to probe internal circuit nodes using tiny whisker probes.

Meanwhile, a competitor has designed a pin-compatible equivalent device with the same architecture, but this device includes DfT for analog signal observability (Figure 16.21). By a curious coincidence, this device has exactly the same problems as those discovered by the first group of engineers. Since there are multiple test points in the circuit, the test engineer is able to measure the output of the DAC directly and discovers that it is actually about 0.2 dB too high. The distortion is absent at the input to the power amplifier, proving that the power amplifier is probably introducing the distortion. After further investigation, it is discovered that the power amplifier is introducing a gain error of –0.7 dB, which explains why the total channel gain error is –0.5 dB (0.2–0.7 dB). Since the power amplifier has been shown to be the cause of distortion, the other design engineers concentrate on other tasks while the power amp designer corrects the distortion and gain problems.

Figure 16.20. Audio interface without DfT.

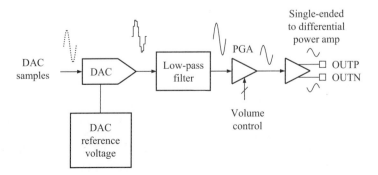

Figure 16.21. Audio interface with analog test accessibility.

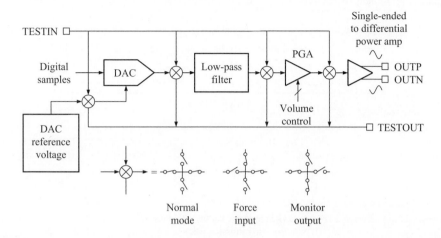

Next, the DAC designer asks the test engineer to measure the DAC reference voltage level using the test bus. The level is 0.19 dB too high, which explains most of the 0.2 dB DAC gain error. The DC reference design engineer then discovers a parasitic resistance that explains the gain reference error. After a design adjustment to correct the parasitic resistance and power amplifier transistor sizes, the second pass design works perfectly. Since the second company's product is first to market, the DfT-enabled design wins in the marketplace over the non-DfT design.

We might be tempted to simply sprinkle test pads throughout the design to achieve the same diagnostic capabilities as analog observability DfT. Test pads require less silicon area than T-switches, along with an analog test bus snaking its way around a die. While test pads are generally a good idea, they can only be accessed using whisker probes that must be positioned by hand under a microscope. This is a time-consuming process that obviously cannot be applied to thousands of units during normal production. It is critical to be able to collect large amounts of characterization using an ATE tester to find correlation between errors in the various internal signals of the DUT.

It is important to realize that breaking a mixed-signal device into sections and testing the pieces is a necessary but insufficient means of guaranteeing system specifications like gain and distortion. Unlike digital circuits, analog circuits do not always behave as a sum of the individual parts; so a divide-and-conquer approach will not always allow system-level specifications to be ignored. Certainly to a first degree, a mixed-signal system behaves as a sum of its parts. Unfortunately, there are many subtle interactions between the various analog and digital circuit blocks that may make the whole behave differently than the individual pieces would indicate. The test engineer should always be prepared to measure both the system-level performance of the whole device and (b) the performance of the individual circuit blocks.

16.7.3 Analog Test Buses, T-Switches, and Bypass Modes

Once a design and test team decide to add analog observability DfT into a new product, the exact method of test point insertion must be chosen. One of the more common ways to provide access to internal analog signals is through analog test buses, such as the ones in Figures 16.18 and 16.21. Using one or more analog test buses, the ATE test program can gain access to internal nodes by opening and closing the appropriate transmission gates or other switching structures. The 1149.4

Figure 16.22. Analog test bus DfT: (a) Capacitive coupling between signals caused by parasitic capacitance. (b) Analog switch implemented using a CMOS transmission gate.

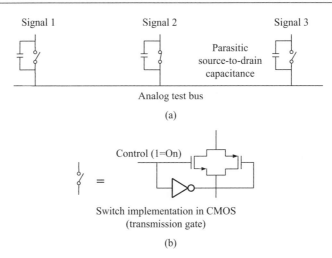

Signal 1   Signal 2   Signal 3

Parasitic
source-to-drain
capacitance

Analog test bus

(a)

Control (1=On)

Switch implementation in CMOS
(transmission gate)

(b)

standard provides for just such internal test access through the AB1 and AB2 analog bus lines. In the 1149.4 standard, the appropriate switches are opened and closed using digital control bits injected through the 1149.1 test access port. In ad hoc architectures, the switches may be closed by any of a variety of means, including test-specific register bits or negative logic levels on normally positive digital input pins (V_{DD} and ground for digital signals, negative voltage to enable test modes). The possibilities for switch control are virtually endless. The more important considerations for this type of test access architecture is the nature of the switches themselves.

One of the biggest problems introduced by analog test buses is the danger of crosstalk between all the observed nodes. Figure 16.22a shows a test bus capable of accessing three internal DUT nodes. The most common CMOS structure for implementing an analog transmission gate is a back-to-back P-channel and N-channel CMOS transistor pair. As shown in Figure 16.22b, an inverter provides complementary control signals to the two transistors so that they are both on or off at the same time. The reason that both an N-channel transistor and a P-channel transistor is required is that the pair allow a larger range of voltages to pass through the transmission gate. If only a P-channel transistor were used, then signals near V_{DD} would not pass due to the transfer characteristics of the P-channel transistor. Likewise, signals near ground would not pass through an N-channel transistor alone.

At first glance, the circuit in Figure 16.22 does not appear to have any problems. But if we realize that the switches are implemented as CMOS transmission gates with parasitic capacitance from drain to source, we see a problem. There is an AC signal path from each signal to the others through the drain-to-source capacitance of each transmission gate. The capacitive coupling path exists even when the transmission gates are all turned off during normal mode. The circuit may suffer from crosstalk problems or it may even break into oscillations due to feedback from one node to a previous node. At low frequencies and low output impedances, this may not pose a problem. Nevertheless, it is a risky approach.

One common solution to the crosstalk problem is the use of a T-switch configuration (Figure 16.23). A T-switch consists of three switches: two in series and one providing a ground to the midpoint of the series connected switches. The grounding switch is closed any time the series

Figure 16.23. T-switches prevent crosstalk between signals.

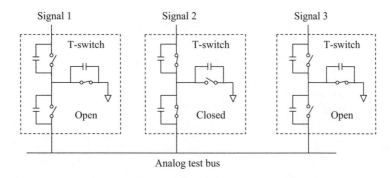

Analog test bus

switches are opened, thereby shunting any potential crosstalk signals to ground. The only problem is that the resulting switch always presents a small capacitive load to ground on the node to be sensed. Fortunately, in many cases the extra load capacitance is entirely acceptable.

Another possible means of implementing analog test accessibility in CMOS circuits is to simply power down a circuit block to provide a bypass path through its circuitry. In this type of scheme, each circuit block can be isolated by powering down some or all of the other blocks in the signal path. The advantage of this approach is that it adds no extra loading or crosstalk paths to the circuit in normal operation. Unfortunately, some circuits can be bypassed by disconnecting their power while others cannot.

16.7.4 Separation of Analog and Digital Blocks

One of the most important forms of mixed-signal DfT is the separation of analog and digital circuit blocks using test modes. The power supply current example of Figure 16.2 is a prime example of this form of DfT. The separation of analog and digital circuits provides several advantages including lower test time and better control over analog circuits so they can be more easily characterized. Let us look at another example of this form of DfT: a digitally controlled automatic gain control (AGC).

A purely analog AGC (Figure 16.24) includes a variable gain amplifier whose gain is automatically adjusted until its output reaches a desired peak amplitude. The peak output level is typically sensed with a peak detector and window comparator to determine whether the amplifier's gain is too high or too low. AGC circuits are commonly found in microphone amplifier circuits. The purpose of the AGC is to maintain a constant voice signal level no matter how loudly or softly the person is speaking. An analog AGC uses the output of the window comparator to adjust a control voltage, which, in turn, adjusts the gain of the amplifier.

A digitally controlled AGC (Figure 16.25) is similar in function to an analog AGC, but it uses digital logic in the feedback path to adjust the amplifier's gain. The digital logic adjusts the gain of a programmable gain amplifier (PGA) up or down depending on the outputs from the window comparator. Since the feedback path is implemented using digital logic, many of the characteristics of the AGC can be adjusted with values written to control registers. Characteristics such as attack time and decay time can be controlled by adjusting the digital logic's response to the window comparator's output. For example, if the decay time is set to be very long, the digital logic would increase the volume of the PGA very slowly after the microphone signal amplitude decreases.

Figure 16.24. Analog automatic gain control (AGC).

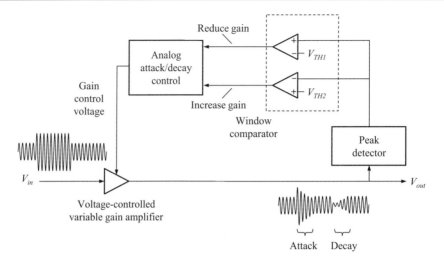

Figure 16.25. Digitally controlled AGC.

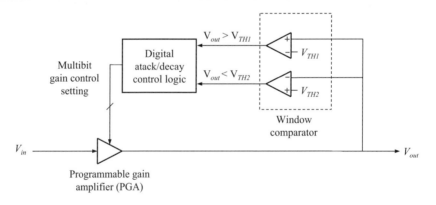

The digital logic in such a circuit can provide hundreds of combinations of attack and decay time. It would be extremely inefficient to measure all the combinations of attack and decay using a variable input signal and observing the gain changes in the PGA to make sure the digital logic increases and decreases volume levels at the correct times. Instead, the digital logic should be separated from the analog circuits to break the AGC into three parts: a PGA, a window comparator, and a digital logic block. One way to break the circuit into pieces is to insert boundary scan cells between the digital logic and the analog cells, as shown in Figure 16.26. The scan cells form what is known as a *scan collar*. The scan collar allows a convenient method to inject the digital stimulus to the feedback logic and to measure its response, as well as to stimulate and measure the digital interfaces to the analog cells. Of course, full scan of the digital logic is an even better approach. Ad hoc register/MUX based isolation circuits are also usable.

Figure 16.26. Digitally controlled AGC with scan-based DfT.

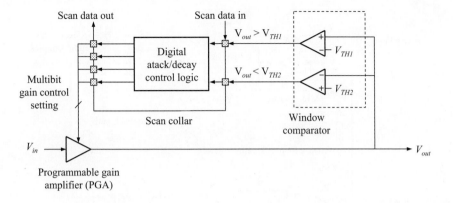

Figure 16.27. AGC attack/decay gain envelope.

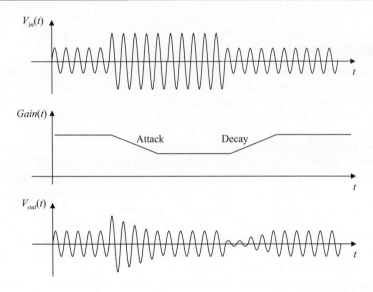

Once the circuits are broken into pieces, the digital logic can be tested using traditional digital methods to guarantee its quality. The remainder of the AGC is a simple PGA and a window comparator, each of which is very easy to test in isolation. The DfT circuits must of course allow the PGA gain to be set directly without the feedback logic, and it must allow direct analog and digital access to the window comparator. Characterization may prove that the AGC does not behave as a sum of pieces; thus the test engineer should also be prepared to perform standard AGC tests, such as measuring the output signal envelope (gain vs. time) with a stepped input (Figure 16.27). This is one case where the sum of the individual pieces will most likely reflect the operation of the whole, because the analog circuits are so simple in nature. A divide-and-conquer approach often works well for circuits that are primarily digital with simple analog interfaces.

Figure 16.28. Analog and digital loopback paths.

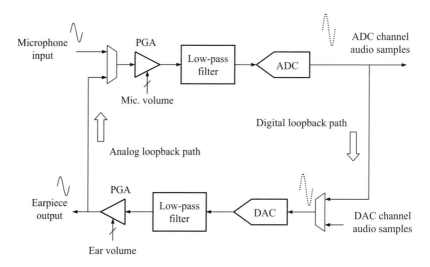

16.7.5 Loopback Modes

Another common mixed-signal DfT approach is the use of digital or analog loopback modes. Figure 16.28 shows the transmit and receive channels from a cellular telephone voice-band interface circuit. This example shows both analog loopback and digital loopback. In analog loopback, the device is placed into a mode that loops the analog DAC channel output (earphone signal) back into the ADC channel input (microphone signal).

In analog loopback mode, it is possible to transmit digitized signals such as sine waves into the DAC channel's digital input and capture digitized samples from the ADC channel's digital output. All circuits in the DAC and ADC channels must work properly for the captured signal to appear at the correct signal amplitude and with low noise and distortion. Loopback mode thus provides a quick and inexpensive way to perform a gross functional test involving most DUT circuits. Any major defect in any of the channel circuits can be detected this way. Analog loopback mode is often used by the original equipment manufacturer (OEM) customer to implement system-level diagnostics that allow the end product (cellular telephone, modem, etc.) to test its own functionality. Loopback DfT thus allows system-level BIST.

Digital loopback mode is similar to analog loopback mode, except that the digital output from the ADC channel is looped back to the digital input of the DAC channel. The test stimulus is a sine wave or other analog signal. The output can be captured and analyzed using a digitizer or other analog measurement instrument.

16.7.6 Precharging Circuits and AC Coupling Shorts

One of the simplest and most effective ways to reduce test time is to provide shorting paths to reduce long RC charging times. Such precharging circuits are useful in a number of different circuit configurations. DC reference decoupling capacitors and DC blocking capacitors are two good examples where shorting paths can be used to bring the DUT to a quiescent state quickly.

Figure 16.29 shows a DC V_{MID} reference circuit with a 1-µF decoupling capacitor located on the device interface board (DIB). This circuit includes a test mode that temporarily reconfigures the buffer amplifier connections to quickly precharge the decoupling capacitor. Except for the

Figure 16.29. Decoupling capacitor precharge circuit.

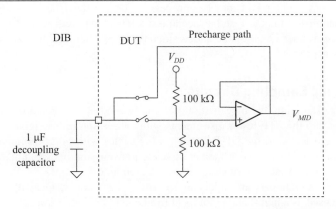

Figure 16.30. Quick-charge switches reduce settling time.

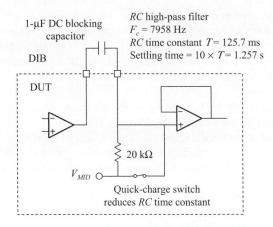

offset of the amplifier, the precharge voltage will exactly match the voltage at the midpoint of the two resistors in the voltage divider, even if the resistors are mismatched. Once the circuit is returned to the normal mode, the decoupling capacitor only has to charge to compensate for the few millivolts of buffer amplifier offset.

Reduction of settling time can also be useful in DC blocking circuits such as the *RC* high-pass filter configuration illustrated in Figure 16.30. One possible way to reduce settling time is to temporarily short the far end of the capacitor to ground (or V_{MID}, depending on the circuit configuration). Yet another possible solution to this problem is to provide a complete bypass of the blocking capacitor during testing, eliminating the *RC* time constant altogether. However, this technique assumes that there will be no clipping problems or impedance matching problems between the two circuits with the *RC* high-pass filter stage removed.

Sometimes, these precharging techniques can be implemented on the ATE device interface board. In these cases, on-chip DfT may be unnecessary. Other times, the nodes are not accessible from the external pins of the DUT and a special test mode is required. As usual, the reduction in test time must be balanced against the added silicon area to determine the cost effectiveness of this type of DfT approach.

16.7.7 On-Chip Sampling Circuits

As DUT signals extend into the megahertz range and beyond, it may become difficult to get the signal under test into the ATE tester instruments without corrupting the signal. Most ATE tester instruments are connected to the DUT through 50-Ω cables with 50-Ω termination resistors at the far end. To the extent that a DUT cannot drive the transmission lines and other reactive loads presented by the tester, a special test structure may be required.

One such test structure is an undersampled strobe comparator, located on-chip.[23-28] Such comparators are used as a high-bandwidth local front end to an undersampling ADC. The timing circuits, successive approximation register (SAR), and SAR DAC are located off-chip to save silicon cost. Since the comparator of the ADC is located on-chip, it is capable of sensing very high-frequency signals. This is an example of a DfT structure that does not necessarily provide lower cost, but may provide the capability to measure signals that otherwise could not be tested in a production environment.

Another similar structure is an on-chip sample-and-hold circuit. Such a circuit can be used in an undersampling mode to down-convert very high-frequency signals on the DUT to lower frequency signals that can easily be measured by an ATE tester. The only drawback to sample-and-hold circuits is that they draw instantaneous currents from the signal under test while charging, thereby introducing current spikes into the signal under test. An undersampled comparator does not suffer from this problem.

Yet another approach to the high-speed signal measurement problem is the use of an on-chip flash ADC, assuming one already exists in the design. This technique is particularly effective in circuits such as hard disk drive PRML channels in which a high-bandwidth ADC is already present on the DUT. The only caveat to this approach and the others presented in this section is that the frequency response of the on-chip sensing element needs to be calibrated on a device-by-device basis for maximum absolute accuracy.

16.7.8 PLL Testability Circuits

The phase-locked loop (PLL) can be one of the more difficult mixed-signal circuit blocks to test, depending on the tightness of the specification limits. Some of the more common specification requirements are settling (lock) time, jitter, center frequency, and frequency range. Lock time and center frequency measurements can be made easier by breaking the analog feedback path of the PLL using a test mode (Figure 16.31). By inserting a midscale voltage into the voltage-controlled oscillator (VCO), it is possible to directly measure the center frequency. Similarly, the maximum and minimum frequencies can be measured by forcing full-scale and minus full-scale voltages into the VCO and observing the VCO's output frequency.

Settling time can be measured by observing the VCO's input voltage during normal operating mode with the analog feedback path enabled for normal operation. Settling time is typically measured by applying a reference frequency that abruptly changes from one frequency to another. The VCO input voltage settles at roughly the same rate that the output frequency settles. Since it is easier to digitize a time-varying voltage than to measure a time-varying frequency on most testers, the VCO input voltage represents a useful test node.

Figure 16.31. PLL DfT structures.

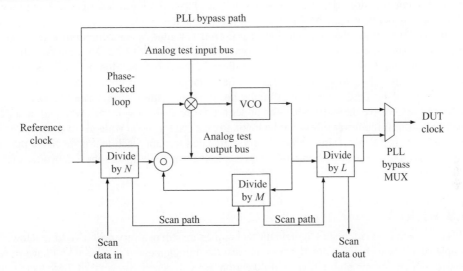

The jitter measurements specified in most PLL circuits can be difficult to measure, especially when they approach the picosecond range. DfT circuits[29–31] have proven useful in the measurement of PLL jitter and other specified parameters.

PLLs are often used as part of a frequency multiplier circuit, which includes a digital divider in the PLL feedback path. If the PLL circuit also includes one or more digital divider blocks, these should be isolated from the analog portions of the PLL to facilitate thorough testing of the divider logic without performing frequency measurements. Scan-based isolation techniques are well suited to this task. Since frequency measurements are usually much more time-consuming on ATE testers than simple digital pattern tests, this type of testability can be useful in reducing test time.

As mentioned in Section 16.5.3, it is critical that PLLs provide a bypass mode so that the tester can inject clock signals into the DUT without the using the PLL. This allows the tester to drive the master clock or other clock input to the DUT in a direct manner.

One final PLL testability signal that has proven useful is the LOCK signal, which tells the tester that the PLL believes it has stabilized to a final frequency. This signal is often needed by the system-level application as well. The LOCK signal allows the system microcontroller to verify that PLL frequency lock has occurred.

16.7.9 DAC and ADC Converters

One of the most time-consuming tests for ADC and DAC converters is all-codes linearity testing. Integral nonlinearity (INL) and differential nonlinearity (DNL) are tests that require each code to be measured with a high degree of accuracy. The large number of measurements associated with a brute force all-codes test such as this leads to very long test times. Each bit of resolution requires twice as much test time; thus a 12-bit converter takes at least 16 times as long to test as an 8-bit converter.

Also contributing to long test times for ADCs and DACs is slow conversion speed. A dual-slope ADC that takes 100 ms per conversion cannot be economically tested for INL and DNL (see Section 7.3.2). Dual-slope ADCs should only be used when absolutely necessary. If a faster successive approximation ADC can be used, it will lead to much lower test times even though its high speed may not be needed by the system-level application.

One DfT technique that leads to lower converter test time is segmentation of the DAC or ADC. This is one of those "easier said than done" suggestions. If a 12-bit DAC can be segmented into two 6-bit DACs, the test time will be significantly lower (see Section 6.3.7). This approach works very nicely as long as the performance of the 12-bit DAC can be reconstructed as a weighted sum of the two 6-bit DACs. The real difficulty lies in superposition errors. The fine DAC (lower 6 bits) may not behave the same when the upper 6-bits are all zero as it does when all upper bits are all ones. But if the lower 6 bits can be made to produce the same voltage curve regardless of the upper bits, then superposition allows for a segmented test approach that is far faster than the 12-bit all-codes test. The design engineer who can find a way to design a robust converter such as this can reduce testing costs significantly.

16.8 RF DFT

RF DfT techniques in contrast to digital DfT techniques are still in a very primitive form. However, a couple of adhoc techniques for RF circuits have been implemented in silicon and have proven effective in high volume production. In this section, we will describe several of these techniques. It is interesting to note that these DfT techniques were the result of a concentrated effort between design and test engineers having a common goal of reducing test costs.

16.8.1 RF Loopback Test

Loopback techniques are commonly used to test ADCs and DACs that reside on the same die. While such approaches do not replace individual ADC and DAC functional tests, it does provide an effective means to screen out gross defects at relatively low cost. For RF transceiver circuits, a similar technique can be applied.[33–36] In the simplest case, the output of the transmit path is connected to the input of the receive path, and both paths can be tested by supplying an analog input stimulus signal and measuring a low frequency output signal, as shown in Figure 16.32. This approach is able to achieve considerable test time reduction, since diagnosis can be made through

Figure 16.32. Loopback test structure for RF Transceiver.

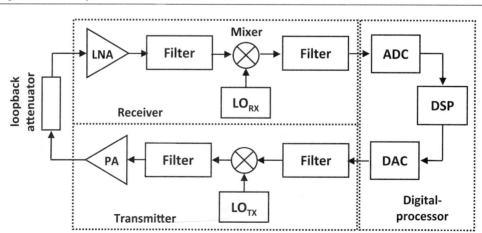

the use of spectral analysis of the transmitted signal. For transceivers with integrated ADC and DAC, this loopback test can also be done in the digital domain with an integrated digital core. This same digital core can also be implemented as a full build-in self-test (BIST). The advantage of this loopback test technique is the simple implementation effort and its flexibility, because the test can be implemented in software and therefore does not depend on the technology of the transceiver. The disadvantage of the loopback technique is mainly related to the fact that the transceiver is tested as a whole, without access to individual circuit blocks or block parameter. The typical measurement parameter for this transceiver loopback test is the BER (see also Chapter 13), which is a good measure for the signal integrity of the device.

16.8.2 RF BIT and BIST

DfT techniques, like built in test (BIT) or built in self test (BIST), are in a very early stage of development for RF circuits. Nonetheless, an implementation of a BIT-or, for that matter, a BIST technique-can provide significant cost saving and improved fault coverage but, in addition, enables factory calibration procedures of critical device settings to be set before and after deployment. This capability can even allow a periodic recalibration during device operation in the end-user system.

Since the BIT and BIST techniques depend strongly on the individual devices, we will show several examples of an RF DFT used in a production environment. Not only are these approaches dependent on the DUT, they are also dependent on the resources available from the ATE. The goal of any BIT technique will be the optimization of the test with respect to an optimal usage of the given ATE resources.

One of the BIT techniques discussed in the open research literature is the idea of implementing a power sensor into the DUT to measure a power proportional to a DC voltage with a low-cost ATE[35,36,38] as depicted in Figure 16.33. Each power sensor can be implemented as peak or RMS sensor. All power sensors are broadband so that they can't select the power dependency as a function of frequency. For the most part, these RMS power sensors operate as a power meter, with

Figure 16.33. Transceiver front-end with loopback and RMS detectors.[35]

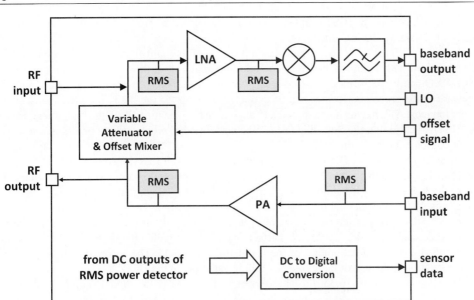

Figure 16.34. RF BIST structure for digital radio processor (DRP™).

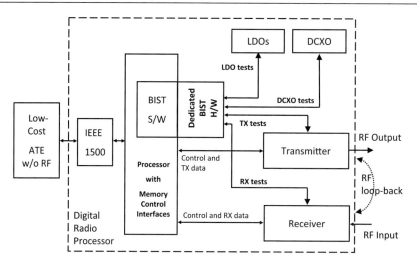

some compromise on dynamic range and accuracy. The dependency of the power gain on process variations must also be incorporated into each implementation. One proposal is to use two or more sensors and measure the difference between input and output power (gain test), assuming that both sensors will have the same shift in performance through a process change.

The BIT test technique using power sensors can be implemented in conjunction with the previously discussed loopback technique, which improves the uncertainty of the loopback test technique, as shown in Figure 16.32. Using power sensors as shown in Figure 16.33, the actual transmit power and the receiver input power can be comparedat multiple test points. By comparing these readings, it can be made sure that a weak transmit performance is not over-shadowed by a strong receive path, or vice versa. This test technique is limited in the detection of nonlinear parameters degradation, however.

Other examples of BIST implementation examples are discussed in the literature.[39-41] Figure 16.34 shows the implementation of a BIST structure used in a Bluetooth (BT) and GSM transceiver SoC. This BIST technique takes advantage of the built-in processor used during normal operation. With additional hardware modules, internal tests of the receiver, transmitter, LDOs, and DCXO can be performed utilizing a software module. The software can be loaded into the firmware of the processor. Results can be extracted off-chip via the IEEE 1500 test bus standard with a low-cost ATE having some digital and analog instrumentation. With the high integration of the BIST structure into the SoC, this test can be performed at probe, at final production test, and at any convenient time in the field.

Another example of a BIST is for the all-digital-PLL (ADPLL) shown in Figure 16.35. Here the ADPLL operates in a digitally synchronous fixed-point phase domain.[41] Since the implementation of the ADPLL is, by definition, digital, the hooks for BIST are digital as well and are designed in the same VHDL design environment. The BIST structure targets two different entities of analog nature: The first target is the oscillator noise, which is not only useful for detecting fabrication defects that results in a degraded noise performance in the digitally controlled oscillator (DCO), but can also serve for a calibration of the oscillator in the field. The second target of the BIST implementation is the test of the frequency tuning varactors in the DCO.

As shown in Figure 16.35, the phase error signal $\phi_E[k]$ is the digital output of the phase detector and is the numerical difference between the reference and variable phase generated by the time-to-digital converter (TDC). This digital signal is internal to the ADPLL and, hence, can

Figure 16.35. ADPLL-based polar transmitter for wireless applications.

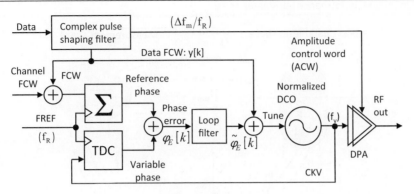

be analyzed with the on-chip processor. By analyzing the phase error signal, defects in the DCO, such as an increase in noise power, can be observed. Typically, the DCO tracking bank consists of digitally biased varactors that are responsible for the frequency tracking and modulation of the output signal. A test needs to be developed to verify that all varactors are functional and that the specified frequency range of the DCO is covered. The BIST technique for the ADPLL shown in Figure 16.35 is based on (a) scanning through the entire varactor bank and (b) testing each of the varactor one at a time. The testing of each varactor individually is based on the (a) ability to address it directly using a digital control word and (b) the ability to detect the effect of the change in frequency by monitoring the phase error signal $\phi_E[k]$. To improve the measurement accuracy of this test by improving the SNR, the BIST is designed so that it periodically toggles the tested capacitance at a rate that allows the accumulation of detected energy at the output of the phase detector. This allows the measurement to be sufficiently reliable for the verification of the functionality of the varactor bank.

The example of the BIST technique just described demonstrates that a well-implemented test structure can not only minimize test costs, but can also be used to improve test coverage. It is essential that the design and test engineers work together from the very beginning of the project with a mindset of developing a manufacturable device that meets the desired electrical specifications. Test costs and test coverage should be a secondary objective in the overall design scheme. The above examples should demonstrate that it is possible to implement BIT and BIST strategies for RF devices, however, at this time it is being driven on an individual basis by the motivation and knowledge of the development team, rather than by established techniques and tools such as those used for digital originated devices.

16.8.3 Correlation-Based Test

A correlation based test, or indirect test, is a test in which the specified parameter is not the one being measured, but another value, for which a correlation can be established. In general, a correlation-based test can be implemented by establishing a correlation between two or more specified parameters, or between a newly designed test with a special stimulus and an adequate measurement. A typical representative for the first class of correlation-based test is the DC to RF correlation-based test. For the second class, the AC to RF correlation based test or alternate test[43] is a typical example. A compromise between both is the multi-V_{dd} testing of RF circuits.[42] The goal of these correlation-based test methods is the implementation of a test which does not require expensive RF equipment in the ATE and which allows a higher multisite test implementation to lower the test cost.

Figure 16.36. Mapping of pass (P) and fail (F) devices from the process parameter space to the measurement and performance parameter space.

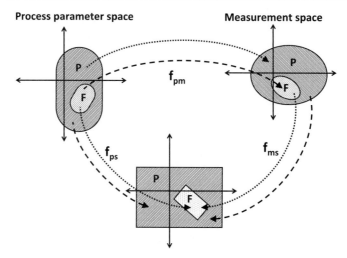

DC-to-RF Correlation

A typical test program will first test continuity and leakage currents at all test insertions, then the bias currents for multiple device modes, eventually even for some internal test modes, before testing a set of RF parameters. Every experienced test and product engineer will have seen some correlation of the bias current to one or more of the RF parameters. The question to be answered is whether a strong correlation can be established as follows: All devices failing the RF tests will also fail one or more DC tests, and those devices that pass the RF test also pass the DC tests. This is depicted in Figure 16.36 where three separate spaces are identified: process parameter space, performance parameter space, and measurement space. Also shown are two separate trajectories: one relating passing devices as they are mapped between spaces, and another relating failed devices. An in-depth correlation analysis needs to be done, for both the passing and the failing devices, before guaranteeing that no fails will be misclassified, but also that there is no yield loss due to misclassification of passing units. The difficulty that the test engineer will experience with this approach is collecting enough evidence to support the correlation in the presence of reasonable levels of manufacturing variation. To date, no acceptable methodology has been identified to assist in identifying these correlations.

To improve the fault coverage, a test method has been published which varies the DC supply voltages V_{DD} and measures the bias currents, as shown in Figure 16.37.[42] This test technique basically consists of increasing the power supply in discrete steps, so that a current signature is generated. When the power supply is ramped up, all transistors in the circuits pass through the regions shown in Figure 16.37, starting with the subthreshold (region A), linear (region B), and saturation (region C). The advantage of a transition across multiple regions is that defects can be detected with distinct accuracy in each of the operating regions. This variation of the bias current is easy to implement and provides data for identifying the current signature of a failed device. A pass/fail criterion can then be established with an in-depth correlation analysis. This method can be extended by ramping some of the power supplies while keeping others at their nominal value, or even by ramping up a DC voltage at one or more RF terminals. These additional tests will increase

Figure 16.37. Power supply ramp with corresponding quiescent current signature.

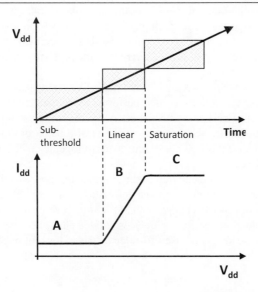

Figure 16.38. Alternate test framework.

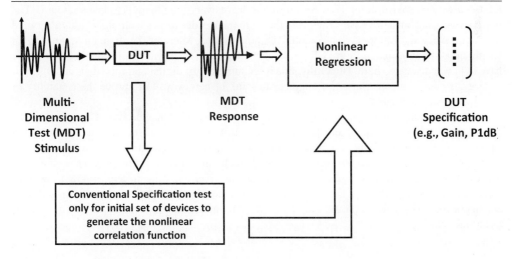

the current signature and thus increase the probability of having a defined fail signature for multiple fails modes. This method is often used a for probe test on the wafer level for a fast screen, but supplemented at the final test, with a full-performance specification verification test strategy.

Analog-to-RF Correlation

As a means to improve the probability of a correlation over and above that predicted by a DC-to-RF correlation, an alternate test methodology[43,44] has been proposed. The principle framework for this test methodology is shown in Figure 16.38. In comparison to traditional full-performance specification-based testing, which uses a nominal test stimulus and measures the nominal specified device response, the alternate test methodology approach replaces the sequential nature of many different specification tests with a single excitation waveform applied to the DUT, while allowing the response signature to be correlated into a single specification. As a result, test time is reduced, and it is conjectured that it might even reduce the performance requirements of the ATE instrumentation.

The alternate test is typically executed in two steps. During the learning cycle, the correlation between the DUT response and the specified performance parameter is established and verified by measuring both the DUT responses for the alternate test stimulus and that resulting from the nominal specified stimulus. The second step is the test of a batch of devices. In this phase, the alternate stimulus and its established correlation is used together with the standard DC tests to screen for defective devices.

The difficulty with the implementation of the alternate test approach is the need to generate a complex test stimulus and to establish a robust correlation to the set of passed and failed devices under reasonable levels of process change. This difficulty prevents this method from being used on a wide range of production test for RF devices. Nonetheless, because this is a relatively new approach, the interested reader is advised to refer to the aforementioned articles to gain more insight into this approach.

16.9 SUMMARY

This chapter has attempted to introduce all the major areas of DfT for mixed-signal circuits, at least in a very cursory fashion. However, trying to learn everything there is to know about DfT is like taking a sip from a fire hose. Obviously there are many sides to this seemingly straightforward topic, and we have only looked at some of the possibilities. Many of the dramatic mixed-signal testing breakthroughs in the future will probably be based on totally new concepts in DfT that we have not even discussed here. Hopefully, this brief summary of digital and mixed-signal DfT will inspire the design engineer and test engineer to think of new ways to improve the testability of their mixed-signal circuits.

Often the design engineers or test engineers get confused as to the purpose of a particular DfT technique. The design engineer may be thinking of ways to evaluate internal circuit nodes to allow easier design debug, while the test engineer may be primarily concerned with test time reduction. There are at least five different objectives of DfT, including lower test cost, higher product and process quality, ease in design diagnostics and characterization, ease in test program development, and enhanced diagnostic capabilities in the customer's system-level application. The design engineers and test engineers should keep in mind what purpose or purposes each DfT structure or technique serves.

PROBLEMS

16.1. List five advantages of DfT.

16.2. Why are robust circuits more testable than circuits whose performance is near failure?

16.3. Why does product cycle time affect profit margins?

16.4. What is the primary purpose of JTAG 1149.1 boundary scan? What is the purpose of the JTAG TAP controller? How can we force the JTAG TAP controller into its reset state regardless of its initial state? What is the maximum number of TCK clock cycles required to accomplish the reset?

16.5. If a digital circuit cannot be reset, what is the likely effect on test time? What is the likely effect on test development cycle time?

16.6. A DSL modem board is fabricated using 5 JTAG 1149.1 compatible devices. Sketch the interconnection of the TAP controllers for this board. How many JTAG interface signals must be connected to the board-level automated tester?

16.7. A design contains 325 flip-flops. Assuming we use a full-scan design methodology with a single scan chain, how many flip-flops would be in the scan chain? If we exercise the scan clock at 10 MHz, how fast can we supply arbitrary parallel test vectors to the DUT circuits under test? How would multiple scan chains affect our maximum test vector rate?

16.8. Gate oxide integrity (GOI) failures result in small leakage currents between the gate of a CMOS transistor and its channel. How might we quickly detect GOI failures in a complex digital design (sketch a transistor-level diagram showing a detected failure)?

16.9. The amplifier of Figure 16.29 has 10 mV of DC offset. The nominal value of V_{MID} is 1.50 V. Assuming a CMOS transmission gate resistance of 2 kΩ, how long does it take for the decoupling capacitor to charge from 0 V to its final value ± 1 mV? After precharging the decoupling capacitor and reversing the state of the two switches, how long should we wait for the V_{MID} voltage to restabilize to within 1 mV of its final value? (Assume that the settling time of the amplifier is negligible.) Compare the total settling time of the circuit in Figure 16.29 with one lacking DfT precharging switches.

16.10. Using MATLAB or similar programming language, simulate the data sequence produced by the LFSR pseudorandom number generator in Figure 16.9. Starting with an initial preset value (seed) of 1, produce a listing of the values produced by the random number generator. How many values are produced before the sequence repeats?

16.11. Using MATLAB or similar programming language, apply the LFSR output values of the previous problem to a 4-bit adder (circuit under test) with carry output. Apply bits D7 through D4 of the LFSR to the A input of the adder and bits D3-D0 to the B input. Apply the output of the adder, including the carry bit, to an 8-bit checksum adder (an 8-bit running sum adder without carry). Reset the 8-bit checksum adder to 0 and apply the pseudo-random sequence to the 4-bit adder and checksum adder through one cycle of the random number sequence. What value does the checksum adder produce? Simulate a defect by forcing the 4-bit adder D1 output to logic LO. What value (signature) is produced by the checksum adder? What type of circuit would we add to a BILBO controller to evaluate the functionality of the adder based on the checksum signature value?

16.12. Design a scannable 4-bit binary counter with synchronous active high reset and ripple carry output using the scannable flip-flop illustrated in Figure 16.8a. Sketch the CLK, SCAN, and SD inputs required to shift the vector 1101 to the QD-QA outputs, latch the response of the combinational logic, and shift out the response (reset the counter to 0000 before performing the scan sequence). Sketch the output at SQ as the scan sequence is performed.

16.13. A 1-kΩ resistor is connected between an output of DUT 1 and an input of DUT 2. Both DUTs are IEEE Std. 1149.4 compatible. Propose a test methodology to verify the connectivity of the resistor to both DUT pins and to verify its value using the AB1 and AB2 analog measurement buses of the 1149.4 analog boundary module. Would your test methodology detect a short to ground on either side of the resistor?

16.14. Would the clock generator circuit shown in Figure 16.39 be well-suited to ATE testing? Why or why not? How might it be improved using DfT (sketch your DfT proposal)?

Figure 16.39. Clock generator circuit for Problem 16.14.

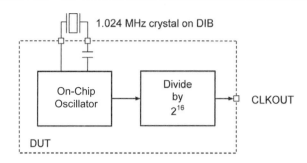

16.15. Propose a calibration procedure for the ADC/DAC BIST scheme of Figure 16.19, assuming we wish to measure absolute gain and frequency response of the DAC and the ADC at the following frequencies: 1, 5, 9, and 13 kHz. You may use only one external signal as a calibration standard, although you may assume that it is error-free.

16.16. In the circuit illustrated in Figure 16.21, the DAC reference voltage is found to be defective, making the channel unusable. We still want to characterize the performance of the remaining circuit blocks while the DC reference block is redesigned. How can we accomplish this characterization using the existing DfT structure?

16.17. What voltages are used for DC-to-RF correlation DfT and what kind of effects can be stimulated with the different voltages?

16.18. Discuss the difficulties when implementing a "learning cycle" into a volume production flow when implementing a correlation based test technique.

16.19. What are the advantages when testing a RF device in the digital domain?

REFERENCES

1. M. Abramovici, M. A. Breuer and A. D. Friedman, *Digital Systems Testing and Testable Design*, revised printing, IEEE Press, New York, January 1998, ISBN 0780310624.

2. E. B. Eichelberger, E. Lindbloom, J. A. Waicukauski, and T. W. Williams, *Structured Logic Testing*, Prentice Hall Series in Computer Engineering, E. J. McCluskey, ed., Prentice Hall, Englewood Cliffs, NJ, 1991.

3. F. C. Wang, *Digital Circuit Testing: A Guide to DfT, ATVG, and Other Techniques*, Academic Press, New York, August 1991, ISBN 0127345809.

4. P. H. Bardell, W. H. McAnney and J. Savir, *Built-in Test for VLSI, Pseudorandom Techniques*, John Wiley & Sons, New York, 1987.

5. L. Avra and E. J. McCluskey, "Synthesizing for Scan Dependence in Built-in Self-Testable Designs," Proceedings of the IEEE International Test Conference, pp. 734–743, 1993.

6. S. Hellebrand et al., Generation of vector patterns through reseeding of multiple-polynomial linear feedback shift registers, in *Proceedings of the IEEE International Test Conference*, 1992, pp. 120–129.

7. M. F. Toner and G. W. Roberts, Towards built-in self-test for SNR testing of a mixed-signal IC, in *IEEE International Symposium on Circuits and Systems*, Chicago, May 1993, pp. 1599–1602.

8. E. Hawrysh and G. W. Roberts, An integration of memory-based analog signal generation into current DfT architectures, in *Proceedings of the IEEE International Test Conference*, 1996, pp. 528–537.

 9. S. Sunter and N. Nagi, A simplified polynomial-fitting algorithm for DAC and ADC BIST, in *Proceedings of the IEEE International Test Conference*, 1997, pp. 389–395.

10. M. Slamani, B. Kaminska, and G. Quesnel, an integrated approach for analog circuit testing with a minimum number of detected parameters, in *Proceedings of the IEEE International Test Conference*, 1994, pp. 631–640.

11. G. W. Roberts and A. K. Lu, *Analog Signal Generation for Built-In-Self-Test of Mixed-Signal Integrated Circuits*, Kluwer Academic Publishers, Boston, April 1995, ISBN 0792395646.

12. C. Dislis, J. H. Dick, I. D. Dear, and A. P. Ambler, *Test Economics and Design for Testability*, 1st edition, Prentice Hall, Englewood Cliffs, NJ, January 1995, ISBN 0131089943.

13. C. M. Maunder, ed., *IEEE Std 1149.1–1993a, Standard Test Access Port and Boundary-Scan Architecture*, IEEE Standards Board, New York, October 1993, ISBN 1559373504.

14. B. Wilkins, ed., *IEEE Std 1149.4–1999, IEEE Standard for a Mixed-Signal Test Bus,* IEEE Standards Board, New York, March 2000, ISBN 0738117552.

15. A. J. van de Goor, *Testing Semiconductor Memories: Theory and Practice*, John Wiley & Sons, Chichester, UK, 1991.

16. P. Mazumder and K. Chakraborty, *Testing and Testable Design of High-Density Random-Access Memories, Frontiers in Electronic Testing*, Kluwer Academic Publishers, Boston, June 1996, ISBN 0792397827.

17. A. J. van de Goor, The Implementation of Pseudorandom Memory Tests on Commercial Memory Testers, in *Proceedings of the IEEE International Test Conference*, 1997, pp. 226–235.

18. K. Parker, J. McDermid, and S. Oresjo, structure and metrology for an analog testability bus,in *Proceedings of the IEEE International Test Conference*, 1993, pp. 309–317.

19. G. Devarayanadurg, P. Goteti, and M. Soma, hierarchy based statistical fault simulation of mixed-signal ICs, in *Proceedings of the IEEE International Test Conference*, 1996, pp. 521–527.

20. R. Voorakaranam et al., Hierarchical specification-driven analog fault modeling for efficient fault simulation and diagnosis, in *Proceedings of the IEEE International Test Conference*, 1997, pp. 903–912.

21. C. Y. Pan, K. T. Cheng, Fault macromodeling for analog/mixed-signal circuits, in *Proceedings of the IEEE International Test Conference*, 1997, pp. 913–922.

22. M. Soma et al., *Analog and Mixed-Signal Test*, B. Vinnakota, ed., Prentice Hall, Englewood Cliffs, NJ, April 1998, ISBN 0137863101.

23. K. Lofstrum, Early capture for boundary scan timing measurements, in *Proceedings of the IEEE International Test Conference*, 1996, pp. 417–422.

24. M. Burns, Undersampling digitizer with a sampling circuit positioned on an integrated circuit, U.S. Patent No. 5,578,935.

25. A. Hajjar and G.W. Roberts, A high speed and area efficient on-chip analog waveform extractor, in *Proceedings, IEEE International Test Conference*, Washington, D.C., October 1998, pp. 688–697.

26. M. Hafed, N. Abaskharoun, and G. W. Roberts, A stand-alone integrated test core for time and frequency domain measurements, in *Proceedings, IEEE International Test Conference*, Atlantic City, NJ, pp. 1031–1040, October 2000.

27. M. Safi-Harb and G. W. Roberts, 70-GHz effective sampling time-base on-chip oscilloscope in CMOS, in *IEEE Journal of Solid-State Circuits*, **42** (8), August 2007, pp. 1743–1757.

28. H. Noguchi, N. Yoshida, H. Uchida, M. Ozaki, S. Kanemitsu, and S. Wadal, A 40Gb/s CDR with adaptive decision-point control using eye-opening-monitor feedback, *Digest of the IEEE International Symposium on Solid-State Circuits Conference*, February 2008, pp. 228–229.

29. B. Veillette and G.W. Roberts, On-chip measurement of the jitter transfer function of charge-pump phase-locked loops, in *Proc. IEEE International Test Conference*, Washington, D.C., November 1997, pp. 776–785.

30. S. Sunter and A. Roy, BIST for phase-locked loops in digital applications, in *Proceedings, IEEE International Test Conference*, Atlantic City, NJ, September 1999, pp. 532–540.

31. D. Fischette, R. DeSantis and J. Lee, An on-chip all-digital measurement circuit to characterize phase-locked loop response in 45-nm SOI, in *Proceedings of the IEEE Custom Integrated Circuits Conference*, Sept. 2009, pp. 19-3.1–19.3-4.

32. B. R. Veillette and G. W. Roberts, A built-in self-test strategy for wireless communication systems, in *Proceedings of the IEEE International Test Conference*, Washington, D.C., October 1995, pp. 930–939.

33. M. Negreiros et al., An improved RF loopback for test time reduction, in *Proceedings of the Design, Automation and Test in Europe*, 2006, pp. 646–651.

34. J-S. Yoon, and W. R. Eisenstadt, Embedded loopback test for RF ICs, in *IEEE Transactions on Instrumentation and Measurement*, **54**(5), 2005.

35. G. Srinivasan, F. Taenzler and A. Chatterjee, Loopback DFT for low-cost test of single-VCO-based wireless transceivers, in *Design & Test of Computers, IEEE*, **25**(2), 2008, pp. 150–159.

36. M. Onabajo et al., Strategic test cost reduction with on-chip measurement circuitry for RF transceiver front-ends—an overview, in *49th IEEE International Midwest Symposium on Circuits and Systems*, Vol. 2, 2006, pp. 643–647.

37. S. Bhattacharya and A. Chatterjee, Use of embedded sensors for builtin-test RF circuits, in *Proceedings of the IEEE International Test Conf.*, 2004, pp. 801–809.

38. A. Valdes-Garcia, R. Venkatasubramanian, R. Srinivasan, J. Silva-Martinez, and E. Sanchez-Sinencio, A CMOS RF RMS detector for built-in testing of wireless transceivers, in *Proceedings of the 23rd IEEE VLSI Test Symposium*, 2005, pp. 249–254.

39. I. Bashir et al., Built-in self testing (BIST) of RF performance in a system-on-chip (SoC), in *Proceedings of the 2005 IEEE Dallas/CAS Workshop: Architectures, Circuits and Implementations of SoC (DCAS'05)*, October 2005, pp. 215–218.

40. O. Eliezer et al., "Built-in self testing of a DRP-based GSM transmitter, in *Proceedings of the 2007 IEEE Radio Frequency Integrated Circuits Symposium*, 2007, pp 339–342.

41. R. B. Staszewski, et al., All-digital TX frequency synthesizer and discrete-time receiver for Bluetooth Radio In 130-Nm CMOS, in *IEEE Journal of Solid-State Circuits*, **39**(12), 2004, pp. 2278–2291.

42. E. Silva1, J. Pineda de Gyvez, and G. Gronthoud, Functional vs. multi-VDD testing of RF circuits, in *Proceedings of the IEEE International Test Conference*, 2005.

43. P. Variyam and A. Chatterjee, Enhancing test effectiveness for analog circuits using synthesized measurements, in *Proceedings of the 16th VTS*, 1998, pp. 132–137.

44. S. Bhattacharya, A. Halder, G. Srinivasan, and A. Chatterjee, Alternate testing of RF transceivers using optimized test stimulus for accurate prediction of system specifications, in *Journal of Electronic Testing: Theory and Applications*, **21**(3), 2005, pp. 323–339.

Gaussian Cumulative Distribution Function Values $\Phi(z)$

z	$\Phi(z)$	z	$\Phi(z)$	z	$\Phi(z)$	z	$\Phi(z)$
−3.0	0.0013	−1.5	0.0668	0.0	0.5	1.5	0.9332
−2.9	0.0019	−1.4	0.0808	0.1	0.5398	1.6	0.9452
−2.8	0.0026	−1.3	0.0968	0.2	0.5793	1.7	0.9554
−2.7	0.0035	−1.2	0.1151	0.3	0.6179	1.8	0.9641
−2.7	0.0047	−1.1	0.1357	0.4	0.6554	1.9	0.9713
−2.5	0.0062	−1.0	0.1587	0.5	0.6915	2.0	0.9772
−2.4	0.0082	−0.9	0.1841	0.6	0.7257	2.1	0.9821
−2.3	0.0107	−0.8	0.2119	0.7	0.758	2.2	0.9861
−2.2	0.0139	−0.7	0.242	0.8	0.7881	2.3	0.9893
−2.1	0.0179	−0.6	0.2743	0.9	0.8159	2.4	0.9918
−2.0	0.0228	−0.5	0.3085	1.0	0.8413	2.5	0.9938
−1.9	0.0287	−0.4	0.3446	1.1	0.8643	2.6	0.9953
−1.8	0.0359	−0.3	0.3821	1.2	0.8849	2.7	0.9965
−1.7	0.0446	−0.2	0.4207	1.3	0.9032	2.8	0.9974
−1.6	0.0548	−0.1	0.4602	1.4	0.9192	2.9	0.9981

APPENDIX B

Some Inverse Values of the Gaussian CDF Function $\Phi(z)$

x	$\Phi^{-1}(X)$	$\Phi^{-1}(1-X)$	x	$\Phi^{-1}(1-X)$	$\Phi^{-1}(1-X)$	x	$\Phi^{-1}(X)$	$\Phi^{-1}(1-X)$
10^{-1}	−1.2815	1.2815	2×10^{-1}	−0.8416	0.8416	4×10^{-1}	−0.2533	0.2533
10^{-2}	−2.3263	2.3263	2×10^{-2}	−2.0537	2.0537	4×10^{-2}	−1.7507	1.7507
10^{-3}	−3.0902	3.0902	2×10^{-3}	−2.8782	2.8782	4×10^{-3}	−2.6521	2.6521
10^{-4}	−3.7190	3.7190	2×10^{-4}	−3.5401	3.5401	4×10^{-4}	−3.3528	3.3528
10^{-5}	−4.2649	4.2649	2×10^{-5}	−4.1075	4.1075	4×10^{-5}	−3.9444	3.9444
10^{-6}	−4.7534	4.7534	2×10^{-6}	−4.6114	4.6114	4×10^{-6}	−4.4652	4.4652
10^{-7}	−5.1993	5.1993	2×10^{-7}	−5.0690	5.0690	4×10^{-7}	−4.9354	4.9354
10^{-8}	−5.6120	5.6120	2×10^{-8}	−5.4909	5.4909	4×10^{-8}	−5.3671	5.3671
10^{-9}	−5.9978	5.9978	2×10^{-9}	−5.8842	5.8842	4×10^{-9}	−5.7685	5.7685
10^{-10}	−6.3613	6.3613	2×10^{-10}	−6.2540	6.2540	4×10^{-10}	−6.1449	6.1449
10^{-11}	−6.706	6.707	2×10^{-11}	−6.6041	6.6041	4×10^{-11}	−6.5006	6.5006
10^{-12}	−7.034	7.034	2×10^{-12}	−6.9372	6.9372	4×10^{-12}	−6.8385	6.8385
10^{-13}	−7.348	7.348	2×10^{-13}	−7.2556	7.2556	4×10^{-13}	−7.1612	7.1612
10^{-14}	−7.651	7.651	2×10^{-14}	−7.5610	7.5610	4×10^{-14}	−7.4703	7.4703

ANSWERS TO PROBLEMS

CHAPTER 1

1.1. Operational amplifiers, active filters, comparators, voltage regulators, analog mixers, analog switches, and transistors. **1.2.** Comparators, analog switches, PGAs, AGCs, ADCs, DACs, PLLs, and switched capacitor filters. **1.3.** Programmable gain amplifier (PGA). **1.4.** Digital-to-analog converter (DAC). **1.5.** Analog-to-digital converter (ADC). **1.6.** Digital signal processor (DSP). **1.7.** Automatic gain control (AGC). **1.8.** Short between metal traces (i.e., blocked etch). Bonus answer: Can also cause a gap in the trace, if using a negative process. **1.9.** A clean room eliminates particles in the air, thus reducing the particulate defects such as those in Figure 1.8. **1.10.** Wafer probe testing, bond wire attachment into lead frames, plastic encapsulation (injection molding), lead trimming, final testing. **1.11.** Two devices do not represent a good statistical sample from which to draw conclusions. Need data from hundreds or thousands of devices. **1.12.** Mainframe, test head, user computer, system computer, DIB. **1.13.** Provides a temporary electrical interface between the ATE tester and the DUT. Also provides DUT-specific circuits such as load circuits and buffer amplifiers. **1.14.** Wafer prober. **1.15.** Requires proactive test engineering, better communications between test and design engineering, coordinated engineering effort. **1.16.** Allows design engineers and test engineers to agree upon an appropriate set of tests. Also serves as test program documentation. **1.17.** Time to market, accuracy/repeatability/correlation, electromechanical fixturing, economics of production testing (test economics). **1.18.** We have to test a total of 5,555,555 devices to get 5 million good ones since we have a 90% yield (yield = ratio of good devices to total devices tested). For the good devices, the time saved is 1.5 s times 5 million devices, or 7.5 million seconds of reduced test time per year. Multiplying this by 3 cents per second, we get a total savings of $225,000 per year in reduced testing costs for the good devices. Multiplying the 0.5-s test time reduction for the bad devices by 555,555 devices per year, we see an additional savings of 3 cents per second times 0.5 s times 555,555 devices, or $8333 per year. Thus the total test cost savings is $233,333 per year. **1.19.** The after-tax profit margin is 12.8% (20% times 100%–36%). Therefore, we would have to ship $233,333/12.8% = 1.823 million dollars worth of product to equal the extra profit offered by the reduced test time. Thus, we have to ship approximately one million extra devices to get the same incremental profit as we get from the 1.5 second test time reduction for this device. Obviously, reducing test time can have as high an impact on profits than selling and shipping millions of extra devices!

CHAPTER 2

2.1. System computers, DC sources, DC meters, relay control lines, relay matrix lines, time measurement hardware, arbitrary waveform generators, waveform digitizers, clocking and synchronization sources, and a digital subsystem for generating and evaluating digital patterns and signals. **2.2.** It improves DC measurement repeatability by removing high frequency noise from the signal under test. **2.3.** A PGA placed before the meter's ADC allows proper ranging of the instrument to minimize the effects of the ADC's quantization error. **2.4.** Using the two-measurement approach, we obtain two readings on the ±5 V-range that have errors of ±5 mV, giving a total differential error of ±10 mV. Using the meter's sample-and-difference mode, the 1-V range can be selected, giving a worst-case error of ±1 mV.Using the two-pass (normal) measurement mode, we have to set the meter to the 5-V range to accommodate the 3.5-V common-mode offset. This gives an error of 0.01% of 5 V, or 0.5 mV. The worst-case error is twice this amount, or 1 mV, since we are subtracting two measurements. Using the sample-and-difference mode, the meter range can be set to the 1-V range. The single measurement has a worst-case error of 0.01% of 1 V, or 0.1 mV. The sample-and-difference mode is much more accurate. **2.5.** Kelvin connections compensate for the *IR* voltage drops in the high force line and current return line of a DC source. **2.6.** A local relay can provide a low-noise ground to the DUT inputs. (Alternate answer: A local relay can provide local connections to load circuits, buffer amplifiers, and other sensitive DIB circuits.) **2.7.** Flyback diodes prevent inductive kickback caused by the relay coil from damaging the drive circuitry. **2.8.** A digital pattern generates digital 1/0 waveforms and high/low comparisons. A digital signal contains waveform information such as samples of a sine wave. **2.9.** Source memory stores digital signal samples and supplies them to a mixed-signal circuit such as a DAC. **2.10.** Capture memory captures and stores digital signal samples from a DUT circuit such as an ADC. **2.11.** Formatting and timing information are combined with one/zero information to reduce the amount of digital pattern memory required to produce a particular digital waveform. It also reduces the required vector (bit cell) rate for complex patterns (see Figure 2.11). Finally, it gives us better control of edge placement. **2.12.** The NRZ waveform could be produced using clocked digital logic operating at 2 MHz. The RZ waveform would require clocked digital logic operating at a period of 100 ns, or 10 MHz. If the stop time for the RZ formatted waveform had to be delayed to 901 ns, we would need digital logic operating at a clock rate of 1 GHz. **2.13.** The CW source and RMS voltmeter are only able to measure a single frequency during each measurement, leading to long test times compared to DSP-based testing. Also, the RMS voltmeter can't distinguish between the DUT's signal and its distortion and noise. **2.14.** The low-pass filter is used to reconstruct, or smooth, the stepped waveform from the AWG's DAC output. **2.15.** The PGA sets the measurement range, reducting the effects of the digitizer's ADC quantization error. **2.16.** Distributed DSP processing reduces test time by splitting the processing task among several processors that perform the mathematical operations in parallel.

CHAPTER 3

3.1. 6.1%. **3.2.** 10.8%. **3.3.** 8.75 V. **3.4.** 5.78 V. **3.5.** 540 Ω. **3.6.** 1000 Ω. **3.7.** 105 Ω. **3.8.** 200 Ω. **3.9.** Output offset = 5 V; input offset = 0.477 V. **3.10.** 0.95 V. **3.11.** 50%. **3.12.** $V_{OS,SE}{}^{+} = -0.1$ V; $V_{OS,SE}{}^{-}$ = 0.2 V; $V_{OS,DE}$ = -0.3 V; $V_{OS,CM}$ = 0.05 V. **3.13.**0.122 V. **3.14.** 7.56 V/V; 17.57 dB. **3.15** V_{OUT}(2.0 V) = 8 V; V_{OUT}(3.0 V) = 11.5 V; gain = 3.5 V/V = 10.88 dB. **3.16.**V_{OUT}(2.0 V) = 5.4 V; V_{OUT}(1.0 V) = 2.85 V; gain = 2.55 V/V = 8.13 dB. **3.17.** G_{ol} = -15392.3 V/V = 83.7 dB. **3.18.** 1.34 mV. **3.19.** 3.0 V. **3.20.** PSS = 0.04 V/V; PSSR = 0.0041 V/V. **3.21.** -85.7 dB. **3.22.** 56.2 μV. **3.23.** 2.606 V. **3.24.** 2.020 V. **3.25.** 2.590 V. **3.26.** 2.490 V. **3.27.** 6 iterations of linear search; 15 iterations of binary search. **3.28.** 9 iterations of linear search; summary of 9 iteration beginning with input of 0.0 V; 9th iteration: Input = 1.820287463, Output = 1.999961500.

CHAPTER 4

4.1. μ = 1.591, σ = 0.0493, σ^2 = 0.0024. **4.2.** (a) 3, (b) 0.0473, (c) 0.99987, (d) 0.9502, (e)

$$F(x)=\begin{cases} 0, & x \le 0 \\ -e^{-3x}+1 & x > 0 \end{cases}$$ **4.3.** max. error< 0.4%. **4.4.** (a) Φ = 0.001349, Φ_{app} = 0.001347, error

= 0.19%. (b) Φ = 0.02872, Φ_{app} = 0.028573, error = 0.49%. (c) Φ = 0.2879, Φ_{app} = 0.2843, error = 1.17%. (d) Φ = 0.4052, Φ_{app} = 0.4016, error = 0.87%. (e) Φ = 0.5, Φ_{app} = 0.5, error = 0%. (f) Φ = 0.4642, Φ_{app} = 0.4621, error = 0.431%. (g) Φ = 0.567, Φ_{app} = 0.5705, error = –0.54%. (h) Φ = 0.999, Φ_{app} = 0.9987, error = –0.00026%. (i) Φ = 0.999, Φ_{app} = 0.999, error = –1.2×10⁻⁸%. (j) Φ = 2.86×10⁻⁷, Φ_{app} = 2.86×10⁻⁷, error = 0.044%. **4.6.** (a) 0.498650, (b) 0, (c) 1.0, (d) 0.8399, (e) 0.9331, (f) 0.2023, (g) 0.9973, (h) 0.0027. **4.7.** (a) 1.60, (b) 0.800, (c) –0.800, (d) 0.800, (e) 0.800. **4.8.** (a) 1.281, (b) –0.5244, (c) 0.5244, (d) 0.5244. **4.9.** μ = 0.0372, σ = 0.0256. **4.10.** $P(23 \leq X \leq 33)$ = 0.1; μ = 50, σ = 28.87. **4.11.** μ = 0.0392, σ = 0.022. **4.12.** $P(X < –0.2)$ = 0.03881, $P(0 < X)$ = 0.01551. **4.13.** $P(X < 9.8)$ = 0.0569, $P(10.0 < X < 10.5)$ = 0.56506. **4.14.** $P(X < 0.70\mathrm{e}{-4})$ = 0.0507, $P(0.70\mathrm{e}{-4} < X < 0.140\mathrm{e}{-3})$ = 0.9044, $P(0.140\mathrm{e}{-3} < X)$ = 0.0478. **4.15.** μ = 0.977, σ =

0.0132, $g(v) = 30.26e^{-2876.87(v-0.977)^2}$. **4.16.** μ = 0.0944, σ = 0.055, $g(v) = 7.23e^{-164.2(v-0.0944)^2}$. **4.17.** μ = 0.0944, σ = 0.055, $g(v) = \begin{cases} 5.231, & -0.00119 \leq v \leq 0.1899 \\ 0, & \text{otherwise} \end{cases}$.

4.20. # code hits at code 50 is 68. **4.21.** # code hits at code 50 is 22. **4.22.** μ = 0, σ = 6.06, $F(x) = \dfrac{1}{21} \sum_{k=-10}^{10} u(x-k)$. **4.23.** μ = 550, σ = 260.1, $F(x) = \dfrac{1}{901} \sum_{k=100}^{1000} u(x-k)$. **4.24.** μ = 0.533, σ = 0.221, $F(x) = \begin{cases} 0, & x \leq 0 \\ 2x^2 - x^4, & x \leq 1 \\ 1, & 1 < x \end{cases}$. **4.25.** 0.0374. **4.26.** 0.0022. **4.27.**

$$f(x) = \frac{625}{1296}\delta(x) + \frac{125}{324}\delta(x-1) + \frac{25}{216}\delta(x-2) + \frac{5}{324}\delta(x-3) + \frac{1}{1296}\delta(x-4).$$

4.28. μ = 10, $P(\text{Errors} = 1)$ = 4.5×10⁻⁴. **4.29.** μ = 0.1, $P(\text{Errors} = 1)$ = 0.090. **4.30.** $P(0 < X)$ = 0.9999999; $P(2 < X)$ = 0.921; $P(X < 3)$ = 0.671. **4.31.** $P(0 < X)$ = 0.503; $P(2 < X)$ = 0.057; $P(X < –3)$ = 0.036. **4.32.** $g(v) = 132.9e^{-55555.5(v+0.005)^2}$. **4.33.** $g(v) = 39.89e^{-0.08(500v-1)^2} + 39.89e^{-0.08(500v+1)^2}$. **4.34.** No closed form. Use numerical method to evaluate.

4.35. $g(x) = \dfrac{1}{\sqrt{2\pi}\sqrt{\sigma_1^2 + 4\sigma_2^2}} e^{-\frac{1}{2}\frac{1}{\sigma_1^2 + 4\sigma_2^2}\left(4x^2 - 4x\mu_1 + \mu_1^2 + 4\mu_2^2 + 8x\mu_2 - 4\mu_1\mu_2\right)}$ **4.36.** $g(x) = 0.066e^{-0.0138(x+2.5)^2}$.

4.37. $g(x) = 0.1596e^{-0.08(x^2+x-0.25)}$.

CHAPTER 5

5.1. 400. **5.2.** 150 ppm. **5.3.** μ = 55.6, error = 0.6 mV. **5.4.** 0.003%. **5.5.** #bits = 16.6, Range setting = 125 mV. **5.6.** –0.322 V < Measurement < 0.324 V. **5.7.** $66.49e^{-13888.9(x-10.1)^2}$. **5.8.** $664.9e^{-1.3888×10^6(x-10.1)^2}$. **5.9.** $664.9e^{-1.3888×10^6(x-0.1)^2}$. **5.10.** 0.0982 V/V < measurement error < 0.1018 V/V. **5.11.** V_{CAL} = –0.0379 + 0.926 × VM. **5.12.** 1.2583 V. **5.13.** Calibration factors: 0.970@1 kHz, 0.984@ 2 kHz, 0.800@3 kHz; Programmed voltage levels: 0.515@1 kHz, 0.559@2 kHz, 0.625@3 kHz. **5.14.** Gain = 1.155 V/V, Offset = –0.135 V. **5.15.** V_{CAL} = 1.961 V, Error = –0.169. **5.16.** RMS noise voltage for 10 μF = 0.5 μV; RMS Noise Voltage for 2.2 μF = 1.1 μV. **5.17.** Settling time for 10 μF = 62.1 ms; Settling time for 2.2 μF = 13.6 ms. **5.18.** Bandwidth needs to be reduced by a factor of 25; test time increases by 25 times. **5.19.** Here we see that increasing the yield through more test time has created a excess profit (Benefit = 0.06625000000 N). This benefit scale with the number of devices tested. **5.20.** Average 11 sample sets. **5.22.** 3σ: UTL = 6.35 V/V, LTL = 5.65 V/V; For 6 σ guardbands, average sample set 9 times. **5.23.** UTL = 6.69 mV, LTL = –6.69 mV. **5.24.** UTL = 9.5 mV, LTL = –9.5 mV. **5.25.** USL = 6.4 V/V, LSL = 5.9 V/V. **5.26.** If 6σ guardbands required, one must average the measurement 7 times. **5.27.** (a) 3, (b) 1, (c) 0. **5.29.** 0.0175 V. **5.30.** 1.82 %. **5.31.** 76.2 %. **5.32.** Because the C_{pk} = 0.44 is less than 1.5, the lot will not meet six-sigma quality standards.

CHAPTER 6

6.2. 5.46 V**6.3.** 0 V. **6.4.** code = [–16., 2.0], [–15., 2.06], [–14., 2.13], [–13., 2.19], [–12., 2.26], [–11., 2.32], [–10., 2.39], [–9., 2.45], [–8., 2.52], [–7., 2.58], [–6., 2.64], [–5., 2.71], [–4., 2.77], [–3., 2.84], [–2., 2.90], [–1., 2.97], [0., 3.03], [1., 3.10], [2., 3.16], [3., 3.23], [4., 3.29], [5., 3.35], [6., 3.42], [7., 3.48], [8., 3.55], [9., 3.61], [10., 3.68], [11., 3.74], [12., 3.81], [13., 3.87], [14., 3.94], [15., 4.00]. **6.5.** V_{LSB} = 1.333 V; OUT = [–8 –10.], [–7, –8.66], [–6, –7.33], [–5, –6.00], [–4, –4.66], [–3, –3.335], [–2, –2.00], [–1, –0.666], [0, 0.666], [1, 2.00], [2, 3.333], [3, 4.66], [4, 6.0], [5, 7.33], [6, 8.66], [7, 10.0]. **6.6.** V_{LSB} = 0.156 V; OUT = [0, 0.], [1, .156], [2, .312], [3, .468], [4, .624], [5, .780], [6, .936], [7, 1.09], [8, 1.25], [9, 1.40], [10, 1.56], [11, 1.72], [12, 1.87], [13, 2.03], [14, 2.18], [15, 2.34], [16, 2.50], [17, 2.65], [18, 2.81], [19, 2.96], [20, 3.12], [21, 3.28], [22, 3.43], [23, 3.59], [24, 3.74], [25, 3.90], [26, 4.06], [27, 4.21], [28, 4.37], [29, 4.52], [30, 4.68], [31, 4.84]. **6.7.** (a) G_{DAC} = 0.365 V/V, offset = –1.72 mV, ΔG = –8.64%, Offset error = –1.7 mV. (b) V_{LSB} = 0.365 V. (c) Absolute error = [0., 0.127], [1., –0.203], [2., –0.227], [3., –0.438], [4., –0.192], [5., –0.493], [6., –0.630], [7., –0.658], [8., –0.904], [9., –0.712], [10., –0.986], [11., –0.959], [12., –1.12], [13., –1.34], [14., –1.37], [15., –1.37] (d) Yes the DAC is monotonic. (e) DNL = –0.236, 0.0702, –0.109, 0.334, –0.196, –0.0539, 0.0792, –0.154, 0.290, –0.186, 0.112, –0.0561, –0.143, 0.0696, 0.117. As max(|DNL|) < ½ LSB, DAC passes test. **6.8.** (a) G_{DAC} = 0.365 V/V, offset = 46.5 mV, ΔG = –9.03 %, offset error = 46.5 mV. (b) V_{LSB} = 0.364 V. (c) Absolute Error = [0., 0.128], [1., –0.203], [2., –0.228], [3., –0.440], [4., –0.192], [5., –0.494], [6., –0.632], [7., –0.659], [8., –0.907], [9., –0.714], [10., –0.989], [11., –0.962], [12., –1.13], [13., –1.35], [14., –1.37], [15., –1.37]. (d) Yes the DAC is monotonic. (e) DNL = –0.231, 0.07, –0.113, 0.35, –0.203, –0.038, 0.07, –0.148, 0.29, –0.176, 0.13, –0.066, –0.121, 0.07, 0.10. As max(|DNL|) < ½ LSB, DAC passes test. **6.9.** (a) G_{DAC} = 0.119 V/V, offset = –0.962V, ΔG = –10.5 %, Offset error = –6.8 mV. (b) V_{LSB} = 0.119 V. (c) Absolute error = [0., .807], [1., .471], [2., .975], [3., .160], [4., 1.21], [5., .118], [6., .185], [7., .151], [8., –.400], [9., .176], [10., –.261], [11., .471], [12., –.807], [13., –.504], [14., –.975], [15., –1.01], (d) Yes the DAC is monotonic. (e) DNL = –0.218, 0.62, –0.697, 1.17, –0.975, 0.18, 0.08, –0.467, 0.76, –0.345, 0.82, –1.10, 0.39, –0.378, 0.14. As max(|DNL|) > ½ LSB, DAC does not pass the test. **6.10.** (a) G_{DAC} = 0.119 V/V, offset = –0.974 V, ΔG = –10.8%, Offset Error = –0.974 V. (b) V_{LSB} = 0.119 V. (c) Absolute error = [0., 0.807], [1., 0.471], [2., 0.975], [3., 0.160], [4., 1.21], [5., 0.118], [6., 0.185], [7., 0.151], [8., –0.400], [9., 0.176], [10., –0.261], [11., 0.471], [12., –0.807], [13., –0.504], [14., –0.975], [15., –1.01]; (d) Yes the DAC is monotonic. (e) DNL = –0.218, 0.62, –0.697, 1.17, –0.975, 0.18, 0.08, –0.467, 0.76, –0.345, 0.82, –1.10, 0.39, –0.378, 0.14; As max(|DNL|) > ½ LSB, DAC does not pass the test. **6.11.** Yes the DAC passes the DNL test. **6.12.** Yes the DAC passes the DNL test. **6.14.** W_0 = 0.194, W_1 = 0.327, W_2 = 0.652, W_3 = 1.30, W_4 = 2.61. **6.15.** W_0 = 0.105, W_1 = 0.208, W_2 = 0.413, W_3 = 0.828. **6.17.** Case A: W_0 = 0.0552, W_1 = 0.1207, W_2 = 0.2414, W_3 = 0.4788; poor data fit; major carrier test does not work.Case B: W_0 = 0.0552, W_1 = 0.1207, W_2 = 0.2414, W_3 = 0.4788; good data fit; major carrier test does work. **6.18.** $t_{settling}$ = 20.65 ns, $t_{risetime}$ = 1.22 ns, OS = 59.3 %, E_{glitch} = 0.544 ns-V. **6.19.** $t_{settling}$ = 22.0 ns, $t_{risetime}$ = 1.64 ns.

CHAPTER 7

7.1. (a) $P(V < 40 \text{ mV}) = 0.5318$, (b) $P(V > 10 \text{ mV}) = 0.4207$, (c) $P(-10 \text{ mV} < V < 40 \text{ mV}) = 0.3674$. **7.2.** (a) $P(V < 40 \text{ mV}) = 0.4522$, (b) $P(V > 10 \text{ mV}) = 0.5$, (c) $P(-10 \text{ mV} < V < 40 \text{ mV}) = 0.3812$. **7.3.** 0.0507. **7.4.** $P(\text{Code} = 325) = 0.2525$; #Code 325 = 101.0. **7.5.** 21.28 mV RMS. **7.6.** #Code

115 = 79.33, #Code 116 = 266.4, #Code 117 = 154.3. **7.7.** V_{LSB} = 206.7 mV; Code width (V) = 0.2125, 0.1417, 0.2125, 0.2480, 0.2480, 0.1771; Code edge (V) = 0.010, 0.2226, 0.3643, 0.5768, 0.8248, 1.073, 1.250. **7.8.** V_{LSB} = 320.3 mV; Code width (V) = 0.339, 0.293, 0.270, 0.225, 0.270, 0.270, 0.315, 0.315, 0.293, 0.339, 0.429, 0.362, 0.315, 0.451; Code edge (V) = 0.075, 0.414, 0.707, 0.977, 1.202, 1.472, 1.742, 2.057, 2.372, 2.665, 3.004, 3.433, 3.795, 4.110, 4.561. **7.9.** V_{LSB} = 0.5155, measurement accuracy = ±17.7 mV. **7.10.** $T_{R,min}$ = 0.98 ms @ 6 hits per code; $T_{R,min}$ = 16.3 ms @ 100 hits per code. **7.11.** ΔV = 1.66 mV; N = 6000. **7.12.** V_{LSB} = 66.8 mV, Code width (V) = 0.0671, 0.0663, 0.0670, 0.0667, 0.0664, 0.0674, 0.0672, 0.0673, 0.0664, 0.0674, 0.0667, 0.0667, 0.0666, 0.0669; Code edge (V) = 0.014, 0.0811, 0.147, 0.214, 0.281, 0.347, 0.415, 0.482, 0.549, 0.616, 0.683, 0.750, 0.817, 0.883, 0.950. **7.13.** V_{LSB} = 708.6 mV, Code width (V) = 0.702, 0.610, 0.812, 0.608, 0.802, 0.599, 0.800, 0.604, 0.800, 0.593, 0.787, 0.587, 0.775, .803; Code edge (V) = 0.020, 0.722, 1.33, 2.14, 2.75, 3.56, 4.15, 4.95, 5.56, 6.36, 6.95, 7.74, 8.32, 9.10, 9.90. **7.14.** $V_{IN,DC}$ = 0.4924 V; $V_{IN,min}$ = –0.0315 V; $V_{IN,max}$ = 1.016 V. **7.15.** $V_{IN,DC}$ = 5.099 V; $V_{IN,min}$ = –0.4020 V; $V_{IN,max}$ = 10.60 V. **7.16.** $F_{T,min}$ = 194.28 @ 10 hits, $F_{T,min}$ = 19.42 @ 100 hits, N = 1,286,797, T_T = 51.5 ms. **7.17.** V_{RES} = 41.9 μV, N = 375,000. **7.18.** DNL = 0.061, –0.0099, –0.1513, –0.2221, –0.1513, –0.1513, –0.0099, –0.0099, –0.0806, 0.061, 0.132, 0.132, –0.0099, 0.414; INL = 0, 0.061, 0.0511, –0.1002, –0.3223, –0.4736, –0.6249, –0.6348, –0.6447, –0.7253, –0.6643, –0.5323, –0.4003, –0.4102, 0; bestfit DNL = 0.0909, 0.0199, –0.1214, –0.1922, –0.1214, –0.1214, 0.0199, 0.0200, –0.0507, 0.0909, 0.1619, 0.1618, 0.0200, 0.4401; bestfit INL: = 0.1821, 0.2730, 0.2929, 0.1715, –0.0207, –0.1421, –0.2635, –0.2436, –0.2236, –0.2743, –0.1834, –0.0215, 0.1403, 0.1603, 0.6004. **7.19.** DNL = 0.056, –0.0852, –0.1555, –0.2963, –0.1555, –0.1555, –0.0148, –0.0148, –0.0852, 0.056, 0.337, 0.126, –0.0148, 0.407; INL = 0, 0.056, –0.0292, –0.1847, –0.4810, –0.6365, –0.7920, –0.8068, –0.8216, –0.9068, –0.8508, –0.5138, –0.3878, –0.4026, 0; bestfit DNL = 0.0829, –0.0584, –0.1287, –0.2694, –0.1287, –0.1287, 0.0120, 0.0121, –0.0584, 0.0828, 0.3639, 0.1528, 0.0120, 0.4294; bestfit INL = 0.2896, 0.3725, 0.3141, 0.1854, –0.0840, –0.2127, –0.3414, –0.3294, –0.3173, –0.3757, –0.2929, 0.0710, 0.2238, 0.2358, 0.6652. **7.20.** DNL = 0.125, –0.0624, 0.125, –0.2500, 0.125, –0.0624; INL = 0, 0.125, 0.0626, 0.1876, –0.0624, 0.0626, 0.; bestfit DNL = 0.1339, –0.05349, 0.1339, –0.2411, 0.1339, –0.05368; bestfit INL = –0.07147, 0.06245, 0.00896, 0.1429, –0.09820, 0.03572, –0.01796.

CHAPTER 8

8.1. (a) F_t = 1/32, UTP = 32, F_f = 1/32. (b) F_t = 13/64, UTP = 64, F_f = 1/64. (c) F_t = 5/64, UTP = 64, F_f = 1/64. (d) F_t = 33/64, UTP = 64, F_f = 1/64. (e) F_t = 65/128, UTP = 128, F_f = 1/128. (f) F_t = 0.078125, UTP = 32, F_f = 1/32. **8.3.** (a) F_t = =250, UTP = 0.004, F_f = 250. (b) F_t = 1625, UTP = 0.008, F_f = 125. (c) F_t = 625, UTP = 0.008, F_f = 125. (d) F_t = 4125, UTP = 0.008, F_f = 125. (e) F_t = 4062, UTP = 0.0160, F_f = 62.5. (f) F_t = 625, UTP = 0.004, F_f = 250. **8.7.** V_{LSB} = 0.723 mV, Noise = 0.2115 mV. **8.8.** 3.29 mV. **8.9.** 261.9 mV, 309.5 mV. **8.10.** 238.1 mV, 285.7 mV. **8.11.** 22.21 μV. **8.12.** 33.0 ns. **8.13.** 1.24 ps. **8.14.** 2.04 MHz. **8.15.** 5.2 bits (ADC), 2.74 bits (DAC). **8.16.** 1.8 mV. **8.17.** 25.4 mV. **8.18.** μ = 0 V and σ = 8.0 mV; μ = 0 V and σ = 2.0 mV. **8.19.** F_t = 19.53 Hz, UTP = 51.2 ms. **8.20.** M = 16; M = 15; 16 cycles are coherent in UTP. **8.21.** 62, 120, 190, 250, 310, 370, 430, 500. **8.26.** frequencies (in Hz): 970, 2010, 3030, 3990; no harmonic or intermodulation collisions. **8.27.** (a) No collisions, (b) no collisions, (c) collisions. **8.28.** 195.3125 kHz. **8.29.** 2.980277714 to 51.200 GHz.

CHAPTER 9

9.1. (a) $v(t) = 6.3662 \sin(2000\pi t) + 2.1221 \sin(6000\pi t) + 1.2732 \sin(10000\pi t)$

(b) $v(t) = -6.3662 \cos(2000\pi t) + 2.1221 \cos(6000\pi t) - 1.2732 \cos(10000\pi t)$

(c) $v(t) = -3.1831\sin(2000\pi t) - 1.5915\sin(4000\pi t) - 1.0610\sin(6000\pi t)$
$\qquad -0.7958\sin(8000\pi t) - 0.6366\sin(10000\pi t)$

(d) $v(t) = 2.5 + 2.0264\cos(2000\pi t) + 0.22521\cos(6000\pi t) + 0.0811\cos(10000\pi t)$

9.2. $v(t) = \dfrac{2}{\pi}\sin(\pi t) - \dfrac{1}{\pi}\sin(2\pi t) + \dfrac{2}{3\pi}\sin(3\pi t) - \dfrac{1}{2\pi}\sin(4\pi t) + \dfrac{2}{5\pi}\sin(5\pi t)$

9.3. $v(t) = -\dfrac{2}{\pi}\cos\left(\pi t + \dfrac{1}{2}\pi\right) + \dfrac{1}{\pi}\cos\left(2\pi t + \dfrac{1}{2}\pi\right) - \dfrac{2}{3\pi}\cos\left(3\pi t + \dfrac{1}{2}\pi\right)$
$\qquad + \dfrac{1}{2\pi}\cos\left(4\pi t + \dfrac{1}{2}\pi\right) - \dfrac{2}{5\pi}\cos\left(5\pi t + \dfrac{1}{2}\pi\right)$

9.4. $v(t) = \dfrac{4}{\pi}\sin(\pi t) + \dfrac{4}{3\pi}\sin(3\pi t) + \dfrac{4}{5\pi}\sin(5\pi t)$

9.5. $v[n] = 0.3667 + 0.8647\cos\left(\dfrac{\pi}{3}n + 1.317\right) + 2.025\cos\left(\dfrac{2\pi}{3}n - 2.684\right) + 1.333\cos(\pi n)$

9.6. $v[n] = 1.445 + 1.419\cos\left(\dfrac{\pi}{3}n + 2.750\right) + 2.074\cos\left(\dfrac{2\pi}{3}n + 1.300\right) + 0.6875\cos(\pi n - 3.141)$

9.10. $v[n] = 0.3667 + 0.8647\cos\left(\dfrac{\pi}{3}n + 1.317\right) + 2.025\cos\left(\dfrac{2\pi}{3}n - 2.684\right) + 1.333\cos(\pi n)$

9.11. $v[n] = 2.0 + 5.2345\cos\left(\dfrac{\pi}{2}n - 0.8124\right) + 1.5811\cos\left(\dfrac{3\pi}{4}n + 0.3218\right)$

9.15. (a)

$$v[n] = 0.25 + 0.5\cos\left(\dfrac{\pi}{5}n\right) + 0.1\sin\left(\dfrac{\pi}{5}n\right) + 2.1\cos\left(\dfrac{2\pi}{5}n\right) - 0.9\cos\left(\dfrac{3\pi}{5}n\right) - 0.1\sin\left(\dfrac{3\pi}{5}n\right)$$

(c) $v[n] = 0.25 + (0.25 - j0.05)\,e^{j\pi n/5} + 1.05e^{j2\pi n/5} + (-0.45 + j0.05)\,e^{j3\pi n/5}$
$\qquad + (-0.45 - j0.05)\,e^{j7\pi n/5} + 1.05e^{j8\pi n/5} + (0.25 + j0.05)\,e^{j9\pi n/5}$

9.16.

(a) $v[n] = 0.4\cos\left(\dfrac{\pi}{5}n\right) + 0.8\sin\left(\dfrac{\pi}{5}n\right) + 0.5\cos\left(\dfrac{3\pi}{5}n\right) - 0.5\sin\left(\dfrac{3\pi}{5}n\right)$

(c) $v[n] = (0.2 - j0.4)\,e^{j\pi n/5} + (0.25 + j0.25)\,e^{j3\pi n/5} + (0.25 - j0.25)\,e^{j7\pi n/5} + (0.2 + j0.4)\,e^{j9\pi n/5}$

9.17. $c = [985, 516, 358, 282, 240, 216, 204, 100]$
$\qquad \phi(rads) = [1.2, -1.71, -2.01, -2.25, -2.48, -2.71, -2.92, -3.14]$

9.20. Rectangular window: PG = 1.0, ε = 1.0, ENBW = 1.0; Hanning window: PG = 0.666, ε = 0.6123, ENBW = 1.5; Blackman window: PG = 0.42, ε = 0.5519, ENBW = 1.73; Kasier (β = 10) Window: PG = 0.31, ε = 0.3980, ENBW = 3.19. **9.21.** $A[N = 64, rms]$ = 0.707, error = 0.27% in ±10% BW; $A[N = 512, rms]$ = 0.7063, error = 1.0% in ±10% BW; $A[N = 1024, rms]$ = 0.7071, error = 2.7% in ±10% BW; $A[N = 8192, RMS]$ = 0.7071, error = 1.0% in ±10% BW; Increasing BW, error will decrease. **9.22.** $A[N = 64, RMS]$ = 0.7070, error = 0.27% in ±10% BW; $A[N = 512, RMS]$

= 0.7063, error = 1.0% in ±10% BW; $A[N = 1024, RMS] = 0.7071$, error = 2.7% in ±10% BW; $A[N = 8192, RMS] = 0.7071$, error = 1.0% in ±10% BW; Increasing BW, error will decrease. **9.23.** $\Delta f = 7812.5$ Hz (N = 128); $\Delta f = 122.07$ Hz (N = 8192). **9.24.** Rectagular: $A[N = 8192, rms] = 0.7071$, error = 0.011% in ±10% BW; Balckman: $A[N = 8192, rms] = 0.7071$, error~0% in ±10% BW; **9.29.** 0.9487 V. **9.30.** 0.2372 V. **9.31.** (a) [1.9500, 1.2400, −1.6500, −0.7400, 2.7500, −0.7400, −2.0500, 1.2400];(b) [1.5800, 0.8200, 1.8000, 0.8450, 0.4170, 1.1800, 0.1980, 1.1600];(c) [0.9000, 0.1640, 0.7740, 1.3400, −0.1740, −0.9000, −0.1640, −0.7740, −1.3400, 0.1740]. **9.32.** same solution as 10.31. **9.33.** $v[n]= 0.00998+ 0.1069\cos\left(\dfrac{\pi}{8}n+1.2566\right) + 0.00056\sin\left(\dfrac{5\pi}{8}n-2.66\right)$. **9.34.** 12.2 µV.

CHAPTER 10

10.1. (a) 6.019 dBV (b) −12.8 dBm (c) −20.0 dBV (d) 0.495 V. **10.2.** 0.714 V, 8.05 degrees, 0.505 V, −5.93 dBV. **10.3.** 0.707 V, 135.0 degrees, 0.5 V, −6.02 dBV. **10.4.** (a) 0.99 V/V, (b) $2+ 0.2V_{IN} + 0.03V_{IN}^2$, (c) $a_1 + 2a_2V_{IN} + 3a_3V_{IN}^2 + \cdots + na_nV_{IN}^{n-1}$, (d) $\dfrac{4}{1+ V_{IN}^2}$. **10.5.** (a) 1.9125 V/V, −0.388 dB. 1.86 V/V, −0.63 dB (b) 0.622 dB, 0.370 dB, 0.194 dB, 0. dB, −0.159 dB. **10.6.** $F_t = 9406.25$ Hz, 2.1 V/V, 6.44 dB. **10.7.** 500, 850, 860, 8500, 8600, 12000, 16000. **10.8.** (a) 0.1 V (b) 0.0489 V. **10.9.** 0.365 V. **10.10.** Absolute gain V/V = 1.0, 0.99, 0.90, 0.96, 1.0, 0.75, 0.46, 0.23; Relative Gain dB : = 0., −0.088, −0.92, −0.36, 0., −2.4, −6.8, −13.0. **10.11.** 19,698.27 Hz. **10.12.** (a) X_{IN} (Vrms) = [0.70, 0.70, 0.70, 0.70, 0.70, 0.70, 0.70, 0.70, 0.70]; X_{OUT} (Vrms) = [0.70, 0.67, 0.66, 0.70, 0.67, 0.45e−2, 0.22e−1, 0.17e−1, 0].(b) phase$_{IN}$ (degrees) = [16., 86., 160., 38., 130., 58., −58., −74., −140.]; phase$_{OUT}$ (degrees) = [16., 120., −140., 140., 30., −86., 3.5, 0.15, 45.].(c)Absolute gain (V/V) = [1.0, 0.96, 0.94, 1.0, 0.96, 0.64e−2, 0.31e−1, 0.24e−1, 0].(d) Relative gain (dB) = [0., 0., −.18, 0., 0., −44., −30., −32., −360.].(e) Phase unwrapped (degrees) = [0., 30., 61., 110., 250., 220., 61., 77., 180.].(f) Group Delay (s) = [1.7×10^{-7}, 1.7×10^{-7}, 2.7×10^{-7}, 7.8×10^{-7}, -1.7×10^{-7}, -8.8×10^{-7}, 8.9×10^{-7}, 5.7×10^{-7}].Group Delay Distortion = [2.2×10^{-7}, 2.2×10^{-7}, 3.2×10^{-7}, 8.3×10^{-7}, -1.1×10^{-7}, -8.3×10^{-7}, 1.4×10^{-7}, 6.2×10^{-7}]. **10.13.** (a) 34.8 dB. (b) 54.8 dB. (c) 34.8 dB. (d) 12.9 dB. **10.14.** $\dfrac{S}{H_3} = \dfrac{4a_1+ 3a_3A^2}{a_3A^2}$. **10.15.** Second-order IM frequencies: {0, 800, 1100, 1900, 2700, 3400, 4500, 6100, 7200, 7900, 8000, 8700, 9500, 9800, 11400, 11700, 12500, 13200, 14000, 14800, 15100, 17000}; Third-order IM frequencies: {200, 500, 600, 1000, 1400, 2400, 2900, 4000, 4300, 4700, 4800, 5100, 5500, 5800, 5900, 6600, 6700, 7400, 7700, 8200, 8500, 9200, 9300, 10000, 10100, 10400, 10800, 11100, 11200, 11600, 11900, 12000, 12700, 13000, 13500, 13800, 14500, 14600, 14900, 15300, 15400, 15700, 16100, 16400, 16500, 16900, 17200, 18000, 18300, 18880, 19100, 19800, 19900, 20200, 20600, 21000, 21400, 21700, 22200, 22500, 23300, 23600, 24400, 25500}. **10.17.** 45.5 dB. **10.18.** −37.3 dB, −38.7 dB. **10.19.** PSR (dB) = [−80.00000000, −60.00000000, −57.72113296]; PSRR(dB) = [−86.02059992, −72.04119982, −69.97681008]. **10.20.** XLR (dB)=[−63.18758762,−65.12578788,−61.60396270]; XRL(dB)=[−65.12578788,−64.10273744, −59.10518798]. **10.21.** (a) 64.8 µV (b) 80.7 dB (c) 1.4 µV/sqrt(Hz). **10.22.** 8.6 µV/sqrt(Hz). **10.23.** 0.346 V **10.24.** −52.0 dBV. **10.25.** (a) −36.9 dB, relative to FS. (b) −32.0 dBm0. **10.26.** (a) −70.9 dB, relative to FS. (b) 16.0 dBmC0. **10.27.** 63.9 dB.

CHAPTER 11

11.1. $N_{ADC} = 4096$, $N_{DAC} = 1024$. **11.2.** $F_{s,DAC} = 17,875$ Hz, $F_{s,DIG} = 32,832$ Hz. **11.3.** $F_{s,AWG} = 14,658$ Hz, $F_{s,ADC} = 14,658$ Hz. **11.4.** $F_{s,AWG} = 64,000$ Hz, $N_{ADC} = 256$. **11.5.** 1.886 V, CF = 1.337. **11.6.** 0.221 V (in-band), 0.158 V (first-image), 0.0651 V, 0.0582 V. **11.7.** 1.76 V (in-band), 0.00260 V (first-image). **11.8.** F_{images} = {7000, 17000, 31000, 41000, 55000, 65000, 79000} **11.9.** F_{images} = {9000, 15000, 33000, 39000, 57000, 63000, 81000}. **11.10.** first case: $F_{images\ of\ 55\ kHz}$ = {7000, 17000, 31000, 41000, 65000}; $F_{images\ of\ 63\ kHz}$ = {9000, 15000, 33000, 39000, 57000}; No overlap. 2nd case: $F_{images\ of\ 65\ kHz}$ = {7000., 17000., 31000., 41000., 55000.}; Here an image of 65 kHz overlaps with the tone at 55 kHz. **11.11.** M_{alias} = 9 (coherent), $T_{s,eff}$ = 1.21×10^{-11}s. **11.12.** M_{alias} = 80 (coherent), $T_{s,eff}$ = 2.41×10^{-11}s. **11.13.** M_{alias} = 9.6 (noncoherent), $T_{s,eff}$ = 3.08×10^{-11} s. **11.14.** 83.84 GHz. **11.15.** (a) 2 (b) 1 (c) 1.2 (d) 1009. **11.19.** 0.0191 (M = 1), 0 (M = 8). **11.20.** 0.8103 V peak. **11.21.** $G_{INSTRINSIC}$ = 1.01358 V/V. **11.22.** $G_{INSTRINSIC}$ = 1.01303 V/V. **11.23.** G_{DAC} = 17.1 mV/bit, ΔG_{DAC} = 3.28 dB. **11.24.** G_{ADC} = 203.5 bits/V, ΔG_{ADC} = –0.047 dB. **11.25.** –53.5 dB. **11.26.** 3000 Hz. **11.27.** 15th Harmonic: bin 367, 5734.375 Hz. 23rd Harmonic: bin 393, 6140.625 Hz. **11.28.** No frequency overlaps. **11.29.** G_{ADC} = 52.33 bits/V, PSSR = –32.4 dB. **11.30.** SNR = 83.0 dB, ENOB = 13.49 bits.

CHAPTER 12

12.1. WLAN (2.4–2.5GHz & 5.5GHz); WiMax (3.5 GHz), BT (2.4GHz), GPS (1.5 GHz). **12.2.** Dynamic range, power range, signals can be described more efficient with wave characteristic rather than voltage. **12.3.** –174dBm/Hz, –90dBm, 20–30dBm, 60dBm. **12.4.** A scalar measure is described by the magnitude only, a vector by magnitude and phase (or real and imaginary). **12.5** 0.316 m. **12.6.** lump equivalent; λ = 0.22 m .**12.7.** The wavelength will be shortened by $1/\sqrt{\varepsilon_{reff}}$. **12.8.** Multiple sinusoidal waves (signals) can be added for any given time to a resulting waveform (signal), but these signals can be treated as individual waves or signals. **12.9.** All circuits with an imaginary part not equal to zero (e.g. capacitive or inductive circuits). **12.10.** The power is the area under the product of voltage and time by time or

$$P = \frac{1}{nT_0} \int_0^{nTo} V_{peak} \sin\left(\frac{2\pi}{T_0}t\right) \cdot I_{peak} \sin\left(\frac{2\pi}{T_0}t + \varphi\right) dt$$

12.11. $\text{Re}[P] = 0.0313\cos(0.2\pi) + 0.0313\cos(0.2\pi)\cos(4\times10^{10}\pi t)$

$\text{Im}[P] = -0.0313\sin(0.2\pi)\sin(4\times10^{10}\pi t)$. **12.12.** 0.0253 W. **12.13.** 25 W; 43.98 dBm. **12.14.** The crest factor is a measure for the peak-to-average power ratio. While the test calculation is often done with respect to average power, the peak power must be observed and tracked as it can compress a subsystem of an RF system (e.g. the input stage of the RF-ATE) is not properly taken into account. **12.15.** 6.34 W. **12.16.** 15.05 dB. **12.17.** Noise: 11dB; UMTS: 15dB; OFDM signals:

11dB. **12.18.** $P|_{dBm} = 10 \cdot \log_{10}\left(\frac{P}{1\ mW}\right)$ dBm . **12.19.** Power: 3dB; voltage: 6dB. **12.20.** $Z_S = Z_L^*$.

12.21. $Z_S = Z_L$. **12.22.** P_L = 0.11 W, P_A = 0.13 W, P_S = 0.32 W. **12.23.** P_L = 0.21 W, P_A = 0.25 W, P_S = 0.32 W. **12.24.** The insertion loss is the ratio of power received at a load before inserting a two-port network into a system to the power received at the load after insertion of the two-port network. The insertion loss does not include loss due to source and load matching since this is a loss that is already present before the insertion of the two-port network was made. The transducer loss is the ratio of the power delivered to a load to the power produced (maximal available) by the source. It is equal to the sum of the attenuation of the two port network and the corresponding mismatch loss. **12.25.** 0.928 dB. **12.26.** 0.757 dB. **12.27.** Dissipative loss is loss caused by any reactive element in an RF system (in contrast to reflection or radiation loss). **12.28.** –141.5 dBm/Hz.**12.29.** Any test equipment with a measurement bandwidth less than 1 Hz and an appropriate noise figure.**12.30.** The system noise figure can be calculated with the Friis equation $F_{sys} = F_1 + \dfrac{F_2 - 1}{G_1} + \dfrac{F_3 - 1}{G_1 G_2} + \cdots + \dfrac{F_n - 1}{G_1 G_2 \cdots G_{n-1}}$, it should be obvious fromt his expression that the total system noise figure can be minimized when the first gain stage has a low noise figure and high gain. **12.31.** 33.84 dB. **12.32.** 5.42 dB. **12.33.** Phase noise in the frequency domain is equivalent to short-term random fluctuations in the phase of a waveform. It can be measured as a power ratio or as jitter in time domain. **12.34.** S_{11} is the input reflection coefficient and describes the ratio of the power reflected from a port to the power of the incident wave. It is a measure for the efficiency of the input match. S_{22} is the output reflection coefficient and describes the ratio of the power reflected from the output port to the power of the incident wave. It is a measure for the efficiency of the output match. S_{21} is the forward transmission gain and describes the power transmitted through a network compared to the input power. S_{12} is the reverse transmission gain or isolation. It describes the power ratio of the power transmitted from the output to the input of a two port to the input power at the output. **12.35.** S_{21} and S_{12} are equal for passive reciprocal networks. **12.36.** The VSWR or voltage standing wave ratio is the ratio of the amplitude of a partial standing wave at an maximum node to the amplitude at the minimum node in an RF transmission line caused by reflections. The reflection coefficient describes the amplitude of a reflected wave relative to the incident wave. The reflection coefficient is a complex number. The return loss is the loss of signal power resulting from the reflections caused by impedance mismatch A high return loss is desirable as it results in a lower insertion loss. The insertion loss is the loss of signal power resulting from reflections, radiation and resistive losses when connecting a load to source with an two-port network (e.g., a transmission line with matching components). The mismatch loss is the amount of power that will not be available to the output due to impedance mismatches and reflections. With an ideal impedance match, the mismatch loss is the amount on power which can be gained. **12.37.** $\Gamma = 0.0086 e^{j0.9817}, RL = 41.3$ dB, $ML = 0.0003$ dB. **12.38.** $\rho = 0.184, RL = 14.7$ dB. **12.39.** The mismatch uncertainty is caused by the scalar description of the reflections and is the estimation of the maximal mismatch loss. The mismatch uncertainty is a measure of the power range a system can receive when the matching conditions are only known as scalar values at impedance mismatches. Mismatch uncertainty is caused by standing waves on lines and the associated variance on voltage across the line. The mismatch uncertainty is often the most significant measurement error. **12.40.** 0.016 dB $\leq ML \leq 0.251$ dB; $MU = 0.234$ dB. **12.41.** 0.67 dB. **12.42.** 0.255 dB. **12.43.** $RL_{IN} = 21.2$ dB; $RL_{OUT} = 18.0$ dB; $RL_{METER} = 21.0$ dB, 0.007 dB $\leq ML_{IN} \leq 0.197$ dB; $MU_{IN} = 0.19$ dB, 0.006 dB $\leq ML_{OUT} \leq 0.20$ dB; $MU_{OUT} = 0.19$ dB.

12.44. –3.0 dB. **12.45.** 0.125.

12.46. $IL = 3.9$ dB, $\rho_{IN} = 0.3, \rho_{OUT} = 0.9, Z_{IN} = 27.5 - j6.2\ \Omega, Z_{OUT} = 5.25 - j49.7\ \Omega$.

12.47. The information of a frequency modulated signal is in the frequency, while the amplitude can be constant. Noise only impacts the amplitude. **12.48.** The information of an AM or FM signal is in one single sideband of the transmitted signal. By removing the carrier and other sidebands, all transmitted signal power can be used for one sideband containing all information. However, for demodulation the carrier will need to be recreated. **12.49.** GFSK is a filtered FSK, which limits the required bandwidth.

CHAPTER 13

13.1. See Figure 13.1; mixer, tracking filter, detector, local oscillator, sweep oscillator, display. **13.2.** Amplifier, mixer, local oscillator, filter, ADC. **13.3.** The spectrum analyzer is using a tracking filter to limit the measurement bandwidth which lowers the minimum detectable power. **13.4.** The amplifiers or ADC are working above their compression point in the non-linear range, harmonics will be generated in the ATE system. In addition the calibration of the ATE is valid only for the linear range, so the displayed power level will also be off. **13.5.** The noise in the measurement path will increase, thus limiting the minimal detectable power. **13.6.** The signal spectrum will spread as shown in Figure 13.4 adding noise components to the natural phase noise of the signal. **13.7.** 18.115 MHz. **13.8.** 2.482568 GHz. **13.9.** 23.4 dB. **13.10.** $11.9 \text{ dB} \le G_A \le 14.2 \text{ dB}$. **13.11.** 0.273 dB. **13.12.** $109.6 - j13.0 \ \Omega$. **13.13.** 4.48 dB. **13.14.** $17.0 - j37.2 \ \Omega$. **13.15.** $G_A = 19.7 \text{ dB}$, $Z_{IN} = 28.3 - j5.8 \ \Omega$. **13.16.** The IP3 is a measure for the nonlinearity of the system. As higher the IP3 value is as higher power can be measured without adding harmonic signals to the measurement signal. **13.17.** 2nd order response has the slope of 2; third-order response has a slope of 3. The slope is caused by the trigonometric identities used when multiplying the signals. **13.18.** 24 dBm. **13.19.** $G = 14 \text{ dB}$, $\text{IIP}_3 = 3 \text{ dBm}$, $\text{OIP}_3 = 17 \text{ dBm}$. **13.20.** Binary search, interpolation method, and modulation methods with step function and continuous power sweep. **13.21.** The modulation methods capture the output signal for various input power level with a single capture, which is faster and has a better accuracy than measuring multiple individual power settings. **13.22.** Both results are quite close. $P1dB = -2.32 \text{ dBm}$ using the linear approximation and P1dBm $= -2.35 \text{ dBm}$ using the polynomial fit approach. **13.23.** 0.8652 dBm. **13.24.** Failed 1dB compression test. **13.25.** Y-factor and cold noise method. **13.26.** The noise source is terminating the DUT. Any change in the source impedance will change the mismatch loss, and thus the noise power injected into the DUT. **13.27.** 3.1 dB. **13.28.** 5.03. **13.29.** -81.4 dBm. **13.30.** $£(f) = \dfrac{2.51}{2.51^2 \pi^2 + f^2} \quad \text{rad}^2/\text{Hz}$

13.31. -78.9 dBc/Hc. **13.32.** The ATE should source a reference frequency of 89.8437500 MHz for a sampling frequency of 10.00084926 GHz. The offset frequency will appear at 1.2208 MHz. **13.33.** -110.6 dBc/Hz. **13.34.** As shown in Figure 13.23, the measurement path of a vector signal analyzer consists of a mixer with local oscillator, gain and attenuator stages, and filter providing a low frequency captured signal to the analysis block, which can be built as DSP, or implemented with software. The hardware is identical to the ATE shown in Figure 13.2. **13.35.** All major parameters of a transceiver impact the EVM value, especially noise, phase noise, nonlinearity, phase in-balance.

13.36. $M(t) = \sqrt{\sin^2\left(\pi 10^{10} \cdot t + \dfrac{\pi}{12}\right) + \sin^2\left(\pi 10^{10} \cdot t + \dfrac{2\pi}{5}\right)}$, $\phi(t) = \tan^{-1}\left[\dfrac{\sin\left(\pi 10^{10} \cdot t + \dfrac{2\pi}{5}\right)}{\sin\left(\pi 10^{10} \cdot t + \dfrac{\pi}{12}\right)}\right]$

and phase angle difference $= -57$ degrees.

13.37. $M(t) = \left[-\cos\left(1.8 \times 10^9 \pi t + \dfrac{5\pi}{14}\right) - \sin\left(1.8 \times 10^9 \pi t\right)\right]$
$+ j\left[\cos\left(1.8 \times 10^9 \pi t + \dfrac{\pi}{8}\right) - \cos\left(1.8 \times 10^9 \pi t\right)\right]$

CHAPTER 14

14.1. $\mu = 0.0015$, $\sigma = 0.000353$, pp = 0.001.

14.2. $J_{PER}[n] = 0.001\sin\left(\dfrac{7\pi}{512}n\right) - 0.001\sin\left[\dfrac{7\pi}{512}(n-1)\right]$

$J_{CYC}[n] = 0.001\sin\left(\dfrac{7\pi}{512}n\right) - 0.002\sin\left[\dfrac{7\pi}{512}(n-1)\right] + 0.001\sin\left[\dfrac{7\pi}{512}(n-2)\right]$

14.6. 17.1 V/sqrt(Hz). **14.7.** 2.42 mV. **14.8.** –117.0 dBc/Hz. **14.9.** –117.0 dBc/Hz

14.10. $S_p(f) = \dfrac{2.5\times 10^{-26}}{\pi^2\left(10^6 + f^2\right)}\ \dfrac{\text{s}}{\text{Hz}}$ $S_p(f) = \dfrac{2.5\times 10^{-6}}{\pi^2\left(10^6 + f^2\right)}\ \dfrac{\text{UI}}{\text{Hz}}$

14.13. (a) $\dfrac{1}{\sigma\sqrt{2\pi}}e^{-\frac{1}{2}\frac{(t-a)^2}{\sigma^2}}$ (b) $\dfrac{1}{\sigma\sqrt{2\pi}}e^{-\frac{1}{2}\frac{(t-a-\mu)^2}{\sigma^2}}$

(c) $\dfrac{1}{4}\delta(t-b+a) + \dfrac{1}{4}\delta(t-b-a) + \dfrac{1}{4}\delta(t+b+a) + \dfrac{1}{4}\delta(t+b-a)$

(d) $\dfrac{1}{8}\delta(t-b+a) + \dfrac{1}{8}\delta(t-b-a) + \dfrac{1}{8}\delta(t+b+2a) + \dfrac{1}{8}\delta(t-b-2a) + \dfrac{1}{8}\delta(t+b+a)$

$\quad + \dfrac{1}{8}\delta(t+b-a) + \dfrac{1}{8}\delta(t+b+2a) + \dfrac{1}{8}\delta(t+b-2a)$

14.14. $1.994\times 10^8 e^{-5\times 10^{17}t^2} + 1.994\times 10^8 e^{-50(10^8 t-1)^2}$

14.15. $9.97\times 10^7 e^{-0.5(10^9 t-11)^2} + 9.97\times 10^7 e^{-0.5(10^9 t-9)^2} + 9.97\times 10^7 e^{-0.5(10^9 t-1)^2} + 9.97\times 10^7 e^{-0.5(10^9 t+1)^2}$

14.17. $P_e(V_{TH} = 0.6) = 9.865\times 10^{-10}$, $P_e(V_{TH} = 0.75) = 1.2\times 10^{-6}$. **14.18.** $P_e(V_{TH} = 0.0) = 31.6\times 10^{-6}$.

14.19. $P_e(t_{TH} = 2.5\times 10^{-10}) = 2.86\times 10^{-7}$; $P_e(t_{TH} = 4\times 10^{-10}) = 0.0113$; **14.20.** $\dfrac{2}{3}\delta(t) + \dfrac{1}{3}\delta(t-10^{-9})$

14.21. 7.17×10^{-8}. **14.22.** 3.29×10^{-7}. **14.23.** $2.701\times 10^{-10} \le \text{BER} \le 5.498\times 10^{-10}$; TestTime=166.67s. **14.24.** 10. **14.25.** NE = 1, 0.499×10^{-3}; NE = 2, 0.276×10^{-2}; NE = 10, 0.5830×10^{-2}. **14.27.** 1.8×10^{13}, 2.002 days. **14.28.** $N_{T,min} = 2.01\times 10^{12}$; $N_{T,max} = 4.77\times 10^{11}$; $T_{T,min} = 2014.4$ s' $T_{T,max} = 477.1$ s. **14.30.** 31. **14.32.** 4095, seed = 0000,0000,0000,0001; yes periodic. **14.33.** #1: 11001011 1001011100101; #2: 10111001011 1001011100; same pattern just time shifted with respect to one another. **14.34.** $V_{logic0} = -0.5819$ V; $V_{logic1} = 0.5819$ V; noise0 = 0.0803 V; Noise1 = 0.0803 V. BER($V_{TH} = 10$ mV) $= 3.18\times 10^{-13}$. **14.35.** V_{TH}(BER = 10^{-6}) = 0.9300 V; V_{TH}(BER = 10^{-8}) = 1.088 V; test time = 0.469 s. **14.36.** 2.41×10^{-8}. **14.37.** 1.026×10^{-10}. **14.38.** $2.33\times 10^{-10} <$ BER $< 10^{-16}$. **14.39.** 3.11×10^{-12}. **14.40.** 2.44×10^{-11}. **14.42.** 1.44×10^{-9}. **14.43.** 1.56×10^{-9}. **14.44.** 1.33×10^{-11}. **14.45.** 2.56×10^{-11}. **14.46.** DJ = 0, RJ = 4×10^{-12}, TJ = 5.39×10^{-11}. **14.47.** DJ = 2.2×10^{-11}, RJ = 5×10^{-11}, TJ = 8.96×10^{-10}. **14.48.** DJ = 2.3×10^{-11}, RJ = 6×10^{-12}, TJ = 1.03×10^{-10}. **14.50.** BER = 1.2×10^{-13}; yes the system performs with a BER less than 10^{-12}. **14.52.** Bins, 0, 33, 66, 99, ..., 33k, ..., 495. RMS level ranges from –17 dB to –24 dB. **14.53.** PJ = 0.379, DDJ = 0.1544 , BUJ = 0.103.

14.54. $pdf_G = \dfrac{1.0 \times 10^{11}\sqrt{2}}{\sqrt{\pi}} e^{-2 \times 10^{22} t^2}$

$$pdf_{SJ} = \begin{cases} \dfrac{1}{\pi\sqrt{\left(1.0 \times 10^{-22}\right)^2 - t^2}}, & |t| \le 1.0 \times 10^{-11} \\ \\ 0, & \text{otherwise} \end{cases}$$

$$pdf_{DCD} = \frac{1}{2}\delta\left(t + 1.5 \times 10^{-12}\right) + \frac{1}{2}\delta\left(t - 1.5 \times 10^{-12}\right)$$

$$pdf_{ISI} = \frac{1}{2}\delta\left(t + 3.0 \times 10^{-12}\right) + \frac{1}{2}\delta\left(t - 3.0 \times 10^{-12}\right)$$

14.55. $1.995 \times 10^{10} e^{-0.005(2 \times 10^{12} t - 9)^2} + 1.995 \times 10^{10} e^{-0.005(2 \times 10^{12} t - 3)^2} + 1.995 \times 10^{10} e^{-0.005(2 \times 10^{12} t + 3)^2}$

$$+ 1.995 \times 10^{10} e^{-0.005(2 \times 10^{12} t + 9)^2}$$

14.56. $M_J = 3$, $M_C = 8 \times 10^5$, $N = 5 \times 10^6$.

CHAPTER 15

15.1. 446.3 mΩ, HF = LF = ILoad = 0.56 A, HS = LS = 0 A, V_{drop} = 249 mV per line, V_{OUT} = 3.3V + 2 × 249 mV = 3.80 V. **15.2.** 201 nH/m, 15.4 nH total, −0.001%; 150 nH/m, 11.5 nH total, 0.00004%. **15.3.** 83 pF/m, 2.5 pF total; 120 pF/m, 3.6 pF total.

15.4. $|H(f)| = \left[\dfrac{1}{\sqrt{1 + \left(\dfrac{f}{F_C}\right)^2}}\right]$ V/V and $\angle H(f) = -\dfrac{180}{\pi}\tan^{-1}\left(\dfrac{f}{F_C}\right)$ degrees

where $F_C = \dfrac{1}{2\pi RC}$ Hz $= \dfrac{1}{2\pi \times 50 \times 10^3 \times 3.6 \times 10^{-12}}$.

15.5. 45 pF/m, 9.2 pF total; 50 pF/m, 10.3 pF total. **15.6.** 50 pF/m, 14.7 pF + 35 pF = 49.7 pF total, gain = 0.141 = −17.02 dB, buffer amplifier is needed. Using refined estimate: 110 pF/m, 32.4 pF + 35 pF = 67.4 pF total, gain = 0.104 = −19.63 dB, buffer amplifier is needed. **15.7.** 110 pF/m, 400 nH/m, Z_o = 60Ω; To lower the characteristic impedance, we would make the spacing smaller or widen the trace. Changing the length has no effect on characteristic impedance in an ideal transmission line. **15.8.** v_{signal} = 1.5× 10^8 m/s = 0.5c, T_d = 1.3 ns, 22.5 pF. **15.9.** L = 7.53 m, or 294 in., much longer than the stripline length. We can use a lumped element model for this line. **15.10.** 100 % (cable length = one wavelength), 16.7 pF, 360 degrees. **15.11.** F_c = 12.7 MHz. Max. attenuation is 0.00005 dB; We should not need a buffer amplifier unless the output becomes unstable when loaded with 125 pF. **15.12.** The LC network will consist of a shunt C of 1.43 pF and a series L of 56.5 nH.**15.13.** The values of the shunt-serces C network is 0.38 pF and 0.69 pF. **15.14.** Two possible solutions: (a) Θ_d = 143.6 degrees, Θ_{OS} = 57.7 degrees, $Y_{in,stub}$ = j 0.03169 S; (b) Θ_d = 91.9 degrees, Θ_{OS} = 122.3 degrees, $Y_{in,stub}$ = $-j$ 0.03169 S. **15.15.** Θ_d = 87.3 degrees, Θ_{OS} = 20.7 degrees, $Y_{in,stub}$ = $-j$ 2.64. **15.16.** 60.2 MHz. **15.17.** 106. **15.18.** 0.38 degrees; We can reduce this error using a pair of buffer amplifiers placed near the DUT to isolate the DUT outputs from the tester capacitance. Of course we need to calibrate the phase mismatch between the DIB buffer amplifiers.

15.19. $v(t) = 5.0 \text{ V} - \dfrac{100 \text{ mV}}{2} \times \dfrac{4.7 \text{ k}\Omega}{1 \text{ k}\Omega} \sin(2\pi ft) = 5.0 \text{ V} - 235 \text{ mV} \times \sin(2\pi \times 2400 \ t)$

15.20. $v(t) = -235 \text{ mV} \times \sin(2\pi \times 2400 \ t)$. **15.21.** Power dissipation is caused by a resistive component associated with the line impedance—for example, the ohmic resistance of the copper trace forming a microstrip line. To reduce the power dissipation, the ohmic resistance needs to be minimized (use a wide and thick trace).

CHAPTER 16

16.1. Lower cost of test, increased fault coverage/improved process control, diagnostics and characterization, ease of test program development, and system-level diagnostics. **16.2.** Robust circuits can tolerate more measurement error without failing test limits. Since measurement accuracy generally comes at the expense of longer test time (averaging, for example), robust circuits with wide design margins can be tested more economically. **16.3.** Delayed time to market results in lower unit prices due to more competition. Lower unit prices lead to smaller profit margins. **16.4.** The IEEE Std. 1149.1 test interface is primarily designed for board-level, chip-to-chip interconnect testing. The TAP controller provides a standard, consistent interface to the scan circuits of the 1149.1 boundary scan circuits. Hold TMS at logic 1 while clocking TCK. No more than five TCK clock cycles are required. **16.5.** Test time will increase due to the tester's match mode search process. Test development time will likely increase due to match mode code development. **16.6.** Only four lines are required: TMS, TCK, TDI, and TDO **16.7.** 325 flip flops (all flip flops are in the scan chain). 10 MHz/(325 + 1). (we need one clock cycle to capture circuit response between parallel vectors). Multiple scan chains allow parallel testing, which will reduce the time require to scan data in and out. **16.8.** IDDQ testing. **16.9.** Using Eq. (6.38), $t_s = \ln(1 \text{ mV}/1.5 \text{ V}) \times$ 2 k$\Omega \times$ 1μF = 15 ms. After precharging, R increases to 50 kΩ (two 100 kΩ resistors in parallel). To recover from the worst-case op-amp offset of 10 mV takes an additional settling time of $t_s =$ $\ln(1 \text{ mV} / 10 \text{ mV}) \times 50 \text{ k}\Omega \times 1\mu\text{F} = 115$ ms, for a total settling time of 130 ms. (Notice that the op-amp's offset causes most of the settling time.) Without DfT, the circuit would normally take t_s $= \ln(1 \text{ mV}/1.5 \text{ V}) \times 50 \text{ k}\Omega \times 1\mu\text{F} = 366$ ms. **16.13.** Force 1 V at the AB1 line and connect DUT 1's signal pin to the AB1 line. Connect DUT2's signal line to the internal ground connection through the ABM ground switch. Measure the voltages at the two signal pins using AB2 and subtract to get V_{drop}. Measure the current, I, supplied into AB1. Use $V_{drop} = IR$ to calculate the value of R. This method would not catch shorts to ground, but repeating the process from DUT 2 to DUT 1 will catch shorts. **16.14.** A reference multitone at a known amplitude at 1, 5, 9, and 13 kHz is applied to the input of the ADC. The DSP measures the ADC signal level and calculates the ADC gain. Then the test mux is switched to connect the DAC output to the ADC. The DAC is set to produce a multitone at the same frequencies. Its output is measured using the ADC. The ADC gain error at each frequency is removed through an on-chip calibration routine. This gives the DAC gain at each frequency. **16.15.** Place the first switch network into the "Force input" mode and apply a reference voltage to the DAC using the TESTIN analog bus. **16.16.** No, the circuit has a very long divide time. Also, it can only be clocked through a coupling capacitor. To improve the design, provide a bypass mode around the divider to allow observation of the oscillator output frequency in it's normal mode of operation. Also, provide a bypass path to clock the digital divider directly so it can be driven without the timing shift produced by the capacitor. Finally, use partitioning to split the divider into subcircuits that will count faster, or better yet, use a full-scan methodology for the divider to allow very fast testing. **16.17.** The voltage range includes the subthreshold, linear, and saturation to activate the active circuits in multiple conditions (see Figure 18.41). **16.18.** For the learning cycle a test cell with RF instrumentation as well as the test stimulus for correlation based test is needed for a subset of units. This will require a different setup than the production setup if the goal is to run production on the cheapest (with the minimum set of instrumentation) ATE. Additional cycles are required starting a lot in production when first establishing the correlation. **16.19.** There are two advantages, one of which is that a digital channel of the ATE is cheaper and more digital channels are typically available per tester. In addition, well-established BIST techniques like scan can be utilized to implement an efficient test. A more practical advantage is that, in general, digital tests are more robust in production when considering the impact of contact resistance on the test result, mismatch uncertainty due to discontinuities, calibration, and so on. (see also Chapters 12 and 13).

A